Unsöld · Baschek The New Cosmos

Albrecht Unsöld Bodo Baschek

The New Cosmos

Translated by
William D. Brewer

Fourth Completely Revised Edition

With 242 Figures

Springer-Verlag
Berlin Heidelberg New York London Paris
Tokyo Hong Kong Barcelona Budapest

Professor Dr. *Albrecht Unsöld*

Institut für Theoretische Physik und Sternwarte, Universität Kiel, Olshausenstrasse 40,
W-2300 Kiel 1, Fed. Rep. of Germany

Professor Dr. *Bodo Baschek*

Institut für Theoretische Astrophysik, Universität Heidelberg, Im Neuenheimer Feld 561,
W-6900 Heidelberg 1, Fed. Rep. of Germany

Translator:

Professor *William D. Brewer,* Ph. D.

Fachbereich Physik, Freie Universität Berlin, Arnimallee 14,
W-1000 Berlin 33, Fed. Rep. of Germany

Cover: Optical image of the radio galaxy NGC 5128 = Centaurus A. Recorded by the Anglo-Australian telescope. (© Anglo-Australian Telescope Board, photography by David Malin)

Frontispiece: The galaxy group (NGC 6769-71) in Pavo with indications of gravitational interaction between the galaxies: faint haloes and connecting arcs, deformation of spiral arms and central regions. (By S. Laustsen with the 3.6 m-telescope of the European Southern Observatory)

Title of the original German edition:
A. Unsöld, B. Baschek: *Der neue Kosmos*. Fünfte, überarbeitete und erweiterte Auflage
© Springer-Verlag Berlin Heidelberg 1967, 1974, 1981, 1988, 1991
ISBN 3-540-53757-0 5. Auflage Springer-Verlag Berlin Heidelberg New York
ISBN 0-387-53757-0 5th Edition Springer-Verlag New York Berlin Heidelberg

---

ISBN 3-540-52593-9 4. Auflage Springer-Verlag Berlin Heidelberg New York
ISBN 0-387-52593-9 4th Edition Springer-Verlag New York Berlin Heidelberg

ISBN 3-540-90886-2 3. Auflage Springer-Verlag Berlin Heidelberg New York
ISBN 0-387-90886-2 3rd Edition Springer-Verlag New York Berlin Heidelberg

Library of Congress Cataloging-in-Publication Data. Unsöld, Albrecht, 1905- [Neue Kosmos. English] The new cosmos / Albrecht Unsöld, Bodo Baschek. – 4th completely rev. ed. p. cm. Translation of: Der neue Kosmos. Includes bibliographical references and index. ISBN 0-387-52593-9. 1. Astronomy. I. Baschek, B., 1935-. II. Title. QB43.2.U5713 1991 520–dc20 91-26110

Printed in Germany

Typesetting: K + V Fotosatz, W-6124 Beerfelden
Printing: Druckhaus Beltz, W-6944 Hemsbach/Bergstr.
Binding: J. Schäffer GmbH & Co. KG, W-6718 Grünstadt
54/3140-543210 – Printed on acid-free paper

In Memoriam

M.G.J. Minnaert

(12.2.1893 – 26.10.1970)

Preface
to the Fourth English Edition

Rapid progress in the field of astronomy has continued undiminished since the publication of the previous edition. New observational techniques made possible by space travel, the development of highly sensitive photodetectors, and the use of powerful computers both for data collection and treatment and for the theoretical interpretation of observations have all contributed decisively to this progress. Although many subfields of astronomy still present only an incomplete picture, it nevertheless seems desirable to us to attempt, in this thoroughly revised new edition, to give a cogent description of observational methods, theoretical fundamentals, and results from the *entire* field of astronomy.

In our opinion, the chief goal of this book is still to provide an *introduction* to the whole of astronomy and astrophysics for students of astronomy and other natural sciences, professionals from neighboring fields, and amateur astronomers, corresponding to the current state of the art in astronomical research and enabling the reader to gain access to the more advanced astronomical literature.

The enormous development and multiplication of astronomical knowledge has necessarily resulted in a marked increase in the length of this edition as compared to the previous one. While the treatment of traditional subjects could be only modestly condensed, new topics, e.g. in connection with the Solar System, the great variety of astronomical instruments and observational methods, nuclear evolution of stars, the complex physical state of interstellar matter, galaxies and clusters of galaxies, and the activity phenomena in galactic cores, as well as modern cosmology and its relation to elementary particle theories, have added considerable material to this edition.

The tested arrangement of the material from previous editions has been retained in the present book, aside from a few minor modifications and a finer subdivision of some sections, intended to improve their clarity. Insofar as astronomical units are not employed, we have for the first time used the international system of units (SI) for all the numerical values quoted in the book; in addition, their values in the Gaussian-cgs system previously used are also given for the most part. In an appendix, we have collected the more important relations between the two systems of units.

In the preparation of previous editions of the "New Cosmos", A. Unsöld was pleased to have the tireless support of his colleagues at Kiel. Critical readings of the manuscript and numerous helpful suggestions were provided by B. Baschek, T. Gehren, H. Holweger, D. Reimers, E. Richter, and V. Weidemann; the production of the manuscripts was assisted by Ms. A. Wagner, Ms. G. Mangelsen, and Ms. G. Hebeler. He expresses his heartfelt thanks to all of these people once again on the occasion of his retirement from active research.

A major portion of the present new edition was prepared by B. Baschek, who bears the responsibility for the book, in particular for its newer parts. In all phases of our work on this edition, colleagues too numerous to name here have helped us generously with advice and support, by providing new material and with constructive criticism on particular sections. Our Heidelberg colleagues J. Köppen, R. Wehrse, and P. Hauschildt helped us willingly by their critical reading of the entire manuscript. An important contribution to the book was made by Ms. G. Höhns, Ms. B. Hoffmann, and in particular Ms. M. Wolf, who made many of the drawings, through their tireless work in helping to prepare the manuscript. We express our sincere gratitude to all of them.

The present English version of the "New Cosmos" is a translation of the Fifth German edition of 1991, which itself represents a modernized and improved version of the completely revised Fourth German edition that appeared in 1988. Compared to this edition, important astronomical results from the past two years have been added, for example some of the information obtained from the fly-by of Voyager 2 past the Neptune system, and all numerical data were updated to their current values. Needless to say, all the errors which came to our attention have been corrected and inconsistencies eliminated. Furthermore, some "selected exercises" have been given; their solution is intended to contribute to the understanding of the material by the interested reader. Also, a few new colored illustrations were added to the book.

Following the appearance of the Fourth German edition, we received a number of communications containing critical remarks and suggestions, which were for the most part taken into account in the present version of the book. We thank all those colleagues who have made helpful suggestions, in particular Prof. D. Schlüter, Kiel, and Prof. J. Oxenius, Brussels, for their careful critical reading of the text.

Special thanks are owed to Prof. W. D. Brewer, Berlin, for his excellent translation of the text, for constructive suggestions, and for continued collegiality. We thank the Springer-Verlag, especially Dr. H. Lotsch and Mr. C.-D. Bachem, for their agreeable cooperation during the preparation and production of this book.

Kiel and Heidelberg,
February 1991

Albrecht Unsöld
Bodo Baschek

From the Author's Preface to the First Edition

This book may serve to present the modern view of the universe to a large number of readers whose over-full professional commitments leave them no time for the study of larger monographs . . . Such a book should not be too compendious. Accordingly, the author has been at pains to allow the leading ideas of the various domains of astronomical investigation to stand out plainly in their scientific and historical settings; the introductory chapters of the three parts of the book, in the framework of historical surveys, should assist the general review. With that in mind, the title was chosen following Alexander von Humboldt's well known book *"Kosmos, Entwurf einer physischen Weltbeschreibung"* (1827 – 1859). On the other hand, particular results – which admittedly first lend color to the picture – are often simply stated without attempting any thorough justification.

The reader seeking further information will find guidance in the Bibliography. This makes no pretensions to completeness or historical balance. References in the text or in captions for the figures, by quoting authors and years, make it possible for the reader to trace the relevant publications through the standard abstracting journals.

I wish to thank my colleagues V. Weidemann, E. Richter and B. Baschek for their critical reading of the book and for much helpful counsel, and H. Holweger for his tireless collaboration with the proofs. Similarly, my thanks are due to Miss Antje Wagner for the careful preparation of the typescript.

Kiel, April 1966 *Albrecht Unsöld*

Contents

1. Introduction

Astronomy, the study of the stars and other celestial objects, is one of the exact sciences. It deals with the quantitative investigation of the cosmos and physical laws which govern it: with the motions, the structures, the formation, and the evolution of the various celestial bodies.

Astronomy is among the oldest of the sciences. The earliest human cultures made use of their knowledge of celestial phenomena and collected astronomical data in order to establish a calendar, measure time, and as an aid to navigation. This early astronomy was often closely interwoven with magical, mythological, religious, and philosophical ideas.

The study of the cosmos in the modern sense, however, dates back only to the ancient Greeks: the determination of distances on the Earth and of positions of the celestial bodies in the sky, together with knowledge of geometry, led to the first realistic estimates of the sizes and distances of the objects in outer space. The complex orbits of the Sun, the Moon, and the planets were described in a mathematical, kinematical picture, which allowed the calculation of the positions of the planets in advance. Greek astronomy attained its zenith, and experienced its swan song, in the impressive work of Ptolemy, about 150 A. D. The name of the science, *astronomy*, is quite appropriately derived from the Greek word "*αστηϱ*" = star or "*αστϱον*" = constellation or heavenly body.

At the beginning of the modern period, in the 16th and 17th centuries, the Copernican view of the universe became generally accepted. Celestial mechanics received its foundation in Newton's Theory of Gravitation in the 17th century and was completed mathematically in the period immediately following. Major progress in astronomical research was made in this period, on the one hand through the introduction of new concepts and theoretical approaches, and on the other through observations of new celestial phenomena. The latter were made possible by the development of new instruments. The invention of the telescope at the beginning of the 17th century led to a nearly unimaginable increase in the scope of astronomical knowledge. Later, new eras in astronomical research were opened up by the development of photography, of the spectrograph, the radio telescope, and of space travel, allowing observations to be made over the entire range of the electromagnetic spectrum.

In the 19th and particularly in the 20th centuries, physics has played a decisive role in the elucidation of astronomical phenomena; astrophysics has steadily increased in importance over "classical astronomy". There is an extremely fruitful interaction between astrophysics/astronomy and physics: on the one hand, astronomy can be considered to be the physics of the cosmos, and there is hardly a discipline in physics which does not find application in modern astronomy; on the other hand, the cosmos with its often extreme states of matter offers the opportunity to study physical processes under conditions which are unattainable in the laboratory. Along with physics, and of course mathematics, applications of chemistry and the Earth and biological sciences are also of importance in astronomy.

Among the sciences, astronomy is unique in that no experiments can be carried out on the distant celestial objects; astronomers must content themselves with *observations*. "Diagnosis from a distance", and in particular the quantitative analysis of radiation from the cosmos over the widest possible spectral range, thus play a central role in astronomical research.

The rapid development of many branches of astronomy has continued up to the present time. With this revised edition of *The New Cosmos*, we have tried to keep pace with the rapid expansion of astronomical knowledge while maintaining our goal of providing a comprehensive – and comprehensible – introductory survey of the *whole* field of astronomy. We have placed emphasis on observations of the manifold objects and phenomena in the cosmos, as well as on the basic ideas which provide the foundation for the various fields within the discipline. We have combined description of the observations directly with the theoretical approaches to their elucidation. Particular results, as well as information from physics and the other natural sciences which are required for the understanding of astronomical phenomena, are, however, often simply stated without detailed explanations. The complete bibliography, together with a list of important reference works, journals, etc., is intended to help the reader to gain access to the more detailed and specialized literature.

We begin our study of the cosmos, its structure and its laws, "at home" by considering our *Solar System* in Chap. 2. This chapter, like the three following chapters, starts with a historical summary which is intended to give the reader an overview of the subject. We first become acquainted with observations of the heavens and with the motions of the Earth, the Sun, and the Moon, and introduce celestial coordinates and sidereal time. The apparent motions of the planets and other objects are then explained in the framework of the Newtonian Theory of Gravitation. Before considering the planets and other objects in the solar system in detail, we give a summary of the development of space research, which has contributed enormously in recent years to knowledge of our planetary system. Chapter 2 ends with a discussion of the individual planets, their moons, and other smaller bodies such as comets and meteorites.

Prior to taking up the topic of the Sun and other stars, it is appropriate to describe the basic principles of astronomical *observation methods*, and we do this in Chap. 3. An impressive arsenal of telescopes and detectors is available to today's astronomer; with them, from the Earth or from space vehicles, he can investigate the radiation emitted by celestial bodies over the entire range of the electromagnetic spectrum, from the radio and microwave regions through the infrared, the visible, and the ultraviolet to the realm of highly energetic radiations, the X-rays and gamma rays. The use of computers provides an essential tool for the modern astronomer in these observations.

Chapter 4 is devoted to *stars*, which we first treat as individual objects. We give an overview of the different types of stars such as those of the main sequence, giants and supergiants, white dwarfs and neutron stars, as well as the great variety of variable stars (Cepheids, magnetic stars, novas, supernovas, pulsars, ...), and become acquainted with their distances, magnitudes, colors, temperatures, luminosities, and masses. In this chapter, the *Sun* plays a particularly important role: on the one hand, as the nearest star, it offers us the possibility of making incomparably more detailed observations than of any other star; on the other, its properties are those of an "average" star, and their study thus yields important information about the physical state of stars in general. The treatment of the *physics of individual stars* occupies an important place in Chap. 4. In particular, the theory of radiation and atomic spectroscopy provide the basis for the quantitative investigation of the radiation and the spectra of the Sun and other stars, and for the understanding of the physical-chemical structure of their outer layers, the stellar atmospheres. Understanding of the mechanism of energy release by thermonuclear reactions and by gravitation is of decisive importance for the study of stellar interiors, their structures and evolution. The formation and evolution of stars and their late stages (white dwarf, neutron star or black hole) are considered in the following chapter, together with the treatment of stellar clusters.

In Chap. 5, we take up *stellar systems* and the macroscopic structure of the universe. Making use of our knowledge of individual stars and their distances from the Earth, we first develop a picture of our own Milky Way galaxy, to which the Sun belongs together with about $100 \cdot 10^9 = 10^{11}$ other stars. We discuss the distribution and the motions of the stars and star clusters and then treat interstellar matter: dust and gas clouds of low density. After making the acquaintance of methods for the determination of the enormous distances in intergalactic space, we turn to other galaxies, among which we find a variety of types: spiral and elliptical galaxies, infrared galaxies, radio galaxies, and the distant quasars. In the centers of many galaxies, we observe an "activity" involving the appearance of extremely large amounts of energy, whose origins are still a mystery.

Galaxies, as a rule, belong to larger systems, called clusters of galaxies. These are, in turn, ordered in clusters of clusters of galaxies, the superclusters, which finally form a "lattice" enclosing large areas of empty intergalactic space and defining the macroscopic structure of the universe. Like individual stars, the galaxies and the clusters of galaxies evolve with the passage of time. The mutual gravitational influence of the galaxies plays an important role in their development.

At the conclusion of Chap. 5, we consider the *universe as a whole*, its content of matter, radiation, and energy, and its structure and evolution throughout the expansion which has taken place over the roughly $2 \cdot 10^{10}$ years from the "Big Bang" to the present time.

After pressing out to the far reaches of the cosmos, we finally return to our Solar System. In Chap. 6, we take up the problems of the *formation and evolution of the Sun and the planets* and give particular attention to the development of the Earth and of life on Earth.

2. Classical Astronomy. The Solar System

We shall begin our treatment of astronomy with a historical overview of classical astronomy (Sect. 2.1) from ancient times up through the introduction of the heliocentric system by Nicholas Copernicus, Tycho Brahe, Johannes Kepler, Galileo Galilei and their contemporaries, and the founding of celestial mechanics by Isaac Newton at the close of the 17th century. We then turn in Sect. 2.2 to the description of motions on the *celestial sphere* and the coordinate systems used to describe the positions of objects in the sky. In Sect. 2.3, we treat the motions of the *Earth*, its rotation around its own axis and its revolution around the Sun, which make themselves evident as motions on the celestial sphere; in this section, we also take up the sidereal time scale. Following these preparatory remarks, we make the acquaintance of the objects of our *Solar System*, beginning in Sect. 2.4 with the Moon, its motions, its phases, and with lunar and solar eclipses. In Sect. 2.5, we survey the motions of the planets, the comets, and other bodies in the Solar System, and the methods for determining distances between them. After reviewing the fundamentals of mechanics and the theory of gravitation, we discuss in Sect. 2.6 some of their applications to celestial mechanics and, in Sect. 2.7, to the calculation of the orbits of artificial satellites and space probes. Section 2.7 also contains a brief chronology of space-research missions. Finally, we treat the physical structure of the objects in the Solar System: the planets, their moons, and the asteroids in Sect. 2.8; and the comets, meteors, and meteorites in Sect. 2.9.

2.1 Humanity and the Stars – Observing and Thinking

A Historical Introduction to Classical Astronomy

Unaffected by the evolution and the activities of mankind, the objects in the heavens have moved along their paths for millenia. The starry skies have thus always been a symbol of the "Other" – of Nature, of deities – the antithesis of the "Self" with its world of inner experience, striving and activity. The history of astronomy is at the same time one of the most exciting chapters in the history of human thought. Again and again, there is an interaction between the appearance of new *forms of thought* and the discovery of new *phenomena*, the latter often with the aid of newly-developed *instruments*.

We cannot treat here the great achievements of the ancient Middle Eastern peoples, the Sumerians, Babylonians, Assyrians, and the Egyptians; nor do we have the space to describe the astronomy of the the Far Eastern cultures in China, Japan, and India, which was highly developed by the standards of the time.

The concept of the *universe* and its investigation in the modern sense dates back to the ancient Greeks, who were the first to dare to shake off the fetters of black magic and mythology and, aided by their enormously flexible language, to adopt forms of thinking which allowed them, bit by bit, to "comprehend" the phenomena of the cosmos.

How bold were the ideas of the pre-Socratic Greeks! Thales of Milet, about 600 B. C., had already clearly understood that the Earth is round, and that the Moon is illuminated by the Sun, and he predicted the Solar eclipse of the year 585 B. C. But is it not just as important that he attempted to reduce understanding of the entire universe to a *single* principle, that of "water"?

The little that we know of Pythagoras (in the middle of the 6th century B. C.) and of his school seems surprisingly modern. The spherical shapes of the Earth, the Sun, and the Moon, the Earth's rotation, and the revolution of at least the inner planets, Venus and Mercury, were already known to the Pythagorans.

After the collapse of the Greek states, *Alexandria* became the center of ancient science; there, the quantitative investigation of the heavens made rapid progress with the aid of systematic measurements. The numerical results are less important for us today than the happy realization that the great Greek astronomers made the bold leap of applying the laws of *geometry* to the cosmos! Aristarchus of Samos, who lived in the first half of the 3rd century B.C., attempted to compare the *distances* of the Earth to the Sun and the Earth to the Moon with the *diameters* of the three bodies by making the assumption that when the Moon is in its first and third quarter, the triangle Sun-Moon-Earth makes a right angle at the Moon. In addition to carrying out these first quantitative estimates of dimensions *in* space, Aristarchus was the first to teach the *heliocentric system* and to recognize its important consequence that the distances to the fixed stars must be incomparably greater than that from the Earth to the Sun. How far he was ahead of his time with these discoveries can be seen from the fact that by the following generation, they had already been forgotten. Soon after Aristarchus' important achievements, Eratosthenes carried out the first measurement of a degree of arc on the Earth's surface, between Alexandria and Syene: he compared the difference in latitude between the two places with their distance along a much-traveled caravan route, and thereby determined the circumference and diameter of the Earth fairly precisely. However, the greatest observer of ancient times was Hipparchus (about 150 B.C.), whose *stellar catalog* was still nearly unsurpassed in accuracy in the 16th century A.D. Even though the means at his disposal naturally did not allow him to make significantly better determinations of the basic dimensions of the Solar System, he was able to make the important discovery of *precession*, i.e. the yearly shift of the equinoxes and thus the difference between the tropical and the sidereal years.

The theory of *planetary motion*, which we shall treat next, was necessarily limited in Greek astronomy to a problem in *geometry* and *kinematics*. Gradual improvements and extensions of observations on the one hand, and new mathematical approaches on the other, formed the basis for the attempts of Philolaus, Eudoxus, Heracleides, Appollonius, and others to describe the observed motions of the planets; their attempts employed the superposition of ever more complicated circular motions. Ancient astronomy and planetary theory attained its final development much later, in the work of Claudius Ptolemy, who wrote his 13-volume Handbook of Astronomy (Mathematics), *Μαθηματικης Συνταξεως βιβλια ιγ*, in Alexandria about 150 A.D. His "Syntax" later acquired the adjective *μεγιστη*, "greatest", from which the arabic title *Almagest* is derived. The Almagest is based, to a large extent, on the observations and research of Hipparchus, but Ptolemy also added much new material, particularly in the theory of planetary motion. At this point, we need only sketch the outlines of Ptolemy's geocentric system: the Earth rests at the midpoint of the universe. The motions of the Sun and the Moon in the sky may be represented fairly simply by circular orbits. The planetary motions have been described by Ptolemy using the *theory of epicycles*: each planet moves on a circle, the so-called epicycle, whose nonmaterial center moves around the Earth on a second circle, the deferent. We shall not delve further into the refinements of this system involving additional, in some cases eccentric circular orbits, etc. The common features and differences to the Copernican heliocentric system will be discussed in Sect. 2.6, where the latter is treated in detail. The intellectual posture of the Almagest clearly shows the influence of Aristotelian philosophy, or rather of *Aristotelianism*. Its modes of thought, originally the tools of vital research, had long since hardened into the dogmas of a rigid school; this was the principal reason for the remarkable historical durability of the Ptolemaic world-system.

We cannot go into detail here about how, following the decline of the academy in Alexandria, first the Nestorian Christians in Syria and later the Arabs in Bhagdad took over and continued the work of Ptolemy.

Translations and commentaries on the Almagest were the basic sources of the first Western textbook on astronomy, the *Tractatus de Sphaera* of Ioannes de Sacrobosco, a native of England who taught at the University of Paris until his death in the year 1256. The *Sphaera* was issued again and again, and often commentated; it was still "the" text for teaching astronomy in Galileo's time, three centuries later.

The intellectual basis of the new thinking was provided in part by the conquest of Constantinople by the Turks in 1453: thereafter, numerous scientific works from antiquity were made accessible to the West by Byzantine scholars. For example, some very fragmentary texts concerning the heliocentric system of the ancients clearly made a strong impression on Copernicus. The result was a turning-away from the rigid doctrine of the Aristotelians in favor of the much more lively and flexible thinking of the schools of Pythagoras and Plato. The "platonic" idea that the process of understanding the universe consists of a progressive adaptation of our inner world of concepts and ways of thinking to the more and more precisely-studied outer world of phenomena has become the hallmark of modern research from Cusanus through Kepler to Niels Bohr. Finally, with the blossoming of a practical approach to life exemplified by the rise of crafts and trades, the question was no longer "What did Aristotle say?", but rather "How can you do this ...?".

In the 15th century, a completely new spirit in science and in life arose, at first in Italy and soon thereafter in the North as well. The sententious meditations of Cardinal Nicholas Cusanus (1401–1464) have only today begun to be properly appreciated. It is fascinating to see how his ideas about the infinity of the universe and about quantitative scientific research arose from religious or theological considerations. Near the end of the century (1492), the discovery of America by Christopher Columbus added the classic expression "il mondo e poco" to the new spirit. A few years later, Nicolas Copernicus (1473–1543) founded the *heliocentric system*.

About 1510, Copernicus sent a letter to several noted astronomers of his time; it was rediscovered only in 1877, and was entitled "Nicolai Copernici de Hypothesibus Motuum Caelestium A Se Constitutis Commentariolus". It foreshadowed the greatest part of the results which were later published in his major work, "De Revolutionibus Orbium Coelestium Libri VI", which appeared in Nuremberg in 1543, the year of his death.

Copernicus maintained the idea of the "perfection of circular motion" which had formed the basis for astronomical thought throughout antiquity and the Middle Ages; he never considered the possibility of another form of motion.

It was Johannes Kepler (1571–1630) who, starting from the Pythagorian-Platonic traditions, was able to break through to a more general point of view. Making use of the observations of Tycho Brahe (1546–1601), which were vastly more precise than any that had preceded them, he discovered his three *Laws of Planetary Motion*. Kepler derived his first two laws from an enormously tedious trigonometric calculation of the motions of Mars reported by T. Brahe in his "Astronomia Nova, Seu Physica Coelestis Tradita Commentariis de Motibus Stellae Martis Ex Observationibus G. V. Tychonis Brahe" (Prague, 1609). The third law was reported in his "Harmonices Mundi Libri V" (1619). We can only briefly mention Kepler's ground-breaking works on optics, his Keplerian telescope, his Rudolphinian Tables (1627), and numerous other achievements.

About the same time, the Italian Galileo Galilei (1564–1642) directed the *telescope* which he had built in 1609 to the heavens and discovered, in rapid succession: the "maria", the craters, and other mountain formations on the Moon; the numerous stars of the Pleiades and the Hyades; the four largest moons of Jupiter and their free orbits around the planet; the first indication of the rings of Saturn; and sunspots. His "Galileis Sidereus Nuncius" (1610), in which he describes the discoveries with his telescope, the "Dialogo Delli Due Massimi Sistemi Del Mondo, Tolemaico, e Copernico" (1632), and the "Discorsi E Dimonstrazioni Matematiche Intorno A Due Nuove Scienze Attenenti Alla Meccanica Ed ai Movimenti Locali" (1638), which was written after his condemnation by the Inquisition and contained the beginnings of theoretical mechanics, are masterworks not only in the scientific sense but also as works of art. The observations with the telescope, Tycho Brahe's observation of the supernova of 1572 and that of 1604 by Kepler and Galileo, and finally the appearance of several comets required what was perhaps the most essential scientific insight of the time: that, in contrast to the opinion of the Aristotelians, there is *no fundamental difference* between cosmic and earthly matter and that *the same natural laws hold in the realms of astronomy and of terrestrial physics* (this had already been recognized by the ancient Greeks in the case of the laws of geometry). This leap of thought, whose difficulty only becomes clear to us when we look back at Copernicus, gave impetus to the enormous upswing of scientific research at the beginning of the 17th century. W. Gilbert's investigations into electricity and magnetism, Otto v. Guericke's experiments with vacuum pumps and electrification machines, and much more, were stimulated by the revolution in the astronomical worldview.

We have no space here to pay tribute to the many observers and theoreticians who developed the new astronomy, among whom such important thinkers as Hevelius, Huygens, and Halley are particularly prominent.

An entirely new era of natural science began with Isaac Newton (1642–1727). His major work, "Philosophiae Naturalis Principia Mathematica" (1687), begins by placing theoretical mechanics on a firm basis using the *calculus of infinitesimals* ("fluxions"), which he developed for this purpose. Its connection with the *law of gravitation* explains Kepler's Laws and in one stroke provides the justification for the whole of terrestrial and *celestial mechanics*. In the area of optics, he invented the reflecting telescope and investigated the interference phenomena known as "Newton's Rings". Almost casually, he developed the basic approaches leading to numerous branches of theoretical physics.

Only the "Princeps Mathematicorum", Carl Friedrich Gauss (1777–1855), is of comparable importance; to him, astronomy owes the theory of *orbit calculation*, important contributions to *celestial mechanics* and advanced geodesics as well as the method of Least Squares. Never again has a mathematician shown such a combination of intuition in the choice of new areas of research and of facility in solving particular problems.

Again, this is not the place to pay tribute to the great theoreticians of celestial mechanics, from Euler to Lagrange and Laplace to Poincaré; and we can only mention the great observers, such as W. and J. Herschel, F. W. Bessel, and F. G. W. and O. W. Struve, when we describe their discoveries. At the conclusion of this overview, we quote just one historical date: the measurement of the first trigonometric *stellar parallaxes*, and thus the distance to particular stars, by F. W. Bessel (61 Cygni), F. G. W. Struve (Vega), and T. Henderson (α Centauri) in the year 1838. This remarkable achievement of astronomical observational technique was, in the last analysis, a critical step towards the modern exploration of space.

2.2 The Celestial Sphere. Astronomical Coordinate Systems. Geographic Latitude and Longitude

Since antiquity, human imagination has combined the easily-recognized groups of stars into *constellations* (Fig. 2.2.1). In the northern sky, the Great Bear (ursa major, the Big Dipper) is readily seen. We can find the Pole Star (Polaris) by extending the line joining the two

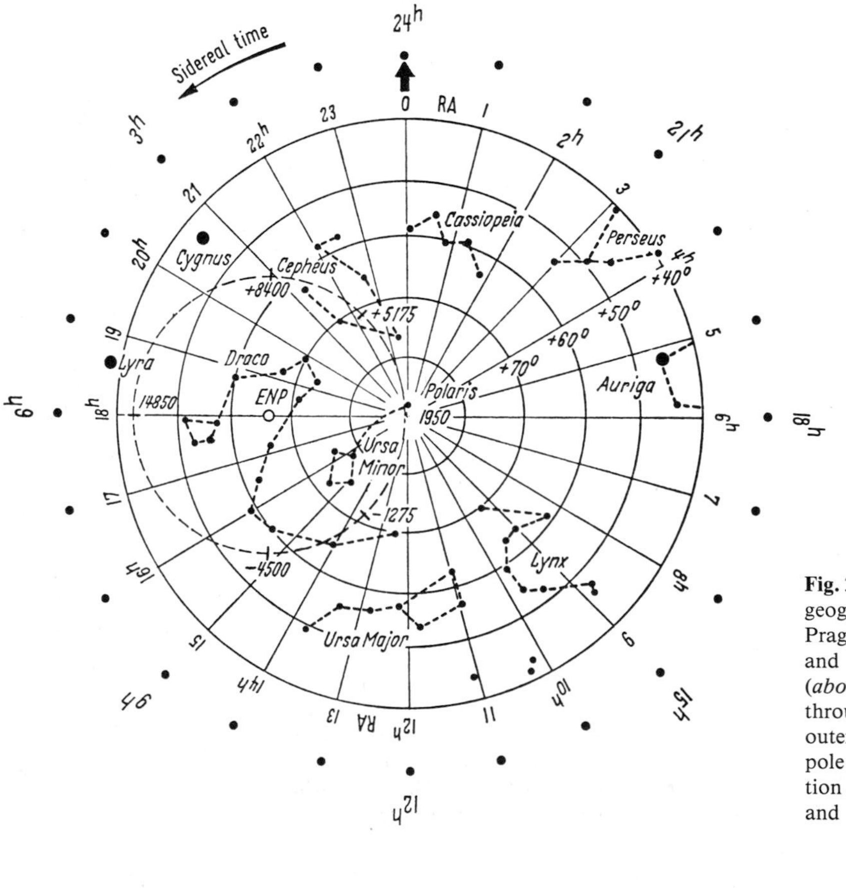

Fig. 2.2.1. Circumpolar stars as seen from a location with a geographic latitude of $\phi = +50°$ (about that of Frankfurt or Prague). The coordinate lines give the right ascension *RA* and the declination $\delta(+40°$ to $+90°)$. The indicator arrow (*above*) moves with the stars, and its extension passes through the Aries point; it shows the sidereal time on the outer dial. Precession: the celestial pole circles about the pole of the ecliptic ENP once every 25700 years. The location of the celestial north pole is indicated for several past and future dates

brightest stars of the Big Dipper until it is about five times longer. Continuing about the same distance past Polaris (which is the brightest star in the Little Dipper or Small Bear), we see the "W" of Cassiopeia. Using a sky globe or a star map, we can readily find the other constellations. In his "Uranometria Nova" (1603), J. Bayer named the stars in each constellation $\alpha, \beta, \gamma \ldots$, as a rule in the order of decreasing brightness. Besides these Greek letters, we also use the numbering system of the "Historia Coelestis Britannica" (1725), compiled by the first Astronomer Royal, J. Flamsteed. The Latin names of the constellations are usually abbreviated to 3 letters (Appendix A.2).

In addition, we define the following features on the *celestial sphere* (in mathematical terms, the infinitely distant sphere on which the stars seem to be projected (Fig. 2.2.2):

a) The *horizon* with the directions North, West, South, and East;
b) vertically above our position the *zenith*, directly under us the *nadir*;
c) the curve which passes through the zenith, the nadir, the celestial pole, and the north and south points is the *meridian*; and
d) the curve which is perpendicular to the meridian and the horizon, passing through the zenith and the east and west points, is the *prime vertical.*

In the coordinate system defined by these features, we denote the momentary position of a star by giving two angles (Fig. 2.2.2): (a) the *azimuth* is measured along the horizon in the direction SWNE, starting sometimes from the S- and sometimes from the N-point; (b) the *altitude* is 90° − the angle to the zenith.

The celestial sphere appears to rotate once each day around the *celestial axis* (passing through the celestial North and South Poles). The *celestial equatorial plane* is perpendicular to this axis. The *position* of a star (Fig. 2.2.3) at a given time on the celestial sphere, imagined to be infinitely distant, is also described by the *declination* δ, which is positive from the equator to the North Pole and negative from the equator to the South Pole, and by the *hour angle* t, which is measured from the meridian in the direction of the diurnal motion, i.e. towards W.

In the course of a day, a star therefore traces out a circle on the sphere (Fig. 2.2.3); its plane is parallel to the plane of the celestial equator. On the meridian, the

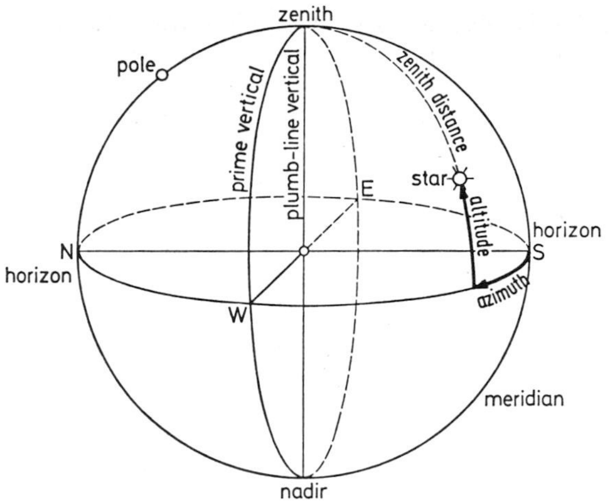

Fig. 2.2.2. The celestial sphere. The horizon with north, east, south, and west points. The (celestial) meridian passes through the north point, the (celestial) pole, zenith, south point, and the nadir. Coordinates: altitude and azimuth

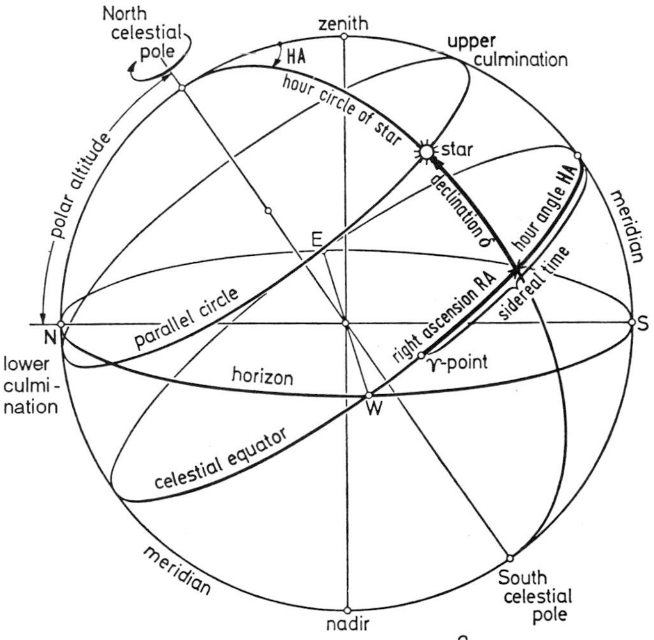

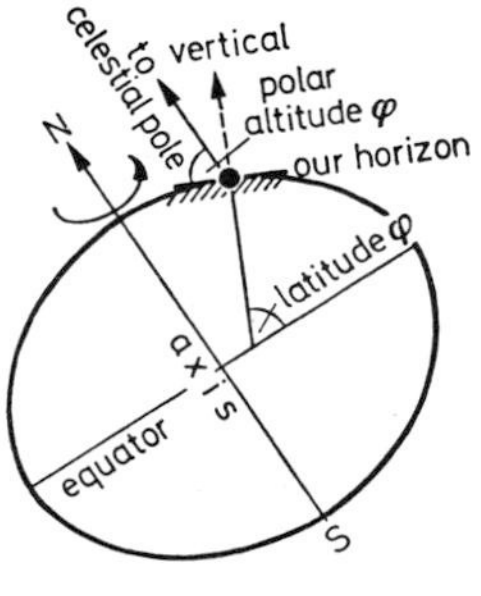

Fig. 2.2.3. Celestial coordinates: right ascension RA and declination δ. The hour angle t = sidereal time minus the right ascension RA. Lower right: the Earth (polar flattening exaggerated). Polar altitude = geographic latitude ϕ

greatest height reached by a star is its *upper culmination*, and the least height is its *lower culmination.*

We also mark the *Aries Point* ♈ on the celestial equator; we shall deal with it in the following section. It marks the point reached by the sun on the vernal equinox (March 21), on which the day and the night are equally long. The hour angle of the Aries point defines the *sidereal time* τ. If we now imagine the clock hand drawn in Fig. 2.2.1 (which intersects the celestial equator at the Aries point) to rotate with the stars, it will indicate the (sidereal) time on the clock dial around the outside of the drawing.

We are now in a position to determine the coordinates of a celestial object on the sphere independently of the time of day: we call the arc of the equator from the Aries point to the hour-circle of a star the *Right Ascension* RA of that star. It is quoted in hours, minutes, and seconds. 24 h (hora) correspond to 360°, or

$$1\,\mathrm{h} = 15^\circ \qquad 1^\circ = 4\,\mathrm{min}$$
$$1\,\mathrm{min} = 15' \qquad 1' = 4\,\mathrm{s}$$
$$1\,\mathrm{s} = 15'' \,.$$

From Fig. 2.2.3, one can readily read off the relation:

$$\text{Hour angle } t = \text{sidereal time } \tau \text{ minus right ascension RA} \,. \qquad (2.2.1)$$

The declination δ, our second stellar coordinate, has already been defined.

If we now wish to train a telescope on a particular star, planet, etc., we look up its right ascension RA and declination δ in a star catalog, read the time from a sidereal clock, and adjust the setting circles of the instrument to the angle hour t calculated from (2.2.1) and to the declination (+ north, − south). The particularly precisely determined positions of the so-called *fundamental stars* (especially for determinations of the time, see below) are to be found, along with those of the Sun, the Moon, the planets, etc. in the astronomical yearbooks or ephemerides; the most important of these is the Astronomical Almanac.

The Copernican system attributes the apparent rotation of the celestial sphere to the fact that the Earth rotates about its axis once every 24 h of sidereal time. The horizon is defined by a plane tangent to the Earth at the location of the observer; more precisely, by an infinite water surface at the observer's altitude. The zenith or vertical is the direction of a plumb-bob perpendicular to this plane, i.e. the direction of the local acceleration of gravity (including the centrifugal acceleration caused by the Earth's rotation). The *polar altitude* (the altitude of the celestial pole above the horizon) is given from Fig. 2.2.3 by the *geographic latitude* ϕ (the angle between the vertical and the Earth's equatorial plane); it can be readily measured as the average of the altitudes of the Pole Star or a circumpolar star at the upper and lower culminations.

The *geographic longitude* l corresponds to the hour angle. If the hour angle of the same object is measured *simultaneously* at Greenwich (zero meridian, $l_G = 0°$) and, for example, in New York, the difference gives the geographic longitude of New York, l_N. The determination of the latitude requires only a simple angle measurement, while that of a longitude necessitates a precise time measurement at two places. In earlier times, the "time markers" were taken from the motions of the Moon or of one of the moons of Jupiter. The introduction of the "seaworthy" chronometer by J. Harrison (ca. 1760–65) brought a great improvement, as did the later transmission of time signals by telegraph and still later by radio.

A few further facts: at a location having the latitude ϕ, a star of declination δ reaches an altitude of $h_{max} = 90° - |\phi - \delta|$ at its upper culmination and $h_{min} = -90° + |\phi + \delta|$ at its lower culmination. Stars with $\delta > 90° - \phi$ always remain above the horizon (circumpolar stars); those with $\delta < -(90° - \phi)$ never rise above the horizon.

In measuring stellar altitudes, we must take the refraction of light in the Earth's atmosphere into account. The apparent shift of a star (the apparent minus the true altitude) is termed the *refraction*. For average atmospheric temperature and pressure, the refraction Δh of a star at altitude h is summarized in the following table:

h =	0°	5°	10°	20°	40°	60°	90°
Δh =	34′50″	9′45″	5′16″	2′37″	1′09″	33″	0″ .

The refraction decreases slightly for increasing temperature and for decreasing atmospheric pressure, for example, in a low-pressure zone or in the mountains.

2.3 The Motions of the Earth. Seasons and the Zodiac. Time: The Day, the Year, and the Calendar

We now consider the *orbital motion* or *revolution* of the Earth around the Sun in the Copernican sense, and then the daily *rotation* of the Earth about its own axis, as well as the motions of the axis itself. We first place ourselves in the position of an observer in space. In Sect. 2.6, we shall derive Newton's theory of the motions of the Earth and the planets starting from his principles of mechanics and law of gravitation.

The apparent annual motion of the Sun in the sky was attributed by Copernicus to the revolution of the Earth around the Sun on a (nearly) circular orbit. The plane of the Earth's orbit intersects the celestial sphere as a great circle called the *ecliptic* (Fig. 2.3.1). This makes an angle of 23°27′ with the celestial equator, the *obliquity of the ecliptic*. This means that the Earth's axis retains its direction in space relative to the fixed stars during its annual revolution around the Sun; it forms an angle of 90°−23°27′ = 66°33′ with the Earth's orbital plane.

A brief summary will suffice to explain the *seasons* (Figs. 2.3.1, 2), starting with the Northern Hemisphere.

In the Northern Hemisphere, the Sun reaches its maximum altitude (midday altitude) at a geographical latitude ϕ on the 21st of June (the first day of summer or summer solstice), $h = 90° - |23°27' - \phi|$. On the 22nd of December (winter solstice), it has its lowest midday altitude, $h = 90° - \phi - 23°27'$. It can reach the zenith at latitudes up to $\phi = +23°27'$, the Tropic of Cancer. North of the Arctic Circle, $\phi \geq 90° - 23°27' = 66°33'$, the Sun remains below the horizon around the winter solstice; near the summer solstice, the "midnight sun" acts as a circumpolar star.

In the Southern Hemisphere, summer corresponds to winter in the Northern Hemisphere, the Tropic of Capricorn to the Tropic of Cancer, etc.

The *zodiac* is the term for a band in the sky on each side of the ecliptic. Since ancient times, it has been divided into 12 equal "signs of the zodiac" (Fig. 2.3.2).

It is often expedient for calculating the motions of the Earth and the planets to use a coordinate system oriented on the ecliptic and its poles. The (ecliptical) longitude is measured along the ecliptic starting from the Aries or ♈ point, like the right ascension in the direction of the annual motion of the Sun. The (ecliptical) latitude is measured, analogously to the declination, in a direction

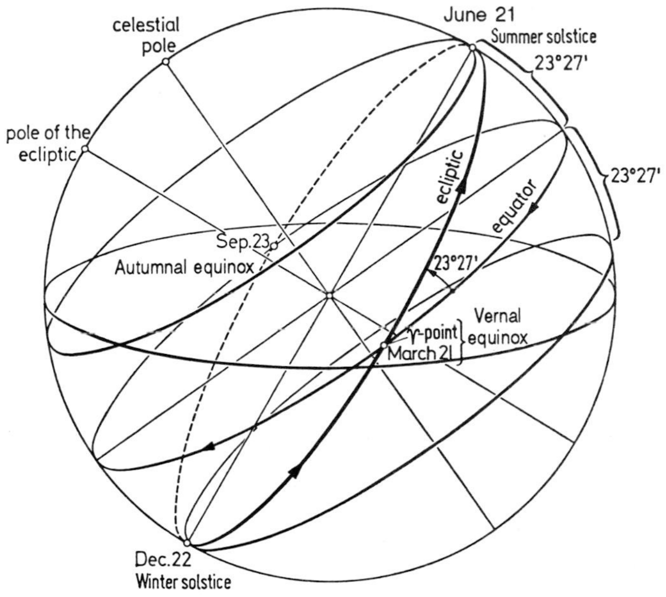

Fig. 2.3.1. Annual (apparent) motions of the Sun among the stars. The Ecliptic. The seasons

Starting date of Season	Name	Coordinates of the Sun: Right ascension RA [h]	Declination δ	The Sun enters the constellation
March 21 Spring	Vernal equinox [a]	0	0°	Aries
June 21 Summer	Summer solstice	6	+23°27′	Cancer
September 23 Autumn	Autumnal equinox [a]	12	0°	Libra
December 22 Winter	Winter solstice	18	−23°27′	Capricorn

[a] On these dates, the day and night arcs of the Sun are equal and each correspond to 12 hours

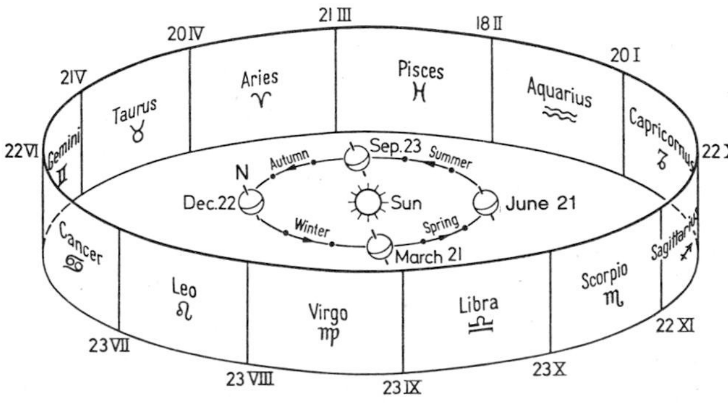

Fig. 2.3.2. The orbit of the Earth around the Sun. The seasons. The zodiac and the signs of the zodiac. The Earth is at perihelion (closest approach to the Sun) on the 2nd of January, and at aphelion (farthest distance from the Sun) on the 2nd of July

perpendicular to the ecliptic. The *ecliptical coordinates* in the sky must naturally not be confused with the similarly-named geographical coordinates!

The apparent annual motion of the Sun contains irregularities which were already known to ancient astronomers; they were recognized by J. Kepler as consequences of his first two laws of planetary motion, which we shall treat in more detail in Sect. 2.5.1:

Kepler's 1st Law: The planets move on elliptical orbits, with the Sun at one (common) focus;
Kepler's 2nd Law: The radius vector of a planet sweeps out equal areas in equal times;
Kepler's 3rd Law: The squares of the orbital periods of two planets are in the ratio of the cubes of their orbital semimajor axes.

The quantities needed for the geometric definition of the Earth's orbit or that of another planet around the Sun are shown in Fig. 2.3.3: we first note the semimajor axis a. The distance from a focus to the midpoint is denoted by $a \cdot e$, where the pure number e is called the eccentricity of the orbit. At the *perihelion*, the closest approach to the Sun, the distance of the Earth or other planet from the Sun is $r_{min} = a(1-e)$; at the *aphelion*, the point furthest from the Sun, $r_{max} = a(1+e)$. The diurnal motion of the Sun in the sky, i.e. the angle passed through by the radius vector of the Earth each day, is found from Kepler's 2nd law to obey $(r_{max}/r_{min})^2 = [(1+e)/(1-e)]^2$. The corresponding apparent diameters

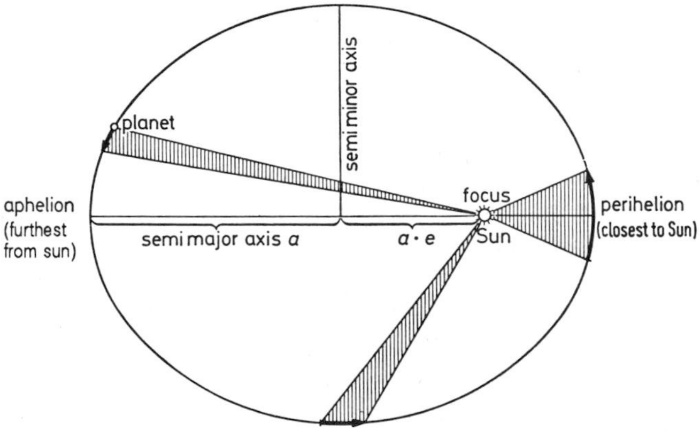

Fig. 2.3.3. The orbital ellipse of a planet. The semimajor axis is a, and the distance from the midpoint to a focus (the Sun) is $a \cdot e$, where e is the eccentricity. (The actual eccentricity of the orbits of the planets is much smaller than shown here)

of the Sun's disc have the ratio $(1+e)/(1-e)$. Measurements yield an eccentricity for the Earth's orbit of $e = 0.0167$. At present, the Earth passes through its perihelion about January 2nd. The approximate coincidence of this date with the beginning of the year is purely accidental.

Hipparchus had already discovered that the ♈ point is not fixed with respect to the celestial equator, but rather moves forward by about 50″ each year. This has led to the advancing of the ♈ point from the constellation Aries in ancient times to the constellation Pisces today. The *precession* of the equinoxes described is due to the fact that the celestial pole rotates about the fixed pole of the ecliptic on a circle of radius 23°27′ with a period of 25 700 years (Fig. 2.2.1); or, expressed differently: every 25 700 years, the Earth's axis of rotation traces out a cone with an opening angle of 23°27′, centered around the axis of the Earth's orbit.

This precession shifts the position relative to the stars of the celestial coordinate system in which we measure the right ascension RA (or α) and the declination δ; therefore, in quoting the positions of individual stars or in star catalogs, the *equinox* for which RA and δ are measured, must always be specified. The stellar positions also change because of the stars' proper motion (Sect. 5.2.2), so that the *epoch* of the observations must also be given, allowing them to be referred to the same time. Table 2.3.1 includes the various corrections which are to be applied to the RA (depending on the values of RA and δ) and to δ (depending only on RA) to take account of the precession during a 10-year period.

Superimposed on the precession, which has a period of 25 700 years, is a superficially similar motion having a period of 19 years, the *nutation*. Finally, the axis of the Earth's rotation fluctuates relative to the body axis of the Earth itself by about ±0.2″; analysis of this motion shows it to have a periodic part, with the so-called Chandler Period (433 days) as well as an annual contribution and an irregular part. The resulting fluctuations in the polar altitude are continuously checked at a series of observation stations. We shall return to the explanation of the various motions of the Earth's axis in Sect. 2.6.

For now, we continue examining the problem of the *measurement of time*. Our daily life depends to a considerable extent on the position of the Sun. Thus, the

$$\text{true solar time} = \text{hour angle of the Sun}$$

was first defined. This is the time indicated by a simple

Table 2.3.1. Precession for 10 years

a) Δ (RA) in minutes of time (+ = increase, − = decrease)

RA for northern objects [h]	6	7 5	8 4	9 3	10 2	11 1	12 0	13 23	14 22	15 21	16 20	17 19	18	
80°	+1.77	+1.73	+1.60	+1.40	+1.14	+0.84	+0.51	+0.19	−0.12	−0.38	−0.58	−0.70	−0.75	80°
70°	1.12	1.10	1.04	0.94	0.82	0.67	0.51	0.35	+0.21	+0.08	−0.02	−0.08	−0.10	70°
60°	0.898	0.885	0.846	0.785	0.705	0.612	0.512	0.412	+0.319	+0.240	+0.178	+0.140	+0.126	60°
50°	0.778	0.768	0.742	0.700	0.645	0.581	0.512	0.444	+0.380	+0.324	+0.282	+0.256	+0.247	50°
$\lvert\delta\rvert$ 40°	0.699	0.693	0.674	0.644	0.606	0.560	0.512	0.464	+0.419	+0.380	+0.350	+0.332	+0.335	40°
30°	0.641	0.636	0.624	0.603	0.576	0.546	0.512	0.479	+0.448	+0.421	+0.401	+0.388	+0.384	30°
20°	0.593	0.590	0.582	0.570	0.553	0.533	0.512	0.491	+0.472	+0.455	+0.442	+0.434	+0.431	20°
10°	0.552	0.550	0.546	0.540	0.532	0.522	0.512	0.502	+0.492	+0.484	+0.478	+0.476	+0.473	10°
0°	+0.512	+0.512	+0.512	+0.512	+0.512	+0.512	+0.512	+0.512	+0.512	+0.512	+0.512	+0.512	+0.512	0°
RA for southern objects [h]	18	19 17	20 16	21 15	22 14	23 13	0 12	1 11	2 10	3 9	4 8	5 7	6	

b) $\Delta\delta$ in minutes of arc (+ = increase in δ, in the southern sky thus decreasing absolute values $|\delta|$!)

RA [h]	0 24	1 23	2 22	3 21	4 20	5 19	6 18	7 17	8 16	9 15	10 14	11 13	12
$\Delta\delta$	+3.34′	+3.23′	+2.89′	+2.36′	+1.67′	+0.86′	0.0′	−0.86′	−1.67′	−2.36′	−2.89′	3.23′	−3.34′

sundial; 12:00 h corresponds to the upper culmination of the Sun. However, due to the nonuniformity of the Earth's orbital motion around the Sun (Kepler's 2nd law) and to the inclination of the ecliptic, this time varies throughout the year. The *mean solar time* was therefore introduced; it is based on a fictitious "mean Sun", which passes through the equator uniformly in the same time required by the true Sun in its passage around the ecliptic. The hour angle of this fictitious Sun defines the mean solar time. The difference

$$\text{true solar time} - \text{mean solar time} = \text{equation of time}$$

is thus composed of two contributions, which derive from the eccentricity of the Earth's orbit and from the inclination of the ecliptic, respectively. Their extrema are

	Feb. 12	May 14	July 26	Nov. 4
Equation of time	−14.3 min	+3.7 min	−6.4 min	+16.4 min .

The mean solar time is different for each meridian. As a simplification for transportation and communication, it has been agreed upon to use the local time of a *particular* meridian within a suitably defined "time zone", for example, Greenwich Mean Time (GMT) in England and Western Europe, Central European Time (CET) in Central Europe, Eastern Standard Time (EST) on the East Coast of the USA and Canada, etc.

For scientific purposes, for example, astronomical or geophysical measurements at stations which may be spread around the globe, the same time is used *everywhere*, namely

$$\text{world time or universal time (UT)} = \text{mean solar time at the Greenwich meridian} .$$

This time is measured in 24-hour days, starting with 0:00 h at midnight. For example, 12:00 h corresponds to 07:00 h EST or 13:00 h CET.

For astronomical observations, the relation between mean solar time and sidereal time is required. The "mean Sun" moves relative to the ♈ point by 360° or 24 h from west to east in the course of a year (365 d). The mean solar day is thus 24 h/365 d or 3 min 56 s longer than a sidereal day. A sidereal clock gains about 2 h per month relative to a "normal" UT clock. To make this clearer, we list the *sidereal time* for several dates at 0:00 h local time (midnight). This is given by the hour angle of the ♈ point or the right ascension of stars which cross the meridian at midnight (and are thus favorable for observation):

0:00 h local time	January 1	April 1	July 1	October 1
Sidereal time or RA at the meridian	6.7 h	12.6 h	18.6 h	0.6 h

The suitable unit for long periods of time is the *year.* We define:

one *sidereal year* = 365.25637 mean solar days

as the time between two passages of the Sun through the same point in the sky (from sidus: "star"); it is thus the true orbital period of the Earth.

One *tropical year* = 365.24220 mean solar days

is the time between two passages of the Sun through the Aries point ♈ or beginning of Spring (from τροπεῖν: "to turn"). Since this point moves 50.3″ to the west each year, the tropical year is correspondingly shorter than the sidereal year. The seasons and the calendar are calculated relative to the tropical year.

Since it is preferable for practical reasons that each year comprise an integer number of days, in everyday life we use

the *civil year*
$= 365.2425 = 365 + \frac{1}{4} - \frac{3}{400}$ mean solar days .

It corresponds to the prescription for leap years given by the *Gregorian calendar,* introduced in 1582 by Pope Gregory XIII. Every 3 years with 365 days are followed by a leap year with 366 days, except for the whole centuries which are *not* divisible by 400. We cannot discuss here the earlier Julian calendar introduced by Julius Caesar in 45 B.C., nor other problems of chronology which are interesting from a cultural-historical point of view.

To simplify chronological calculations dealing with long periods of time, and in particular for observations and ephemerides of variable stars, etc., it is desirable to avoid the variation in the lengths of years and months. Following a suggestion of J. Scaliger (1582), the *Julian days* are simply counted in an unbroken succession. Each Julian day begins at 12:00 h UT (mean Greenwich midday). The beginning of Julian day 0 was fixed at 12:00 h UT on January 1st in the year 4713 B.C. On January 1, 1990, the Julian day no. 2447893 began at 12:00 h UT.

Independently of the detailed definition, we use the symbols yr for "year" and d for "day".

Astronomical time reckoning was long based on the assumed uniformity of the Earth's rotation. The basic physical principle underlying terrestrial time measurements was already recognized by Ch. Huygens ("Horologium Oscillatorium", 1673): every clock consists of an oscillatory mechanism which is isolated as far as possible from its surroundings (a pendulum, clock movement, etc.) and held in motion by a driving mechanism (weight, spring, etc.) with the least possible feedback. The *pendulum chronometer,* steadily improved over the years, was for three centuries one of the most important instruments in every astronomical observatory. The *quartz clock*, considerably less susceptible to disturbances, consists of an oscillating piezoelectric quartz crystal, which is kept in motion by a loosely-coupled electrical oscillator circuit. The pinnacle of metrological precision, however, was attained in recent times by the *atomic clock*, which uses the oscillation frequency of cesium atoms (the isotope ^{133}Cs) in the vapor phase to measure time. The frequency corresponds to the transition between the two energetically lowest hyperfine levels. In simple terms, it is the frequency with which the nuclear spin precesses relative to the angular momentum vector of the rest of the atom. Excitation of the transitions, transformation to a lower frequency, and the display of the result are all performed electronically. The extreme accuracy of atomic clocks, which attain a relative accuracy of their oscillation frequency of about 10^{-14}, is the basis for various fundamental measurements and observations in physics and astronomy.

The comparison of astronomical observations with groups of quartz clocks and later with atomic clocks has shown that the rotational period of the Earth is not constant, but rather exhibits fluctuations of the order of a millisecond, in part with an annual period, in part aperiodic; these are related to changes in the mass distribution in the Earth.

The *second* was *defined* in 1967 as 1 s = the time required for 9192631770 oscillations of the radiation from the transition between the lowest two hyperfine structure levels of the ground state of ^{133}Cs; it is a base unit of the international system of units (SI).

The atomic-clock measurements of time by standards institutes are combined into International Atomic Time (TAI). From it, Coordinated Universal Time (UTC) is derived as an approximation to universal time as corrected only for the variations in the Earth's axis; the latter is not uniform and is termed UT1. It is a measure of the true rotation of the Earth about its axis and differs from

UTC by the correction

$$\Delta UT = UT1 - UTC \ ,$$

which is promulgated regularly by the Bureau International de l'Heure (BIH) and other similar institutions.

In everyday life, for navigation, geophysics, etc., the measurement of time is still necessarily based on the Earth's rotation; therefore, UTC is "switched" by one whole second ahead or back whenever the magnitude of Δ UT approaches 1 s.

Independently of the progress in the physical measurement of time it has been discovered that the motions of the planets and the Sun, or of the Earth and, in particular, the Moon, exhibit small, common variations over long periods of time relative to the ephemerides calculated according to the laws of Newtonian mechanics and gravitation theory. On the one hand, there is a secular (i.e., progressive) increase in the length of the day, which is caused by the braking effect of tidal friction (Sect. 2.6). Another contribution to the variations shows no such obvious origins. However, comparison of the variations for different objects forces us to attribute them to deviations of "astronomical time", which is based on the Earth's rotation, from "physical time", based on Newton's laws. Because of this empirical fact, it was decided in 1950 to base all astronomical ephemerides on a time scale derived from the basic laws of physics; the latter is termed *Ephemeris Time* (ET). The small corrections, Ephemeris Time minus Universal Time, are found for the most part by very accurate observations of the motion of the Moon. They can only be determined retrospectively; for most predictive purposes, they can be extrapolated with sufficient accuracy. In 1956, the ephemeris second was defined as the 31 556 925.9747th part of tropical year 1900.

Ten years later, it was decided to relate the ephemeris second to the atomic time unit. This, however, disturbs the internal consistency of the system of ET. In 1984, ephemeris time was replaced by *Dynamical Time* (TD). The questions of a complete amalgamation of the atomic time system, UTC, and dynamical time, and furthermore of taking the location of the measurement in the sense of general relativity theory (Sect. 4.12.9) into account (i.e., the influence of the gravitational potential) will require additional experiments and international negotiations.

2.4 The Moon. Eclipses of the Moon and the Sun

The *Moon* appears to us as a disc in the sky of mean diameter 31′; it is thus just the same apparent size as the Sun. Its distance from the Earth is small enough to be determined by triangulation from two widely-separated points on the ground (e.g., on the same meridian). Astronomers refer to the angle subtended by the equatorial radius of the Earth, seen from the Moon, as the *equatorial horizontal parallax* of the Moon. It has a mean value equal to 3422.6″. Since the Earth's radius is known to have the value 6378 km, one can calculate from these two numbers the average distance of the Moon from the center of the Earth:

$$60.3 \text{ Earth radii} = 384\,400 \text{ km}$$

and therefore the Moon's radius:

$$0.272 \text{ Earth radii} = 1738 \text{ km} \ .$$

We shall take up the physical structure of the Earth and the Moon in Sect. 2.8. First, we consider the Moon's orbit and its motions from the viewpoint of an observer.

The Moon orbits around the Earth, in the same direction as the Earth around the Sun, in one *sidereal month* = 27.32 d; that is, after that time it has returned to the same point in the heavens.

The origin of the *phases of the Moon* is illustrated in Fig. 2.4.1. Their period, the *synodic month* = 29.53 d

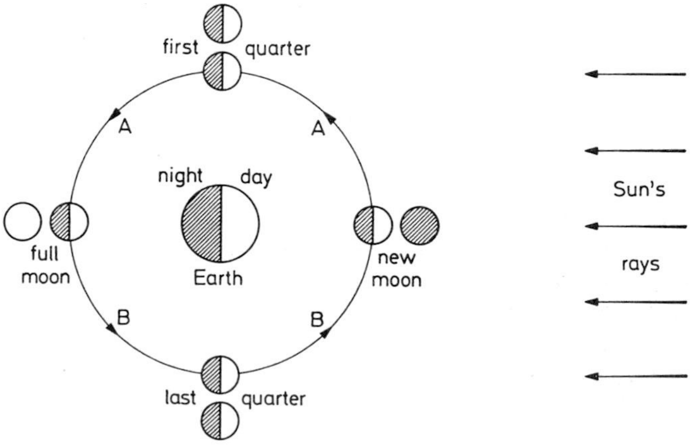

Fig. 2.4.1. The phases of the Moon; the Sun is at the right. The outer pictures indicate the Moon as seen from the Earth: waning Moon A, waxing Moon B

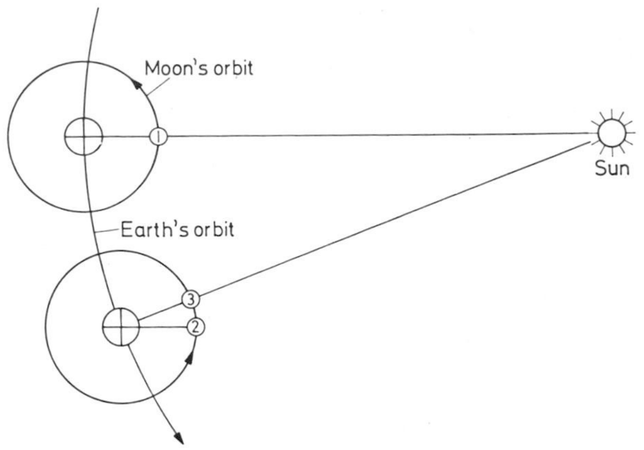

Fig. 2.4.2. The synodic month (1→3) is longer than the sidereal month (1→2), because the Earth moves onwards in its orbit in the meantime

(1→3 in Fig. 2.4.2), is the time after which the Moon returns to exactly the same position relative to the Sun, and is longer than the sidereal month (1→2 in Fig. 2.4.2). The Moon moves in an easterly direction relative to the Sun by 360°/29.53 = 12.2° each day, and relative to the stars, by 360°/27.32 = 13.2°.

The difference between the sidereal and synodic daily motion of the Moon is equal to the daily motion of the Sun, i.e. 360°/365 ≃ 1°. This becomes immediately clear if we consider that the daily motion is nothing other than the *angular velocity* in astronomical units. We could just as well write

$$\frac{1}{\text{sidereal month}} - \frac{1}{\text{sidereal year}} = \frac{1}{\text{synodic month}} .$$

More precisely, the *orbit of the Moon* around the Earth is an ellipse with eccentricity $e = 0.055$. The point in the orbit where the Moon is closest to the Earth (analogous to the perihelion in the Earth's orbit around the Sun) is called the *perigee*, and the most distant point is the *apogee*. The plane of the Moon's orbit is inclined relative to the Earth's orbit (the plane of the ecliptic) by an angle $i = 5.1°$. The Moon crosses over the ecliptic from the south to the north at ascending *nodes*, and passes "below" the ecliptic (for observers in the Northern Hemisphere!) at descending nodes.

As a result of the perturbation (gravitational attraction) caused by the Sun and the planets, the Moon's orbit also includes the following motions:

1) The perigee rotates "directly" around the Earth in the plane of the Moon's orbit, i.e. in the same sense as the revolution of the Earth around the Sun, with a period of 8.85 yr.
2) The nodes of the Moon's orbit, or the *line of nodes*, in which the orbit of the Moon crosses the Earth's orbital plane, has a retrograde motion in the ecliptic, i.e. in the opposite sense from the Earth's revolution, with a period of 18.61 yr, the so-called *period of nutation*.

This "regression of the lunar nodes" furthermore causes a corresponding "nodding" of the Earth by a maximum of 9″; this is the nutation of the Earth's axis mentioned previously.

The average time between two successive passages of the Moon through the same node is called the *draconitic month* = 27.2122 d. It is important for the prediction of eclipses (see below).

If we were to observe the orbits of the Moon and the Earth around the Sun from a spaceship, we would see, in agreement with a simple calculation, that the Moon's orbit is always concave as seen from the Sun (Fig. 2.4.3).

Let us now consider the *rotation* of the Moon and its other motions around its center of gravity. These can be determined very precisely by observing the motion of a sharply defined crater or similar feature on the lunar disc.

The fact that the Moon always shows us more or less the same face is because the lunar rotational period is equal to the orbital period, i.e. equal to a sidereal month. The two periods apparently were equalized by tidal interactions (Sect. 2.6.6) between the Moon and the Earth.

Careful observation shows, however, that the "face" of the Moon wobbles somewhat. The so-called *geometric librations* of the Moon have the following causes:

1) The equator and the orbital plane of the Moon form an angle of ≃6.7°; the latitudinal libration caused by this is equal to about ±6.7°.

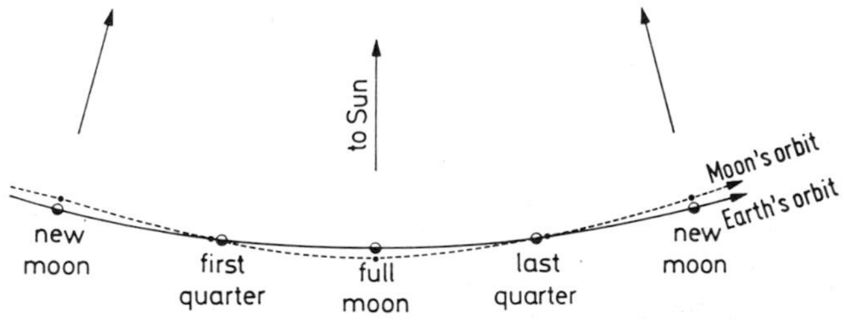

Fig. 2.4.3. The orbits of the Earth and Moon around the Sun

2) The rotation of the Moon is uniform (following Newton's law of inertia), but its revolution, from Kepler's second law taking the eccentricity of its orbit into account, is not; this causes a longitudinal libration of about $\pm 7.6°$.
3) The equatorial radius of the Earth appears to subtend an angle of 57′ from the Moon, the lunar horizontal parallax; the daily rotation of the Earth thus causes a diurnal libration.

Furthermore, there is the considerably smaller *physical libration*, which is due to the fact that the Moon is not quite spherical in shape and therefore performs small oscillations in the gravitational field (mainly that of the Earth).

All together, the librations have the effect that we can observe 59% of the Moon's surface from the Earth.

Having studied the motions of the Sun, the Earth, and the Moon, let us turn to the impressive spectacle of the lunar and solar eclipses!

A *lunar eclipse* occurs when the full Moon is covered by the shadow of the Earth. As with shadows on the Earth, we distinguish between the central part of the shadow, the umbra, and the surrounding half-shadow, the penumbra. If the Moon is completely in the umbra of the Earth, we speak of a total eclipse; if only a part of the Moon's disc is in the umbra, we have a partial eclipse. From the known geometrical facts we can calculate that a lunar eclipse can last for at most 3 h 40 min, while totality lasts for at most 1 h 40 min. Because the Sun's light is absorbed and scattered by the Earth's atmosphere more strongly at the blue end of the visible spectrum than at the red end, the outer edge of the penumbra on the Moon is not sharp and that of the umbra is also noticeably fuzzy. Furthermore, the penumbra and to a lesser extent the umbra seem to have a reddish-coppery color.

If the new Moon passes in front of the Sun, a *solar eclipse* occurs (Fig. 2.4.4). It can be partial or total. If the apparent diameter of the Moon is smaller than that of the Sun, we will observe only a ring-shaped eclipse when the Moon's shadow is centered on the Sun. In a partial eclipse, an observer on the Earth is in the penumbra of the Moon; in a total eclipse, the observer is in the umbra. In the case of a ring-shaped eclipse, the vertex of the Moon's shadow cone is between the Moon and the observer.

Total eclipses are particularly important for astrophysical observations of the outer layers of the Sun and the nearby interplanetary material; the bright sunlight is then completely blocked off outside the Earth's atmosphere.

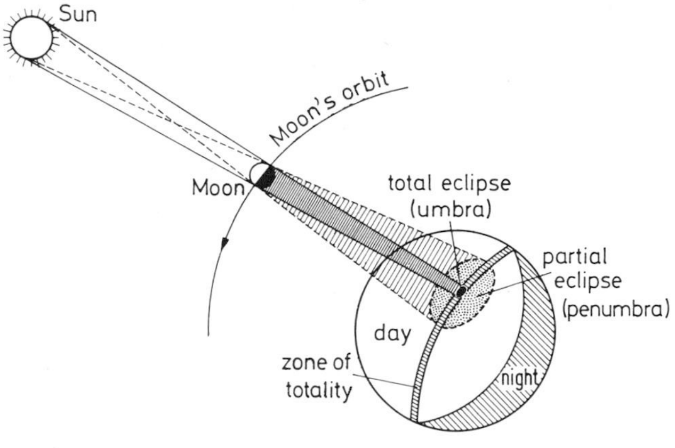

Fig. 2.4.4. An eclipse of the Sun (shown schematically). The Moon moves from W to E across the Sun's disk. In the umbra, a total eclipse is observed; in the penumbra, a partial eclipse

Relative to the Sun, the Moon moves through an angle of 0.51″ per second in the sky, in agreement with the length of the synodic month. This corresponds to a distance of 370 km on the Sun. Observations of eclipses carried out with good time resolution therefore yield an angular resolution which is generally better than that of available telescopes.

Stellar occultations by the Moon, which, like solar eclipses, must be predicted for each location in particular, are also very sharply defined in time, since the Moon has no atmosphere. They are important for checking the orbit of the Moon, for determining the fluctuations in the Earth's rotation, or for establishing the ephemeris time scale. Since the Moon moves 0.55″ per second relative to the stars, photometric measurements of these occultations with good time resolution may, in favorable cases, yield the angular diameters of the tiny stellar discs. Still more important for radio-astronomical observations with high angular resolution are occultations of radio-emitting objects by the Moon.

It was already known to the ancient eastern cultures that eclipses of the Sun and Moon (in the following, we shall refer for short simply to "eclipses") succeed each other with a period of about 18 yr 11 d, the so-called *Saros cycle*. This cycle is based upon the fact that an eclipse can occur only when the Sun *and* the Moon are rather close to a node in the lunar orbit. The time which the Sun requires to return to a particular lunar node is, due to the regression of the nodes, slightly less than a tropical year, namely 346.62 d; this time is called an *eclipse year*. As we may readily verify, the Saros cycle corresponds to an integral number of eclipse years:

223 synodic months = 6585.32 d

and

19 eclipse years = 6585.78 d ;

furthermore,

229 anomalous months = 6585.54 d
(from perigee to perigee, 27.555 d) .

Thus, an eclipse configuration indeed repeats itself with good accuracy after 18 yr 11.33 d. In one year, as one can show by considering the orbits of the Earth and the Moon, taking their diameters into account, there can be a maximum of 3 lunar eclipses and 5 solar eclipses. At a particular location, lunar eclipses, which can be seen from a whole hemisphere of the Earth, are relatively frequent; a total solar eclipse is, in contrast, very rare.

2.5 Orbital Motions and Distances in the Planetary System

The planets known since ancient times (with their time-honored symbols), Mercury ☿, Venus ♀, Mars ♂, Jupiter ♃, and Saturn ♄, have fascinated people again and again; their motions in the sky often appeared erratic but were found, step by step, to obey regular laws.

In Sect. 2.1, we briefly summarized the efforts made in ancient times to explain the motions of the planets. Here, we immediately adopt the *heliocentric* point of view, as developed by N. Copernicus in 1543. We shall furthermore drop the insistance on circular orbits which Copernicus had retained as a last vestige of Aristotelianism and make use of J. Kepler's elliptical orbits and his three laws of planetary motion (1609 and 1619). We thus place ourselves at the threshold of modern mathematical-physical thinking, which took on a clear form through the work of Galileo (1564–1642) and was consolidated into the beginnings of classical mechanics and gravitation theory in Newton's Principia (1687).

In Sect. 2.5.1, we describe the planets and their orbits and define the orbital parameters necessary to fully specify their motions. In Sect. 2.5.2 we then summarize the orbits of the comets and meteors. Finally, in Sect. 2.5.3, we discuss the determination of the Earth-Sun distance, the fundamental "Astronomical Unit" (AU), as well as the Doppler effect which results from the motion of the Earth, and the aberration of light.

We shall defer the discussion of the physical structure of the planets and other objects in the Solar System to Sects. 2.8, 9.

2.5.1 Planetary Motions and Orbital Elements

The origin of the *direct* (west–east) and of the *retrograde* (east–west) motions of the planets is explained in Fig. 2.5.1, taking Mars as an example.

Referring to Fig. 2.5.2, we first consider the motion around the Sun of an *inner* planet, e.g. Venus, as seen from our more slowly revolving Earth.

The planet is closest to us at its *lower conjunction*. It then moves away from the Sun in the sky and, as Morning Star, reaches its greatest westerly *elongation* of 48°. At the *upper* conjunction, Venus is at its greatest distance from the Earth and is closest to the Sun as seen in the sky. It then again moves away from the Sun and reaches its greatest easterly elongation of 48° as the Evening Star. The ratio of the orbital radii of Venus and the Earth is established by the maximum elongation of ±48° (for Mercury, ±28°). The *phases* of Venus, which are readily recognized in Fig. 2.5.2, and the corresponding changes in its apparent diameter (9.9″ to 64.5″) were immediately discovered by Galileo with his telescope; they prove that the Sun is also at the center of the true orbit of Venus. The planet reaches its maximum apparent brightness, as

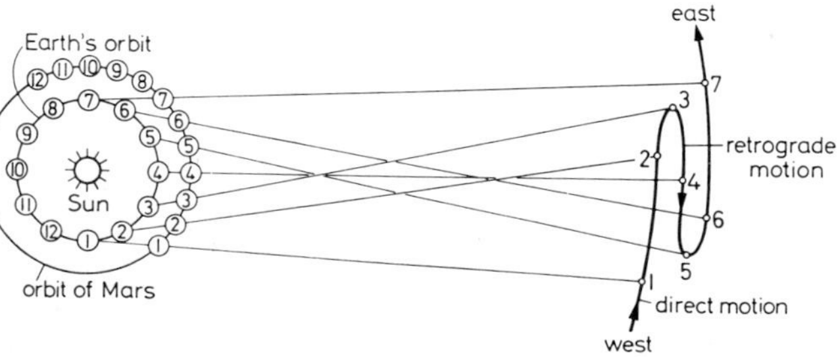

Fig. 2.5.1. Direct (west-east) and retrograde (east-west) motions of the planet Mars. The positions of the Earth and Mars on their orbits are numbered from month to month. At 4, Mars is in apposition to the Sun; it is overtaken here by the Earth and thereafter shows retrograde motion. At this time, it is closest to the Earth and most readily observed. The orbit of Mars is inclined relative to that of the Earth, i.e. the plane of the ecliptic, by 1.9°

can be seen from Fig. 2.5.2, in the neighborhood of its greatest elongations. In the lower conjunction, Venus (and Mercury) can pass in *front* of the Sun. These transits of Venus were previously of interest for the determination of the distance to the Sun or of the solar parallax.

An *outer planet*, for example Mars (Fig. 2.5.3), is nearest to us at its *opposition*; it then has its culmination at midnight true local time, has its largest apparent diameter, and is most favorable to observe. When it is near the Sun in the sky, it is said to be in *conjunction*. The outer planets do not go through the full cycle of *phases* from "full" to "new". The angle between the Earth and the Sun, as seen from the planet, is called the phase angle ϕ. The fraction of the planet's hemisphere which faces the Earth and is dark is thus $\phi/180°$. The phase angle of an outer planet passes through a maximum at the *quadratures*, i.e. when the planet and the Sun form an angle of 90° in the sky. The largest phase angle of Mars is 47°, that of Jupiter is only 12°.

The true time taken by a planet to revolve around the Sun is termed its *sidereal period*. The *synodic period* is the time it requires for a revolution in the sky relative to the Sun, i.e. the time between two successive, corresponding conjunctions. In analogy to the Moon, the following relations hold for the planets (subtraction of the angular velocities!):

$$\frac{1}{\text{synodic period}} = \left|\frac{1}{\text{sidereal period}} - \frac{1}{\text{sidereal period of the Earth}}\right| . \qquad (2.5.1)$$

For example, the sidereal period of Mars is found from the observed synodic period of 780 d and the length of the sidereal year (365 d), to be 687 d.

J. Kepler was the first to derive the true form of *Mars' orbit*, by combining pairs of observations of Mars which were taken at intervals equal to Mars' sidereal period, that is when the planet had returned to the same point on its orbit. Thus he could localize Mars from two points on the Earth's orbit separated by 687 d; since the latter was sufficiently well-known, he was able to trace out the true orbit of Mars. Two fortunate circumstances, namely that the conic sections had been thoroughly investigated by Apollonius of Pergae; and that of the then-known planets (except Mercury, which is difficult to observe), Mars has the greatest orbital eccentricity, $e = 0.093$, made it possible for Kepler to arrive at his first two laws of planetary motion. He discovered the third law only after 10 further years, guided by the unshakeable conviction

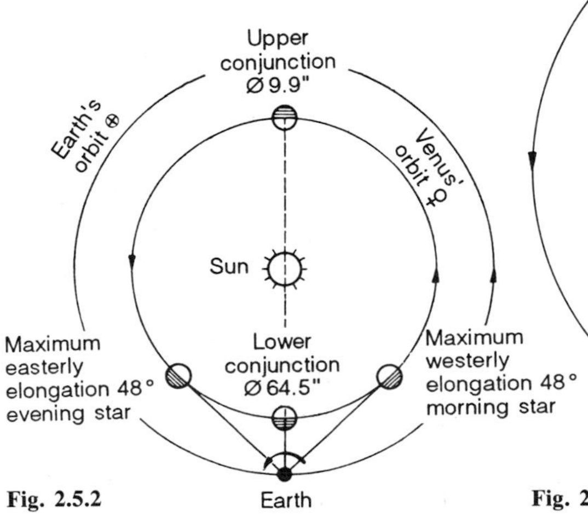

Fig. 2.5.2

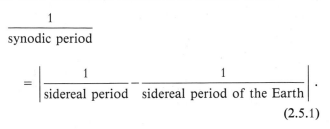

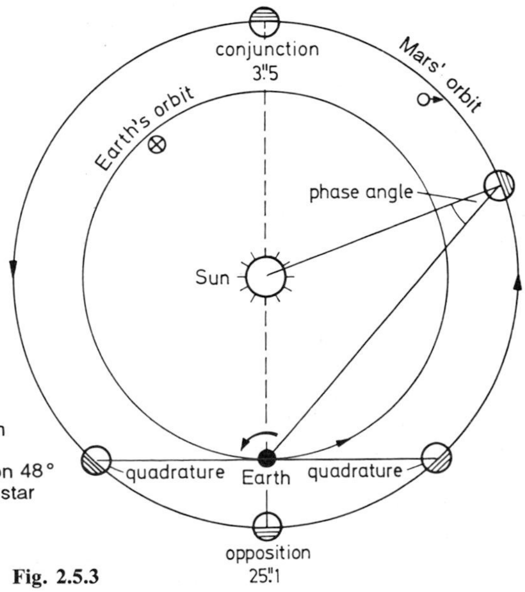

Fig. 2.5.3

Fig. 2.5.2. The orbit and phases of Venus, an inner planet. The elongation of Venus in the sky cannot exceed ±48° (Mercury, ±28°). The phases are similar to those of the Moon. The maximum brightness occurs near the maximum elongation

Fig. 2.5.3. The orbit and phases of Mars, an outer planet. The maximum brightness and largest angular diameter of 25.1″ occur when it is in opposition

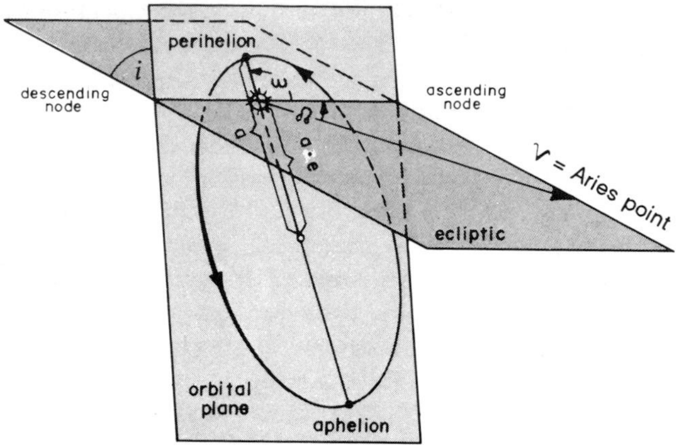

Fig. 2.5.4. The orbital elements of a planet or comet

that a "universal harmony" must somehow express itself in the orbits of the planets.

The complete description of the orbit of a planet or comet (Sect. 2.5.2) around the Sun requires the *orbital elements* defined in Fig. 2.5.4:

1) The semimajor axis a. It is measured either in terms of the semimajor axis of the Earth's orbit, 1 *Astronomical Unit* (AU), or in kilometers.
2) The eccentricity e [distance at perihelion: $a(1-e)$; distance at aphelion: $a(1+e)$].
3) The inclination i of the orbital plane relative to the ecliptic.
4) The length of the ascending node ☊ (angle from the Aries point ♈ to the ascending node).
5) The distance ω of the perihelion from the node (angle from the ascending node to the perihelion). The sum of the two angles ☊ and ω, where the first is measured in the ecliptic and the second in the orbital plane, is called the perihelion length, $\tilde{\omega}$.
6) The period P (sidereal period, measured in tropical years) or the mean daily motion n (in degrees or arc seconds per day).
7) The epoch E or the time of the passage through the perihelion T.

The orbital elements a and e determine the size and shape of the orbit (Fig. 2.3.3), i and ☊ fix the orbital plane, and ω determines the position of the orbit within the plane. The motion along the orbit depends upon P and T; the period P is, in fact, determined by the value of a, from Kepler's third law, apart from small corrections (see below).

Table 2.5.1 contains the orbital elements of the planets which are of interest to us here. In addition to the planets known in ancient times, we have included Ceres, the brightest of the asteroids or planetoids, as well as Uranus, Neptune, and Pluto.

Of these last planets, *Uranus* was discovered fortuitously in 1781 by W. Herschel. J. Kepler had already suspected the presence of an object in the gap between the orbits of Mars and Jupiter (Fig. 2.5.5). The first asteroid, *Ceres*, was discovered in this region on January 1st, 1801 by G. Piazzi in Palermo; however, it was "lost" again in mid-February as it neared the Sun. By October of the same year, the 24-year-old C.F. Gauss had calculated its orbit and ephemerides, so that it could be

Table 2.5.1. Some orbital elements of the planets (Epoch 1980.0)

Name		Symbol	Sidereal period [yr]	Semimajor axis of orbit [AU]	Semimajor axis of orbit [10^6 km]	Eccentricity e	Inclination i to ecliptic	Average orbital velocity v [km s^{-1}]
inner planets	Mercury	☿	0.241	0.387	57.9	0.206	7.0°	47.9
	Venus	♀	0.615	0.723	108.2	0.007	3.4°	35.0
	Earth	⊕	1.000	1.000	149.6	0.017	–	29.8
outer planets	Mars	♂	1.881	1.524	227.9	0.093	1.8°	24.1
	Minor planets:	Ceres	4.601	2.766	413.5	0.077	10.6°	17.9
	Jupiter	♃	11.87	5.205	779	0.048	1.3°	13.1
	Saturn	♄	29.63	9.576	1432	0.056	2.5°	9.6
	Uranus	♅	84.67	19.28	2884	0.050	0.8°	6.8
	Neptune	♆	165.5	30.14	4509	0.008	1.8°	5.4
	Pluto	♇	251.9	39.88	5966	0.253	17.1°	4.7

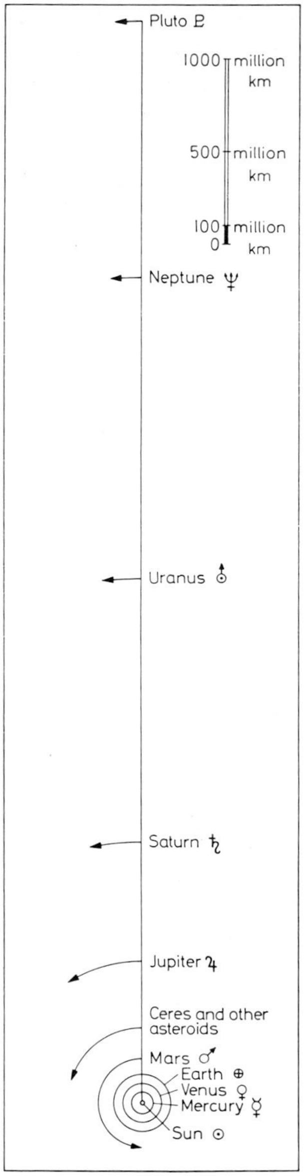

Fig. 2.5.5. Mean orbital radii of the planets (semimajor axes a). The arcs correspond to the average motion per year; in one year, Venus circles the Sun 1.62 times and Mercury circles it 4.15 times

re-discovered by F. Zach. Following this mathematical achievement, C. F. Gauss in 1809 solved the general problem of orbit determination, i.e. the derivation of all the orbital elements of a planet, a comet, etc. from three complete observations. Today, several thousand asteroids are known, for the most part between the orbits of Mars and Jupiter (Sect. 2.8.6).

Perturbations of the orbit of Uranus led J. C. Adams and J. J. Leverrier to postulate the existence of a planet with a still longer orbital period; they calculated its orbit and ephemerides. J. G. Galle discovered *Neptune* close to the calculated position in 1846.

Similarly, perturbations of the orbits of Uranus and Neptune led to the postulate of a transneptunian planet. The search for this planet, which went on for many years and in which P. Lowell († 1916) played a major role, was finally successful in 1930, when C. Tombaugh at the Lowell Observatory discovered *Pluto* as a faint star of magnitude 14.9. Its orbital parameters, with an eccentricity $e = 0.25$ and an inclination of 17.1°, are quite different from those of the other planets. Pluto spends some time within the orbit of Neptune, and its orbital period is in "resonance" (3:2) with that of Neptune. Pluto reached perihelion in 1989.

2.5.2 Comets and Meteors

Our planetary system also includes the *comets* and *meteors* ("shooting stars").

The *comets* (Fig. 2.9.1) are named for the year of their perihelion and numbered in the order of their passage through perihelion in that year; the name of the discoverer is often included, too. (A provisional name is given according to the year of discovery, with lower-case letters indicating the order of discovery in that year). In ancient times and in the Middle Ages, the comets were relegated to the Earth's upper atmosphere, in accordance with the doctrine of the immutability of the heavenly regions. The first proof that this doctrine was incorrect, was given by Tycho Brahe's precise observations of the comets of 1577 and 1585, from which he derived parallaxes that showed that the comet of 1577, for example, must be at least six times more distant from the Earth than the Moon. Newton recognized that the comets move on elongated ellipses or parabolas around the Sun, i.e. on conic sections with eccentricities equal to or a little smaller than 1. His contemporary E. Halley improved the method of determining their orbits and in 1705 was able to show that *Halley's comet* of 1682, which bears his name, must have a period of 76 yr. From Kepler's third law, the semimajor axis of its orbit is equal to $2 \cdot 76^{2/3} = 36$ AU, i.e. its aphelion lies somewhere outside the orbit of Neptune. Halley's orbit calculation also demonstrated that the bright comet of 1682 was identical with those of 1531 and 1607, and he could predict its return in 1758. All together, 29 appearances of Halley's comet have been witnessed since the year 240 B.C. The most recent perihelion of Halley's comet took place on February 9th, 1986.

In searching for comets, observations in the *far infrared*, which are sensitive to the thermal radiation emitted by the comet's dust, can be very useful. Thus, the first infrared observation satellite, IRAS, was able to discover six new comets in the single year 1983, while they were still extremely faint in the visible region.

The *orbits* of the roughly 1300 known comets fall into two groups:

a) *Comets of long orbital period* with periods of revolution between 10^2 and 10^6 years, and perihelia in the range of 1 AU (high probability of discovery): the inclinations i of their orbits are randomly distributed; direct and retrograde motions are about equally probable. The eccentricities e are slightly smaller than or nearly equal to 1, so that their orbits are long, thin ellipses or, as a limiting case, parabolas. Hyperbolic orbits, with $e > 1$, are only rarely produced by perturbations from the major planets. Since the velocities v of these comets are quite small at large distances from the Sun, it is likely that they originate from a cloud which accompanies the Sun on its path in the Milky Way. It is estimated that this "Oort Cloud" (with a diameter of about 50000 AU) contains 10^{12} comets, whose total mass is however equal to only about 10 times the mass of the Earth.

b) *Comets of short orbital period*, with periods less than 200 years (denoted by the letter P and the name of the discoverer): they move for the most part in elliptical orbits with small inclinations i (mean inclination of this group $\bar{i} \simeq 20°$). Nearly half of these comets have their aphelia in the range 5–6 AU, i.e. in the neighborhood of Jupiter's orbit ($a_{♃} \simeq 5.2$ AU). The mean values of the orbital data correspond to $a \simeq 3.6$ AU and $e \simeq 0.56$. This "Jupiter family" of comets evidently arose through capture of long-period comets by the planet Jupiter (similar comet families associated with other planets have not been identified with certainty). Comets decay in time by breaking up and by vaporization of cometary matter, so that the swarm of short-period comets must be constantly replenished by capture of new objects.

The shortest known orbital period is that of Encke's comet (1789 I). The comets Schwassmann-Wachmann 1925 II ($a = 6.1$ AU, $e = 0.11$) and Oterma 1942 VII ($a = 4.0$ AU, $e = 0.14$) have nearly circular orbits.

The comet IRAS-Araki-Alcock (1983 d) is worthy of mention; it was discovered using the IRAS satellite on the basis of its far-infrared radiation and passed the Earth at a distance of only 0.032 AU, i.e. about 12 times the distance to the Moon.

The showers of "*falling stars*" or *meteors* which appear to emanate from a particular point in the sky (their "radiant", similar to a vanishing point in perspective drawings) on certain days each year are, as indicated e. g. by their periodicity, simply the debris of comets, whose orbits crossed or nearly crossed that of the Earth. In some cases, the cometary matter seems to be fairly well concentrated along the orbit, so that especially lively meteor showers are observed with the corresponding period: an example is the famous case of the Leonids (radiant *RA* 152°, δ +22°), which were observed by Alexander v. Humboldt from South America in 1799, and can be attributed to the comet 1866 I, with a 33 year period. In addition, there are sporadic meteors which show no recognizable periodicity. The fact that "falling stars" are really small objects from space, which enter the Earth's atmosphere and are heated to incandescence by it, was first demonstrated in 1798 by two students in Göttingen, Brandes and Benzenberg. They made observations of meteors from two sufficiently distant points and calculated the altitude of their tracks. Earlier, in 1794, E. F. F. Chladni had shown that *meteorites* are just meteoric material (from larger meteors) which has reached the Earth's surface.

Comets or meteors whose orbits were originally *hyperbolic*, i.e. objects which have entered the Solar System from outer space, have not been observed.

We shall return to the physical structure of the planets, their atmospheres and their moons, as well as that of comets, meteors, and meteorites in Sects. 2.8.9.

2.5.3 Distance Determination. The Doppler Effect and Aberration of Light

We need to turn to the important question of how the *distance* from the Earth to the Sun (or, more precisely, the semimajor axis of the Earth's orbit, which we have defined as the astronomical unit, 1 AU) can be measured in established units such as kilometers. Astronomers prefer to refer to the *solar parallax* $\pi_{\odot}$, i.e. the angle which the equatorial radius of the Earth, R_E (known from geodetic measurements to be 6378 km) subtends when seen from the center of the Sun. The solar parallax is too small for direct measurement, as can be done in the case of the lunar parallax. Therefore, the first step is to determine the distance to a planet or asteroid whose orbit brings it sufficiently near to the Earth, by making observations from

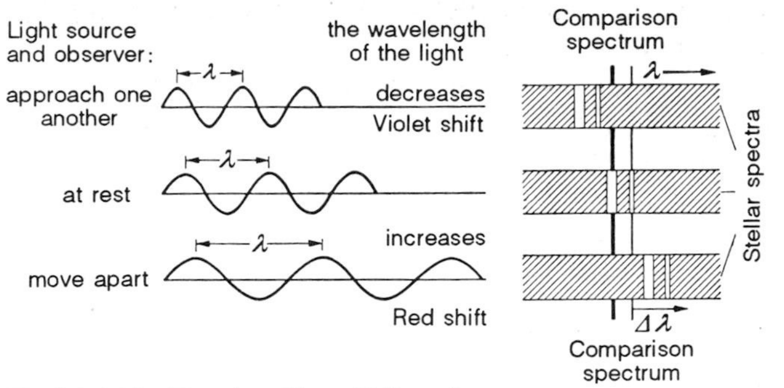

Fig. 2.5.6. The Doppler effect, $\Delta\lambda/\lambda = v/c$

several observatories in both the Northern and Southern Hemispheres. In former times, Mars at opposition or Venus at its lower conjunction were used; more recently, extensive series of observations of the oppositions of the asteroid Eros, which is more favorable for such determinations, have been carried out. These astronomical methods of distance determination have recently acquired serious competition from *radar techniques*, which allow the precise determination not only of the distance to the Moon, but also that to Venus and Mars by measurement of the round-trip travel time of reflected radiofrequency signals, combined with the velocity of light c which is known from terrestrial experiments. Combining the individual measurements allows the calculation of the radius of the Earth's orbit using Kepler's 3rd law; the details involve difficult celestial-mechanical calculations.

Instead of the radius of the Earth's orbit, the orbital *velocity* of the Earth can be determined in km/s by using the *Doppler effect* (Fig. 2.5.6): when a radiation (light) source moves relative to an observer with a radial velocity (the velocity component in the direction of the line of sight between source and observer) v, the wavelength λ of the radiation (or its frequency $\nu = c/\lambda$) appears to be shifted by

$$\Delta\lambda = \frac{\lambda v}{c} \quad \text{or} \quad \Delta\nu = \frac{\nu v}{c} . \tag{2.5.2}$$

A source whose relative motion is away from the observer (by definition a positive velocity) produces an increase in the wavelength λ, i.e. a red shift of the spectral lines and a reduction in the frequency ν, and *vice versa*.

In practice, either the radial velocity of a fixed star relative to the Earth is followed over most of a year, using the Doppler shift in its spectral lines, or else the relative velocity between the Earth and, e.g., Venus is determined from the frequency shift of reflected radar signals (reflection from a moving mirror yields twice the frequency shift quoted above).

A similar consideration formed the basis of the historically important first measurement of the velocity of light by O. Römer in 1675: he determined the *frequencies of revolution* of the larger moons of Jupiter from their transits behind the planet's disc. When the Earth is moving away from Jupiter, these frequencies appear to be reduced, due to the finite propagation velocity c of light; when it is moving towards the planet, the frequencies are apparently increased. Starting from the contemporary value of the solar parallax, Römer obtained a relatively good numerical value for the velocity of light. The fact that he anticipated Doppler's principle (2.5.2) by nearly two hundred years before its first spectroscopic application is seldom mentioned in texts on astronomy and physics.

Another effect which is due to the finite velocity of light is the *aberration* of light (Fig. 2.5.7), which was

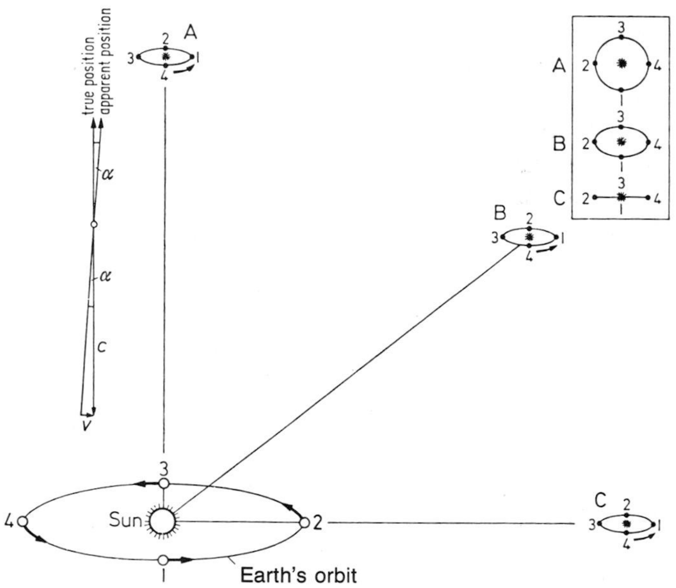

Fig. 2.5.7. The aberration of light. The light from the stars seems to be deflected in the direction of the velocity vector of the Earth (*left*) by an angle v/c, where v is the component of the Earth's velocity perpendicular to the light propagation direction and c is the velocity of light. A star at the pole thus describes a circle of radius equal to the ratio (Earth's velocity/velocity of light): $\alpha = 20.49''$; in the ecliptic, it moves on a line of maximum extension $\pm\alpha$; and in between, it describes an ellipse (*indicated at the upper right*). ($*$) shows the true position of the star; an observer looking to the right sees the star at intervals of $\frac{1}{4}$ year in the positions 1-2-3-4

discovered by J. Bradley in 1728 while he was attempting to measure the parallax of fixed stars. When a star is observed which would be perpendicular to the Earth's orbit for an observer at rest, the Earth's (and the observer's) orbital motion makes it necessary to incline the telescope by a small angle in the direction of the Earth's orbital velocity v in order to see the star; this aberration angle is v/c. As a result, a star at the celestial pole appears to describe a small circle in the course of a year (Fig. 2.5.7, upper right); in the ecliptic it seems to move back and forth on a straight line, and in between, it moves on an ellipse. The usual analogy is that of an astronomer who is walking rapidly through rain which is coming straight down, and must incline his umbrella forwards in order to stay dry.

This demonstration, like our elementary derivation of the Doppler effect, is imperfect; it neglects the principle of the constancy of the velocity of light in all frames of motion, independently of the motion of the source, which was demonstrated by the Michelson-Morley experiment. A consistent explanation of all the effects of order v/c in that experiment, and especially of those of order $(v/c)^2$, is given by Einstein's special relativity theory (1905).

Finally, we give a summary of the numerical values of the quantities discussed and their various relationships:

Earth's equatorial radius	$R_E = 6.378 \cdot 10^6$ m
Solar parallax (equatorial horizontal parallax)	$\pi_\odot = 8.794''$
Astronomical unit [AU] = semimajor axis of the Earth's orbit	$A = R_E/\pi_\odot$ $= 1.496 \cdot 10^{11}$ m
Velocity of light	$c = 2.9979 \cdot 10^8$ m/s
Light propagation time for 1 AU	$\tau_A = A/c = 499.0$ s
Mean orbital velocity of the Earth	$\bar{v} = 2.98 \cdot 10^4$ m/s
Rotational velocity of the Earth at the equator	$v_E = 465$ m/s
Aberration constant (Epoch 2000)	$\kappa = \bar{v}/c = 9.94 \cdot 10^{-5}$ $\triangleq 20.496''$

2.6 Mechanics and the Theory of Gravitation

Following the tedious and even dangerous beginnings made by G. Galilei and J. Kepler, I. Newton in his *Principia* (1687) gave the first complete treatment of the mechanics of terrestrial and extraterrestrial systems. Combining it with his law of gravitation, in the same work he derived Kepler's laws and many other observed regularities in the motions of objects in the Solar System. It is not surprising that the further development of celestial mechanics remained an important field for the great mathematicians and astronomers for nearly two hundred years.

We begin our treatment here by stating the most important concepts and laws of *mechanics* and the *theory of gravitation* for later reference: Newton's laws of motion (Sect. 2.6.1), conservation of momentum (Sect. 2.6.2), conservation of angular momentum (Sect. 2.6.3), conservation of energy and the virial theorem (Sect. 2.6.4), and finally Newton's law of universal gravitation, from which we immediately derive Kepler's laws (Sect. 2.6.5). We then treat the origin of tides (Sect. 2.6.6) and make the connection, from the modern point of view, between the Ptolemaic and the Copernican world-systems. The application of celestial mechanics to artificial satellites and space probes will be treated in Sect. 2.7.

2.6.1 Newton's Laws of Motion

We formulate Newton's three *basic laws of mechanics* in modern language:

I. A body remains in a state of rest or moves with constant velocity in a straight line as long as it is not subject to an external force (law of inertia).

We denote *velocities* in direction and magnitude by a vector (arrow)[1] $\boldsymbol{v}$; in the same way, we represent the *force* $\boldsymbol{F}$. For the addition and subtraction of these quantities, we use the vector parallelogram: the two vectors are represented by their components in a rectilinear coordinate system x, y, z, i.e. by their projections on the corresponding axes; thus, for example, $\boldsymbol{v} = \{v_x, v_y, v_z\}$. . . . If a moving body has the mass m, the vector mass times velocity

$$\boldsymbol{p} = m\boldsymbol{v} \tag{2.6.1}$$

is defined as its momentum. This important concept allows the formulation of the *law of momentum*:

II. The rate of change with time of the momentum of a body is proportional to the magnitude of the external force which acts on it, and is in the direction of that force.

[1] Vectors are denoted in print by boldface characters.

Mathematically formulated, we write for *one* body (t being the time):

$$\frac{d\boldsymbol{p}}{dt} = \frac{d}{dt}(m\boldsymbol{v}) = \boldsymbol{F} \ . \tag{2.6.2}$$

Law I is clearly just a special case for $\boldsymbol{F} = 0$ of law II. We can interpret the velocity $\boldsymbol{v}$ as the rate of change of the position vector $\boldsymbol{r}$ with components in x, y, and z, and we thus write $\boldsymbol{v} = d\boldsymbol{r}/dt$ and, for m = constant, also

$$m\frac{d^2\boldsymbol{r}}{dt^2} = \boldsymbol{F} \ . \tag{2.6.3}$$

This formulation (force = mass×acceleration) is however only valid for constant masses, while (2.6.2) also remains valid within special relativity theory, where the mass depends on the velocity through the formula $m = m_0/\sqrt{1-(v^2/c^2)}$ ($m_0 = m_{v=0}$ being termed the rest mass). If we consider N objects, denoted by indices $k = 1, 2, 3 \ldots N$, then (2.6.2) corresponds to the N vector equations or $3N$ component equations:

$$\frac{d\boldsymbol{p}_k}{dt} = \frac{d}{dt}(m_k\boldsymbol{v}_k) = \boldsymbol{F}_k \ . \tag{2.6.4}$$

Newton's final law deals with the interactions of two bodies; it states that:

III. The forces which two bodies exert on one another have equal magnitudes and opposite directions.

If $\boldsymbol{F}_{ik}$ is the force which body i exerts on body k, we thus have

$$\boldsymbol{F}_{ik} = -\boldsymbol{F}_{ki} \ , \tag{2.6.5}$$

the law of *action and reaction.*

As a simple example of Newton's laws of motion, we consider a mass m (Fig. 2.6.1) which moves at the end of a string of length r in a horizontal circle with the constant velocity $v = |\boldsymbol{v}|$. (The magnitude of a vector quantity, i.e. the "length of the arrow", is generally denoted by absolute value signs, e.g. $|\boldsymbol{v}|$, or the corresponding Roman letter v). The angular velocity $d\phi/dt$ (angle ϕ in radians) is then equal to v/r. If we trace successive velocity vectors $\boldsymbol{v}$ using the same starting point, thus drawing a so-called hodograph, we can immediately see that the acceleration $|d\boldsymbol{v}/dt|$ is equal to $(v/r)v$ and points towards the center of the circular path. We thus obtain the law of *centrifugal*

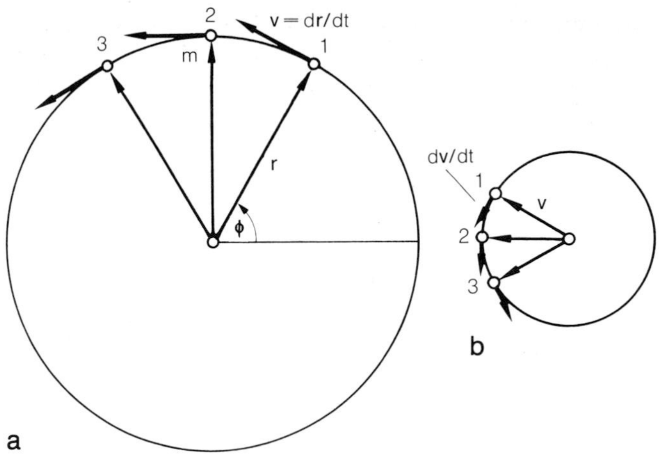

Fig. 2.6.1 a,b. Calculation of centrifugal force. (**a**) A point mass m on a circular orbit, with the position vector $\boldsymbol{r}$ at the times $1, 2, 3, \ldots$ and the velocity vector $\boldsymbol{v} = d\boldsymbol{r}/dt$ tangential to the orbit. The magnitude of $\boldsymbol{v}$ is $v = r \cdot d\phi/dt$. (**b**) Hodograph. The velocity vector $\boldsymbol{v}$ at the times $1, 2, 3, \ldots$. The acceleration vector $d\boldsymbol{v}/dt$ is in the direction of the tangent to the hodograph and is therefore parallel to $-\boldsymbol{r}$. The magnitude of the acceleration is $v \cdot d\phi/dt = v^2/r$

force which was derived by C. Huygens even before Newton:

$$F = mv^2/r \ . \tag{2.6.6}$$

Newton's 3rd law states that the string pulls on its anchor at the center with the *same* force that pulls the object towards the center of its circular orbit.

When the details of the internal structure of a sufficiently small object of mass m are not important in mechanics, we speak of a *point mass*. In the theory of planetary motion, for example, we can consider the Earth to be a point mass.

From Newton's three laws for the motion of individual point masses, we go on to the equations of motion for a *system* of point masses. From them, we derive the three conservation laws of mechanics, which we shall often use later in this book.

2.6.2 The Conservation of Linear Momentum

In a system of point masses m_k, which we denumerate by the indices i or k ($i, k = 1, 2, 3 \ldots N$), we distinguish between internal forces $\boldsymbol{F}_{ik}$ which, for example, mass i exerts on mass k, and external forces $\boldsymbol{F}_k^{(e)}$, which are exerted on the mass k from "outside". This point mass obeys the equation of motion:

$$\frac{d\boldsymbol{p}_k}{dt} = \boldsymbol{F}_k^{(e)} + \sum_{i=1}^{N} \boldsymbol{F}_{ik} \; . \qquad (2.6.7)$$

Summing[2] over all values of k, we obtain with (2.6.5)

$$\frac{d}{dt} \sum \boldsymbol{p}_k = \sum \boldsymbol{F}_k^{(e)} \; .$$

If we consider the N masses as *one* system, its total momentum is

$$\boldsymbol{P} = \sum \boldsymbol{p}_k \qquad (2.6.8\,\mathrm{a})$$

and the total external force on the system is

$$\boldsymbol{F} = \sum \boldsymbol{F}_k^{(e)} \; . \qquad (2.6.8\,\mathrm{b})$$

The *equation of motion* then becomes, analogously to that for a single point mass,

$$\frac{d}{dt}\boldsymbol{P} = \boldsymbol{F} \; . \qquad (2.6.9)$$

When no external forces act ($\boldsymbol{F} = 0$), the system exhibits the *law of conservation of momentum*:

$$\boldsymbol{P} = \sum \boldsymbol{p}_k = \text{constant} \; . \qquad (2.6.10)$$

The meaning of (2.6.9,10) can perhaps be more intuitively clarified if we define the center of gravity S of our system with the overall mass

$$M = \sum m_k \qquad (2.6.11)$$

by its position vector $\boldsymbol{R}$ according to the equation

$$M\boldsymbol{R} = \sum m_k \boldsymbol{r}_k \; . \qquad (2.6.12)$$

With this, (2.6.9) is converted into the equation of motion of the center of gravity,

$$M\frac{d^2\boldsymbol{R}}{dt^2} = \boldsymbol{F} \; , \qquad (2.6.13)$$

in analogy to that of a single point mass. We can see from this last equation that in the case of no net external force, $\boldsymbol{F} = 0$, the center of gravity must exhibit (following Newton's 1st law) a straight-line inertial motion with a constant velocity, $d\boldsymbol{R}/dt$ = constant.

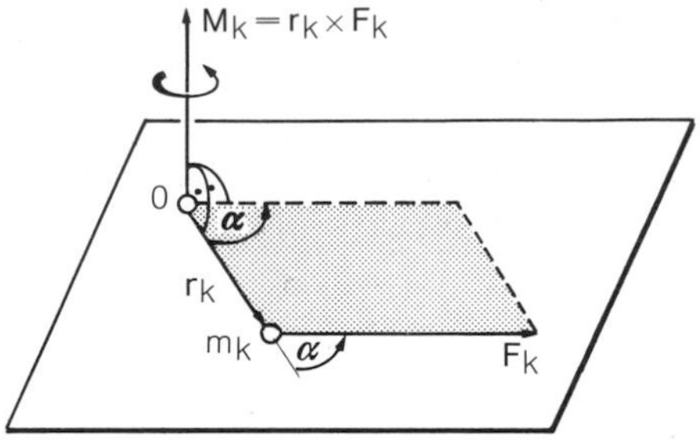

Fig. 2.6.2. The torque $\boldsymbol{M}_k = \boldsymbol{r}_k \times \boldsymbol{F}_k$. The absolute magnitude of $\boldsymbol{M}_k$ is $|\boldsymbol{r}_k| \cdot |\boldsymbol{F}_k| \cdot \sin\alpha$, i.e. equal to the area of the parallelogram spanned by $\boldsymbol{r}_k$ and $\boldsymbol{F}_k$

2.6.3 Conservation of Angular Momentum

We first consider a point mass m_k (Fig. 2.6.2), which is free to rotate about a fixed point 0 on a lever arm of length $\boldsymbol{r}_k$. A force $\boldsymbol{F}_k$ acts on m_k. This force "tries" to rotate the mass around an axis through 0 and perpendicular to the plane containing $\boldsymbol{r}_k$ and $\boldsymbol{F}_k$; only the "tangential" component of the force $|\boldsymbol{F}_k| \sin\alpha$ is effective in producing such a rotation, where α is the angle between $\boldsymbol{r}_k$ and $\boldsymbol{F}_k$. The quantity [lever arm $|\boldsymbol{r}_k|$ times effective force component $|\boldsymbol{F}_k| \sin\alpha$], drawn as a vector perpendicular to the plane containing $\boldsymbol{r}_k$ and $\boldsymbol{F}_k$ is, in mathematical terms, the vector product $\boldsymbol{r}_k \times \boldsymbol{F}_k$, and in physical terms, it is the moment of the force $\boldsymbol{F}_k$ about 0, also known as the *torque*, $\boldsymbol{M}_k = \boldsymbol{r}_k \times \boldsymbol{F}_k$.[3] The torque is the equivalent of the force for rotational motion; similarly, we can define a quantity corresponding to the linear momentum $\boldsymbol{p}_k = m\boldsymbol{v}_k$, the *angular momentum* $\boldsymbol{L}_k = \boldsymbol{r}_k \times \boldsymbol{p}_k = \boldsymbol{r}_k \times m\boldsymbol{v}_k$.

We now multiply Newton's equation of motion, (2.6.4), vectorially from the left by $\boldsymbol{r}_k$:

$$\boldsymbol{r}_k \times \frac{d}{dt} m_k \boldsymbol{v}_k = \boldsymbol{r}_k \times \boldsymbol{F}_k \; ,$$

or[4]

[2] In the following sections, all summation signs $\sum$, when not otherwise noted, imply a summation from $k = 1$ to N.

[3] The vector product is defined so that a r.h. screw which is being turned from $\boldsymbol{r}_k$ to $\boldsymbol{F}_k$ will bore itself in the direction of $\boldsymbol{M}_k$.

[4] We initially obtain

$$\frac{d}{dt}(\boldsymbol{r}_k \times m_k \boldsymbol{v}_k) = \frac{d\boldsymbol{r}_k}{dt} \times m_k \boldsymbol{v}_k + \boldsymbol{r}_k \times \frac{d}{dt}(m_k \boldsymbol{v}_k) \; .$$

However, in the first term, $d\boldsymbol{r}_k/dt = \boldsymbol{v}_k$. The vector product of the two parallel vectors $\boldsymbol{v}_k$ and $m_k \boldsymbol{v}_k$ also vanishes, since its magnitude is equal to the area between the factor vectors.

$$\frac{d}{dt}(\boldsymbol{r}_k \times m_k \boldsymbol{v}_k) = \boldsymbol{r}_k \times \boldsymbol{F}_k \ . \tag{2.6.14}$$

If we have a system of point masses, we define the total torque of all internal and external forces relative to the fixed point 0 as

$$\boldsymbol{M} = \sum \boldsymbol{r}_k \times \boldsymbol{F}_k = \sum \boldsymbol{r}_k \times \left(\boldsymbol{F}_k^{(e)} + \sum_{i=1}^{N} \boldsymbol{F}_{ik} \right) \tag{2.6.15a}$$

and the total angular momentum by

$$\boldsymbol{L} = \sum \boldsymbol{r}_k \times \boldsymbol{p}_k = \sum \boldsymbol{r}_k \times m_k \boldsymbol{v}_k \ . \tag{2.6.15b}$$

The equation of motion then becomes

$$\frac{d\boldsymbol{L}}{dt} = \boldsymbol{M} \ , \tag{2.6.16}$$

i.e., the time derivative of the angular momentum vector is equal to the combined torque due to all the forces.

If only *central forces* act in our system, such as gravitational forces, for example, which act along the line connecting two point masses, then the contribution to $\boldsymbol{M}$ from internal forces vanishes and the right side of (2.6.16) contains only the torque due to external forces, $\boldsymbol{M}^{(e)}$.

If, furthermore, no external forces are present or if the resultant torque is zero, then $d\boldsymbol{L}/dt = 0$ and the important theorem of *conservation of angular momentum* holds:

$$\boldsymbol{L} = \sum \boldsymbol{r}_k \times m_k \boldsymbol{v}_k = \text{constant} \ . \tag{2.6.17}$$

2.6.4 Conservation of Energy and the Virial Theorem

If a point mass m_k moves under the influence of a force $\boldsymbol{F}_k$ along a differential path element $d\boldsymbol{r}_k$, the latter making an angle α with the force $\boldsymbol{F}_k$, then the net *work*

$$dA = |\boldsymbol{F}_k| \, |d\boldsymbol{r}_k| \cos\alpha = \boldsymbol{F}_k \cdot d\boldsymbol{r}_k \tag{2.6.18}$$

is performed. The scalar product of the two vectors is denoted by a dot "·". If we calculate the work performed on passing along a finite section of an orbit, 1→2, using the Newtonian equation of motion (2.6.4) and $\boldsymbol{v}_k = d\boldsymbol{r}_k/dt$, we find

$$\int_1^2 \boldsymbol{F}_k \cdot d\boldsymbol{r}_k = \int_1^2 \frac{d}{dt}(m_k \boldsymbol{v}_k) \cdot d\boldsymbol{r}_k = \frac{1}{2} m_k \boldsymbol{v}_k^2 \Big|_1^2 \ . \tag{2.6.19}$$

The quantity $m_k \boldsymbol{v}_k^2/2$, or its sum over several point masses, is called the *kinetic energy* E_{kin}. Furthermore, if $\sum \boldsymbol{F}_k \cdot d\boldsymbol{r}_k$ is a complete differential,

$$-dE_{\text{pot}} = -\sum \left(\frac{\partial E_{\text{pot}}}{\partial x_k} dx_k + \frac{\partial E_{\text{pot}}}{\partial y_k} dy_k + \frac{\partial E_{\text{pot}}}{\partial z_k} dz_k \right) , \tag{2.6.20}$$

i.e. if the work performed by the forces, $\sum \int \boldsymbol{F}_k \cdot d\boldsymbol{r}_k$, is independent of the actual orbits of the point masses and depends only on the initial and final positions, then the sum E of kinetic energy E_{kin} and *potential energy* E_{pot} is constant. The important theorem of *conservation of energy* is then valid

$$E = E_{\text{kin}} + E_{\text{pot}} = \text{constant} \ . \tag{2.6.21}$$

In a system of point masses with gravitational forces, as in many other cases, E_{kin} depends only on the velocities and E_{pot} only on the positions of the point masses.

An important basis for the understanding of many problems, not only in celestial mechanics but also in the structure and evolution of stars and stellar systems, is the *virial theorem* of R. Clausius (1870). In an isolated system of point masses, we consider the time variation of the quantity $\sum \boldsymbol{p}_k \cdot \boldsymbol{r}_k$. By differentiating, we find

$$\frac{d}{dt} \sum \boldsymbol{p}_k \cdot \boldsymbol{r}_k = \sum \frac{d\boldsymbol{p}_k}{dt} \cdot \boldsymbol{r}_k + \sum \boldsymbol{p}_k \cdot \boldsymbol{v}_k \ , \tag{2.6.22}$$

or, using the equation of motion (2.6.4) and the kinetic energy defined above,

$$\frac{d}{dt} \sum \boldsymbol{p}_k \cdot \boldsymbol{r}_k = \sum \boldsymbol{F}_k \cdot \boldsymbol{r}_k + 2E_{\text{kin}} \ . \tag{2.6.23}$$

If we now average this expression over a sufficiently long time, the mean value of the left side is equal to zero (as long as $\boldsymbol{r}_k$ and $\boldsymbol{p}_k$ are finite for all the point masses, and therefore $\sum \boldsymbol{p}_k \cdot \boldsymbol{r}_k$ remains bounded), and we obtain the virial theorem:

$$\overline{E_{\text{kin}}} = -\frac{1}{2} \overline{\sum \boldsymbol{F}_k \cdot \boldsymbol{r}_k} \ . \tag{2.6.24}$$

In order to evaluate the *virial*, $\sum \boldsymbol{F}_k \cdot \boldsymbol{r}_k$, we must know the forces $\boldsymbol{F}_k$. In the important case that the point masses are held together by forces $\propto r^{-2}$, in particular by *gravitational forces* (Sect. 2.6.5), the virial is equal to the time-averaged potential energy of the system E_{pot}.[5] The virial theorem in this case thus states that the total energy E is distributed between the kinetic energy and the potential energy in such a way that, *averaged over time*,

$$\overline{E_{\text{kin}}} = -\tfrac{1}{2}\overline{E_{\text{pot}}} \tag{2.6.25}$$

or, using (2.6.21),

$$\overline{E_{\text{kin}}} = -E \ . \tag{2.6.26}$$

2.6.5 The Law of Gravitation. Celestial Mechanics

In order to obtain a theory of celestial motions, Newton had to add his *law of universal gravitation* (ca. 1665) to his basic laws of mechanics:

Two point masses m_i and m_k at a distance r attract each other along the line joining them with a force of magnitude

$$F = G\frac{m_i m_k}{r^2} \ . \tag{2.6.27}$$

By carrying out an integration, which we shall not repeat here, Newton first demonstrated that exactly the same law (2.6.27) holds for the attraction of two spherical mass distributions of finite size (e. g., the Sun, a planet, etc.) as for the corresponding point masses. He then verified the law of gravitation by assuming that free fall near the Earth's surface (Galileo) and the lunar orbit are both dominated by the gravitational attraction of the Earth:

The *acceleration* (force/mass) for a free fall can be determined in experiments with falling objects, or, more precisely, with a pendulum. On the equator, its numerical value is 9.78 m/s^2, or, taking into account the centrifugal acceleration of the Earth's rotation, 0.034 m/s^2, it is equal to

$$g_{\oplus} = 9.81 \text{ m/s}^2 \ . \tag{2.6.28}$$

On the other hand, the Moon moves on its circular orbit of radius r with a velocity $v = 2\pi r/T$ ($T = 1$ sidereal month) and thus experiences the acceleration (Fig. 2.6.1):

$$g_{☾} = \frac{v^2}{r} = \frac{4\pi^2 r}{T^2} = 2.72\cdot 10^{-3} \text{ m/s}^2 \tag{2.6.29}$$

where $r = 384\,400$ km $= 3.844\cdot 10^8$ m, and $T = 27.32$ d $= 27.32\times 86\,400$ s.

The accelerations $g_{\oplus}$ and $g_{☾}$ are, in fact, related as the inverse squares of the radii of the Earth, R, and the lunar orbit, r, i.e.,

$$g_{\oplus} : g_{☾} = \frac{1}{R^2} : \frac{1}{r^2} = 3620 \ . \tag{2.6.30}$$

The numerical value of the *universal gravitation constant* G appears here only as a product with the initially likewise unknown mass of the Earth $\mathcal{M}$. Similarly, in other astronomical problems, G occurs only as a product with the mass of the attracting celestial object. Thus G can, for fundamental reasons, *not* be determined by astronomical measurements; instead, it must be found from terrestrial experiments.

The first attempt at such a measurement was that of N. Maskelyne in 1774, who investigated the deflection of a plumb-bob by a large mass (a mountain). In 1798, H. Cavendish used a rotational balance, and P. v. Jolly performed measurements with a suitable beam balance in 1881. The result of modern determinations is

$$G = (6.673 \pm 0.001)\cdot 10^{-11} \text{ m}^3\text{s}^{-2}\text{kg}^{-1} \ . \tag{2.6.31}$$

Since the acceleration of gravity on the surface of the Earth (whose rotation and polar flattening will be neglected for the moment) is related to its mass $\mathcal{M}$ and radius R by the equation $g = G\mathcal{M}/R^2$, which we have already used, we can now calculate the Earth's mass $\mathcal{M}$ and *average density* $\bar{\varrho} = \mathcal{M}/(4\pi/3)R^3$, obtaining:

---

[5] *Proof:* From (2.6.20), the force acting on the point mass m_k has the components

$$\boldsymbol{F}_k = \left\{-\frac{\partial E_{\text{pot}}}{\partial x_k}, -\frac{\partial E_{\text{pot}}}{\partial y_k}, -\frac{\partial E_{\text{pot}}}{\partial z_k}\right\} \ .$$

One now shows that for the sum in the virial,

$$\sum \boldsymbol{F}_k\cdot\boldsymbol{r}_k = -\sum\left(x_k\frac{\partial E_{\text{pot}}}{\partial x_k} + y_k\frac{\partial E_{\text{pot}}}{\partial y_k} + z_k\frac{\partial E_{\text{pot}}}{\partial z_k}\right) = E_{\text{pot}}$$

is valid. E_{pot} is namely composed of terms of the type

$$\frac{1}{r_{kl}} = [(x_k - x_l)^2 + (y_k - y_l)^2 + (z_k - z_l)^2]^{-1/2} \ ,$$

from which we find

$$\frac{\partial}{\partial x_k}\left(\frac{1}{r_{kl}}\right) = -(x_k - x_l)\,r_{kl}^{-3} \ .$$

Our conjecture is thereby verified.

$$\mathcal{M} = 5.97 \cdot 10^{24}\ \mathrm{kg}$$

and (2.6.32)

$$\bar{\varrho} = 5500\ \mathrm{kg/m^3} = 5.5\ \mathrm{g/cm^3}\ .$$

We shall return to a discussion of the geophysical significance of these numbers.

First, we go back to Newton's *Principia* and derive Kepler's laws from the basic equations of mechanics and the law of gravitation, in order to gain a deeper understanding of the laws themselves and of the numerical values which occur in them.

The mass of the Sun is clearly very much greater than that of any of the planets, so we start by treating the Sun as fixed and calculate the radius vectors $\boldsymbol{r}$ or $r = |\boldsymbol{r}|$ of the planets from the center of the Sun. The mutual attraction of the planets among themselves, their *perturbations*, will initially be left out of the calculations.

The motion of *one* planet around the Sun is governed by the central force $G\mathcal{M}m/r^2$, where $\mathcal{M}$ is the mass of the Sun, and $m(\ll \mathcal{M})$ is that of the planet. Conservation of angular momentum thus holds, i.e. the angular momentum vector of the planet

$$\boldsymbol{L} = \boldsymbol{r} \times m\boldsymbol{v} \tag{2.6.33}$$

(where $\boldsymbol{v}$ is the planet's orbital velocity vector) is constant in magnitude and direction. That means that $\boldsymbol{r}$ and $\boldsymbol{v}$ remain in the same plane $\perp \boldsymbol{L}$, the spatially-fixed *orbital plane* of the planet. The magnitude of $|\boldsymbol{r} \times \boldsymbol{v}| = rv \sin \alpha$ (Fig. 2.6.3) is twice the area which is swept out by the radius vector $\boldsymbol{r}$ of the planet per unit time. Conservation of angular momentum is therefore identical with the statement that each planet follows an orbit in a fixed plane with constant areal velocity (Kepler's 2nd and part of his 1st laws).

We shall not reproduce here the somewhat tedious calculations which show that the orbit of a point mass (planet, comet, etc.) under the influence of a central force

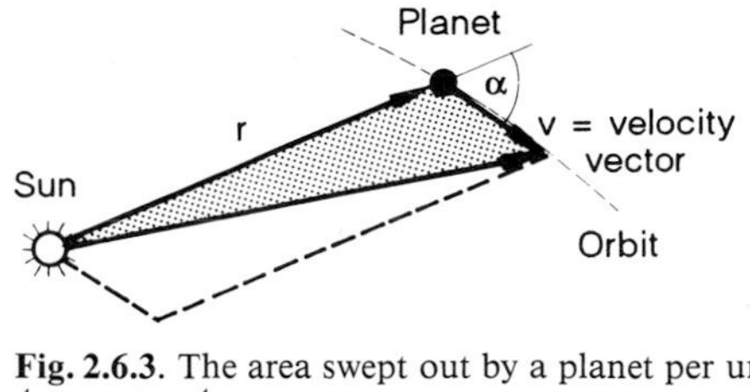

Fig. 2.6.3. The area swept out by a planet per unit time within its orbit, $\frac{1}{2}|\boldsymbol{r} \times \boldsymbol{v}| = \frac{1}{2} rv \sin \alpha$

proportional to r^{-2} must be a *conic section*, i.e. a circle (eccentricity $e = 0$), an ellipse ($0 < e < 1$), a parabola ($e = 1$) or a hyperbola ($e > 1$) with the Sun in one focus (Kepler's first law).

However, we will write down the *law of energy conservation* for planetary motion: if we imagine a point mass m which was initially at rest at a distance r from the Sun (mass $\mathcal{M}$) under the influence of the gravitational force $G(\mathcal{M}m/r^2)$ to be moved out to infinite distance ($r \to \infty$), the work required is equal to its original potential energy $E_{\mathrm{pot}}(r)$, i.e.

$$E_{\mathrm{pot}}(r) = -G \int_r^\infty \frac{\mathcal{M}m}{r^2}\, dr = -G \frac{\mathcal{M}m}{r}\ . \tag{2.6.34}$$

Since no work is performed when the motion is tangential, only when it is radial, we can readily see that this expression is independent of the path along which the mass moves (the path of integration). The potential energy per unit mass, $\phi(r) = -G\mathcal{M}/r$ is called the *potential* at a distance r from the Sun. It is one of the fundamental concepts of celestial mechanics and of theoretical physics.

The total energy $E = E_{\mathrm{kin}} + E_{\mathrm{pot}}$ of a planet

$$E = \frac{1}{2} m v^2 - G \frac{\mathcal{M}m}{r} \tag{2.6.35}$$

thus remains constant in time. We can see again from this that the orbital velocity increases in going from the aphelion to the perihelion.

We shall carry out the calculation of planetary motion and the derivation of Kepler's 3rd law only for *circular orbits.*[6] However, in view of future applications, we shall not restrict the mass of the planet to be very much smaller than the mass of the Sun. We therefore consider the motion of two masses around their common center of gravity S, and, on the other hand, the relative motion of the two masses with relation to, for example, the larger one. Let m_1 and m_2 be the masses and a_1 and a_2 their respective distances from the center of gravity; then $a = a_1 + a_2$ is their mutual distance (Fig. 2.6.4). Then, using the definition of the center of gravity, we have

$$a_1 : a_2 : a = m_2 : m_1 : (m_1 + m_2)\ , \quad \text{or}$$

$$m_1 a_1 = m_2 a_2 = \left(\frac{m_1 m_2}{m_1 + m_2} \right) a\ . \tag{2.6.36}$$

[6] The general calculation can be found in any text on classical mechanics.

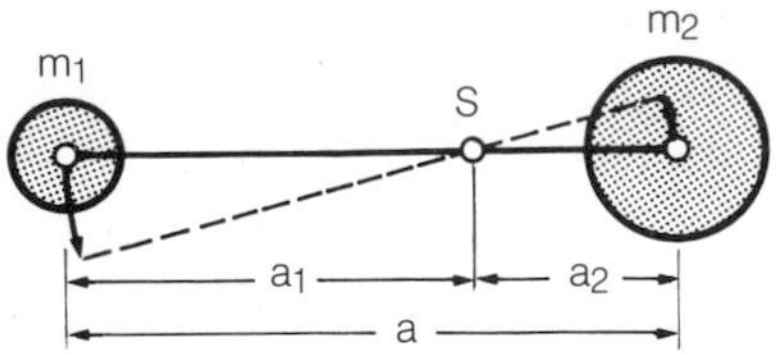

Fig. 2.6.4. The motion of the masses m_1 and m_2 about their common center of gravity S; $m_1 a_1 = m_2 a_2$

For each of the two masses, the force of gravitational attraction, $Gm_1 m_2/a^2$, must be balanced by the centrifugal force. Calling the orbital period of the system T, we find for m_1

$$\frac{m_1 v_1^2}{a_1} = \left(\frac{2\pi}{T}\right)^2 m_1 a_1 \ . \tag{2.6.37}$$

Because of action = reaction [as expressed by (2.6.36)]; a similar equation and the same value of T naturally hold for m_2. Applying this equation once more, together with (2.6.36), we obtain after rearranging

$$\frac{a^3}{T^2} = \frac{G}{4\pi^2}(m_1 + m_2) \ . \tag{2.6.38}$$

If we relax the requirement of a circular orbit (we shall not carry out the explicit calculation here), the masses m_1 and m_2 are found to move on similar conic sections around the center of gravity S; the relative orbit is also a corresponding conic section. Instead of orbital radii, we then have *orbital semimajor axes*, which are likewise denoted by a_1, a_2, and a. We thus obtain a generalized version of Kepler's 3rd law (2.6.38). In the Solar System, as we shall see, the mass of e.g. the largest planet, Jupiter, is only about 1/1000 the mass of the Sun. To this accuracy, we can therefore set the expression $m_1 + m_2$ on the right-hand side equal to the Sun's mass, $\mathcal{M}_\odot$. By inserting the numerical values for the orbit of the Earth or another planet in (2.6.38), we obtain the solar mass. Its apparent disc radius of 16′, together with a, gives the solar radius, $R_\odot$. From $(4\pi/3)R^3\bar{\varrho} = \mathcal{M}$, we finally obtain the mean density of the Sun, $\bar{\varrho}_\odot$. The numerical values are:

$$\begin{aligned} &\text{Mass} && \mathcal{M}_\odot = 1.989 \cdot 10^{30}\,\text{kg} \ , \\ &\text{Radius} && R_\odot = 6.960 \cdot 10^{8}\,\text{m} \ , \\ &\text{Density} && \bar{\varrho}_\odot = 1409\,\text{kg/m}^3 \ . \end{aligned} \tag{2.6.39}$$

The uncertainty in these values is in each case about ± one unit in the last place.

In a similar way, we calculate the *masses of the planets* (Table 2.8.1) by applying Kepler's 3rd law (2.6.38) to the orbits of their moons. If the planet has no moons, then mutual perturbations of the planets must be used; this is, of course, much more difficult. Space travel has opened up further possibilities of determining planetary masses by observing the orbital motions of artificial satellites and space probes.

It is instructive to consider the "Kepler problem" again from the point of view of energy conservation (2.6.35): for a *circular orbit*, the centrifugal force is equal to the attractive force of the two masses, and therefore $mv^2/r = G\mathcal{M}m/r^2$ or

$$\frac{mv^2}{2} = \frac{1}{2} G \frac{\mathcal{M}m}{r} \ . \tag{2.6.40}$$

Thus, the kinetic energy E_{kin} is equal to $-\frac{1}{2}$ times the potential energy,

$$E_{\text{kin}} = -\tfrac{1}{2} E_{\text{pot}} \ . \tag{2.6.41}$$

This is a special case of the virial theorem (2.6.25) for the periodic motion of a single point mass attracted by a central force.

If, on the other hand, we consider a *parabolic orbit* (e.g., a nonperiodic comet), we find that at infinity, both the kinetic and the potential energy are zero. From $E = 0$ it follows from (2.6.35) that

$$\frac{mv^2}{2} = G\frac{\mathcal{M}m}{r} \quad \text{or} \quad E_{\text{kin}} = -E_{\text{pot}} \ . \tag{2.6.42}$$

At the same distance from the Sun, the parabolic velocity, i.e. v on a parabolic orbit, is larger by a factor of $\sqrt{2}$ than the velocity on a circular orbit. For example, if the mean velocity of the Earth is 30 km/s, that of a comet or meteor swarm which meets us on a parabolic orbit would be $30\sqrt{2} = 42.4$ km/s.

Besides the Kepler problem, Newton solved numerous other problems in celestial mechanics.

We shall first consider *precession*, at least in outline form. The "wobble" of the Earth's axis around the pole of the ecliptic is a result of the same physical phenomenon as the wobbling of a top under the influence of gravity: the equitorial bulge of the Earth is pulled into the plane of the ecliptic by the Sun and the Moon, whose

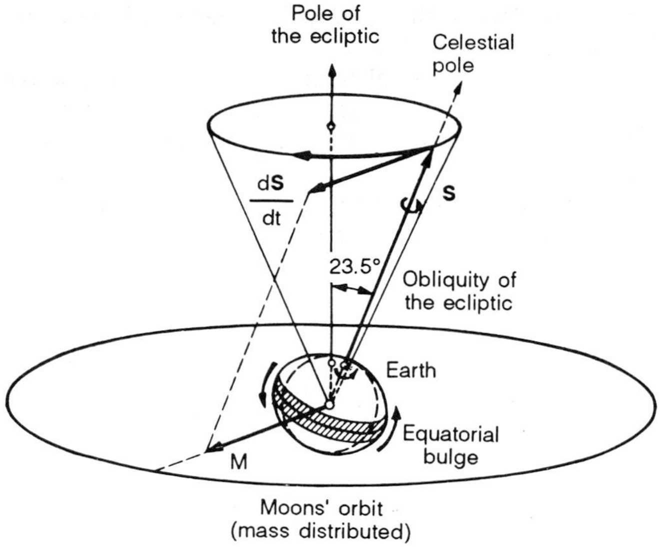

Fig. 2.6.5. Lunar precession

mass we can imagine to be distributed around their orbits over the long precession period of the Earth's axis (the Moon has only a small orbital inclination). The angular momentum vector $\boldsymbol{S}$ ("spin") of the Earth's rotation, which is essentially parallel to the rotation axis, reacts to this torque $\boldsymbol{M}$ (Fig. 2.6.5) according to (2.6.16). The rate of change in $\boldsymbol{S}$, given by $d\boldsymbol{S}/dt = \boldsymbol{M}$, produces the circling motion of $\boldsymbol{S}$, i.e. of the Earth's axis on a cone of constant vertex angle, as can be seen in Fig. 2.6.5. Numerical calculation yields the correct period for the lunisolar precession (Fig. 2.2.1).

For later applications, we give the intrinsic angular momentum (spin) of a *rigid* rotating body, in which all the mass points have the same angular velocity $\boldsymbol{\omega}$ ($\omega = |\boldsymbol{\omega}| = 2\pi/P_{\mathrm{rot}}$; P_{rot} is the period of rotation):

$$\boldsymbol{S} = \sum_k \boldsymbol{r}_k \times m_k \boldsymbol{v}_k = \sum_k \boldsymbol{r}_k \times m_k (\boldsymbol{\omega} \times \boldsymbol{r}_k) = J\boldsymbol{\omega} \ , \quad (2.6.43)$$

where

$$J = \sum_k m_k r_k^2 \quad (2.6.44)$$

is the *moment of inertia* of the body relative to the particular axis of rotation. For a homogeneous sphere of mass $\mathcal{M}$ and radius, we have

$$J = \tfrac{2}{5} \mathcal{M} R^2 \ . \quad (2.6.45)$$

2.6.6 The Tides

Next, we turn briefly to the old problem of the tides. Galilei became involved in controversies with his contemporaries over a wholly incorrect theory of the 12-hour alternation of ebb and flow (the tides having been known to Mediterranean peoples only by hearsay); they were a major contributor leading to his unfortunate trial by the Inquisition. It was again Newton who developed the elements of a *static* theory of the tides (Fig. 2.6.6).

An acceleration vector (equal to the difference between the gravitational and the centrifugal accelerations) acts, for example on a water droplet in the ocean, as a result of the mutual motion of the Earth and the Moon around their common center of gravity. It points *upwards* at the upper and the lower culmination of the Moon, i.e. at the passages of the Moon through the meridian in the north and the south. The water will therefore be lifted at those times; we observe high water. Two water "peaks" and two "valleys" circle the Earth continually, following the apparent motion of the Moon and thus getting 51 minutes later each day (1 lunar day = 24 h 51 min).

The tidal acceleration a_T on the surface of the Earth (at the Earth's radius R) at, for example, the lower culmination (point A in Fig. 2.6.6) is the *difference* between the accelerations of gravity due to the attractive mass $\mathcal{M}$ acting at the Earth's center (at a distance r from $\mathcal{M}$) and that at the point A (at a distance of $r+R$)

$$a_{\mathrm{T}} = \frac{G\mathcal{M}}{r^2} - \frac{G\mathcal{M}}{(r+R)^2} \simeq \frac{2G\mathcal{M}}{r^3} R \ . \quad (2.6.46)$$

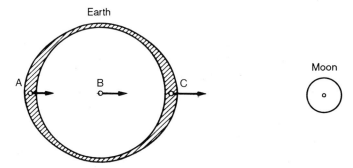

Fig. 2.6.6. The static theory of the tides. The acceleration relative to the Moon of the three points indicated corresponds to their differing distances from it: A, lower culmination of the Moon, $a - a_{\mathrm{T}}$; B, center of the Earth, a; C, upper culmination of the Moon, $a + a_{\mathrm{T}}$. The (rigid) Earth has, as a whole, the acceleration a. Therefore, at the points A and C, an acceleration a_{T} is in excess, and it produces a high water level at both these points

The tidal force is therefore proportional to r^{-3} and is a stronger function of the distance than the gravitational force. For this reason, the tidal force due to the Sun is only about half as great as that from the Moon, although the Sun's mass is enormously greater. At the new Moon and full Moon, the tidal forces of the Moon and the Sun act together and produce a spring tide; in the first and last quarters of the Moon, they oppose each other and we have a neaptide. This static theory, in fact, only explains the rough features of tidal phenomena; the *dynamic* theory of tides investigates the forced oscillations which are produced in the various oceans by tidal forces with the different periods of the solar and lunar motions. The prediction of the tides according to G. Darwin's theory is based essentially on a harmonic (Fourier) analysis and synthesis using these frequencies. Tidal friction in straits produces a braking effect on the Earth's rotation, as we mentioned in Sect. 2.3, and thus an *increase in the length of the day.* According to the law of conservation of angular momentum (2.6.17), the angular momentum of rotation lost by the Earth must be transferred to the orbital motion of the Moon. Since, from Kepler's 3rd law, the angular momentum per unit mass is proportional to the square root of the orbital radius, the Moon must be gradually moving away from the Earth.

2.6.7 The Ptolemaic and the Copernican Worldviews

Before leaving the theory of planetary motions, we will consider the decisive change which was made in going from the Ptolemaic to the Copernican worldviews, looking back from our modern standpoint (Fig. 2.6.7).

Viewed from the Sun (heliocentric system), let us denote the position vector of a planet by $\boldsymbol{r}_P$ and that of the Earth by $\boldsymbol{r}_E$. Then as seen from the Earth (geocentric system), the position of the planet is given by the difference vector

$$\boldsymbol{R} = \boldsymbol{r}_P - \boldsymbol{r}_E \ . \tag{2.6.47}$$

We turn once more to the *geocentric* view of Ptolemy: (a) for the *outer* planets, e.g. Mars, we begin by first drawing the vector $\boldsymbol{r}_P$ from the *Earth* and letting it rotate as it did before around the Sun. Following (2.6.47), we then add the vector $-\boldsymbol{r}_E$ (the position vector of the Sun as seen from the Earth) and thus obtain the position vector of the planet, $\boldsymbol{R}$, as seen from the Earth. The imaginary circle which $\boldsymbol{r}_P$ describes with its sidereal period around the Earth is the Ptolemaic deferent. The other circle around

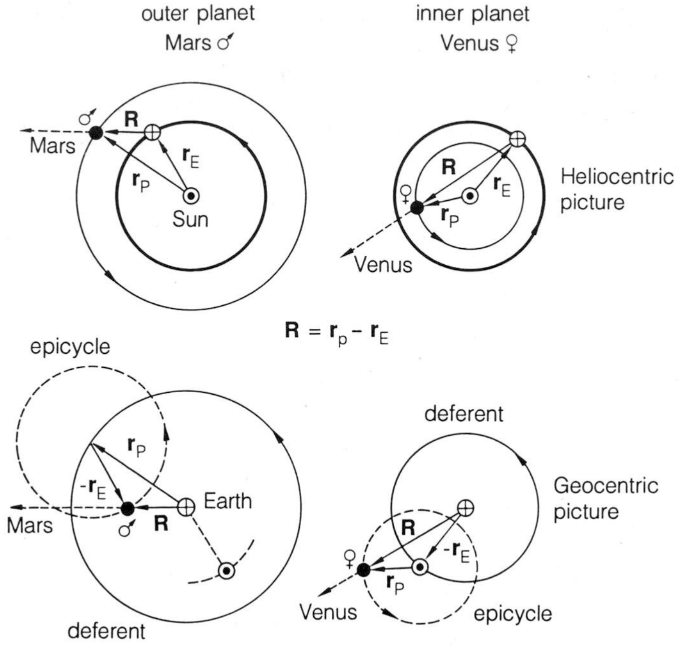

Fig. 2.6.7. The motion of an outer planet (Mars) and an inner planet (Venus) on the celestial sphere, represented in heliocentric and in geocentric pictures (⊙ Sun, ⊕ Earth). The dashed arrow indicates in each case the position of the planet in the sky. Correspondingly, for

	outer planets	inner planets
Deferent	r_P	$-r_E$
Epicycle	$-r_E$	r_P

the point $\boldsymbol{r}_P$ which is described by the planet at the end of the vector $-\boldsymbol{r}_E$ with the sidereal period of the Earth is the Ptolemaic epicycle. (b) For the *inner* planets, it appeared more reasonable to the Alexandrians to first let the larger vector circle the Earth with a period of one sidereal year as the deferent, and then to allow the smaller vector $\boldsymbol{r}_P$ to circle the point $-\boldsymbol{r}_E$ with the sidereal period of the planet as the epicycle.

So far, the geocentric constructions still correspond exactly to the equation $\boldsymbol{R} = \boldsymbol{r}_P - \boldsymbol{r}_E$. Scheme *b*, applied to *all* the planets, would represent the system of T. Brahe.

In fact, however, we have not yet completed the transition to the Ptolemaic system: as long as only the positions of the planets in the sky, i.e. only their directions but not their distances, could be measured, only the direction, not the magnitude, of the vector $\boldsymbol{R}$ was of importance. One could therefore draw the vector $\boldsymbol{R}$ on a different scale for each planet. This means that the vectors

$$\boldsymbol{R}'_{\mathrm{P}} = A_{\mathrm{P}} \boldsymbol{R}_{\mathrm{P}} \ , \tag{2.6.48}$$

with a fixed but purely arbitrary factor A_P for each planet, give a completely satisfactory representation of the motions of the planets in the sky in the Ptolemaic system.

Now we can see clearly what was lost in our stepwise return from the Copernican to the Ptolemaic system:

1) The change of coordinate system means relinquishing a simple mechanical explanation.
2) The scale factors A_P leave the positions of the planets in the celestial sphere unchanged, but the mutual relations of the planetary positions among themselves are lost.
3) The fact that in the Ptolemaic system the annual period, corresponding to the motion of r_E, is introduced *independently* for *each* planet, is another indication of the 'clumsiness' of the ancient worldview.

However, it is important to recognize that a purely *kinematic* consideration of the planetary motions in the sky did not allow a decision to be made between the ancient and the modern views. Only Galileo's observations with his telescope (1609) led to progress: (i) Jupiter, with its freely orbiting moons, could be seen as a 'model' of the Copernican solar system. (ii) The phases of Venus are determined by the relative positions of the Sun, Earth, and Venus. The smallness of the phase angle, e.g., for Jupiter also supports Copernicus. Even the very *idea* of a celestial *mechanics* presupposes, as we should keep in mind, that the basic *similarity* of cosmic and terrestrial matter and its physics be understood.

2.7 Artificial Satellites and Spacecraft. Space Research

As a conclusion to our discussion of celestial mechanics we shall take a brief look at the orbits of artificial satellites and spacecraft. Here, we will consider only the gravitational field of the Earth, limit ourselves to circular orbits, and neglect the braking effect of the atmosphere.

A satellite in a circular orbit near the Earth (Earth's radius $R = 6378$ km) must have, according to (2.6.40), a velocity near $v_0 = 7.9$ km/s and an orbital period near $T_0 = 84.4$ min. For a larger orbit with a radius r, from Kepler's 3rd law the velocity must be $v = v_0(r/R)^{-1/2}$ and the period must be $T = T_0(r/R)^{3/2}$. Of particular importance is the fact that the period becomes equal to one sidereal day for $r = 6.6$ R. A geostationary satellite at this distance then "stands" at an altitude of nearly 36000 km above a fixed spot on the Earth's surface.

In order to allow a spacecraft which lacks its own drive motor to escape from the gravitational field of the Earth (alone) and travel to infinite distances, it must start with at least the parabolic or escape velocity $v_0\sqrt{2} = 11.2$ km/s.

Jules Verne, in his great novel "From the Earth to the Moon" (1865)[7], suggested a solution to this problem using a gigantic cannon. This would not work, however, since the initial velocity of a cannon shell cannot be much greater than the velocity of sound in the gases from the explosive, which is too small.

Higher velocities can be reached by using *rockets*. We shall first deal with the mechanics of rocket propulsion by considering a rocket without the influence of gravity (i.e. on a horizontal test ramp or in space) and without air resistance. Let the mass of the rocket's hull and other parts plus fuel at time t be given by $m(t)$. The change in $m(t)$ per unit time, corresponding to the mass of combustion products expelled from the rocket per unit time, is dm/dt. If we now call the expulsion velocity of the combustion gases v_E, then a momentum equal to $-(dm/dt)\,v_E$ will be transferred to the rocket per unit time. Considering the acceleration of the rocket from the point of view of an astronaut moving along with it, we obtain the Newtonian equation of motion:

$$m(t)\frac{dv}{dt} = -\frac{dm}{dt}v_E$$

or

$$\frac{dv}{v_E} = -\frac{dm}{m(t)} \ . \tag{2.7.1}$$

By integrating and using the initial condition $v = 0$ and $m = m_0$ at $t = 0$, we find the *rocket equation*

$$v = v_E \ln \frac{m_0}{m(t)} \ . \tag{2.7.2}$$

[7] Our futurologists should pale with envy on reading Jules Verne's predictions: his launching point was only 150 km from Cape Canaveral. To observe the projectile, a reflecting telescope of about 200″ (!) is constructed; one of the first tests is the complete resolution of the Crab Nebula!

If we had taken the (homogeneous) gravitational field in the neighborhood of the Earth into account, for a vertical takeoff we would have had on the right the additional well-known term $-gt$.

Taking the relatively favorable values $v_E = 4$ km/s and $m_0/m = 10$ on burnout, we find a final velocity (without air resistance!) for our single-stage rocket of $v =$ 9.2 km/s. For real space travel, it is thus necessary to use *multi-stage* rockets, whose basic principle quickly becomes clear on repeatedly applying the rocket equation (2.7.2).

In the 2nd World War, Germany constructed the V-2 rocket, which could attain heights of nearly 200 km. After the war, in the USA the V-2 and improved rockets were employed for the investigation of the highest layers of the atmosphere (the ozone layer and the ionosphere) and of short-wavelength solar radiation ($\lambda < 285$ nm). Today, research using simple rockets is still of importance, since with a relatively modest investment it allows us to study the spectral regions which are strongly absorbed by the atmosphere: the far ultraviolet, X-rays, and gamma radiation from cosmic objects; and, on the other side of the visible spectrum, the infrared out to the mm-wave region and finally radio-frequencies which are reflected by the ionosphere, with wavelengths from about 30 m to 1 km. The observation time is, to be sure, limited to a few minutes.

In 1957, Soviet researchers succeeded in placing the first *artificial satellite*, Sputnik 1, into an orbit with a minimum altitude of 225 km and a maximum altitude of 950 km. Since then, thousands of satellites have been launched for purposes of research and communication. Of particular importance for lengthy series of astronomical observations outside the Earth's atmosphere is an excellent stabilization of the satellites to less than 1″, which was first attained by the OSO and OAO series launched by NASA (the National Aeronautics and Space Administration of the USA). Some satellites which have been of importance for astronomical research are listed in Table 2.7.1. Their instrumentation will be described in part in Chap. 3; we shall discuss the results obtained with these satellites and telemetered back to Earth in connection with the objects observed.

The first *manned* space flight was ventured upon in 1961 by Yuri Gagarin, who circled the Earth in the space vehicle Vostok 1. In 1973/74, three different teams carried out astronomical, biological, and technical experiments in the first space laboratory, Skylab.

Since the early 1980's, the launching of satellites and space research missions by NASA has been carried out

Table 2.7.1. Some astronomical satellites

Launch	Name		Applications (wavelength or energy range)
1962/75	OSO-Series	Orbiting Solar Observatory	Solar UV, X-rays, gamma rays
1970	SAS-1 = UHURU	Small Astronomical Satellite (one of the Explorer series)	X-rays
1972	OAO-3 = Copernicus	Orbiting Astronomical Observatory	Stellar UV radiation with high resolution (95 – 156 nm)
1975	COS-B		Gamma rays (50 – 5000 MeV)
1978	IUE	International Ultraviolet Explorer	UV spectroscopy (115 – 320 nm)
1978	HEAO-2 = Einstein	High Energy Astronomical Observatory	X-rays (first imaging telescope) (0.1 – 3 keV)
1983	IRAS	Infrared Astronomical Satellite	Infrared (12 – 100 μm)
1983	EXOSAT	European X-Ray Observatory	X-rays (0.04 – 40 keV)
1987	Ginga	(Japanese) Galaxy	X-rays (1.5 – 30 keV)
1989	HIPPARCOS	High Precision Parallax Collecting Satellite	Stellar parallaxes
1989	COBE	Cosmic Background Explorer	Microwave radiation (0.1 – 10 mm)
1990	HST	Hubble Space Telescope	2.4 m telescope for optical and ultraviolet range
1990	ROSAT	Röntgen Satellite	83 cm X-ray telescope (0.1 – 2 keV)
1991	GRO	Gamma Ray Observatory	Gamma rays (0.02 – 30000 MeV)

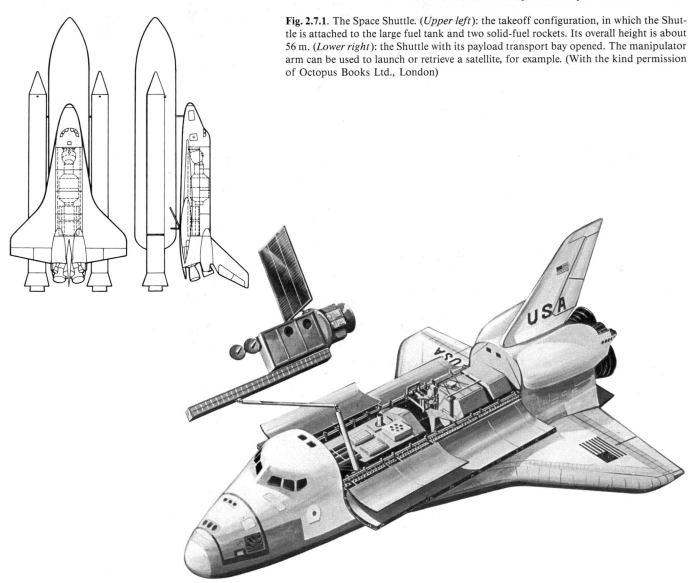

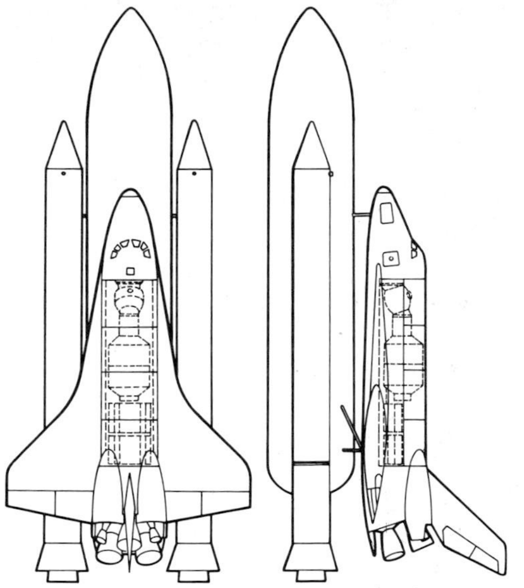

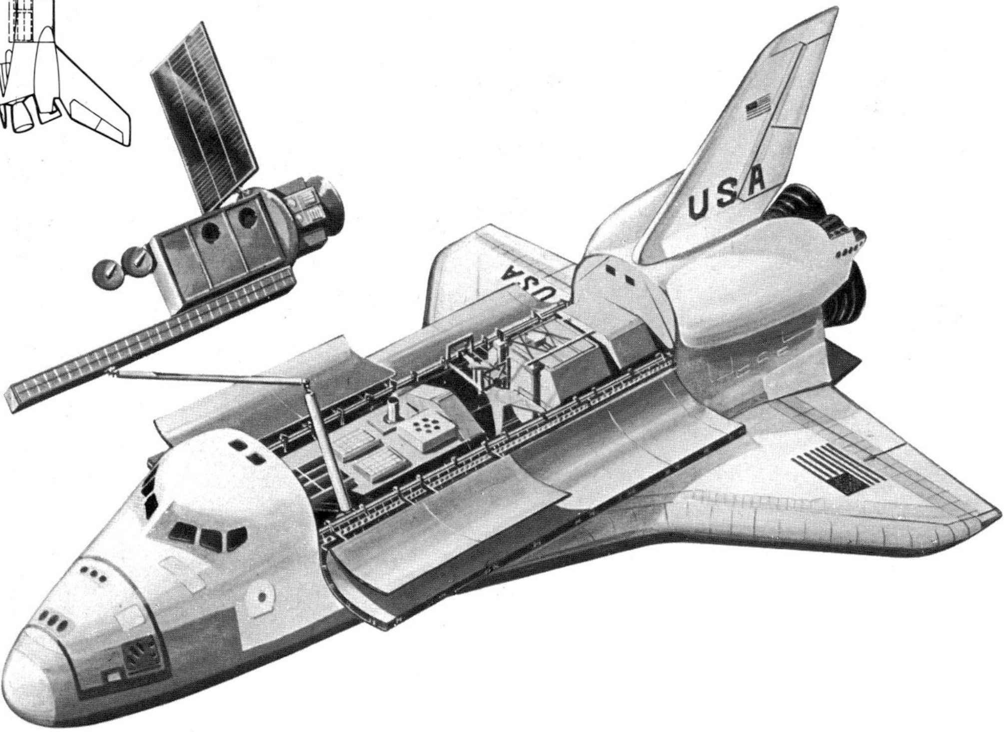

Fig. 2.7.1. The Space Shuttle. (*Upper left*): the takeoff configuration, in which the Shuttle is attached to the large fuel tank and two solid-fuel rockets. Its overall height is about 56 m. (*Lower right*): the Shuttle with its payload transport bay opened. The manipulator arm can be used to launch or retrieve a satellite, for example. (With the kind permission of Octopus Books Ltd., London)

increasingly by using the manned, reusable Space Shuttle instead of multistage launch vehicles. The explosion and loss of the Space Shuttle "Challenger" in 1986 represented a considerable setback for the space program. The Shuttles (Fig. 2.7.1) are fully maneuverable and can land on a runway.

The vertical takeoff system consists of two solid-fuel rockets whose motors can be recovered and reused, as well as a large, non-reusable fuel tank and the actual Shuttle (orbiter), which resembles an airplane. Depending on the payload, orbits of maximum altitude from 200 to 1000 km can be reached. Satellites such as the geostationary satellites mentioned above, which require orbits of higher altitude, employ an additional rocket motor. Several satellites can be launched on one Shuttle flight, containers with instrumentation may be set out and retrieved, and repairs and maintenance may be performed on satellites. The Shuttle itself can serve as an orbiting laboratory (Spacelab) and its use may permit the construction of future large space stations.

The investigation of celestial objects outside the Earth using space vehicles naturally began with the Moon, our nearest neighbor in space. In 1959, the Soviet probe Luna 1 approached the Moon to within 5000 km, Luna 2 made a "hard" landing on its surface, and Luna 3 transmitted the first pictures of the back side of the Moon. Figure 2.7.2 illustrates the unmanned flight of one of the American Ranger space probes to the Moon (1961–65), with a hard landing. A new era in the exploration of the Moon began in 1966 with the successful soft landing of Luna 9 and the placement of Luna 10 in a Lunar orbit, making it the Moon's first artificial satellite.

Preparations for *landing men on the Moon*, a difficult and costly undertaking, were carried out by NASA in the period from 1966–69. Possible landing sites were investigated by various Lunar Orbiter and soft-landed Surveyor probes, and manned test flights with Lunar orbiting as well as a number of simulation experiments were carried out. The manned landing itself was accomplished using the following principle: a three-stage rocket is used to reach the Moon. While one of the astronauts circles the Moon a number of times in the command unit, the two others land on the surface in an auxiliary craft (LM: Lunar Module). After completing their mission there they take off with the aid of a rocket motor, couple their LM to the third stage of the main rocket, and start on the return journey after discarding the now-useless LM. The main difficulties in landing result from the heating during reentry into the Earth's atmosphere (heat shield!) and the temporary interruption of radio communication because of ionization of the air.

The first landing on the Moon was made in 1969 with Apollo 11 by the astronauts N. Armstrong, M. Collins, and E. Aldrin in the Mare Tranquillitatis (Fig. 2.7.3). They left behind a research station, containing among other things a seismometer, and brought back 22 kg of lunar rocks and loose material. Through 1972, there were five further landings on the Moon as part of the Apollo project.

Soviet researchers have in the meantime developed the techniques of unmanned, automated or remote-control exploration of the Moon. In 1970/76, they brought back rock samples from the Moon using *unmanned* spacecraft of the Luna series. The remote-controlled lunar vehicles Lunachod 1 and 2 explored further regions of the lunar landscape.

Along with numerous observations concerning the structure and history of the Lunar surface, the Apollo flights also introduced a novel method of studying the celestial mechanics of the *Earth-Moon system*, the

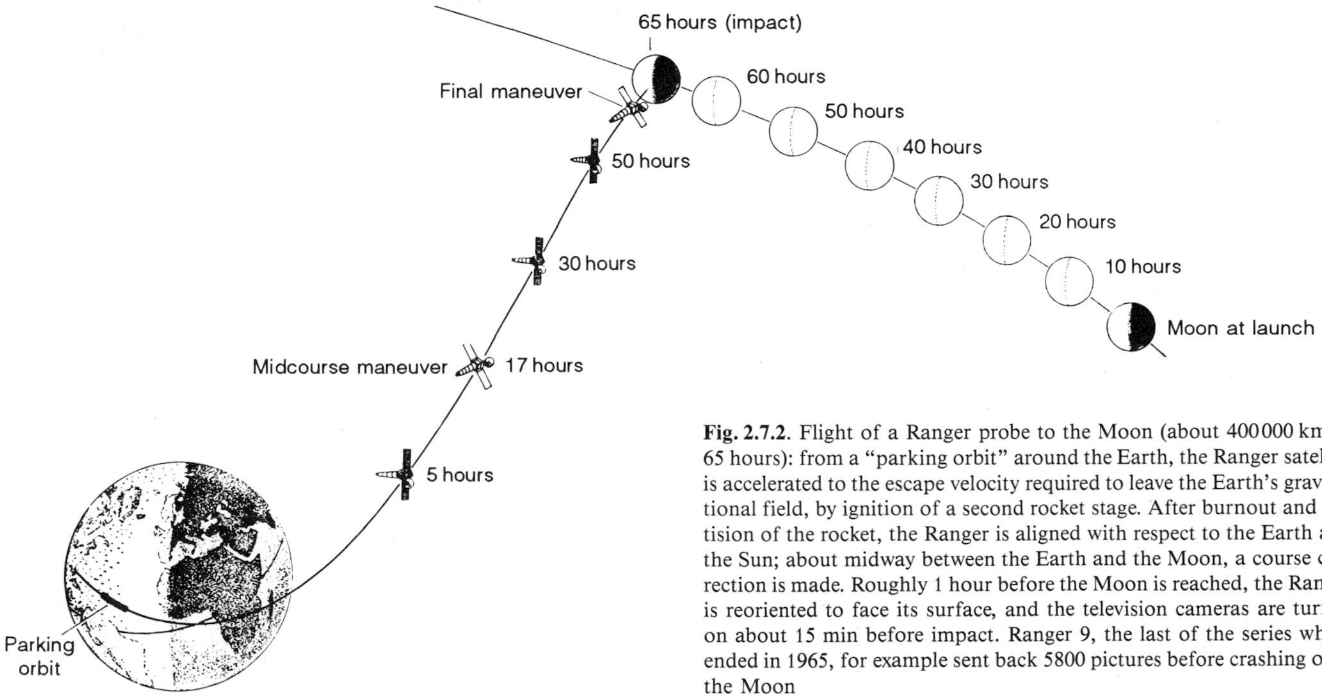

Fig. 2.7.2. Flight of a Ranger probe to the Moon (about 400000 km in 65 hours): from a "parking orbit" around the Earth, the Ranger satellite is accelerated to the escape velocity required to leave the Earth's gravitational field, by ignition of a second rocket stage. After burnout and jettision of the rocket, the Ranger is aligned with respect to the Earth and the Sun; about midway between the Earth and the Moon, a course correction is made. Roughly 1 hour before the Moon is reached, the Ranger is reoriented to face its surface, and the television cameras are turned on about 15 min before impact. Ranger 9, the last of the series which ended in 1965, for example sent back 5800 pictures before crashing onto the Moon

Fig. 2.7.3. The Lunar Module of Apollo 11 has landed in the Mare Tranquillitatis. Astronaut E. Aldrin is setting up the seismic station. Many small craters can be recognized in the foreground; pieces of rock are strewn about on the ground, and the shoes of the astronauts' space suits leave sharp imprints

"Moon-Cat's-eye", or, to use its official name, the "Laser Ranging Retro-Reflector": a glass or quartz prism, having the form of the cut-off corner of a cube, has the property that an incident beam of light is returned exactly along its incoming direction due to reflection from all three of the cubic faces. Reflectors of this type were set up on the Moon. Using a large reflecting telescope, intense laser beams can be sent to the Moon and, from their transit times for the round trip, the distance can be determined with a precision of about 15 cm. Using this technique, the accuracy of various important parameters of the Earth-Moon system can be improved by several orders of magnitude.

The investigation of the *Solar System* using unmanned space probes began in the 1960's with flights to Venus and Mars. The American flights initially made no attempt at landing, but rather gathered data from "fly-by's" of the probes in the series Mariner and Pioneer as close as possible to the planets; in contrast, several of the Soviet space probes in the Venera and Mars series were launched with the goal of placing instrumentation capsules on the planetary surfaces. We unfortunately cannot describe in detail here the great variety of instrumentation used in the space probes, which includes cameras for direct photography, spectrometers, radiometers, and detectors for particles and magnetic fields, and other instruments.

The first successful fly-by of Venus was that of Mariner 2 in 1962; that of Mars was by Mariner 4 in 1965. Mariner 10 investigated Venus in 1974 and then continued on to Mercury, after its orbit was suitably redirected by the passage through the gravitational field of Venus. This fly-by technique ("slingshot technique") was later employed for the missions of the Voyager probes to the outer planets, for example using the gravitational field of Jupiter (Fig. 2.7.4).

The atmosphere of *Venus* was first studied *in situ* in 1967 from a capsule ejected from the space probe Venera 4 and lowered to the surface by a parachute. In 1970, the first of a series of landings on the surface of Venus was successfully carried out by Venera 7. For longer series of observations, artificial satellites (orbiters) were placed in orbit around the planet. For example, Pioneer Venus 1 has been carrying out a cartographic investigation of the surface of Venus since 1978, and Magellan since 1990, using radar mapping techniques.

Orbiters have also been employed for the study of *Mars*: Mariner 9, Mars 2 and 3 (1971), Mars 5 (1973), and most recently Viking 1 and 2 (1976). While the instrument capsule landed by Mars 3 transmitted data for only 20 s, the two Vikings have provided us with a large amount of information on the details of the Martian surface.

Space missions to the *major planets* began in 1972 with the launch of Pioneer 10, which reached Jupiter in 1973, and that same year with Pioneer 11, which flew past Jupiter in 1974 and also investigated Saturn during a close approach in 1979. The two space probes Voyager 1 and 2, launched in 1977, have carried out close-up investigations of the Jupiter and Saturn systems; Voyager 2 also reached Uranus in 1986 and finally Neptune in 1989 (Fig. 2.7.4).

The *interplanetary medium* in the neighborhood of the Sun was studied in 1974–76 by the two "Sun probes" Helios 1 and 2, whose orbits attained perihelia of about 0.3 AU.

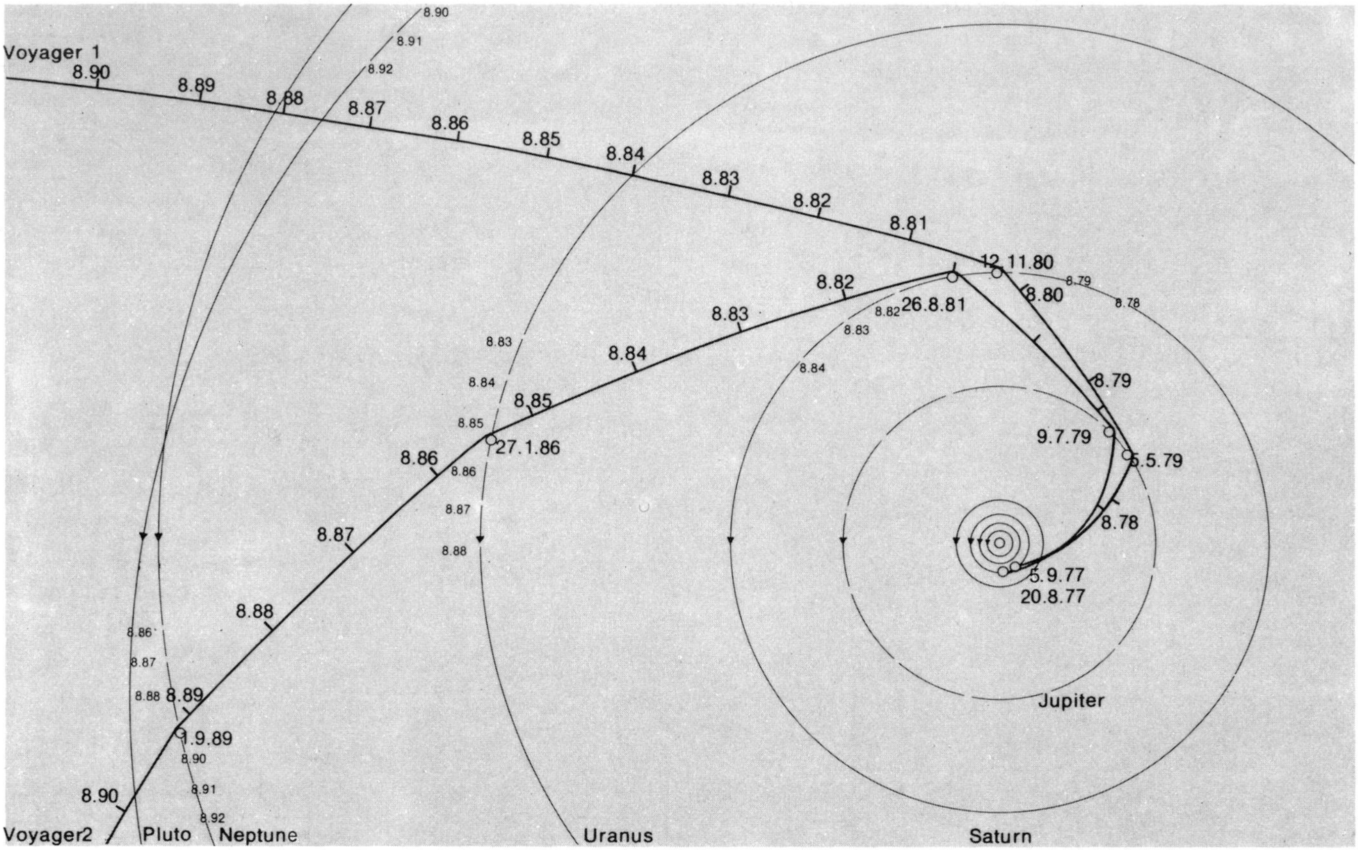

Fig. 2.7.4. The orbits of the space probes Voyager 1 (launched on Sept. 5, 1977) and Voyager 2 (August 20, 1977) to the major planets. The gravitational field of Jupiter was used to produce the required redirection of the orbit to allow a near passage to Saturn (in 1980 and 1981; "slingshot technique"). The favorable constellation of the planets allowed Voyager 2 to pass near Uranus in 1986, and near Neptune in 1989. (G. Hunt and P. Moore, 1983; with the kind permission of the Herder-Verlag, Freiburg)

Following the passage in 1985 of the International Cometary Explorer through the tail of Comet Giacobi-Zinner 1986, the return of *Halley's Comet* in the same year offered an opportunity for no less than five space probes with a variety of instrumentation to investigate the cometary gas and dust *in situ* and to photograph its head from close up: the two Japanese probes Sakigake and Suisei approached Halley to within $7 \cdot 10^6$ and $1.5 \cdot 10^5$ km, respectively; the Soviet probes Vega 1 and 2 first flew by Venus (Vega = Venus-Halley, from Russian *Ve*nera-*Ga*llei) and then approached Halley's Comet to within 8000 to 9000 km; and the European probe Giotto passed the cometary head at a distance of only 600 km.

The results of the manned and unmanned flights to the Moon as well as those of the interesting missions to other objects within the Solar System will be discussed together with the detailed descriptions of the objects concerned.

2.8 The Physical Structure of the Planets and Their Moons

The study of the planets and their satellites has developed in recent years under the influence of space research into one of the most interesting but also most difficult branches of astrophysics. In particular, understanding the observations has required all the available resources of physical chemistry and the geological sciences (geology, mineralogy, etc.). In Sect. 2.8.1, we shall begin with some remarks concerning experimental methods, although some of the necessary instruments cannot be treated before we come to Chap. 3. We then discuss some general theoretical viewpoints, the energy balance of the planets (Sect. 2.8.2), their interior structure and stability (Sect. 2.8.3), and the properties of their atmospheres (Sect. 2.8.4).

Proceeding from these basics, we then treat the Earth and the Moon, which are the members of the Solar system which have been studied in most detail, and the earthlike planets Mercury, Venus, and Mars (Sect. 2.8.5); finally, we discuss the asteroids (Sect. 2.8.6). All these celestial bodies have mean densities $\bar{\varrho}$ in the range from 3000 to 5500 kg/m^3, corresponding roughly to those of terrestrial rocks or metals. We then turn to the the major planets Jupiter, Saturn, Uranus, and Neptune, which have completely different structures, with mean densities $\bar{\varrho}$ in the range 700 to 1600 kg/m^3, corresponding roughly to those of liquefied gases in the laboratory (Sect. 2.8.7). Finally, we treat Pluto, which is much smaller than the other outer planets (Sect. 2.8.8).

It is only reasonable to treat the formation and evolution of our Solar System (cosmogony) in connection with stellar evolution, in Chap. 6.

2.8.1 Ways of Studying the Planets and Their Satellites

We need make no further remarks about methods of determining the apparent and true diameters of the planets or their masses, and the mean densities $\bar{\varrho}$ which are calculated from those data. Numerical values (for the following sections as well) are collected in Table 2.8.1; Fig. 2.8.1 gives some feeling for the true sizes of the planets and the Sun.

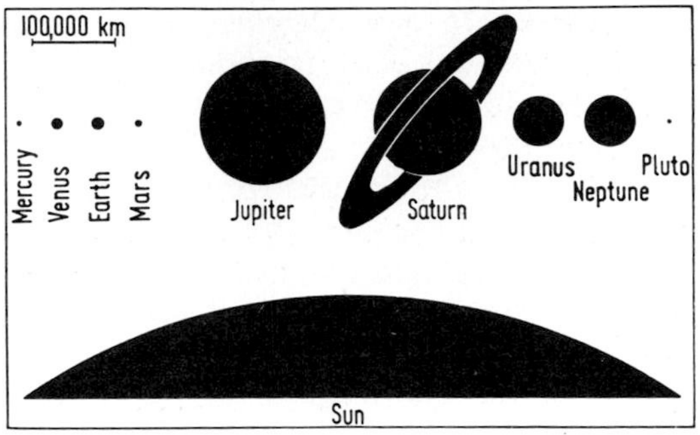

Fig. 2.8.1. The actual relative sizes of the planets and the Sun

The *rotational periods* can be determined by observation of any sufficiently permanent surface phenomena. Another possibility is to use the Doppler effect of the Fraunhofer lines in reflected sunlight or the absorption lines of the atmosphere itself.

Radar methods allow us to discern ring-shaped zones around the midpoint of the planetary disk of nearby planets, by making use of the different propagation times of the radar waves. If the planet rotates, sectors parallel to the projection of the axis of rotation can also be differentiated according to the Doppler shift of the reflected waves. Using the 300 m paraboloid antenna near Arecibo, Puerto Rico in 1964/65, the first clearcut measurements

Table 2.8.1. Physical properties of the planets and the Moon. Radius in units of $R_\oplus = 6378.1$ km, mass in units of $\mathcal{M}_\oplus = 5.97 \cdot 10^{24}$ kg. Overall acceleration g contains the centrifugal acceleration

	Equatorial radius $R/R_\oplus$	Mass $\mathcal{M}/\mathcal{M}_\oplus$	Mean density $\bar{\varrho}$ [g cm^{-3} $= 10^3$ kg m^{-3}]	Sidereal rotation period [d]	Inclination of equator to orbital plane	Overall acceleration at the equator g [m s^{-2}]	Global effective temperature $\bar{T}_{eff}$ [K]
Mercury	0.38	0.055	5.43	58.65	2°	3.7	443
Venus	0.95	0.82	5.24	243.0[a]	3°	8.9	230
Earth	1.00	1.00	5.52	0.997	23.5°	9.8	255
Moon	0.27	0.012	3.34	27.32	6.7°	1.6	274
Mars	0.53	0.11	3.93	1.03	23.9°	3.7	216
Jupiter	11.2	317.8	1.33	0.41	3.1°	23.2	125
Saturn	9.45	95.1	0.70	0.45	26.7°	9.3	94
Uranus	4.02	14.6	1.27	0.72	98°	(8.4)	58
Neptune	3.89	17.1	1.64	0.67	29°	(10.9)	58
Pluto	0.18	(0.002)	(2.1)	6.39	122°	(0.7)	(42)

[a] retrograde rotation

of the *rotation of Venus and Mercury* were finally achieved, and were able even to localize individual craters and mountains on the surface of Venus through the completely opaque cloud layer.

Conclusions about the *distribution of mass* in the interiors of the planets and satellites can be reached by measuring the polar flattening due to their rotation, from their precession in the gravitational fields of other masses (Sect. 2.6.5), as well as from precise measurements of the gravitational potential or acceleration of gravity at their surfaces and (using artificial satellites or space probes) some distance out from the surface. In the case of the Earth and the Moon, investigations of the propagation of seismic waves also give information on the depth dependence of the elastic constants and the densities.

The *reflectivity* is described by quoting the *albedo*, which is defined as the ratio of the intensity of sunlight reflected or scattered in all directions to that of the incident light. Its magnitude and wavelength dependence give information on the nature of the surface, particularly in the case of the planetoids and moons. More detailed information can be obtained from the brightness as a function of the phase angle, or the surface brightness as a function of the angles of incidence and of reflection, as well as the *polarization* of the reflected light, when such data are available.

Some gases, at least, can be identified in the *spectrum* of a planet by terrestrial observations, by utilizing their absorption bands, which occur in addition to the Fraunhofer lines of the reflected solar spectrum. By making the observations outside the Earth's atmosphere, e.g., from a satellite, we can avoid interference from the terrestrial bands due to H_2O, CO_2, O_3, etc., and can also identify other components of the planetary atmosphere by means of spectral lines in the ultraviolet and the infrared.

We obtain information about the *temperatures* in the atmospheres (or on the surface of the planet or moon when the atmosphere is sufficiently transparent) by determining the intensity of the emitted thermal radiation in the infrared or mm- to dm-wave regions of the radio spectrum. However, a theoretical model of the atmosphere is necessary to decide which layers correspond to the measured temperatures.

The many possibilities of *space travel* for investigating the Solar System, such as near approaches of space probes to the planets and their satellites, manned and unmanned landings with *in situ* measurements, or artificial satellites around other planets for longer observations, have already been discussed in Sect. 2.7; there we also gave a summary of the most important missions to the Moon and the planets thus far.

Before we turn to the individual planets and their satellites, we shall first present some theoretical considerations about their energy balances, their interior and atmospheric structures, and their stabilities.

2.8.2 The Global Energy Balance of the Planets

The *input of radiant energy* of the Sun at a distance of 1 AU is given by the solar constant, $S = 1.37\,\mathrm{kW/m^2}$ (Sect. 4.3.2); for a planet with an orbital radius r, it is

$$S(r) = S\left(\frac{r}{1\,\mathrm{AU}}\right)^{-2} . \qquad (2.8.1)$$

The planet (of radius R) absorbs $\pi R^2(1-A)S(r)$ of this power, where A is its mean albedo (mean reflectivity). For the Earth, $A \simeq 0.3$, whereby a major part is caused by the clouds (whose reflectivity $\simeq 0.5$) which cover on the average about 50% of the surface.

Radiant losses occur for the most part in the infrared; for a "black" planet at temperature T, they would be given by the Stefan-Boltzmann radiation law as the surface area $4\pi R^2 \times \sigma T^4$, where the radiation constant $\sigma = 5.67 \cdot 10^{-8}\,\mathrm{W\,m^{-2}K^{-4}}$ (Sect. 4.2.3).

For objects like the planets, which are not black bodies, a *global effective temperature* $\bar{T}_{\mathrm{eff}}$ is *defined* by the Stefan-Boltzmann law.

Since the radiant energy input together with the heat input Q from internal energy sources just balances the radiant losses, we have

$$\pi R^2(1-A)S(r) + 4\pi R^2 Q = 4\pi R^2 \sigma \bar{T}_{\mathrm{eff}} . \qquad (2.8.2)$$

For the earthlike planets, the internal energy sources may be neglected in comparison to the radiant energy input from the Sun; in the case of the Earth, the mean heat input, which is predominantly due to the energy released by the decay of radioactive elements in the crust, is only $Q = 0.06\,\mathrm{W/m^2} \simeq 10^{-4}\,S$. On the other hand, infrared measurements for the major planets Jupiter, Saturn, and Neptune indicate radiant losses which are 2 to 3.5 times greater than the absorbed solar radiation:

	Jupiter	Saturn	Neptune
$\dfrac{4\sigma \bar{T}_{\rm eff}}{(1-A)S(r)}$	1.9(±0.2)	≃ 3.5	2.4(±1)

This energy is due to the release of gravitational energy or to heat remaining from the time of the formation of the planets. Whether or not Uranus also has an internal energy source is still not completely clear: in contrast to the infrared observations, the temperature distribution on the planetary surface seems to indicate an internal source.

The global effective temperatures are listed in Table 2.8.1. As we shall see in the discussions of the individual planets, the actual temperatures which occur in the atmospheres and on the planetary surfaces differ considerably from $\bar{T}_{\rm eff}$. On the one hand, rotation and atmospheric flow play a decisive role in equalizing the temperatures in the day and night portions of the planet; on the other, a strong dependence of atmospheric transparency on the wavelength of the radiation can lead to the well-known "greenhouse effect" or to selective heating of particular layers, for example of the ozone layer in the Earth's atmosphere.

2.8.3 Interior Structure and Stability

The pressure distribution in the interior of a planet or moon (or also of a star; see Sect. 4.12.1) is determined by its *hydrostatic equilibrium*. If we consider a volume element with the base area dA and height dr at a distance r from the center, then its mass $\varrho(r)dA\,dr$ (ϱ: density) will be attracted by all the masses which are closer to the center. That part of the mass of a planet $\mathcal{M}(r)$ which is located within a sphere of radius r is given by

$$\mathcal{M}(r) = \int_0^r \varrho(r')4\pi r'^2 dr' \quad \text{or} \quad \frac{d\mathcal{M}(r)}{dr} = 4\pi r^2 \varrho(r) \; . \tag{2.8.3}$$

$\mathcal{M}(r)$ produces a gravitational acceleration at its surface according to Newton's law of gravitation:

$$g(r) = \frac{G\mathcal{M}(r)}{r^2} \; . \tag{2.8.4}$$

Within our volume element, the pressure p thus changes by the amount

$$\begin{aligned} -dp\,dA &= \varrho(r)dA\,dr \times g(r) \\ \text{Force} \quad &= \text{Mass} \times \text{Acceleration} \; . \end{aligned} \tag{2.8.5}$$

The *hydrostatic equation* for the planet is thus given by

$$\frac{dp}{dr} = -\varrho(r)g(r) = -\varrho(r)\frac{G\mathcal{M}(r)}{r^2} \; . \tag{2.8.6}$$

For the special case of a *homogeneous sphere* with $\varrho(r) = \bar{\varrho} =$ const., we can readily integrate (2.8.6) and thus obtain an estimate of the pressure $p_{\rm c}$ at the center of the planet. Let $\mathcal{M}$ be the total mass of the planet and R its radius; then from (2.8.6) we find, using $\mathcal{M}(r)/\mathcal{M} = (r/R)^3$:

$$p_{\rm c} = \int_R^0 \frac{dp}{dr}dr = \bar{\varrho}\int_0^R \frac{G\mathcal{M}}{r^2}\left(\frac{r}{R}\right)^3 dr = \frac{1}{2}\bar{\varrho}\frac{G\mathcal{M}}{R} \; . \tag{2.8.7}$$

It can be shown that a homogeneous mass distribution gives a minimum value for $p_{\rm c}$, as long as we only consider the case that $\varrho(r)$ increases monotonically towards the center.

For the pressure at the center of the Earth, we obtain from (2.8.7) and (2.6.32) the estimate $p_{\rm c} \simeq 1.7\cdot 10^{11}$ Pa $\simeq$ 1.7 Mbar[8], which is about a factor of two smaller than the actual value.

In the general case, we shall need the *equation of state* of the material, $p = p$ (ϱ, T; chemical composition), in order to find a solution of the hydrostatic equation, and we can therefore only obtain the pressure or density distribution if the temperature profile $T(r)$ is known. This quantity is determined by energy transport. In the interior of an earthlike planet, the equation of state is only weakly dependent on T, so that here $\bar{\varrho}$ and $\mathcal{M}$, together with information about the mass distribution (Sect. 2.8.1) and the chemical composition, practically determine the planet's internal structure. The pressures deep in the interior of a planet exceed those which are at present obtainable in the laboratory; the equation of state must therefore be derived from theoretical considerations.

We now consider the *stability* of a satellite with respect to tidal forces. A satellite which orbits around its central body at a distance r will tend to be pulled apart by the tidal forces of the latter (Sect. 2.6.6) and, if it approaches too closely, will disintegrate or will never be formed. If

[8] 1 Pa (Pascal) = 1 N·m^{-2} = 1 kg·m^{-1}s^{-2} is the SI unit of pressure. 1 Pa = 10 dyn/cm^2 = 10^{-5} bar.

the satellite is not too small, we can neglect its internal cohesive forces and can readily estimate the critical tidal force. We assume for the central body (planet) the mass $\mathcal{M}$, radius R, and mean density $\bar{\varrho}$, and correspondingly for the satellite $\mathcal{M}_S$, R_S, and $\bar{\varrho}_S$. Now we can estimate the mutual attraction of the parts of the satellite by replacing it by two masses $\simeq \mathcal{M}_S/2$ at a distance R_S, i.e.:

$$\frac{G\mathcal{M}_S\mathcal{M}_S}{4R_S^2} . \tag{2.8.8}$$

On the other hand, the tidal force which pulls the two fictitious masses $\mathcal{M}_S/2$ apart is, according to (2.6.46), equal to $G\mathcal{M}\mathcal{M}_S R_S/r^3$. Centrifugal (i.e. inertial) terms are of the same order of magnitude as these forces. The condition for stability of the satellite is therefore

$$\frac{G\mathcal{M}_S\mathcal{M}_S}{4R_S^2} \geq cG\frac{\mathcal{M}\mathcal{M}_S}{r^3}R_S , \tag{2.8.9}$$

where c is a constant of the order of one. If we now use the fact that for the satellite $\mathcal{M}_S = (4\pi/3)\bar{\varrho}_S R_S^3$, and correspondingly for the planet, $\mathcal{M} = (4\pi/3)\bar{\varrho}R^3$, we obtain

$$\frac{r}{R} \geq (4c)^{1/3}\left(\frac{\bar{\varrho}}{\bar{\varrho}_S}\right)^{1/3} . \tag{2.8.10}$$

A more precise calculation due to E. Roche (1850) yields (for a satellite which rotates synchronously, as does our Moon) the *stability limit*:

$$\frac{r}{R} \geq 2.44\left(\frac{\bar{\varrho}}{\bar{\varrho}_S}\right)^{1/3} . \tag{2.8.11}$$

A (large) satellite with the same density as its central body is thus not " allowed" to approach the latter closer than 2.44 planetary radii.

For *smaller* satellites, in contrast, it was remarked by H. Jeffreys (1947) that the internal cohesive forces must be considered. These are larger than the intrinsic gravitational forces for objects with radii $R_S \leq R_0$, where R_0 is determined by the *tensile strength* ζ, i.e. by the maximum force per initial cross-section which can pull on the body without tearing it apart. In the framework of our estimate of the attractive force (2.8.8), we find:

$$\zeta R_0^2 \simeq \frac{G\mathcal{M}_S^2}{4R_0^2} = \frac{1}{4}\left(\frac{4\pi}{3}\right)^2 G\bar{\varrho}_S^2 R_0^4 \tag{2.8.12}$$

or, up to a factor of the order of one,

$$R_0 \simeq \sqrt{\frac{\zeta}{G\bar{\varrho}_S^2}} . \tag{2.8.13}$$

For stone, the tensile strength is $\zeta \simeq 10^8$ Pa; for ice, it is about $2\cdot 10^7$ Pa; and for loose material, for example corresponding to carbonaceous chondrites (Sect. 2.9.2), it is $\leq 10^6$ Pa. Therefore, for a density $\bar{\varrho}_S \simeq 3000\ \mathrm{kg\,m^{-3}}$, R_0 lies in the range 30 ... 300 km. From these considerations, we see that the hydrostatic equation (2.8.6) represents a good approximation for the structure of planets and satellites which are not too small. Finally, for smaller satellites, we can estimate the stability limit by comparing ζR_0^2 with the tidal force $G\mathcal{M}\mathcal{M}_S R_S/r^3$. We readily see that this limit lies closer to the central body than *Roche's limit* (2.8.11) which applies to larger satellites dominated by their intrinsic gravitation. Precise calculations, which also take into account the dynamics of the process of disintegration, yield $r/R \simeq 1.4$.

2.8.4 The Structure of Planetary Atmospheres

The pressure distribution in a planet's atmosphere is, like the pressure distribution in its interior, governed by the hydrostatic equation (2.8.6). If the thickness of the atmosphere is small compared to the planetary radius R, we can take the gravitational acceleration (2.8.4) to be constant, $g = G\mathcal{M}/R^2$. Introducing the altitude $h = r - R$, we find

$$\frac{dp}{dh} = -g\varrho(h) . \tag{2.8.14}$$

The pressure p is related to the density ϱ (or the particle density n), the temperature T, and the mean molecular mass $\bar{\mu}$ by the equation of state of an ideal gas:

$$p = \varrho\frac{kT}{\bar{\mu}m_\mathrm{u}} = \varrho\frac{\mathcal{R}T}{M} = nkT . \tag{2.8.15}$$

$k = 1.38\cdot 10^{-23}$ J/K is the Boltzmann constant, $\mathcal{R} = 8.31\ \mathrm{J\cdot K^{-1}\cdot mol^{-1}}$ is the universal gas constant, $m_u = 1.66\cdot 10^{-27}$ kg is the atomic mass constant, and M [kg/mol] is the molecular mass. Substitution in (2.8.14) gives

$$\frac{dp}{p} = -\frac{g\bar{\mu}m_\mathrm{u}}{kT}dh = -\frac{dh}{H} \tag{2.8.16}$$

with the equivalent height or scale height

$$H = \frac{kT}{g\bar{\mu}m_u} \; . \tag{2.8.17}$$

If H is constant, we can easily integrate and obtain the barometric formula, which is usable for moderate altitudes:

$$\ln(p) - \ln(p_0) = -\frac{h}{H} \quad \text{or} \quad p = p_0 e^{-h/H} \; , \tag{2.8.18}$$

where p_0 is the pressure at the surface or at the initial height $h = 0$. Quite generally, it follows from (2.8.16) for the region between two levels at h_1 and h_2 with the pressures p_1 and p_2:

$$\ln(p_2) - \ln(p_1) = -\int_{h_1}^{h_2} \frac{dh}{H(h)} \; . \tag{2.8.19}$$

As a result of (2.8.16), the structure of a planetary atmosphere thus depends on (a) the gravitational acceleration g; (b) the mean molecular mass $\bar{\mu}$, i.e. the chemical composition and possibly also the dissociation and ionization of the atmospheric gases; and (c) the temperature distribution $T(h)$. This last quantity is determined by the mechanisms of *energy transport*, that is the input and outflow of thermal energy into each layer from h to $h + dh$, which result from convection and radiation (and in individual layers also from heat conduction).

We can obtain the temperature gradient in a *convective* atmosphere, in which hot matter rises adiabatically (without heat exchange with its environment) and cooler matter falls, by logarithmic differentiation with respect to h of the well-known adiabatic equation:

$$T \propto p^{1-1/\gamma} \; , \tag{2.8.20}$$

where $\gamma = c_p/c_v$ is the ratio of the specfic heats at constant pressure and constant volume. We obtain

$$\frac{1}{T}\frac{dT}{dh} = \left(1 - \frac{1}{\gamma}\right)\frac{1}{p}\frac{dp}{dh} \; . \tag{2.8.21}$$

If we now use the hydrostatic equation (2.8.14) together with the equation of state (2.8.15) and the relation $c_p - c_v = k/(\bar{\mu}m_u)$, we find the *adiabatic* temperature gradient

$$\frac{dT}{dh} = -\frac{g}{c_p} \; . \tag{2.8.22}$$

In the lower, convectively unstable part of the Earth's atmosphere, the troposphere, we obtain for dry air ($c_p = 1005\ \mathrm{J \cdot kg^{-1} K^{-1}}$), using $g = 9.81\ \mathrm{m/s^2}$, a gradient equal to 9.8 K/km; for moist air, the latent heat which is liberated on condensation of moisture leads to a gradient which is only half as large. The measured mean temperature decrease with increasing altitude is 6.5 K/km.

We will investigate the other limit, that of an atmosphere with energy transport by *radiation*, in Sect. 4.8.1, in connection with radiative energy transport in *stellar* atmospheres. The relations derived there are in principle also applicable to planetary atmospheres. In a planetary atmosphere with surface heating due to unabsorbed solar radiation in the visible region, we observe a temperature increase, the so-called *greenhouse effect*, when the reradiated energy in the infrared spectral region is strongly absorbed by the atmosphere.

The general case of an interaction between convective and radiative energy transport cannot be treated here, nor can the atmospheric currents and winds, which are due to the variable input of solar energy depending on the time of day and the season. We shall limit ourselves to reporting some particular results in the descriptions of the individual planets.

First, we ask the question as to whether a planet or satellite can retain its *own atmosphere*. The molecules of a gas with mass m and temperature T have, according to the kinetic theory of gases, a most-probable velocity given by

$$\bar{v} = \sqrt{\frac{2kT}{m}} \; , \tag{2.8.23}$$

where k is again Boltzmann's constant. According to (2.6.42), a molecule having the velocity v can escape from a celestial body of mass $\mathcal{M}$ and radius R when $v^2/2 \geq (G\mathcal{M})/R$. Taking into account the Maxwell-Boltzmann probability distribution of molecular velocities $v > \bar{v}$, we can understand that, e.g., Mercury, our Moon, and most of the satellites in the Solar System can have practically no atmospheres, and that, on the other hand, Saturn's largest moon, Titan, at the temperature calculated from (2.8.2), could, in fact, retain its atmosphere for a very long time.

In the outermost layers of a planetary atmosphere, the *exosphere*, there are only infrequent collisions between the gas particles owing to the low density. The electrically *neutral* particles thus move practically on Keplerian orbits in the gravitational field of the planet; the exosphere cannot be described by the hydrostatic equation. For the motion of *charged* particles, which are formed in the upper atmosphere by ionization processes mainly due to solar UV radiation, the planet's magnetic field is the determining factor. The extent of the *magnetosphere* is determined by the interaction between the planetary magnetic field and the plasma flowing out from the solar wind (Sect. 4.10.7).

2.8.5 The Earth and the Moon. The Earthlike Planets and the Asteroids

Inside the asteroid belt – which practically divides the Solar System into two physically differing zones – all the planets have masses less than 1 Earth mass and mean densities between 3900 and 5500 kg/m^3. They evidently consist essentially of solid matter. Their atmospheres are chemically oxidizing; they contain O_2, CO_2, H_2O, N_2 . . . . It thus seems justified to collect them under the name "earthlike planets". We shall examine our Earth somewhat more carefully, as a paradigm for the other planets; of course, we cannot attempt to give a summary of all of geophysics here. Knowledge of our Moon has made a quantum leap within a few years, thanks to the successes of space travel. Mercury, Venus, and Mars have also been examined at close distance in recent years.

a) Internal Structures

As a result of its rotation, the *Earth* is to a good approximation a *flattened ellipsoid of rotation*, the so-called terrestrial spheroid, with

$$\begin{aligned} &\text{equatorial radius} && a && = 6378.1\ \text{km} \ , \\ &\text{polar radius} && b && = 6356.8\ \text{km} \ , \quad \text{and} \\ &\text{flattening ratio} && \frac{a-b}{b} && = \frac{1}{298} \ . \end{aligned}$$

Flattening and centrifugal force have the effect that the gravitational acceleration at the equator is 1/189 weaker than at the poles. *Mars*, with its somewhat slower rotation, has a flattening of 1/171. Precise measurements of the gravitational potential show that both planets have a "pear shape"; for the Earth, the largest deviation from the spheroidal shape is only 17 m, while for Mars it is nearly 2 km. *Venus*, as a result of its extremely slow rotation, exhibits no deviation from a spherical shape.

The *Moon*, due to its rotation which is synchronous with its orbital period around the Earth, is elongated along the axis connecting its midpoint with that of the Earth; its geometrical midpoint lies about 2 km further from Earth than its center of gravity. *Mercury*, owing to the (3 : 2) resonance of its rotation with the (eccentric) orbital motion, is subject to constant deformation by tidal forces.

The *mean densities* of the planets and the Moon have already been discussed (Table 2.8.1). The density of the *Earth's crust* (granite, basalt) is 2600 to 3000 kg/m^3, and similar densities have been found for the crusts of the earthlike planets.

The *moments of inertia* about the rotational axis, $J = \alpha \mathcal{M} R^2$ ($\mathcal{M}$: mass, R: equatorial radius of the planet), which have been obtained from observations of the precession or determinations of the gravitational field using artificial satellites, indicate that the density increases with depth. In contrast to a homogeneous sphere for which $\alpha = 0.40$, the value found for the Earth is $\alpha = 0.33$, and for Mars $\alpha = 0.38$; the value for the Moon, $\alpha = 0.39$, differs only slightly from the homogeneous case.

More precise information about the density variations in the interior of a planet is available from the propagation of *seismic waves*. During an earthquake, longitudinal and transverse elastic waves are generated at the epicenter, which is relatively near to the surface of the planet. These propagate through the interior of the planet and will be refracted, reflected, and will interact with each other depending on the depth dependence of the elastic constants and the density. By careful studies of the propagation of seismic waves around 1906, E. Wiechert in Göttingen found that the Earth's interior contains several *surfaces of discontinuity*, at which the elastic constants and the density ϱ change suddenly. The *Earth's crust* has a thickness of about 30 to 40 km under flat regions, increasing to up to 70 km under recently-formed mountain ranges. Beneath the oceans, it decreases down to $\simeq$ 10 km. Its lower limit is the *Mohorovicic discontinuity* or "Moho". Below it, down to 2800 km depth, is the *mantle*, with a density ϱ = 3300 to 5700 kg/m^3; it probably consists mainly of silicates. From 2800 km to 6370 km depth, we have the *core*, with 10000 to 12000 kg/m^3. In it, there is no propagation of transverse waves. In this sense, we can consider the outer part of the core to be liquid, but with an enormous viscosity. Thorough investigations, however, indicate that the *inner core* below 5000 km depth is again solid.

The chemo-mineralogical *composition* of the Earth's core can be only indirectly determined. Laboratory experiments at high pressures and temperatures lead to a phase diagram which is compatible with the geophysical data if the assumption is made that the core, like iron meteorites, consists in the main of Fe and Ni, along with small amounts of S. The *pressure* increase towards the center can be calculated with sufficient accuracy from the hydrostatic equation (2.8.6). The increase in the *temperature* with increasing depth can be measured in deep wells; one obtains a geothermic depth dependence of up to 30 K/km. The temperature distribution at great depths is determined by the heat production due to decay of the radioisotopes ^{238}U, ^{232}Th, and, to a lesser extent, ^{40}K; and, on the other hand, by the slow transport of heat to the exterior by the thermal conductivity and convection of the magma. The temperature of the Earth's core must be at least several thousand degrees, but certainly ≤ 10000 K.

Our knowledge of the interiors of *Mercury*, *Venus*, and *Mars* is considerably less certain than that of the Earth. Model calculations yield on the whole a similar variation of the density (Fig. 2.8.2). While the Fe-Ni core makes up about 30% of the total mass of the Earth and Venus, it forms a noticeably smaller core of 15% in the case of Mars, and is considerably larger at 60% for Mercury. (However, the internal structure of Mercury can also be represented using a nearly homogeneous model).

The mean density of the *Moon*, 3340 kg/m^3, is noticeably similar to that of the Earth's mantle; the average composition of the Moon cannot be the same as that of the Earth. Some information about its internal structure was obtained from the seismometers set up during the Apollo missions. Surprisingly, the seismic waves caused on the Moon by the impact of meteorites or the crash of the discarded Lunar lander could be tracked for up to an hour. The extremely small damping of moonquake waves in comparison to earthquake waves can be attributed to a difference in scattering or to particular dissipation mechanisms. The Moon probably has *no* core of Fe and Ni; the upper limit for the possible radius of such a core is about 380 km.

b) Radioactive Dating. The Earth's History

The methods of radioactive dating give us the most accurate information about the age of the Earth, of the Moon and of meteorites, as well as about the length of the various geologic eras. The radioactive decay of the

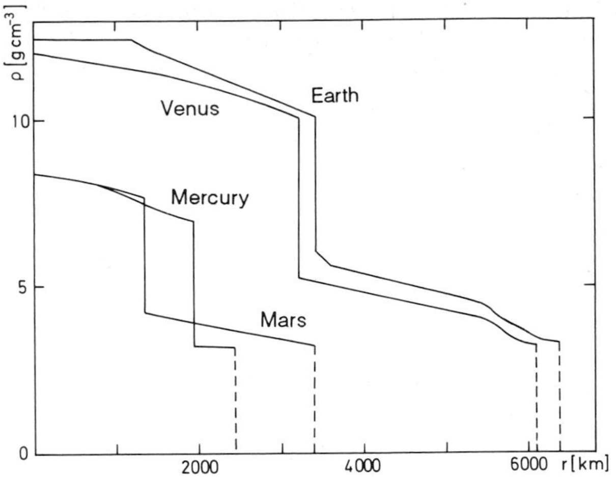

Fig. 2.8.2. The dependence of the density $\varrho(r)$ on distance from the centers of the earthlike planets Mercury, Venus, Earth, and Mars. ϱ is in [g cm^3 = 10^3 kg m^{-3}]

following sufficiently long-lived isotopes is employed (we list only the stable end products of the whole decay chain in each case):

		Half-life T [yr]
^{147}Sm	$\rightarrow$ $^{143}Nd + {}^4He$	$1.06 \cdot 10^{11}$
^{187}Re	$\rightarrow$ $^{187}Os + e^-$	$4.4 \cdot 10^{10}$
^{87}Rb	$\rightarrow$ $^{87}Sr + e^-$	$4.88 \cdot 10^{10}$
^{232}Th	$\rightarrow$ $^{208}Pb + 6\,{}^4He$	$1.40 \cdot 10^{10}$
^{238}U	$\rightarrow$ $^{206}Pb + 8\,{}^4He$	$4.47 \cdot 10^9$
^{235}U	$\rightarrow$ $^{207}Pb + 7\,{}^4He$	$7.04 \cdot 10^8$
^{40}K	$\rightarrow$ $^{40}Ca + e^-$	$1.28 \cdot 10^9$
	$\searrow$ ^{40}Ar (K capture)	$1.1 \cdot 10^{10}$

In all the methods, the ratio of *end* product to *parent* isotope is determined. This ratio was initially ($t = 0$) equal to zero, and after a time t it is given by $2^{t/T} - 1$.

The oldest rocks in the Earth's crust have an age between 3.7 and $3.9 \cdot 10^9$ yr. The age of the Earth since the last mixing – or separation – of its material (it is by no means clear what the mechanism of this process may have been) can be determined by investigating the abundance distribution of the lead isotopes in uranium-free lead minerals or by finding its original value in the "archaic lead" from iron meteorites, which contain practically no uranium or thorium. The relatively exact value obtained for the age of the Earth and of the oldest meteorites, and thus for the *age of the Solar System*, is

$$(4.53 \pm 0.02) \cdot 10^9 \text{ yr} . \tag{2.8.24}$$

The oldest rocks on the Moon likewise are dated at about $4.5 \cdot 10^9$ yr, i.e. the Moon was formed at the same time as the Earth, within the accuracy of the measurements.

The most important *geological periods* of the Earth's history are briefly characterized and their absolute datings are summarized in Table 2.8.2.

c) Magnetic Fields. Planetary Tectonics

The *Earth* possesses a magnetic field, which corresponds roughly to a dipole field with a magnetic flux density at the equator of $3.1 \cdot 10^{-5}$ Tesla (T) or 0.31 Gauss (G). We can characterize a dipole by its *magnetic moment* $\boldsymbol{M}$, which points in the direction of the dipole axis and from which we can obtain the vector of the *magnetic induction* or *magnetic flux density* $\boldsymbol{B}$ by taking the gradient:

$$\boldsymbol{B} = -\nabla \frac{\boldsymbol{M} \cdot \boldsymbol{r}}{r^3} = -\nabla \frac{M \sin \lambda}{r^2} \quad . \qquad (2.8.25)$$

Here, λ is the magnetic latitude; the field strength at the poles ($\lambda = 90°$) is $B_p = 2M/R^3$ (R: radius of the planet), and at the equator ($\lambda = 0°$) it is $\frac{1}{2}B_p$. The radial component of $\boldsymbol{B}$ is $B_r = B_p \sin \lambda$, the component in the direction of λ is $B_\lambda = -\frac{1}{2}B_p \cos \lambda$, and the azimuthal component is $B_\phi = 0$. At large distances from the origin, B decreases as r^{-3}.

In the case of the *geomagnetic field*, $\boldsymbol{M}$ forms an angle of $\alpha = 11.5°$ with the axis of rotation of the Earth and the dipole axis is shifted from the center of the Earth by $450\ \mathrm{km} \simeq 0.07 R_\oplus$. The magnetic moment of the Earth's magnetic field is $8 \cdot 10^{15}\ \mathrm{T \cdot m^3}$ or $8 \cdot 10^{25}\ \mathrm{G \cdot cm^3}$; it is currently decreasing by 0.05% per year, and the dipole axis precesses with about $0.04°\ \mathrm{yr}^{-1}$ (a period of about 9000 yr?).

In the case of *Mercury*, the space probe Mariner 10 surprisingly determined a magnetic field with a moment of $5 \cdot 10^{12}\ \mathrm{T \cdot m^3}$ and $\alpha \simeq 11°$. In contrast, *Venus* and *Mars* probably have no planetary magnetic fields. (The extremely small fields which were measured are likely due to magnetized plasma from the solar wind striking their surfaces.) Our own *Moon* also has no measureable dipole field ($M < 10^9\ \mathrm{T \cdot m^3}$); the weak remanent magnetization of some lunar rocks, corresponding to a few percent of the Earth's field, is an indication of an *earlier* lunar magnetic field.

The magnetic field of the *Earth* and its *secular variation* (the rapid variations caused by the Sun will be treated later) can be interpreted as follows, according to W. M. Elsasser and E. Bullard: the liquid matter in the outer regions of the Earth's core forms large vortices driven by convection and heat exchange as mentioned above. If traces of a magnetic field are present in such a vortex of conducting material, they can be amplified, as in the self-exciting dynamo invented by W. v. Siemens. The details of such dynamos in the Earth's interior are not yet completely clear; however, the dynamo theory offers the only possibility of explaining the terrestrial magnetic field, its secular variations, and its periodic reversals (see below).

The geomagnetic field in past epochs can be determined thanks to the circumstance that when certain minerals are formed, the magnetic field which happens to be present at the time is, so to speak, frozen in to them (P. M. S. Blackett, S. K. Runcorn, and others). Such paleomagnetic determinations have shown that the magnetic field vectors in previous times can best be brought into an orderly scheme by applying the hypothesis of *continental drift*, originally deduced from the arrangement of the continents on the globe by A. Wegener (1912). If the reasonable assumption is made that the Earth had essentially a magnetic dipole field in earlier times, we can reconstruct the relative positions of the continents in previous geologic epochs from the paleomagnetic measurements, and can show, in good agreement with a large number of geological and paleontological observations, how the present continents were formed bit by bit in the course of the Earth's history. For example, the Atlantic ocean had its beginnings about $1.2 \cdot 10^8$ yr ago, i.e. in the Jurassic to Cretaceous periods, as a narrow trench, similar to the Red Sea today. On the average, America and Europe have moved apart by a few centimeters per year in the intervening time (Fig. 2.8.3). In the 1950's, paleomagnetic investigations thus provided strong evidence in favor of the continental drift theory, which had previously been the subject of controversy for many years; then, in the 1960's, they provided an insight into the basic mechanisms of the continental motion: it had been noticed from the paleomagnetic measurements that directly adjacent layers of minerals often indicate diametrically opposed directions for the magnetic field. Was this to be interpreted as a spontaneous remagnetization of the rocks (an effect which is known in physics) or as a reversal of the entire terrestrial magnetic field? Detailed investigations of precisely dated series of layers showed that the latter was the case! This no longer seems so surprising if one considers that in the self-exciting dynamo, the direction of the current is determined by the random

Table 2.8.2. Earth's history

Period Age in 10^6 yr before present	Times of principal mountain formations	Development of flora	Development of fauna	Epochs	First appearance and disappearance	Period Age in 10^6 yr before present
Quaternary		Later angiosperms	Cenozoic	Snails; Mussels Mammals		Quaternary
2					First humans →	2
Tertiary	Alpine	Neophytic				Tertiary
65		Earliest angiosperm		Goniatites Ammonites; Dinosaurs	Extinction of the ammonites and dinosaurs →	65
Cretaceous					First angiosperms →	Cretaceous
144				Brachiopods	First birds →	144
Jurassic		Later gymnosperms	Mesozoic			Jurassic
213					First mammals →	213
Triassic		Mesophytic				Triassic
248		Earliest gymnosperms			Extinction of many paleozoic animals →	248
Permian						Permian
286	Variscan	Later pteridophytes			First reptiles →	286
Carboni-ferous						Carboni-ferous
360		Palaeophytic		Armored fish	First amphibians → First spermatophytes →	360
Devonian			Palaeozoic			Devonian
408		Earliest pteridophytes		Trilobites; Graptolites		408
Silurian	Caledonian				First vascular plants →	Silurian
438						438
Ordovician		Eophytic (Age of algae)			First vertebrates →	Ordovician
505						505
Cambrian					Development of many phyla of invertebrates →	Cambrian
590(?)						590(?)
Precambrian	Assyntic Algomian Laurentian		Proterozoic Archean Eozoic	Precambrian		Precambrian
	oldest rocks: 3800×10^6 yr				Oldest traces of life (stromatolites): 3500×10^6 yr →	

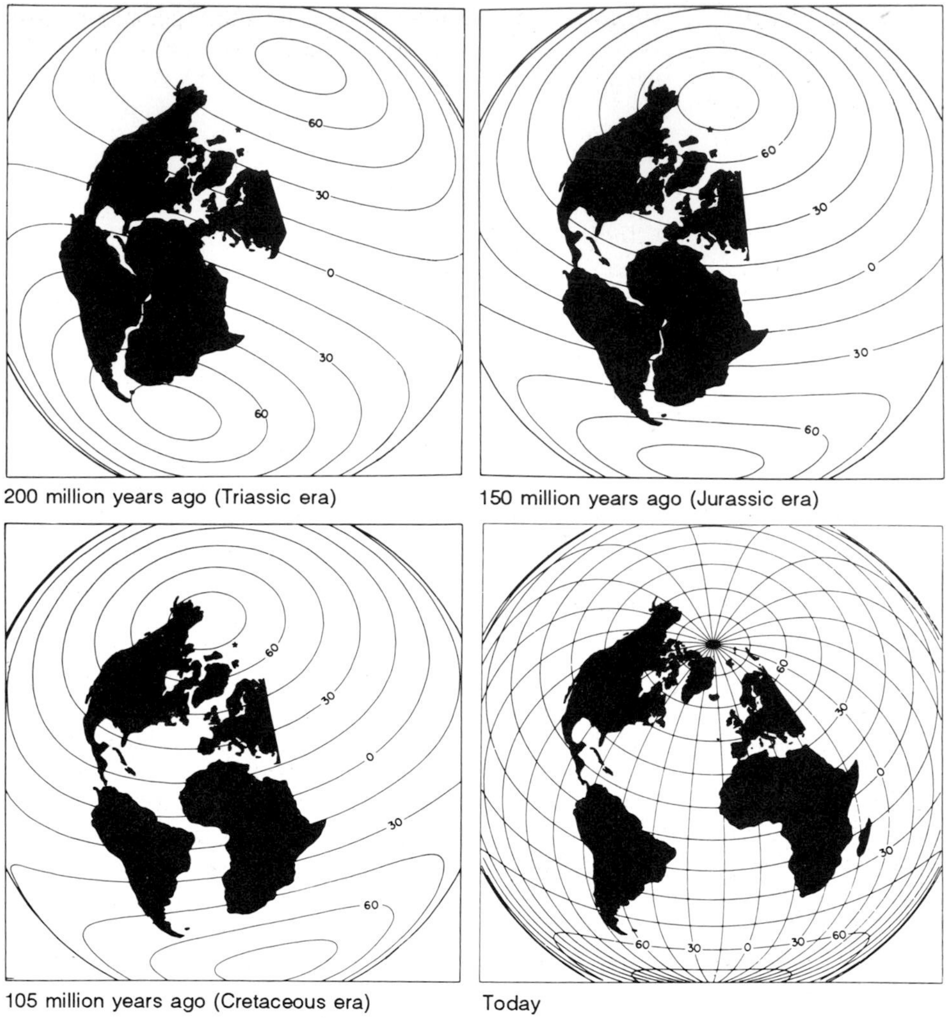

Fig. 2.8.3. The continental drift. The positions of the continents around the Atlantic ocean (in relation to North America) at different times. The continents move apart with velocities of 2.5 to 4 cm per year. (With the kind permission of the Wissenschaftliche Verlagsgesellschaft, Stuttgart)

weak magnetic fields present when the machine is started up. After it had become clear that the geomagnetic field changes its sign in irregular periods of several hundred thousand years, it was discovered by H. H. Hess in 1962, by F. J. Vine and D. H. Matthews in 1963, and by others, that the floor of the Atlantic ocean shows a whole series of strips with approximately north-south orientation and alternating directions of magnetization. Their magnetic dating indicates that the ocean floor has been spreading for about $1.2 \cdot 10^8$ yr, starting from a mid-atlantic ridge about halfway between America and Europe and moving out on both sides, pushing the two continents apart. This fruitful theory of *ocean floor spreading* was soon complemented through the recognition by H. H. Hess, T. Wilson and others (ca. 1965) that in the course of all these processes, large slabs or plates in the lithosphere (i.e. the upper portion of the Earth's crust) are shifted in one piece or are broken along fault lines (this is the origin of tectonic earthquakes); we thus refer to *plate tectonics*.

The driving force behind all of these geologic occurrences is, as had been assumed by A. Holmes as early as 1928, clearly *convection currents*, especially in the upper portion of the mantle; they well up along the mid-oceanic ridges. Here, fresh crust material and fresh plates are formed and forced apart on both sides. Where the oceanic plates are pushed back down into the mantle, deep oceanic trenches are formed, and behind them, due to compression of the continental crust, young mountain ranges are produced by folding. These and many other basic geological processes have such a simple explanation! The necessary energy is provided by the radioactive decay of minerals within the Earth; from the geothermal temperature profile and the thermal conductivity of the rocks, it can be estimated that the entire Earth obtains about 10^{21} J/yr of *thermal energy* in this way. About 0.1% of this energy goes to produce earthquakes. If our "thermodynamic machine" produces mechanical energy with an efficiency of even $\simeq 1\%$, this would be quite sufficient to drive the convection in the mantle. Geologic observations furthermore support the idea that the tectonic activity of the Earth has been subject to strong quantitative and qualitative changes in the course of time, with maxima at 0.35, 1.1, 1.8, and $2.7 \cdot 10^9$ yr ago. S.K. Runcorn has attempted to explain these periods of increased tectonic activity in terms of flow transitions, in each case from a particular convection flow mode to a more complex one with a larger number of convection cells.

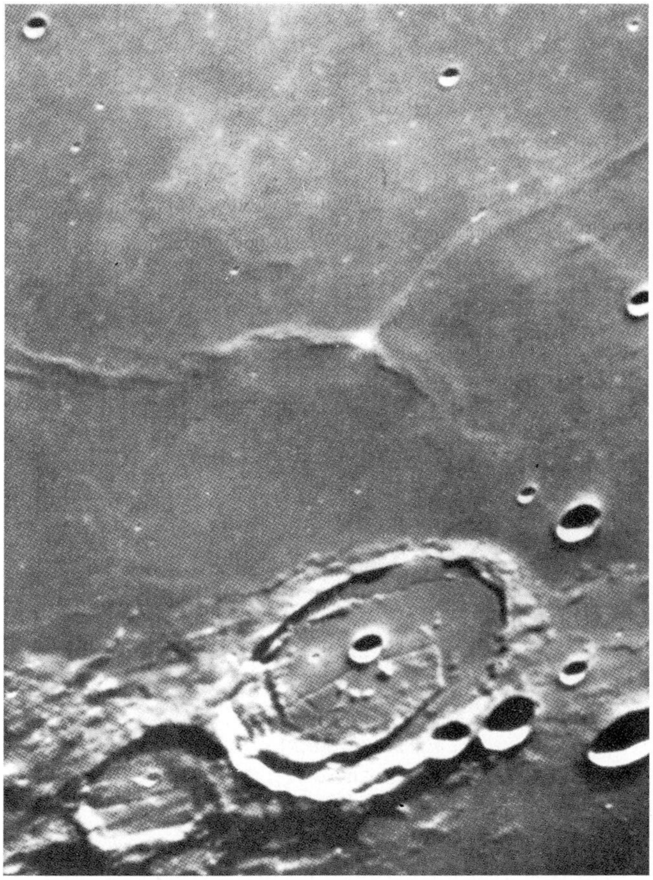

Fig. 2.8.4. The Moon. At the rim of the Mare Serenitatis (*above*), the crater Posidonius (*lower right*) with a diameter of 100 km, and the smaller crater Chacornac (*lower left*) may be seen. In the Mare Serenitatis, there is a richly structured mountain range of $\simeq 180$ m altitude. Numerous small craters can be seen scattered over the landscape. This photograph was made with the 120″ reflector at the Lick Observatory on March 25, 1962

Owing to the continual formation of new, hot lithospheric material and its subsequent cooling, plate tectonics or ocean floor spreading represents the most important process for cooling the Earth's interior.

We cannot go into the details of oceanic physics here, especially since within the Solar System, oceans are unique to the Earth.

Still very little is known about matter flow, convection, etc. in the interiors of the earthlike planets. On the small planets *Mercury* and *Mars*, as well as on the *Moon*, all tectonic and volcanic activity has now ceased; heat transport from the interiors is accomplished by the thermal conductivity of the lithospheres. It is astonishing that Mercury can generate a magnetic field. Its metallic core is, to be sure, relatively large (Fig. 2.8.2); but its rotational velocity, which plays an important role in driving the dynamo, is much slower than e.g., that of Mars, which has no magnetic field of its own.

In the case of *Venus*, there are no clear topographical indications for plate tectonics, although its inner structure and energy sources are quite similar to those of the Earth. Venus also has no magnetic field. It is possible that differences in the thickness of the crust and its temperature and thermal conductivity, as well as the slow rotation, are responsible for the lack of plate tectonics and of a magnetic field on Venus. Loss of heat from the interior of Venus is probably mostly due to the thermal conductivity of the crust and to cooling of hot magma, which reaches the surface in several volcanic regions.

We shall now turn to the description of the surfaces and the surface-forming processes on the earthlike planets and their moons, beginning with our own Moon.

d) The Lunar Surface

Selenographic investigations with increasingly larger telescopes (Fig. 2.8.4), with space probes and finally through the landings of the Apollo astronauts have given us detailed information on the formations of the lunar surface (Fig. 2.8.5).

There can now be no more doubt that all of the circular formations, from the enormous maria (terming them "seas" has only historical significance) – Mare Imbrium has a diameter of 1150 km and a depth of 20 km –

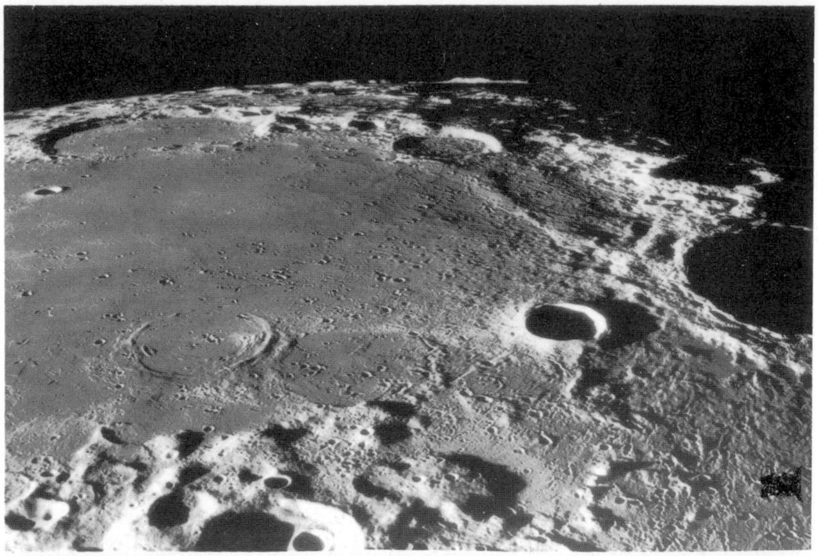

Fig. 2.8.5. The Mare Nectaris with the craters Theophilus, Mädler, and Daguerre. The photo was taken from the command module of Apollo 11 at an altitude of 100 km. In the Mare Nectaris, two "ghost craters" can be seen in the foreground, mostly covered over with lava; otherwise, there are only numerous small craters

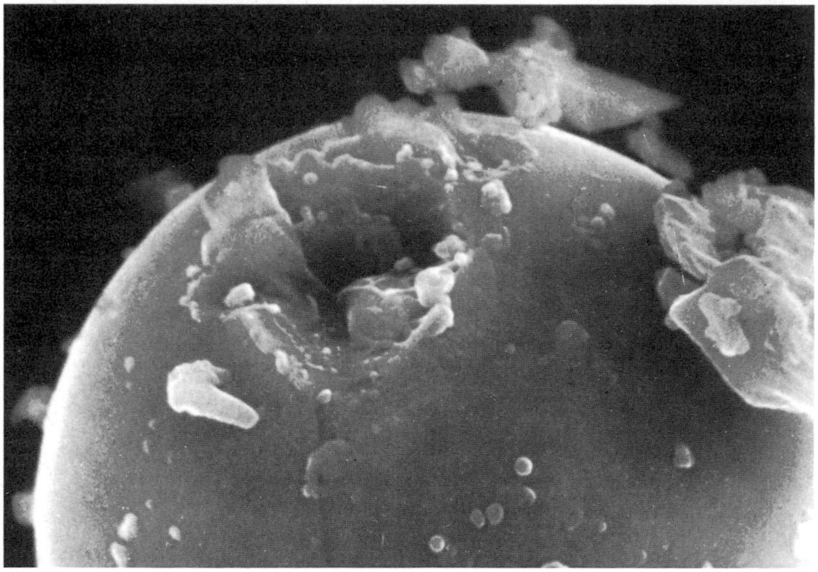

Fig. 2.8.6. In lunar dust (collected by Apollo 11), glasslike spherules are found; they were formed from stone liquified by meteorite impacts. This electron microscope image by E. Brüche and E. Dick (1970) shows one of these spherules which has a diameter of 0.017 mm; the impact of a micrometeorite has produced a "microcrater" on its surface

through the craters and down to microscopic holes of a few micrometers diameter in smooth surfaces (Fig. 2.8.6), were caused by the impact of meteor-like bodies with planetary velocities. The depth of these *impact craters* varies with the diameter in the same way as in terrestrial explosion craters. The floors of the maria and of many larger craters were evidently later flooded by liquid lava. After it had hardened, the smooth surface was again covered with smaller craters.

The nature of the meandering *rills* was for a long time one of the major riddles of lunar research. The Apollo 15 astronauts were able to go to the so-called Hadley Rill and to observe its steep walls clearly from a short distance. It is probably a lava channel, which was originally "roofed over" to a considerable extent. Similar lava channels or tubes can be found in terrestrial volcanoes. Other, longer rills may be explained as cracks in cooling lava.

E. Pettit and S. B. Nicholson determined by means of infrared measurements that the shadow's edge during an eclipse of the Moon changes its temperature only very slowly, and thus they could estimate the low thermal conductivity of the surface material. In fact, the surface of the Moon is covered to a large extent with fine dust and loose fragments of stone (regoliths), which were produced by the impacts of the meteorites.

Many of the larger craters are surrounded by a system of bright *rays*, readily seen even with a small telescope. They evidently consist of material which was thrown out when the crater was formed.

The highest mountains are to be found in the bright, crater-pocked highlands, the *Terra regions*. Their height is limited by the breaking strength of the material and is of the same order as those on the Earth, as was already determined by Galileo, who observed the lengths of their shadows along the terminator or day-night boundary on the Moon.

The *Moon rocks* brought back by the astronauts from the maria are filled with bubble-like hollows, similar to terrestrial lavas which have hardened under low pressure. They are thus *igneous* rocks, i.e. they were formed by hardening from the melt. The *Moon dust* mentioned above was formed by pulverizing of such rocks. The *breccias* consist of dust and other small particles which have become clumped together. Mineralogically, the lunar rocks are roughly similar to terrestrial basalts. In Fig. 2.8.7, we compare the frequency of occurrence of some important elements in lunar material with those in terrestrial basalts and in eucrites (Sect. 2.9.2), i.e. the basalt-like achondrites and type 1 carbonaceous chondrites, meteorites which are the most similar to solar matter.

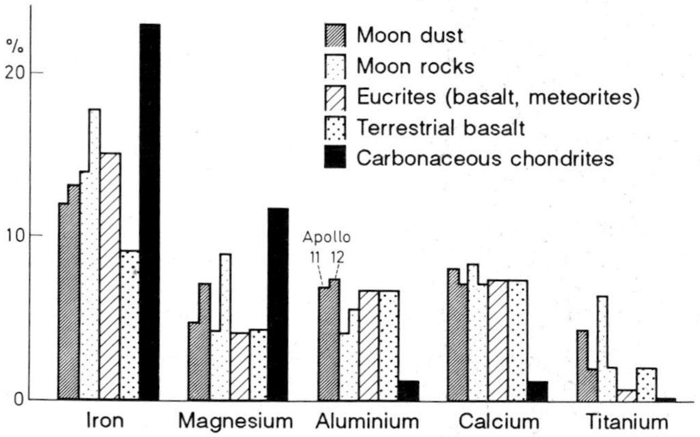

Fig. 2.8.7. The abundance distribution of the elements iron, magnesium, aluminum, calcium, and titanium (in weight percent) from samples of lunar material (*left half*: Apollo 11, Mare Tranquillitatis; *right half*: Apollo 12, Oceanus Procellarum), from Eucrites, i.e. basalt-like achondritic meteorites, and from carbonaceous chondrites of type 1. In the last-named meteorites, the abundance distribution of all non-volatile elements corresponds to that of original solar matter

This comparison and other studies of rarer (so-called trace) elements show that some elements are enriched in lunar rocks compared to solar matter by up to $\simeq 100$-fold, while the elements which bind to Fe (siderophilic elements), Ni, Co, Cu, ..., are reduced by factors of up to $\simeq 100$.

While the age of the mountain ranges goes back to $4.5 \cdot 10^9$ yr, the Mare Tranquillitatis is found to be only $3.7 \cdot 10^9$ yr old, and the Mare Imbrium $3.9 \cdot 10^9$ yr. From ca. 3.3 to $4.0 \cdot 10^9$ years ago, the maria were partially filled with basalt lava, which welled up from deeper regions. Their formation was thus completed only about 10^9 yr *after* that of the Moon itself. These data, together with the statistics of the lunar craters and their "covering over", indicate that the cosmic bombardment was initially extremely strong and then decreased in the course of the first $\simeq 10^9$ yr of the Earth's and the Moon's history, at first rapidly and later much more gradually.

The great plains of the maria, the "drowned" craters and some remarkable hills on the Moon however demonstrate, as already mentioned, that in addition to meteoric impacts, also volcanic processes, i.e. the melting of rocks and outflows of basaltic lava, must have played an important role in the formation of the lunar surface.

e) The Surfaces of the Earthlike Planets

Mercury is very difficult to observe, since it is never more than $\pm 28°$ away from the Sun. Radar measurements together with older visual observations have shown that the rotational period of Mercury is not, as was previously believed, equal to its period of revolution (88 d); instead, it is 58.65 d, or exactly 2/3 of the period of revolution.

The space probe Mariner 10, which flew past Mercury three times in 1974/75, transmitted numerous high-quality pictures of nearly 50% of the planetary surface with a resolution comparable to that with which the Moon can be seen from the Earth. Mercury's surface, like that of the Moon, is thickly covered with craters. The largest, the *Caloris Basin*, has a diameter of 1300 km, similar to the Mare Imbrium. Other structures (plains, terrae, ...) also exhibit great similarity to the Moon's surface. Because of Mercury's stronger gravitational field, the matter ejected from impact craters is not thrown as far as on the Moon.

The surface formations on Mercury and the Moon have remained essentially intact because tectonic and volcanic activity ceased early on both bodies and, due to the lack of an atmosphere, no weathering and erosion

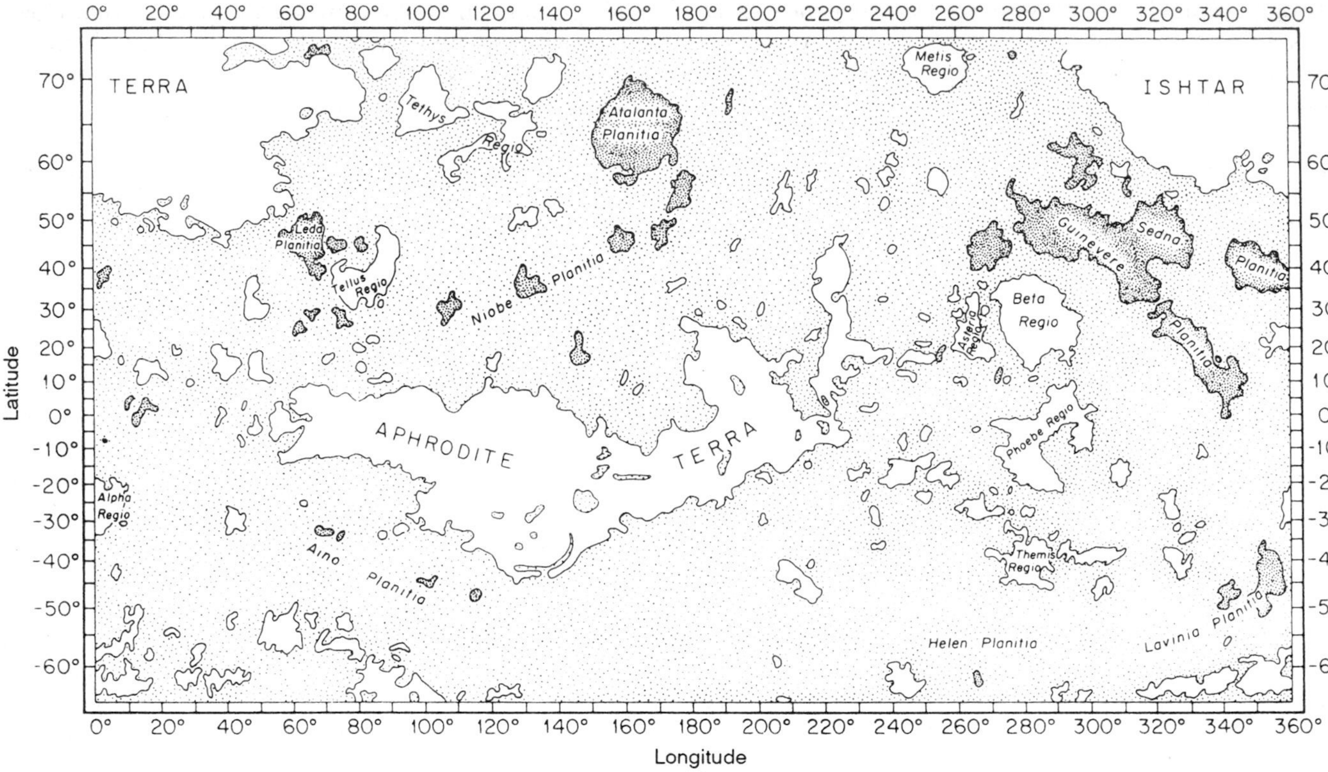

Fig. 2.8.8. The topography of the surface of Venus in a Mercator projection, from radar observations by the Pioneer Venus Orbiter. The dotted areas have hights within 1 km from the average altitude level (plains); the light areas are highlands above 1 km altitude; and the dark areas are lowlands below 1 km. G.E. McGill: Nature **296**, 14 (1982). (Reproduced by permission of Macmillan Magazines Ltd., London, and of the author)

have occurred. Currently, "weathering" can only take place through the energetic protons of the solar wind, through the impacts of micrometeorites, which produce craters of 1 to 20 mm diameter, and through the strong temperature variations, particularly on Mercury (600 K on the day side compared to 100 K on the night side).

The surface of *Venus* cannot be observed visually, because it is completely covered by a thick layer of clouds. Its exploration became possible only with the development of high-sensitivity radar technology and through the soft landings of space probes of the Soviet Venera series.

The radar investigations led to the surprising result that the rotation of Venus is *retrograde*, with a sidereal period of 243.0 d. While it was possible to localize some individual mountains on the surface of Venus with earth-bound radar, the topographic structure was later surveyed by the Pioneer Venus orbiter, which has been orbiting the planet in a strongly elliptical orbit since 1978, with a horizontal resolution of ≥ 30 km and a vertical resolution which can be as low as several hundred meters. About 70% of the surface is taken up by *rolling plains*, whose altitude differs by less than 1 km from the mean level, i.e. from the radius of the planet (Fig. 2.8.8). A number of scattered circular formations could be impact craters (or also calderas of volcanic origin). About 20% of the surface is depressed by ≤ 1.5 km; some of these *lowlands* are roughly circular in form with diameters of several hundred km and are reminiscent of the old, large impact craters or basins on the Moon or Mercury. The remaining 10% is occupied by *highlands* (terrae), which are comparable in size to terrestrial continents. *Ishtar Terra* contains an extended plateau 3 to 4 km high, with high, steep mountains around the edges, including the *Maxwell Montes*, which at 11 km is the highest point on Venus. *Beta Regio* apparently contains large shield volcanoes.

Venera 15 and 16, which have been circling the planet since 1983, have yielded radar images with a (horizontal) resolution of 1 to 2 km, greatly improved compared to Pioneer, on which e.g., Ishtar Terra can be seen to contain flat circular structures with diameters of ≥ 100 km (of volcanic origin?).

Especially the high mountain ranges and the long trenches in the highlands are similar to terrestrial structures which have been formed by plate tectonics. However, on Venus, no global system of the apparently tectonic formations is found on the rims of the plates, as is characteristic of the Earth. The crust of Venus has probably not broken up into plates. Beta Regio and Aphrodite Terra are extended *volcanic* areas, which are presumably still active today and in which the hot magma from the interior of the planet gives up its heat to the exterior. Here also, the *electrical storms* concentrate; they can be observed through their low-frequency radio emissions. Considerably more details of the surface structure than yielded by radar observations have been obtained thanks to the pictures and the analyses of the surface material sent back by several soft-landed space probes, first in 1972 by Venera 8, then in 1975 by Venera 9 and 10 with the first pictures of low resolution, and finally in 1982 by Venera 13 and 14, which transmitted color pictures permitting the recognition of structures of a few mm diameter. In the brief time intervals (2 h and 1 h) during which these last-named vehicles survived the inhospitable surface conditions on Venus, with temperatures around 740 K and pressures around 90 bar, among other things rock samples were obtained with a drill and chemically analyzed in the interior of the vehicle using X-ray fluorescence following irradiation by radioactive ^{55}Fe and ^{238}Pu. At both landing places, on the edge of Beta Regio at an altitude of 2 km (Venera 13) and 960 km distant in the lowlands (Venera 14), we find rocks of volcanic origin, quite similar to the basalts which are widespread on the terrestrial ocean floors and the lunar maria.

Since 1990, Magellan has been circling Venus as an artificial satellite in an orbit ranging from 290 to 8000 km altitude, with the goal of mapping practically the whole surface at a resolution of about 120 m using radar techniques.

In contrast to Venus, on *Mars* the surface is readily observable owing to the thin atmosphere. The lively reddish color of the planet is due to a decrease of its reflectivity in the short-wavelength region of the visible spectrum. This, along with polarimetric measurements by A. Dollfus, indicates iron oxides. Visual observations already permit a multiplicity of structures to be seen. However, the long-popular *canals of Mars* have been found to be due to a physiological-optical contrast phenomenon: our eyes have the tendency to connect outstanding points and vertices by lines (think of the constellations!). With the telescope, one recognizes two white *polar caps*, which are reduced in size in the Martian summer and increase during the winter.

The first television images from Mariner 4 (1965) showed numerous craters on the surface of Mars, whose diameters ranged from the limit of resolution at a few km up to $\simeq 120$ km. These craters correspond closely to those on our Moon. Mariner 9, which circled Mars from 1971 on, delivered images with a considerably improved resolution. They showed that the surface was shaped not only by meteorite impacts, but also to a large extent by volcanic activity (volcanic craters, shield volcanoes, calderas), by tectonics, by erosion (Fig. 2.8.9), and by the deposit of minerals.

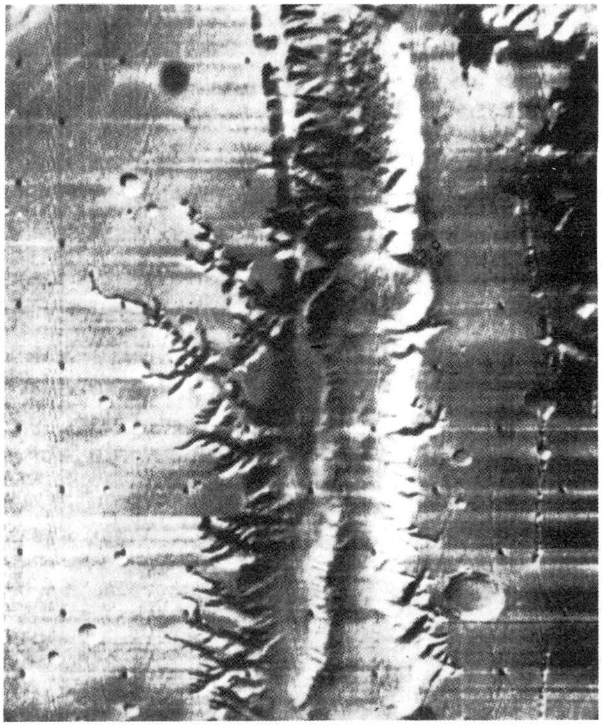

Fig. 2.8.9. This Mariner 9 photograph (taken January 12, 1972) of an area of about $500 \cdot 380$ km^2 on the surface of Mars shows, along with several meteorite craters, a part of the over 2500 km long canyon system of the Coprate region. This canyon was probably formed through a complex interaction of tectonics and erosion. The row of small craters on the right may perhaps be interpreted as having a volcanic origin

Fig. 2.8.10. This close-up photo of the surface of Mars from the Viking 1 Lander (July 22, 1976) shows numerous shards of stone ranging from a few cm to several m in size. The ground between the stones is covered by fine-grained material, which is to some extent piled up in the "wind shadows" of the stones. The lower edge of the picture is approximately 4 m away from the camera, and the horizon is at about 3000 m; the width of the large boulder on the horizon is roughly 4 m

Since the summer of 1976, the two Viking Orbiters 1 and 2 have been circling the planet and have yielded a large number of pictures of its surface, confirming the enormous variety of surface structures. The image quality is better than of the earlier pictures from Mariner 9, whose sharpness was reduced by a screen of dust that was stirred up by a great storm and settled again only after some months. Each of the Orbiters released a probe (Viking Lander) to make a soft landing on the surface and investigate more closely two regions of the northern hemisphere at 23° and 48° latitude, separated by 180° longitude (Chryse Planitia and Utopia Planitia).

On Mars, we find a large-scale *asymmetry* in the surface formations corresponding to its "pear shape": the northern hemisphere is dominated by lower-lying plains with few impact craters, while the southern hemisphere shows a very high density of craters, comparable to that of the terrae on the Moon. The Tharsis region near the equator is impressive, with several great shield volcanoes of more than 20 km height. Olympus Mons has a diameter of 600 km at its base and rises to 26 km altitude; it has a caldera 80 km in diameter.

All the surface contours on Mars are somewhat rounded in comparison to those on the Moon as a result of erosion. The complex system of canyons and twisted crevices, reminiscent of dried-up river beds, indicates that in earlier times there were probably great floods. Today, water is found only in the form of ice and as traces of water vapor, not as a liquid; a large amount of water may be bound up in the minerals of the crust.

In spite of the numerous sandstorms, wind erosion probably does not now play an important role in forming the surface of Mars, since even very old formations can be found which still retain sharp features. Only around the polar caps do we find deposits of basaltic sand of ≤ 100 m depth left by the wind (transported from equatorial regions) as well as a belt of dunes (around the northern polar cap only).

The relatively flat, yellow-brown colored *landing places* of the two Viking Landers remind us of a stony desert on

Earth, with their many small stones and fine-grained, windblown sand (Fig. 2.8.10). During the Martian winter, a thin white layer of frost can sometimes be seen. X-ray fluorescence spectra of samples collected at the two landing places indicate iron-containing clays and hydroxides, along with sulfates and carbonates; furthermore, about 1 wt.% of water is bound in the surface rocks. Directly over the *northern* polar cap in the summer, a surprisingly high temperature of 205 K was measured, which, together with the low atmospheric pressure of only about 7 mbar near the surface, excludes the existence of CO_2 ice in equilibrium with the atmosphere. Therefore, in contradiction to earlier observations, the main constituent of the polar caps must be *water ice*, which is covered by precipitated CO_2 ice depending on the season. (The southern polar cap, in contrast, probably consists of CO_2 ice.)

The temperature on the surface is found from the radiation temperatures in the radiofrequency region to be 210 K on the average, with variations between $\simeq 180$ and 300 K, in good agreement with theoretical values.

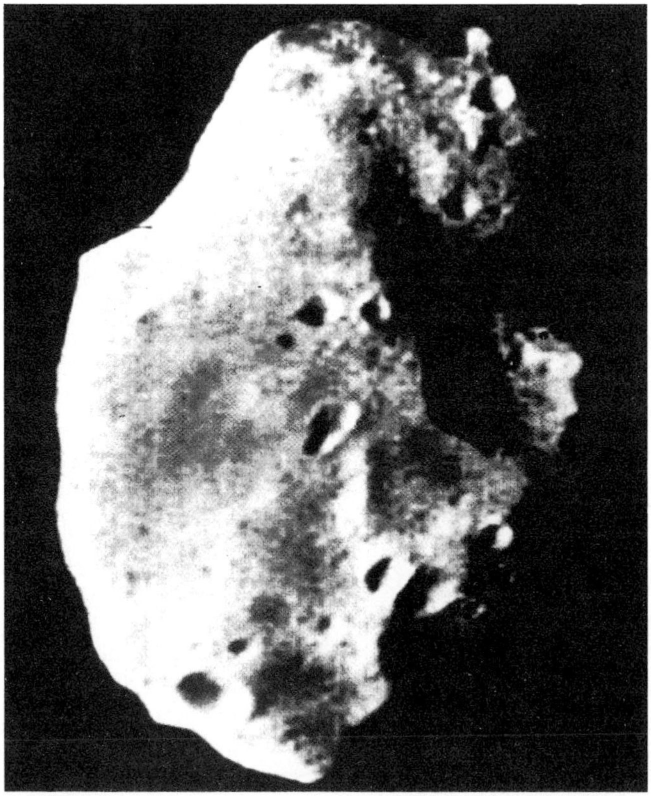

Fig. 2.8.11. A Mariner 9 photograph of Mars' moon Phobos. The diameter of this irregularly-shaped satellite is about 20 km. It is covered with impact craters having diameters up to 5.3 km

Both the Viking Landers carried out several experiments to detect the presence of *life* on Mars. Surface samples were heated to 800 K and the liquids driven off were analyzed with gas chromatography and mass spectrometry. The result was negative, since aside from CO_2 and H_2O, no fragments of organic molecules were detected. On the other hand, the microbiological experiments, which were supposed to react to gas exchange, metabolism and carbon assimilation, have not permitted a clear-cut distinction to be made as yet between biological and chemical activity.

In 1877, A. Hall discovered the tiny moons of Mars, *Phobos* (Fig. 2.8.11) and *Deimos*. The period of revolution of Phobos, 7 h 39 min, is considerably shorter than the rotational period of the planet. Mariner 9 photographs show the traces of strong bombardment by meteorites on the two moons, which are 20 km and 12 km in diameter, respectively.

Although we have met widely differing surface structures on the earthlike planets and their moons, they may all be understood as the results of the same *basic surface forming processes*. For all the objects in the Solar System which have formed solid crusts, the strong bombardment by meteorites, especially in the first 10^9 yr after their formation, is the most important process. The result is surfaces covered with *impact craters*, which we find all the way out to the satellites of the Uranus and Neptune systems (Sect. 2.8.7).

In the cases of Mercury, Earth's Moon, and the two small moons of Mars, this original surface structure is for the most part still present, since essentially no processes have occurred which would have destroyed it. In contrast, on Venus, the Earth, and Mars, the crust was reformed on the one hand by *tectonic* and *volcanic* effects, and on the other by *weathering* and *erosion*. While on the Earth, which remains geologically very active, the crust is modified into a variety of forms on a planetary scale mainly by plate tectonics and the associated volcanic activity, on Venus and Mars mostly local tectonic processes such as the rise and fall of mountain ranges, as well as volcanic activity, play a dominant role. On planets with atmospheres and constituents which have condensed from them, we have a variety of *leveling processes*, mechanical and chemical weathering as well as transport and deposit by wind, water, or glaciers. On the Earth, erosion by water (dissolution of minerals, cracking by frost) is dominant; on Venus, owing to the high temperatures and pressures, chemical weathering is probably the most important process. On Mars, we have found indications of water erosion in the past and of wind erosion.

f) The Atmospheres of the Earthlike Planets

Mercury and Earth's *Moon* have no atmospheres in the strict sense. They merely have *exospheres* (Sect. 2.8.4) with very low concentrations of particles, on the order of $10^9\,m^{-3}$ on the day side and $10^{11}\,m^{-3}$ on the night side; they result mainly from an equilibrium between supply of particles by the solar wind and loss as a result of thermal motion.

In contrast, *Venus*, *Earth*, and *Mars* have atmospheres of differing densities (Table 2.8.3). We have already discussed the basics of their global energy balances in Sect. 2.8.4. The *surface temperatures* of Venus, and to a lesser extent of the Earth, are as a result of the "greenhouse effect" *higher* than the effective temperature, while on Mars with its very thin atmosphere they more closely correspond to the equilibrium value (2.8.2).

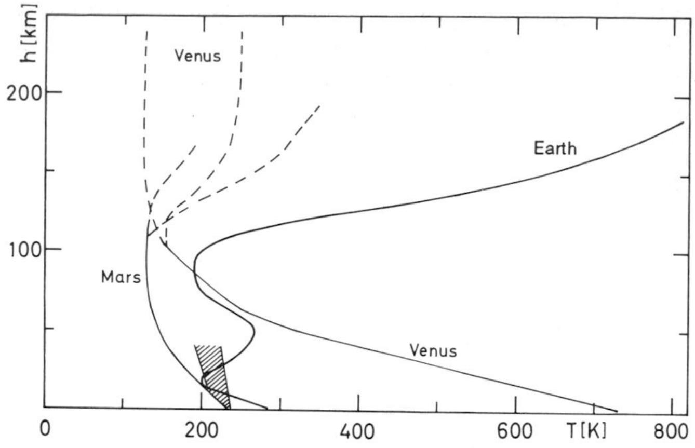

Fig. 2.8.12. The mean temperature as a function of altitude in the atmospheres of the Earth, Venus, and Mars. In the case of Venus, there are large temperature differences in the thermosphere between the day and night sides of the planet. The atmosphere of Mars is subject to wide temperature variations; the (smoothed) measurements of the Viking Lander are shown. During major dust storms, the temperature in the lower part of its atmosphere increases (*shaded region*)

Table 2.8.3. Atmospheres of the earthlike planets

	Venus	Earth	Mars
"Solar constant" $S(r)$ [$kW\,m^{-2}$]	2.6	1.4	0.6
Average Albedo A	0.7	0.3	0.2
Effective temperature $\bar{T}_{eff}$ [K] (2.8.2)	230	255	216
Surface temperature T_0 [K]	735	220 – 310	145 – 245
Surface pressure p_0 [10^5 Pa = 1 bar]	90	1	$7\cdot10^{-3}$
Relative pressure variations $\Delta p/p$	$\leq 10^{-3}$(?)	$\simeq 0.01$	0.1
Troposphere:			
Equivalent height $H = \dfrac{kT}{g\bar{\mu}m_u}$ [km], (2.8.17)	14	8	10
Altitude of the Tropopause [km]	60	10	15
Average temperature gradient [$K\,km^{-1}$]	8	6.5	3[a]
Chemical composition (relative mass fraction)			
CO_2	**0.96**	$3\cdot10^{-4}$	**0.95**
N_2	0.03	**0.78**	0.03
O_2	$7\cdot10^{-5}$	**0.21**	$1\cdot10^{-3}$
CO	$2\cdot10^{-5}$	$1\cdot10^{-6}$	$7\cdot10^{-4}$
H_2O	$\simeq 1\cdot10^{-3}$	$(1-28)\cdot10^{-3}$	$3\cdot10^{-4}$[a]
Ar	$7\cdot10^{-5}$	$9\cdot10^{-3}$	0.02
Average molecular mass $\bar{\mu}$	43.4	29.0	43.5

[a] wide variations

The *chemical composition* (Table 2.8.3) of the atmospheres of Venus and Mars, which consist mainly of CO_2, is very similar, while the Earth's atmosphere is distinguished by a high proportion of N_2 and O_2. The Earth probably also previously had an atmosphere similar to that of Venus and Mars, formed by volcanic activity and by outgassing. Through condensation of water into oceans, in which CO_2 dissolves and can then react with silicate minerals to form carbonates, as well as through the presence of living organisms (photosynthesis of oxygen by plants), however, its composition was fundamentally changed (Sect. 6.4).

In particular, with a view to comparisons with the other planets, we first consider briefly the structure of the *Earth's* atmosphere. Its pressure profile is, according to (2.8.19), determined in the main by the temperature profile $T(h)$. The latter is fixed by the mechanisms of energy transport, i.e. the input and output of thermal energy into each layer between h and $h+dh$. In the Earth's atmosphere (Fig. 2.8.12), the removal of absorbed solar energy in the lowest layer, the *troposphere*, takes place mainly through convection, which leads to a uniform decrease of the temperature on going upwards. Above the so-called *tropopause*, at an altitude of roughly 10 km, radiation becomes the dominant mechanism of energy transport and we arrive at the *stratosphere*, which is almost isothermal. What is now important is the decisive mechanism of absorption of solar radiation on the one hand, and its reradiation into space at longer wavelengths on the other. At an altitude of 25 km we find a warm layer in connection with the formation of ozone, O_3, up to the *stratopause* at an altitude of about 60 km. In the *mesosphere* which lies above it, CO_2 radiates energy in the infrared, while warming through absorption by O_3 no longer occurs, so that a brief decrease in temperature results. Above the *mesopause* at about 80 km altitude, in the *thermosphere*, the temperature rises to about 1000 K (night side) or 2000 K (day side) as a result of the dissociation and ionization of the atmospheric gases N_2 and O_2 by solar UV radiation.

Photoionization produces the electrically conducting layers of the *ionosphere* (with the maximum electron density of the D layer at about 90 km, of the E layer at about 115 km, and of the F layer at 300–400 km altitude). These layers allow the transmission of electromagnetic waves of relatively long wavelengths completely around the globe.

The recombination of electrons and ions in the E layer produces the emission lines and bands called the *airglow*.

Up to the so-called *turbopause* at 120 km altitude, the atmosphere is well mixed; above it, the different gaseous constituents separate by diffusion. For example, above about 300 km, atomic oxygen O is predominant, and above about 2000–3000 km, atomic hydrogen H is mainly present; it can be observed from satellites through the strong Lyman α emission at $\lambda = 121.6$ nm.

From the *exosphere* (above about 500 km altitude), particles from the atmosphere can escape into space (Sect. 2.8.4). The terrestrial magnetic field is very important for the dynamics of the ionized constituents. The extent of the *magnetosphere* is determined by interactions with the solar wind on the side of the Earth directed towards the Sun. At the subsolar point, its boundary, the *magnetopause*, is at about 10 earth-radii; on the night side, a plasma trail extends to about 1000 earth-radii. At 1.6 and 3.5 earth-radii, we find higher concentrations of charged particles in the "*radiation belts*" discovered in 1958 by J. A. van Allen.

For the global *energy balance* of the Earth's atmosphere, the *clouds* (consisting of H_2O) play a decisive role. On the average, they cover about 50% of the surface; a typical cloud layer has an albedo of about 0.5. The details of both the input of solar radiation and the albedo of Earth's atmosphere and its surface depend on the latitude. The excess of absorbed radiation in the equatorial zones as opposed to the polar regions provides the driving force for atmospheric and oceanic currents. We cannot delve further here into the dynamics of planetary atmospheres and weather, a complex subject in itself.

The discovery in 1932 by W. S. Adams and T. Dunham based on infrared spectra that *Venus* possesses a thick atmosphere rich in CO_2 was followed by determinations of the radiation temperatures in the cm- and dm-wavelength spectral regions, and later the fly-bys of the Mariner satellites, the parachute drops of instruments by the Soviet Venera probes, and observations by the American Pioneer Venus Orbiter; the last two investigations in particular yielded considerable information about the planet's atmosphere. This dense atmosphere, with a pressure of about 90 bar and a temperature of 735 K at the surface, consists almost entirely of CO_2 and N_2 (Table 2.8.3). The traces of HCl and HF which have been detected in the lower atmosphere (as well as the drops of sulfuric acid in the clouds) probably result from chemical reactions of the atmosphere with surface minerals. Due to its high density, the lower portion of Venus' atmosphere exhibits only very weak pressure and temperature variations and low wind velocities (≤ 1 m/s at the surface).

In the upper part of the troposphere, between about 50 and 70 km altitude at temperatures of 370 to 220 K, there are dense *cloud layers* which completely block optical observations of the planet's surface and prevent all but a few percent of the incident sunlight from reaching it, owing to scattering. Both above and below the cloud cover there are in addition layers of haze about 20 km thick. Observations with ultraviolet light reveal high-contrast structures in the cloud layer (Fig. 2.8.13), which are between 10 and 1000 km in size and persist for several days; dark V- or Y-shaped formations are particularly noticeable. The clouds consist mainly of H_2SO_4 droplets a few μm in diameter with a concentration of several 10^8 droplets per m^3; in addition, there are 10 to 15 μm diam. particles, which apparently consist of solid or liquid sulfur. The sulfuric acid and its dissociation product SO_2, together with the CO_2 and H_2O in the atmosphere, are presumably responsible for the strong greenhouse effect on Venus.

In order to investigate the horizontal structure of Venus' atmosphere before their mission to Halley's comet, in 1985 Vega 1 and 2 dropped weather balloons into the cloud layer; at 54 km altitude, the wind velocity of 70 m/s is considerably greater than the rotational velocity of the planet, 1.8 m/s.

The *atmosphere of Venus* has no stratosphere; the mesosphere follows directly above the tropopause or the

Fig. 2.8.13. A photograph of the cloud structures of Venus from an altitude of 65000 km taken with the cloud photopolarimeter of the Pioneer Venus Orbiter (January 18, 1979, during the 45th orbit of the artificial satellite). The bright cloud bands near the poles, which are higher than the adjacent background, are set off clearly from the other formations. The north pole is at the top of the picture

cloud layer, followed by the thermosphere (Fig. 2.8.12). The exosphere begins at an altitude of about 160 km. With temperatures in the range 120 to 250 K, it is relatively cool in comparison to the upper atmosphere of the Earth. The turbopause is at about 140 km altitude; above 150 km, atomic oxygen is predominant, and above 250 km, helium and hydrogen. The ionosphere has its maximum around 140 km with ion densities of several $10^{11}\,m^{-3}$, comparable to the Earth's E-layer. Its main constituent up to 200 km altitude is the molecule-ion O_2^+, and higher up O^+.

The *upper atmosphere* of Venus shows diurnal and annual variations. Furthermore, clearcut thermal structures have been observed, which are to some extent not yet understood. For example, in the mesosphere at around 90 km altitude, the poles are warmest and the subsolar point coolest; at the mesopause, there is a relative maximum in temperature in the middle of the night side. In addition, infrared observations near both poles show long structures ($\simeq$ 4000 km) which are up to 35 K warmer than their surroundings and rotate in the retrograde sense with a period of about 3 d.

The *atmosphere of Mars*, very thin in comparison to those of Venus and Earth, exhibits extensive pressure and temperature variations. Also characteristic are violent *sandstorms* on a planetary scale. The atmosphere reacts sensitively to fluctuations in the absorption of solar radiation, which in turn is strongly dependent on the albedo of the polar caps and of the dust which is stirred up by storms. In particular, the temperature gradient in the lower atmosphere decreases strongly with increasing dust content (Fig. 2.8.12).

Spectroscopic observations and the mass-spectroscopic measurements of the Viking Landers indicate that Mars' atmosphere, which is well mixed up to an altitude of 120 km, consists mainly of CO_2 and N_2 with some other constituents present in minimal concentrations (Table 2.8.3). Water vapor plays a special role; it is present only in trace quantities and is subject to very strong local and seasonal variations. (Under the present conditions on Mars, free H_2O can exist as a stable phase only in the form of ice or vapor, however not as a liquid.) In spite of its low concentration, the water vapor in the atmosphere is probably nearly saturated and plays an essential part in the formation of the clouds which have been observed. The *weather on Mars* is characterized by various types of thin ice clouds, surface fog, and by diurnally and annually varying winds and dust storms.

The short-wavelength "air glow" from the upper atmosphere of Mars contains the Lyman α emission line of hydrogen; also, atomic oxygen and carbon, in addition to CO_2^+ and CO, have been detected.

2.8.6 Asteroids or Planetoids

Today, we have catalogued several thousand asteroids, most of them between Mars and Jupiter, whose data are collected in the Tucson Revised Index of Asteroid Data (TRIAD). The total number of asteroids is estimated to lie between 10^4 and 10^6. Objects with known orbits receive a number and a name.

The excentricities of their *orbits* have a maximum frequency of occurrence at $e \simeq 0.14$, and the inclinations of the orbits at $i \simeq 10°$; both these values differ considerably from those of the comets (Sect. 2.5.2). The semimajor axes a are mainly in the range between 1.8 and 5.2 AU. Following Hirayama and others, we distinguish several *families* of asteroids with similar orbital parameters. The distribution of orbits according to their orbital periods shows several clusters and gaps (named for D. Kirkwood), which correspond to integral ratios with the orbital period of Jupiter; for example, we find a well-developed gap at a ratio of 3:1 (i.e. with a semimajor axis of 2.5 AU), and a cluster in the group of the *Trojans* at the so-called libration point (1:1) of Jupiter's orbit.

433 Eros, for example, has unusual orbital parameters, with $e = 0.23$; at opposition, it approaches the Earth to within 0.15 AU and thus permits excellent determinations of the solar parallax. Hermes, whose orbit is not exactly known, approached the Earth in 1937 to less than 0.005 AU. 1566 Icarus ($e = 0.87$, $a = 1.08$ AU) approaches the Sun at perihelion more closely than Mercury; 2060 Chiron, by contrast, remains for the most part between the orbits of Saturn and Uranus ($a = 13.7$ AU). With $i = 64°$, Tantalus exhibits an unusually large orbital inclination.

The *masses* of the three largest asteroids can be determined from the perturbations which they cause to the orbits of other asteroids (J. Schubart, 1973). Recent values are $1.2 \cdot 10^{21}$ kg $= 2.0 \cdot 10^{-4} \mathcal{M}_\oplus = 6.0 \cdot 10^{-10} \mathcal{M}_\odot$ for 1 Ceres, $3.5 \cdot 10^{-5} \mathcal{M}_\oplus$ for 2 Pallas, and $4.5 \cdot 10^{-5} \mathcal{M}_\oplus$ for 4 Vesta. The overall mass of all the asteroids is estimated to be $5 \cdot 10^{-4}$ times the mass of the Earth or 2.5 times the mass of Ceres.

The angular *diameter* can be measured directly for only the largest and brightest asteroids. For example, 1 Ceres is found to have a diameter of 940 km, 2 Pallas of 538 km, and 4 Vesta of 576 km, with uncertainties of a few tens of kilometers; their mean densities are thus of the order of 3000 kg/m^3, somewhat smaller than those of the earthlike planets. For the smaller asteroids, diameters to 0.4 km can be estimated from their brightness or polarization of reflected light by assuming a reflectivity curve.

Many asteroids show a periodic variation of brightness, indicating their *rotational periods*. Remarkably, they all lie between 3 and 17 h. Brightness variations of 0.1 to 0.3 mag are typical; however, 433 Eros, for example, has a stronger fluctuation with an amplitude of 1.5 mag and a period of 5.3 h, and 1620 Geographus has an even larger amplitude of 2.0 mag with a 5.2 h period. The light curves of many asteroids indicate that they have elongated, irregular shapes.

Spectrophotometric measurements of the reflected light in the optical and infrared regions allow us to identify various surface characteristics. About two thirds of the asteroids belong to the so-called *type C*, with a flat spectrum that shows weak structure, and a low albedo (0.03 to 0.08); they are thus probably related to the carbonaceous chondrites (Sect. 2.9.2).

With this we close our rather incomplete overview of the earthlike planets and turn to the large planets, which, as their masses and mean densities already demonstrate, are of quite different character (Table 2.8.1). We will treat Jupiter in detail, as a representative example, but we must be more brief in describing Saturn, Uranus, Neptune, and Pluto.

2.8.7 The Major Planets

a) Jupiter

Jupiter, the largest and most massive of the planets (1/1047 solar masses), exhibits a dense atmosphere on observation from the Earth, with clear *stripes* parallel to the equator (Fig. 2.8.14a), similar to the wind belts on the

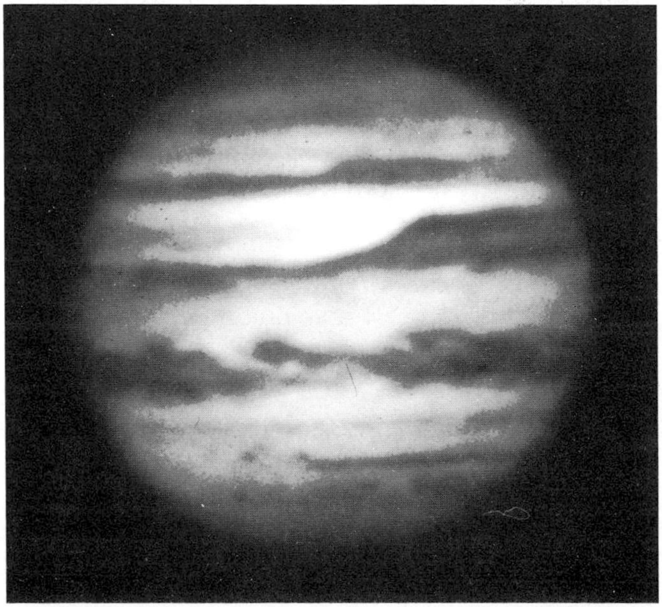

Fig. 2.8.14a. Jupiter. An image made by B. Lyot and H. Camichel with the 60 cm refractor on the Pic du Midi

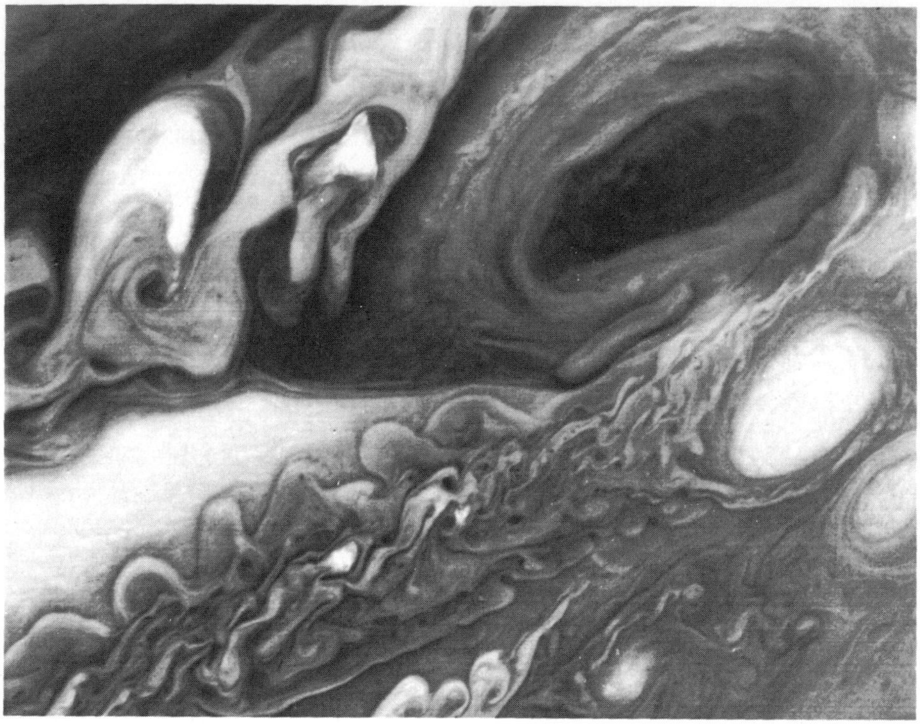

Fig. 2.8.14b. This Voyager 1 photograph of Jupiter (March 1st, 1979) from a distance of $4.3 \cdot 10^6$ km shows the region immediately to the southeast of the Great Red Spot, with a vortex-like flow field around one of the "white ovals". The smallest structures which can be distinguished are about 80 km in size

Earth. The so-called *great red spot*, discovered in 1665 by G. Cassini, is a large oval structure which has persisted for a long time.

The first close-up pictures and measurements were provided in 1973 by the space probe Pioneer 10, which flew by Jupiter with a closest approach of 130 000 km ($\simeq 2$ Jupiter radii). In 1979, the two probes Voyager 1 and Voyager 2 transmitted impressive high-resolution images of the flow patterns on the planet (Fig. 2.8.14b): we see among other things cloud bands, convection cells, jetstreams, vortices, white and brown ovals of varying size, and circulation systems. Neighboring structures can have relative velocities of over 100 m/s. In addition to long-lived formations, for example the Red Spot, we can observe changes in large structures on a time scale of a few days.

As early as 1932, R. Wildt identified the strong absorption bands in the spectrum of Jupiter with higher overtones of the molecular vibrations of *methane*, CH_4, and of *ammonia*, NH_3; this represented a turning point in the investigation of the major planets. In 1951, G. Herzberg succeeded in detecting some infrared band lines of the hydrogen molecule, H_2, which are weak due to their small quadrupole transition probabilities. During a stellar occultation by the disk of Jupiter in 1971, some insight was gained into the layer structure of the planetary atmosphere, and from the equivalent altitude (Sect. 2.8.17), its main components could be concluded to be *hydrogen* and *helium*. Furthermore, especially by means of infrared spectroscopy, numerous *trace compounds* were discovered, such as ethane, C_2H_6, ethyne (acetylene), C_2H_2, water, H_2O, hydrocyanic acid, HCN, phosphine, PH_3, germane, GeH_4, and the deuterated molecules HD and CH_2D.

In the infrared, one can observe into layers at up to 225 K. In this region, the transparency of the atmosphere is more and more limited by *clouds*; only occasionally, through gaps in the cloud layer, can deeper layers at up to 280 K (and a pressure of 5 bar) be seen. The highest cloud layer (at about 150 K) consists presumably of NH_3 crystals; the deeper layers are probably NH_4SH and H_2O (and also H_2S?), while the explanation of the red and brown colors requires that trace compounds (free radicals?) also be present. The weather on Jupiter appears to be much more complex than in the Earth's atmosphere, particularly because of the occurrence of chemical reactions.

A detailed analysis of all the observations together with the basic premises of a theory of the inner planetary structure yields the following *model*: Jupiter consists for the most part of *unchanged solar material*, with hydrogen and helium in the (atomic) ratio He/H $\simeq 0.1$ being the most abundant elements. In spite of the high pressures and densities, the hydrogen in the planetary interior remains for the most part *liquid*, owing to the relatively high temperatures (≤ 30000 K). At 0.77 Jupiter radii, i.e. at a pressure of $3 \cdot 10^{11}$ Pa and a density of 1000 kg/m^3, a phase transition takes place from liquid molecular hydrogen, H_2, to liquid *metallic* hydrogen. Jupiter in all probability has a *solid core*, which contains about 4% of its mass or about 14 Earth masses, and consists of a mixture of *stones* (SiO_2, MgO, FeO, and FeS) and *ices* (CH_4, NH_3, H_2S, and H_2O). This core could have served in the early phase of the formation of the Solar System as a "condensation nucleus" for the hydrogen- and helium-rich solar material. The models give a density for the core of $\geq 2 \cdot 10^4$ kg/m^3 and a pressure of about 10^{13} Pa. The excess of thermal radiation from Jupiter compared to the solar irradation has already been discussed in Sect. 2.8.2.

The *radio-frequency emission* of Jupiter is thermal at wavelengths $\lambda \leq 1$ cm, with a radiation temperature of 120 K, corresponding approximately to that of the tropopause. In the decimeter-wave region, the intensity of the radio emission increases and, due to its partial polarization, must have its origin in nonthermal synchrotron radiation in the planetary magnetic field (Sect. 5.6.1). In the meter wavelength region, there are in addition sharply localized radiation *bursts* from well-defined sources on the planetary disk.

The measurements of the Pioneer and Voyager probes indicate that Jupiter has a dipole-like *magnetic field* with about $4 \cdot 10^{-4}$ Tesla or 4 Gauss at the equator and a dipole moment (Sect. 2.8.25) of $1.5 \cdot 10^{20}$ T·m^3 ($1.5 \cdot 10^{30}$ G·cm^3), whose axis is tilted away from the planet's axis of rotation by about 10°. Jupiter's *magnetosphere*, similar to the Van Allen belts of the Earth, traps enormous numbers of energetic electrons and protons, as well as thermal plasma. At high altitudes in the Lyman region, one can observe the emission lines of hydrogen and helium, a sort of airglow.

In 1610, Galileo discovered the four brightest *satellites* J1 to J4 (previously denoted as I through IV). With increasingly larger telescopes, a total of 12 have been discovered from the Earth; the number of known satellites has now increased to 16 as a result of the Voyager missions (Table 2.8.4).

The circular orbits ($e \leq 0.01$) of the eight inner satellites are all nearly in the equatorial plane of the planet (regular orbits). The outer satellites, in contrast, have larger eccentricities (0.13 to 0.38) and orbital inclinations, with the four outermost showing a retrograde motion. They are presumably asteroids which have been captured by Jupiter. The eight outermost satellites are small and dark; for two of them (J6 and J7), colors and albedos have been measured which are similar to those of type *C* asteroids (Sect. 2.8.6).

Table 2.8.4. Jupiter's moons. Distance *a* from the center of the planet in units of Jupiter's radius $R_J = 71400$ km, and the moons' radii *R* (or semiaxes) in [km]

a/R_J	Name	R	a/R_J	Name	R
1.80	J15 Adrastea	$\simeq 20$	156	J13 Leda	$\simeq 5$
1.80	J16 Metis	$\simeq 20$	161	J6 Himalia	$\simeq 90$
2.55	J5 Amalthea	135·85·75	164	J10 Lysithea	$\simeq 10$
3.11	J14 Thebe	$\simeq 40$	165	J7 Elara	$\simeq 40$
5.9	J1 Io	1816	291	J12 Ananke	$\simeq 10$
9.5	J2 Europa	1563	314	J11 Carme	$\simeq 15$
15.1	J3 Ganymede	2638	327	J8 Pasiphae	$\simeq 20$
26.6	J4 Callisto	2410	333	J9 Sinope	$\simeq 15$

Amalthea, the innermost known moon until the Voyager missions (E.E. Barnard, 1892), is a reddish, elongated and irregularly-shaped body, whose orbit lies just outside the Roche stability limit (2.8.11). A *Ring system* was unexpectedly discovered by the Voyager probes; it consists of a flat (≤ 30 km), bright ring with a radius of about 1.8 Jupiter radii, which is embedded in a disk and a weak halo, and of a much weaker ring which extends to about 3 Jupiter radii. At the outer edge of the bright ring, two small moons (J15 and J16) move co-orbitally, i.e. practically on the same orbit, with periods of revolution of only 7 h 8 min.

The radii and masses or mean densities of the four *Galilean satellites* are rather exactly known:

Moon	J1 Io	J2 Europa	J3 Ganymede	J4 Callisto
Mass	4.7	2.6	7.8	5.6 [$10^{-5}\mathcal{M}_J$]
Average Density	3550	3040	1930	1830 [kg m^{-3}]

While Io and Europa ($\bar{\varrho} \geq 3000$ kg/m^3) are predominantly composed of silicates, the mean densities of Ganymede and Callisto indicate a mixture of ice and silicates (about 1:1), which is pressed together by gravity. Ganymede is the largest moon in the Solar System.

The surface of Io, the innermost of the four large moons, shows spots and is sprinkled with varying, mainly

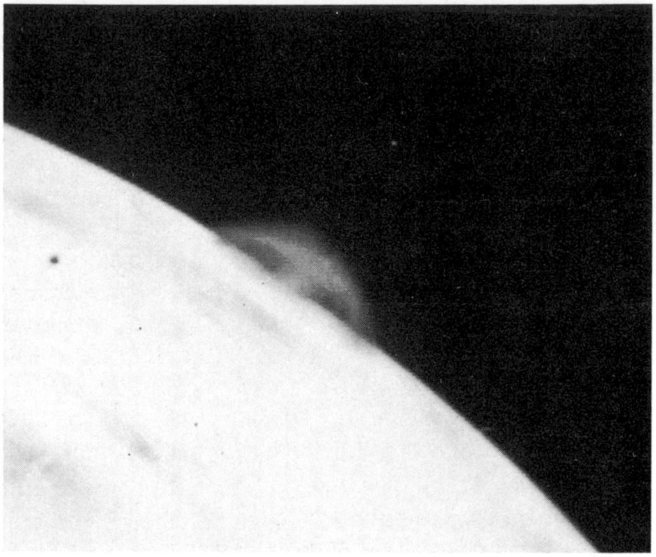

Fig. 2.8.15. In this Voyager 1 picture (March 4, 1979) of Jupiter's moon Io, an eruption of one of the active "volcanoes" (Prometheus) can be readily discerned at the rim of the moon; it rises to more than 100 km above the surface

gray-yellow colors; its rocks are rich in sodium-potassium compounds and especially in sulfur and sulfur compounds. In a few mountain ranges, the crust rises up to about 10 km altitude. Particularly noticeable is the large number of calderas with diameters of ≥ 200 km, indicating dead volcanoes, as is the rarity of impact craters. The most surprising discovery of the Voyager probes was the strong *surface activity* on Io: all together, nine active "volcanoes" were observed, which eject sulfur- and oxygen-containing gases with velocities up to 1 km/s nearly 300 km high in eruptions lasting several hours (Fig. 2.8.15). This vulcanism continually changes Io's surface and makes the lack of impact craters understandable. A few dark spots, which are about 150 K hotter than their surroundings, can be explained as being recently-solidified lava. The energy source for this activity is probably heating of the interior by strong *tidal forces*, which are generated because Io's orbit is forced to be somewhat eccentric due to resonances with the orbital periods of Europa and Ganymede.

Io has its own thin *atmosphere* (SO_2), about 120 km thick, and an ionosphere (S, O, and Na ions) up to about 700 km altitude, which are constantly renewed from the active surface, since interactions with Jupiter's magnetosphere create a toroidal tube of plasma with particle densities of $\geq 2 \cdot 10^9\,\mathrm{m}^{-3}$ which encloses Io's entire orbit. The interactions with the magnetosphere have a strong influence on the radio bursts from Jupiter, also.

The next Galilean moon, *Europa*, also shows only few impact craters. Its bright, icy surface is covered by a web of dark, interlacing lines. It is astonishingly smooth (altitude differences ≤ 100 m); the lines seem to be painted on. In contrast, the two outermost major moons of Jupiter, *Ganymede* and *Callisto*, are covered with impact craters like our Moon or Mercury.

b) Saturn

Saturn (Fig. 2.8.16) is to a large extent similar to Jupiter. The rings of Saturn were discovered in 1659 by Chr. Huygens with the telescope he constructed (Galileo had already seen indications of them); he also discovered the brightest moon of Saturn, Titan. J.E. Keeler's measurement of the rotational velocity of Saturn's rings in 1895 using the Doppler effect of the reflected sunlight shows that the various zones of the rings revolve at different rates corresponding to Kepler's 3rd law, and thus they consist of small particles.

A quantum leap in our knowledge of the planet, its ring system, and its moons was provided by the fly-bys of the space probes Pioneer 11 and especially of Voyagers 1 and 2, which reached Saturn in 1979, 1980, and 1981 following their exploration of the Jupiter system.

The *spectrum* of Saturn, the chemical composition of its *atmosphere*, and the cloud structure and flow patterns are all similar to those of Jupiter. In the case of Saturn, the stripes which run parallel to the equator are broader and reach out to higher latitudes. The structures appear more washed out due to a more dense layer of haze above the clouds. The temperature of the upper cloud layer,

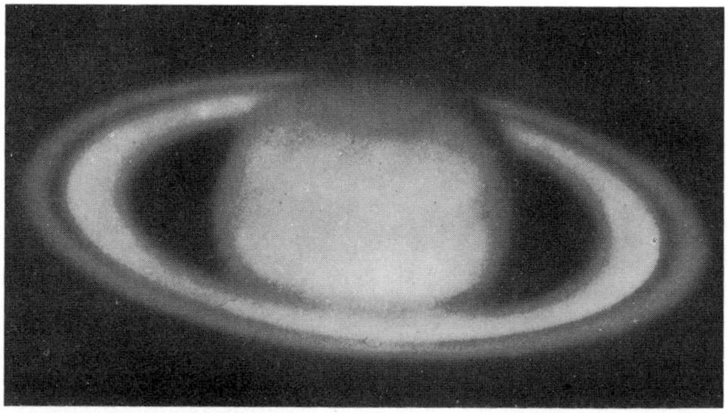

Fig. 2.8.16. Saturn. A photograph by H. Camichel using the 60 cm refractor on the Pic du Midi

consisting of NH_3 crystals, is about 110 K, and that of the tropopause is about 80 K.

More precise analyses show that the abundance of *helium* in the atmosphere of Saturn is noticeably less than in the case of Jupiter. Apparently a separation of hydrogen and helium occurs on the planet, which is somewhat less massive and cooler than Jupiter; the helium sinks down to lower atmospheric layers. The energy released in this process probably makes a major contribution to the thermal radiation, which in Saturn's case also exceeds the incoming radiation from the Sun (Sect. 2.8.2).

Similar to the case of Jupiter, model calculations for the inner structure indicate a solid ice-silicate core of about 16 Earth masses.

Like Jupiter, Saturn has a *magnetosphere*, extending out to from 20 to 40 Saturn radii. Corresponding to the dipole moment of $4.6 \cdot 10^{18}\,\mathrm{T \cdot m^3}$ or $4.6 \cdot 10^{28}\,\mathrm{G \cdot cm^3}$, the magnetic field strength at the equator is $2 \cdot 10^{-5}$ T or 0.2 G, similar to that of the Earth. It is of interest for the dynamo theory of the generation of planetary magnetic fields that the direction of Saturn's dipole moment is almost precisely (<1°) parallel to its axis of rotation. Finally, Saturn is also surrounded by *radiation belts* containing energetic electrons and protons, with intensity maxima at roughly 7 and 4 Saturn radii from the center of the planet; they evidently contain structure due to the inner moons.

The number of known *satellites* of Saturn has increased to (at least) 17 following the two Voyager missions (Table 2.8.5). Here, too, the orbital eccentricities and inclinations of the inner satellites out to and including Titan are small, while those of the outer moons are considerably larger. S 9 Phoebe, the darkest outer moon, has a retrograde orbit.

The largest moon of Saturn, S 6 *Titan*, is the only moon in the Solar System to have its own *atmosphere*. It came as a surprise when in 1944 G. P. Kuiper discovered

Table 2.8.5. Saturn's moons and ring system. Distance r from the center of the planet and orbital semimajor axis a in units of Saturn's radius $R_s = 6 \cdot 10^4$ km, and the moons' radii R (or semiaxes) in [km]. Roche's stability limit (2.8.11) equals $3.0\,R_s$ (for an average satellite density of $1300\,\mathrm{kg\,m^{-3}}$)

r/R_s	Ring	Division	a/R_s	Moon	R [km]
1.1	D				
1.24	C				
1.50		Maxwell			
1.53	B				
1.95		Cassini			
2.03	A	Encke			
2.28		Pioneer	2.28	S 15 Atlas	20·10·?
2.32	F		2.31	S 16 Prometheus	70·50·40
			2.35	S 17 Pandora	55·45·35
			2.51	S 11 Epimetheus	70·60·50
			2.51	S 10 Janus	110·100·80
≃2.8	G		3.1	S 1 Mimas	196
3.5	E		4.0	S 2 Enceladus	255
			4.9	S 3 Tethys	530
			4.9	S 13 Telesto	17·14·13
≃5.0			4.9	S 14 Calypso	17·11·11
			6.3	S 4 Dione	560
			6.3	S 12 Helene	18·16·15
			8.7	S 5 Rhea	765
			20.3	S 6 Titan	2575
			24.6	S 7 Hyperion	205·130·110
			59	S 8 Iapetus	730
			215	S 9 Phoebe	110

absorption bands of methane, CH_4, in its spectrum, similar to those from Saturn itself; later, L. Trafton found bands attributable to H_2. According to the estimate in Sect. 2.8.4, the gravitational field of Titan is, indeed, sufficiently strong to hold an atmosphere.

Voyager 1 approached Titan to within 4000 km and was in particular able to investigate its atmosphere spectroscopically; previously, there had been diverse opinions about its composition. Surprisingly, it was found to contain mostly molecular nitrogen, N_2, with a relative abundance of 90%. Also, in addition to methane, CH_4 ($\simeq 3\%$) and H_2 ($\simeq 0.2\%$), which had long been known, a number of trace constituents were observed in the infrared spectrum, especially hydrocarbons such as ethane, C_2H_6, propane, C_3H_8, ethyne (acetylene), C_2H_2, and ethene, C_2H_4, as well as cyano compounds. In order to explain the mean molecular mass of 28.6, known from observations of occultations, it must be presumed that the second most abundant constituent is argon, with about 10% relative abundance. Dense layers of haze block the view down to the lower levels of the atmosphere and the surface of Titan. Model calculations yield a pressure at the surface of 1.6 bar ($1.6 \cdot 10^5$ Pa) and a surface temperature of 94 K. The density of the atmosphere is thus of the same order as that of Earth's atmosphere. Under these conditions, methane should be present in liquid form. Perhaps methane (together with other hydrocarbons) plays a similar role to that of water on the Earth, forming seas, clouds, and rain.

The mean density of Titan is 1940 kg m^{-3}, similar to that of Jupiter's moon Ganymede; it indicates here, also, an inner structure composed of ice and silicates in roughly equal proportions.

The remaining seven *major moons*, S1 Mimas, S2 Enceladus, S3 Tethys, S4 Dione, S5 Rhea, S7 Hyperion, and S8 Iapetus, have icy surfaces covered by numerous impact craters. The density of craters is of the same magnitude as on the Moon, i.e. the rate of impacts of chunks of stone in the first 10^9 years of the Solar System (Sects. 6.1, 2) was comparable out at Saturn's orbit to that near the Earth. From the density of 1200 to 1400 kg/m^3, the interior of these moons probably also consists of an ice-silicate mixture.

Within the orbit of Mimas, there are two *minor moons*, S11 Epimetheus and S10 Janus, sharing nearly the same orbit; apparently they exchange their relative positions during their periodic close approaches. An additional small moon (S12) has the same orbit as Dione, maintaining a constant angular distance to the larger satellite. On the orbit of Tethys there are actually two minor moons, S13 Telesto and S14 Calypso, at the socalled Lagrangian points. The *accompanying moons* or *"shephard moons"* in some of Saturn's rings are particularly interesting (see below).

Direct images and observations of occultations by the Voyager probes have given us the following picture of the *ring system* (Table 2.8.5): the rough structure in the radial direction consists of seven ring zones. In addition to the three brighter rings, which have long been known (denoted from the outermost to the innermost as A, B, and C, with the readily-visible Cassini division between A and B), and the narrow F-ring discovered by Pioneer 11, there are three weaker rings. The D-ring is inside the C-ring and reaches from it to near the planetary surface; the other two (G and E) are further out, near the orbits of the moons Mimas and Enceladus, respectively. The ring system is extremely thin (< 3 km) in a direction perpendicular to the equatorial plane, and its total mass is probably not more than 10^{-5} of Saturn's mass.

The high spatial resolution of the Voyager images provided some very surprising results concerning the fine structure of the ring system (Fig. 2.8.17): in the ring zones, several hundred to thousand ringlets can be observed, consisting of thin bright and dark regions with sharp boundaries and apparently irregular spacings. These ringlets are in some cases only a few hundred meters in width. Even within the Cassini division, we find a series of thin rings. The deviations from a circular shape in some of the ringlets are of interest, as are the nearly radial structures ("spokes") in the B-ring, which appear dark in backward-scattered light; they persist for about one rotational period of the planet. The F-ring, which is only about 100 km across, exhibits an unusual, nearly unchanging structure; it consists of a few twisted strands, in which thickenings and kinks can be seen. This ring is accompanied on each side by a small, irregularly-shaped moon (S16 and S17) of about 100 km diameter. A small accompanying moon (S15 Atlas) has also been found near the sharp outer edge of the A-ring. The accompanying moons, along with resonances with the or-

Fig. 2.8.17. A Voyager 1 photograph of the ring system of Saturn from a distance of $8 \cdot 10^6$ km, taken in 1980. The long-known broad ring zones A, B, and C with the Cassini and Encke divisions, as well as the thin F ring with its inner accompanying satellite (arrow) are indicated. The picture shows nearly 100 individual rings, which in turn consist of further narrower rings, as shown by images with better resolution from Voyager 2; thus for Saturn all together, at least 100000 rings are estimated to exist. (From G. Briggs and F. Taylor, 1982)

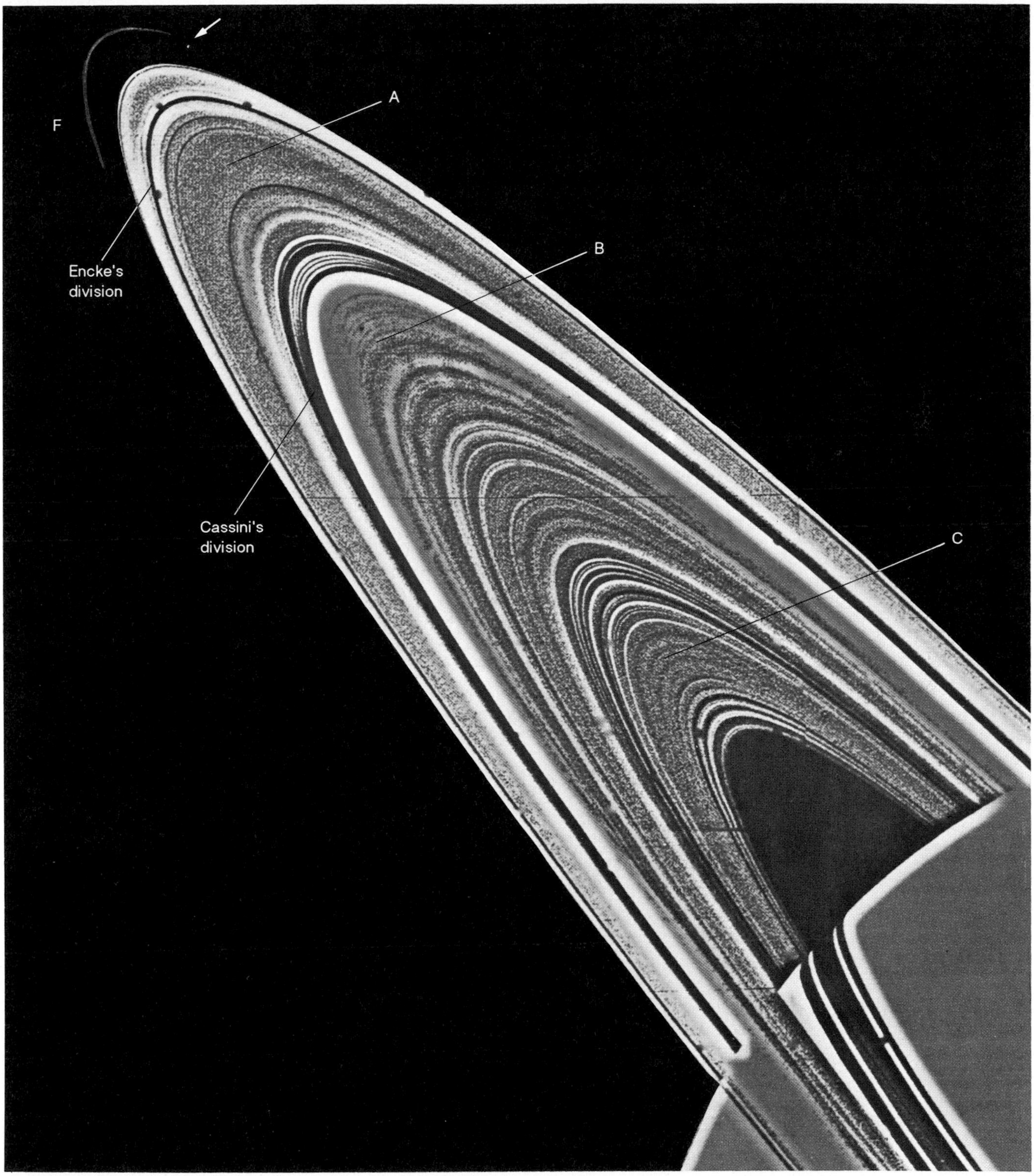
F
A
Encke's division
B
Cassini's division
C

bital periods of the major moons, no doubt play an important role in forming the structure of the rings. The "spoke" phenomenon is possibly caused by the interaction of charged dust particles with the magnetosphere of Saturn.

We mention finally that a large satellite, according to E. Roche (Sect. 2.8.3), could not exist within three Saturn radii from the center of the planet owing to the strong tidal forces.

Table 2.8.6. The five major moons of Uranus. Distance a from the center of the planet in units of Uranus' radius $R_U = 25600$ km, and moons' radii R in [km]

a/R_u	Name	R	Discovery
5.1	U5 Miranda	240	G.P. Kuiper (1948)
7.6	U1 Ariel	660	W. Lassell (1851)
10.5	U2 Umbriel	510	W. Lassell (1851)
17.2	U3 Titania	780	W. Herschel (1787)
23.1	U4 Oberon	780	W. Herschel (1787)

c) Uranus

As seen from the Earth, the disk of Uranus hardly shows any recognizable details.

The five large, relatively dark moons which were known before the Voyager missions are listed in Table 2.8.6. They orbit Uranus practically in its equatorial plane on almost circular, prograde orbits. Except for the smallest, U5 Miranda, they are comparable in size to e.g., Saturn's moon Rhea. Their densities are of the order of 1600 kg/m^3, so that their interiors probably consist of an ice-silicate mixture.

The discovery during observations of a stellar occultation by Uranus in 1977 of a system of nine narrow, *dark rings* came as a complete surprise; they were seen as a series of short, sharp eclipses of the star before and after the expected actual occultation by the planet.

Finally, in 1986, the fly-by of Voyager 2 brought a plethora of new information, as it had in the cases of the Jupiter and Saturn systems, also; it included the discovery of 10 new, small moons and an additional ring.

The now-known *rotational period* of Uranus, 17.2 h, is consistent with an *inner structure* having a silicate/iron core (of roughly 7 Earth masses?), surrounded by a mantle of ice (H_2O, CH_4, and NH_3) and a massive gas shell (H_2, He).

The cool *atmosphere* of the planet ($T \simeq 60$ K) is dominated by molecular hydrogen, H_2. The helium abundance of about 12%, known from occultations observed in the radio-frequency region, corresponds to the solar mixture; methane, CH_4, is relatively abundant in the atmosphere, but ammonia, NH_3, is amazingly rare.

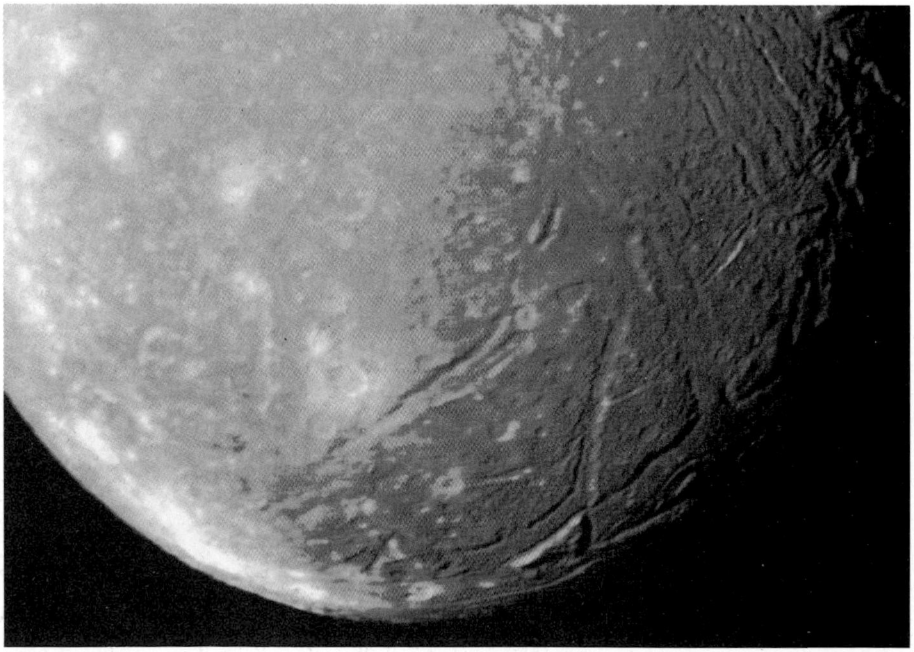

Fig. 2.8.18. A picture of Uranus' moon U1 Ariel taken by Voyager 2 from a distance of 170000 km. Along with many craters, the surface shows well-defined rills and valleys

At a greater depth (corresponding to a pressure of 1.6 bar), there is a layer of CH_4 clouds. Uranus is surrounded by an extended corona of neutral hydrogen; on the sunward side, an intense emission from hydrogen molecules ("electroglow"), whose excitation mechanism is still not understood, is observed in the upper atmosphere.

Uranus, like Jupiter and Saturn, has a magnetic field and a *magnetosphere* and emits radio-frequency radiation. The magnetic field at the equator is $2.3 \cdot 10^{-5}$ T or 0.23 G; the magnetosphere has an extent of about 20 Uranus radii ($R_U = 25\,600$ km) and forms a long plasma tail on the side away from the sun. The magnetosphere of Uranus has a structure and dynamics which are unusual among the planets, as a result of the large angle of about 59° between the magnetic dipole axis and the planetary rotation axis, the latter being nearly in the orbital plane of the planet. (Only Neptune has a similarly large angle of inclination).

The *ring system* consists of 10 very narrow, extremely dark rings (albedo $\simeq 0.04$) in the equatorial plane at distances between 1.6 and 1.95 R_U from the center of the planet. The outermost, widest ring (ε-ring) is between 20 and 100 km across and is accompanied on either edge by a small shephard moon; it contains unusually large chunks of stone (≥ 1 m). The remaining rings are of the order of only 10 km across; additional accompanying moons were not discovered. As in the case of Saturn, some of the Voyager images of Uranus also show finer ring structures and dust-like matter outside the 10 "main rings".

The small, very dark moons newly discovered by Voyager 2 all lie between 2.1 and 3.4 R_U, i.e. within the orbit of U5 Miranda. The *surfaces* of the five previously-known major moons, which are covered with ice and dark matter, show in the Voyager pictures not only the expected large number of craters, but also (except for U2 Umbriel) complex geological structures such as rills, valleys, steps, dislocations, etc. (Fig. 2.8.18). These indicate an earlier tectonic activity.

d) Neptune

The size, mass, and spectrum of Neptune, and therefore its structure, are to a great extent similar to those of Uranus.

Our knowledge of this planet has been increased considerably as a result of the fly-by of the space probe Voyager 2, which approached the planetary surface to within 5000 km in 1989. The bluish *atmosphere*, which consists mainly of hydrogen, helium, and methane, shows numerous structures (Fig. 2.8.19). Besides bands and small spots, a "*great dark spot*" is observed 22° south of the equator, a long-lived vortex analogous to the Red Spot on Jupiter. This large spot has a rotation period of 18.3 h, and thus moves with a velocity of 0.3 km s^{-1} in a direction opposite to the rotation of the planet (rotational period 16.1 h). About 50 km above the blue cloud cover, bright streamers of cirrus clouds are visible. The flow fields in Neptune's atmosphere, surprisingly strong in comparison to those of Uranus, are probably due to the large internal energy of the planet, which exceeds the energy input from solar radiation by a factor of 2 to 3 (Sect. 2.8.2).

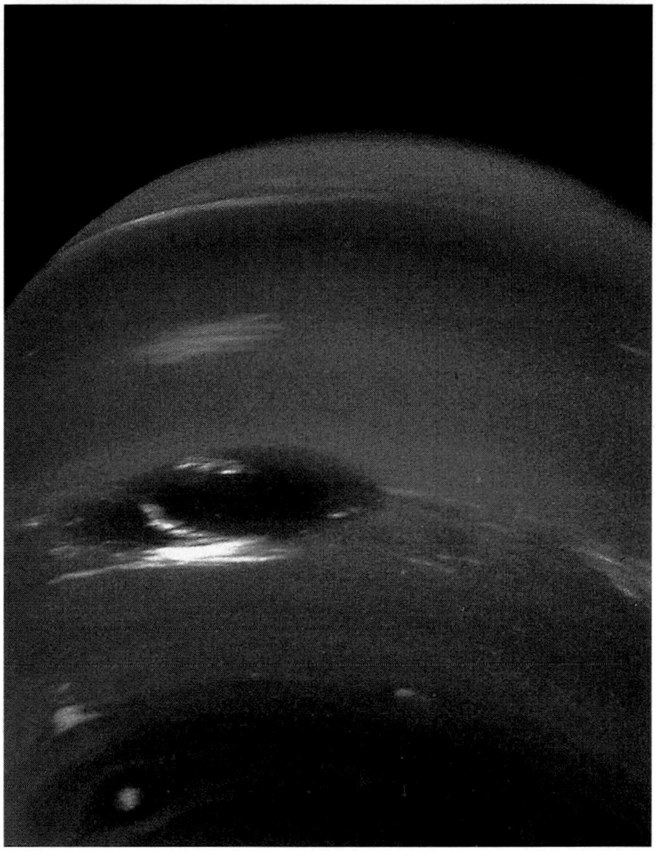

Fig. 2.8.19. An image of Neptune transmitted by Voyager 2 (August 1989). The blue atmosphere shows only a faint banded structure. On the left at a latitude of 22° south is the "Great Dark Spot" with cirrus clouds in its vicinity; somewhat further south is a small dark spot with a light central region

Neptune also possesses a dipole-like *magnetic field*, which is to be sure weaker than those of the other major

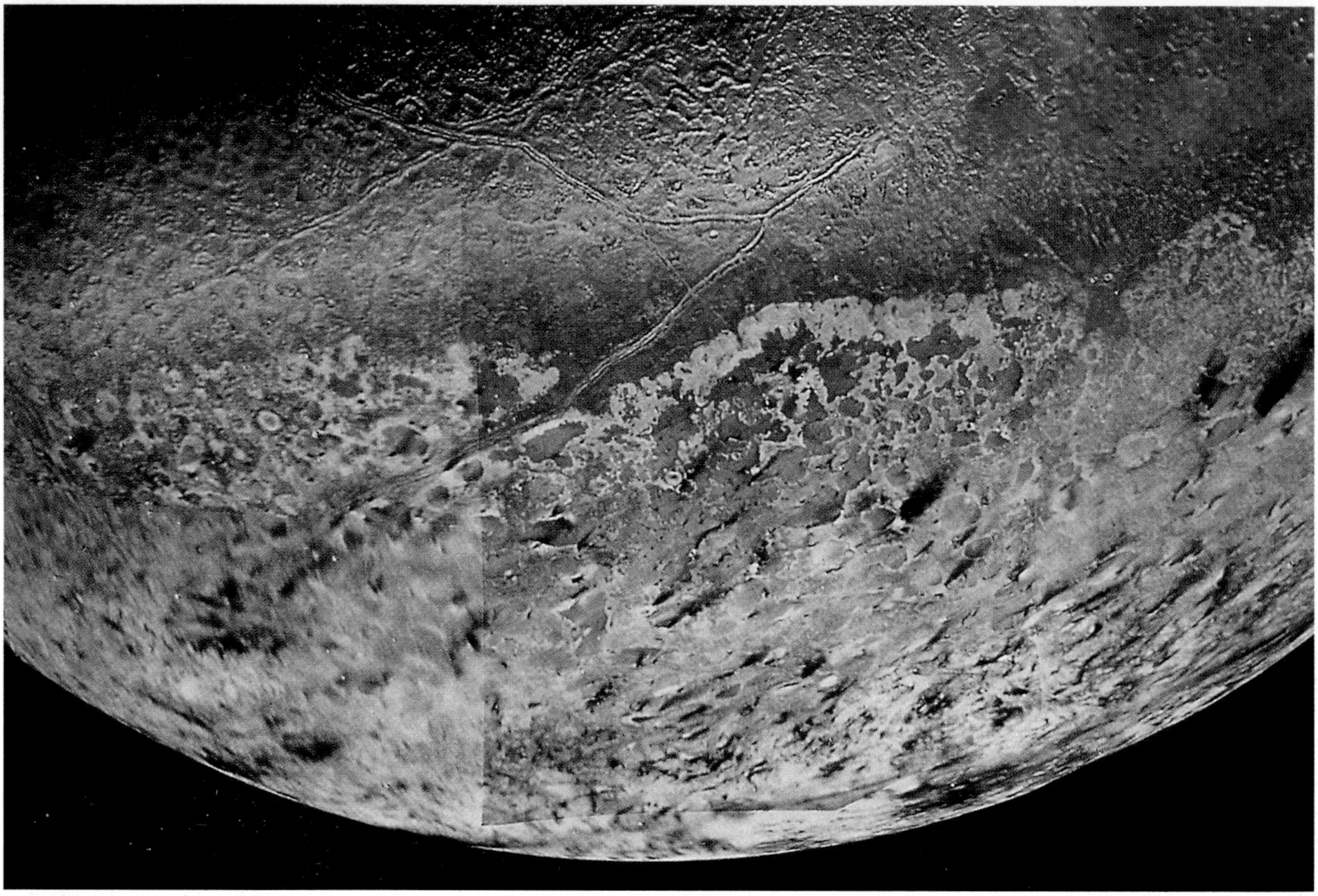

Fig. 2.8.20. A picture of Neptune's moon N1, Triton, taken by Voyager 2 (August 1989). This picture, made up of several superimposed close-up images, shows the surface structures in the neighborhood of the south pole

planets. Its axis is inclined to the rotation axis of the planet by 47°, a large angle comparable to that of Uranus. The magnetosphere of Neptune contains relatively low particle densities.

Before the Voyager mission, two moons of Neptune were known: N1 *Triton*, which is nearly as large as Earth's Moon, with a radius of 1360 km, and which orbits the planet with a retrograde motion at a distance of 14.3 planetary radii ($R_N = 24800$ km); and the much smaller N2 *Nereid*, with a radius of 200 km and a strongly elliptical orbit with a semimajor axis of 227 R_N. Both moons have large orbital inclinations (20° and 30°). Furthermore, observations of stellar occultations by Neptune in 1984/85 gave indications of an irregularly-shaped ring or ring fragments at a distance of about 3 R_N.

Voyager 2 discovered six additional moons, observed a complex ring system, and transmitted details of the surface of Triton, which the probe passed at a distance of 38000 km. The six new *satellites* orbit the planet on circular paths at between 1.9 and 4.7 R_N distance, all nearly in the equatorial plane. They are, like Nereid, irregularly-shaped, dark objects with radii between 25 and 210 km.

The *ring system* consists of two narrow, sharply bounded rings at distances of 2.1 and 2.6 R_N, one diffuse ring lying within the orbits of the moons, and an extended disk of finely divided dust particles. The outer sharp ring contains large, irregular concentrated regions, which were probably the origin of the stellar occultations observed from the Earth.

The large, reddish moon N1 *Triton* exhibits numerous surface structures (Fig. 2.8.20): along with rough terrain and long cracks, we find "frozen lakes" with terraced banks; still unexplained flat, dark areas with bright edge regions; and dark spots over 50 km long. These last features may consist of dust, which was blown out by eruptions of "geysers" (of liquid nitrogen from deeper layers?). The small number of impact craters on Triton indicates a geologically young surface. The area around the south pole is at present covered with frozen methane and nitrogen; because of its high albedo, the temperature here is extremely low, 38 K.

The atmosphere of Triton is very thin (pressure at the surface about 15 µbar) and contains mainly nitrogen with traces of methane. Occasionally, isolated clouds and haze are observed.

2.8.8 Pluto

This faint ($m_v \simeq 15$ mag) outermost planet of our Solar System is difficult to observe. In the course of a series of astronomical observations at the US Naval observatory in 1978, J. W. Christy found a systematically-occurring elongation of the planetary disk, whose analysis indicated the presence of a moon; it was later observed directly using speckle interferometry. P1 *Charon*, about 2 mag fainter than Pluto, has an orbital period of 6.39 d, synchronous with the rotation of Pluto. Its mean distance from the planet is only 0.9" or 20000 km. As a result of the discovery of this satellite, the mass and thus the mean density of Pluto can be much more reliably determined than was previously possible from the perturbations of Neptune. In particular, the observation of a series of mutual occultations of Pluto and Charon in the 1980's made it possible to determine the parameters of the system more precisely. (Such a favorable inclination of the orbital plane relative to the Earth occurs only twice during a revolution of the planet around the Sun.) The mass of the total system is about 1/400 Earth masses, of which about 10% are attributable to Charon. The mass ratio of the satellite to the planet is by far the largest in the Solar System, and the Pluto-Charon system can be thought of as a double planet. Pluto's radius is about 1150 km. Its mean density is thus found to be about 2100 kg/m³, i.e. somewhat larger than that of Saturn.

Spectrophotometric investigations indicate that Pluto, like Neptune's moon Triton, is covered with *methane ice*, which, depending on the distance to the Sun, sublimes strongly, forming a thin atmosphere of CH_4. The surface temperature of Pluto (near its perihelion) is around 60 K.

2.9 The Physical Properties of the Comets, Meteors, and Meteorites. The Interplanetary Medium

Having already made the acquaintance of the asteroids or planetoids and the satellites of the planets, we now turn to the remaining *"small objects"* of our Solar System, first dealing with the comets (Sect. 2.9.1), then with meteors and meteorites (Sect. 2.9.2), and finally with the interplanetary dust (Sect. 2.9.3), which represents the continuation of the meteorites to still smaller particle sizes.

2.9.1 Comets

Photographs taken with a suitable exposure time (Fig. 2.9.1) show that a comet consists of a (seldom re-

Fig. 2.9.1. Comet Mrkos, 1957 d. Photograph made with the Mount Palomar Schmidt camera, 1957 August 23.18. Above, we see the extended, richly structured type I or plasma tail; below, the thicker, nearly featureless type II or dust tail

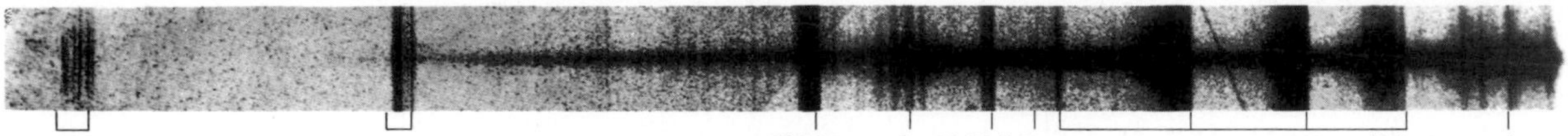

Fig. 2.9.2. A spectrum of the head of comet 1941 I Cunningham (at a distance of 0.87 AU from the Sun). In the center is the continuous spectrum of reflected sunlight; further out are the emission bands of the molecules OH, NH, CN, C_2, C_3, ... (*) indicates the skyglow line [OI] at 630 nm

cognizable) *nucleus* having a diameter of often only a few kilometers. It is surrounded by the *coma* which is like a diffuse, misty shroud and often takes the form of a series of parabolic shells or rays stretching out from the head. The nucleus and the coma together are called the *head* of the comet; its diameter is in the range $2 \cdot 10^4$ to $2 \cdot 10^5$ km. Roughly within the orbit of Mars, comets develop the well-known *tail*, which, in its visible portion, can attain a length of 10^7 and sometimes even $1.5 \cdot 10^8$ km = 1 AU.

The brighter comets can be observed in the ultraviolet region of the spectrum from satellites. It is found that the head is surrounded by a *halo* out to a distance of several 10^7 km, consisting of atomic hydrogen which radiates strongly in the Lα line at $\lambda = 121.6$ nm.

The *spectrum of the comet's head* in the optical region shows in part reflected sunlight (Fig. 2.9.2), whose intensity distribution indicates scattering from dust particles with diameters of the order of the wavelength of visible light ($\simeq 0.6$ μm). In addition, there are emission bands from numerous molecules and radicals such as CN, CH, C_2, C_3, NH, NH_2, and OH, and from radical ions such as CO^+, CH^+, OH^+, N_2^+, CO_2^+, and H_2O^+. Near the Sun, the spectral lines of atoms like [OI], Na, Ca, Cr, Fe, and Ni also occur.

In the infrared, we can observe the silicate structures at $\lambda = 10$ and 18 μm (Sect. 5.3.1) and, especially, the thermal radiation of the dust particles in the comet. In the microwave region, in addition to OH and CH which were already known from optical spectroscopy, hydrogen cyanide, HCN, hydrogen sulfide, H_2S, methyl cyanide, CH_3CN, and water, H_2O were discovered. In the ultraviolet, in addition to the atoms and molecules known from other spectral regions, as well as H for example C, C^+, O, S, S_2, CO, CS, and CN^+ have been found.

The hydrogen halo as well as the fact that all of the molecules in comets are composed of the cosmically-abundant light elements H, C, N, and O indicate that comets consist essentially of *solar matter.*

The *spectra of comets' tails* show for the most part molecular or radical *ions*: N_2^+, CO^+, OH^+, CH^+, CN^+, CO_2^+, and H_2O^+.

The characteristic shapes and motions of comets' tails require the assumption of a repulsive force originating from the Sun, whose magnitude exceeds that of the gravitational force by a large factor.

The broad *diffuse and curved tails* (Type II) consist mainly of small *dust particles* (≤ 1 μm). For such particles, the radiation pressure (each absorbed or scattered light quantum $h\nu$ transfers a momentum $h\nu/c$) can indeed reach a multiple of the gravitational force, as required by the observations.

The *narrow, elongated tails* (Type I), in contrast, consist for the most part of molecular ions, as indicated by their spectra. The calculated radiation pressure is in this case not sufficient to explain the observed large ratio of radiation acceleration to gravitational acceleration. According to L. Biermann (1951), these *plasma tails* are, instead, blown away from the Sun by an always-present corpuscular radiation, the *solar wind* (Sect. 2.9.3). At a distance equal to that of the Earth's orbit, the solar wind consists of a stream of ionized hydrogen particles, i.e. protons and electrons, with a density of 10^6 to 10^7 particles per m^3 and a velocity of about 500 km/s. Thus, the often (but not always) observed influence of solar activity on comets can be understood.

Halley's comet, which is periodic (Sect. 2.5.2), was investigated from close up in 1986 by several space probes. The European probe Giotto passed through the inner coma and approached the nucleus to within 600 km; using mass spectrometers and impact detectors, gas and dust particles could be detected *in situ*, and close-up photographs of the nucleus were obtained.

Based on the spectrographic and direct observations, we can construct the following picture of the *development of a comet*:

At a large distance (≥ 5 AU) from the Sun, only the *nucleus* of roughly 1 to 10 km diameter is present; its mass is in the range 10^{12} to 10^{15} kg. Its composition is not well known. According to F. Whipple (1950), it can be considered as a "dirty snowball", consisting of a mixture of ice (mainly H_2O, but also CH_4, NH_3, ...) and small grains of silicates and nickel-iron, similar to meteorites (Sect. 2.9.2). The nucleus of Halley's comet has an elongated, nonsymmetric shape, similar to a potato or a peanut, with dimensions of about 15 km for the long axis and 7 to 10 km for the short axis. Its surface has an irregular structure and is very dark (albedo about 0.02 to 0.04).

As the nucleus approaches the Sun, substances such as H_2O, CH_4, NH_3 etc. evaporate and begin to form the *coma*. These *mother molecules* become dissociated and ionized by solar radiation and by interactions with the solar wind, and stream away with velocities of the order of 1 km/s. Through multiple chemical reactions, other particles are formed in the outer coma and are excited to fluorescence by the solar radiation. The dust particles embedded in the nucleus are likewise released into the coma during this evaporation process; in the case of Halley's comet, an extremely irregular release of dust from the surface of the nucleus, in the form of jets, was observed. The smaller dust particles (≤ 1 μm) are then driven away from the Sun by radiation pressure.

The gas in the outer coma is carried along by the solar wind and forms the *plasma tail*. On the side facing the Sun, a shock wave forms as a result of the braking of the solar wind by the coma of the comet; in the case of Halley's comet, for example, it was about 10^6 km from the nucleus. The molecules and radicals in the tail are further ionized by energetic solar radiation, while recombination of the positive ions with electrons is rare owing to the low particle density. Therefore, ionic spectral lines dominate in the spectra of comets' tails.

2.9.2 Meteors and Meteorites

The *meteors* or *falling stars* represent only a portion of all of the small bodies of our Solar System. A distinction is sometimes made between a *meteor*, which is a brief, luminous trail in the heavens, ranging from "telescopic meteors" to fireballs which shine as bright as day, and the body which causes it, the (small) *meteoroid* or the (larger) *meteorite*. Since energy conservation requires that heavenly bodies on roughly parabolic orbits in the neighborhood of the Earth have a velocity of 42 km/s, and on the other hand, the orbital velocity of the Earth itself is 30 km/s, depending on the direction of approach (morning or evening), relative velocities between 12 and 72 km/s can be reached. On entering the Earth's atmosphere, the objects are heated. In the case of larger objects, the heat cannot penetrate sufficiently rapidly to the interior, and the surface forms melt-pits and burns off; such objects reach the ground as *meteorites*. The largest known meteorite is the Hoba West, in Southwest Africa, with a mass of about 60 tons. It must have required very large masses to make some of the *meteoritic craters* on the Earth (as on the Moon and other bodies in the Solar System). For example, the well-known crater of Canyon Diablo in Arizona has a diameter of about 1300 m and (today) a depth of 174 m. According to the geologic evidence, it must have been formed about 20000 years ago by the impact of an iron meteorite of ca. two million tons mass. The Nördlinger Ries in southern Germany, with a diameter of about 25 km, is probably also a meteoritic impact crater, which was formed $15 \cdot 10^6$ years ago in the Tertiary Era.

Small meteorites burn up in the atmosphere as "falling stars" at an altitude of about 100 km. On their paths through the upper atmosphere, they ionize a tube-shaped region of air. Large meteor showers thus contribute to the ionosphere, along with the so-called anomalous E-layer at about 100 km altitude. Furthermore, such a luminous, ionized cylinder emits electromagnetic waves like a wire, mostly at right angles to its own axis; this is the basis for the enormous impetus which has been given to the study of meteors by *radar technology*. On the radar screen, only the larger objects themselves are visible, but the direction perpendicular to the "ion-tube" is readily determined even for meteors which are well below the level of visual detectability. Velocities can also be measured with radar techniques. Their decisive advantage compared to visual observation is the fact that they are independent of clouds and the time of day, so that they eliminate the falsification of statistical investigations by such influences.

The most precise information about somewhat brighter meteors is obtained from photographic observations using wide-angle cameras of strong light gathering power, when possible from two stations a suitable distance apart. In this way, one can determine the precise spatial position of the orbits. Rotating sectors ("choppers") interrupt the image of the orbit and allow the calculation of the orbital velocity. The intensity of the streak records the optical magnitude and its often rapid temporal variations.

Since the air resistance varies as the cross-sectional area $\propto$ (diameter)^2, but the gravitational force as the mass $\propto$ (diameter)^3, it is readily seen that for smaller and smaller particles, air resistance becomes so dominant that they never begin to burn up and simply drift slowly to the ground, undamaged. These *micrometeorites* are smaller than a few hundred μm. Using suitable collection apparatus, they can be found in large quantities on the ground and also in deep-sea mud; a difficulty is naturally to distinguish them from terrestrial dirt. In recent times, research with satellites and space probes has yielded important results concerning micrometeorites and the interplanetary dust (Sect. 2.9.3). Further information has been obtained from the study of microcraters on the Moon (Sect. 2.8.5 d).

The meteorites, being the only cosmic matter which is directly accessible on the Earth, have been carefully investigated, at first using mineralographic and petrographic methods, and more recently for traces of radioactive elements and anomalous isotopes. It was originally believed that the chemical analyses of a large number of meteorites by V. M. Goldschmidt and the Noddacks represented the *"cosmic abundance distribution"* of the elements; today, these data are used in conjunction with the quantitative analysis of the Sun to give information on the early history of the meteoritic material and of our Solar System.

A first classification of the meteorites divides them into *iron meteorites* and *stone meteorites*. The former have densities of about 7800 kg/m^3, and their Fe-Ni crystals are arranged in characteristic Widmannstätten etch patterns (Fig. 2.9.3) which rule out the possibility of confusion with terrestrial iron.

The stone meteorites have densities of about 3400 kg/m^3, and can themselves be divided into two subclasses: the frequently-occurring *chondrites*, characterized by silicate grains of mm-dimensions (chondrules), and the rarer *achondrites*. The finer classification is shown in Fig. 2.9.4. The *carbonaceous chondrites* of type C1 have chemical compositions corresponding essentially to unmodified solar matter (Table 4.9.1); only the noble gases and other readily volatile elements are less abundant or lacking. The matrix in which the chondrules of carbonaceous chondrites are embedded contains many organic compounds, e.g., amino acids and complex ring compounds, which are *not*, as occasionally presumed, of biogenic origin (Sect. 6.2).

The matrix of the carbonaceous chondrites C1 must have been formed at temperatures below about 360 K. The formation of the (older) chondrites and in particular the separation of metals (Fe, Ni, . . .) and silicates implies complicated separation processes which are only partially

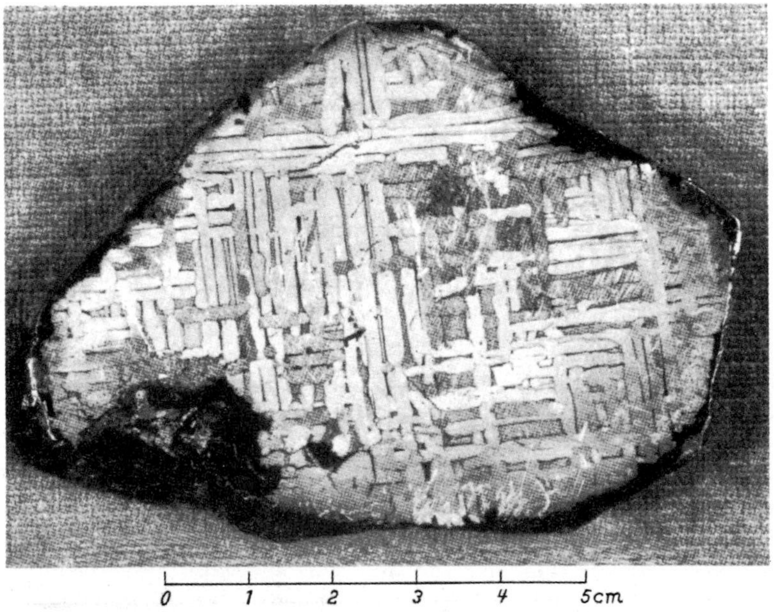

Fig. 2.9.3. The iron meteorite Toluca, named for the place where it was found. This polished and etched cut surface exhibits Widmannstätten figures, which are due to Fe-Ni crystal layers of kamazite (7% Ni) and taenite (high nickel content) which fit together parallel to the four pairs of surfaces of the octahedra; such meteorites are termed octahedrites

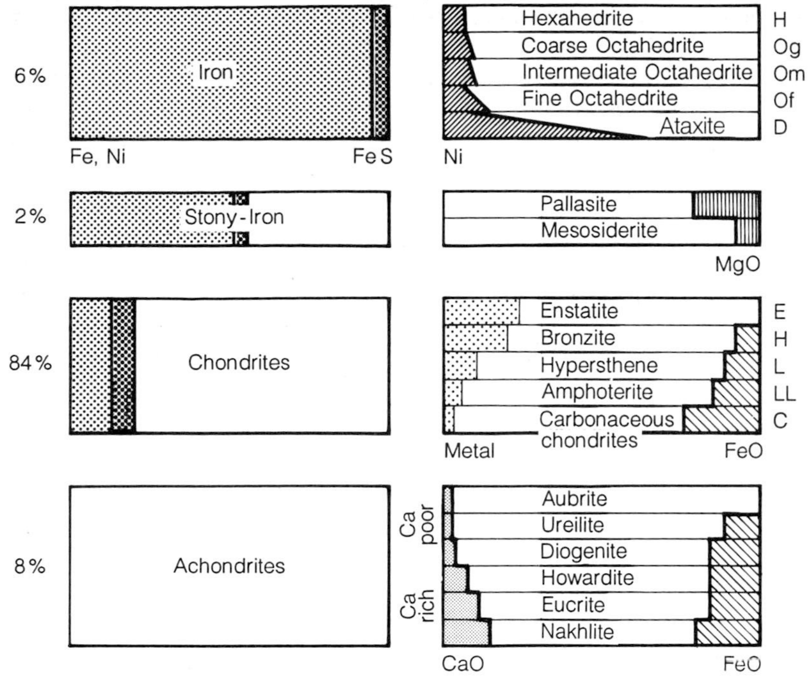

Fig. 2.9.4. The division of meteorites into four main classes (*left*) according to the ratio of metals (*shaded*) to silicates (*white*). A more detailed division is carried out according to various chemical and structural criteria (*right*). At the far left, the percentage of occurrences is indicated. (E. Anders, 1969)

understood. H.C. Urey (1952), then later J.W. Larimer, E. Anders, and others have calculated the successive formation of various chemical compounds and minerals depending on the pressure and temperature. According to these results, the chondrites required formation temperatures of $\simeq$ 500 to 700 K. We shall return to the subject of the formation of meteorites in connection with the development of the Solar System in Sect. 6.2.

The discovery of small diamonds in certain meteorites caused considerable consternation. However, it is now clear that they were not formed under high static pressure, as on the Earth, but rather under the influence of shock waves resulting from the collision of two meteorites in space.

The *tectites* or glass meteorites are greenish-black objects a few centimeters across which are composed of silicate-rich glass (70–80% SiO_2, density near 2400 kg/m^3), whose often rounded or conical shape shows that they must have passed through the air at high speeds in the molten state. Tectites are found only in certain areas, e.g., the *moldavites* in Bohemia. W. Gentner showed that several of these groups can be associated with meteoritic craters of the same age and in neighboring regions; for example, the moldavites are associated with the Nördlinger Ries mentioned above.

Radioactive dating (Sect. 2.8.5b) yields a *maximum age* for the meteorites which is the same as that of the Earth within the error limits, namely $4.5 \cdot 10^9$ years. On the other hand, it can be determined how long a meteorite was bombarded by energetic protons from the (constant) cosmic radiation background in space. Their main component penetrates about 1 m into matter and produces various stable and radioactive isotopes by spallation of heavy nuclei. The *irradiation age* can be found from the abundance ratios of these product isotopes. The irradiation ages of the tough iron meteorites are several 10^8 to 10^9 years, while those of the much more fragile stone meteorites are only 10^6 to $4 \cdot 10^7$ years. These ages are a measure of the elapsed time since the object in question was broken off from a larger body by a collision in space.

Some parts of certain meteorites contain encapsulated *noble gases*. Their isotopic abundance distribution indicates that they, like the noble gases in Moon dust and in the surfaces of Moon rocks, are for the most part products of the *solar wind*. Investigation of the *xenon* contained in meteorites furthermore showed that in some cases, those Xe isotopes were most abundant which are (as detailed study shows) decay products of two relatively short-lived and thus no longer present radioisotopes of iodine and plutonium, ^{129}I (halflife $1.6 \cdot 10^7$ yr) and ^{244}Pu

(halflife $8.3 \cdot 10^7$ yr). These investigations, which are very complex in their details, indicate that all meteorites, including the carbonaceous chondrites, were formed within at most the last $\simeq 10^7$ years.

Isotopic anomalies are found in the inclusions in some meteorites which cannot be explained either by radioactive decay processes or by fractionation or spallation due to cosmic rays. For example, the C3 chondrite *Allende* exhibits an enrichment of ^{16}O up to 5% relative to the usual oxygen isotopic ratios, a large amount of the isotope ^{26}Mg, which results from the radioactive decay of short-lived ^{26}Al (halflife $7.4 \cdot 10^5$ yr), and additional less striking anomalies. Here, we are probably dealing with remains of "presolar matter", whose composition of nuclides dates from a time before the formation of the Solar System.

We still know very little about the *origin* of the meteors and meteorites, and their relation to the comets and the asteroids (planetoids). From orbital information, especially from radar measurements, we can conclude that a portion of the meteors is of *cometary* origin; another portion, the sporadic meteors, travels on statistically-distributed elliptical orbits with eccentricities $\simeq 1$. Hyperbolic orbits or the corresponding velocities do not occur. On the other hand, a quantitative comparison of the reflectivity over the whole range from the ultraviolet into the infrared of various types of meteorites with that of the planetoids shows that there must be some relationship, at least of the larger meteorites, to the *asteroids*. We can find asteroids ranging in type from that of the iron meteorites to that of the carbonaceous chondrites. A considerable portion of the meteorites must therefore have its origin in the asteroid belt, and have been formed there as debris from collisions between larger objects.

2.9.3 Interplanetary Matter

Going from the meteors and meteorites to still smaller particle sizes, i.e. particle radii below 100 μm or masses below 10^{-8} kg, we find the *interplanetary dust*.

The reflection and scattering of sunlight from interplanetary particles with radii of 10 to 80 μm gives rise to the *zodiacal light*, which can be observed as a conical-shaped brightness in the sky in the region of the zodiac, in Spring shortly after sunset in the west, or in Autumn just before sunrise in the east. Opposite to the Sun, one can observe the faint *gegenschein* (counterglow). In total eclipses of the Sun, the strong forward scattering by interplanetary dust (Tyndall scattering) gives rise to a continuation of the zodiacal light in the neighborhood of the Sun as an outer part of the solar corona, the so-called *F* or *Fraunhofer corona*. It is called that because its spectrum, like that of the zodiacal light, contains the dark Fraunhofer lines of the solar spectrum. In both cases, the scattered light is partially polarized.

The *intrinsic radiation* of the dust particles contributing to the zodiacal light, in the far infrared region, was observed from the IRAS satellite. Three narrow emission regions are clearly visible above the background radiation: the brightest coincides with the ecliptic, while the other two are at a distance of about 10° above and below the ecliptic. They correspond to bunches of dust at about 2.5 AU distance from the Sun, i.e. in the asteroid belt (Sect. 2.8.6), and they could have been formed as debris from collisions between asteroids (with orbital inclinations $i \simeq 10°$).

In recent times, it has been possible to *directly* collect and analyze interplanetary dust particles and micrometeorites with satellites and space probes, and from the tiny (diameter ≤ 1 mm) impact craters on the Moon.

Figure 2.9.5 gives a summary of the total *flux* of particles of varying mass in the neighborhood of the Earth. In total, a mass of about $4 \cdot 10^4$ kg falls onto the earth daily, mostly in the form of micrometeorites.

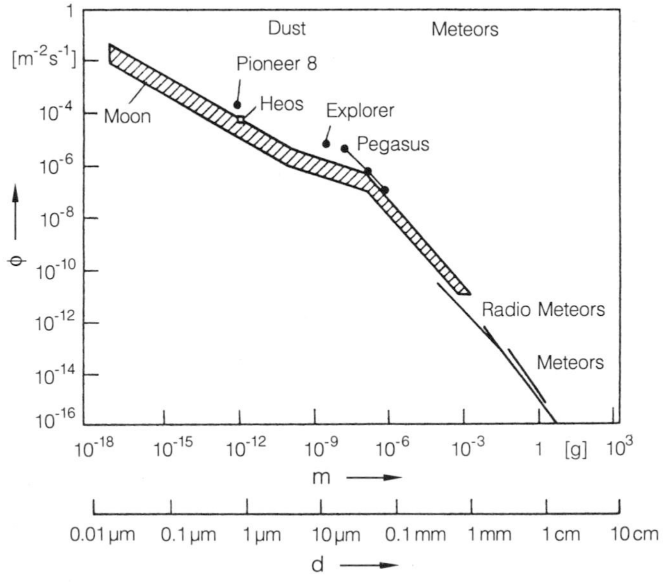

Fig. 2.9.5. The cumulative particle flux $\phi(>m)$ at 1 AU distance from the Sun as a function of the mass m or diameter d of the particles. $\phi(>m)$ is the total number of particles with masses larger than m per m^2 and s. Values for the smaller particles are taken from satellite and space-probe observations as well as from the analysis of microcraters on the Moon's surface. (H. Fechtig et al., 1981)

The comets are probably the most important source of interplanetary dust, which is released from them when they are near the Sun. Smaller dust particles are continually produced due to collisions of larger particles and by evaporation of larger objects in the neighborhood of the Sun. The radiation pressure of the Sun drives sufficiently small particles (diameter $\leq 1\ \mu m$) out of the Solar System; these so-called β-meteoroids, on hyperbolic orbits, have been observed from space probes.

In addition to dust, the interplanetary medium includes on the one hand the magnetic plasma of the *solar wind* (Sect. 4.10.7), which at a distance of 1 AU from the Sun has a mean particle density of about $10^7\ m^{-3}$ and a magnetic field of about $6 \cdot 10^{-9}$ T, and streams outwards with an average velocity of 470 km/s. Furthermore, in interplanetary space we also find the highly energetic particles of *cosmic radiation*, both of solar origin and from the galaxy (Sect. 5.3.8).

3. Astronomical and Astrophysical Instruments

Having come to know classical astronomy and our Solar System, we have reached what is perhaps the most suitable point to consider the more important astronomical and astrophysical instruments and experimental methods all together, before turning to the astrophysics of the Sun and the stars.

In conjunction with a historical introduction, we shall first give an overview of the electromagnetic spectrum and related basic concepts (Sect. 3.1). We then describe telescopes and detectors for the optical and ultraviolet spectral regions (Sect. 3.2), and for the infrared and radio-frequency regions (Sect. 3.3). Following a treatment in Sect. 3.4.1 of the most important physical processes in the interactions of high-energy photons and particles with matter, e.g. in a detector, we can deal with instruments for the observation of cosmic rays (Sect. 3.4.2), gamma rays (Sect. 3.4.3), and X-rays (Sect. 3.5). To conclude, we discuss briefly the role of electronic computers in modern astronomical research.

3.1 The Development of Astronomical Observation Methods

A Historical Introduction and Overview of the Electromagnetic Spectrum

The great advances in research are often associated with the invention or introduction of new types of *instruments*. The telescope, the clock, the photographic plate, photometer, spectrograph, and finally the whole arsenal of modern electronics and space travel each are associated with an epoch of astronomical research. However, equally important – and we should not forget this – is the creation of new *concepts* and approaches for the analysis of the observations. Brilliant scientific attainments are indeed always based upon a combination of the formulation of new concepts and of instrumental developments, which only together can achieve an advance into previously unknown realms of Nature. One is tempted to say with Simon Stevin (1548–1620) "Wonder en is gheen wonder".

The invention of the *telescope* (G. Galilei 1609; J. Kepler 1611) opened a new era for astronomy with previously unsuspected observational possibilities, due to the enormous increase in magnification and light-gathering power. The accessible spectral region of *visual* observations is limited by the sensitivity of the human eye; visible light occupies only a small portion of the electromagnetic spectrum, from about 400 to 750 nm wavelength, from violet through blue, green, and yellow to red. Only in the 19th century, after the invention of the *photographic plate*, did a light detector for astronomical observations become available which, on the one hand, stores and "integrates over" the incident light, and on the other, possesses sensitivity beyond the visible spectral region.

The wavelength region which is accessible for astronomical observations from the Earth's surface is limited by the transmission of the *Earth's atmosphere* (Fig. 3.1.1). The "*optical window*" includes the near ultraviolet and the near infrared, in addition to the visible region. On the short-wavelength end, it is limited by the absorption of atmospheric ozone, O_3, near $\lambda = 300$ nm; on the long-wavelength end, by the absorption of water vapor, H_2O, at about $\lambda = 1$ µm. Out to about 20 µm, observations are possible in several narrow windows. Only in the *radio-frequency region*, above $\lambda = (1-5)$ mm, does the atmosphere again become transparent. Although the propagation of long-wavelength electromagnetic waves in free space was discovered in 1888 by H. Hertz, available radio receivers were for a long time not sufficiently sensitive to detect radio emissions from cosmic sources. Not until 1932 were the radio emissions of the Milky Way accidentally discovered by K.G. Jansky, while he was searching for disturbances of radio transmissions due to electrical storms.

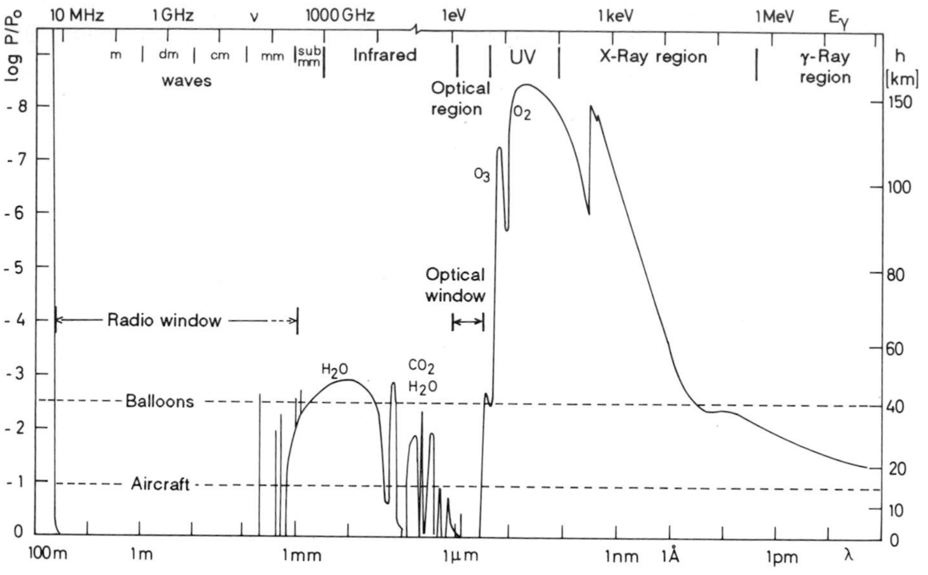

Fig. 3.1.1. The transmission of the Earth's atmosphere for electromagnetic radiation as a function of the wavelength λ (*lower scale*) or the frequency ν and the corresponding photon energy $E_\gamma = h\nu$ (*upper scale*). The altitude h [km] (and the corresponding pressure P in units of the pressure at ground level, $P_0 \simeq 1$ bar $= 10^5$ Pa) at which the intensity of the incident radiation is reduced to one-half its initial value is shown, and the maximum altitude for observations from aircraft and balloons is also indicated

Before we turn to the development of radio astronomy and the techniques for making observations in other spectral regions, it is appropriate to summarize some basic concepts of *electromagnetic radiation.*

Electromagnetic waves propagate *in vacuum* at the speed of light ($c = 2.998 \cdot 10^8\ \mathrm{m\,s^{-1}}$) (and in matter with an index of refraction n at the velocity c/n). They are determined by the time variation of the electric field vector $\boldsymbol{E}$ or, via the Maxwell equations, by the related magnetic field vector $\boldsymbol{H}$. For a given direction of propagation, they are characterized by their period of oscillation or their frequency ν, by the phase of the oscillation, by the plane of oscillation (polarization), and by the amplitude of $\boldsymbol{E}$ or of $\boldsymbol{H}$. The following relation holds for the frequency ν and the wavelength λ (strictly speaking, only in vacuum):

$$c = \nu\lambda \ . \tag{3.1.1}$$

The unit of frequency is 1 Hz (Hertz) $= 1\ \mathrm{s^{-1}}$; wavelengths are quoted in suitable multiples of the meter, depending on the spectral region, e.g. 1 nm $= 10^{-9}$ m for the visible region. In addition, special units such as 1 Å (Ångström) $= 0.1$ nm are also used. Besides ν and λ, the circular frequency $\omega = 2\pi\nu = 2\pi c/\lambda$, the wavenumber $\tilde{\nu} = 1/\lambda = \nu/c$, and the circular wavenumber $k = 2\pi\tilde{\nu}$ are also commonly quoted.

The amplitudes of the electric and magnetic fields are related in magnitude by:

$$E = \sqrt{\frac{\mu_0}{\varepsilon_0}}\, H = cB \ . \tag{3.1.2}$$

The magnetic flux density $\boldsymbol{B}$ in vacuum is given by $\boldsymbol{B} = \mu_0 \boldsymbol{H}$. The electric field constant (permittivity constant of vacuum), $\varepsilon_0 = 8.854 \cdot 10^{-12}\ \mathrm{A \cdot s \cdot V^{-1} \cdot m^{-1}}$, and the magnetic field constant (permeability constant of vacuum), $\mu_0 = 4\pi \cdot 10^{-7}\ \mathrm{V \cdot s \cdot A^{-1} \cdot m^{-1}}$, depend on the speed of light via the relation $\varepsilon_0 \mu_0 c^2 = 1$.

The energy density of an electromagnetic wave in vacuum is:

$$w = \frac{1}{2}(\varepsilon_0 \boldsymbol{E}^2 + \mu_0 \boldsymbol{H}^2) = \frac{1}{2}\left(\varepsilon_0 \boldsymbol{E}^2 + \frac{\boldsymbol{B}^2}{\mu_0}\right) , \tag{3.1.3}$$

and the vector of the energy-current density (Poynting vector) is defined as:

$$\boldsymbol{S} = \boldsymbol{E} \times \boldsymbol{H} \ . \tag{3.1.4}$$

In the *international system of units* (SI), E is measured in [V/m], H in [A/m], B in [T = Tesla], w in [J/m^3], and S in [W/m^2]. For visible electromagnetic radiation, i.e. *light*, there are special SI units derived from the base unit of light intensity, 1 cd (Candela), which corresponds to the radiated power in a unit solid angle [W/sr].

In the *Gaussian unit system*, the relations corresponding to (3.1.2–4) are

$$E = H = B \ , \tag{3.1.2a}$$

$$w = \frac{1}{8\pi}(E^2+H^2) = \frac{1}{8\pi}(E^2+B^2) \ , \tag{3.1.3a}$$

and

$$S = \frac{c}{4\pi} E \times H \ , \tag{3.1.4a}$$

where E is measured in electrostatic units, H in [Oe = Oersted], B in [G = Gauss], all three having the dimensions [$\mathrm{cm}^{-1/2}\,\mathrm{g}^{1/2}\,\mathrm{s}^{-1}$]; w is expressed in [$\mathrm{erg/cm}^3$] and S in [$\mathrm{erg\,s}^{-1}\,\mathrm{cm}^{-2}$].

In general, electromagnetic radiation from cosmic sources is an *incoherent* superposition of wavelengths of differing frequencies and polarization directions, and the rapid oscillations themselves are not of interest, but only *average values* over many periods of oscillation, for example, the average of the energy-current density S, proportional to E^2 or to H^2. We shall describe the basics of radiation theory such as intensity, radiation current density, etc. in more detail in Sect. 4.2.

Especially at high frequencies, it is often expedient to consider electromagnetic radiation not as a wave, but as a stream of *photons*, and to characterize it not by ν or λ, but by the *photon energy*[1]

$$E = h\nu = \hbar\omega = \frac{hc}{\lambda} \ ; \tag{3.1.5}$$

here, $h = 2\pi\hbar = 6.626 \cdot 10^{-34}\,\mathrm{J \cdot s} = 6.626 \cdot 10^{-27}\,\mathrm{erg \cdot s}$ is Planck's constant. An often-used unit is the *electron volt*, $1\,\mathrm{eV} = 1.602 \cdot 10^{-19}\,\mathrm{J}$, which corresponds to a frequency of $2.418 \cdot 10^{14}\,\mathrm{Hz}$ and whose equivalent wavelength is given by

$$E\,[\mathrm{eV}] = \frac{1.240 \cdot 10^{-6}}{\lambda\,[\mathrm{m}]} = \frac{1240}{\lambda\,[\mathrm{nm}]} \ . \tag{3.1.6}$$

We now turn again to the radio-frequency region, in which *radio astronomy* has opened up completely new realms for observation. The "radio window" of the Earth's atmosphere is limited on the short-wavelength end at $\lambda \simeq 1$ to 5 mm or $\nu \simeq 300$ to 60 GHz by the absorption of atmospheric oxygen, and on the long-wavelength end at $\lambda \simeq 50$ m or $\nu \simeq 6$ MHz by reflection from the ionosphere.

Following the discovery by K.G. Jansky in 1932 of the radio-frequency emissions of the Milky Way in the meter wavelength region ($\lambda = 12$ to 14 m), G. Reber from 1939 on carried out a survey of the heavens at $\lambda = 1.8$ m using a 9.5 m parabolic mirror (in the garden of his house!). During the 2nd World War, in 1942, J.S. Hey and J. Southworth detected the radio emissions of the perturbed and of the quiet Sun, using receivers which had in the meantime been improved for use in radar apparatus. In 1951, various researchers in Holland, the USA, and Australia almost simultaneously discovered the 21 cm line of interstellar hydrogen, which had been predicted by H.C. van de Hulst; its Doppler effect immediately offered enormous possibilities for investigating the motions of interstellar matter in our Milky Way galaxy and in other cosmic formations. Since the 1950's, an almost explosive development of radio astronomy has occurred, owing to the construction of individual telescopes and of a wide variety of antenna systems based on the principles of interferometry and of aperture synthesis (M. Ryle) which have had better and better *angular resolution*, as well as to the increasing detection sensitivity, especially through low-noise *amplifiers* (Sect. 5.1). Using Very Long Baseline Interferometry with transcontinental base lengths, an angular resolution of better than 10^{-4} seconds of arc has now been attained, by far exceeding the precision of optical observations. Since about 1970, radio-astronomical observations have been extended to the millimeter and recently to the submillimeter regions as a result of progress in amplifier technology, in particular.

Observations *outside the atmospheric envelope* of the Earth using rockets (whose instrument package can be returned to the ground) or satellites and space probes (whose data are returned to Earth telemetrically) have allowed astrophysics, particularly solar research, to extend to all the spectral regions which would otherwise be completely absorbed by the Earth's atmosphere: the medium and far *infrared* between 20 and 350 μm, which is absorbed by atmospheric water vapor and oxygen bands; the *far ultraviolet* past the transmission limit of atmospheric ozone at $\lambda = 285$ nm; the contiguous *Lyman region*, where absorption is mainly due to atmospheric oxygen, O_2; then the *X-ray* and finally the *gamma ray* regions. Although we can investigate the solar spectrum continuously from wavelengths in the radio-frequency region out to the X-ray region, for galactic and extragalactic research we must keep in mind the Lyman continuum of the *interstellar* hydrogen atoms (Sect. 5.3). Their ab-

[1] The conventional symbol E is not to be confused with the magnitude of the electric field $\boldsymbol{E}$.

sorption sets in strongly at 91.2 nm and allows no view until we reach the X-ray region beyond about $\lambda = 1$ nm.

At the beginning of "space astronomy", immediately after the end of the 2nd World War, observations were made using V2 rockets which were developed in Germany during the war, and later with stabilized research rockets: in 1946, H. Friedman and his group obtained the first ultraviolet spectrum, and in 1948, T.R. Burnight recorded the first X-ray image of the *Sun*. Following the launching of the first artificial satellite Sputnik 1 (1957), a rapid development in the investigation of the Solar System and in astronomical observation using satellites and space probes occurred, which we have to some extent already described in Sect. 2.7.

From the 1960's on, beginning with the Orbiting Solar Observatory (OSO) and Orbiting Astronomical Observatory (OAO), satellites with sufficient positional stability ($\leq 1''$) to permit longer series of observations of the Sun and other cosmic sources in the ultraviolet, X-ray, and gamma-ray regions became available. In 1962, (nonsolar) *X-ray astronomy* began with the accidental discovery of the strong source Sco X-1 by R. Giacconi and coworkers using a rocket-carried instrument; in 1970, the first X-ray satellite, UHURU, began its survey of the skies. In the 1970's, *gamma-ray astronomy* also attained a sufficient angular resolution to be able to resolve individual cosmic sources. Astronomy with observation of high-energy photons has developed in a few years into one of the most interesting areas of research, making use of an arsenal of new and unusual instruments.

In the infrared, the in general weak sources must be detected against a background of thermal radiation from the instrument itself and from the Earth's atmosphere. The development of cooled semiconductor detectors since the 1960's has produced a rapid improvement in our ability to observe infrared sources from stratospheric balloons and aircraft, as well as from the first infrared satellite observatory (IRAS) launched in 1983.

Observations from space also offer clear advantages in the *optical region*, owing to the fact that the angular resolution is no longer limited by air motions. The improvement of the resolution and of the detection limit for optical astronomy expected from the 2.4 m Hubble Space Telescope in the 1990's has not been achieved to the extent planned, at least in the initial phase following the telescope's launching, owing to an error in the optical system. The rapid development of new, high-sensitivity detectors after about 1970 has also increasingly allowed earthbound optical observation access to extremely faint light sources.

To end our introductory overview, we shall mention several astronomical observation methods which are not based on the detection of electromagnetic radiation. In 1912, V.F. Hess had already discovered *cosmic radiation* in the course of a balloon ascent; today, it represents an important research area of *high-energy astronomy*. While the secondary particles resulting from interactions of the primary energetic protons and heavier nuclei in the upper atmosphere can be investigated on the ground, the primary radiations themselves must be observed from outside the atmosphere. The search for *gravitational radiation* from cosmic sources has thus far remained unsuccessful. The interesting area of *neutrino astronomy*, whose early development was due to R. Davis Jr. (1964), is at present still limited to the Sun and to unusual events such as the explosion of the supernova SN 1987 A in the nearby Large Magellanic Cloud.

3.2 Telescopes and Detectors in the Optical and Ultraviolet Regions

From the instrumental viewpoint, there is no difference in principle between the ultraviolet and the optical spectral regions, if we exclude the extreme ultraviolet ($\lambda \leq 110$ nm). Instruments intended for ultraviolet astronomy must, however, be designed for observations from space, since absorption by atmospheric ozone prevents observations from the ground for $\lambda \leq 300$ nm (Fig. 3.1.1). In the laboratory, measurements at $\lambda \leq 200$ nm (vacuum ultraviolet) must be carried out in vacuum owing to absorption by the air. Below $\lambda \simeq 110$ nm, the reflectivity of the usual mirror materials decreases sharply, so that here, as in the X-ray spectral region (Sect. 3.5), systems with grazing incidence mirrors must be used.

To begin, in Sect. 3.2.1 we treat the most important types of telescopes, the refractor and the reflector, with their focussing systems, as well as the Schmidt mirror. In the following Sect. 3.2.2, we discuss some basic concepts such as resolving power, light-gathering power, and the principle of the interferometer. Section 3.2.3 gives an overview of detectors used for observing the radiation, from the photographic plate to modern high-sensitivity semiconductor detectors. As dispersing element for the spectral decomposition of light, either a prism or, more usually, a diffraction grating is used; the basic principles of the spectrograph will be treated in Sect. 3.2.4. In Sect. 3.2.5, we introduce some telescopic instruments intended for observations from space, and describe the prospects for constructing earthbound telescopes having light-collecting mirror surfaces with effective diameters of more than 10 m.

3.2.1 Conventional Telescopes

The principles of Galileo's telescope (1609) and Kepler's telescope (1611) are recalled in Fig. 3.2.1; in both, the magnifying power is determined by the ratio of the focal lengths of the objective and the ocular lenses. Galileo's arrangement yields an upright image and therefore became the prototype of opera glasses and binoculars. Kepler's telescope, in contrast, casts a real image in the focal plane, which for visual observations is viewed through the ocular, using it like a magnifying glass. This permits a reticle to be inserted in the common focal plane of the objective and the ocular and can thus be used to fix angles precisely, e.g. on the meridian circle. If the reticle is extended to include a *micrometer scale*, among other things the relative positions of binary stars can be determined.

The disturbing colored borders (chromatic aberration) in the images of telescopes with simple lenses were eliminated by J. Dollond and others about 1758 by the invention of *achromatic lenses*. An achromatic positive lens consists of a convex lens (positive lens) made of crown glass, whose dispersion compared to its refracting power

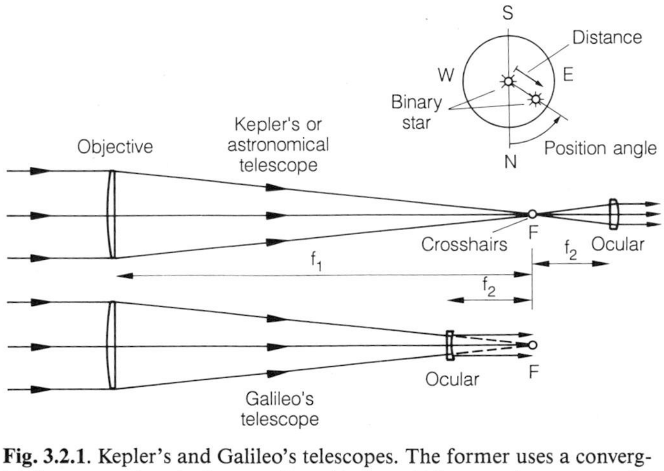

Fig. 3.2.1. Kepler's and Galileo's telescopes. The former uses a converging lens as ocular, the latter a diverging lens. F denotes the common focal point of the objective and the ocular lenses. The magnification V is equal to the ratio of the focal lengths of the objective, f_1, and of the ocular, f_2; in this figure, a ratio $V = f_1/f_2 = 5$ has been chosen. In Kepler's telescope, a crosshair or a micrometer scale can be inserted in the focal plane at F to allow the quantitative observation of binary stars. (The curvature of the lenses is exaggerated in the drawing)

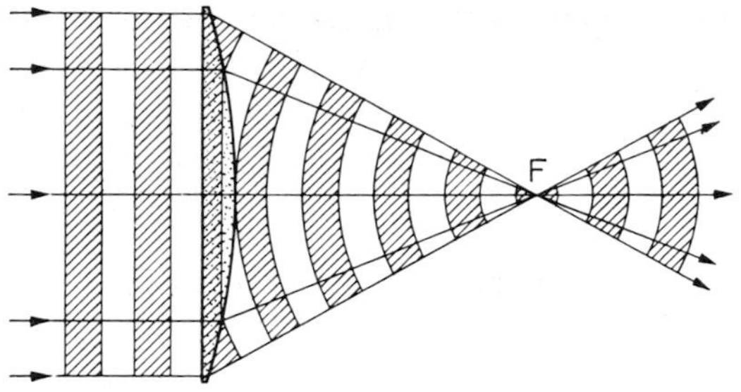

Fig. 3.2.2. Image formation by a plane-convex lens. Wavefronts of plane-wave light from a star are incident at the far left; the light rays are perpendicular to the wavefronts. The velocity of light in the glass of the lens is a factor n smaller than in vacuum (n = index of refraction); as a result, the wavefronts are bent into spherical surfaces which converge on the focal point F and then diverge behind it

is relatively small, combined with a concave lens (negative lens) made of flint glass, whose dispersion is large compared to its refracting power. With an objective consisting of two lenses, it is possible to eliminate the change of the focal length f with λ (i.e. to make $df/d\lambda = 0$) at only *one* wavelength λ_0. For a *visual* objective lens, λ_0 is chosen to be about 529 nm, corresponding to the wavelength of maximum sensitivity of the eye; for a *photographic* objective lens, in contrast, λ_0 is taken to be about 425 nm, the wavelength of maximum sensitivity of a photographic blue plate.

Let us examine in more detail the imaging of a region in the sky by a telescope, for example onto a photographic plate in the focal plane of the objective lens! The task of converting the incident plane wave coming from "infinity" into a convergent spherical wave is accomplished by the lens, making use of the fact that in glass (index of refraction $n \geq 1$), the light moves at a speed which is n times slower and therefore the wavelength is n times shorter than in vacuum. As a result, the wavefront is delayed in the middle of the objective lens relative to its outer edges (Fig. 3.2.2).

What the *lens telescope* or *refractor* does by inserting layers of material of $n > 1$ of differing thickness into the optical path, is accomplished in the *mirror telescope* or *reflector* by means of a concave mirror (I. Newton, about 1670). This arrangement has the *a priori* advantage of possessing no chromatic aberrations. A *spherical mirror* (Fig. 3.2.3 a) focusses a beam of axial rays at a focal length f equal to half the radius of curvature R, as can be seen by simple geometric considerations. Rays which are more distant from the optical axis are focussed at

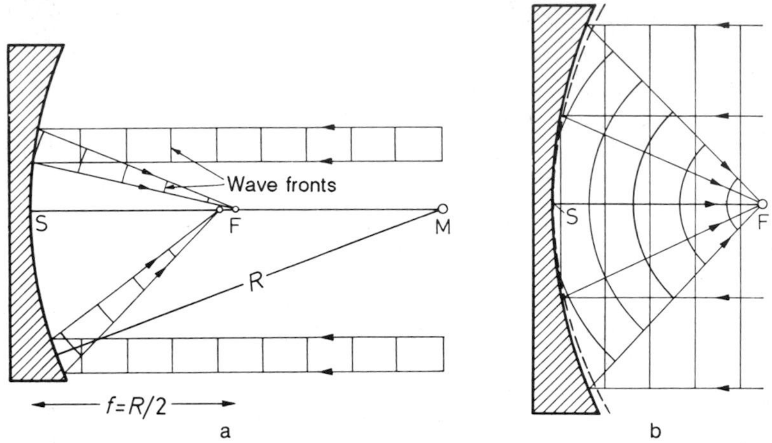

Fig. 3.2.3. **(a)** A spherical mirror. A bundle of rays near the optical axis (*upper bundle*) converges at the focal point *F*, whose distance from the crown of the mirror, *S*, corresponds to the focal length $f = R/2$, with *R* the radius of curvature of the mirror. A bundle which is further from the optical axis (*lower bundle*) converges on a point closer to *S*: this gives rise to spherical aberration. The planar wavefronts incident from the right are converted into spherical wavefronts on reflection by the mirror. **(b)** A parabolic mirror combines all the incident rays which are parallel to the optical axis precisely at the focal point *F*, i.e. the plane wave which is incident parallel to the axis is converted into a single convergent spherical wave. The conformal sphere (– – –) has the same curvature at the center point *S* as the paraboloid

somewhat shorter distances from the center of the mirror; this imaging error is called *spherical aberration*. The exact focussing of a beam of axial rays in a *single* focal point is accomplished by a *paraboloid* mirror (Fig. 3.2.3 b); this can be seen immediately by considering the paraboloid as a limiting case of an ellipsoid, whose right focal point has moved away to infinity. Unfortunately, however, a paraboloid mirror yields a good image only in the immediate neighborhood of the optical axis. At a larger *aperture ratio* D/f (D = mirror diameter; focal or *F*-ratio f/D), the usable diameter of the image field is very small, owing to the imaging errors from obliquely incident rays which increase rapidly as one moves away from the axis.

The *mount* of a telescope has the task of allowing it to follow the diurnal motion of the Earth with sufficient precision. The commonly-used equatorial mount therefore has a 'right ascension' axis which is parallel to the axis of the Earth and is driven by a sidereal clock, and perpendicular to it a 'declination' axis; both of them are equipped with corresponding divided scales for reading the celestial position.

The azimuthal mount, with a vertical and a horizontal axis, permits a compact telescope construction with uniform axis loading and is convenient for large, heavy instruments. The problem of the necessary *nonuniform* motion about *both* axes can now be readily solved by a computer-controlled drive system. Azimuthal mounts have long been used for large radiotelescopes but are not widespread in optical astronomy.

Refractors, with aperture ratios in the range 1 : 20 to 1 : 10, are usually equipped with the so-called Fraunhofer or *German Mount*, as for example the largest instrument of this type, the refractor at the Yerkes Observatory of the University of Chicago (Fig. 3.2.4), with an objective of 1 m diameter and 19.4 m focal length, which was completed in 1897. Large refractors have hardly been constructed since the turn of the century. For special applications, e.g. in position astronomy or for the visual observation of binary star systems, however, lens telescopes are still preferred today to reflecting telescopes.

Reflectors are usually constructed with an aperture ratio of 1 : 5 to 1 : 2.5 and equipped either with one of the various types of *fork mounts* (the declination axis passes through the center of gravity of the tube and is reduced to two bosses on either side of the latter), or with an English Mount, in which the north and south bearings of the long right ascension axis rest on separate pillars. The largest reflectors (with the primary mirror in a *single* piece) are at present the Hale telescope on Mount Palomar, California, placed in service in 1948, with a diameter of 200″ or 5 m and a focal length of 16.8 m for the primary mirror (Fig. 3.2.5), and, since 1976, the (azimuthally mounted) 6 m reflector of the Soviet Special Astrophysical Observatory in Zelenchuk (Caucasus Mountains) whose primary mirror has a 24 m focal length.

Since about 1970, the number of conventional reflecting telescopes of more than 3 m diameter has increased sharply. Instruments in this class in the *Northern Hemisphere* are at the Lick Observatory on Mount Hamilton in California (3.1 m), at the Kitt Peak National Observatory of the USA in Arizona (4.0 m), on Mauna Kea in the Canada-France-Hawaii 3.6 m telescope, at the German-Spanish Astronomical Center in Calar Alto in Southern Spain (3.5 m), and at the European Observatory on the Roque de los Muchachos on the island of La

Fig. 3.2.4. The 40″ (1 m) refractor of the Yerkes Observatory; it has a Fraunhofer or German mount

Fig. 3.2.5. The 200″ (5 m) Hale reflector on Mount Palomar

Palme (4.2 m). In the *Southern Hemisphere*, they include those at the Cerro Tololo Inter-American Observatory in Chile (4.0 m), at the European Southern Observatory (ESO) on La Silla in Chile (3.6 m), and at the Anglo-Australian Observatory at Siding Spring in Australia (3.9 m).

In contrast to lens telescopes, the image quality of reflectors is influenced strongly by temperature variations, unless materials of very small thermal expansion coefficients are used. *Glasses*, e.g. Pyrex, from which the 5 m Palomar mirror was constructed, have a linear thermal expansion coefficient $\alpha \simeq 30 \cdot 10^{-7}\,K^{-1}$; quartz has $\alpha \simeq 6 \cdot 10^{-7}\,K^{-1}$. The ability since about 1965 to manufacture and work large blocks of high-quality *glass ceramics*, e.g. Zerodur, has permitted significant progress. This mixture of an amorphous glass and a crystalline ceramic component with opposing thermal properties exhibits no thermal expansion to speak of ($\alpha \leq 10^{-7}\,K^{-1}$). The *reflecting* surface consists of an aluminum layer of only 100 to 200 nm thickness which is vacuum-evaporated onto the mirror.

With many reflectors (Fig. 3.2.6), one can observe either in the *primary focus* of the main mirror or, by means of a plane secondary mirror which reflects the beam out the side of the tube, in the *Newtonian focus*. It is also possible to attach a convex mirror in front of the primary focus and to produce the image at the *Cassegrain focus* (*F*-ratio 1 : 20 to 1 : 10) behind an opening in the center of the main mirror. In newer telescopes, the *Ritchey-Chrétien system* (1 : 10 to 1 : 7) is often used; here, *both* mirrors in the Cassegrain system are replaced by mirrors with hyperboloid-like forms in order to obtain a larger field of view (0.5°) and a coma-free image.[2] In both the

[2] Coma: an image aberration in which rays not parallel to the optical axis are not focussed at the same point, so that the image of a point source is drawn out into a "comet's tail" shape.

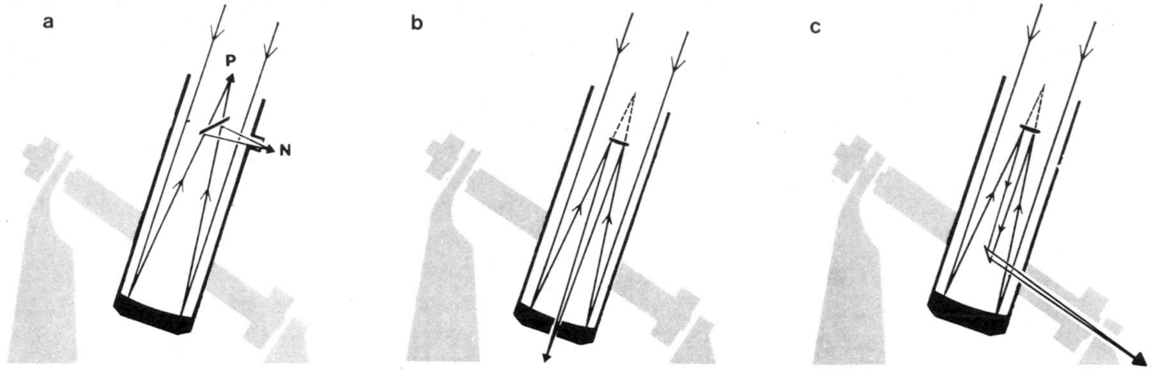

Fig. 3.2.6 a – c. Different focus arrangements, using the example of the 2.5 m Hooker reflector on Mount Wilson, the first of the large modern telescopes (completed in 1917), which was only recently decomissioned. **(a)** Primary (P) and Newtonian (N) focus (*f*-ratio 1 : 5); **(b)** Cassegrain focus (1 : 16); **(c)** Coudé focus (1 : 30)

Cassegrain and the Ritchey-Chrétien systems, boring through the primary mirror can be avoided and the image can be directed out of the tube by a plane mirror to the *Nasmyth focus*. Finally, by means of a complicated arrangement of mirrors, the light can be directed through the hollow pole axis and the image of a star can be projected for example onto the entrance slit of a *fixed* large spectrograph in the *coudé focus* (1 : 45 to 1 : 30).

The desire of astronomers for a telescope with a large field of view *and* a large aperture ratio (light-gathering power) was fulfilled by the ingenious invention of the *Schmidt camera*. B. Schmidt (1930/31) first noticed that a spherical mirror of radius R will focus small beams of parallel rays, which are incident from any direction but pass roughly through the neighborhood of the spherical center of curvature, onto a concentric sphere of radius $R/2$ – corresponding to the well-known focal length of the spherical mirror, $f = R/2$. With a small aperture ratio, one can thus obtain a good image on a *curved plate* over a *large angular region*, if, in addition to the spherical mirror, an entrance iris diaphragm is placed at the center of curvature, i.e. at a distance equal to twice the focal length from the mirror (Fig. 3.2.7). If a high light-gathering power is also desired, so that the entrance iris must be opened wide, spherical aberration becomes apparent as a smearing out of the stellar images. This was eliminated by Schmidt, who placed a thin, aspherically-ground *correction plate* in the entrance iris, compensating the differences in optical path by corresponding thicknesses of glass and a small shift in the focal plane. These path differences are seen in Fig. 3.2.3 b to correspond to the distance between the paraboloid and the surface of the conformal sphere. Due to the smallness of these differences, correction is possible simultaneously for a large range of incident angles and without introducing disturbing chromatic aberrations.

Schmidt telescopes, as a rule, have *F*-ratios of 1 : 3.5 to 1 : 2.5, but can be constructed with ratios down to 1 : 0.3. The 48″ Mount Palomar Schmidt telescope (1 : 2.5) has a correction plate of diameter 48″ = 122 cm. To avoid vignetting, the spherical mirror must have a larger diameter of 183 cm. This instrument was used for the famous Palomar Observatory Sky Survey; ca. 900 image fields of $7° \cdot 7°$, each taken with a blue plate and a red plate having detection limits of 21 mag and 20 mag, respectively, cover

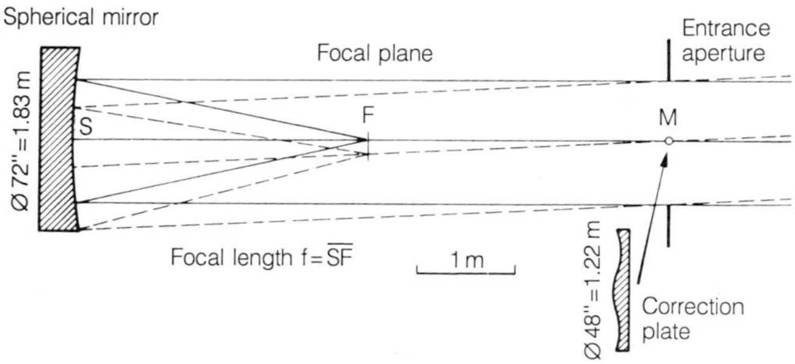

Fig. 3.2.7. A Schmidt mirror. Bernhard Schmidt based his instrument on a spherical mirror with a radius of curvature $R = MS$. Bundles of parallel rays, even those incident at a considerable angle from the optical axis, are collimated by the entrance aperture at the center point M of the mirror and thus are combined into a spherical wavefront around M with a radius $R/2 = MF$; this is termed the focal surface. The focal length is then $f = FS = R/2$. To remove the spherical aberration, an aspherically-ground, thin correction plate is placed in the aperture. (The dimensions shown correspond to those of the Mount Palomar 48″ Schmidt telescope)

the entire celestial Northern Hemisphere to a declination of $-32°$.

In the case of the *Maksutsov-Bouwers telescope* (or camera), the Schmidt correction plate for avoiding spherical aberrations is replaced by a large *meniscus lens* having spherical surfaces.

Among the special instruments for *positional astronomy*, we must at least mention the *meridian circle* (O. Römer, 1704). The telescope can be moved about an east-west axis in the meridian. The right ascension is determined by the time of transit of a star through the meridian (using cross-hairs in the focal plane). The horizontal cross-hair in the field of view, together with the divided circle attached to the axis, allow the determination of the culmination altitude and thus of the declination of the star. Modern position determinations attain a precision of a few hundredths of a second of arc. On a divided circle with 1 m radius, 0.1 second of arc corresponds to 0.5 μm!

3.2.2 Resolving Power and Light-Gathering Power. Optical Interferometers

We shall now attempt to develop a clear picture of the usefulness of different telescopes for various applications! The visual observer is primarily interested in the *magnification*. This is, as stated above, simply equal to the ratio of focal lengths of the objective and ocular. A limit to imaging smaller and smaller sources is, however, set by the diffraction of light at the entrance aperture of the telescope. The smallest angular distance between two stars, e.g. binary stars, which can still just be separated in the image, is termed the *resolving power.* A square aperture of side D (which is simpler to treat than a circular aperture) produces a diffraction image from the parallel rays of light arriving from a star; it is bright in the center, and has on either side a dark band resulting from interference, where the light excitation from the two halves of the aperture (Fresnel zones) cancels. From Fig. 3.2.8, this corresponds to an angle (in radians) of λ/D. For a *circle* of radius D, we find for the angular spacing

$$\phi = 1.22\frac{\lambda}{D}\ , \tag{3.2.1}$$

in which the diffraction disks of two stars half cover each other and can just still be separated. Equation (3.2.1) thus yields the *resolving power* of the telescope. If ϕ is calculated in seconds of arc and the diameter of the tele-

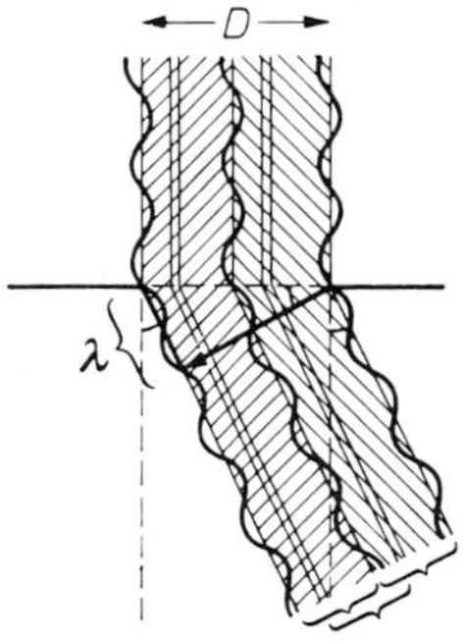

Fig. 3.2.8. Diffraction of light by a slit or rectangle of width D. When two bundles of rays spaced a distance $D/2$ apart have a path length difference equal to one-half wavelength, i.e. $\lambda/2$, then the first diffraction minimum results from interference; it occurs at an angle of deflection given by λ/D

scope is in meters, we obtain for $\lambda = 550$ nm a theoretical resolving power of

$$\phi = \frac{0.14''}{D\,[\mathrm{m}]}\ , \tag{3.2.2}$$

i.e., for example, for a 5 m mirror, $\phi = 0.03''$. In the "oldest optical instrument", the *human eye*, the resolving power for a pupil diameter of several mm is found to be of the order of $1'$.

With a focal length f, (3.2.1) corresponds to a *linear* extension of the diffraction disk in the focal plane of $l = f \tan\phi$ or, for small angles,

$$l = 1.22\,\lambda\frac{f}{D} \tag{3.2.3}$$

where D/f is the aperture ratio. For $\lambda = 550$ nm, l is then equal to $0.67 f/D$ [μm].

Longer photographic exposures made from the ground are subject to a smearing of point images of the order of $1''$ due to *seeing* (scintillation), so that even a small telescope of somewhat more than 10 cm diameter reaches the maximum angular resolution permitted by the atmosphere.

The *aperture* or the *diameter D* of the telescope mirror determines in the first instance the *collecting surface* ($\propto D^2$) for the light from a star, a nebula, etc. The collected light or energy flux is imaged onto the receptor in the focal plane; the intensity or "speed" I is a measure of how strongly the light is concentrated on the receptor surface.[3] The light from a distant star, i.e. that of a "*point*

[3] Precisely defined concepts relating to the propagation of radiation will be treated in Sect. 4.2.1.

source", whose diameter is small compared to that of its diffraction disk, is spread in any case over an area given by (3.2.3), so that its intensity is at the most equal to:

$$I \simeq \left(\frac{D}{\lambda}\right)^2 \left(\frac{D}{f}\right)^2 . \qquad (3.2.4)$$

For a given ratio D/f, I thus increases proportionally to D^2. If the diameter δ of the smallest resolving element of the receptor is now chosen to be larger than the dimension l of the diffraction disk, the intensity of the point source depends only on D^2, not on the F-ratio.

On the other hand, for an *extended source*, e.g. a comet or a gaseous nebula, with an angular diameter whose image ϕf is larger than the diffraction disk or the receptor resolution, the intensity is proportional to $(D/f)^2$. This is also valid for the case that the disk caused by seeing ($\phi \simeq 1''$) is larger than the diffraction disk. The power of an instrument for determining surface brightness therefore depends, like that of a camera, in the first instance only on the F-ratio, but secondarily on absorption and reflection losses in the optical system. In both respects, the Schmidt mirrors and their variants are unmatched.

A much more complex question is that of how faint a *star* can still be detected. The *limiting magnitude* of a telescope is clearly determined by the criterion of whether the small stellar disk – its size given by scintillation, diffraction, photographic-plate graininess, detector resolution, etc. – can still be distinguished from the background due to airglow and other disturbances. Thus, fainter stars can initially be seen by using an instrument of larger diameter; for a given aperture, a smaller F-ratio, i.e. a longer focal length, is advantageous. In practice, the exposure times must also be considered!

As we have seen, the theoretical resolving power of a telescope is determined by the *interference* of edge rays. A somewhat improved resolution was obtained by A.A. Michelson using his *stellar interferometer*, by placing two slits at a spacing D in front of the objective (Fig. 3.2.9 a). A "pointlike" star then yields a system of interference fringes with the angular spacing:

$$\phi = \frac{n\lambda}{D} \quad (n = 0, 1, 2, 3 \ldots) . \qquad (3.2.5)$$

If a binary star is observed with this instrument, with its two components separated by an angular distance y along the axis of the two slits, then the interference fringes from the two stars are superposed; the brightest fringes are obtained when $y = n\lambda/D$. In between, the interference

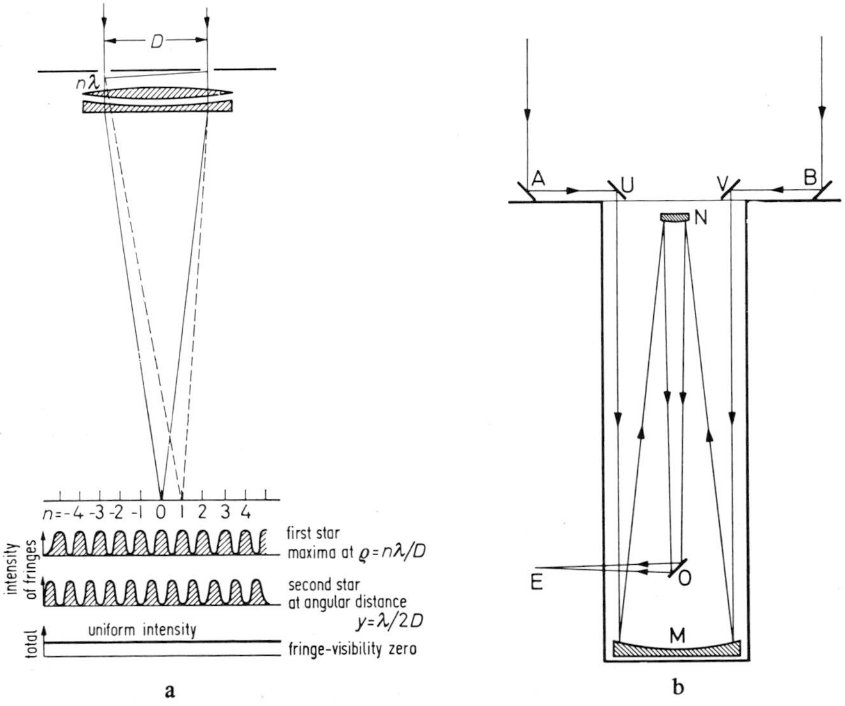

Fig. 3.2.9. (a) A stellar interferometer as developed by A.A. Michelson. A "pointlike" star produces a system of interference fringes whose distances from the optical axis are $\phi = n \cdot \lambda/D$ ($n = 0, \pm 1, \pm 2, \ldots$). The fringes from two (equally bright) stars superpose and give a constant intensity at angular distances $y = 1/2\lambda/D$, $3/2\lambda/D, \ldots$, i.e. zero visibility. **(b)** The 6.1 m Michelson interferometer of the Mount Wilson Observatory. A steel beam above the aperture of the 2.5 m reflector carries the inner two 45° mirrors U and V, which are fixed, and the outer, movable mirrors A and B. The distance AB (maximum 6 m) corresponds to the slit width D in **(a)**. Observations are made visually at the Cassegrain focus E

fringes become undetectable, when the components of the binary star are equally bright; otherwise, they pass through a minimum. Conversely, if one slowly moves the two slits apart, the interference fringes are most readily observable for $y = 0, \lambda/D, 2\lambda/D, \ldots$ and least visible for $y = \lambda/2D, 3\lambda/2D, \ldots$.

When a disk of (angular) diameter y' is observed, it is, as can be shown by an exact calculation, nearly equivalent to two bright points at a spacing $y = 0.41\,y'$, and the first visibility minimum is obtained at a slit distance D_0 corresponding to $0.41\,y' = \lambda/2D_0$ or

$$y = 1.22\frac{\lambda}{D_0} . \tag{3.2.6}$$

Compared to using a normal telescope, it at first may seem that little is to be gained [D_0 is $\leq D$ in (3.2.1)], but in fact the criterion of *visibility* of the interference fringes is less dependent on seeing than, for example, a measurement with a cross-hair micrometer. Michelson and others were thus first able to measure the diameter of Jupiter's moons, closely-spaced binary stars, etc.

Later, however, Michelson inserted a system of mirrors like that in binoculars in front of the mirror of the 2.5 m reflector of the Mt. Wilson Observatory, and could thus make $D_0 > R$ (Fig. 3.2.9b). It then became possible in the 1920's to measure the diameters of several red giant stars directly (the largest are $\simeq 0.04''$).

An important point in the Michelson interferometer is to bring the two rays together with the correct phase. This difficulty, which made the construction of still larger instruments impossible, was overcome by the *correlation interferometer* developed by R. Hanbury Brown, as follows: two concave mirrors at a separation D collect the light of a single star, each one onto a photomultiplier. The correlation of the current fluctuations from the two photomultipliers in a particular frequency range is measured. The strength of this correlation is, as can be shown by theory, related to D and y in *exactly* the same way as the visibility of the interference fringes in the Michelson interferometer.

The correlation interferometer in Narrabri, Australia, uses two 6.5 m mirrors, which are each constructed as a mosaic of hexagonal plane mirrors (the image quality is unimportant here); they are mounted on a circle of diameter $D_0 = 188$ m, on which they can be moved, so that their spacing D can be varied in the range $\leq D_0$. The maximum resolving power corresponding to D_0, for $\lambda = 420$ nm, is, from (3.2.1), equal to $6 \cdot 10^{-4}$ seconds of arc. With this instrument, R. Hanbury Brown and R. Q. Twiss in 1962 determined the diameters of about 30 brighter stars, whereby for each star an integration time of the order of 100 h was required (stable electronics!).

The limitation of resolving power of a ground-based telescope to roughly $1''$ due to seeing can be avoided by applying *speckle interferometry*, developed by *A.* Labeyrie in 1970. In the usual astronomical exposures, the diffuse seeing-disk is caused by turbulent fluctuations, in particular of the index of refraction in the atmosphere. On the other hand, a "momentary exposure", which is obtained in a time shorter than the mean fluctuation time of the turbulence elements (≤ 0.1 s), yields a diffraction image composed of numerous small speckles (Fig. 3.2.10). Although the speckles are spread over an angular range of the order of $1''$, the angular diameter of each individual speckle is on the average equal to λ/D, and thus corresponds to the resolving power of the telescope with aperture D (3.2.1), e.g. $0.03''$ in the visible region for a 4.0 m telescope. Such a *speckle interferogram* contains coded information (more precisely: an autocorrelation function) about the distribution of brightness of the object, e.g., about the ratio of magnitudes and the separation of two binary-star components, down to angular separations of order λ/D, to be sure with a very unfavorable signal/noise ratio. The art of speckle interferometry lies in superimposing 100 to 10^6 images, depending on the brightness of the object, and in then extracting its brightness distribution by Fourier analysis. This technique offers the advantage that it can also be applied to faint objects (≥ 13 mag). In favorable cases, in which a suitable point source lies in the field of view, *speckle holography* can even be used to reconstruct the high-resolution image.

3.2.3 Optical Detectors

Just as important as the telescopes themselves are the means of detecting and measuring the radiation from stars, nebulae, etc.

Visual observation plays a role today only where quick recognition of small angles or fine details near the noise level of scintillation is required, i.e. for example in the observation of visual binary stars.

The *photographic plate* remains one of the most important tools of the astronomer today. Among the four most-often used Kodak plate types, the highest-sensitivity blue plates of emulsion type O have a *sensitivity range* to ≤ 500 nm. For exposures in the visible and the red spec-

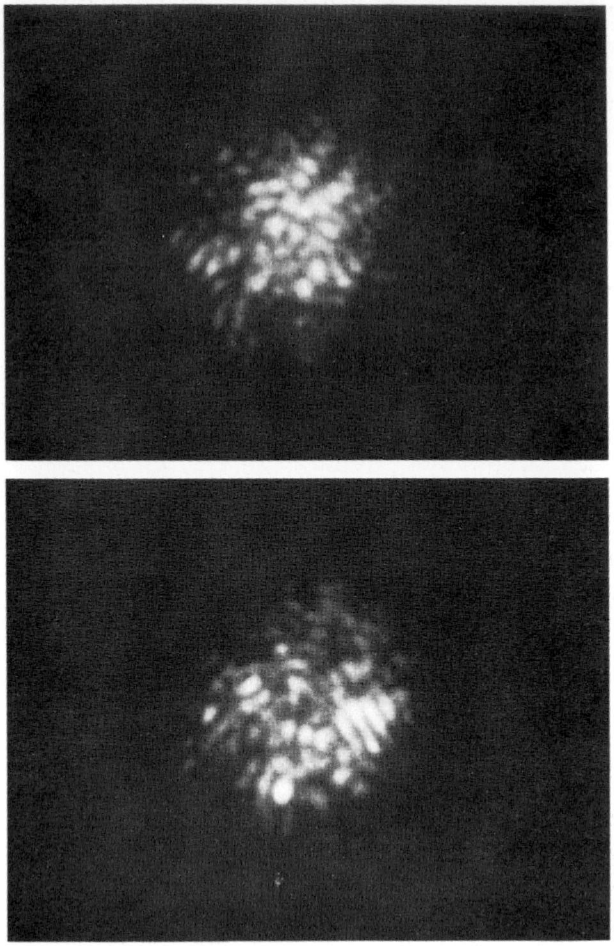

Fig. 3.2.10. Speckle images of α Lyr (Vega). The photographs were made by G. Weigelt using the 1 m mirror on the Hoher List through a 10 nm bandpass interference filter at $\lambda = 510$ nm. Exposure time 0.002 s, time interval between the two pictures 0.016 s. The angular size of the speckle image is about 1″. (Reproduced by permission from *Sterne und Weltraum*)

tral regions, the emulsions J (≤ 550 nm), D (≤ 640 nm), and F (≤ 700 nm), which have been sensitized with organic dyes, are used in conjunction with suitable colored filters. The N and Z plate types for infrared images (≤ 900 nm or ≤ 1.15 μm, respectively) must be hypersensitized, e.g. with ammonia, shortly before exposure.

The (spatial) resolving power of astronomical photographic plates, which is determined by the *grain size* of the emulsion, is in the range between about 5 and 25 μm. Using complex reduction methods, one can determine the position of a star image to within a fraction of one μm.

The *density* S of a plate is defined as $S = -\log I/I_0$, where I/I_0 is the transmission, i.e. the ratio of the intensity I transmitted by the darkened plate to the undarkened intensity I_0 (for an unexposed plate, I/I_0 would be 1 and $S = 0$). The relationship between the total incident radiation intensity $E = It$ (more precisely: the number of photons incident during the exposure time t) and the density is *nonlinear*. The *characteristic curve*, S as a function of $\log E$, increases (above a threshold intensity) at first only slowly; it then becomes linear, the slope of this part of the curve giving the *contrast* of the emulsion; and finally, it reaches a saturation region where it flattens out again. A *single* photographic plate covers only a very limited brightness region (*dynamic* range about 1 : 20, maximum 1 : 100) before everything "goes black".

The characteristic curve must be determined individually and empirically for each plate. For this purpose, calibration or density marks are placed at the edge of the image or the spectrum using, for example, light sources of known intensity. *Photographic photometry* for the determination of stellar magnitudes must therefore be carried out differentially as far as possible. In *iris diaphragm photometry* (used in particular with Schmidt cameras), an iris diaphragm is closed around the stellar image and used to measure its darkening, the calibration in terms of magnitude classes being carried out with the aid of photoelectrically determined brightness sequences (see below). In recent times, photographic images of regions of the sky, and also photographically recorded spectra, have been evaluated automatically using *microdensitometers* with a short step length, under program control. The data are then immediately available in a computer for further reduction. The attainable accuracy of photographically recorded magnitudes is in the range 5 to 10%.

Photoelectric photometry has used, since the 1950's, *photomultiplier* or *electron multiplier* tubes with suitable amplifiers. The incident photons release electrons from the *photocathode*, a thin layer of, for example, an alkali metal evaporated onto glass, as long as their energies are greater than the work function for electron emission (outer photoelectric effect). The electrons are then accelerated onto additional electrodes (dynodes) where they cause secondary electron emission, and thus amplify the initial electron current in an avalanche process, giving an amplification ratio of 10^6 to 10^7. The dark current, which represents background noise, can be reduced by cooling the electrodes. The spectral sensitivity can be adjusted for measurements from the ultraviolet to the near infrared by a suitable choice of photocathode material.

It is important to be aware of the relative advantages and disadvantages of photographic and photoelectric stellar photometry, and thus of their expedient use in combination: photoelectric methods cover a similar spectral region to that of the photographic plates, but have a greater stability and accuracy, and react in a *linear* fashion over a larger region of incident light intensity. Their *quantum yields*, i.e. the fraction of incident photons which gives a signal at the anode, are in the range 20 to 30% in the blue, while photographic emulsions reach at best 2 to 4%. The photomultiplier is a *single-channel* detector, which collects all the electrons onto one anode; image details are thus lost; the spatial resolution is determined by the area of the photocathode. Therefore, an observer must measure each individual star photoelectrically; in contrast, *one* photographic exposure registers an enormous number of stars. As a result, the following applications are in general employed for the two methods: *photoelectric* photometry for magnitude scales, precise magnitudes, color indices etc.; *photographic* photometry for intensity measurements of large numbers of stars (star clusters!) and for surveying for particular types of stars, etc., using photoelectrically determined scales. Recently, photometry with two-dimensional CCD detectors (see below) has increased in importance.

The wish to record more and more distant galaxies and other faint objects as well as their spectra with acceptable exposure times led as early as the 1930's to the development of the first image converters by A. Lallemand and others. The goal is to combine the advantages of the photographic plate with its extremely high image resolution, and the favorable properties of the photoelectric detector, i.e. linearity, a large dynamic range with ready amplification, and higher quantum yield. To this end, many small photoelectric detectors are combined in a one- or two-dimensional array to give a *multichannel* detector. Progress in microelectronics and solid state physics and their applications to computer manufacture, to television technology, and for military uses (night-vision apparatus, reconnaissance ...) has led since about 1970 to an enormous development and a rapidly increasing application of various kinds of high-sensitivity photoelectric detectors in optical astronomy. In the following section, we introduce some of the most important of the newer types of detectors, without making a claim to completeness.

In the *image intensifier* or *image converter*, electrons are released from a photocathode by the incident light and are accelerated, but then, unlike the photomultiplier, they are not all collected by one anode, but rather, using electron optics (applied magnetic or electrostatic fields, combined with glass fibers to make the image plane conform to the shape of the electrodes), they are imaged onto a fluorescent screen. This image, having a 50 to 100 times increased intensity compared to the original incident light, is then photographed directly from the screen, e.g. using a blue-sensitive plate. By placing several amplifier stages in series (Fig. 3.2.11), the amplification can be increased to around 10^6 to 10^7, although with reduced image quality.

In the *electronographic camera* (A. Lallemand and others), the focussed electrons are detected not by a fluorescent screen, but rather directly by using their effect on a film with a special nuclear emulsion which is sensitive to charged particles (Sect. 3.4.2). The difficulty of manipulating the film *inside* the evacuated multiplier tube is avoided in the *Spectracon* (J. D. McGee, 1971) by accelerating the electrons to ca. 40 keV and letting them pass out of the tube through a mica plate of about 5 μm thickness, which serves as an exit window. The image is then produced in a film which is placed *outside* the window.

The *microchannel plate* consists of a thin bundle of parallel glass capillaries ("microchannels", between 1 and 15 μm diameter and about 1 mm long) which is placed directly behind a photocathode. The photoelectrons are accelerated in the microchannels by a voltage applied to electrodes evaporated on the front and back surfaces of the plate, and impact on the inner channel walls, causing the emission of a cascade of secondary electrons from a thin semiconducting layer applied there. The microchannel plate thus exhibits the high avalanche amplification of a photomultiplier, but at the same time maintains spatial resolution by guiding the electrons within the chan-

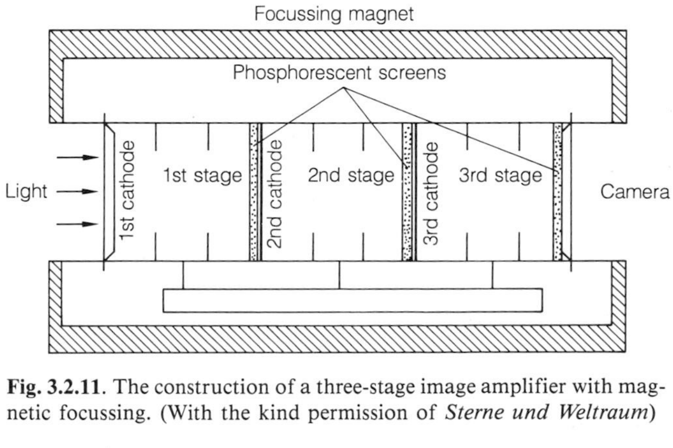

Fig. 3.2.11. The construction of a three-stage image amplifier with magnetic focussing. (With the kind permission of *Sterne und Weltraum*)

nels. The image at the anode can be further amplified and enhanced.

Among *television* imaging systems, especially the *Vidicons*, of Type SIT (Silicon Intensified Target) and Type SEC (Secondary Electron Conduction) have been developed for astronomical applications in conjunction with image intensifiers. In the imaging section of the television camera, the photoelectrons are initially accelerated and focussed onto a target which consists of a thin foil; the optical image is thus directly converted into an electrical charge distribution, which is then amplified within the target. For example, in the SEC-Vidicon, secondary electrons are emitted by a thin alkali-metal layer and "sucked up" onto a metallic signal plate by a low voltage, so that in the target, a moderately amplified (50 to 100 times), positively charged image is formed. This image is finally sampled by an electron beam, similar to that in a television tube, in the output section, and, after additional amplification, it is read out, for example into a computer for further treatment. The amplification factor of an SIT-Vidicon is somewhat higher (several 10^3); however, for long exposures, the target must be cooled (to $-60°$ C).

In *photon-counting detectors*, for example the Image Photon Counting System (IPCS) developed by A. Boksenberg and coworkers, the amplification is raised to such a high value by a multistage image converter that signals can be recognized on the fluorescent screen which are due to *single* incident photons. A television camera views this screen and serves as a spatially-resolving "intermediate memory"; it transmits the image to a computer, where it is then digitally reconstructed.

In a *semiconductor detector* such as the *silicon diode*, the incident photons create *electron-hole pairs* within the solid semiconductor material, if they have sufficient energy to lift electrons from the valence band into the conduction band (inner photoelectric effect). The free charges can then be collected in potential wells on the boundary layer between p- and n-doped silicon by an externally-applied voltage, and thus build up a charge image during long exposure times. The dark current due to electrons released by thermal lattice vibrations must be suppressed by cooling to below -100 °C. Silicon diodes have especially good quantum yields in the red and infrared: at $\lambda = 600$ to 700 nm, they reach values around 80%. The disadvantage that an avalanche amplification is not readily possible, in contrast to devices using the outer photoelectric effect, is offset to a considerable degree by the high quantum yields and the storage ability of semiconductor detectors. The small size of photodiodes permits arraying a large number of them in one- or two-dimensional structures (diode arrays) with electronic control circuits on a single chip.

The silicon diode responds not only to light, but also to (sufficiently energetic) *electrons*. This fact is made use of in the *Digicon* (E.A. Beaver and C.E. McIlwain, 1971), in which the electrons emitted from a photocathode, after acceleration in a vacuum tube, strike a (one-dimensional) diode array which serves as detector. Imaging of the photocathode on the diode array is accomplished by electron optics. Each electron, after acceleration to about 20 keV, produces several 10^3 electron-hole pairs in the diode (energy of formation several eV).

To an increasing extent since the mid-1970's, the *CCD* (*C*harge-*C*oupled *D*evice) has found application in optical astronomy. It is a two-dimensional array of extremely small semiconductor detectors, invented at Bell Laboratories in 1970 by W.S. Boyle and G.E. Smith and developed for television applications. A CCD (Fig. 3.2.12) consists of a thin n-p-doped silicon platelet (chip), on which, separated by a thin insulating silicon oxide layer, a two-dimensional arrangement of small electrodes is placed. These electrodes, together with the elec-

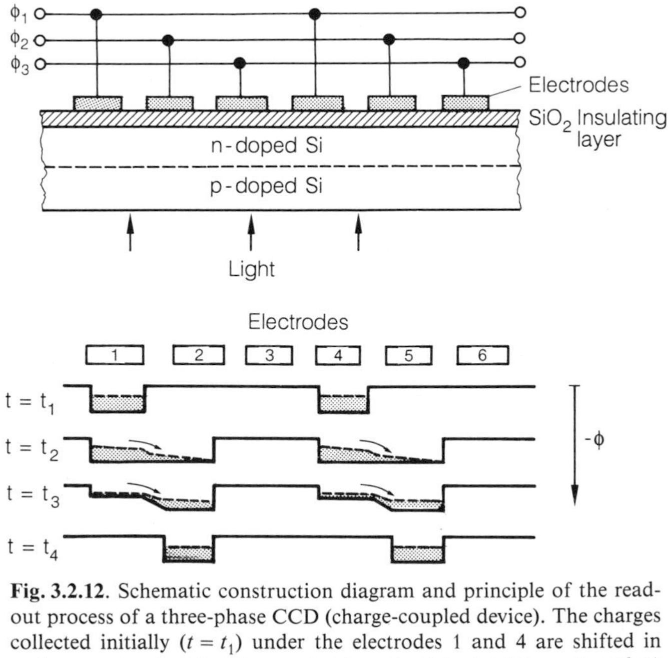

Fig. 3.2.12. Schematic construction diagram and principle of the read-out process of a three-phase CCD (charge-coupled device). The charges collected initially ($t = t_1$) under the electrodes 1 and 4 are shifted in such a way by stepwise changes in the potentials ϕ (at times t_2 and t_3) that they are located under electrodes 2 and 5 at time t_4. (Reproduced by permission from *Sterne und Weltraum*)

tronic circuitry, define the smallest independent receptors or image elements (pixels, from *pic*ture *el*ements), which have roughly 30 μm · 30 μm areas. During the exposure, the electrons released, proportional to the number of incident photons, are collected in the potential well of each corresponding pixel. After completion of the exposure, control circuitry carries out a *charge-coupled readout process* by suitable changes in the potentials; in this process, the charge distributions of the pixels are moved to the edge of the picture, line by line, and are read into a computer memory via an amplifier. The readout frequency can be very high (≥ 1 MHz), so that the production of new charges during the readout process remains insignificant. In the computer, several charge-images can be digitally combined to give the final picture. CCD chip technology is currently undergoing rapid development; at present, chips with over 1000 · 1000 pixels in an area of several cm^2 are in use.

The impressive success of CCD cameras in recent years for the observation of extremely faint objects is based in particular on their favorable quantum yields and the large range of linear response. With respect to the number of image elements, the CCD is still far behind the photographic plate (a Schmidt camera image contains of the order of 10^9 pixels).

3.2.4 Spectrographs

A precise analysis of the spectral distribution of the light from cosmic sources is the task of *spectroscopy*. We shall here consider only its experimental aspects, by (somewhat artificially) deferring the corresponding basic concepts and applications to later sections.

The decomposition of light into its wavelength components requires a prism or a grating as the *dispersive element* of a spectrograph. The telescope here has the task only of focussing the light of a cosmic source, e.g. a star, onto the entrance slit of the spectrograph. Spectrographs of high light-gathering power for the investigation of faint objects are placed at the primary focus of the mirror. Large spectrographs are mounted independently in a separate room behind the coudé focus. The Cassegrain focus has the advantage that relatively large spectrographs can be mounted there, rigidly attached to the telescope, avoiding light losses by reflection from additional mirrors.

In a *grating spectrograph* (Fig. 3.2.12), the *collimator* initially selects parallel rays of light and directs them to the grating. The focal length of the collimator mirror is chosen to be as large as allowed by the dimensions of the available grating; its aperture ratio should be equal to that of the telescope, in any case no smaller (full illumination of the grating!). The spectrally-decomposed light is then imaged in the *spectrograph camera*, for which the grating serves as entrance aperture. These cameras are frequently constructed as *Schmidt cameras* (Fig. 3.2.7), whose advantages we have already seen: large field of view, low absorption and reflection losses, small curvature of the image field, and negligible chromatic aberrations. It is often expedient to combine the grating and the camera mirror into a *concave grating*.

A *diffraction grating* (line grating) yields intensity maxima (e.g., as a reflection grating), i.e. spectral lines, when the optical path difference between neighboring rays is given by (Fig. 3.2.13)

$$a(\sin\phi - \sin\phi_0) = n\lambda \quad (n = \pm 1, \pm 2, \ldots) \ . \tag{3.2.7}$$

Here, a is the spacing of the parallel lines of the grating, (the grating constant), ϕ_0 is the fixed angle of incidence onto the grating, ϕ is the angle of the reflected rays, and n is the spectral order. Neighboring orders are separated by an angle $\sin\phi = \lambda/a$ from one another. Overlapping orders, which would complicate the spectra (for example, the 2nd order at $\lambda = 400$ nm is coincident with the 1st order at 800 nm) must be separated out using color filters together with the spectral response of the detector.

The angular dispersion of the grating, is, according to (3.2.7), *independent* of the wavelength:

$$\frac{d\phi}{d\lambda} = \frac{n}{a\cos\phi} \ , \tag{3.2.8}$$

an advantage compared to a prism with its rather nonlinear dispersion. When the spectrograph camera has a focal length f, (3.2.8) corresponds to a linear dispersion given by

$$\frac{dx}{d\lambda} = f\frac{d\phi}{d\lambda} \ . \tag{3.2.9}$$

The *spectral resolving power*

$$A = \frac{\lambda}{\Delta\lambda} \ , \tag{3.2.10}$$

is defined for a grating by the spacing $\Delta\lambda$ for which two sharp spectral lines are just recognizably separated from

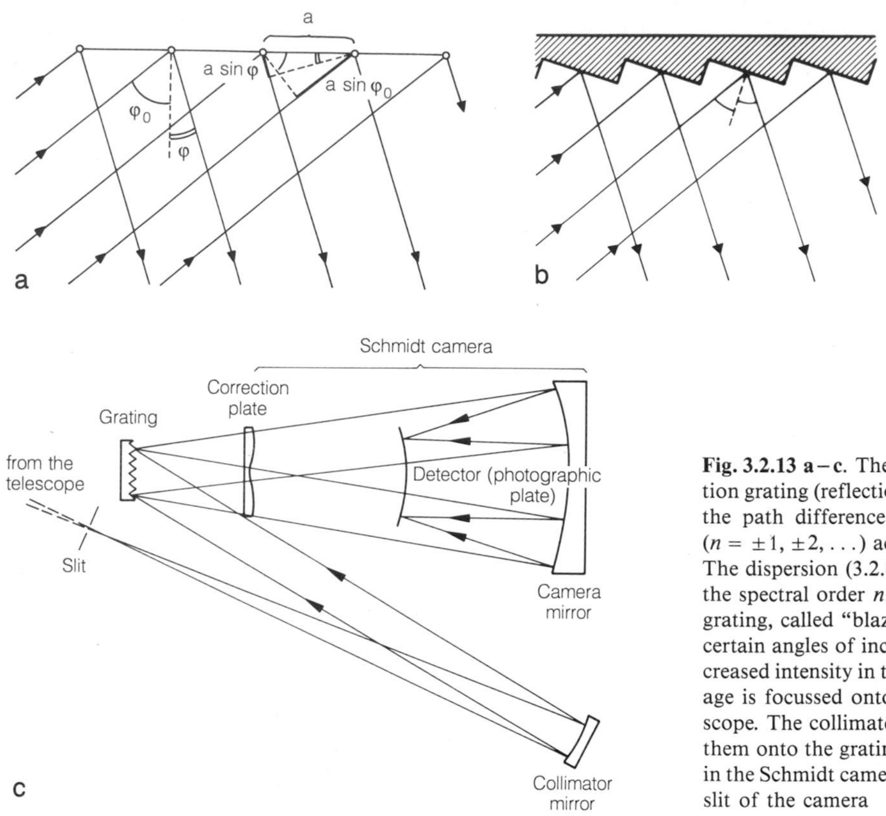

Fig. 3.2.13 a–c. The grating spectrograph. **(a)** Principle of the diffraction grating (reflection grating). Interference maxima are obtained when the path difference of neighboring rays, $a(\sin\phi - \sin\phi_0)$ equals $n\lambda$ ($n = \pm 1, \pm 2, \ldots$) according to (3.2.7), where a is the grating constant. The dispersion (3.2.8) and resolving power (3.2.12) are proportional to the spectral order n. **(b)** By vapor deposition of e.g. aluminum on the grating, called "blazing", specular reflection from the grating steps for certain angles of incidence and emission can be obtained, giving an increased intensity in the spectra. **(c)** A coudé spectrograph. The star's image is focussed onto the entrance slit of the spectrograph by the telescope. The collimator mirror makes the light rays parallel and reflects them onto the grating. The spectrally decomposed light is then imaged in the Schmidt camera; the grating serves simultaneously as the entrance slit of the camera

each other. For a grating of dimension D, according to the theory of diffraction, (3.2.1), $\Delta\phi = \lambda/D$ is the smallest resolvable angle, so that (for a small exit angle ϕ), with (3.2.8), we have

$$\Delta\lambda = \frac{a}{n}\Delta\phi = \frac{\lambda}{n}\frac{a}{D} . \quad (3.2.11)$$

If we now introduce the total number of grating lines, $N = D/a$, we find:

$$A = nN , \quad (3.2.12)$$

i.e. the resolving power depends only on the number of lines and the order.

In astronomical spectroscopy, diffraction effects can be for the most part neglected, since $\Delta\lambda$ is limited by the spatial resolution of the detector, i.e. by the grain size of the photographic emulsion or the image element of the diode array (Sect. 3.2.3). The image of the entrance slit produced by the collimator and the camera optics of the spectrograph should, for reasons of economy, be adapted to this resolution. If one wishes, for example, to obtain spectra, e.g., of stars of a particular limiting magnitude with a telescope of a given size in a particular exposure time (exposure times >5 h are inconvenient in practice), then the camera focal length and the dispersion (i.e., the number of lines) of the grating are predetermined.

Today, practically only *grating spectrographs* are in use, since it is now possible to manufacture the lines of the grating with the required profile so that only a particular reflection angle ("blaze angle"), and thus a particular order n, shows a strong light intensity. With blaze angles $\leq 25°$, a spectral resolving power A in the range 500 to 5000 in low orders with Cassegrain spectrographs having grating sizes D up to about 15 cm can be obtained; using coudé spectrographs with gratings of about 50 cm size, the resolving power can be up to 10^5.

The most important accessories to a spectrograph are (a) a light source with a known spectrum as *wavelength standard* (e.g. a neon-iron hollow cathode lamp), which is imaged above or below the spectrum being measured and, in particular, allows the determination of Doppler shifts; and (b) for photographic spectroscopy an auxiliary optical path, which allows the recording of continuous spectra of sequentially variable intensities from a lamp with *known* intensity properties. These allow the characteristic curve of the plate for each wavelength (Sect. 3.2.3) to be registered, and thus permit the photometric evaluation of the spectrum.

In recent times, along with "classical" grating spectrographs, the newer *Echelle spectrographs* have found increasing application. They permit a high spectral resolution ($A = nN \leq 10^5$) with an *Echelle grating* (stepped grating) in *high* orders ($n \simeq 10 \ldots 1000$) at a large blaze angle (ca. 65°). The overlapping orders are separated at *right angles* to the dispersion direction of the Echelle grating by a second grating which need have only a limited resolving power and can be made as a concave grating to serve simultaneously as the camera mirror, so that altogether a "two-dimensionally" arrayed spectrum is obtained (Sect. 3.2.5).

For *spectral surveys* of whole starfields, the *objective prism* is used, i.e. a prism is placed in front of the telescope at the point of minimal diversion, giving the spectrum of each star on the photographic plate in the focal plane. For example, the Henry Draper Catalog was prepared in this manner by B. E. C. Pickering and A. Cannon at the Harvard observatory; in addition to position and magnitude, it lists the spectral type of about a quarter of a million stars. Instead of an objective prism, an *objective grating* may also be used.

3.2.5 Space Telescopes and Large Earth-Based Telescopes

In order to observe cosmic sources in the ultraviolet ($\lambda \leq 300$ nm), we have seen that the only possibility is to employ a telescope outside the Earth's atmosphere. But even in the visible region, a space telescope offers notable advantages, which must, to be sure, be weighed against the cost of placing an instrument in space. The limitation on resolution due to seeing no longer applies, and the background brightness is reduced by the amount due to the atmosphere, so that for the same mirror diameter, a space telescope can detect fainter objects than an Earth-based instrument. Furthermore, the useful duty cycle of a space telescope is greater, since there, observations can be carried out day and night, independently of weather conditions.

Following initial rocket experiments with very limited observation times, the Copernicus satellite (OAO-3) with an 82 cm telescope offered from 1972 to 1981 the first opportunity to carry out high-resolution *spectroscopy in the ultraviolet* in the range $\lambda \geq 95$ nm as a long series of measurements. The "conventional" spectrograph, in which the spectrum is sampled stepwise by a photomultiplier, limited these spectroscopic observations to relatively bright stars.

The IUE Satellite (International Ultraviolet Explorer), launched in 1978, carries a 45 cm telescope of the Ritchey-Chrétien design, with two spectrographs in its Cassegrain focus, one for the spectral region $\lambda = 115-195$ nm and one for the region $\lambda = 190-320$ nm. In each of these regions, either a spectrum of low resolution (3.2.10), using a concave grating (with $\lambda/\Delta\lambda \simeq$ a few 100),

Fig. 3.2.14. The ultraviolet spectrum of the planetary nebula NGC 3242 in Hydra. The photograph was made by the *International Ultraviolet Explorer* in the range $\lambda = 115-200$ nm (J. Köppen and R. Wehrse, 1979), and shows a section of an échelle spectrum from the 66th to the 125th orders with a resolution $\lambda/\Delta\lambda = 1.2 \cdot 10^4$, corresponding to $\Delta\lambda \simeq 0.01$ nm. Each order band covers a region of about 2.5 nm. The emission lines of the nebula, of which the strongest are the C III doublet at $\lambda = 190.7/190.9$ nm (*left*) and the He II line at $\lambda = 164.0$ nm (*center*) are superposed on the continuum from the central star. At the upper right, the resonance line of H I Lα, $\lambda = 121.5$ nm, is visible on the one hand as a broad absorption band due to the interstellar medium, and on the other as emission from the geocorona. In the lower part of the picture, the track of a spurious particle may be seen

or else a high-resolution spectrum using an Echelle grating (with $\lambda/\Delta\lambda = 1.2 \cdot 10^4$, see Fig. 3.2.14) can be recorded. The detectors are SEC television cameras (Sect. 3.2.3), the ultraviolet light first being transformed into visible light by an image converter (CsTe cathode with a MgF_2 entrance window) and guided to the photocathode of the television camera tube using fiber optics. After an exposure, the image, consisting of $768 \cdot 768$ pixels, is read out by an electron beam, the video signal is digitized (with 256 discrete intensity levels), and is transmitted to a computer at the ground station, together with information for calibration, image reconstruction, etc. The nearly geostationary, elliptical orbit of the IUE, with an apogee of about 46000 km, permits continuous contact with the ground station. By the use of *two-dimensional*, more sensitive detectors, a considerably weaker limiting magnitude is attained than in the Copernicus satellite, in spite of the smaller telescope. With exposure times of the order of 2 h, high-resolution spectra from 10th magnitude stars can be recorded.

The spectral region *below* 91.2 nm has still been only slightly investigated, because the strong absorption in the Lyman continuum by the *interstellar medium* limits possible observations to our immediate neighborhood in the Milky Way Galaxy. However, the Apollo-Soyuz mission (1975) demonstrated that earlier estimates were too pessimistic and that the inhomogeneous distribution of the interstellar gas would permit a view out to greater distances (up to about 100 pc) in several "holes". The spectrometers with objective gratings on board the space probes Voyager 1 and 2 cover the region down to 50 nm with a resolution of 2.5 nm. The interstellar medium becomes transparent again only in the X-ray region.

With NASA's *2.4 m Hubble Space Telescope* (HST), a large instrument for the optical and ultraviolet spectral regions is available for the first time, for a number of years from 1990 onwards. The telescope, with a Ritchey-Chrétien optical configuration, was expected to attain nearly the diffraction-limited angular resolution (3.2.1), with $\lesssim 0.1''$ at $\lambda = 633$ nm in the Cassegrain focus (aperture ratio 1:24). Due to an error in the optical system, however, this goal was not initially reached; as a result of spherical aberration (Sect. 3.2.1), only about 15% of the light is concentrated within a disk of $0.07''$ diameter. The remaining light is distributed over a disk of $\gtrsim 1''$ diameter. In the focal plane of the Space Telescope, the following instruments can be selected for use: (a) a *wide-angle camera* with a maximum field of view of $2.7' \cdot 2.7'$, covered by four CCD detectors in a mosaic, each with $800 \cdot 800$ pixels. Its spectral sensitivity ranges from the ultraviolet into the near infrared (115 nm – 1.1 μm). (b) A *camera for faint objects* with a more narrow field of view and more limited spectral range (120 – 600 nm), in which, with a view to image reconstruction, detectors with a very high spatial resolution of $0.008''$ to $0.044''$ are used (image intensifiers with television camera tubes capable of single-photon detection). (c) A *low-resolution spectrograph*, with $\lambda/\Delta\lambda = 250 \ldots 1300$ in the range 115 – 700 nm, using a Digicon detector (a linear diode array with 512 elements, cf. Sect. 3.2.3), for investigation of faint objects; and (d) a *high-resolution spectrograph* for the ultraviolet (110 – 330 nm), with $\lambda/\Delta\lambda = 2 \cdot 10^3 \ldots 10^5$, also using Digicon detectors; it is illustrated in Fig. 3.2.15.

Parallel to the design of unmanned space telescopes on satellites, and the planning of observation from a manned space station, the development of *large ground-based optical telescopes* with diameters in the range 10 to 25 m continues. Now that optical detectors have reached

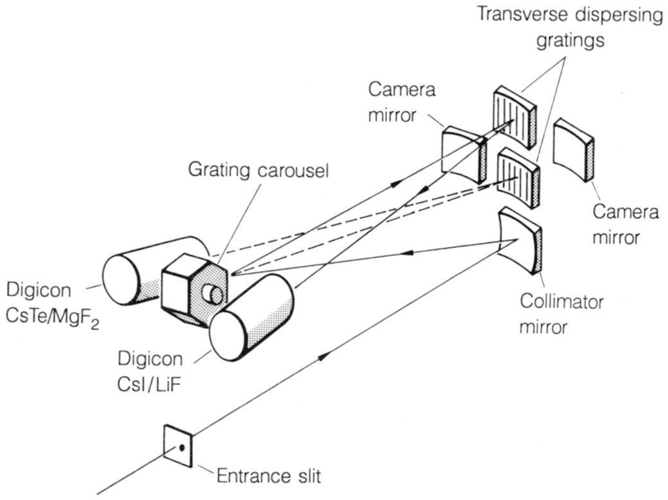

Fig. 3.2.15. The high-resolution spectrograph of the Hubble Space Telescope. The parallel rays of light from the collimator mirror are first spectrally decomposed by one of the six plane gratings, which are mounted on a carousel so they can be interchanged. For the highest resolution, $\lambda/\Delta\lambda = 10^5$ (the optical path shown), an échelle grating is used, followed by a transverse spectral decomposition using a concave grating which serves simultaneously as the camera mirror. Two digicons with a one-dimensional arrangement of 512 diodes serve as detectors, with either a CsI cathode and LiF window ($\lambda = 105 - 170$ nm), or with CsTe/MgF_2 (170 – 320 nm), depending on the wavelength range. Spectra of lower resolution ($\lambda/\Delta\lambda = 2 \cdot 10^3$ or $2 \cdot 10^4$) are produced by the remaining five plane gratings on the carousel and, depending on the spectral region, are focussed onto the corresponding digicon by one of the two camera mirrors. (From J. Brandt et al., 1982)

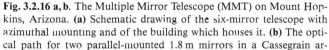

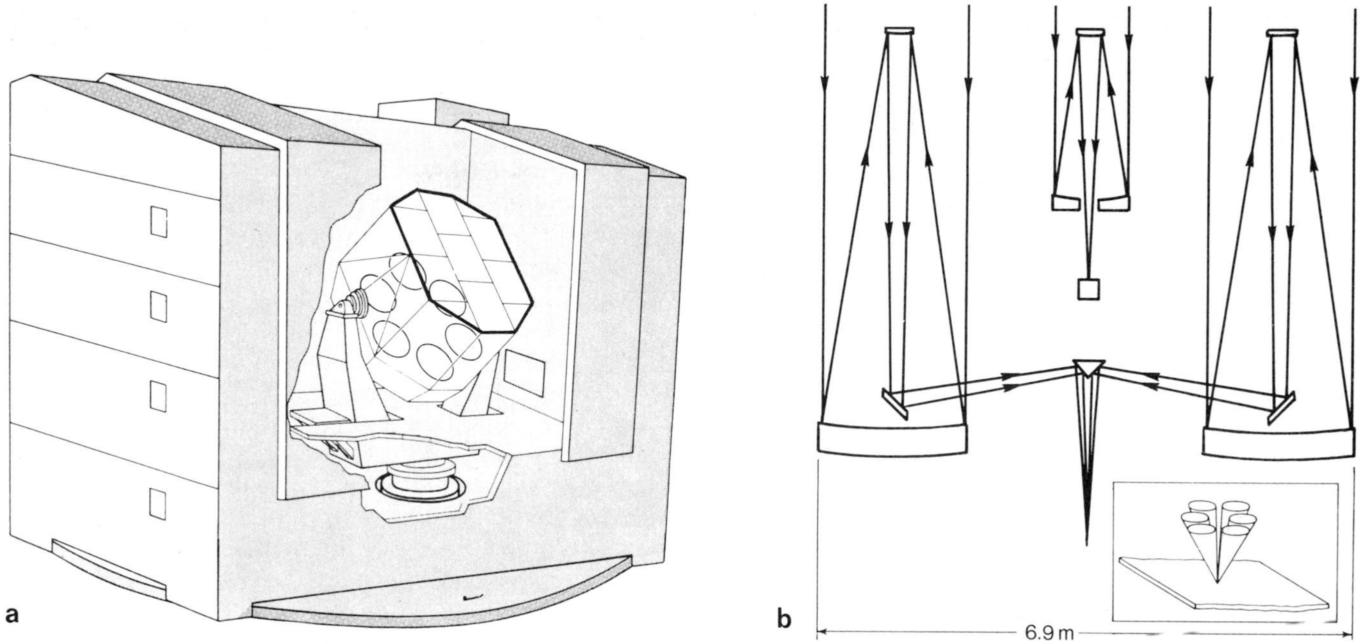

Fig. 3.2.16 a, b. The Multiple Mirror Telescope (MMT) on Mount Hopkins, Arizona. **(a)** Schematic drawing of the six-mirror telescope with azimuthal mounting and of the building which houses it. **(b)** The optical path for two parallel-mounted 1.8 m mirrors in a Cassegrain arrangement with a common focus. The central axis contains a 0.76 m guide telescope used for controlling the motion of the MMT and for adjusting the main mirrors. (Reproduced with permission of the American Institute of Physics, New York, and of the authors)

a high sensitivity near to the theoretical limit, an increase in the power of optical telescopes is mainly a question of increasing the size of their light-gathering areas. In particular, spectroscopic investigations, which are not very strongly influenced by atmospheric scintillation, can be extended to extremely faint objects with all the advantages of a ground-based telescope: ready access and the possibility of using more complex instruments, combined with the relatively low cost compared to space telescopes.

The construction of telescopes of conventional design, with a rigid primary mirror having a high-precision surface ($\leq \lambda/10$) as a *single* piece, has currently reached its limits with mirror diameters of about 4 to 6 m, due to the increasing technical problems (weight!) and cost of further increases in mirror diameter. The development of future large telescopes therefore is tending on the one hand in the direction of lighter and thus *thinner* mirrors, and on the other towards the construction of the primary reflecting element as a mosaic of *several* individual mirrors. Since, however, thin mirrors are readily deformed, the principle of *active* or *adaptive optics* must be applied: the shape and alignment of the reflecting surfaces must be continually controlled and adjusted by an automatic regulation system connected to a powerful computer. The various technical possibilities include, among others, a single, thin giant mirror in connection with active optics, or the division of a large paraboloid reflecting surface into a mosaic of smaller segments, each of which is actively adjusted. Furthermore, the optical paths of a number of telescopes arranged in an array can be combined at a common focal point, or finally, several telescopes can be connected together with a common mount and focal point.

As the precursor of a new generation of large telescopes, the *Multiple Mirror Telescope* (MMT) developed using the last-mentioned principle by the University of Arizona and the Smithsonian Astrophysical Observatory, on Mount Hopkins in Arizona, has been used successfully for optical and infrared observations since 1979 (Fig. 3.2.16). Six similar mirrors, each with 1.8 m diameter and an aperture ratio 1 : 2.7, are mounted together around a common axis; the light from the individual mirrors is combined by secondary mirrors at a quasi-Cassegrain focus. In this way, a light-collecting surface is available which corresponds to a conventional telescope of $\sqrt{6} \cdot 1.8\ \mathrm{m} = 4.4\ \mathrm{m}$ diameter, while the angular resolution remains limited by the dimensions of the individual mirrors. Using an additional guide telescope, the alignment of the individual telescopes is controlled by laser beams. The common azimuthal mounting allows a very compact construction: the MMT is housed in the thermally insulated part of a rectangular building of only 17 m height and 20 m · 13 m floor space. The entire building, weighing "only" 450 t, can be rotated in the azimuthal direction.

Additional large telescopes designed according to various technical concepts are under construction or in the planning stages at several places; it remains to be seen which solution will prove to be the best.

3.3 Telescopes and Detectors for the Radio-Frequency and Infrared Spectral Regions

The second major region of transparency of the Earth's atmosphere, besides the "optical window", is the "radio-frequency window"; on the long-wavelength end, it is bounded by reflection from the ionosphere, mainly in the F-layer (Sect. 2.8.5 f.). The longest wavelength which is transmitted varies strongly, depending on fluctuations in the electron density of the ionosphere, between about $\lambda = 12$ and 100 m (i.e. between 24 and 3 MHz). At the short-wavelength end of the region, below about $\lambda = 5$ mm, absorption by atmospheric oxygen and water vapor limits observability more and more, until finally in the submillimeter range below $\lambda = 0.35$ mm (860 GHz), ground-based astronomy is no longer possible. Below about 20 μm, the atmosphere again becomes transparent as we approach the optical window, at first however only in narrow regions between the absorption bands of water vapor (Fig. 3.3.1).

We shall begin with a description of the most important types of radio telescopes (Sect. 3.3.1). In order to obtain high angular resolution in spite of the long wavelengths, both parabolic mirrors with the largest possible diameters, and especially interferometers are employed, the latter consisting of many individual telescopes at a considerable distance from one another. In Sect. 3.3.2, we give a very short summary of the receivers and spectrometers used for the radio-frequency region. Finally, in Sect. 3.3.3, we briefly describe instruments for the infrared spectral region, which lies between the radio-frequen-

Fig. 3.3.1. The 100 m radio telescope of the Max Planck Institute for Radio Astronomy, Bonn, at Effelsberg in the Eifel mountains

cy and the optical regions and requires observation techniques combining elements of those used in the two neighboring regions. It is essential for the observation of astronomical sources in the infrared to suppress the intense thermal radiation background from the Earth and from the telescope itself.

3.3.1 Radiotelescopes

The weak radio signals from space are initially collected by an *antenna*, which should have the largest effective area possible and good directional characteristics (angular resolution); they are then passed to a *receiver* or *radiometer* for amplification and rectification, and on to a *detector*, so that finally their intensity can, for example, be traced by a strip-chart recorder or stored in digital form in a computer for further analysis. The radiation which is collected by the radiometer can also be decomposed into its frequency components with good frequency resolution in a *spectrometer*, or it can be analyzed in terms of its state of polarization in a *polarimeter*.

The most flexible type of antenna, with good directional characteristics over a large frequency range, is the *parabolic mirror*, which, so long as the wavelength to be collected is not too short, can be made of reflecting sheet metal or wire mesh (mesh diameter $\leq\lambda/10$). The precision of the reflecting surface determines the shortest wavelength which can be observed. The radiation is collected at the focal point by a *feeder antenna* (a horn or, for $\lambda \geq 20$ cm, a shielded dipole) and input to the receiver via a transmission line. Analogously to an optical telescope, the signal can be taken up by the feeder antenna at the primary focus or, via a secondary reflecting surface, at a secondary focus, e.g. the Cassegrain focus (Fig. 3.2.6). The latter arrangement offers the possibility of placing several different receivers behind the primary mirror for parallel use.

The *resolving power* of a mirror of diameter D in the radio-frequency region is also given by (3.2.1), as long as $\lambda \ll D$. It is usual in radioastronomy to quote the *beam width* as the angle between the points in the directional characteristic curve at which the energy sensitivity has decreased from its maximum value by one-half (HPBW: Half Power-Beam Width). The beam width

$$\phi = 1.03\,\frac{\lambda}{D} \qquad (3.3.1)$$

(in radians) differs only slightly from (3.2.1) in terms of the numerical coefficient.

In the *cm- and lower dm-wavelength region*, several large radiotelescopes with diameters over 60 m have been available for some time, beginning in the 1950's with the construction of the 76 m telescope at Jodrell Bank near Manchester in England. The currently largest fully-directable radiotelescope, the azimuthally mounted 100 m parabolic mirror of the Bonn Max-Planck-Institute for Radioastronomy at Effelsberg in the Eifel Mountains (finished in 1972; Fig. 3.3.1), can be utilized for observations from about 50 cm to 6 mm wavelength (0.6 to 50 GHz).

The effective direction of observation of the *fixed* 305 m spherical mirror in a valley at Arecibo, Puerto Rico can be varied within a limited range by changing the phase of detection of the incident waves at the focus of the mirror (rotation of the feeder antenna).

The angular resolution of even these largest radiotelescopes is poor compared to that of Galileo's first optical telescope! According to (3.3.1), for a 100 m telescope at $\lambda = 50$ cm, it is only about 18′. It is therefore understandable that radio astronomers quite early developed instruments using the interferometer principle for the *meter wavelength* region, in order to obtain better resolution.

The *radiointerferometer* of M. Ryle, which corresponds precisely to Michelson's stellar interferometer (Sect. 3.2.2), attains a high resolving power. In it, the signals from two radiotelescopes are combined, retaining *phase information*, and then further amplified. In contrast to the optical region, the phase of radio-frequency waves can be transmitted over several kilometers by cable and up to several tens of kilometers by a radio link.

The principles of the linear diffraction grating and the two-dimensional grid grating (for a fixed wavelength) have also been applied to antenna technology in order to obtain a high angular resolution with *multielement interferometers*. In the *Mills Cross* (B. Y. Mills, 1953), two long, cylindrically-parabolic antennas, each of which has a planar directional characteristic (i.e., good angular resolution in only *one* dimension), are mounted in a cross-shaped arrangement and are alternately connected to the receiver in phase and then with a $\lambda/2$ phase shift in one of the transmission lines. Taking the difference of the two signals yields a synthetic, rod-shaped power beam in the center of the cross, with an angular resolution of the order of λ/D, where D is here the length of the crossarms. Thus, for example, with the crossed antenna in Molonglo, Australia, whose arm lengths are 1.6 km, an angular resolution at $\lambda = 73.5$ cm (408 MHz) of $1.4' \cdot 1.4'$ is obtained.

The principle of the *correlation interferometer*, first applied in radioastronomy by R. Hanbury Brown and R.Q. Twiss, has already been described for its "optical" version.

It has furthermore been shown by M. Ryle how the principles of *aperture synthesis* can be applied to use the information on amplitude and phase obtained *successively* from *several* small antennas at suitable, preselected positions, rather than that collected by a single large instrument during a particular time interval. For this purpose, the individual antennas, usually in an X-, T-, or Y-shaped arrangement, are moved with respect to each other, and the Earth's rotation is also used, to change their relative distances (projected onto a sphere). The angular resolution according to (3.3.1) corresponds to that of a single antenna whose aperture would be equal to the area covered in the course of time by all the small antennas taken together.

Among the large aperture-synthesis telescopes for the cm and dm ranges, we mention the 5 km Synthesis-Cross at Cambridge, England, which consists of 8 fixed and 4 movable individual 13 m-telescopes; the Synthesis-Radiotelescope at Westerbork, in the Netherlands, with 12 fixed and 2 movable 25 m mirrors in an east-west arrangement with a maximum baseline of 1.6 km; and the Very Large Array (VLA) at Socorro, New Mexico, USA, with 27 movable 25 m-telescopes in a Y-shaped arrangement, the lengths of the three arms being up to 21 km (Fig. 3.3.2). With these antenna systems, angular resolutions of a few 0.1″ to 1″ are obtained.

In order to attain a resolution much better than 1″, since 1970 the technique of Very Long Baseline Interferometry (VLBI) has been developed: two radiotelescopes which are very far apart, up to distances of order of the Earth's diameter, are used. They detect the radiation from the same source at exactly the same frequency independently of one another as a function of time, and thus of the effective baseline. The signals (i.e. the intermediate frequencies, see below) are recorded digitally on videotape; they are accompanied by extremely precise time markers from two atomic clocks, so that their interference patterns can later be analyzed with *known phases* in a computer (correlator). Instrumental phase fluctuations and disturbances in the atmosphere, which are different for the two sites at the large distances used, can be nearly eliminated by suitable combination of the signals from three or more telescopes (arranged along a "closure curve"). Thus, using a worldwide network of the largest radiotelescopes with D on the order of $\geq 10^4$ km, the obtainable resolving power and positional precision according to (3.3.1) is down to 0.0001″, i.e. about 10^4 times that achievable in optical measurements, which is limited by the seeing! Using an additional telescope on a satellite in a strongly elliptical orbit, the angular resolution could be further improved to several 10^{-5} seconds of arc.

Fig. 3.3.2. The Very Large Array (VLA) in Socorro, New Mexico (USA). This is a radio synthesis-telescope with 27 movable 25 m dia. mirrors in a Y-shaped arrangement (maximum arm lengths 21, 21, and 19 km). (With the kind permission of the American Institute of Physics, New York, and of the author)

In the *mm and sub-millimeter wavelength region*, the resolution becomes more favorable, according to (3.3.1), but the requirements for surface precision of the mirrors become considerably more demanding, especially considering thermal deformations due to differential heating during the day-and-night cycle. Besides steel, carbon-fiber strengthened plastics with a reflecting foil surface are used in this region. Among the largest telescopes for the mm region are the 45 m mirror at Nobeyama, Japan, built in 1985 and usable down to the 3 mm range; and the French-German 30 m mirror of the IRAM (Institut de Radio Astronomie Millimetrique) on the Pico Veleta near Granada, Spain, for wavelengths down to 1.3 mm. As with longer wavelengths, interferometry and Very Long Baseline Interferometry (VLBI) are carried out in the mm region.

Techniques for observation in the sub-millimeter region are undergoing rapid development. The currently existing large instruments for the wavelength range up to 0.3 mm include the 15 m mirror on La Silla, Chile (SEST = Sweden-ESO Sub-mm Telescope) and on Mauna Kea, Hawaii (James Clerk Maxwell Telescope), as well as the IRAM array consisting of three 15 m mirrors on the Plateau de Bure in the French Alps.

3.3.2 Receivers and Spectrometers for the Radio-Frequency Region

We cannot treat amplifier technology here, but we shall mention at least a few principles and the most important types of amplifiers.

The in general weak, incoherent radio-frequency radiation from outer space ("noise") has to be detected and measured in the receiver or radiometer against the background of antenna noise and that of the receiver components, as well as the background radiation from the Earth's atmosphere in the case of shorter wavelengths ($\lambda \leq 1$ cm).

In the receiver, a *noise power* (H. Nyquist) is produced by the *thermal* motion of electrons in every (ohmic) resistance at temperature T in the frequency bandwidth $\Delta\nu$

$$W_{\mathrm{R}} = kT\Delta\nu \ , \qquad (3.3.2)$$

where $k = 1.38 \cdot 10^{-23}\,\mathrm{J\,K^{-1}}$ is Boltzmann's constant. In addition, there is "shot noise" and "semiconductor noise" in some components such as transistors, diodes, etc. The overall noise in the receiver, W_{E} ($> W_{\mathrm{R}}$) can, analogously to (3.3.2), be represented by an effective *receiver noise temperature* T_{E}

$$W_{\mathrm{E}} = kT_{\mathrm{E}}\Delta\nu \ , \qquad (3.3.3)$$

where T_{E} is higher than the physical temperature T. The main contribution to W_{E} or T_{E} comes in the *first* amplifier stage and the components before it, so that particularly *low-noise* amplifiers are used for this function. Since the contribution due to thermal noise is proportional to T, W_{R} and thus W_{E} can frequently be considerably reduced by *cooling*.

The *total noise power* W of the receiver system is the sum of antenna noise, W_{A}, and the receiver noise W_{E} itself:

$$W = W_{\mathrm{A}} + W_{\mathrm{E}} = k(T_{\mathrm{A}} + T_{\mathrm{E}})\Delta\nu \ , \qquad (3.3.4)$$

where W_{A}, again analogously to (3.3.2), can be expressed in terms of the *antenna temperature* T_{A}. In the ideal case, W_{A} comprises only the noise power of the cosmic object being observed; for an extended source which fills the power beam of the antenna, T_{A} would then be equal to the brightness temperature T_{B} of the source (5.6.1). In practice, W_{A} also includes a noise contribution from the antenna itself, and, for $\lambda \leq 1$ cm, radiation from the atmosphere and possibly from the ground, or for $\lambda \geq 30$ cm, the background radiation from our Milky Way galaxy (nonthermal synchrotron radiation, Sect. 5.6.1).

The *limit of sensitivity*, i.e. the weakest radiation power ΔW which can still be detected above the overall noise power W, is given for a measurement in a frequency bandwidth $\Delta\nu$ with an integration time τ by

$$\Delta W = \frac{W}{\sqrt{\tau\Delta\nu}} \ , \qquad (3.3.5)$$

since in this case, the number of independent measured values is $N = \tau\Delta\nu$ and the relative accuracy, according to the law of statistical fluctuations, is $\Delta W/W = 1/\sqrt{N}$.

For the detection of weak cosmic signals it is thus important to have a stable amplifier (constant amplification) over the integration time τ. Furthermore, W_{E} must be kept as low as possible, since all contributions to W are of stochastic nature and, in principle, cannot be separated from one another. The frequency bandwidth is determined to a large extent by the problem at hand (a continuum measurement or observation of a single spectral line).

Fluctuations in the amplification can be eliminated to a large extent by the method of R.H. Dicke (1946), employing a *Dicke or modulation receiver*, in which the receiver input is periodically switched between the antenna and a resistance at a fixed temperature, and the *difference signal* is then analyzed. If the resistor is replaced by a second antenna, which is pointed at a neighboring area of the sky, the noise contributions from celestial background and atmospheric radiations can be eliminated by taking the difference between the signals from the two antennas (beam switching). Another possibility for eliminating the background noise is to shift the radiotelescope mirror at a low "wobble frequency" (≤ 0.1 Hz) between the source of interest and a neighboring area, and again to use only the difference signal (on-off technique).

Following the first amplifier stage, the high-frequency radio signal is usually converted to a (low) *intermediate frequency* (difference frequency) in a *superheterodyne* receiver using a *mixer* to combine with a fixed frequency from a stable *oscillator*; the intermediate frequency signal is then further analyzed.

In the case of *dm- and meter-waves* ($\lambda \geq 30$ cm or $\nu \leq 1$ GHz), the nonthermal background noise radiation from our own galaxy dominates all other noise contributions; its brightness temperature increases from a few 10 K to well above 1000 K with increasing λ. Thus, in this range, there is no need for particularly low-noise amplifiers, and *conventional* amplifiers from radio technology, e.g. silicon transistor amplifiers and, more recently, (uncooled) field effect transistors ($T_E \simeq 300$ K) are used.

In the *cm-wave region* (1 to 30 cm, 30 to 1 GHz), the celestial background is "cold" (radiation temperature 3 to 10 K: microwave background radiation, Sect. 5.9.4), so that it is worth the effort to use *low-noise* amplifiers. With *parametric amplifiers* or *field effect transistors* cooled to about 20 K, noise temperatures T_E of several 10 K are obtained; with *masers* cooled to still lower temperatures ($T \simeq 4$ K), one obtains $T_E < 10$ K and thus an enormous increase in the precision of measurements of faint radio sources, to be sure at the price of reduced bandwidth.

Going into the *mm-wave region*, absorption by the Earth's atmosphere increases rapidly, and the atmospheric temperature soon reaches about 300 K. For the longer-mm-wavelength region, parametric amplifiers and maser amplifiers are used. Below about $\lambda = 3$ mm ($\nu > 100$ GHz), there are currently no low-noise amplifiers available. Here, the signal is fed *directly* into a mixer and reduced to an intermediate frequency in the 1 GHz range without preamplification; the latter can then be amplified conventionally. With cooled Schottky mixers ($T \simeq 20$ K) at 100 GHz, $T_E \geq 200$ K is obtained; with very low-temperature ($T \leq 4$ K) Josephson mixers or SIS mixers (Superconductor-Insulator-Superconductor), $T_E \simeq 100$ K.

Receiver technology for the observationally difficult *submillimeter wavelength region* ($\nu \geq 300$ GHz) is currently undergoing rapid development; it uses "quasioptical" components in heterodyne receivers, based on a "hybridization" of waveguide techniques as used for microwaves, and the methods of geometric optics as used in the infrared and optical regions.

For the investigation of radio-frequency *spectral lines*, as for example the 21 cm line of neutral hydrogen or the many molecular lines in the cm- and mm-regions, the broadband signal from the receiver is decomposed into its component frequencies in a spectrometer and their intensities are determined. In a *filter spectrometer*, the original frequency band (in some casses after conversion to another frequency range using the heterodyne principle) is divided up by a large number (several hundred) of narrow bandpass filters with fixed center frequencies. The signals from the individual spectrometer channels are then analyzed in parallel. In the *acusto-optical spectrometer*, the electromagnetic waves are first converted to ultrasonic oscillations using the piezoelectric effect. These then produce standing waves in a (transparent) crystal or a liquid, corresponding to fluctuations in the density and the index of refraction of the material, and thus forming a "diffraction grating" which is irradiated with monochromatic light. The spectral energy distribution of the radio signal is thus finally converted into an optical diffraction image, whose intensity distribution can be measured, e.g. with a diode array.

3.3.3 Observation Methods in the Infrared

Observations from the ground can be made only in the *near* infrared ($0.8\ \mu m \leq \lambda \leq 1.2\ \mu m$) and in limited "windows" through the absorption by atmospheric water vapor in the *mid* infrared ($1.2\ \mu m \leq \lambda \leq 20\ \mu m$); owing to the strong decrease in water vapor content of the air with increasing altitude, they are best carried out from high mountains (≥ 4 km). Astronomical observations in the *far* infrared ($\lambda \geq 20\ \mu m$), in contrast, have to be made from aircraft (altitude range ≤ 15 km) and balloons (≤ 40 km) or from rockets and satellites. Only in the very-far infrared or submillimeter region ($\lambda \geq 350\ \mu m =$

0.35 mm) does the atmosphere again become transparent in several windows (Fig. 3.1.1).

In the entire infrared range, optical *telescopes* can in principle be used (Sect. 3.2.1). However, since at $T \simeq$ 300 K the background radiation from not only the atmosphere but also from the telescope parts themselves is strong in the infrared, this bothersome thermal radiation has to be reduced as far as possible, by limiting the field of view, by keeping to a minimum the number of reflecting surfaces, and by cooling the telescope and the detectors, among other methods. In order to detect the faint infrared radiation from cosmic sources against the background of intense, variable emissions from the Earth's atmosphere and from the sky, a difference signal between the source and a neighboring area of the heavens is obtained by periodic "wobbling" as in the radio-frequency region (for example, by wobbling the secondary mirror in a Cassegrain arrangement at about 10 Hz; this is termed chopping or beam switching).

Along with temporarily "modified" optical telescopes, especially constructed *infrared telescopes* are employed, for example, the 3.2 m mirror of NASA and the British 3.8 m telescope, both on Mauna Kea (Hawaii). Longer observations from altitudes of 12 to 15 km have been possible since 1974 with the *Kuiper Airborne Observatory*, an aircraft equipped with a 91 cm mirror for infrared astronomy. The first *infrared satellite*, IRAS (Infrared Astronomical Satellite) carried out a sky survey in 1983 with a 57 cm telescope in four broad wavelength regions between 12 and 100 μm with an angular resolution on the order of 1′. Using a tank filled with liquid helium, the entire telescope was maintained for 10 months at a temperature $\lesssim 10$ K, and the focal plane with its detectors at a still lower temperature of ≤ 2 K, until the cooling agent was spent. IRAS found about 250000 infrared "point sources", among them many in distant galaxies whose light is primarily in the infrared, and made some surprising discoveries (among others new comets, interstellar dust clouds, and a dust ring around Vega).

As *detectors* for the *near* infrared region ($\lambda \leq 1.2$ μm), photographic plates, photomultipliers, and image converters, as in the optical region, but with special cooled photocathodes (AgOCs or GaAsP) are used; silicon photodiodes are also employed. Out to about $\lambda = 5.5$ μm, indium antimonide (InSb) photodiodes cooled with liquid nitrogen at 77 K or liquid helium at 4.2 K are used; lead sulfide cells are also employed to about 4 μm. In the *mid* and *far* infrared, cooled semiconductor detectors predominate; they can also be used in CCD cameras. In a *photoconductive detector*, the absorption of a light quantum in a semiconductor creates an electron-hole pair, by exciting an electron from the valence band through the energy band gap which separates it from the conduction band. The briefly increased electrical conductivity (until recombination of the pairs occurs) is then used for detection. Compared to the corresponding optical detectors, the photon energies in the infrared are considerably lower, so that the detector material must have a small band gap. In pure semiconductors such as germanium or silicon, the band gap is of the order of 1 eV, corresponding to $\lambda \simeq 1.2$ μm. Doping with suitable materials, however, creates "impurity states" within the band gap, with energy spacings down to about 0.01 eV, so that the response limit of the photoconductor is extended to much longer wavelengths. The semiconductor must be cooled to a temperature at which these energy states are not populated thermally, i.e. cooling to lower temperatures is necessary as the limiting wavelength increases. By doping silicon with In, Ga, As, or Sb or germanium with Cu, and cooling to below about 10 K, detectors for infrared radiation out to the 20 to 30 μm range can be fabricated; germanium detectors doped with Ga and cooled to liquid helium temperatures (≤ 4 K) are sensitive even as far out as 120 μm.

In the entire infrared region, *thermal detectors* are also used: here, the absorbed radiation is converted into thermal energy in a crystal. In the *bolometer*, the strong temperature dependence of the electrical resistivity of a suitably doped, cooled semiconductor crystal is used to detect the infrared radiation. At liquid ^{4}He or liquid ^{3}He temperatures (4.2 or 0.3 K, respectively), a gallium-doped *germanium bolometer* has a sensitive range from 1 to about 1000 μm.

For *spectroscopy* in the infrared, besides cooled prisms and gratings of medium resolving power, high-resolution *Fourier spectrometers* are also employed. The *heterodyne spectrometers* developed for the microwave region (Sect. 3.3.2) permit spectroscopy of line sources in the far infrared for wavelengths above ca. 100 μm.

3.4 Instruments for High-Energy Astronomy

From the long-wavelength region of the electromagnetic spectrum, we now jump to the region of the shortest wavelengths, or the highest frequencies and energies: to the *gamma-ray* and *X-ray* regions. Reflection, e.g. from the surface of a mirror, is as a result of the strong penetrating power of these high-energy photons possible only for soft X-rays (energies below a few keV) at grazing incidence, i.e. at angles $\leq 1°$ to the surface. At higher energies, the detection of the photons is carried out with instruments developed in high-energy physics for counting energetic particles, making use of the *energetic electrons* released from the detector material by the photons.

Along with high-energy electromagnetic radiation, the radiation consisting of *energetic particles* has become increasingly important: the *solar wind* blows solar matter, i.e. mainly high-energy hydrogen and helium ions and electrons, with velocities of about 200 to 1000 km s^{-1} into space. Protons of (on the average) about 500 km s^{-1} have an energy of 1 keV. The range of even higher energies is taken up by *cosmic radiation*, which reaches us with energies of up to 10^{10} eV from outbursts on the Sun, and in addition from the Milky Way (and possibly from other galaxies) with energies of a few 10^8 up to 10^{20} eV; it continually bombards the Earth's atmosphere (Sect. 5.3.8).

Before briefly discussing the most important types of particle detectors and telescopes constructed with them, we first summarize in Sect. 3.4.1 those physical processes involving the interactions of energetic photons and particles with matter which are necessary for understanding not only the instruments themselves, but also the propagation, e.g. of gamma rays or cosmic rays, from their sources to their final detection.

3.4.1 The Interactions of Energetic Photons and Particles with Matter

The energy of photons in the optical and ultraviolet regions suffices only to eject electrons from the outer (valence) shells of atoms or ions; in contrast, *X-ray photoionization*, i.e. at energies above about 100 eV or wavelengths below about 10 nm, proceeds mainly by emission of electrons from the *inner* shells (core electrons).

Also in contrast to optical spectroscopy, in the X-ray region, except from highly ionized atoms, absorption lines (i.e., transitions between two bound energy states) are seldom observed, since the less strongly-bound states are in general occupied by electrons. Instead, *ionization edges* are characteristic of X-ray absorption spectra; at an edge, the absorption cross section increases abruptly at a frequency ν_n or a wavelength λ_n corresponding to the binding energy of an electron in a shell of principal quantum number n:

$$h\nu_n = \frac{hc}{\lambda_n} = E_n \;, \qquad (3.4.1)$$

and then decreases towards higher energies at a rate proportional to E^{-3} or λ^3. E_n is proportional to Z^2/n^2, where Z is the nuclear charge (atomic number) (4.7.3).[4]

The proportionality between $\sqrt{\nu}$ and Z for corresponding edges of different elements in the X-ray region was first discovered by H. Moseley in 1913.

For the innermost electron of the so-called *K shell*, the ionization energy is given by

$$E_K = 13.6\, Z_{\rm eff}^2\,[\rm eV] \;, \qquad (3.4.2)$$

where $Z_{\rm eff} = Z - s$ is the effective nuclear charge, taking account of the shielding s due to the other electrons. For example, the binding energy in the K shell of hydrogen ($Z = 1$) is 13.6 eV; for oxygen ($Z = 8$), it is 530 eV; for silicon ($Z = 14$), it is 1.84 keV; and for iron ($Z = 26$), it is 7.11 keV. We shall not discuss the fine structure of the edges here; they reflect the different "subshells" of the atom. In Fig. 3.4.1, we show as an example the absorption

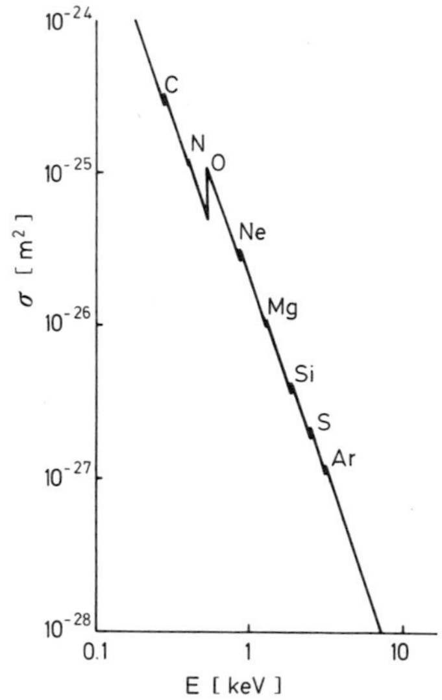

Fig. 3.4.1. Photoionization in the X-ray region for the element mixture in the interstellar medium: the dependence of the absorption cross-section σ per hydrogen particle on the photon energy $E = h\nu$. The K edges of the more abundant elements are indicated; the strongest edge at $E = 0.53$ keV is due to oxygen. The absorption coefficient is $\kappa = \sigma \cdot n_{\rm H}$, where $n_{\rm H}$ is the particle density of hydrogen (H). (Reproduced by permission of the Cambridge University Press, Cambridge, England)

[4] In Section 4.7.1,3, we shall discuss the basic principles of atomic spectroscopy and optical transitions, which are complex in comparison to transitions in the X-ray region.

cross sections in the X-ray region for matter of a composition corresponding to that of the interstellar medium (Sect. 5.3).

We shall describe the interaction of a photon or also of another particle with an atom or ion by means of the energy- or frequency-dependent *interaction cross section* σ, which has the dimensions of an area. If the particle density is n atoms or ions per unit volume, then the absorption coefficient (per unit volume) is given by $\kappa = n\sigma$. The *reaction rate r*, i.e. the number of interaction processes per unit time, is then

$$r = n\sigma v = \kappa v \ , \tag{3.4.3}$$

where v is the relative velocity of the reactants, which in the case of photons is equal to the velocity of light, c. Finally, the *mean free path l*, the distance after which a reaction occurs on the average, is given by

$$l = (n\sigma)^{-1} = \kappa^{-1} \ . \tag{3.4.4}$$

Instead of l, we can use the mass column ϱl, expressed, for example, in kg m^{-2} (ϱ: mass density), which the particles pass through on the average before undergoing an interaction process: $\varrho l = \varrho\kappa^{-1} = \kappa_M^{-1}$, the last equation defining the *mass absorption coefficient* $\kappa_M = \kappa/\varrho$, (4.2.15).

As long as the energy $h\nu$ of the photons is small relative to the rest energy of an electron, $m_0c^2 = 511$ keV ($m_0 = 9.11 \cdot 10^{-31}$ kg is the rest mass of the electron), they give up their energy mainly through photoionization processes. At higher energies, i.e. in the *gamma-ray region*, another process becomes important: *Compton scattering* of photons on electrons, $\gamma + e^- \rightarrow \gamma + e^-$. The energy loss on scattering corresponds to a change in wavelength by the amount:

$$\Delta\lambda = \frac{2h}{m_0 c}\sin^2\frac{\phi}{2} \ , \tag{3.4.5}$$

where ϕ is the scattering angle of the photon. $\Lambda = h/(m_0c) = 2.43 \cdot 10^{-12}$ m is the Compton wavelength of the electron. It is, by the way, the wavelength corresponding to the rest energy of the electron ($m_0c^2 = h\nu = hc/\Lambda$).

The cross section for Compton scattering in the case of low photon energies, $h\nu \ll m_0c^2$, is given by the *Thomson scattering coefficient*:

$$\sigma_T = \frac{8}{3}\pi r_0^2 = \frac{8\pi}{3}\left(\frac{e^2}{4\pi\varepsilon_0 m_0 c^2}\right)^2 = 6.65 \cdot 10^{-29}\,\mathrm{m}^2 \ , \tag{3.4.6}$$

independently of the energy; r_0 is the classical electron radius, and ε_0 is the permittivity constant of vacuum, equal to $8.85 \cdot 10^{-12}$ As$\cdot$V^{-1} m^{-1}.

Above $h\nu \simeq m_0c^2$, σ decreases from σ_T roughly in inverse proportion to the photon energy $h\nu$, as described by the *Klein-Nishina scattering formula*; in the limit $h\nu \gg m_0c^2$, it is given by

$$\sigma = \sigma_T \frac{3}{8}\frac{m_0c^2}{h\nu}\left(\ln\frac{2h\nu}{m_0c^2} + \frac{1}{2}\right) \ . \tag{3.4.7}$$

If the energy of the photon becomes greater than $2m_0c^2 \simeq 1.02$ MeV, then *pair production* in the Coulomb field of a charge Z, $\gamma \rightarrow e^+ + e^-$, becomes possible. The cross section for this process is of the order of

$$\sigma(e^\pm) \simeq \alpha Z^2 \sigma_T \ , \tag{3.4.8}$$

where $\alpha = 1/137.04$ is the fine structure constant. Above several tens of MeV, $\sigma(e^\pm)$ thus becomes larger than the cross section for Compton scattering and pair production is then the dominant process in the interaction of energetic photons with matter. The electron-positron pair takes on the direction of the gamma quantum which produced it, within an angle of the order of $m_0c^2/h\nu$.

Now, what are the most important interaction processes when energetic (charged) *particles* pass through matter? Like the photons, both protons and heavy nuclei as well as electrons *ionize* the atoms of the material and lose their energy thereby. They readily destroy the weak bonds within molecules and crystal lattices. For electrons, in addition to ionization, an important energy loss mechanism is *Bremsstrahlung*, an electromagnetic radiation emitted as a result of their (braking) acceleration by the charges of ions and electrons in the material. Finally, there are *nuclear processes*, when energetic protons or heavy nuclei strike atomic nuclei in the material.

The *energy loss due to ionization* of a particle of rest mass m_0 and charge z with energy E or velocity v on passing a distance x through a material of density ϱ (with atoms having charge Z and atomic mass A) is given by

$$\frac{dE}{d(\varrho x)} = -\frac{z^2}{v^2}f(v)\frac{Z}{A} \ , \tag{3.4.9}$$

where $f(v)$ is a weakly varying function[5] of v. The properties of the ionized material thus hardly enter (Z/A!); however, the energy loss is on the one hand proportional to the square of the charge of the ionizing particle, and on the other, for energies $E \simeq m_0c^2$, it is inversely proportional to v^2 or to its energy E. At higher energies, $dE/d(\varrho x)$ increases only slowly with increasing E. The *minimum* energy loss, at $E \simeq m_0c^2$, is $0.2z^2$ MeV on a path corresponding to $1\ \mathrm{kg/m^2}$. If the electrons emitted have sufficiently high energies, they can initiate secondary ionization processes.

While energetic protons and heavy nuclei maintain their original directions of motion practically unaffected, the lighter electrons are strongly deflected.

The cross section for *Bremsstrahlung* or *free-free radiation* (Sect. 4.7.1) of a *relativistic* electron ($E \gg m_0c^2$, $v \simeq c$) in the Coulomb field of a charge Z depends for one thing on its energy E, and for another, it is proportional to $\alpha Z^2 \sigma_T$, where σ_T is again the Thomson cross section and $\alpha = 1/137.04$ is the fine structure constant; this is similar to the formula for pair production by a gamma quantum (3.4.8). The corresponding energy loss rate $-dE/dt$ (3.4.3) is proportional to $\alpha Z^2 \sigma_T E c n$, with $n \propto \varrho/A$ being the particle density (A: atomic mass). If we again relate the energy loss to the mass column passed through, $d(\varrho x) = \varrho c\, dt$, we find

$$\frac{dE}{d(\varrho x)} = -\frac{E}{\xi_0} \qquad (3.4.10)$$

with the "*radiation length*" $\xi_0 \propto A(\alpha \sigma_T Z^2)^{-1}$.

For example, for air, ξ_0 is $365\ \mathrm{kg\,m^{-2}}$. Along a path of length ξ_0, according to (3.4.10), the initial energy E of the electron decreases by a factor $e^{-1} \simeq 0.37$. Depending on the material, above 10 to 100 MeV the Bremsstrahlung losses dominate losses due to ionization.

For ions, energy loss via Bremsstrahlung is unimportant, since its cross section is inversely proportional to the square of the particle mass.

When a relativistic electron passes through matter, there is a high probability that roughly half of its energy will be converted through Bremsstrahlung into one or two gamma quanta. The radiation is emitted preferentially within an angular spread of m_0c^2/E along the initial direction of motion. Since a high-energy gamma quantum mainly produces electron-positron pairs in the field of a charge Z (see above), and since both the resulting electron and the positron emit further Bremsstrahlung quanta, an *electromagnetic cascade* occurs in the material:

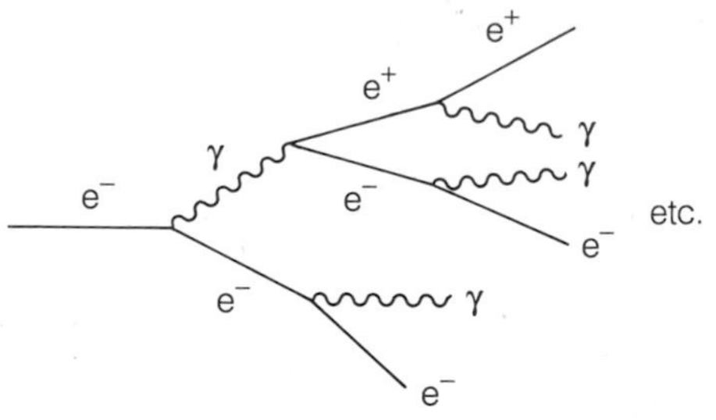

This cascade finally runs down when the energies of the gamma quanta are no longer sufficient for pair production.

If a particle moves through a medium with a velocity v which is *higher* than the velocity of light c/n in that medium (with index of refraction n), it will emit *Cherenkov radiation* at a particular angle to its direction of motion, determined by v and n. This radiation can be observed in the visible region (and recently, using wavelength converting scintillators, in the UV).

An energetic *proton* or a heavier nucleus can react with the *nucleons* in the nuclei of matter which it penetrates, if it approaches them to within a distance $\leq 10^{-15} A^{1/3}$ m (A: atomic mass), the range of the nuclear force. In this process, 40 to 50% of its energy is converted into a variety of particles, especially into π *mesons*, but also to individual nucleons, hyperons, etc. The charged particles lose most of their energy by ionization of the material; the secondary nucleons etc. can, however, react with additional nuclei if their energy is sufficiently high, leading to a *nuclear cascade*. The mean free path with respect to to strong interactions, e.g., for a cosmic-ray particle in the Earth's atmosphere, corresponds to about $100\ \mathrm{kg\,m^{-2}}$, only about a hundredth of the whole atmosphere. The *neutral pions*, π^0, decay after a mean lifetime of only about $1.8 \cdot 10^{-16}$ s into two gamma quanta, which for their part initiate electromagnetic cascades. The charged pions, $\pi^{\pm}$, decay in a few 10^{-8} s into muons, $\mu^{\pm}$, which have high penetrating power, as well as into neutrinos and antineutrinos.

---

[5] The relation between velocity and total energy (including the rest energy m_0c^2) is $E = \gamma m_0 c^2$, where $\gamma = [1-(v^2/c^2)]^{-1/2}$ is the Lorentz factor. The kinetic energy is $E_{\mathrm{kin}} = (\gamma - 1) m_0 c^2$. For nonrelativistic particles ($v \ll c$), γ becomes $\simeq 1 + (v/c)^2/2 \simeq 1$, $E \simeq m_0c^2 + m_0v^2/2$, and $E_{\mathrm{kin}} = m_0 v^2/2$; for extreme relativistic particles ($\gamma \gg 1$), we have $E \simeq E_{\mathrm{kin}} \gg m_0c^2$.

3.4.2 Particle Detectors and Telescopes for Cosmic Radiation

The orbit of an energetic particle can be directly observed in a *bubble chamber*. The ions which are produced along its track act as nucleation points for vaporization in a superheated liquid with boiling hysteresis, so that the path becomes visible as a string of small bubbles of vapor. The particle's energy can be derived from the degree of ionization, i.e. the intensity of the track; a magnetic field can be applied to affect the particle orbits and allow the determination of their charge and mass. Bubble chambers are, however, not suitable for use in satellites or space vehicles due to their size and complexity.

The *cloud chamber*, in use since the 1930's, in which the tracks of particles are made visible as condensation trails in a supersaturated vapor generated by rapid expansion, is now of only historical interest.

In *plastic detectors*, the tracks of strongly ionizing particles can be detected using the permanent damage to the material which they produce. Since the damaged areas have an increased chemical reactivity, the particle tracks can be made visible by etching.

In *nuclear emulsions*, which consist of a gelatine layer containing suspended AgBr crystals much like a conventional photographic emulsion, the crystals are first activated by secondary electrons. After development, the tracks of ionizing particles become visible as a result of precipitation of silver grains. The density of grains at each point is proportional to the energy loss dE/dx.

A *scintillation detector* converts the energy of charged particles or energetic photons to *visible light flashes* via secondary electrons released in a suitable crystal or liquid; the yield is a few percent, and the light flashes are detected using photomultipliers. The intensity of the current pulse from the photomultiplier is proportional to the energy loss of the primary particle. Materials for scintillation detectors are NaI and CsI crystals, thallium doped NaI, and also organic liquids and plastics.

In order to register energetic particles above a particular threshold energy or velocity, *Cherenkov detectors* are used; they are made of a transparent, solid dielectric material such as plexiglas, or employ a gas. The weak Cherenkov light in the optical region can once again be detected by photomultipliers. In the so-called RICH (Ring Imaging Cherenkov) detector, the UV Cherenkov radiation from highly energetic particles is first converted to visible light by a suitable scintillator, then detected, for example by a CCD camera.

Individual charged particles can be detected by a *gas counter tube*, in which they ionize the gas charge, e.g. Ne or Xe, along their paths. A high voltage is applied between the outer metal wall of the tube and a thin wire along its axis, giving rise to an electron avalanche due to ionization of the gas, which can be detected as a current pulse on the wire electrode. In the *gas proportional counter*, the voltage is set so that the total number of electrons detected is directly proportional to the number of electrons produced by the primary incident particles. Gas proportional counters are currently not so much used as particle detectors, but instead as detectors for X-radiation (Sect. 3.5).

By increasing the applied voltage, one can operate a gas counter tube, as in the *Geiger-Müller detector*, in the saturation region, in which each particle ionizes the whole gas and produces a spark. Counters of this type can be made with very small dimensions and assembled into a *spark chamber*. Because of their good positional resolution, spark chambers are suitable for determining the orbits of particles.

In *solid-state detectors*, *electron-hole pairs* are created in a semiconducting material, which may be fabricated from an extremely pure single crystal. The pairs, similarly to the electron-ion pairs in a gas counter, then make the material conducting and can be registered as a voltage pulse after amplification. The disadvantages of the relatively small detector areas and the need for cooling are compensated by the high energy resolution of semiconductor detectors, which is due to the well-defined energy of the electron-hole pairs and to the short stopping ranges of the primary particles in the material.

Now that we have briefly introduced the most important processes by which energetic particles interact with matter, and the numerous types of detectors which are based on them, we may ask how a *particle telescope* for observation of cosmic radiation can be constructed. Since these energetic particles can neither be reflected nor focussed by matter, owing to their high penetrating power, it is necessary to construct a particle telescope in a completely different manner from, for example, an optical telescope. To select a particular type of energetic particles, characterized by their charge z and mass m, at the same time discriminating against other types of particles, and to determine their direction of incidence and energy E, one assembles several detectors having different sensitivities, together with shielding and absorbing elements, into a telescope. Its geometric arrangement, together with its electronic logic, which includes coincidence and anti-

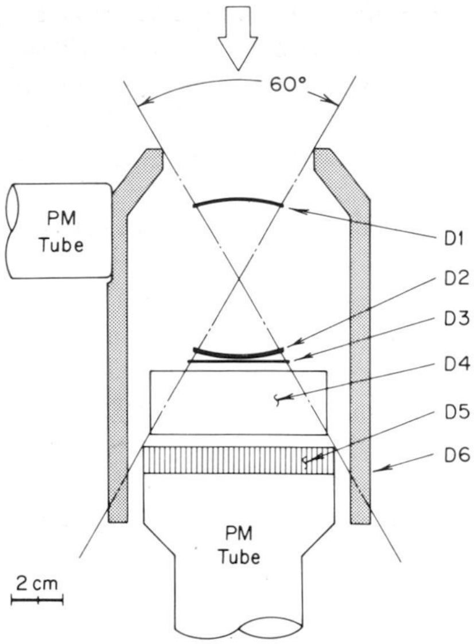

Fig. 3.4.2. A telescope for detecting cosmic-radiation particles. A cross-section through the telescope constructed by the University of Chicago for the IMP-7 satellite (launched 1972), and used to measure beryllium isotopes from cosmic radiation in the energy range around 100 MeV per nucleon. The various detectors D1 to D6, together with logic circuits, detect the type, energy, and direction of incidence of the particles and eliminate signals from secondary and background radiations; the silicon drift counters D1 through D3 determine the angular resolution of the telescope. "Desired" particles must be stopped in the Tl-doped CsI counter D4, and therefore may not produce a signal in the Cherenkov detector D5 or in the plastic scintillator shielding D6 (PM tube = photomultiplier tube). M. Garcia-Munoz et al., Astrophys. J. **217**, 859 (1977). (Reproduced by permission of the University of Chicago Press, the American Astronomical Society, and the authors)

coincidence circuits that operate on the detector signals, are designed to yield the desired information (Fig. 3.4.2, 3). The observable *angular range* of the telescope can, for example, be geometrically defined by the transmission of a particle through two particular detectors; details of particle orbits can be obtained from numerous small detectors arranged, for example, into a spark chamber. The determination of the *charge* z requires the analysis of the energy loss $dE/d(\varrho x)$, e.g. using the degree of ionization in a detector (3.4.9), with as many variously-responding detectors as possible. Finally, in order to find the particle's *mass* m_0, the total kinetic energy $(\gamma-1)\, m_0 c^2$ (γ: Lorentz factor) must be precisely measured. For this purpose, the particle must be stopped in the telescope. This is verified by a "last detector", in which the particular particle is *not* allowed to produce a signal and which is connected via an anticoincidence circuit to the other detectors with their positive detection signals.

The particles of galactic cosmic radiation interact strongly even with the upper layers of the Earth's atmosphere. Thus, complex apparatus can be placed on the ground to investigate the cascades or air showers of secondary particles, from which conclusions can be drawn about the primary radiation. On the other hand, however, the *primary cosmic radiation*, which consists mainly of protons of ≥ 1 GeV energy and, to a lesser extent, of heavier nuclei and electrons, must be observed outside the atmosphere. This places heavy demands on the technology, since in particular for the most energetic parti-

Fig. 3.4.3. A large-area particle telescope for observing cosmic-radiation isotopes at several 100 MeV per nucleon from a stratospheric balloon. The dimensions are roughly 1.4 m · 1.4 m · 1.4 m (a collaboration of the University of Siegen, Germany, with the Goddard Space Flight Center of NASA; photograph by M. Simon)

cles, which have relatively low intensities, large, heavy telescopes must be employed over long periods of time. Satellites and space vehicles are used, as well as stratospheric balloons; it must be remembered, however, that even at an altitude of 50 km, a layer of atmosphere of about 20 kg m^{-2} (Fig. 3.1.1) still remains above a balloon-carried telescope.

3.4.3 Gamma-Ray Telescopes

Gamma rays from space are absorbed in the upper layers of the Earth's atmosphere; in addition, large numbers of gamma quanta are produced by the impact of highly energetic cosmic ray particles on the upper atmosphere. Therefore, astronomy in the gamma-ray region (above a few hundred keV photon energy) can only be carried out from balloons, satellites, or space vehicles. Rockets with limited observation times are not suitable owing to the weak photon flux in the gamma-ray spectral region. Only extremely high-energy gamma rays can be investigated by ground-based instruments using the air showers which they produce in the atmosphere.

In the 1960's, instruments in balloons and satellites such as Explorer XI (1961) and the Orbiting Solar Observatory OSO-3 (1967) were able to detect the gamma radiation from the Sun and our Milky Way galaxy, although only limited angular- and energy-resolutions were attained. For example, the angular resolution of OSO-3 was only about 20°. At the beginning of the 1970's, longer series of observations with sensitive gamma-ray telescopes could be carried out; they first attained an angular resolution worthy of mention. Thus, the surveys performed by the satellites SAS-2 and COS-B (launched in 1972 and 1975) at energies ≥ 100 MeV had an angular resolution of a few degrees and an energy resolution $\Delta E/E$ of about 50% (Fig. 5.3.17). Finally, the instrumentation of the large gamma-ray satellite GRO (*G*amma *R*ay *O*bservatory), which was launched in 1991, can attain an angular resolution near 10′ at an energy resolution of around 20% in the high-energy range(100 to 2000 MeV), and of 4 to 10% at energies of a few MeV.

In principle, the *detection* of a gamma quantum is performed using the particles released by it in the detector, so that all of the multiplicity of *particle detectors* used in high energy physics (Sect. 3.4.2) can be employed for this purpose. At low energies E_γ, below a few tens of MeV, the detection is performed using Compton backscattering electrons; at higher energies, electron-positron pair formation is used (Sect. 3.4.1). In the region of *low energy* gamma rays, the determination of the direction of incidence of the gamma radiation is particularly difficult. It can be accomplished by using, for example, two large-area scintillation detectors spaced some distance one behind the other. The relatively good angular resolution at high energies is based upon the fact that the $e^{\pm}$ pair retains the direction of the gamma quantum which produced it to within an angle of $\simeq mc^2/E_\gamma$ (the rest energy of the electron $mc^2 = 0.511$ MeV).

Cosmic gamma radiation with its low photon fluxes has to be detected against a strong background of secondary gamma radiation generated by cosmic rays in the atmosphere and in the detector itself. Gamma-ray telescopes therefore consist of a suitable geometric arrangement of several types of detectors with differing response probabilities for charged particles and photons, together with electronic logic circuits.

In the case of the *gamma-ray telescope* on COS-B (Fig. 3.4.4), a spark chamber serves as the primary detector; it contains layers of wire mesh and tungsten plates, and is filled with a mixture of neon and ethane gases. The incident gamma quantum creates an electron-positron pair in the tungsten, which then ionizes the gas along its path and is detected on passing through a scintillation or Cherenkov counter. The system of wires in the spark chamber localizes the orbits of the electrons and posi-

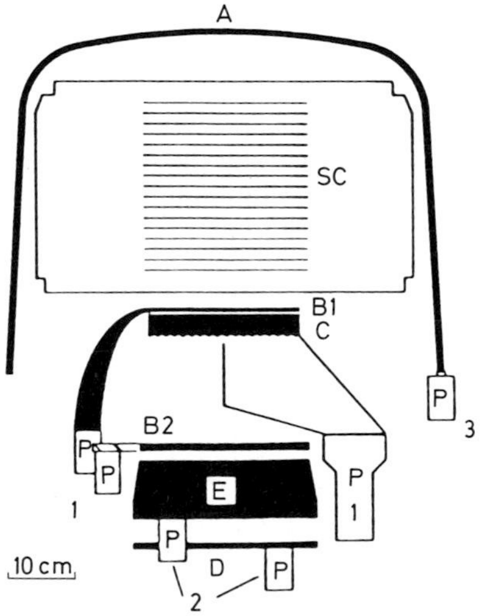

Fig. 3.4.4. Schematic construction of the gamma-ray telescope on the COS-B satellite. SC: wire spark chamber; B1,2: scintillation counters; C: Cherenkov detector; A: anticoincidence shield; D: plastic scintillators; E: CsI crystal; P: photomultipliers. (With the kind permission of *Sterne und Weltraum*)

trons and thus determines the *direction* of the gamma quantum. To this end, a discharge in the ionized gas along the tracks is initiated by a high-voltage pulse triggered from the detection counter, leading to current pulses in the "affected" wires. In order to verify that the electron-positron pair was in fact produced by a gamma quantum and not by a cosmic-ray particle, the spark chamber is surrounded by a large-area scintillation counter which responds to cosmic radiation, but not to gamma radiation. An *anticoincidence circuit* relative to the detection counter eliminates triggering of the spark chamber by the unwanted particles. The *energy* of the gamma quantum is finally determined from the energy of the electron-positron pair, which produces flashes of light upon absorption in an additional detector system consisting of a cesium iodide and a plastic scintillator; their intensity depends on the energy.

The *Gamma-Ray Observatory* (GRO) carries four instruments, which can detect cosmic gamma rays simultaneously in different energy ranges. A detector system consisting of NaI scintillation crystals and photomultipliers which "sees" the *entire* sphere, except for the part covered by the Earth, is used in particular for investigating the short, rapidly varying gamma bursts (Sect. 4.11.6) and other transient sources in a relatively low energy range, ≤ 2 MeV. The resolving time is less than one ms and the positional accuracy (for strong sources) is about 1°. A *spectrometer* is made up of a system of NaI and CsI scintillation counters and provides an energy resolution $\Delta E/E$ of 4 to 12% between 0.1 and 10 MeV for spectroscopic observations at an angular resolution of about 10′. In this energy range, there are many gamma rays from radioactive nuclides, which can give information about nucleosynthesis in the stars (Sect. 5.4.5). At somewhat higher energies (1–30 MeV), an imaging *"Compton telescope"* has an energy resolution near 7% and an angular resolution of about 10′. In this telescope, the gamma quanta which are scattered in liquid scintillators are detected by scintillation crystals mounted at a suitable distance. Finally, for the high energy range (20 MeV–30 GeV), the GRO makes use of an additional telescope consisting of numerous spark chambers and a large NaI scintillation crystal, which can determine the positions of the gamma-ray sources to a precision of 5 to 10′ with a modest energy resolution ($\simeq 20\%$).

At the *highest energies* $\geq 10^{12}$ eV, gamma rays can be detected indirectly from the *ground* by means of their electromagnetic *cascades*, which result from their interactions with the particles of the upper atmosphere. At the very highest energies $\geq 10^{14}$ eV, the large air showers reach the ground and can be registered by the usual particle telescopes. Below 10^{14} eV, the *Cherenkov radiation* from the energetic particles can be detected in the optical region with relatively small telescopes. For example, the Crab-nebula pulsar, the binary star system Cyg X-3, and the radiogalaxy CenA were discovered in the high-energy region in this way. The difficulty of air shower observations is to differentiate the electromagnetic cascades originating with gamma quanta from the much more frequently-occurring "nuclear" cascades due to cosmic ray particles, which contain considerably more muons.

3.5 X-Ray Telescopes and Detectors

Let us now turn to the remaining region of the electromagnetic spectrum, the X-ray region; it includes photon energies from about 0.1 to a few 100 keV, or wavelengths from about 10 to a few 10^{-3} nm. Corresponding to its position between the ultraviolet and the gamma ray regions, the X-ray region can be observed with instruments which occupy an intermediate position between those for the neighboring spectral regions. For photon energies of up to a few keV, *imaging* telescopes with *grazing incidence mirrors* can be constructed; at higher energies, as in the gamma-ray region, detectors and telescope arrangements must be adapted from high-energy physics.

Astronomy in the X-ray region must also be carried out outside the Earth's atmosphere. The first cosmic X-ray sources (aside from the Sun) were discovered in 1962 by R. Giacconi and coworkers using a rocket. Numerous observations of X-ray sources followed, made from rockets and balloons. A complete survey of the heavens was accomplished by the first *X-ray satellite*, UHURU, launched in 1970; it revealed about 340 sources. In the following decade, the number of known X-ray sources increased to well over 1000 as a result of observations made from satellites; of these, many could be identified with astronomical objects of a variety of types. The most surprising discovery of nonsolar X-ray astronomy has been the great variety of *temporal variability* of cosmic sources.

Optical imaging in the X-ray region was first obtained for the Sun using a primitive *pinhole camera*. Then mechanical *collimators* were constructed, behind which, for example, a large-area scintillation counter was placed to detect the X-ray quanta (Fig. 3.5.1). The gyroscopic motion of the rocket or the rotation of the satellite is used

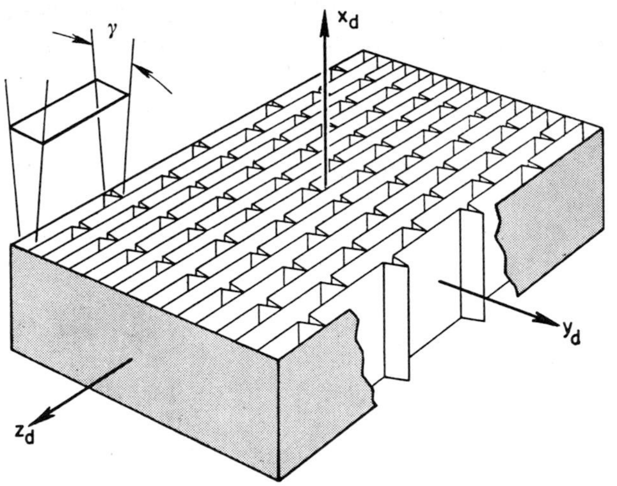

Fig. 3.5.1. A cellular collimator for observing cosmic X-ray sources. The angular resolving power in the γ direction is about 1°. Below the collimator, there would be in the operating configuration a photon counter of the same area

to sample larger areas in the celestial sphere. The moderate angular resolution of the order of 1° can be improved by using *modulation collimators*, in which the X-ray intensity is modulated by motion of, for example, wire meshes.

Only in 1964 did R. Giacconi take up the *X-ray mirror telescope* developed by H. Wolter in Kiel in 1951 (initially developed for X-ray microscopy). It is based on the following considerations: a metal surface, no matter how well polished, gives specular reflection for X-rays only at grazing incidence (the angle between the ray and the surface must be less than a few degrees). A first attempt could thus be made to image distant objects by using a ring-shaped segment of an extended *paraboloid*, covering the unused part of the entrance opening by a round metal plate (Fig. 3.5.2 a). This type of telescope would, however, produce a very poor image due to the enormous aberrations (coma). According to Wolter, these can be considerably reduced by following the first reflection from a paraboloid by a second reflection from a coaxial and confocal *hyperboloid* placed down the optical path (Fig. 3.5.2 b). To increase the radiation-collecting area of the telescope, several of these confocal double mirrors are placed within one another.

The *Einstein Observatory* (HEAO-2 satellite) launched by NASA carried from 1978 to 1981 the first *imaging* Wolter telescope, developed under the leadership of R. Giacconi. It had a 56 cm aperture and was suited for long observation times with high angular resolution (≤2″) in the region from 0.1 to 4 keV; it permitted the discovery of very faint X-ray sources. The European EXOSAT satellite (1983–1986) carried two small 27 cm Wolter telescopes on a highly eccentric orbit (perigee 190000 km), with which observations could be continued for days (without interruption by the Earth's shadow). The satellite ROSAT, which was designed by J. Trümper and coworkers and launched in 1990, will carry out a complete sky survey from a circular orbit at an altitude of 580 km using an 83 cm Wolter telescope. Later, interesting objects will be investigated in more detail. The image detectors are sensitive in the range from 0.1 to 2.5 keV, and observations can be carried out either at reduced energy resolution with an angular resolution of 3″, or in four separate spectral regions ($E/\Delta E \simeq 2.5$) with an angular resolution of 25″. It is expected that the survey will reveal about 10^5 new X-ray sources.

The most important type of *detector* in the X-ray region is the *gas proportional counter*, which is usually filled with Ar or Xe. These can be constructed with a

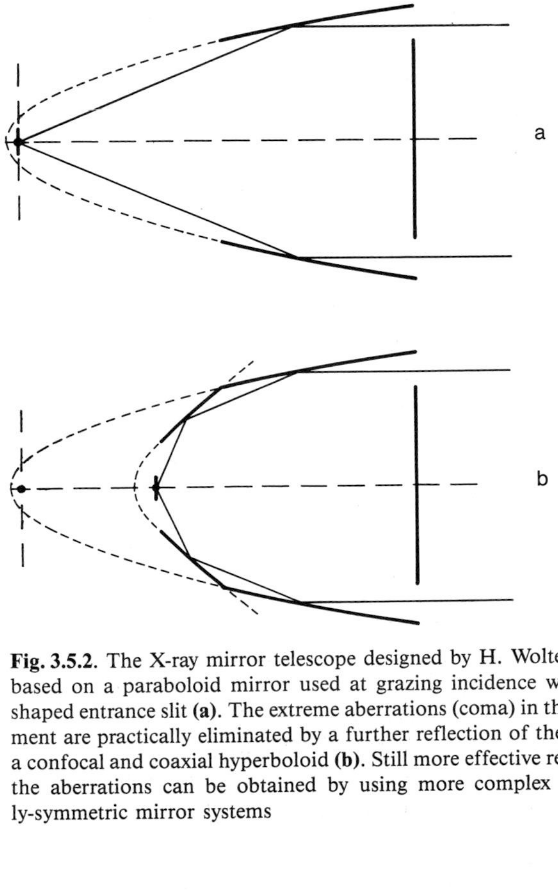

Fig. 3.5.2. The X-ray mirror telescope designed by H. Wolter (1951) is based on a paraboloid mirror used at grazing incidence with a ring-shaped entrance slit **(a)**. The extreme aberrations (coma) in this arrangement are practically eliminated by a further reflection of the rays from a confocal and coaxial hyperboloid **(b)**. Still more effective reduction of the aberrations can be obtained by using more complex rotationally-symmetric mirror systems

large area and give a yield of 30 to 40% up to energies ≤ 20 keV. In order to distinguish ionization in the detector due to X-ray quanta from that due to cosmic ray particles, anticoincidence circuits or electronic pulse shape discrimination are necessary, as in all X-ray detectors. In the gas-filled (Xe) *scintillation proportional counters*, the electrons knocked out by X-ray quanta are accelerated so weakly that they do not ionize any more gas atoms, but instead excite them to light emission; the light is then observed in the optical region using conventional photomultipliers.

Solid-state detectors, made from lithium-doped germanium and cooled by liquid nitrogen or helium, currently have the best energy resolution but can be manufactured with only small sensitive surface areas. To use the imaging properties of a Wolter telescope, *position-sensitive detectors* are required. In the *Imaging Proportional Counter* (IPC), the localization of the X-ray photons is performed by crossed wire arrays together with electronic discrimination methods. Arrangements of channel plates and CCD cameras have recently found application.

The *Sun* is unique as an X-ray source in that the radiation flux is sufficiently high to permit images to be registered on photographic film, such as those taken on the Skylab mission in 1973 (Fig. 4.10.7).

In the hard X-ray region (≥ 10 keV), *scintillation detectors* made of Na(Tl)I or Cs(Na)I are used; the light flashes are detected by photomultipliers.

Discrimination of more or less narrow wavelength ranges in the X-ray region is accomplished with *filters*, making use of the absorption edges of the elements contained in the filter, which are sharply bounded on the long wavelength side (Sect. 3.4.1). The spectral resolution or energy resolution of the detectors themselves is not very high. The most favorable resolution is obtained from *solid state detectors*, which attain 1 to 10% depending on their size and the energy range. In the soft X-ray region, *transmission gratings* or Bragg *crystal spectrometers* can give high resolutions of $\lambda/\Delta\lambda$ from 10^3 to 10^4.

* * *

To complete our treatment of observational methods in the various spectral or energy regions, we should mention the enormous development of *electronic computers* which, one can properly say, have opened a new era of astronomical observation and measurement methods. The rapid developments in semiconductor technology, miniaturization, and low-cost mass production of components have permitted computers of steadily increasing speed and storage capacity to be applied in astronomy to such an extent that today the collection and analysis of observational data would be completely unimaginable without them. We mention here only a few examples, such as the steering of telescopes (azimuthal mounting), the variation of mirror surfaces on the principle of active optics, the combined analysis of enormous amounts of data from very long baseline interferometry to yield high resolution radio images, the logical control circuits for particle and gamma-ray detectors, control of readout of CCD cameras and production of a "digital image" in the computer, or the reduction of "raw data" by either automatic or interactive programs, often of great complexity. Finally, "space astronomy" using satellites and space vehicles has become possible only through the application of powerful computers.

Astronomers have now become accustomed to storing *observational material* for the most part as *numerical fields or arrays* on suitable data storage media (magnetic tapes or disks, semiconductor memories ...). Even the information on photographic plates is increasingly digitized so that it can be directly input into a computer.

Finally, we mention that not only observational astronomy but also *theoretical astrophysics* with its enormously complex model calculations, for example of radiation transport, stellar structures, or magnetohydrodynamic processes, would be unthinkable today without powerful computers having large memories.

The fantastic "efficiency" of computers opens unsuspected possibilities for research on the one hand, but on the other, it presents astronomers with a great responsibility.

4. The Sun and Stars. Astrophysics of Individual Stars

Having considered classical astronomy, our Solar System, and astronomical instruments and observation methods, we now turn to the *astrophysical investigation of the Sun and other stars*: initially, we shall treat the latter as individuals. Stellar evolution, star systems, galaxies, and the cosmos as a whole will be the subject of Chap. 5.

We begin in Sect. 4.1 with some historical remarks in order to give the reader a first insight into solar and stellar astrophysics. We then introduce in Sect. 4.2 the basic concepts of radiation theory, which are of central importance to all of astrophysics. In Sect. 4.3, we deal with the radiation of the *Sun*: with the continuum emitted from its photosphere which contains the dark Fraunhofer lines. Later, in Sect. 4.10, we make the acquaintance of the layers of the solar atmosphere which lie above the photosphere: the chromosphere and the corona, as well as the solar wind and the various manifestations of solar activity. First, however, in Sect. 4.4 we treat the radiation of the *stars*, their basic parameters and their distances; in Sect. 4.5, we deal with their spectra, and in Sect. 4.6 with their masses, along with an overview of binary star systems. The chromospheres, coronas, and activity phenomena of stars are discussed in Sect. 4.11 in connection with a description of the most important types of *variable stars.*

With a view to the quantitative interpretation of solar and stellar spectra, and those of other astronomical objects, we summarize in Sect. 4.7 the basic principles of atomic spectroscopy and of the excitation and ionization of atoms. With this basis, in Sect. 4.8 we can treat the physics of *stellar atmospheres* and in Sect. 4.9 the theory of stellar spectral lines and the quantitative *analysis of stellar spectra*, with the goal of determining the relative abundances of the chemical elements.

We conclude our treatment of the astrophysics of individual stars in Sect. 4.12 with a discussion of the *internal structures* and of energy production in the Sun and other stars.

4.1 Astronomy + Physics = Astrophysics

Historical Introduction

At the conclusion of Sect. 2.1, we mentioned the first measurements of trigonometric stellar parallaxes by F. W. Bessel, T. Henderson, and F. G. W. Struve in 1838. They provided the final confirmation of the Copernican worldview (whose validity was by then hardly doubted). But in addition, they provided a secure foundation for the determination of all cosmic *distances.* Bessel's parallax of 61 Cygni, $p = 0.293''$, indicates that this star is at a distance of $1/p = 3.4$ parsec or 11.1 light years. With this value, we can, for example, compare the luminosity of this star directly with that of the Sun. However, the *trigonometric parallaxes* only became an important method in astrophysics after about 1903, when F. Schlesinger developed their photographic determination to a tool of unbelievable precision ($\simeq 0.01''$).

Information about the *masses* of stars is obtained from *binary star systems.* W. Herschel's observations of Castor (1803) left no doubt that here, two stars were revolving about one another on elliptical orbits under the influence of their mutual attraction. As early as 1782, J. Goodricke had observed the first *eclipsing variable* star, Algol. The application of the variety of information obtained from binary star systems is the work of H.N. Russell and H. Shapley (1912). The first *spectroscopic* binary system (determination of the motion by means of the Doppler effect) was Mizar, whose binary nature was discovered by E.C. Pickering in 1889.

Stellar photometry, the measurement of the apparent brightness of stars, obtained a firm basis in the middle of the 19th century, after its beginnings in the 18th century (Bouguer, 1729; Lambert, 1760; etc.). For one thing, in 1850 N. Pogson introduced the definition of a *magnitude* as corresponding to a decrease in the logarithmic brightness of 0.400, i.e., a brightness ratio of $10^{0.4} = 2.512$. Furthermore, J. C. F. Zöllner in 1861 constructed the first *visual spectrophotometer*, with which even the colors of stars could be determined. About the same time, the great catalogues of stellar magnitudes and positions made knowledge of the stars available on a widespread basis: e.g. in 1852/59, the *Bonner Durchmusterung*, with data for 324000 stars down to about 9.5 mag, was prepared by F. Argelander and coworkers; later followed the *Cordoba Durchmusterung* for the southern hemisphere.

Photographic stellar photometry was founded by K. Schwarzschild with the Göttinger Aktinometrie in 1904/08. He also immediately recognized that the color index (= photographic minus visual brightness) can be used as a measure of the color and thus of the temperature of a star. Soon after the invention of the photoelectric cell, around 1911, H. Rosenberg, P. Guthnick, and J. Stebbins began the development of *photoelectric* photometry. Since the invention of the photomultiplier shortly after the second World War, it has become possible to measure the brightnesses of stars with a precision of a few thousandths of a magnitude in expediently chosen wavelength ranges, such as the system of UBV magnitudes (Ultraviolet, Blue, Visual) of H.L. Johnson and W.W. Morgan (1951). The corresponding color indices can then be derived.

Parallel to stellar photometry, the *spectroscopy* of the Sun and stars was developed. In 1814, J. Fraunhofer discovered the dark lines in the solar spectrum which bear his name. In 1823, with an extremely modest apparatus, he was able to observe similar lines in the spectra of several stars and noted their differences. Modern *astrophysics*, i.e., the investigation of the stars with physical methods, began in 1859, when G. Kirchhoff and R. Bunsen in Heidelberg discovered *spectral analysis* and explained the Fraunhofer lines in the solar spectrum (Fig. 4.1.1). Beginning in 1860, G. Kirchhoff had formulated the basic principles of a theory of radiation, in particular Kirchhoff's Theorem, which deals with the relation between the absorption and emission of radiation in thermal equilibrium. This theorem, together with Doppler's principle ($\Delta\lambda/\lambda = v/c$), represented for forty years the entire theoretical basis of astrophysics. The spectroscopy of the Sun and of stars was at first directed to the following applications:

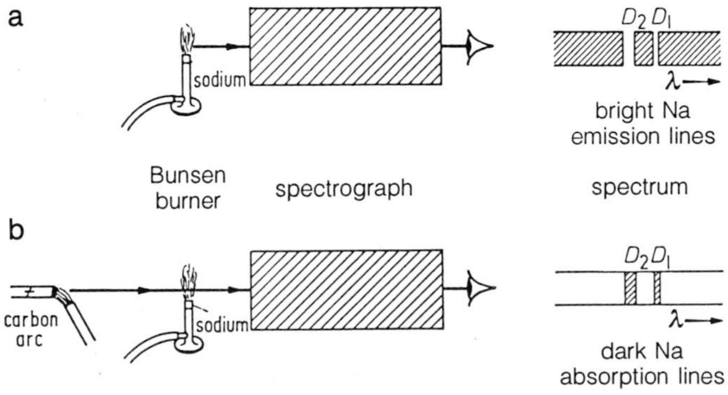

Fig. 4.1.1 a, b. The basic experiment of spectral analysis, after G. Kirchhoff and R. Bunsen, 1859. **(a)** A Bunsen flame to which some sodium (salt) has been added shows bright Na emission lines in its spectrum (*right*). **(b)** When the light of a carbon-arc lamp, which is at a much higher temperature, is passed through the sodium flame, a continuous spectrum is obtained with dark Na absorption lines, similar to the solar spectrum

1) Recording of the spectra and determination of the spectral-line wavelengths of all the elements in the *laboratory*. Identification of the lines from stars and other cosmic light sources (W. Huggins, F.E. Baxandall, N. Lockyer, H. Kayser, C.E. Moore-Sitterly, and many others).

2) Photographic registration and increasingly precise evaluation of the spectra of *stars* (H. Draper, 1872; H.C. Vogel and J. Scheiner, 1890, and others) and of the *Sun* (H.A. Rowland, 1888/89).

3) *Classification* of the stellar spectra, initially in a one dimensional series, essentially as a function of (decreasing) temperature. Following early works by W. Huggins, A. Secchi, H.C. Vogel and others, E.C. Pickering and A. Cannon et al. created the Harvard Classification (starting in 1885) and the Henry Draper Catalogue. Later progress was made as a result of the discovery of the *luminosity* as the second classification parameter, and therewith the determination of *spectroscopic parallaxes* by A. Kohlschütter and W.S. Adams in 1914; much later, under "modern" aspects, the Atlas of Stellar Spectra was compiled by W.W. Morgan, P.C. Keenan, and E. Kellman, using the MKK classification (1943).

4) However, for a long time the main interest of astronomers was held by the determination of the *radial velocities* of stars using Doppler's principle. Following visual attempts by W. Huggins in 1867, the first usable photographic radial velocity measurements were obtained by H.C. Vogel in 1888. The rotation of the Sun had already been verified spectroscopically by that time. The further development of the technique of precise radial velocity determinations was in particular the achievement of W.W. Campbell (1862–1938) at the Lick observatory.

What might be termed the conclusion of this period in astrophysics came with the discovery in 1913 of the

Hertzsprung-Russell diagram. As early as 1905, E. Hertzsprung had recognized the difference between giant and dwarf stars. H.N. Russell drew the well-known diagram based on improved trigonometric parallaxes, with the spectral types on the abscissa and the absolute magnitudes on the ordinate; it showed that most of the stars in our neighborhood fall within the narrow band of the *main sequence* (Fig. 4.5.1), while a smaller number occur in the region of the *giant stars*. Russell at first based a *theory of stellar evolution* on this diagram (a star begins as a red giant, undergoes compression and heating to join the main sequence, then cools along the main sequence), which however had to be abandoned ten years later. We shall take up further developments in this theme in Chap. 5.

What astrophysics most needed at the beginning of the present century was an extension of its physical and theoretical *fundamentals*. The theory of *cavity radiation* or *black-body radiation*, i.e., of the radiation field in thermal equilibrium, begun by G. Kirchhoff in 1860, had been completed in 1900 by M. Planck with the discovery of the *quantum theory* and the law of *spectral energy distribution* for black-body radiation. Astronomers then attempted to apply this theory to the *continuous spectra of stars* in order to estimate their temperatures. However, the brilliant K. Schwarzschild (1873–1916) soon developed one of the future pillars of the theory of stars, his *theory of stationary radiation fields*. He showed in 1906 that in the photosphere of the Sun (i.e., in the layers which emit the main portion of the solar radiation), energy transport takes place from within to the outer layers by means of radiation. Under this assumption, he calculated the increase in temperature with increasing depth, and was able to obtain the correct *center-limb-darkening* relation for the solar disk. Schwarzschild's paper on the solar eclipse of 1905 is a masterpiece of the mutual interaction of observation and theory. In 1914, he investigated radiation exchange theoretically and spectroscopically in the broad H- and K-lines of the solar spectrum. It was evidently clear to him that the further development of his ideas required above all an atomic theory of absorption coefficients, i.e., of the interaction of radiation and matter; he thus accepted with great enthusiasm the quantum theory of atomic structure founded by N. Bohr in 1913. He produced his famous papers on the quantum theory of the Stark effect and on band spectra. Schwarzschild died in 1916, much too soon, at the age of only 43.

The relationship between the theory of radiation equilibrium and the new atomic physics was clarified on the one hand by A.S. Eddington in 1916/26 with his *theory of the internal constitution of the stars*. On the other hand, J. Eggert (1919) and M.N. Saha (1920) sought the key to the interpretation of the spectra of the Sun and of stars in the theory of *thermal ionization and excitation*. The fundamental principles of the modern *theory of stellar atmospheres* and of the solar and stellar spectra thus were rapidly developed. We should mention also the works of R.H Fowler, E.A. Milne, and C.H. Payne on *ionization* in stellar atmospheres (1922/25); the measurement and calculation of *multiplet intensities* by L.S. Ornstein, H.C. Burger, H.B. Dorgelo, as well as R. de L. Kronig, A. Sommerfeld, H. Hönl, and H.N. Russell; and then the important contributions of B. Lindblad, A. Pannekoek, M. Minnaert and others. By 1927, a serious attempt could be made to combine the theory of *radiation transport*, which had been further developed, with the *quantum theory of absorption coefficients*, and to construct a rational theory of solar and stellar spectra (M. Minnaert, O. Struve, A. Unsöld). This made it possible to determine the *chemical composition* of stellar atmospheres from their spectra and thus to study *stellar evolution* empirically in relation to energy production by nuclear processes in the interior of stars (see below).

The theory of *convective flow* in stellar atmospheres and especially in the Sun was founded in 1930 by A. Unsöld with the discovery of the hydrogen convection zone. The connection to hydrodynamics (theory of mixing length) was soon thereafter made by H. Siedentopf and L. Biermann, after S. Rosseland had already pointed out the astrophysical importance of turbulence.

We know that the ionized gases in stellar atmospheres and in other cosmic objects have a high electrical conductivity. T.G. Cowling and H. Alfvén pointed out in the 1940's that *cosmic magnetic fields* might persist for very long times until the currents associated with them are consumed by ohmic dissipation. The magnetic fields and flow fields interact continuously; the basic principles of electrodynamics and of hydrodynamics must be combined into the science of *magnetohydrodynamics*. Empirically, G.E. Hale in 1908 had discovered magnetic fields of up to 4 T in sunspots by observing the Zeeman effect of the Fraunhofer lines (incidentally, he started from physically quite incorrect hypotheses!). The weak magnetic field on the remainder of the solar surface, measured in 1952 by H.W. Babcock, has more recently been found to result from the superposition of numerous, thin flux tubes, each having a magnetic field intensity of a few tenths of a Tesla. The sunspots, the faculae which surround them, prominences, flares, and many other solar phenomena are statistically correlated in the 2×11.5-year cycle of *solar activity*. All of this, of which at present only the least part is understood, belongs to the area of magnetohydrodynamics and plasma physics.

Only by studying the flow phenomena and magnetic fields in the Sun can we arrive at a certain degree of understanding of the structure and excitation of the *corona*, the outermost shell of the Sun with a temperature of 1 to 2×10^6 degrees. Here, very complex processes occur which lead to the production of X-rays and the radiofrequency radiation of the Sun, as well as various types of corpuscular radiation and the solar wind.

The investigation of the chromospheres, coronas, and activity phenomena of other *stars* has become possible through observations in the ultraviolet and X-ray spectral regions made from space. The ultraviolet resonance lines of common elements from hot giant and supergiant stars exhibit evidence in their short-wavelength absorption components of massive *stellar winds* with velocities of several thousand km s^{-1}. The *loss of mass* connected with these stellar winds is orders of magnitude higher than in the Sun and has an important influence on the evolution of these stars. A. Deutsch found already in the 1950's that cool giant stars are surrounded by extended, slowly expanding gas shells; these could later be observed using the infrared emissions of the dust particles and molecular emisson lines in the radio-frequency region.

Since the 1920's, the *theory of the internal structure* of the Sun and stars has developed in parallel with progress in the understanding of the outer, "visible" stellar layers. In relation to the age of the Earth and the Sun of about $4.5 \cdot 10^9$ years, known from radioactivity measurements, J. Perrin and A.S. Eddington suggested already in 1919/20 that the energy which is continuously radiated by the Sun is generated by conversion of hydrogen to helium. In 1938, H. Bethe and C.F. v. Weizsäcker were then able to show, based on the understanding of *nuclear physics* which had in the meantime developed, just *which* thermonuclear reactions would, for example, be able to slowly "burn" hydrogen to helium in the interior of main-sequence stars at temperatures of about 10^7 K.

A "view" into the interior of stars is in general possible only in the case of the *Sun*: the *neutrinos* which are generated in the solar interior by the fusion of hydrogen reach the outside with practically no interactions. Their detection was achieved by R. Davis and coworkers in an experiment which has been running since the mid-1960's; the measured flux still differs from that predicted by theory. Besides the Sun, the only other cosmic source of neutrinos which could thus far be identified has been the supernova SN 1987 A, relatively near to the Earth in the Large Magellanic Cloud. Information about the interior of the Sun is also obtained from "*helioseismology*", in which the various types of *solar oscillations* (R.B. Leighton, 1961; F.L. Deubner, 1975; etc.) are analyzed on the basis of the periodic, large-area components in the photospheric flow patterns.

The observations and theoretical investigations relating to stellar evolution and the associated creation of the chemical elements will be treated in Chap. 5. Here, however, we shall consider the compact *final stages* of stellar evolution following the exhaustion of all energy sources: R.H. Fowler recognized in 1926 that *white dwarfs*, for example the star accompanying Sirius, have a structure which is completely different from that of normal stars; their high densities are due to the Fermi degeneracy of their electrons. A star of, e.g., 0.5 solar masses may shrink down to the size of the Earth. Even its stored thermal energy suffices to supply its weak radiation losses over billions of years. According to S. Chandrasekhar (1931), white dwarfs are stable only below a mass limit of about 1.4 solar masses.

Near the end of the evolution of massive stars, the stellar matter can become even more dense (to about 10^{17} kg m^{-3}), causing the protons and electrons to combine into neutrons. These *neutron stars*, which were predicted already in 1932 by L. Landau, shortly after the discovery of the neutron, were first treated theoretically in the framework of general relativity theory by J.R. Oppenheimer and G.M. Volkoff in 1939. Only in 1967 were they discovered experimentally by A. Hewish using their radio emissions. W. Baade and F. Zwicky had realized in 1934 that the collapse of a star into a neutron star releases sufficient gravitational energy to explain the violent outburst of a *supernova*. Still stronger compression finally leads, according to general relativity, to the so-called *black holes*, whose enormous gravitational fields can prevent the escape even of light quanta.

The discoveries of *X-ray binary stars*, of X-ray and gamma-ray bursters, etc., with their intense, temporally variable emissions, have provided an additional opportunity since the 1970's to observe neutron stars and, possibly, black holes. In a *close* binary star system, the components exchange matter in the course of their evolution. If a compact object, e.g. a neutron star, accretes the matter which is sucked out of its companion star, the gravitational energy which is released can be converted into X- and gamma-radiation. Unusual *high-energy* phenomena are exhibited by the X-ray binary system SS 433, with a pair of strongly focussed "jets" that precess about the rotation axis and have relativistic emission energies, and by Cyg X-3 with its emission of extremely energetic gamma quanta (of roughly 10^{15} eV).

A highly interesting possibility for testing A. Einstein's theory of general relativity is offered by the "binary star pulsar" PSR 1913+16, discovered in 1974. The lengthening of its period corresponds to the predicted energy loss rate through emission of *gravitational waves* as a result of the movements of mass within the system.

Summarizing briefly, we can characterize the development of the *physics of stars* (not including stellar evolution, which will be treated later) as follows: (1) theory of radiation: interactions between radiation and matter; (2) thermodynamics and hydrodynamics of flow processes; (3) magnetohydrodynamics and plasma physics; in parallel (4) nuclear physics and energy generation; matter at high densities; finally, in more recent times: (5) high-energy astrophysics: cosmic corpuscular radiation, X- and gamma-rays, neutrinos, and gravitational waves.

4.2 The Theory of Radiation

In order to understand the radiation fields in the atmospheres and the interiors of stars, and also in planetary atmospheres, in gaseous nebulas, and in interstellar space, we now take up the *fundamentals of the theory of radiation.* In this section, we shall emphasize applications of the theory to radiation whose analysis yields information on the structure and composition of stellar atmospheres.

We begin in Sect. 4.2.1 with the phenomenological introduction of intensity and radiation flux, and later, in Sect. 4.2.4, we treat the radiation density. In Sect. 4.2.2, after an (again phenomenological) description of absorption and emission, we introduce the important radiation transport equation. In Sect. 4.2.3, we then make the acquaintance of the special case of thermodynamic equilibrium and the radiation properties of a cavity radiator or blackbody radiator.

4.2.1 Intensity and Radiation Flux

In the radiation field, about which we initially make no special assumptions, we place a surface element $d\sigma$ having a surface normal $\boldsymbol{n}$ (Fig. 4.2.1); we then consider the radiation energy which passes through $d\sigma$ per unit time at an angle θ to $\boldsymbol{n}$ within a small range of solid angle $d\omega$ (characterized by the directional angles θ and ϕ). By spectral decomposition, we separate out the frequency range ν to $\nu+d\nu$ and write:

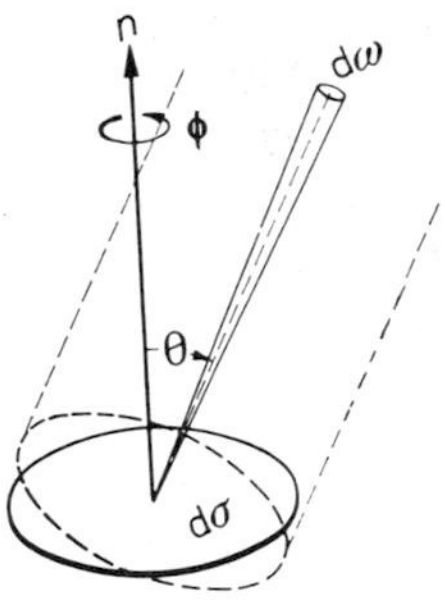

◄ **Fig. 4.2.1.** The definition of radiation intensity

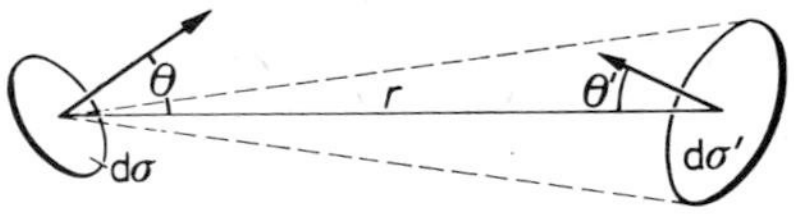

Fig. 4.2.2. Radiation emitted from one surface element $d\sigma$ and observed at a second, $d\sigma'$

$$dE = I_\nu(\theta,\phi)\,d\nu\cos\theta\,d\sigma\,d\omega \ , \tag{4.2.1}$$

where $d\omega = \sin\theta\ d\theta\ d\phi$, and $d\sigma\ \cos\theta$ is the cross-sectional area of the radiation beam.

The *radiation intensity* $I_\nu\ (\theta,\phi)$ is then defined as the amount of energy which passes through a surface normal to the direction (θ,ϕ) per unit solid angle (1 sr = 1 steradian) and unit frequency range (1 Hz) in one second. We can relate the spectral decomposition to the wavelength range instead of the frequency range; from $\nu = c/\lambda$, we have $d\nu = -(c/\lambda^2)d\lambda$. Then from $I_\nu d\nu = -I_\lambda d\lambda$, it follows that

$$I_\lambda = \frac{c}{\lambda^2} I_\nu \ , \tag{4.2.2}$$

or, written symmetrically, $\nu I_\nu = \lambda I_\lambda$.

The intensity of the *total radiation* is obtained by integrating over all frequencies or wavelengths:

$$I = \int_0^\infty I_\nu d\nu = \int_0^\infty I_\lambda d\lambda \ . \tag{4.2.3}$$

As a simple application, we calculate the energy dE radiated per unit time from one surface element $d\sigma$ into a second surface element $d\sigma'$ at a distance r from the first. The surface normals of $d\sigma$ and $d\sigma'$ are supposed to make the angles θ and θ' with the line r connecting them (Fig. 4.2.2). $d\sigma'$ as seen from $d\sigma$ subtends a solid angle $d\omega = \cos\theta' d\sigma'/r^2$; we thus have

$$dE = I_\nu d\nu \cos\theta \, d\sigma \, d\omega$$

or

$$dE = I_\nu d\nu \frac{\cos\theta \, d\sigma \cos\theta' d\sigma'}{r^2} \ . \qquad (4.2.4)$$

On the other hand, $d\sigma$ as seen from $d\sigma'$ subtends a solid angle $d\omega' = \cos\theta \, d\sigma/r^2$. We can thus also write, in a manner symmetrical to the first equation in (4.2.4),

$$dE = I_\nu d\nu \cos\theta' d\sigma' d\omega' \ . \qquad (4.2.5)$$

Thus, the *intensity*, for example, of solar radiation, as defined by Eq. (4.2.1), is *the same* in the immediate vicinity of the Sun and somewhere far out in space.

What is meant in the somewhat fuzzy terms of everyday language by "strength", e.g. of sunshine, is more closely related to the exact concept of radiation flux density or, for short, the *radiation flux*. We define the radiation flux F_ν[1] in the direction $\boldsymbol{n}$ by writing the total energy of radiation with frequency ν which passes through our surface element $d\sigma$ per unit of time (Fig. 4.2.1) as:

$$F_\nu d\sigma = \int_{\theta=0}^{\pi} \int_{\phi=0}^{2\pi} I_\nu(\theta,\phi) \cos\theta \, d\sigma \sin\theta \, d\theta d\phi \ . \qquad (4.2.6)$$

In an *isotropic* radiation field (I_ν independent of θ and ϕ), F_ν is zero. It is often expedient to decompose F_ν into the

Emitted Radiation $(0 \leq \theta \leq \pi/2)$:

$$F_\nu^+ = \int_0^{\pi/2} \int_0^{2\pi} I_\nu \cos\theta \sin\theta \, d\theta d\phi \qquad (4.2.7)$$

and the

Incident Radiation $(\pi/2 \leq \theta \leq \pi)$:

$$F_\nu^- = \int_\pi^{\pi/2} \int_0^{2\pi} I_\nu \cos\theta \sin\theta \, d\theta d\phi \ ,$$

so that

$$F_\nu = F_\nu^+ - F_\nu^- \ .$$

In analogy to (4.2.3), the total radiation flux is defined as:

$$F = \int_0^\infty F_\nu d\nu = \int_0^\infty F_\lambda d\lambda \ . \qquad (4.2.8)$$

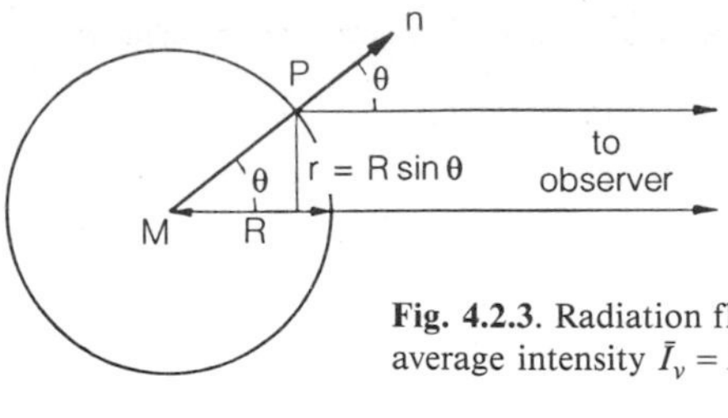

Fig. 4.2.3. Radiation flux F_ν from a star. The average intensity $\bar{I}_\nu = F_\nu/\pi$

We now consider the *radiation emitted by a (spherical) star* (Fig. 4.2.3)! The intensity I_ν of the radiation emitted by the star's atmosphere will depend only on the exit angle θ (calculated with respect to the normal at the particular position), i.e., $I_\nu = I_\nu(\theta)$. The same angle θ occurs once more (Fig. 4.2.3) as the angle between the line connecting the observer with the midpoint M of the star, and the line connecting M with the observed point P. The distance of the point P from M, projected onto a plane perpendicular to the direction of observation, in units of the radius of the star, is thus equal to $\sin\theta$.

The *mean intensity* $\bar{I}_\nu$ of the radiation emitted from the apparent disk of the star (or the Sun) towards the observer is given according to (4.2.1) by

$$\pi R^2 \bar{I}_\nu = \int_0^{\pi/2} \int_0^{2\pi} I_\nu(\theta) \cos\theta R^2 \sin\theta \, d\theta d\phi \ . \qquad (4.2.9)$$

Dividing both sides by R^2 and comparing them with (4.2.7), we see that

$$\bar{I}_\nu = \frac{F_\nu^+}{\pi} \ , \qquad (4.2.10)$$

i.e., the mean radiation intensity of the stellar disk is equal to $1/\pi$ times the radiation flux at the star's surface, F_ν. If the star (of radius R) is at a large distance r from the observer ($\cos\theta \simeq 1$), he sees its disk with a solid angle $d\omega = \pi R^2/r^2$ and therefore detects, as a result of (4.2.7), a radiation flux of

$$f_\nu = \bar{I}_\nu d\omega = F_\nu R^2/r^2 \ . \qquad (4.2.11)$$

The theory of *stellar* spectra must therefore aim at the calculation of the *radiation flux*. The great advantage of *solar* observations, on the other hand, is due to the fact

[1] Frequently, the radiation flux is also denoted as πF_ν. This so-called astrophysical flux F_ν is thus a factor π smaller than F_ν.

Table 4.2.1. Basic terms and units of measurement of radiation

	Radiation quantity		SI units	cgs units
Extended sources	Intensity or surface brightness	I_ν [a]	$\mathrm{W\,m^{-2}\,Hz^{-1}\,sr^{-1}}$	$\mathrm{erg\,s^{-1}\,cm^{-2}\,Hz^{-1}\,sr^{-1}}$
		I	$\mathrm{W\,m^{-2}\,sr^{-1}}$	$\mathrm{erg\,s^{-1}\,cm^{-2}}$
Unresolved (point) sources	Radiation flux density or radiation flux	F_ν, f_ν [a]	$\mathrm{W\,m^{-2}\,Hz^{-1}}$ [b]	$\mathrm{erg\,s^{-1}\,cm^{-2}\,Hz^{-1}}$
		F, f	$\mathrm{W\,m^{-2}}$	$\mathrm{erg\,s^{-1}\,cm^{-2}}$
			1 SI unit $= 10^3$ cgs units	

[a] In the λ scale the corresponding quantities of the radiation field, I_λ, F_λ, f_λ, are usually quoted per μm, nm or Å, according to the particular application. Thus, for example, I_λ is given in $[\mathrm{W\,m^{-2}\,\mu m^{-1}\,sr^{-1}}]$, $[\mathrm{W\,m^{-2}\,nm^{-1}\,sr^{-1}}]$, $[\mathrm{erg\,s^{-1}\,cm^{-2}\,Å^{-1}\,sr^{-1}}]$, etc.
[b] In many cases, particularly in radiastronomy, the jansky ($1\,\mathrm{Jy} = 10^{-26}\,\mathrm{W\,m^{-2}\,Hz^{-1}}$) is used as the unit of flux.

that, as first recognized by K. Schwarzschild, we can there measure I_ν directly as a function of θ.

In general, the radiation flux is the characteristic measurable quantity for the observation of unresolved sources (point sources), and the intensity or surface brightness is that for the observation of area sources. Table 4.2.1 summarizes the units which are currently in common use.

4.2.2 Emission and Absorption. The Radiation Transport Equation

We shall now describe the emission and absorption of radiation, at first also phenomenologically: let a volume element dV emit per unit time into the solid angle $d\omega$ and in the frequency interval $\nu \ldots \nu + d\nu$ the energy

$$j_\nu d\nu dV d\omega \ . \tag{4.2.12}$$

The *emission coefficient* j_ν, in general, depends on the frequency ν as well as on the nature and state of the emitting material (chemical composition, temperature, and pressure); it may also depend on the emission direction. The overall energy output of an isotropically emitting volume element dV per unit time is

$$dV 4\pi \int_0^\infty j_\nu d\nu \ . \tag{4.2.13}$$

We compare this emission to the energy loss by *absorption* which a beam of radiation of intensity I_ν experiences in passing through a layer of matter of thickness ds. This is gven by

$$\frac{dI_\nu}{ds} = -\kappa_\nu I_\nu \ . \tag{4.2.14}$$

The coefficient κ_ν (again dependent on ν, the nature and state of the material, possibly also on the direction) is termed the *absorption coefficient*. In (4.2.14), we related it to a layer of unit thickness, i.e., to a unit volume. Instead, we could consider a layer containing a unit mass of material above a unit surface area; this leads to the mass absorption coefficient $\kappa_{\nu,\mathrm{M}}$. If ϱ is the density of the material, we have $\kappa_\nu = \kappa_{\nu,\mathrm{M}}\varrho$. The absorption coefficient per atom is denoted as the *atomic absorption coefficient* or *absorption cross section* $\kappa_{\nu,\mathrm{at}}$. With a particle density n, we find $\kappa_\nu = \kappa_{\nu,\mathrm{at}} n$. The units are given by

Absorption coefficient $\kappa_\nu\ [\mathrm{m^{-1} = m^2 m^{-3}}]$

Mass absorption coefficient $\kappa_{\nu,\mathrm{M}} = \dfrac{\kappa_\nu}{\varrho}\ [\mathrm{m^2\,kg^{-1}}]$

Atomic absorption coefficient $\kappa_{\nu,\mathrm{at}} = \dfrac{\kappa_\nu}{n}\ [\mathrm{m^2}]$. (4.2.15)

Combining the changes in the intensity of our radiation beam along the path ds due to absorption (4.2.14) and to emission (4.2.12), we obtain the *transport equation for radiation* of frequency ν

$$\frac{dI_\nu}{ds} = -\kappa_\nu I_\nu + j_\nu \ . \tag{4.2.16}$$

If our beam passes through a nonemitting layer of thickness s, we have

$$\frac{dI_\nu}{I_\nu} = -\kappa_\nu ds \tag{4.2.17}$$

and the intensity I_ν of the radiation after passing

through this absorbing layer is related to the intensity $I_{\nu,0}$ of incident radiation by

$$\frac{I_\nu}{I_{\nu,0}} = e^{-\tau_\nu} , \tag{4.2.18}$$

where the dimensionless quantity τ_ν is defined as

$$\tau_\nu = \int_0^s \kappa_\nu ds = \int_0^s \kappa_{\nu,M} \varrho \, ds \tag{4.2.19}$$

and is termed the *optical thickness* of the transmitting layer. For example, a layer of optical thickness $\tau_\nu = 1$ attenuates a beam to $e^{-1} = 0.368$ of its original intensity. In the case that the layer neither emits nor absorbs, i.e., for propagation of radiation in vacuum, we again find that the intensity along the beam stays constant, cf. (4.2.4 and 5).

Using the optical thickness $d\tau_\nu = \kappa_\nu ds$, we can rewrite the radiation transport equation (4.2.16) in the form

$$\frac{dI_\nu}{d\tau_\nu} = -I_\nu + \frac{j_\nu}{\kappa_\nu} = -I_\nu + S_\nu , \tag{4.2.20}$$

where the *source function* has been introduced as the ratio of emission to absorption coefficients:

$$S_\nu = \frac{j_\nu}{\kappa_\nu} . \tag{4.2.21}$$

Our phenomenological approach for the absorption coefficient in the radiation transport equation includes *all* physical processes which decrease the intensity I_ν in the observed direction of radiation. Along with the actual absorption given by (4.2.14), in general we must take into account the *scattering* from particles in the layer (Sect. 4.8.2); it causes radiation to be deflected out of the incident direction with a cross section σ_ν. We then speak of *extinction* with an extinction coefficient $k_\nu = \kappa_\nu + \sigma_\nu$, and the corresponding optical thickness $d\tau_\nu = k_\nu ds$. In a similar manner, the emission coeffcient j_ν in (4.2.16) may also contain a contribution to the intensity from scattering into the observed direction from the layer *ds*.

As an example, we give here the extinction of the radiation of frequency ν by the *Earth's atmosphere* for a star with the zenith distance z (Fig. 4.2.4). It corresponds to an attenuation factor of

$$\frac{I_\nu}{I_{\nu,0}} = e^{-\tau_\nu \sec z} , \tag{4.2.22}$$

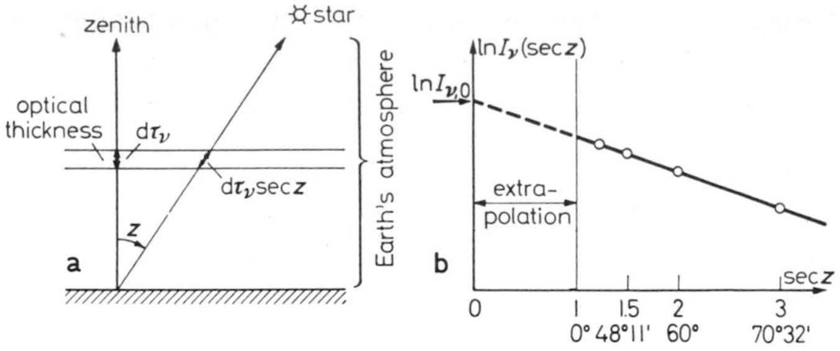

Fig. 4.2.4. **(a)** The atmospheric extinction of the radiation at frequency ν from a star with zenith distance z. **(b)** Extrapolation to $\sec z \to 0$ yields the extraterrestrial radiation intensity $I_{\nu,0}$

where $\sec z = 1/\cos z$ and τ_ν is the optical thickness of the Earth's atmosphere (as measured in a direction perpendicular to the ground) at the frequency ν. The *extraterrestrial intensity* $I_{\nu,0}$ can, by the way, be obtained by linear extrapolation to $\sec z \to 0$ of the values of $\ln I_\nu$ measured at various zenith distances z, as long as τ_ν is not too large.

4.2.3 Thermodynamic Equilibrium and Blackbody Radiation

We meet a particularly clear situation when we consider a radiation field which is in *thermal equilibrium* or *temperature equilibrium* with its surroundings. Such a radiation field can be generated by immersing an (otherwise arbitrary) cavity in a heat bath at the temperature T. In such a bath, all objects have by definition the same temperature; it thus seems justified to speak of *cavity or blackbody radiation at a temperature* T.[2] In an isothermal cavity, every body or every surface element will emit and absorb the same amount of radiation energy per unit time. Starting from this fact, we can now easily show that the intensity I_ν of the cavity radiation is *independent* of the material content and the form of the walls of the cavity and of the direction (isotropy). We need only consider a cavity with two different chambers H_1 and H_2 (Fig. 4.2.5). If $I_{\nu,1} \neq I_{\nu,2}$, we could obtain energy by putting a radiometer (the well-known "light mill" sometimes seen in shop window displays) in the opening between the two cavities; we would then have a perpetual motion machine of the 2nd kind! It is readily shown that $I_{\nu,1} = I_{\nu,2}$ must hold for *all* frequencies, radiation directions, and

[2] We have here assumed that the index of refraction in the cavity is 1. It would be simple to generalize this assumption.

polarization directions by inserting into the opening between H_1 and H_2 sequentially a colored filter, a tube oriented in a particular direction, or a polarizing filter; in each case, as a result of the 2nd Law of Thermodynamics, one cannot construct a perpetual motion machine of the 2nd kind. The intensity I_ν of the cavity radiation is thus a *universal function* $B_\nu(T)$ *of* ν *and* T.

A small hole in the cavity wall takes up all the incident radiation by multiple reflections and absorption in its interior, and thus represents a so-called *black body*. Cavity radiation is therefore also termed *blackbody radiation* or, for short, simply black radiation.

It can be produced and measured by "sampling" a cavity at the temperature T through a sufficiently small opening. The important function $B_\nu(T)$, whose existence was recognized in 1859 by G. Kirchhoff and which was first explicitly calculated in 1900 by M. Planck, we shall call the *Kirchhoff-Planck function*, $B_\nu(T)$. Planck recognized that he could find a function which would agree with measurements of the intensity of cavity radiation only by extending the laws of classical mechanics and thermodynamics to include the well-known requirements of *quantum mechanics*. He thus found the *Planck radiation formula*, which we give in terms of both frequency ν and of wavelength λ:

$$B_\nu(T) = \frac{2h\nu^3}{c^2}\frac{1}{e^{h\nu/kT}-1}$$

or (4.2.23)

$$B_\lambda(T) = \frac{2hc^2}{\lambda^5}\frac{1}{e^{hc/k\lambda T}-1}$$

with the two important limiting cases

$$\frac{h\nu}{kT} \gg 1: \quad B_\nu(T) \simeq \frac{2h\nu^3}{c^2}e^{-h\nu/kT} \quad \text{(Wien's law)} \tag{4.2.24a}$$

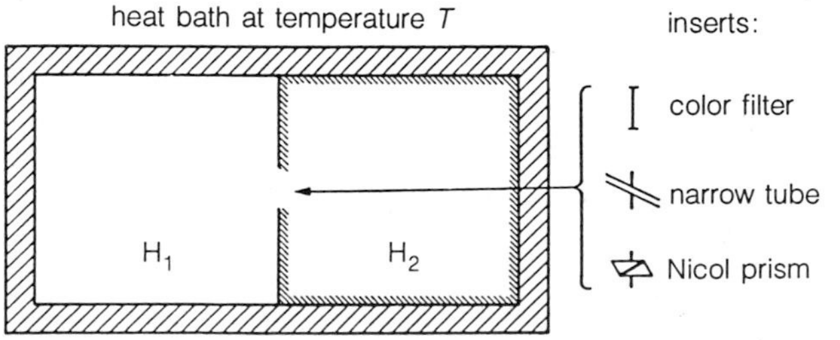

Fig. 4.2.5. Black-body radiation is unpolarized and isotropic. Its intensity is a universal function of ν and T, the Kirchhoff-Planck function $B_\nu(T)$

and

$$\frac{h\nu}{kT} \ll 1: \quad B_\nu(T) \simeq \frac{2\nu^2 kT}{c^2} \quad \text{(Rayleigh-Jeans law)} . \tag{4.2.24b}$$

In the Rayleigh-Jeans radiation law, the quantity h characteristic of quantum mechanics has vanished. For light quanta $h\nu$ whose energy is much smaller than the thermal energy kT, quantum mechanics (quite generally) is transformed into classical mechanics (Bohr's correspondence principle). The radiation constant which occurs in the exponent of the λ-representation (4.2.23) is called c_2; it is given by

$$c_2 = \frac{hc}{k} = 1.439 \cdot 10^{-2}\,\mathrm{m \cdot K} . \tag{4.2.25}$$

The maximum in the Kirchhoff-Planck function $B_\nu(T)$ or $B_\lambda(T)$ is given by *Wien's displacement rule*:

$$\frac{c}{\nu_{\max}} \cdot T = 5.10 \cdot 10^{-3}\,\mathrm{m \cdot K}$$

or (4.2.26)

$$\lambda_{\max} \cdot T = 2.90 \cdot 10^{-3}\,\mathrm{m \cdot K} .$$

The intensity of the *total radiation* of the black body is obtained by integrating (4.2.23) over frequencies, $B(T) = \int_0^\infty B_\nu(T)d\nu$. The *total radiation flux*, i.e., the radiation emitted by a black unit surface into the cavity, is $F^+ = \pi B(T)$. If one carries out the integration over $B_\nu(T)$, the result is the *Stefan-Boltzmann Radiation Law* found experimentally in 1879 by J. Stefan and derived theoretically in 1884 by L. Boltzmann using a calculation of the entropy of cavity radiation:

$$F^+ = \pi B(T) = \sigma T^4 \tag{4.2.27}$$

with the radiation constant

$$\sigma = \frac{2\pi^5 k^4}{15c^2h^3} = 5.67 \cdot 10^{-8}\,\mathrm{W \cdot m^{-2} K^{-4}}$$
$$= 5.67 \cdot 10^{-5}\,\mathrm{erg \cdot s^{-1}\,cm^{-2} K^{-4}} . \tag{4.2.28}$$

In *thermodynamic equilibrum*, i.e., in a cavity at temperature T, the emission and absorption of an arbitrary volume element must be equal. We write them for a volume element of base area $d\sigma$ and height ds in the frequency domain ν within a solid angle element $d\omega$ perpendicular to $d\sigma$. From (4.2.12 and 14), we have for the

$$\text{Emission } s^{-1} = j_\nu d\nu\, d\sigma\, ds\, d\omega \ ,$$

and the

$$\text{Absorption } s^{-1} = \kappa_\nu ds B_\nu(T) d\sigma\, d\omega\, d\nu \ . \tag{4.2.29}$$

Setting the two quantities equal, we obtain *Kirchhoff's Law*:

$$j_\nu = \kappa_\nu B_\nu(T) \ . \tag{4.2.30}$$

This important result means that in a state of thermodynamic equilibrium, the ratio of the emission coefficient j_ν to the absorption coefficient κ_ν is the same universal function $B_\nu(T)$ of ν and T which also describes the intensity of the cavity radiation. For the source function in the radiation transport equation (4.2.20), we then find $S_\nu = j_\nu/\kappa_\nu = B_\nu(T)$.

We shall calculate as an important application the radiation intensity I_ν which is emitted by a *layer of matter* (e.g., the plasma in a gas discharge) at the constant temperature T with a thickness s in the direction of observation of the radiation (Fig 4.2.6):

From a volume element with a unit cross-sectional area and a length in the direction of observation from x to $x+dx$ $(0 \leq x \leq s)$, a contribution to the radiation intensity equal to $\kappa_\nu B_\nu(T)dx = B_\nu(T)dt_\nu$ is emitted per second and sr, where $dt_\nu = \kappa_\nu dx$ is the optical thickness. Before leaving the layer, this contribution is attenuated by the factor $\exp(-t_\nu)$, with $t_\nu(x) = \int_0^x \kappa_\nu dx$. Thus for the whole layer of thickness s, whose optical thickness is $\tau_\nu = t_\nu(s)$, we obtain

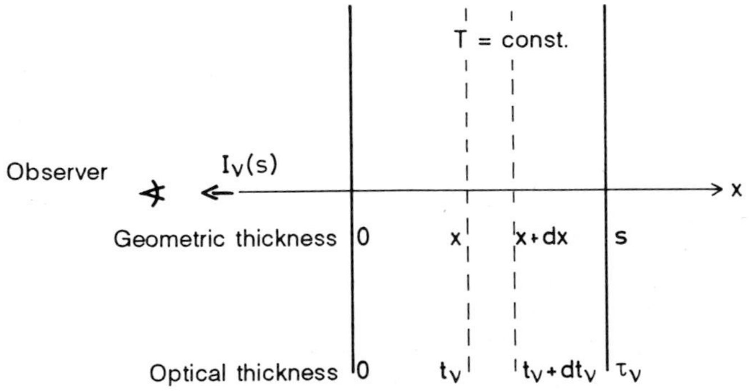

Fig. 4.2.6. Emission from a layer of thickness s at a constant temperature

$$I_\nu(s) = \int_0^{\tau_\nu} B_\nu(T) e^{-t_\nu} dt_\nu \ , \tag{4.2.31}$$

or, since T is supposed to be constant within the layer,

$$I_\nu(s) = B_\nu(T)(1-e^{-\tau_\nu}) \simeq \begin{cases} \tau_\nu B_\nu(T) & \text{for} \quad \tau_\nu \ll 1 \\ B_\nu(t) & \text{for} \quad \tau_\nu \gg 1 \ . \end{cases} \tag{4.2.32}$$

That is, the radiation emitted by an isothermal *optically thin* layer ($\tau_\nu \ll 1$) is equal to its optical thickness times the Kirchhoff-Planck function. The radiation intensity from an *optically thick* layer, in contrast, approaches that of a black body but cannot become greater than it. Applied to a spectral line, this leads to the well-known phenomenon of *self absorption*: although two broad spectral lines from an optically thin layer have an intensity ratio $\simeq \kappa_1 : \kappa_2$, their ratio approaches unity from an optically thick layer. (This subject will be taken up again in Sect. 4.9.)

If the matter in the layer is not in thermodynamic equilibrium, then Kirchhoff's Law cannot be applied to the radiation transport equation, but the solution (4.2.32) remains valid if we replace the Kirchhoff-Planck function by the source function $S_\nu = j_\nu/\kappa_\nu \neq B_\nu(T)$, which is constant within the layer. To be sure, its calculation based on the individual processes of interaction of the radiation with the material in the plasma requires a considerably greater effort (Sect. 4.7.3).

4.2.4 The Radiation Density

As the final fundamental concept of the theory of radiation, we introduce the *radiation density* u_ν; it is the monochromatic energy density of the radiation field in the frequency range ν to $\nu+d\nu$. The relationship of u_ν to the radiation intensity I_ν is found from the following considerations: according to (4.2.1), the radiation energy $d\sigma \int I_\nu d\omega$ passes per unit time through the basal area $d\sigma$ of a cylindrical volume element of height dl, and it spends the time dl/c in the volume element (distance traveled divided by the velocity of light, c). The energy density is therefore

$$u_\nu = \frac{1}{c} \int I_\nu d\omega = \frac{4\pi}{c} J_\nu \ , \tag{4.2.33}$$

in which we denote by

$$J_\nu = \frac{1}{4\pi} \int I_\nu d\omega \qquad (4.2.34)$$

the *mean intensity*, averaged over all solid angles. The radiation density u_ν corresponds to a *photon density*

$$N_\nu = \frac{4\pi}{c} \frac{J_\nu}{h\nu} , \qquad (4.2.35)$$

since the energy of each photon is equal to $h\nu$.

If we integrate (4.2.34) over the whole spectrum, we obtain the *total radiation density*

$$u = \int_0^\infty u_\nu d\nu = \frac{4\pi}{c} J \qquad (4.2.36)$$

and the corresponding photon density

$$N = \frac{4\pi}{c} \int_0^\infty \frac{J_\nu}{h\nu} d\nu . \qquad (4.2.37)$$

For the isotropic radiation field of a *black body*, the intensity and the mean intensity are equal to the Kirchhoff-Planck function (4.2.23), $I_\nu = J_\nu = B_\nu(T)$, and therefore we have

$$u_\nu = \frac{4\pi}{c} B_\nu(T) . \qquad (4.2.38)$$

The total radiation density of a black radiation field is then given by the *Stefan-Boltzmann Law* (4.2.28):

$$u = \frac{4\pi}{c} B(T) = aT^4 \qquad (4.2.39)$$

with the radiation constant

$$a = \frac{4\sigma}{c} = 7.65 \cdot 10^{-16}\ \mathrm{J \cdot m^{-3}\,K^{-4}} . \qquad (4.2.40)$$

4.3 The Sun: Radiation and Velocity Fields in the Photosphere

We briefly summarize once more the data relevant to the Sun; when dealing with the stars, we shall often use them as intuitively apparent *units of measure*.

Making use of the solar parallax of 8.794″, we first obtained the mean distance from the Earth to the Sun, the *astronomical unit*:

$$1\ \mathrm{AU} = 149.6 \cdot 10^6\ \mathrm{km} = 23\,456\ \text{equatorial Earth radii} . \qquad (4.3.1)$$

The corresponding apparent radius of the solar disk is

$$15'59.63'' = 959.63'' = 0.004652\ \mathrm{rad} . \qquad (4.3.2)$$

We obtain from this the radius of the Sun

$$R_\odot = 696\,000\ \mathrm{km} . \qquad (4.3.3)$$

Therefore, on the Sun

$$1'' = 725\ \mathrm{km} . \qquad (4.3.4)$$

Owing to atmospheric scintillation, the limit of resolving power for most observations made from the Earth is about 500 km.

Conversely, an astronomical unit corresponds to 215 solar radii. The flattening of the Sun as a result of its rotation is at the limit of current detectability ($\lesssim 0.01''$).

From Kepler's 3rd law, we obtain the mass of the Sun:

$$\mathcal{M}_\odot = 1.989 \cdot 10^{30}\ \mathrm{kg} . \qquad (4.3.5)$$

With it, we calculate the mean density of the Sun, $\bar{\varrho}_\odot = 1409\ \mathrm{kg\,m^{-3}}$; we readily obtain the gravitational acceleration at the Sun's surface

$$g_\odot = 274\ \mathrm{m\,s^{-2}} \qquad (4.3.6)$$

or a factor of 27.9 times the acceleration of gravity on the Earth's surface.

The *spectrum* of the Sun appears on observation to consist of a *continuum*, if we initially disregard the far ultraviolet and the radiofrequency regions; it contains many dark *absorption lines*, the *Fraunhofer lines*. Those layers of the solar atmosphere which give rise to the continuous radiation and the major portion of the Fraunhofer lines are termed the *photosphere*.

In Sect. 4.3.1, we give an overview of the photospheric spectrum and its intensity curve from the center of the Sun's disk to the edge. Then, in Sect. 4.3.2, we discuss observations of the absolute energy distribution in the solar spectrum and derive the luminosity and the effective temperature of the Sun. We delay the analysis of the Fraunhofer spectrum until Sect. 4.9. However, in

Sect. 4.3.3, we do treat the "grainy" structure of the Sun's surface, the granulation, as well as the currents and oscillations in the photosphere ("solar seismology").

During total solar eclipses, the higher layers of the atmosphere can be observed by themselves at the edge of the solar disk; they do not emit a continuum, but rather practically only the emission lines corresponding to the Fraunhofer spectrum. This layer is called the *chromosphere*; it yields only a small contribution to the intensity of the absorption lines. Above it comes the *corona*, which stretches outwards into the interplanetary medium. We shall deal with these outermost layers of the Sun, and likewise with all the various phenomena of *solar activity* (sunspots, prominences, eruptions, etc.) or, as it is also called, with the *disturbed Sun*, in Sect. 4.10.

4.3.1 The Spectrum of the Photosphere. Center-Limb Variation

Using large grating spectrometers, for example in a tower telescope (Fig. 4.3.1), the solar spectrum can be recorded with extremely high resolution, so that the Fraunhofer lines can be investigated in great detail (Fig. 4.3.2).

The wavelength λ, the identifications, and the intensities of the *Fraunhofer lines* are contained in the tables of Rowland (1895/97) or in their (second) revised edition by C.E. Moore, M.G.J. Minnaert, and J. Houtgast (1966) for the range from 293.5 to 877 nm, as well as in newer compilations, which also include to some extent the neighboring spectral regions. The Utrecht Photometric Atlas of the Solar Spectrum, by M.G.J. Minnaert, G.F.W. Mulders, and J. Houtgast (1940) contains data on the *intensity distribution* in the solar spectrum as determined with the microphotometer; there are also several newer atlases, with in part greater dispersion or resolution and also applied to other spectral regions. The Utrecht measurements and many others refer to the *center* of the solar disk. Towards the limb, the solar disk exhibits a considerable *center-limb darkening*; the spectral lines also show a center-limb variation (relative to the continuum at the same position).

We specify the position of observation on the solar disk (Fig. 4.2.3) by giving its distance from the center in units of the solar radius, $r/R_\odot$, or, for theoretical purposes, by giving the angle of emission θ relative to the normal to the solar surface. We thus have:

$$\frac{r}{R_\odot} = \sin\theta \quad \text{and} \quad \mu = \cos\theta = \sqrt{1-(r/R_\odot)^2} \; . \qquad (4.3.7)$$

The center of the Sun corresponds to $\sin\theta = 0$, $\cos\theta = 1$ and the limb of the Sun corresponds to $\sin\theta = 1$, $\cos\theta = 0$. The *radiation intensity* at the surface of the Sun (optical thickness $\tau_0 = 0$, Sect. 4.8.1) at a distance $r/R_\odot = \sin\theta$ from the center of the disk is denoted by $I_\lambda(0,\theta)$ or $I_\nu(0,\theta)$ referred to a wavelength or a frequency scale, cf. (4.2.2). The *center-limb variation* of the radiation intensity, $I_\lambda(0,\theta)/I_\lambda(0,0)$ is measured by setting the entrance slit of the spectrograph at various positions on the solar image. The main difficulty in this and all other measurements of details on the solar disk lies in the elimination of scattered light from the instrument and the Earth's atmosphere. A summary representation of measurements of the center-limb variation at different wavelengths is given in Fig. 4.3.3. The ratio of the mean intensity of the solar disk, $\bar{I}_\lambda = F_\lambda/\pi$ (4.2.10), to the intensity at the center of the disk, $I_\lambda(0,0)$, can be readily calculated using (4.2.7).

4.3.2 The Absolute Energy Distribution, Luminosity, and Effective Temperature of the Sun

Absolute values for the radiation intensity, for example at the center of the Sun's disk, $I_\lambda(0,0)$, are obtained by reference to a cavity radiator (black body) at a known temperature. The highest fixed point on the temperature scale which has been precisely determined by gas thermometry is the melting point or *solidification point of gold*, T_{Au}. The temperature scale above T_{Au} is based on this fixed point and on pyrometer measurements using Planck's radiation formula or the radiation constant c_2, see (4.2.23,26). The International Practical Temperature Scale of 1968 uses the value $T_{Au} = 1337.58$ K or 1064.43 °C and the radiation constant $c_2 = 1.4388 \cdot 10^{-2}$ m·K. (The zero point of the Celsius scale is at 273.15 K.)

The extinction of the Earth's atmosphere must, of course, be precisely determined at a number of wavelengths (Fig. 4.2.4) and corrected for.

Since the solar spectrum (Figs. 4.3.2, 4) contains many Fraunhofer lines, the best procedure is to measure the intensity $I^m_\lambda(0,0)$ of the spectrum *including* the lines, averaged within sharply-defined wavelength regions of e.g. $\Delta\lambda = 2$ nm (superscript m). The integration over the solar disk using the center-limb variation measured in the same wavelength regions then gives directly the corresponding mean values of the radiation flux, $F^m_\lambda(0)$. Using absolute measurements of this kind, high-resolution spectra of both the center of the disk and of the radiation integrated over the disk can be calibrated.

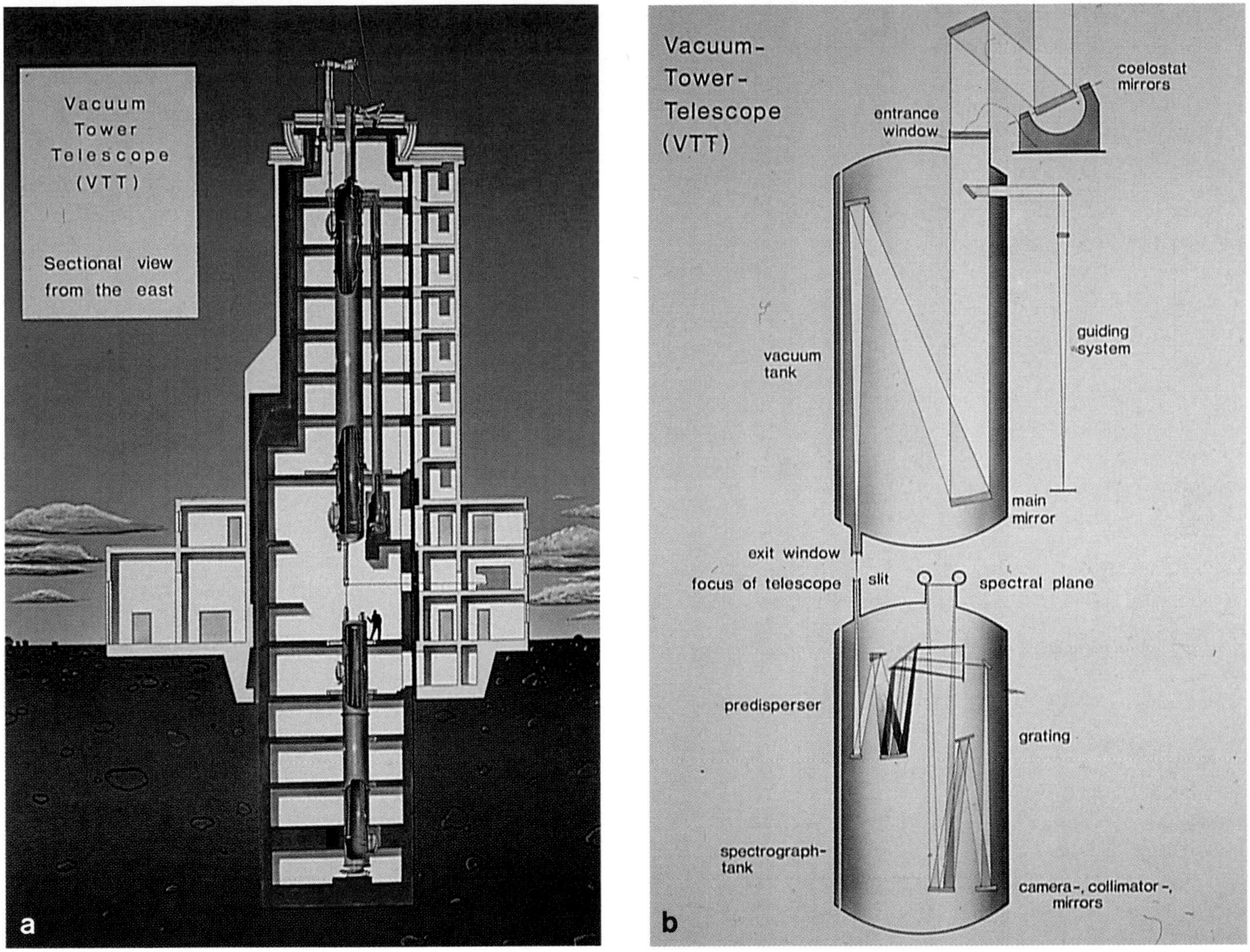

Fig. 4.3.1 a, b. The 60 cm vacuum tower telescope of the German Solar Observatory at the Observatorio del Teide of the Instituto de Astrofisica de Canarias on Tenerife: a cutaway view and a schematic drawing of the optical path. The two computer-controlled coelostat mirrors direct the sunlight vertically downwards through a tube of 2.5 m diameter, which is evacuated to better than 10^{-3} atmospheres to reduce heating; at the base of the tube is the main mirror with a focal length of 46 m. After emerging from the vacuum tube, the light may either be redirected horizontally into laboratories, or vertically into a high-resolution échelle spectrograph, which is located in a chamber 16 m deep below the ground. The building, 38 m high, consists of two towers, one within the other; the inner tower houses the instruments and the outer one protects against vibrations, particularly those due to wind. (From E. H. Schröter and E. Wiehr, 1985)

The *continuum* between the lines, $I_\lambda^c(0,0)$ or $F_\lambda^c(0)$, is also of interest for the physics of the Sun. Its determination in the long-wavelength region $\lambda > 450$ nm can be carried out without difficulties. In the blue and violet spectral regions, however, the lines are so closely spaced that the determination of the "true" continuum becomes problematic below $\lambda = 450$ nm. In this region, only a "local quasi-continuum" can be defined using those places in the spectrum ("windows") which are the least influenced by the lines. In this part of the spectrum, the "true" continuum can be determined only by reference to a well-developed theory.

The results of the most precise absolute measurements and center-limb observations up to the present, by H. Neckel and D. Labs, as well as the newest high-resolution Fourier spectra by J. Brault, are shown in Fig. 4.3.4.

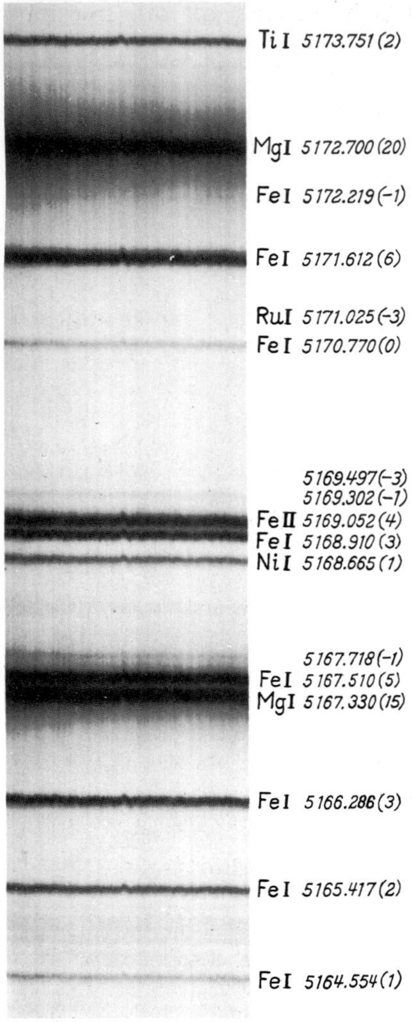

Fig. 4.3.2. A solar spectrum (from the center of the Sun's disk), $\lambda = 516-518$ nm. 5th order of the vacuum grating spectrograph of the McMath-Hulbert Observatory. The wavelength in Å (1 Å = 0.1 nm), the identification, and the (estimated) intensity of the Fraunhofer lines are noted at the right

The fraction η of the radiation absorbed by the lines from the continuum, averaged over well-defined regions $\Delta\lambda$ of about 1 to 10 nm width,

$$\eta_\lambda^{\mathrm{m}} = 1 - I_\lambda^{\mathrm{m}}(0,0)/I_\lambda^{\mathrm{c}}(0,0) \tag{4.3.8}$$

or the corresponding expression for $F_\lambda^{\mathrm{m}}(0)$, is determined from high-resolution spectra by numerical integration. $\eta_\lambda^{\mathrm{m}}$ in the ultraviolet (300 to 400 nm), determined in 1 nm wide regions and relative to the local continuum, lies between 25 and 80%; only above $\lambda > 550$ nm does it drop down to a few percent.

By integration of the radiation flux $F_\lambda(0)$ over all wavelengths, whereby the end regions in the ultraviolet and infrared which are cut off by absorption in the Earth's atmosphere must be corrected using measurements made outside the atmosphere, the *total radiation flux* from the Sun's surface can be obtained:

$$F = \int_0^\infty F_\lambda \, d\lambda = 6.33 \cdot 10^7 \ \mathrm{W\,m^{-2}}$$

$$= 6.33 \cdot 10^{10} \ \mathrm{erg \cdot cm^{-2}\,s^{-1}} \ . \tag{4.3.9}$$

From this value, one readily calculates the total emission of the Sun per unit time, i.e. its *luminosity*:

$$L_\odot = 4\pi R_\odot^2 F$$

$$= 3.85 \cdot 10^{26}\,\mathrm{W} = 3.85 \cdot 10^{33} \ \mathrm{erg\,s^{-1}} \ . \tag{4.3.10}$$

On the other hand, we can obtain the radiation flux S at the distance of the Earth ($r = 1$ AU) according to (4.2.11) by multiplying the mean radiation intensity of the solar disk by its solid angle (as seen from the Earth),

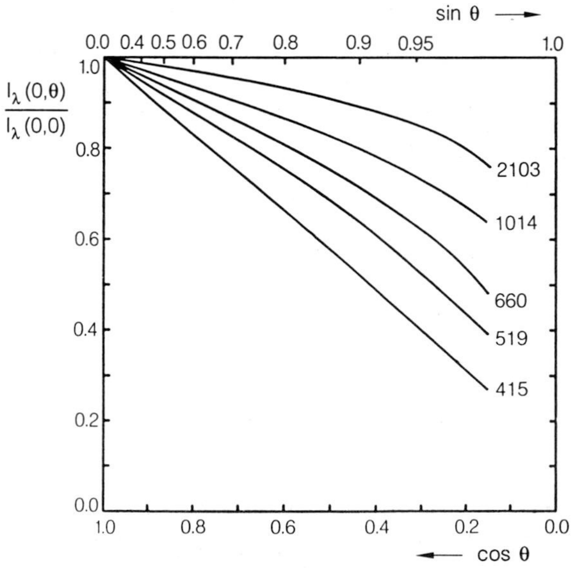

Fig. 4.3.3. The center-limb darkening of the Sun. The ratio of the radiation intensities $I_\lambda(0,\theta)/I_\lambda(0,0)$ is plotted against $\cos\theta$ for several wavelengths λ in [nm] (at which there are no spectral lines) from the blue into the infrared. The non-uniform upper $\sin\theta$ scale gives the distance from the center of the disk in units of the Sun's radius

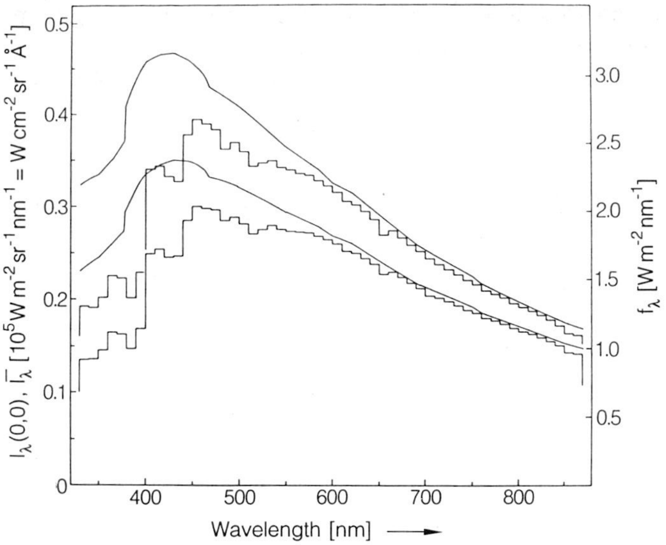

Fig. 4.3.4. The absolute energy distribution in the solar spectrum after H. Neckel and D. Labs (1984). The intensity $I_\lambda(0,0)$ at the center of the disk (*upper curves*) and the intensity $\bar{I}_\lambda = F_\lambda/\pi$ averaged over the disk (*lower curves*). The stepped curves represent intensities averaged over intervals of $\Delta\lambda = 10$ nm, I_λ^m or $\bar{I}_\lambda^m$, including the Fraunhofer lines. The smooth curves show the dependence of the quasi-continuum, I_λ^c or $\bar{I}_\lambda^c$. The right-hand scale shows the radiation flux at the Earth's position ($r = 1$ AU) but outside the atmosphere for the lower curves, $f_\lambda = F_\lambda(R_\odot/r)^2$. Integration of f_λ over all wavelengths yields the solar constant S (4.3.11)

$\pi R_\odot^2/r^2 = 6.800 \cdot 10^{-5}$ sr, or by multiplying the radiation flux F by the attenuation factor $(R_\odot/r)^2$. In this way, we obtain the *solar constant*:

$$S = 1.37 \text{ kW m}^{-2} = 1.37 \cdot 10^6 \text{ erg} \cdot \text{cm}^{-2} \text{ s}^{-1} . \qquad (4.3.11)$$

This important quantity was first determined precisely by K. Ångström (about 1883) and by C. G. Abbot (about 1908), following initial experiments by S. S. Pouillet in 1837. They started by measuring the total solar radiation arriving at the Earth's surface using a "black" receiver, a so-called pyrheliometer. This measurement must then be complemented by relative determinations of the spectrally decomposed radiation, since only then can the atmospheric extinction (Fig. 4.2.4) be eliminated. In recent times, S has been determined from aircraft, rockets, and space vehicles, with smaller and smaller extinction corrections. The combination of *all* these measurements leads to the numerical value quoted in (4.3.11) with an uncertainty of less than ±1%. This solar radiation power is available outside the atmosphere.

Measurements with high relative precision using sensitive radiometers on satellites have shown that the "solar constant" is subject to temporal fluctuations of about 0.1 to 0.3%, which are proportional to the fraction of the surface of the Sun which is covered by sunspots (Sect. 4.10.1).

Next, we shall return once more to consideration of the total radiation flux F and the radiation intensity $I_\lambda(0,0)$ at the surface of the Sun. In order to obtain a rough idea of the temperatures in the solar atmosphere, at first in a rather formal and preliminary way, we interpret the total flux F in terms of the Stefan-Boltzmann radiation law and thus define the *effective temperature* of the Sun:

$$F = \sigma T_{\text{eff}}^4 \; ; \quad T_{\text{eff}} = 5780 \text{ K} . \qquad (4.3.12)$$

Furthermore, we interpret the radiation intensity $I_\lambda(0,0)$ using Planck's radiation law (4.2.23) and thus define the *brightness temperature* T_λ for the center of the solar disk as a function of the wavelength λ. (Analogously, a wavelength-dependent brightness temperature can also be defined by means of the monochromatic radiation flux $F_\lambda(0)$.)

However, the Sun does *not* radiate as a black body (otherwise, for one thing, $I_\lambda(0,\theta)$ would be independent of θ, i.e. the Sun would exhibit no center-limb darkening; for another, the Fraunhofer lines would not occur); thus we cannot interpret T_{eff} and T_λ all too literally. T_{eff} or T_λ will at least be a reasonable approximation to the temperature in *those* layers of the solar atmosphere which give rise to the overall observed radiation or the radiation of wavelength λ, respectively. The effective temperature T_{eff} is furthermore an important characteristic of the solar atmosphere (and, correspondingly, of the atmospheres of stars), since it yields by definition, together with the Stefan-Boltzmann law, the total radiation flux F, i.e. the total energy per unit time which arrives at a unit surface of the Sun from its interior.

The analysis of the solar spectrum, the temperature and density profiles in the photosphere, and the abundances of the chemical elements will be treated in connection with the theory of stellar atmospheres in Sects. 4.8 and 9.

4.3.3 Granulation and Oscillations

If one studies the *solar surface* carefully, it appears (in white light) to have a grainy structure. This *granulation* consists of brighter "granula", whose temperature is 100 to 200 K higher than that of the darker areas in between (Fig. 4.10.2). The granulation pattern changes constantly; photographs taken in series show that the cells have a mean lifetime of about 8 minutes, before they divide or dissolve. The diameters of the granulation cells ranges from the largest of about 3500 km down to the telescopic resolution limit of about 100 km. In photographic solar spectra with good spatial and time resolution, one can distinguish vertical motions of the individual cells with velocities of about 0.5 km s^{-1} from the Doppler shift (2.5.2) of the Fraunhofer lines. Since the entrance slit of the spectrograph admits light from many granulation elements, the absorption lines have a sawtooth-like structure (Fig. 4.3.2).

In addition to the granulation, there is a second network of cells on a larger scale, whose meshes have diameters from 15000 to 40000 km. The flow in this *supergranulation* rises in the center of the cells, with radial velocities of about 0.4 km s^{-1}, and sinks at the edges, with velocities of ≤ 0.2 km s^{-1}. The mean lifetime of the cells (roughly equal to their turnover times) is about 36 h. The supergranulation also reaches into the chromosphere, the layer of the solar atmosphere which lies above the photosphere, and is particularly visible in Ca-II spectroheliograms (Sect. 4.10).

The flow fields of the granulation and supergranulation are closely related to the *magnetic fields* on the Sun. We shall come back to this point in connection with the phenomenon of solar activity in Sect. 4.10.6.

Along with the irregular flow patterns of the granulation, we can see in the Doppler shifts of the Fraunhofer lines nearly periodic, large-area *wave motions* in the photosphere (and the chromosphere). In 1960, R.B. Leighton discovered oscillations in the radial velocities with periods of 5 min or frequencies of $3.3 \cdot 10^{-3}$ Hz and amplitudes from 0.1 to 0.5 km s^{-1}. F.L. Deubner succeeded in resolving these oscillations with respect to frequency and wavelength in 1975, and identified them as a superposition of nonradial *oscillations* of the layers near the surface. Since then, using more refined methods of observation, many further modes of oscillation have been found. The velocity amplitude of the individual modes is very small (≤ 0.2 m s^{-1}); only their stochastic superposition gives amplitudes of up to 500 m s^{-1}. In contrast to pulsating variable stars like the Cepheids (Sect. 4.11.1), on the Sun the higher harmonics with wavelengths $\ll R_\odot$ and very small amplitudes are excited.

An oscillation mode can be interpreted as a standing *acoustic wave* in a particular layer of the Sun, which acts as a resonator. The temperature and density profiles in the Sun determine the propagation conditions of the acoustic waves (velocity, reflection, refraction) and thus define the "walls" of the resonator cavity. Conversely, *helioseismology* can yield information about the *interior* of the Sun, which lies beneath the photosphere and from which no electromagnetic radiation can reach us, by applying a detailed analysis to the complex, variable pattern of oscillation maxima and nodes on the solar *surface* resulting from the superposition of about 10^7 modes of different wavelengths; this is analogous to the analysis of seismic waves on the Earth. A view especially deep into the interior is permitted by radial oscillations of long wavelengths or long periods. These become observable on integration over a large area of the solar surface, whereby oscillations of shorter wavelength are averaged out.

4.4 Magnitudes and Colors, Distances and Radii of the Stars

Having dealt with the Sun, we now turn to the much more distant *stars* and concern ourselves first, in Sect. 4.4.1, with their apparent magnitudes, colors, and spectral energy distributions. In Sect. 4.4.2, we treat the trigonometric measurement of parallaxes and thus the determination of distances to the stars, with which their true or absolute magnitudes can be calculated from the apparent magnitudes. In Sect. 4.4.3, we introduce the bolometric magnitude and the luminosity; finally, we give an overview of stellar radii and the methods for determining them in Sect. 4.4.4.

4.4.1 Apparent Magnitudes, Color Indices, and Energy Distributions

Hipparchus and many of the earlier astronomers had already begun to make catalogues of the brightnesses of stars. *Stellar photometry* in the modern sense dates from the definition of the magnitude scale by N. Pogson in 1850 and the construction of a visual photometer in 1861 by J.C.F. Zöllner; with this instrument, the brightness of a star could be precisely compared with that of an artificial star image by using two Nicol prisms[3].

[3] Two Nicol prisms whose planes of polarization are rotated through an angle α attenuate the transmitted light by a factor $\cos^2\alpha$.

If the radiation fluxes measured from two different stars are in the ratio s_1/s_2, then according to Pogson's definition, their "*apparent magnitudes*" differ by an amount

$$m_1 - m_2 = -2.5 \log (s_1/s_2) \quad [\text{mag}] \tag{4.4.1}$$

or, conversely,

$$s_1/s_2 = 10^{-0.4(m_1 - m_2)} \ . \tag{4.4.2}$$

Thus, a difference of:

$-\Delta m = 1$	2.5	5	10	15	20	[mag]

corresponds to a ratio of the radiation fluxes (or, for short, a brightness ratio) of:

$s_1/s_2 = 2.51$	10	10^2	10^4	10^6	10^8 .

The unit of measure of m is the *magnitude* and is abbreviated as mag or m.

The zero point of the magnitude scale was originally based on the international *pole sequence*, a series of stars in the neighborhood of the pole, whose magnitudes were precisely measured and checked for their constancy. Since brighter stars correspond to smaller and finally to negative magnitudes, it is reasonable to say, for example, "the star α Lyr (Vega) with an apparent (visual) magnitude of 0.14 mag is 1.19 magnitudes *brighter* than α Cyg (Deneb), which has 1.33 mag", or, "α Cyg is 1.19 magnitudes *fainter* than α Lyr".

The early photometric measurements were carried out *visually*. Photographic photometry came into practice in 1904/08, beginning with K. Schwarzschild's Göttinger Aktinometrie, initially using conventional blue-sensitive plates. It was soon discovered that by using plates sensitized in the yellow region and placing a yellow filter in front of them, one could imitate the spectral sensitivity of the human eye. Thus, in addition to visual magnitudes m_v, photographic magnitudes m_{pg} and then photovisual magnitudes m_{pv} could be determined. Today, it is possible to adjust the sensitivity maximum of the apparatus for photographic or photoelectric magnitude determinations to any desired wavelength region by using a suitable combination of plates or photomultipliers with corresponding color or interference filters.

We begin by clarifying our concepts and asking the question, "What do the various magnitudes actually mean?"

Assume that a star of radius R emits at its surface a radiation flux F_λ at the wavelength λ. If it is at a distance r from the Earth, we obtain a radiation flux outside the atmosphere [we shall initially neglect interstellar extinction; see (4.2.11)] equal to

$$f_\lambda = R^2 F_\lambda / r^2 \ . \tag{4.4.3}$$

Now assume that the value which our apparatus indicates for a normal spectrum $f_\lambda \equiv 1$ can be described as a function of wavelength by a sensitivity function E_λ, which is determined by the transmission of the instrument and the sensitivity of the detector. The indicated value is then proportional to the integral:

$$s = \frac{1}{r^2} \int_0^\infty R^2 F_\lambda E_\lambda \, d\lambda \tag{4.4.4}$$

and the apparent magnitude m of our star is given by (with a constant of integration which must be fixed by convention):

$$m = -2.5 \log \frac{1}{r^2} \int_0^\infty R^2 F_\lambda E_\lambda \, d\lambda + \text{constant} \ . \tag{4.4.5}$$

In (4.4.5), we have left out the extinction due to the Earth's atmosphere. Since E_λ in practice takes on reasonably large values only within a limited wavelength range, the *extinction determination* can indeed be carried out as in Fig. 4.2.4 without further problems, as if for monochromatic radiation at the sensitivity maximum of E_λ.

The *visual* magnitudes m_v are, from (4.4.5), thus defined by the sensitivity function of the human eye *times* the transmission of the instrument; the sensitivity maximum, called the isophotic wavelength, lies at about 540 nm, in the green. In a corresponding manner, the sensitivity maximum for the *photographic* magnitudes m_{pg} lies near $\lambda = 420$ nm.

The difference of two apparent magnitudes measured in different wavelength ranges X and Y is termed the

$$\text{color index} = m_X - m_Y \ . \tag{4.4.6}$$

As a standard system for the determination of *stellar magnitudes and colors*, the UBV System developed by H. L. Johnson and W. W. Morgan in 1951 is currently most often used (U: ultraviolet, B: blue, and V: visual); the resulting magnitudes are denoted in brief as:

$$\text{U} = m_U \ , \quad \text{B} = m_B \ , \quad \text{V} = m_V \ , \tag{4.4.7}$$

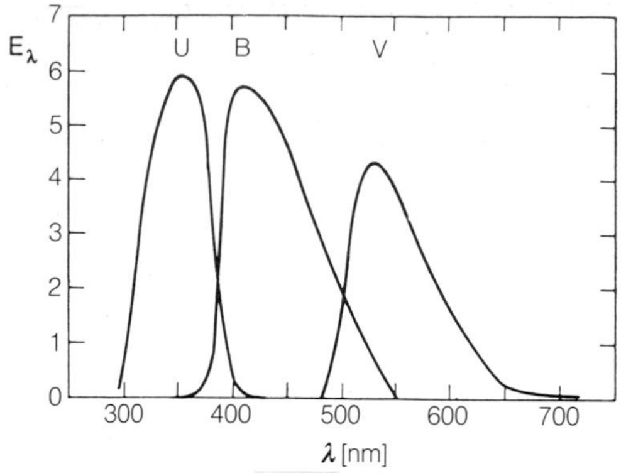

Fig. 4.4.1. Relative sensitivity functions E_λ (referred to a light source with f_λ = const) for UBV photometry. After H. J. Johnson and W. W. Morgan

The corresponding sensitivity functions E_λ, which can be obtained photographically or photoelectrically, are represented graphically in Fig. 4.4.1. Their effective wavelengths for average star colors are:

$$\lambda_U \simeq 365 \text{ nm} \ , \quad \lambda_B \simeq 440 \text{ nm} \ , \quad \lambda_V \simeq 548 \text{ nm} \ . \tag{4.4.8}$$

For hot (blue) or cool (red) stars, these are shifted to shorter or longer wavelengths, respectively.

The three magnitudes U, B, and V are by definition related to each other in such a way (i.e., the three constants in (4.4.5) are chosen in such a way) that for A0V stars (e.g., α Lyr = Vega; Sect. 4.5):

$$\left.\begin{array}{l} U = B = V \ , \\ U-B = B-V = 0 \ . \end{array}\right. \quad \text{(for A0V)} \tag{4.4.9}$$

For practical purposes, including transfer from one instrument to another with a (possibly) somewhat different sensitivity function, the UBV System is referred to a number of precisely measured *standard stars*, whose magnitudes and colors cover a large range.

The *color indices* give a measure of the *energy distribution* in the stellar spectra, as first recognized by K. Schwarzschild; they are thus also a measure of the temperatures of the stellar atmospheres. As a first approximation, we obtain the so-called *color temperature* T_C when we fit the energy distribution in a limited spectral region to Planck's radiation law. In *Wien's approximation*, $c_2/\lambda T \gg 1$, $F \propto \exp(-c_2/\lambda T_C)$; if we limit the integrations in (4.4.5) to the effective wavelengths of the sensitivity functions, we obtain for example:

$$B-V = \frac{2.5 c_2 \log e}{T_C} \left(\frac{1}{\lambda_B} - \frac{1}{\lambda_V} \right) + \text{constant} \ , \tag{4.4.10}$$

or, with $c_2 = 0.014$ m K and (4.4.8),

$$B-V \simeq \frac{0.7 \cdot 10^4}{T_C [K]} + \text{constant} \ . \tag{4.4.11}$$

The numerical values of the color temperatures calculated, e.g., from the color index B−V have no great significance, owing to the considerable deviations of the stars from black body behavior (absorption edges, Fraunhofer lines, ...; see Fig. 4.4.2). The importance of the color indices lies elsewhere: they (or the color temperatures) are related to fundamental parameters, in particular to the *effective temperature* T_{eff} (total radiation flux!), the *surface gravity* g, and the *absolute magnitudes* of the stars (Sect. 4.8), as we shall see from the theory of stellar atmospheres. Since the color indices can be measured photoelectrically with a precision of 0.01 mag without difficulty, it can be expected from (4.4.11) that e.g. around 7000 K, temperature *differences* can be determined with an accuracy of about 1%, which is not obtainable with any other method. The temperatures themselves are of course not nearly so accurately determined.

In addition to the UBV System, the UGR System of W. Becker, with effective wavelengths λ = 366, 463, and 638 nm, and the *six color system* of J. Stebbins, A. E. Whitford, and G. Kron, which reaches from the ultraviolet to the infrared (λ_U = 355 nm, ... λ_I = 1.03 µm), have attained some importance. The mutual interrelation of magnitudes and color indices among the various photometric systems with not too different isophotic wavelengths is calculated with the aid of empirically derived relations which are for the most part linear.

H. L. Johnson and others have extended the UBV System into the red and infrared by adding the following filters and sensitivity-maximum wavelengths, determined essentally by the transmission of the Earth's atmosphere:

	R	I	J	H	K
λ:	0.7	0.9	1.25	1.63	2.2
	L	M	N	Q	
	3.6	5.0	10.6	21	[µm] .

(4.4.12)

If only the brighter stars' colors are to be studied, broadband photometry (with halfwidths of the sensitivity func-

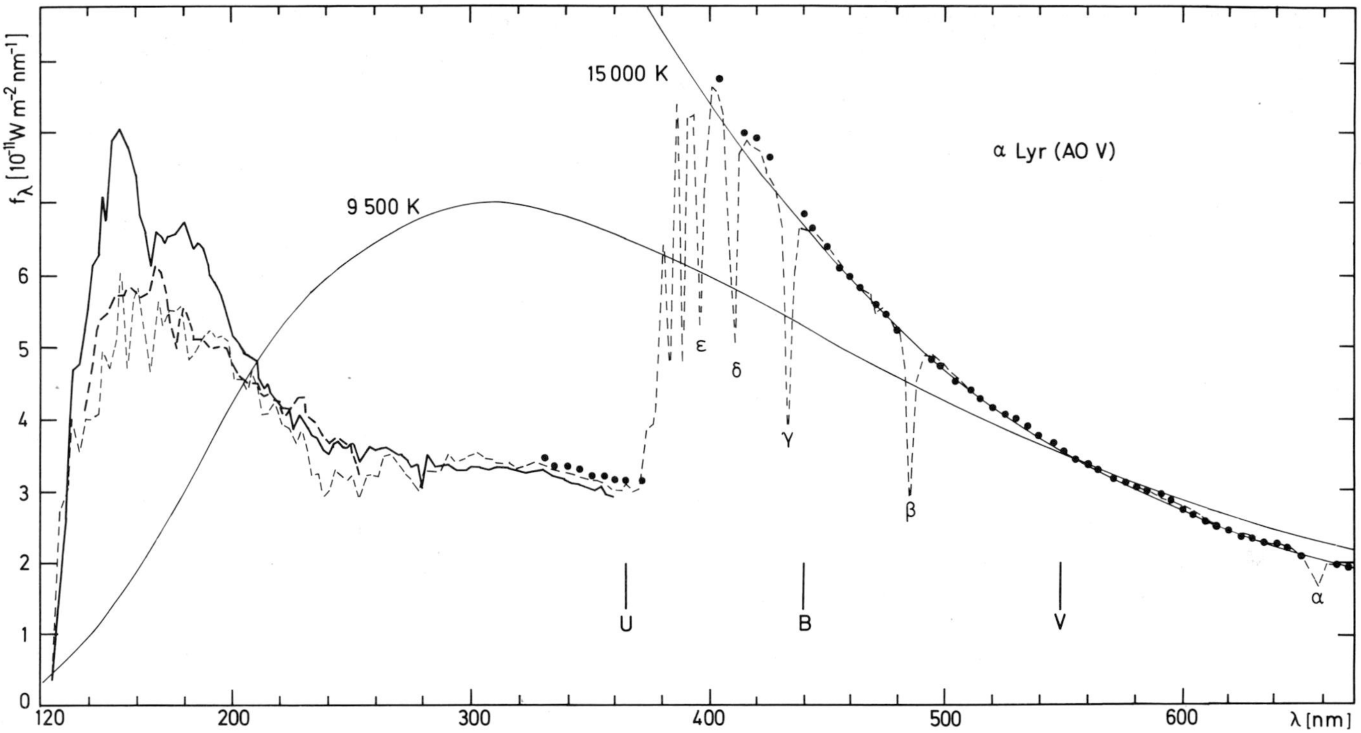

Fig. 4.4.2. The energy distribution in the spectrum of an A0V star, α Lyr (Vega), showing the observed radiation flux f_λ in units of $[10^{-11}\,\mathrm{W\,m^{-2}\,nm^{-1}} = 10^{-9}\,\mathrm{erg\,s^{-1}\,cm^{-2}\,Å^{-1}}]$. *Optical region*: absolute measurements by H. Tüg et al. (1977), corrected for extinction in the Earth's atmosphere, with bandwidths $\Delta\lambda = 1$ nm ($\lambda \leq 570$ nm) or 2 nm, excluding the regions where there are strong hydrogen lines (points ●●●). *Ultraviolet*, $\lambda \leq 350$ nm: satellite observations from TD 1 (C. Jamar et al., 1976), $\Delta\lambda \simeq 4$ nm (– – –) and from Copernicus (A.D. Code and M. Meade, 1979), $\Delta\lambda = 1.2$ or 2.2 nm (———). *For comparison*: the Kirchhoff-Planck function $B_\lambda(T)$, fitted to the data at $\lambda = 555$ nm for 9500 K, corresponding to the color temperature in the blue. Also, a theoretical model (– – –) by R.L. Kurucz (1979) for $T_{\mathrm{eff}} = 9400$ K and $g = 89.1$ m s^{-2}, taking into account absorption lines, and calculated for a spectral resolution of $\Delta\lambda = 2.5$ nm (Sect. 4.8)

tions $\Delta\lambda \geq 50$ nm) is not necessary, and measurements can be carried out with narrow band filters, which offer advantages with respect to a theoretical calibration as a function of the fundamental stellar parameters. The *medium bandwidth photometry* developed by B. Strömgren ($\Delta\lambda \simeq 10-30$ nm) is often used; it employs the following filters:

$$\begin{array}{lccccl} & \mathrm{u} & \mathrm{v} & \mathrm{b} & \mathrm{y} & \\ \lambda: & 350 & 411 & 467 & 547 & [\mathrm{nm}]\;. \end{array} \qquad (4.4.13)$$

The color indices used in this system are u−b and b−y (analogous to U−B and B−V), as well as $c_1 = (\mathrm{u}-\mathrm{b})$-(v−b) and the "metal index" $m_1 = (\mathrm{v}-\mathrm{b})-(\mathrm{b}-\mathrm{y})$.

Furthermore, *narrow band filters* having a halfwidth of a few nm can be employed to investigate individual strong lines or groups of lines in the spectra. For example, Strömgren's uvby System is extended to include the *β index* for brighter stars; this measures the intensity of the hydrogen line Hβ, $\lambda = 486$ nm relative to the neighboring continuum through a filter with $\Delta\lambda \simeq 3$ nm.

For measurements in spectral regions of only a few nm width, photoelectric *spectrophotometry* has become increasingly important since the 1960's. In this method, the narrow spectral regions are not defined by color or interference filters, but rather are defined by a slit or by the width of the detector (for example a photodiode) in the focal plane of a spectrograph. The goal of spectrophotometry, to investigate as much of the entire accessible spectrum as possible, can be accomplished with *one* detector which measures the various parts of the spectrum *sequentially* (scan technique), or with *several* detectors which register the energy distribution in the spectrum by simultaneous measurements at different wavelengths (multichannel spectrophotometry).

Although it is relatively easy to determine *relative* radiation fluxes with spectrophotometry, relating them to carefully investigated standard stars, it is difficult to obtain from these the *absolute* fluxes. The A0V star α Lyr (Vega) serves as the primary photometric standard. We show in Fig. 4.4.2 the absolute measurements of α Lyr by H. Tüg, N.M. White, and G.W. Lockwood (1977), in which the calibration was accomplished by comparison to a cavity radiator (black body) of precisely known temperature. The melting point of platinum at 2042.1 K was one of the fixed points used to determine the temperature scale. The radiation fluxes obtained in this manner should be accurate to 1 to 2%. In the far ultraviolet, there are observations of α Lyr using, for example, the European astronomical satellite TD1, with bandwidths of about 4 nm, and from the Copernicus satellite with 1.2 and 2.2 nm bandwidths. In this range, absolute calibration is very difficult, and the accuracy of the fluxes obtained is probably not better than 10 to 20%.

4.4.2 Distances and Absolute Magnitudes

Due to the orbital motion of the Earth around the Sun, a nearby star describes a small ellipse in the course of a year relative to the more distant, fainter stars (Fig. 4.4.3; not to be confused with the aberration, Fig. 2.5.7!). Its semimajor axis, i.e. the angle which the radius of the Earth's orbit would subtend as seen from the star, is called the heliocentric or annual *parallax p* of the star (from $\pi\alpha\varrho\alpha\lambda\lambda\alpha\xi\iota\varsigma$ = back-and-forth motion).

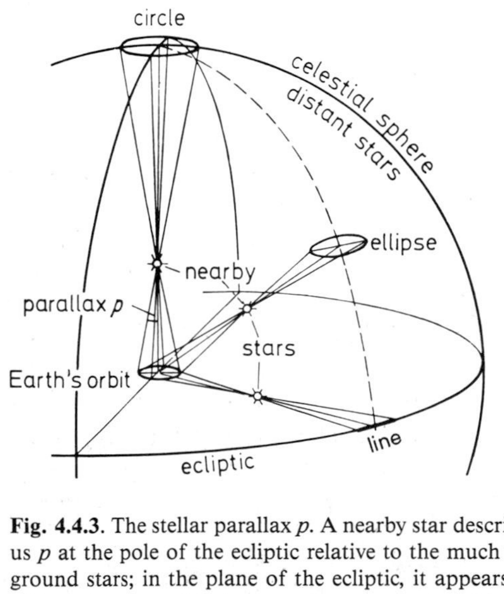

Fig. 4.4.3. The stellar parallax p. A nearby star describes a circle of radius p at the pole of the ecliptic relative to the much more distant background stars; in the plane of the ecliptic, it appears to move on a line of length $\pm p$, and in between it moves on an ellipse

In the year 1838, F.W. Bessel in Königsberg succeeded (with the aid of a Fraunhofer heliometer) in measuring the direct (trigonometric) parallax of the star 61 Cyg, $p = 0.293''$. We should also mention the observations carried out at the same time by F.G.W. Struve in Dorpat and by T. Henderson at the Cape Observatory. The star α Cen, observed by the latter, is, along with its companion Proxima Centauri, our nearest neighbor in interstellar space. Their parallaxes, according to recent measurements, are $0.75''$ and $0.76''$. A fundamental advance was made by F. Schlesinger in 1903, who succeeded in performing photographic determinations of trigonometric parallaxes with a precision of about $\pm 0.01''$.

An improved precision down to a few thousandths of a second of arc is hoped for by using extremely precise position measurements with the European astronomical satellite HIPPARCOS (*Hi*gh *P*recision *Par*allax *Co*llecting *S*atellite); following its launching in 1989, however, this satellite failed to reach the planned altitude; it will probably be able to carry out only a reduced observational program.

The General Catalogue of Trigonometric Stellar Parallaxes by L.F. Jenkins (Yale, 1952 and 1963) and the Catalogue of Bright Stars (D. Hoffleit and C. Jaschek, Yale 1982) are among the most important tools for the astronomer.

The first measurements of stellar parallaxes meant not only a confirmation of the Copernican world view (at that time hardly any longer necessary), but more importantly the first step towards measurements in interstellar space. We define suitable units of measure:

A *distance* of $360 \cdot 60 \cdot 60/2\pi = 206265$ astronomical units or radii of the Earth's orbit corresponds to a parallax of $p = 1''$. This distance is termed 1 parsec (from *par*allax and *sec*ond), abbreviated 1 pc. We thus have:

$$1 \text{ pc} = 3.086 \cdot 10^{16} \text{ m} = 3.26 \text{ light years} , \quad 1 \text{ light year} = 0.946 \cdot 10^{16} \text{ m} = 0.31 \text{ pc} . \tag{4.4.14}$$

This means that light, moving at 300000 km s^{-1}, requires 3.26 years to travel a distance of 1 pc. A parallax p (in seconds of arc) corresponds to a distance of $1/p$ [pc] or $3.26/p$ [light years]. Given the precision of measurement of trigonometric parallaxes, these take us out into interstellar space "only" to distances of at most 100 pc. The *stream or cluster parallaxes*, for groups of stars which "stream" through the Milky Way galaxy with the same velocity vectors, reach out to about 2000 pc; they will be discussed later.

The observed *apparent* brightness of a star (in whatever photometric system) depends, according to (4.4.5), on its true brightness *and* on its distance. We now define the *absolute magnitudes* M of the stars, by moving them from their actual distances r to an imaginary *standard distance* of 10 pc. Following the $1/r^2$ law of photometry [e.g. (4.4.3)], their brightnesses are modified by a factor $(r/10)^2$; in terms of magnitudes, we thus have:

$$m-M = 5\log\frac{r\,[\mathrm{pc}]}{10} = 5\log r\,[\mathrm{pc}]-5$$

$$= -5-5\log p\,[''] \; . \qquad (4.4.15)$$

The quantity $m-M$ is called the *distance modulus*; a modulus $(m-M)$ of:

−5	0	+5	+10	+25	[mag]

corresponds to a distance of: (4.4.16)

1 pc	10 pc	100 pc	10^3 pc = 1 kpc (1 kiloparsec)	10^6 pc = 1 Mpc (1 megaparsec).

In our calculation using the $1/r^2$ law, we initially disregarded *interstellar extinction*. We must recognize that it becomes quite important at distances of more than 10 pc. Then the relation (4.4.15) between the distance modulus and the distance or parallax must be corrected:

$$m-M = 5\log r\,[\mathrm{pc}]-5+A \; , \qquad (4.4.17)$$

where A gives the interstellar extinction in magnitudes. $A = 1$ mag corresponds to an optical thickness τ for extinction in the wavelength region under consideration of about 1, since from (4.2.18), $A = -2.5\log e^{-\tau} = 1.08\tau$.

In principle, absolute magnitudes can be quoted in any photometric system; they are then denoted by the corresponding subscript, for example, M_v = visual absolute magnitude. If no subscript is given, M_v is always to be understood.

We shall now collect the data for the *Sun as a star*, which are important as a point of reference. We need hardly mention that it is metrologically extremely difficult to make a photometric comparison of the Sun with the fainter stars, which have brightnesses at least 10 orders of magnitude less. The *distance modulus* of the Sun is obtained from the definition of the parsec (4.4.2, 15) and is $(m-M)_\odot = -31.57$ mag. With this, in units of magnitude, we obtain:

Apparent magnitude	Color Indices	Absolute magnitude
$U_\odot = -25.85$	$(U-B)_\odot = 0.18$	$M_{U,\odot} = +5.72$
$B_\odot = -26.03$	$(B-V)_\odot = 0.67$	$M_{B,\odot} = +5.54$
$V_\odot = -26.70$		$M_{V,\odot} = +4.87$

(4.4.18)

The limits of error of these data lie in the range of several hundredths of a magnitude.

4.4.3 Bolometric Magnitudes and Luminosities

Along with the radiation in different wavelength ranges, the *total radiation* of the stars is of interest. In analogy to the solar constant, we therefore define the *apparent bolometric magnitude* in the sense of (4.4.5):

$$m_{\mathrm{bol}} = -2.5\log\frac{1}{r^2}\int_0^\infty R^2 F_\lambda\,d\lambda+\mathrm{constant} \; ,$$

or (4.4.19)

$$m_{\mathrm{bol}} = -2.5\log\frac{R^2F}{r^2}+\mathrm{constant} \; ,$$

where

$$F = \sigma T_{\mathrm{eff}}^4 \qquad (4.4.20)$$

again denotes the total radiation flux at the stellar surface. The constant is usually defined in such a manner that the

$$\text{bolometric correction, B.C.} = m_V-m_{\mathrm{bol}} \; , \qquad (4.4.21)$$

corresponding to the color indices, never becomes negative[4]; i.e. its minimum value, which lies at $T_{\mathrm{eff}}\simeq 7000$ K, is set equal to zero. Since the Earth's atmosphere completely absorbs large spectral regions, the bolometric magnitude *cannot* be measured directly from the ground; its name is in fact misleading. It must be determined by making use of measurements from satellites, for example. With the aid of theory, m_{bol} or B.C. can be calculated as a function of other, *measurable* parameters of stellar atmospheres.

[4] A definition with reversed sign can also be found, so that B.C. = $m_{\mathrm{bol}}-m_V$ never becomes positive.

For the Sun, we obtain

$$m_{\text{bol},\odot} = -26.83 \text{ mag} \ ,$$

and (4.4.22)

$$\text{B.C.}_\odot = +0.13 \text{ mag} \ .$$

The *absolute bolometric magnitude* M_{bol} of a star is a measure of its total radiation energy output per unit time, i.e. of its *luminosity* L. The latter quantity is generally referred to the solar luminosity $L_\odot$ as unit of measure. With $(m-M)_\odot = -31.57$ mag, we obtain $M_{\text{bol},\odot} = 4.74$ mag and thus:

$$M_{\text{bol}} = 4.74 - 2.5 \log\left(\frac{L}{L_\odot}\right) \quad [\text{mag}]$$

and (4.4.23)

$$L_\odot = 3.85 \cdot 10^{26}\,\text{W} = 3.85 \cdot 10^{33}\,\text{erg s}^{-1} \ .$$

Since the energy radiated from the stars is generated in their interiors by nuclear processes, the luminosity belongs among the most important starting data for investigations into internal stellar structure.

4.4.4 Stellar Radii

If a star of radius R is located at a distance r, then the very small angle which its radius subtends at this distance is $\alpha = R/r$ in angular units (rad) or $\alpha'' = 206265\ R/r$ in seconds of arc.

Only in the case of the Sun can α be found directly; for several bright stars, *interferometric* methods allow a determination (Sect. 3.2.2). Although measurements with the Michelson interferometer are limited to a very few red giant stars (Sect. 4.5), R. Hanbury Brown and R.Q. Twiss were able to use their correlation interferometer to determine the angular radii of about 30 stars, including dwarfs, to about 0.001″. In recent times, speckle interferometry has also been used to measure the radii of stars.

For stars in the neighborhood of the ecliptic, *lunar occultations* (Sect. 2.4) can be used to measure angular radii, also to about 0.001″. In this case, a photoelectric arrangement with a sufficiently short time constant (≤ 1 ms) registers the intensity fluctuations from Fresnel diffraction of the starlight by the sharp edge of the Moon; these pass over a particular observer in a few tens of ms. The angular radius of the stellar disk can be derived from the contrast of the intensity bands compared to that expected from a point source.

For the majority of stars, the radii must be estimated indirectly using *photometric* methods from their magnitudes or radiation fluxes. From (4.2.11), we obtain the monochromatic radiation flux f_λ to be expected from a star at a distance r:

$$f_\lambda = F_\lambda R^2/r^2 = F_\lambda \alpha^2 \tag{4.4.24}$$

and the total radiation flux (4.4.20):

$$f = F\alpha^2 = \sigma T_{\text{eff}}^4 \alpha^2 \ . \tag{4.4.25}$$

Thus, from the measured radiation fluxes f_λ or f, or using (4.4.5) with the apparent magnitude m measured in one of the photometric standard regions (e.g., U, B, or V), the angular radii can be calculated when the radiation flux F_λ or F at the star's surface, or its effective temperature T_{eff}, are known. These latter quantities can be derived from the color or the energy distribution by making use of the theory of stellar atmospheres.

When greater precision is required in determining the angular radii, the center-limb variation (Sect. 4.3) must be taken into account. Especially in the case of cool giant stars, with their extended atmospheres, the radius depends also on the wavelength or color which is used for the observations.

In order to obtain the *stellar radii* themselves from the angular radii α, we require the distance r or the parallax p, or else the distance modulus or the absolute magnitude (4.4.15), or, because of

$$4\pi r^2 f = 4\pi R^2 F = 4\pi R^2 \sigma T_{\text{eff}}^4 = L \ , \tag{4.4.26}$$

we can use the luminosity L (derived e.g. from the star's spectrum) or the absolute bolometric magnitude M_{bol}.

The largest observed angular radii are found for the red giant stars such as α Ori (Betelgeuze), with $\alpha = 0.024''$, α Sco (Aldeberan), with 0.021″, or o Cet (Mira), with 0.022″. From these values, we derive e.g. for α Ori ($V = 0.9$ mag, $p = 0.017''$, and $T_{\text{eff}} \simeq 3500$ K) a radius $R \simeq 300\ R_\odot \simeq 1.4$ AU. The diameter of this star corresponds to that of Mars' orbit! The radii of the red giants are for the most part somewhat variable.

4.5 Classification of Stellar Spectra; the Hertzsprung-Russell Diagram and the Color-Magnitude Diagram

When, as a result of the discoveries of J. Fraunhofer, G. Kirchhoff, and R. Bunsen, the observation of *stellar spectra* was begun, it quickly became clear that they can, for the most part, be ordered in a *one-parameter* sequence. The correlated changes of the stellar colors or color indices indicate that in doing this, the stars had been ordered according to decreasing *temperature*.

Based on the work of Huggins, Secchi, Vogel, and others, E. C. Pickering and A. Cannon developed the *Harvard Classification* of stellar spectra in the 1880's; it formed the basis of the Henry Draper Catalogue. The series of *spectral classes* ("Harvard types") which are ascribed to the colors of the stars:

$$O - B - A - F - G - K - M \begin{array}{c} \nearrow S \\ \searrow R - N \end{array} \qquad (4.5.1)$$

    blue   yellow   red

resulted from older classification schemes after numerous modifications and simplifications. H. N. Russell's students in Princeton invented the well-known mnemonic for this series: *O Be a Fine Girl, Kiss Me Right Now.*[5]

Between each of these letters, a finer subdivision is indicated by a following number, 0 to 9. For example, a B5 star is between B0 and A0 and has about an equal amount in common with each of these types. The definition of the Harvard sequence is accomplished principally by comparison of the spectra of certain standard stars.

We shall describe the spectral classes with their classification criteria later, in connection with the MK Classification which is in general use today; it has adopted the nomenclature of the Harvard Classification to a large extent.

In the year 1913, H.N. Russell had the happy thought of investigating the relationship between the *spectral type* Sp and the *absolute magnitude* M_v of the stars, by constructing a diagram with the spectral type as abscissa and M_v as the ordinate, and plotting all stars for which the parameters were known with sufficient accuracy. Figure 4.5.1 shows such a diagram, which Russell drew in 1927 using considerably improved observational results, for his textbook; it served for a whole generation as "astronomical Bible".

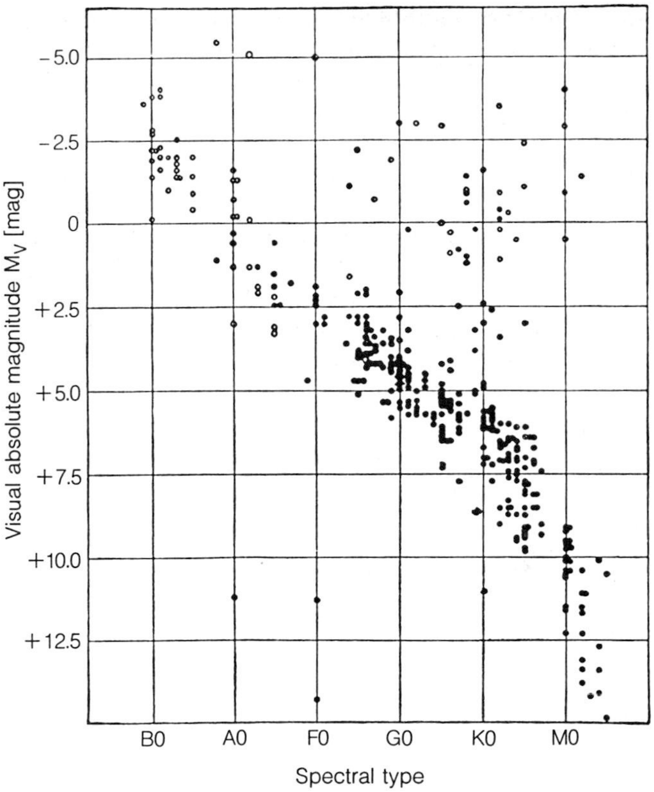

Fig. 4.5.1. Hertzsprung-Russell diagram. The visual absolute magnitude M_v is plotted against spectral type. The Sun corresponds to $M_v = 4.8$ mag and type G2. The points (●) represent stars within 20 pc with reliable parallaxes. For the rarer stars having larger absolute magnitudes (○), along with the trigonometric parallaxes, spectroscopic and cluster parallaxes were also employed

Most stars populate the narrow band of the *main sequence*, which stretches diagonally from the (absolutely) bright blue-white B and A stars (e.g., the belt stars in Orion) through the yellow stars (e.g., the Sun, G2 and $M_v = +4.8$ mag) out to the faint red M stars (e.g., Barnard's star, M5 and $M_v = +13.2$ mag).

On the upper right, we find the group of *giant stars*; in contrast, those stars of the same spectral class which have much smaller luminosities are termed *dwarf stars*. Since, at the same temperatures, the difference in absolute magnitude can only result from a corresponding difference in *stellar radii*, these names seem quite appropriate. The distinction and classification of the giant and dwarf stars dates back to older work (1905) of E. Hertz-

---

[5] (Note, for experts only): S stands for "Smack" (or Sweetheart!).

sprung, so that today we refer to the (Sp, M_V) diagram as a *Hertzsprung-Russell Diagram* (HRD).

Instead of the spectral type Sp, a color index, e.g. B−V, can be plotted; in this way, one obtains a *Color-Magnitude Diagram* (CMD) which is equivalent to the HRD. Figure 4.5.2 shows the CMD (B−V, M_V) with its now very sharply defined main sequence, some yellow giant stars (upper right) and white dwarfs (lower left); it contains individual and cluster stars from our neighborhood with precisely determined parallaxes. Since color indices can be precisely measured even for faint stars, the CMD has become the most important tool of stellar astronomy.

The extremely bright stars along the upper edge of the HRD or CMD are called supergiants. For example, α Cyg (Deneb; A2) has an absolute magnitude $M_V = -7.2$ mag; it thus exceeds the luminosity of the Sun ($M_V = +4.8$ mag) by 12.0 magnitudes, i.e. by a factor of about 60000!

A further clearly recognizable group are the *white dwarfs* at the lower left. Since they have only weak luminosities in spite of their relatively high temperatures, they must be very small; their radii are readily calculated to be barely larger than the Earth's radius. For Sirius'

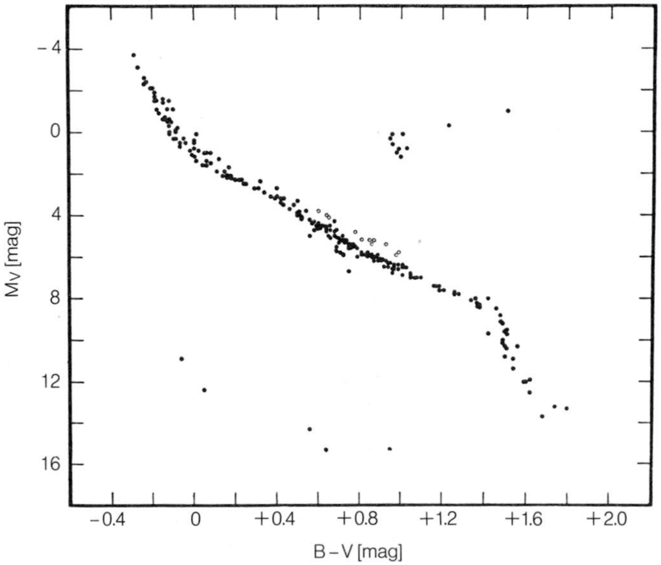

Fig. 4.5.2. The color-magnitude diagram with M_V plotted against B-V according to H.L. Johnson and W.W. Morgan; it contains main sequence stars with trigonometric parallaxes $p \geq 0.10''$ and those from several galactic star clusters with well-known parallaxes, interstellar absorption and reddening. In addition, five white dwarfs (*lower left*) and several yellow giants (*upper right*) are also plotted. The stars from Praesepe which lie above the main sequence are probably binary stars

companion α CMa B and a few similar objects, the masses are also known, so that mean densities of the order of 10^8 to 10^9 kg m^{-3} can be calculated. The internal structure of such stars must therefore be quite different from that of other stars. R.H. Fowler showed in 1926 that in white dwarfs, the matter (more precisely, the electrons) is *degenerate* in the sense of Fermi statistics, in the same way as was demonstrated soon thereafter by W. Pauli and A. Sommerfeld for the conduction electrons in a metal. This means that nearly all the quantum states are completely occupied, taking into account the Pauli Principle, as in the inner shells of a heavy atom.

We shall return to further groups of stars in the HRD, which for the most part are relatively small and specialized, in connection with other topics.

It was noted by E. Hertzsprung in 1905 that stars with sharp hydrogen lines, for example the A2 star α Cyg, are remarkable for their high luminosities. In 1914, W.S. Adams and A. Kohlschütter then showed that the stars of a particular spectral class can be further subdivided, corresponding to their luminosities or *absolute magnitudes* M_V, on the basis of new spectroscopic criteria. Among the absolutely bright stars, for example, the lines from ionized atoms are stronger relative to those from neutral atoms; among the A stars, as mentioned, the hydrogen lines can be used as a criterion for luminosity, etc.

If such a *luminosity criterion* (which can only hold in a particular range of spectral classes!) is calibrated using stars of known absolute magnitude, the resulting calibration curve can be used for the spectroscopic determination of *absolute* magnitudes. If the interstellar absorption (which was completely unknown in 1914!) can be neglected or corrected for, one can, by combining with the known apparent magnitudes of the stars (4.4.17), determine *spectroscopic parallaxes.* We shall have more to say in Chap. 5 about their importance for the investigation of the Milky Way galaxy. Here, we follow the significant insight that the majority of stars can be classified using *two* parameters.

From the Harvard classification, W.W. Morgan and P.C. Keenan developed the two-dimensional MK *Classification*, which is today in general use, and is summarized in "An Atlas of Stellar Spectra. With an Outline of Spectral Classification". Its general principles hold for *any* classification:

1) The classification is based only upon *empirical* criteria, i.e. directly observable absorption and emission phenomena.

2) The observational data are *unified.* In order, on the one hand, to be able to define sufficiently fine spectral criteria, but, on the other, to penetrate far enough into the galaxy, a unified dispersion of $\simeq 125$ Å mm^{-1} is employed, *even* for bright stars.

3) The *transferability* of the classification system to other instruments is guaranteed by a list of suitable *standard stars,* i.e. by direct comparison, *not* by (possibly semitheoretical) descriptions.

4) The classification is done according to *spectral type* Sp and *luminosity class* LC.

A series of standard stars for defining the spectral classes is shown in Fig. 4.5.3. We describe it briefly here, associating the spectral lines used (classification criteria) with the correct chemical elements and ionization states (I: neutral atom, II: singly ionized atom, III: doubly ionized atom, ...). The temperatures quoted correspond *roughly* to the color of the star and are given only to provide an initial orientation (Table 4.5.1).

The luminosity criteria should depend to first order and as sensitively as possible *only* on the luminosity of the star, over the whole range of Sp. W. W. Morgan's *luminosity classes,* LC, at the same time indicate the position of the star in the HRD; they are (with their proper names):

Ia-0	Hypergiants	IV	Subgiants
I	Supergiants	V	Main Sequence (Dwarfs)
II	Bright Giants	VI	Subdwarfs
III	Giants		

As needed, the luminosity classes I to V can be subdivided using the suffixes a, ab, and b. Figure 4.5.3 shows an extract from the Atlas of Stellar Spectra, containing the important spectral classes of the main sequence stars (LC = V) and, for the spectral type A0, the division into luminosity classes I to V (depending on the width of the hydrogen lines).

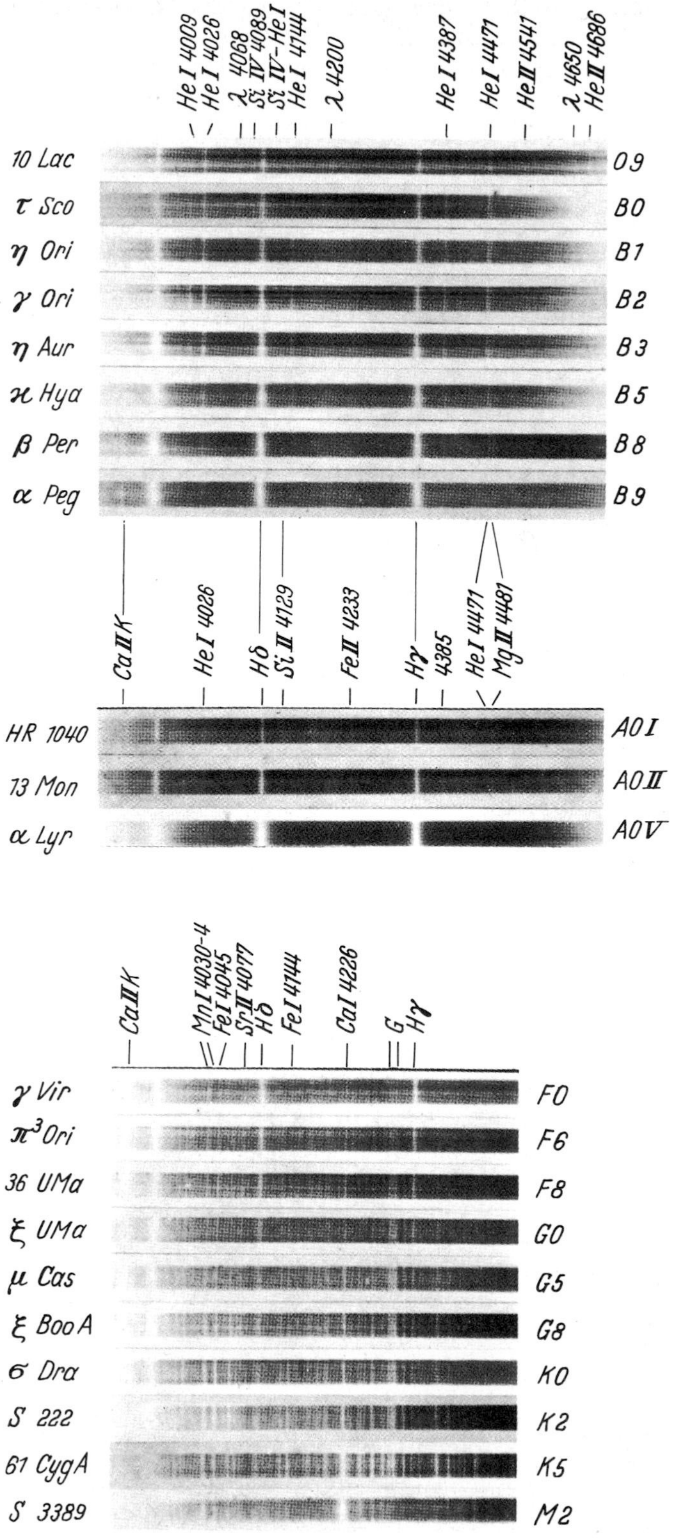

Fig. 4.5.3. The MK classification of stellar spectra, from "An Atlas of Stellar Spectra" by W. W. Morgan, P. C. Keenan, and E. Kellman (1943). Along the main sequence (luminosity class V), above, O9-B9, then A0V and below, F0-M2. In the case of A0, the luminosity classes I (supergiants) and II (bright giants) are also shown, in order to make clear the significance of the absolute magnitudes (spectroscopic parallaxes!). The spectral lines used in the classification are indicated with their identification and wavelength in Å (1 Å = 0.1 nm)

Table 4.5.1. Classification of stellar spectra

Spectral Type	Temperature [K]	Criteria for Classification
O	50000	Lines of highly ionized atoms: He II, Si IV, N III ...; hydrogen H relatively weak; occasionaly emission lines.
B0	25000	He II not present; He I strong; Si III, O II; H stronger.
A0	10000	He I not present; H at maximum; Mg II, Si II strong, Fe II, Ti II weak; Ca II weak.
F0	7600	H weaker; Ca II strong; the ionized metals, e.g. Fe II, Ti II had their maxima at about A5; the neutral metals, e.g. Fe I, Ca I have about the same strength here.
G0	6000	Ca II very strong; neutral metals Fe I etc. strong.
K0	5100	H relatively weak, neutral atomic lines strong; molecular bands.
M0	3600	Neutral atom lines, e.g. Ca I, very strong; TiO bands.
M5	3000	Ca I very strong; TiO bands stronger.
C	3000	Strong CN-, CH-, and C_2-bands; TiO not present. Neutral metals as with K and M.
S	3000	Strong ZrO-, YO-, LaO-bands; neutral atoms as with K and M.

About 90% of all stellar spectra can be accounted for in the MK Classification; those remaining are in part composite spectra of unresolved binary stars, and in part the peculiar spectra of pathological individuals.

Some of the peculiarities of stellar spectra which cannot be taken into account in a two-parameter classification scheme are denoted by the following abbreviations: the suffix n (nebulous) indicates a particularly diffuse appearance of the lines, e denotes emission lines, v means variable spectrum, and p (peculiar) characterizes *any sort of* unusual feature, e.g. an anomalous intensity of the lines of a certain element.

Certain groups of stars are also sometimes denoted by using prefixes; e.g. c stands for especially sharp lines (supergiants, α Cyg Ia or cA2), g for giant stars, d for dwarfs, sd for subdwarfs, and w for white dwarfs. This additional notation should, however, be avoided in the MK Classification.

The *white dwarfs* are denoted by a D written before the spectral class, e.g. DA, DB, DZ and so forth. We mention the *Wolf-Rayet Stars*, which are notable for clear emission lines; they are denoted by WC or WN, depending on whether their spectra contain the lines of carbon or of nitrogen. We shall discuss these and other special groups of stars in another connection in Sect. 5.4.5.

The MK Classification system has been extended and refined numerous times since it was first proposed. We mention the Revised Spectral Atlas for Stars earlier than the Sun, by W. W. Morgan, H. A. Abt, and J. W. Tapscott (1976); the extension to cooler stars (G, K, M, S, and C) by P. C. Keenan and R. C. McNeil (1976 and later), which takes into account the most important abundance anomalies of the giant stars; and the new classification of the stars in the Henry Draper Catalogue by N. Houk (1976/78).

Newer spectral atlases of the MK types are "An Atlas of Objective Prism Spectra" (Michigan 1974) by N. Houk, N. J. Irvine, and D. Rosenbush; the "Bonner Atlas für Objektivprismenspektren" by W. C. Seitter (1970/75); and "An Atlas of Representative Stellar Spectra" by Y. Yamashita, K. Nariai, and Y. Norimoto (Tokyo 1978).

Using observations made from the IUE satellite, a classification system for the early-type spectral types has been established into the *ultraviolet region* ($\lambda = 115$ to 320 nm) with a spectral resolution of about 0.7 nm (A. Heck, D. Egret, M. and C. Jaschek, 1984).

The *relationship* between the parameters Sp and LC of the MK Classification on the one hand, and the color index B−V and absolute magnitude M_V on the other hand, is given in Fig. 4.5.4, using the best currently-available calibrations.

Along with the color-luminosity diagram, the *two color diagram* developed by W. Becker (1942) plays an important role. Here (Fig. 4.5.5), the short-wavelength color index U−B is plotted (downwards) against B−V as the abscissa. For *black body radiators* in this diagram, one obtains approximately a straight line inclined at 45°, as can be readily calculated using (4.4.11) and the corresponding expression for U−B. In Fig. 4.5.5, the relationship between U−B and B−V is plotted for the *main sequence* stars, with the spectral classes and absolute magnitudes indicated.

The great differences between the spectral energy distributions of the stars and of a black-body radiator will be clarified in Sect. 4.8.3, where we make use of the theory of stellar spectra. The applications of the two-color diagram for the determination of interstellar reddening, as well as for identifying particular types of stars, will be discussed in Chap. 5.

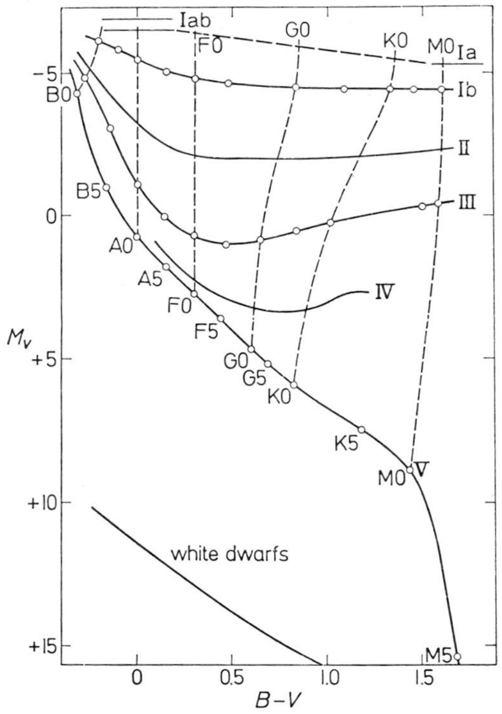

Fig. 4.5.4. Spectral type Sp and luminosity class LC of the MK classification as functions of the color index B-V and the absolute magnitude M_v [mag]

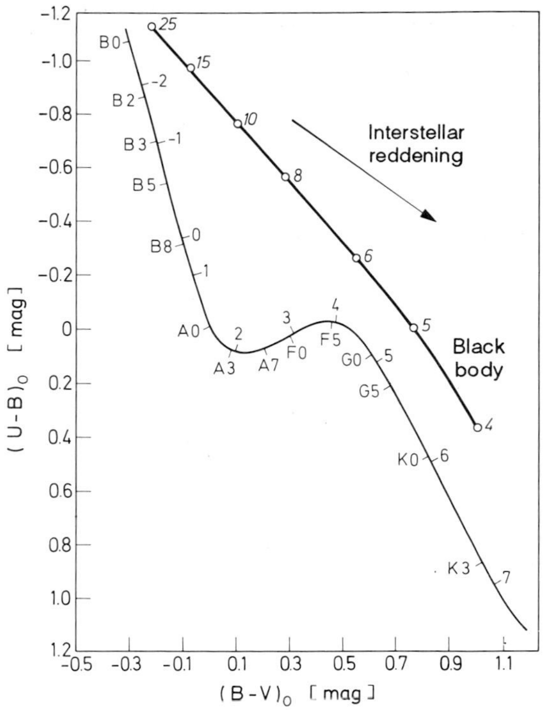

Fig. 4.5.5. A two-color diagram for main sequence stars using the (interstellar-reddening-independent) indices $(U-B)_0$ and $(B-V)_0$ in [mag] (after H.L. Johnson, W.W. Morgan and others). Along the lines, the MK spectral types and the absolute magnitudes of the stars are indicated. Black-body radiators having the $T_{\text{eff}} \cdot 10^3$ K values shown would give the nearly straight line above the main sequence line. Interstellar reddening displaces the data points for stars parallel to the line drawn at the upper right; it refers directly to O stars

4.6 Binary Stars and Stellar Masses

In 1803, F.W. Herschel discovered that α Gem = Castor is a *visual binary star*, whose components move around each other under the influence of their mutual gravitational attraction. The observation of binary stars therefore offers a possibility of extracting the stellar masses $\mathscr{M}$ or at least of reaching quantitative conclusions about them. Since the stellar masses are accessible *only* through their gravitational interactions, the investigation of binary star systems remains of fundamental importance for all of astrophysics today.

We begin with an overview of the types of binary star systems, the visual binary star systems (Sect. 4.6.1), the spectroscopic binary systems and the eclipsing variables (Sect. 4.6.2). In Sect. 4.6.3, we consider briefly the rotational periods of binary star systems and the rotation of stars. The stellar masses, determined in various ways, are summarized in Sect. 4.6.4, where we also give the mass-luminosity relation, which is important for the theory of stellar structure. Particularly interesting are also the *close* binary systems (Sect. 4.6.5), in which exchange of stellar matter can occur between the components, and the few systems which contain a radio pulsar (Sect. 4.6.6).

4.6.1 Visual Binary Stars

We first separate the optical (apparent) pairs using statistical criteria, possibly also taking account of their particular proper motions and radial velocities, from the *physical pairs*, i.e. the true binary stars. The *apparent orbit* of the fainter component (the "companion") around the brighter component is observed with a refractor using a crosshair micrometer (Fig. 3.2.1); its separation (in seconds of arc) and position angle (N 0°–E 90°–S 180° –W 270°) are recorded. If the apparent orbit is plotted, the result is an *ellipse*. If we were to look down onto the orbital plane from a perpendicular direction, we would necessarily find the brighter component at one focus of the orbit. This is, in general, *not* the case, since the orbital plane subtends an *inclination angle* i with the celestial plane (perpendicular to the direction of observation). Conversely, the orbital inclination i can clearly be determined in such a way that the true orbit fulfills Kepler's laws. Let

- a be the semimajor axis of the (relative) true orbit in seconds of arc, and
- p be the parallax of the binary star system in seconds of arc; then
- a/p is the semimajor axis of the true orbit in astronomical units (radii of the Earth's orbit).

If, furthermore, P is the orbital period in years, we can apply *Kepler's Third Law* to obtain the *total mass* $\mathcal{M}_1 + \mathcal{M}_2$ of the two stars (in units of the solar mass)

$$\mathcal{M}_1 + \mathcal{M}_2 = \frac{a^3}{p^3 P^2} \ . \tag{4.6.1}$$

If the motion of the two components has been measured *absolutely* (i.e. relative to the background stars, after subtracting the parallactic and the proper motions), then the semimajor axes a_1 and a_2 of their true orbits about the common center of gravity can be obtained, and from (2.6.36), we find

$$a_1 : a_2 = \mathcal{M}_2 : \mathcal{M}_1 \quad \text{and} \quad a = a_1 + a_2 \ , \tag{4.6.2}$$

so that now the individual masses $\mathcal{M}_1$ and $\mathcal{M}_2$ can be calculated.

In recent times, speckle interferometry (Sect. 3.2.2) has in particular been used to determine the separation ($\geq 0.001''$) of binary star components.

When the fainter component is not directly observable, its presence can still be inferred from the (absolutely measured) motion of the brighter component about the center of gravity (astrometric binary stars). If a_1 is its semimajor axis (again in seconds of arc), we obtain from $a_1/a_2 = \mathcal{M}_2/(\mathcal{M}_1 + \mathcal{M}_2)$ the following relation:

$$(\mathcal{M}_1 + \mathcal{M}_2)\left(\frac{\mathcal{M}_2}{\mathcal{M}_1 + \mathcal{M}_2}\right)^3 = \frac{a_1^3}{p^3 P^2} \ . \tag{4.6.3}$$

In this way, F. W. Bessel found in 1844 from meridian circle observations that Sirius (α CMa, A1 V, $\mathcal{M} = 2.2\,\mathcal{M}_\odot$) must have a "dark" companion. In 1862, A. Clark indeed discovered Sirius B, 9.8 mag fainter. It has an absolute magnitude of only $M_V = 11.2$ mag, although its mass is $0.94\,\mathcal{M}_\odot$. Since the surface temperature of this little star is quite "normal" at about 23000 K, it must be extremely *small* (as already noted). In 1923, F. Bottlinger came to the conclusion "that it is a question of something new here", namely a white dwarf star.

A further interesting application is the precise determination of the orbits of nearby stars in order to search for dark companions, which represent either the transitional phase between stars and planets, or even *planets* themselves. Extensive observations by van de Kamp et al. over a period of years of our second-nearest neighbor, Barnard's star (BD +4° 3561), which is at distance of 1.8 pc and has the spectral type M5 V and mass $0.15\,\mathcal{M}_\odot$, at first indicated two companions having about 0.7 and 0.5 Jupiter masses and orbital periods of 12 and 20 yr, respectively; i.e. a kind of planetary system. Later independent analyses, however, have not confirmed the existence of a periodic motion in the orbit of Barnard's star.

4.6.2 Spectroscopic Binary Stars and Eclipsing Variables

In 1889, E.C. Pickering observed that in the spectrum of Mizar = ζUMa, the lines become double (twice) within a period of $P = 20.54$ d. Mizar was thus shown to be a *spectroscopic binary*. In this particular system, two similar A2 stars move around each other; their angular separation is too small to be resolved telescopically. In other systems, only *one* component can be recognized in the spectrum; the other is evidently too faint. If the radial velocity of one or both components, obtained from the Doppler effect, is plotted against time, a *velocity curve* is obtained. After subtracting the mean or center-of-gravity motion, one can read off the component(s) of the orbital velocity in the direction of observation. From this, the semimajor axis of the orbit itself cannot be determined; however, using methods which we shall not treat in detail here, the quantity $a \sin i$ can be calculated (i is the unknown orbital inclination angle), for component 1 if only it can be observed, otherwise for both components.

If only *one* spectrum is visible, Kepler's 3rd Law and the Center of Gravity theorem immediately yield from (4.6.3):

$$\frac{(a_1 \sin i)^3}{P^2} = (\mathcal{M}_1 + \mathcal{M}_2)\left(\frac{\mathcal{M}_2}{\mathcal{M}_1+\mathcal{M}_2}\right)^3 \sin^3 i$$
$$= \frac{\mathcal{M}_2^3 \sin^3 i}{(\mathcal{M}_1+\mathcal{M}_2)^2} \, . \qquad (4.6.4)$$

The quantity on the right is called the *mass function*. For statistical purposes, one can make use of the fact that the mean value of $\sin^3 i$ over a sphere is equal to 0.59, or, taking the probability of discovery into account, about 2/3. Since $\mathcal{M}_2 < \mathcal{M}_1$, the factor $(\mathcal{M}_2/(\mathcal{M}_1+\mathcal{M}_2))^3$ is in any case $< 1/8$.

When both spectra are visible, one obtains $\mathcal{M}_1^3 \sin^3 i$ and $\mathcal{M}_2^3 \sin^3 i$, and thus the mass ratio $\mathcal{M}_1 : \mathcal{M}_2$.

Of particular interest is the search for planet-like companions based on periodic shifts in the spectral lines from nearby stars. For this search, an extremely high precision of measurement is needed, since for example the variations in the radial velocity of the Sun caused by Jupiter have an amplitude of only $\simeq 0.01$ km s^{-1} with a period of around 12 yr. In recent times, D. W. Latham, B. Campbell, and others, using special techniques, have attained precisions in the range 0.01 to 0.1 km s^{-1}. Thus, for example, in the case of the star HD 114762, which is similar to the Sun and is at a distance of 28 pc, periodic variations in the radical velocity of $\lesssim 0.7$ km s^{-1} with a period of 84 d have been found, indicating a companion with the order of 1 to 10 Jupiter masses.

If the orbital inclination of a spectroscopic binary system is near to 90°, then "eclipses" occur and the system is called an *eclipsing variable*.

The classic example is β Per = Algol, identified by J. Goodricke in 1782, with a period of $P = 2.87$ d.

From the magnitudes of an eclipsing variable, measured over a long period of time, its *period P* is first determined, and then its *light curve*. From the latter (Fig. 4.6.1), the radii of the two stars can be obtained, in units of the radius of the relative orbit, as well as the orbital inclination i. If, in addition, the velocity curve can be determined spectroscopically for one or even both components, then the absolute dimensions of the system and its masses, and thus the mean densities of both stars can be calculated. In favorable cases, even the ellipticity (flattening) and the center-limb darkening of the two stars can be extracted. Thanks to the methods for determination of the elements of eclipsing variables developed by H.N. Russell and H. Shapley, they are today among the most precisely described stars. In *close pairs*

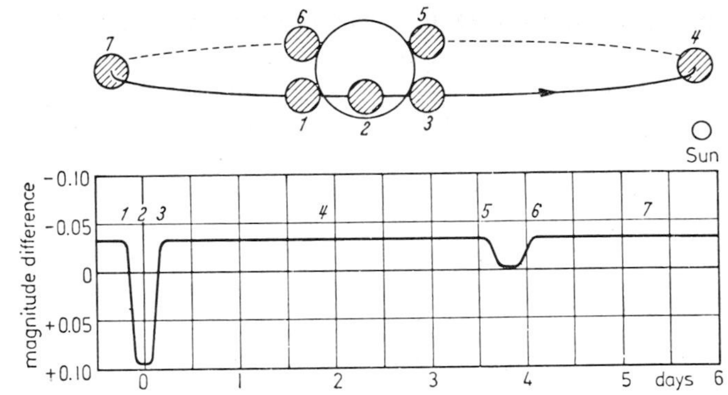

Fig. 4.6.1. Apparent relative orbit and light curve of the eclipsing binary variable IH Cas. Corresponding points on the light curve and the orbit are marked with numbers. The main eclipse of the brighter component by the fainter and smaller one is in this case ring-shaped

(Sect. 4.6.5), the two components also interact physically, as first shown by O. Struve from a careful analysis of the spectra.

4.6.3 The Periods of Binary Systems. Rotation of the Stars

Let us now attempt to give an overview of the general features of binary star systems; a detailed statistical discussion would in any case be of dubious significance, owing to the inevitable non-randomness of the sample (probability of discovery!).

The visual binaries, spectroscopic binaries, and eclipsing variables, which differ only in the manner of their observation, form a continuum, with some overlap. Their periods range from a few hours up to many millenia. Binary stars with short periods mostly have circular orbits; systems of long period prefer large orbital eccentricities. In addition to binary systems, *multiple* systems also occur frequently; they usually contain one or more close pairs. The "binary star" α Gem = Castor discovered by F. W. Herschel consists of *three pairs* A, B, and C, each spectroscopic binaries with periods of 9.21, 2.93, and 0.814 d, respectively. A and B orbit around each other in 420 yr, and Castor C revolves around A+B in several thousand years. In our immediate vicinity within 20 pc, 45%, i.e. nearly half, of the stars are members of binary or multiple systems.

In the spectra of binary stars and eclipsing variables of short period, whose components circle about each other at a close distance, the *Fraunhofer lines* are usually noticeably broad and washed out. This is related to the fact that the two components rotate about each other like a rigid body due to tidal friction, in a manner similar to the Earth-Moon system. The periods of rotation are equal to the orbital (revolution) periods.

If the projection of the equatorial velocity on the line of sight is $v \sin i$, then the Doppler shift at the wavelength λ is $\Delta\lambda = \pm\lambda(v/c)\sin i$. A spectral line which would be sharp in the case of a star at rest now appears to be spread out into a band of width $2\Delta\lambda$, whose profile reflects the brightness distribution of the "stellar disk". If the latter exhibits no center-limb darkening, for example, the line profile is elliptical. A B star of radius $5R_\odot$ which rotates with a period of 1.5 d and has $i = 90°$, for example, has a projected equatorial velocity of $v \sin i = 250$ km s^{-1} and the halfwidth of e.g. the Mg II $\lambda = 448.1$ nm line will then be $\Delta\lambda = \pm 0.37$ nm.

O. Struve and coworkers discovered that there are also *individual stars* in whose spectra all the lines are strongly broadened in this manner and which therefore must *rotate* with equatorial velocities of up to more than 300 km s^{-1}. Like the rapidly rotating binary stars, the rapidly rotating single stars belong quite preferentially to the spectral classes O, B, and A in the upper part of the main sequence. Main sequence stars with spectral classes following F5 V have, in contrast, very small rotational velocities below 10 to 20 km s^{-1}.

We shall return later to the significance of the rotation of single and binary stars and the role of angular momentum in the problems of stellar evolution.

4.6.4 The Stellar Masses

We shall now gain an overview of the masses $\mathscr{M}$ of the stars, obtained from all types of binary star systems. Their numerical values range from about 0.07 $\mathscr{M}_\odot$, the smallest mass found for a "visible" star, to 100 $\mathscr{M}_\odot$, with the majority of stellar masses falling in the region from 0.3 to 3.0 $\mathscr{M}_\odot$. The connection to the other quantities of state of the stars remained unclear until A.S. Eddington in 1924 discovered the *mass-luminosity relation* in connection with his theory of the inner stellar structure. From a modern standpoint, we can understand the essential features as follows: the stars on the main sequence are evidently in analogous stages of development (their energy requirements are supplied by the fusion of hydrogen to helium) and therefore, as a rule, have structures formed "according to the same recipe". A particular mass $\mathscr{M}$ will correspond to internal energy sources of a particular magnitude, which in turn determine the luminosity L of the star. We should thus expect a relationship between the mass $\mathscr{M}$ and the luminosity L, or the absolute bolometric magnitude M_{bol}, of these stars. In fact, as an analysis of all the observational data (Fig. 4.6.2)

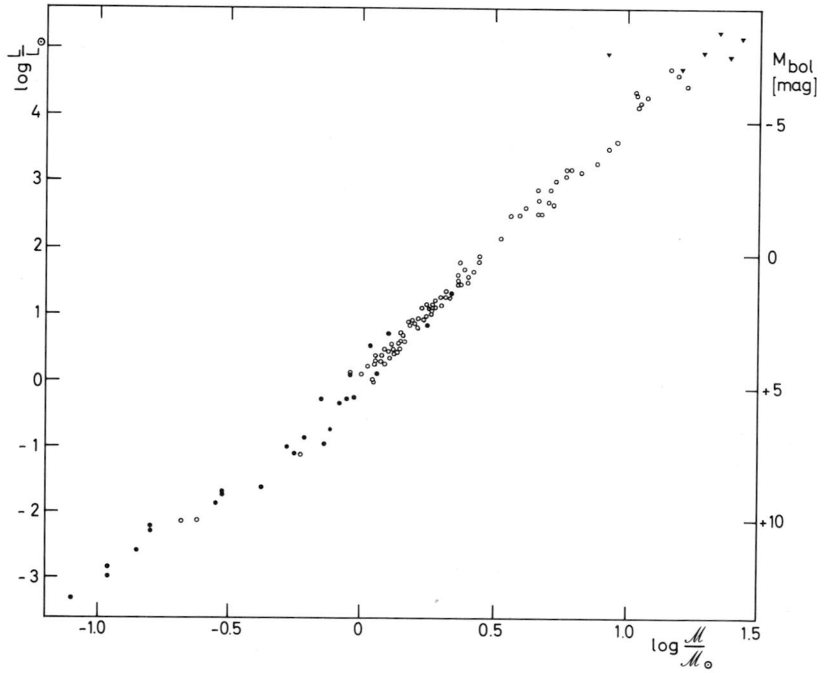

Fig. 4.6.2. Empirical mass-luminosity relation for main sequence stars. The luminosity L or the absolute bolometric magnitude M_{bol} of the stars is plotted as a function of their masses $\mathscr{M}$ (D.M. Popper, 1980). [● visual binary stars; ○ spectroscopic binary stars: optically resolved systems and eclipsing variables (detached systems); ▼ OB eclipsing variables (presumably contact systems)]

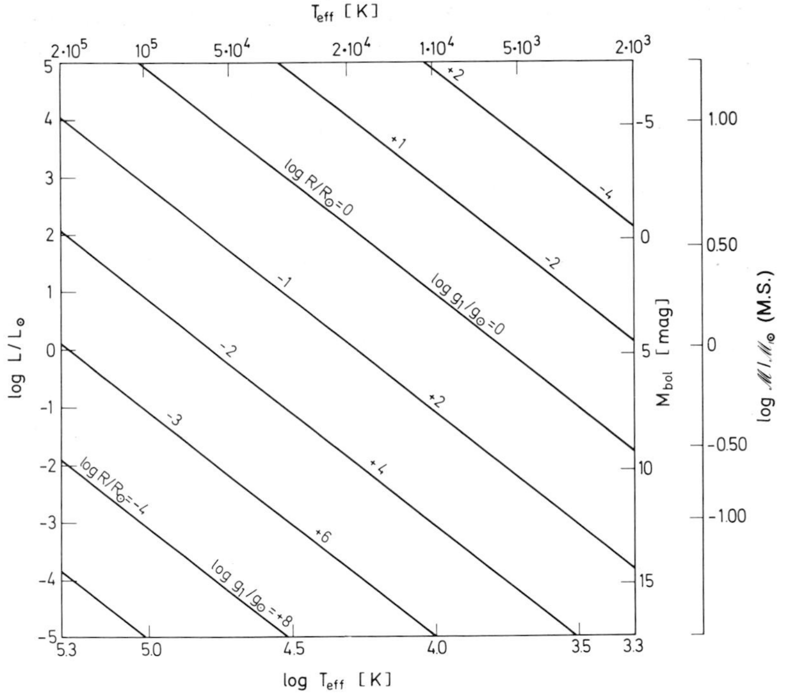

Fig. 4.6.3. The relation between luminosity L or absolute bolometric magnitude M_{bol} (left and right ordinate scales, respectively), effective temperature T_{eff} (abscissas) and stellar radii R and surface gravity g. The latter is given as g_1 for stars of $\mathcal{M} = 1\,\mathcal{M}_\odot$; these numbers are roughly valid for old, evolved stars. For main sequence stars, the mass $\mathcal{M}$ as an approximate function of M_{bol} is indicated on the outer right

shows, such a relation holds for the *main sequence stars*; in the upper mass range, it can be approximated by the empirical formula:

$$\log \frac{L}{L_\odot} = 3.8 \log \frac{\mathcal{M}}{\mathcal{M}_\odot} + 0.8 \ . \tag{4.6.5}$$

The stars which have evolved away from the main sequence (Sects. 5.4.4 and 5) do not obey this relation, as is to be expected; for example, the white dwarfs with an average mass of about $0.5\,\mathcal{M}_\odot$, and especially the neutron stars with masses of the order of $1\,\mathcal{M}_\odot$ (Sect. 4.12.8). For the red giants, for which there are hardly any reliable empirical mass determinations, the mass-luminosity relation is also inapplicable, since they form a heterogeneous group in terms of their evolution. As a statistical average, the masses of the red giants are about $1.1\,\mathcal{M}_\odot$.

Considering the radii R of the stars as given (determined from their absolute magnitudes and, roughly speaking, their temperatures), we can calculate an important quantity for the theory of stellar spectra, the *surface gravity* on the stellar surface:

$$g = \frac{G\mathcal{M}}{R^2} \ . \tag{4.6.6}$$

We find that it has a numerical value which is constant within a factor of 2, $g \simeq 2 \cdot 10^2\ \mathrm{m\,s^{-2}}$ ($\simeq 2 \cdot 10^4\ \mathrm{cm\,s^{-2}}$), for the stars of the main sequence out to the spectral class M2 V.[6] For giants and supergiants, it is considerably smaller (down to about $10^{-2}\ \mathrm{m\,s^{-2}}$), and for white dwarfs, it is much larger (about $10^6\ \mathrm{m\,s^{-2}}$).

In Fig. 4.6.3, we summarize the important results for the relations between the luminosity L, the bolometric magnitude M_{bol}, the effective temperature T_{eff}, the star's radius R, and the surface gravity g; here, we have made use of (4.4.26) and (4.6.6):

$$L = 4\pi R^2 \cdot \sigma T_{\mathrm{eff}}^4 \ , \qquad g = \frac{G\mathcal{M}}{R^2} \ . \tag{4.6.7}$$

L, R, and $\mathcal{M}$ are given in units of $L_\odot = 3.85 \cdot 10^{26}$ W, $R_\odot = 6.96 \cdot 10^8$ m, and $\mathcal{M}_\odot = 1.99 \cdot 10^{30}$ kg. The gravitational acceleration is calculated for a star of mass equal to $1\,\mathcal{M}_\odot$, in units of $g_\odot = 274\ \mathrm{m\,s^{-2}}$, i.e. $g_1/g_\odot$. For stars

[6] It is usual in the theory of stellar spectra to denote the gravitational acceleration as a dimensionless quantity, i.e. simply as $\log g$, where g [$\mathrm{cm\,s^{-2}}$] is actually meant; thus for main sequence stars, $\log g \simeq 4.3$, and for white dwarfs, $\log g \simeq 8$.

on the main sequence and still younger stars, the mass scale on the right is approximately applicable.

As we shall see in Sect. 4.9, the analysis of stellar spectra permits the extraction of the effective temperature T_{eff} and the acceleration of gravity g. It is then possible using (4.6.7) to calculate the *ratio* of mass to luminosity:

$$\frac{\mathcal{M}}{L}=\frac{1}{4\pi G\sigma}\frac{g}{T_{\mathrm{eff}}^4} \quad . \tag{4.6.8}$$

Should one wish to determine $\mathcal{M}$ and L separately, it is necessary either to refer to the theory of internal stellar structure (Sect. 4.12) or to use corresponding empirical data.

4.6.5 Close Binary Star Systems

In the case of *close* pairs of stars, there are for one thing very strong tidal forces acting between the two components, which tend to synchronize their rotational periods with the orbital periods. Furthermore, there is often a direct physical interaction between them; as was first shown by O. Struve in the 1940's and 50's using spectroscopic analyses, there are common gas shells and gas flows from one component to the other. In recent times, it has become clear from, for example, investigations of novas and nova-like variables, and of galactic X-ray sources (Sects. 4.11.5 and 6), that a gas flow often does not enter the companion star directly, but rather, owing to conservation of angular momentum, forms a rotating disk (accretion disk) around it.

The cause of matter exchange in close binary systems is to be found in the changes in stellar radii during the course of the stars' evolution, in particular the enormous increase in radius on approaching the red giant stage (Sect. 5.4.4).

We first consider the *equipotential surface* of a binary star system, whose components are initially still separated. We consider a point at distances r_1 from the mass $\mathcal{M}_1$ and r_2 from $\mathcal{M}_2$; according to (2.6.34), at that point there acts a gravitational potential given by:

$$\Phi_{\mathrm{G}}=-G\left(\frac{\mathcal{M}_1}{r_1}+\frac{\mathcal{M}_2}{r_2}\right) \quad . \tag{4.6.9}$$

If the system rotates with the angular velocity ω, we can represent the centrifugal acceleration $z\omega^2$ (z is the distance from the axis of rotation) by an additional potential $\Phi_z=-z^2\omega^2/2$. On a surface given by

$$\Phi=\Phi_{\mathrm{G}}+\Phi_z=-G\left(\frac{\mathcal{M}_1}{r_1}+\frac{\mathcal{M}_2}{r_2}\right)-\frac{1}{2}z^2\omega^2 \tag{4.6.10}$$

a test mass can be moved freely without performing work; therefore, the surface of e.g. an ocean, or of a celestial body, corresponds to an equipotential surface $\Phi=$ const.

In our binary star system, each component is at first surrounded by "its own" closed equipotential surfaces, out to the point where the first common equipotential surface begins; it surrounds both objects in the shape of an hourglass, and is termed the *Roche surface*. Further out, all the equipotential surfaces surround both objects (Fig. 4.6.4).

If now the object of larger mass, let us say $\mathcal{M}_1$, evolves into a giant star, it can grow out beyond the innermost common equipotential surface. We then have a *semi-detached system*; gas flows from component 1 to component 2, so that the mass ratio can even be reversed.[7]

What can happen in detail will be described in Sect. 5.4.6, after we have treated the various phases of stellar evolution.

If the two components completely fill a common equipotential surface, we speak of a *contact system* or W Ursae Majoris system (Fig. 4.6.4).

4.6.6 Pulsars in Binary Star Systems

A further possibility for determining stellar masses, besides the spectroscopic measurements of the changing radial velocities in binary star systems discussed above (Sect. 4.6.2), is offered by the regular signals sent out at short time intervals by the pulsars (Sect. 4.11.7).

We have already met the frequency shift $\Delta\nu$ from a moving radiation source due to the Doppler effect. The Doppler formula

$$\frac{\Delta\nu}{\nu_0}=\frac{v}{c} \quad , \quad v\ll c \quad , \tag{4.6.11}$$

is applicable not only to electromagnetic waves (or acoustic waves, in which case c is the velocity of sound), but also to *every* regular sequence of signals such as, for example, the "ticking" of a pulsar with a pulse frequency

[7] We denote the originally more massive component in a binary system as the primary component, independently of whether the mass ratio has been reversed in the course of later evolution.

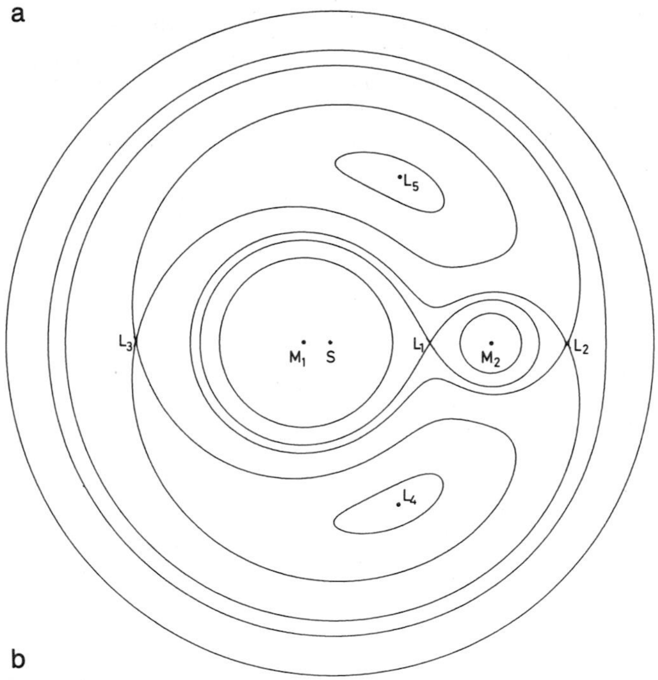

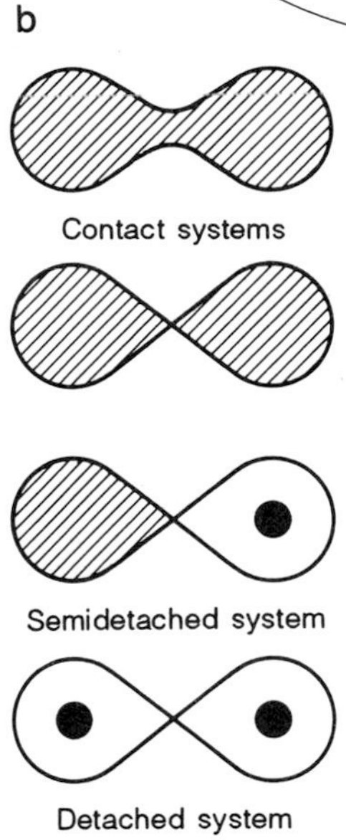

Fig. 4.6.4. **(a)** Geometry of the equipotential surfaces in a close binary star system. The curves of Φ = const according to (4.6.10) in the orbital plane are drawn for a mass ratio $\mathcal{M}_2/\mathcal{M}_1 = 0.17$. The *Roche* limit curve meets itself at the Lagrange point L_1. S is the center of gravity of the system, through which the axis of rotation passes. **(b)** Types of spectroscopic binary systems

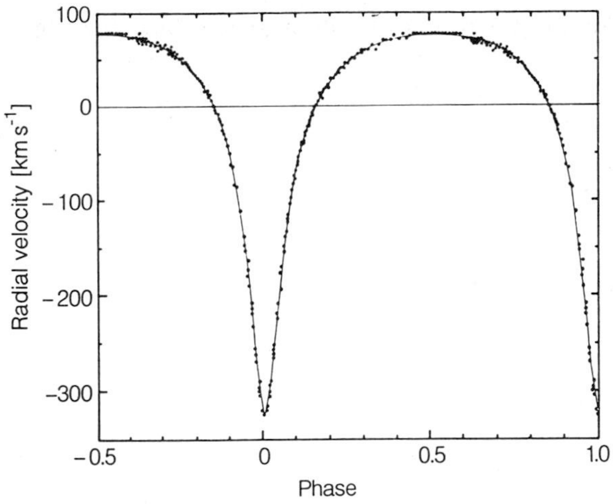

Fig. 4.6.5. The observed radial velocity curve of the binary pulsar PSR 1913+16 over one orbital period. The large orbital eccentricity $e = 0.62$ expresses itself as a noticeable deviation from a sine curve. From R.A. Hulse and J.H. Taylor: Astrophys. J. **195**, L51 (1975). (Reproduced with the kind permission of The University of Chicago Press, The American Astronomical Society, and the authors)

ν_0 or a period $1/\nu_0$. If a pulsar is approaching us with a relative velocity v owing to its motion in a binary star system, the time interval between arriving signals is shortened, or the pulse frequency ν_0 is increased, according to (4.6.11).

The exceptionally high precision with which the arrival of radio pulses can be measured (ca. 1 μs) permits a very precise determination of the orbital elements and, above all, of the masses of the pulsars (neutron stars), whereby we can simply apply the formulas from Sect. 4.6.2. To be sure, the so-called *radio pulsars* (Sect. 4.11.7) are for the most part single stars, but in 1974, R.A. Hulse and J.H. Taylor, in the course of a sky survey with the 300 m radio telescope at Arecibo in Puerto Rico, discovered fluctuations in the pulse frequency of the pulsar PSR 1913+16. This pulsar, which has a period of 0.059 s or a frequency of $\nu_0 = 17$ Hz, exhibits changes in its period of up to 80 μs with a modulation period of 0.323 d; the modulation was interpreted by its discoverers as due to orbital motion in a binary system, with radial velocities of the pulsar between 60 and 330 km s^{-1} (Fig. 4.6.5).

Longer series of observations have yielded unusual orbital elements for this "*binary pulsar*": the semimajor axis of the orbital ellipse of the pulsar, projected onto the sphere, is equal to $a \sin i = 702\,000$ km; with an inclination of $i \simeq 50°$, this corresponds to a semimajor axis of only $1.3\,R_\odot$! The orbital period is $P = 0.32$ d, and the eccentricity of the orbit is $e = 0.62$. The masses, both of the pulsar or neutron star and of its companion, which has still not been clearly identified (also a neutron star?), have been found to be about $1.4\,\mathcal{M}_\odot$. These orbital elements are so extreme that Newton's theory of gravitation is not sufficient to describe them; instead, due to the strong gravitational fields which occur, Einstein's

General Relativity Theory (Sect. 4.12.9) must be employed. Particularly notable is the large *rotation of the periastron* of the orbit, $4.23° \, yr^{-1}$, corresponding to a complete rotation in 85 years. The effect of general relativity is much stronger in this binary-pulsar system than in the analogous precession of the perihelion of Mercury around the Sun, which is 43″ per 100 years, see (4.12.64).

Einstein's gravitational theory predicts that the motions of the masses in a binary system like PSR 1913+16 will give rise to the emission of *gravitational waves* and therefore lead to an energy loss by the system. Indeed, a small systematic *decrease* $\dot{P}$ in the orbital period, $\dot{P}/P = -2.4 \cdot 10^{-12}$, has been observed; it agrees well with the theoretically predicted value and can thus be considered to represent an indirect observation of gravitational radiation.

We know of only a few binary star systems with a radio pulsar: the nine (as of 1990) systems within the galactic disk appear to fall into two classes; one contains more massive stars ($\gtrsim 0.7 \mathcal{M}_{\odot}$) such as PSR 1913+16, and the other includes less massive systems. In addition, there are some binary-star pulsars in globular clusters (Sect. 4.11.7).

The *X-ray pulsars* (Sect. 4.11.6), such as Her X-1, differ from the radio pulsars in that they are, as a rule, members of a *close binary system*. Frequently, in addition to the modulation of the X-ray pulses, the variations in radial velocity can also be observed in the spectrum of the other component. Analysis of the orbits of the known X-ray pulsars yields masses for the neutron stars in the range $1.2 \mathcal{M}_{\odot}$ to $1.6 \mathcal{M}_{\odot}$.

4.7 Spectra and Atoms. Excitation and Ionization

The interpretation of stellar spectra and their classification led to M. N. Saha's theory of thermal excitation and ionization in the year 1920, following important preliminary work by N. Lockyer. This theory is based essentially on the quantum theory of atoms and atomic spectra developed by N. Bohr, A. Sommerfeld and others after 1913. We shall permit ourselves here to recall some fundamentals without giving a complete derivation: the basic concepts of atomic spectroscopy (Sect. 4.7.1); the Boltzmann and the Saha formulas for the excitation and ionization of atoms in thermodynamic equilibrium (Sect. 4.7.2); and the general kinetic equations for excitation and ionization by individual elementary atomic processes when there is no thermodynamic equilibrium (Sect. 4.7.3).

4.7.1 Basic Concepts of Atomic Spectroscopy

The possible energy levels of an atom are graphically represented in an energy term scheme or *Grotrian diagram* (Fig. 4.7.1). We distinguish between:

1) *Discrete negative energy values* ($E<0$) corresponding to the bound or elliptical orbits of the electrons in Bohr's model. Each energy level is characterized by several integral or half-integral quantum numbers, which we initially represent by *one* index n, m, s, etc.

2) *Continuous positive energy values* ($E>0$) corresponding to the free or hyperbolic orbits in Bohr's model. At a large distance from an atom, such an electron has

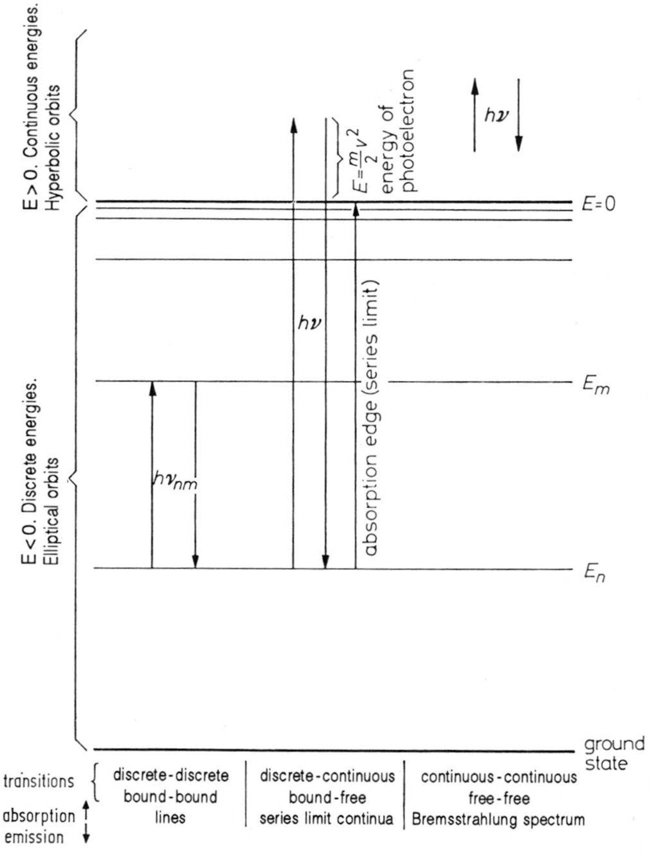

Fig. 4.7.1. Energy level or Grotrian diagram of an atom (schematic)

only kinetic energy, $E_{\text{kin}} = \frac{1}{2} m v^2$, where m is its mass and v its velocity.

When *transitions* between two energy levels E_m and E_n occur, a light quantum of energy $h\nu$ is absorbed (↑) or emitted (↓) ($h = 6.63 \cdot 10^{-34}$ Js $= 6.63 \cdot 10^{-27}$ erg·s, or $\hbar = h/2\pi$ is Planck's constant):

$$h\nu = \hbar\omega = |E_m - E_n| \ . \tag{4.7.1}$$

The frequency ν [s^{-1} or Hz] or the circular frequency ω corresponds to a *wavenumber* (the number of light waves per meter in vacuum), $\tilde{\nu} = \nu/c$ [m^{-1} or Kayser $= \text{cm}^{-1}$] and a *wavelength* $\lambda = 1/\tilde{\nu} = c/\nu$ [m]. Furthermore, the unit 10^{-10} m = 1 Å (one Ångström) is also used. The energy values are often quoted not relative to $E = 0$, but instead relative to the ground state of the atom. Their units are usually not given as J or erg, but rather as cm^{-1} or Kayser; one speaks of the (energy) *terms* and the term scheme of the atom; alternatively, the energy is quoted in units of eV, i.e. electron volts, the energy which an electron gains on passing through a potential difference of 1 Volt. In thermal equilibrium, the energies are always of order kT. In this sense, we write the temperature T in [K] corresponding to an energy E. The following conversion formulas apply:

$$\begin{aligned} 1 \text{ eV} &= 1.602 \cdot 10^{-19} \text{ J} = 1.602 \cdot 10^{-12} \text{ erg} \\ &= 8066 \text{ cm}^{-1} = (1239.9 \text{ nm})^{-1} = (12399 \text{ Å})^{-1} \\ &= 11605 \text{ K} \ . \end{aligned} \tag{4.7.2}$$

The transitions of an atom accompanied by absorption or emission of a light quantum $h\nu$ are quite naturally divided into the following groups:

$E_m < 0$, $E_n < 0$; discrete-discrete or bound-bound transitions with absorption or emission of a *spectral line* whose wavenumber $\tilde{\nu}$ is calculated from the difference of the term values.

$E > 0$, $E_n < 0$; continuous-discrete or free-bound transitions. Continuous-discrete absorption $\nu > \nu_n$ runs up to the series limit or *absorption edge* $h\nu_n = E_n$, causing the emission of a photoelectron with kinetic energy $\frac{1}{2} m v^2 = h\nu - |E_n|$. In the process, the atom is ionized or goes to the next highest ionization state. We denote the spectra of neutral, singly ionized, doubly ionized, etc. atoms, e.g. calcium, by Ca I, Ca II, Ca III, etc.

The inverse process is the capture of a free electron of energy $\frac{1}{2} m v^2$ accompanied by the emission of a light quantum,

$$h\nu = \tfrac{1}{2} m v^2 + |E_n| \ ;$$

it is called *pair recombination*.

$E' > 0$, $E'' > 0$; continuous-continuous or free-free transitions. A light quantum $h\nu = |E' - E''|$ is absorbed or emitted; the free electron gains or loses the corresponding kinetic energy as it passes near the atom or ion.

We shall first concern ourselves with the *discrete* terms ($E_n < 0$) of the atoms or ions.

A particular energy level of an atom or ion with *one* valence electron (the remaining electrons are assumed not to participate in transitions) is described by the following *quantum numbers*:

a) n, the *principal quantum number*. For hydrogen-like orbits (Coulomb field), in the language of Bohr's model, $n^2 a_0/Z$ is the semimajor axis of the orbit; the corresponding energy is:

$$E_n = \frac{1}{4\pi\varepsilon_0} \frac{e^2 Z^2}{2a_0} \frac{1}{n^2} \ , \tag{4.7.3}$$

and the term value is $R_\infty Z^2/n^2$. Here, $a_0 = 5.29 \cdot 10^{-11}$ m $= 0.529$ Å is the first Bohr radius of the hydrogen atom, $\varepsilon_0 = 8.85 \cdot 10^{-12}$ As $\text{V}^{-1}\,\text{m}^{-1}$ is the permittivity constant of vacuum, $R_\infty = 1.097 \cdot 10^7\,\text{m}^{-1}$ is the Rydberg constant, and Z is the effective nuclear charge ($Z = 1$ for a neutral atom, $Z = 2$ for a singly ionized atom, etc.).

b) l is the *angular momentum of the orbital motion* of the electron, measured in the quantum units $\hbar = h/2\pi$. l can take on the integral values $0, 1, 2, \ldots n-1$.

$l = 0$	1	2	3	4	5	refers to an
s	p	d	f	g	h	electron.[8]

c) $s = \pm\frac{1}{2}$ is the *spin* of the electron in the same units.

d) j is the *total angular momentum*, again in units of $\hbar$. It is the vector sum of l and s and has only the two possible values $l \pm \frac{1}{2}$.

An electron with $n = 2$, $l = 1$, and $j = 3/2$ is, for example, referred to as a $2p_{3/2}$ electron.

In atoms or ions with several electrons, the angular momentum vectors are usually coupled as follows (Russell-Saunders or LS coupling):

---

[8] This notation originally referred to the upper (running) term of the series: s: sharp subseries; p: principal series; d: diffuse subseries; f: fundamental series; thereafter alphabetic order.

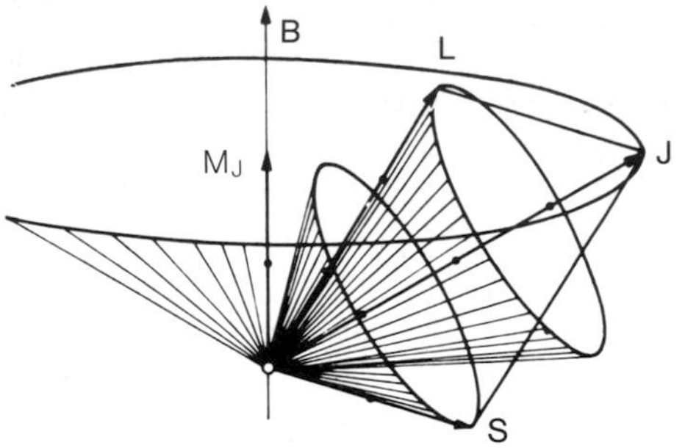

Fig. 4.7.2. Vector diagram for an atom with Russell-Saunders coupling. The vectors of the total orbital angular momentum $\boldsymbol{L}$ and the spin angular momentum $\boldsymbol{S}$ add to give the total angular momentum $\boldsymbol{J}$ (all in units of $\hbar = h/2\pi$). M_J is the component of J in the direction of an applied field $\boldsymbol{B}$. $\boldsymbol{L}$ and $\boldsymbol{S}$ precess about $\boldsymbol{J}$, and $\boldsymbol{J}$ precesses about $\boldsymbol{B}$. The drawing corresponds to an energy level with $L = 3$, $S = 2$, and $J = 3$, i.e. $^5F_3$

The orbital angular momenta add vectorially to give a resultant orbital angular momentum $\boldsymbol{L} = \sum \boldsymbol{l}$, similarly for the spin momenta which give a resultant spin angular momentum $\boldsymbol{S} = \sum \boldsymbol{s}$. The vectors $\boldsymbol{L}$ and $\boldsymbol{S}$ add to give the total angular momentum $\boldsymbol{J}$ (Fig. 4.7.2), which obeys:

$$|L-S| \leq J \leq L+S \ . \tag{4.7.4}$$

L is always an integer; S and J are half-integral or integral for atoms with odd or even numbers of electrons, respectively.

A particular set of values of S and L yields a *term*. As in a one electron system, the orbital angular momentum states are denoted by Roman letters (now capitalized)

$L = 0$	1	2	3	4	5	refers to an
S	P	D	F	G	H	term .

As long as $L \geq S$, the term consists of $r = 2S+1$ energy levels with different J values. The number r is called the *multiplicity* of the term (even for $L < S$) and is written at the upper left of the term symbol; J is written in the lower right as an index, in order to characterize the individual energy levels belonging to the term. A summary of the possible terms of various multiplicities, their energy levels, and the usual notation is given in Table 4.7.1.

In an *applied field*, e.g. a magnetic field, the total angular momentum vector is oriented in such a way that its component M_J parallel to the field is also half-integral or integral. M_J can thus take on the values J, $J-1, \ldots, -J$; i.e. the directional quantization of J yields $2J+1$ possible orientations (Fig. 4.7.2). When the applied field is zero, these $2J+1$ possible levels all have the same energy; the level J is then said to be $(2J+1)$-fold *degenerate*. Furthermore, we divide the terms into two groups, the even or the odd terms, according to their *parity*, depending on whether the arithmetic sum of the l's of the electrons is even or odd. *Odd* terms are denoted by an ^{o} at the upper right of the term symbol.

A transition between two energy levels corresponds to a *spectral line*; the possible transitions between all the levels of a term produce a group of neighboring lines, a so-called *multiplet*. The possible transitions (with emission or absorption of electric dipole radiation, in analogy to the well-known Hertzian dipole radiation) are limited by the followng *selection rules*:

1) There are transitions only between even and odd levels.
2) J changes by only $\Delta J = 0$ or ± 1. The transition $0 \leftrightarrow 0$ is forbidden.

For Russell-Saunders coupling, the additional two rules hold:

3) $\Delta L = 0, \ \pm 1$.
4) $\Delta S = 0$, i.e. no intercombinations (e.g., singlet-triplet).

Russell-Saunders or LS coupling can be recognized for example by the fact that the multplet splittings arising

Table 4.7.1. The terms and J-values of their levels for different quantum numbers L and S in Russell-Saunders coupling

		$S = 0$ $r = 2S+1 = 1$ Singlet	1/2 2 Doublet	1 3 Triplet	3/2 4 Quartet
$L = 0$	S-Term	$J = 0$	$J = 1/2$	$J = 1$	$J = 3/2$
1	P-Term	1	1/2 3/2	0 1 2	1/2 3/2 5/2 (dashed box)
2	D-Term	2	3/2 5/2	1 2 3	1/2 3/2 5/2 7/2
3	F-Term	3	5/2 7/2	2 3 4	3/2 5/2 7/2 9/2

Example: (dashed box) Quartet P-term with the energy levels $^4P_{1/2}$, $^4P_{3/2}$, $^4P_{5/2}$. Statistical weight of the term: $g(^4P) = 4 \cdot 3 = 2+4+6$.

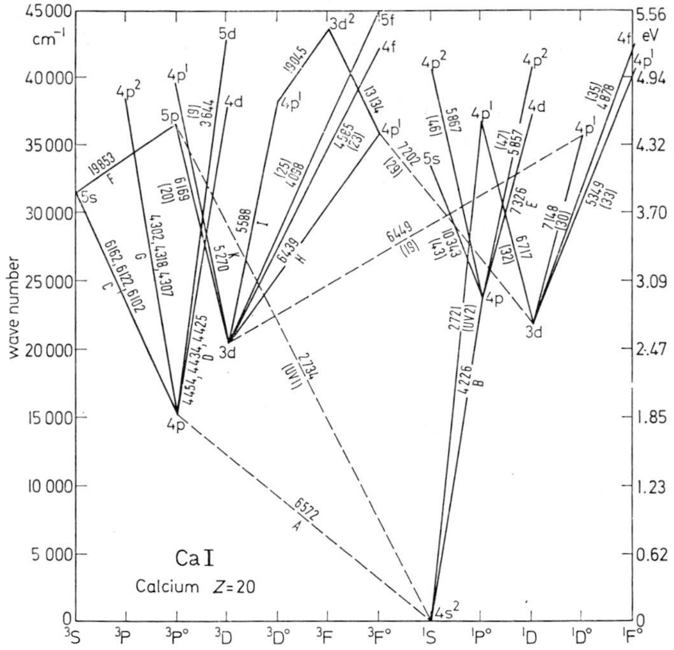

Fig. 4.7.3. A term scheme or Grotrian diagram for the spectrum of neutral calcium, Ca I. The more important multiplets are denoted by their numbers in "A Multiplet Table of Astrophysical Interest" or "An Ultraviolet Multiplet Table" by C. E. Moore, as well as by the wavelengths of their most intense lines in [Å]

from the magnetic interaction between the orbital and spin momenta (magnetic moments) are small in relation to the splittings between neighboring terms or multiplets.

If the selection rules (1) and (2) are *not* fulfilled, there can still be *forbidden transitions* involving electric quadrupole or magnetic dipole radiation with much smaller transition probabilities.

An example of a term scheme or Grotrian diagram is given in Fig. 4.7.3, which shows the term scheme for neutral calcium, Ca I.

The theory of atomic spectra outlined in the previous paragraphs allowed the *classification* of the wavelengths λ or wavenumbers $\tilde{\nu}$ measured in the laboratory for most of the chemical elements and their ionization states. This means that for each measured spectral line, the lower and upper term (usually referred to the ground state) and the term classifications can be identified. For example, Fraunhofer's K line, the strongest line in the visible solar spectrum at $\lambda = 393.37$ nm, corresponds to Ca II $4^2S_{1/2} - 4^2P^o_{3/2}$.

The *intensity* of a line (where we at this point use the word in a qualitative sense without precise definition) will depend essentially on what fraction of the atoms of the element in question are in the *ionization state* and furthermore in the *state of excitation* (energy level) from which the line can be absorbed. The answer to these questions is given by the *Saha theory*, as long as we can assume that the gas is in a state of thermal equilibrium, i.e. that it corresponds sufficiently accurately to the conditions in a closed cavity at the temperature T.

4.7.2 Thermal Excitation and Ionization

We first consider an (ideal) gas at the temperature T, consisting of *neutral* atoms (Fig. 4.7.4). Per unit volume there are:

N atoms in total
N_0 atoms in the ground state 0, and
N_s atoms in an excited state s with the excitation energy χ_s .

We shall soon extend this preliminary notation by adding a second index r, which distinguishes between neutral, singly, doubly ... ionized atoms denoted by $r = 0,1,2...$

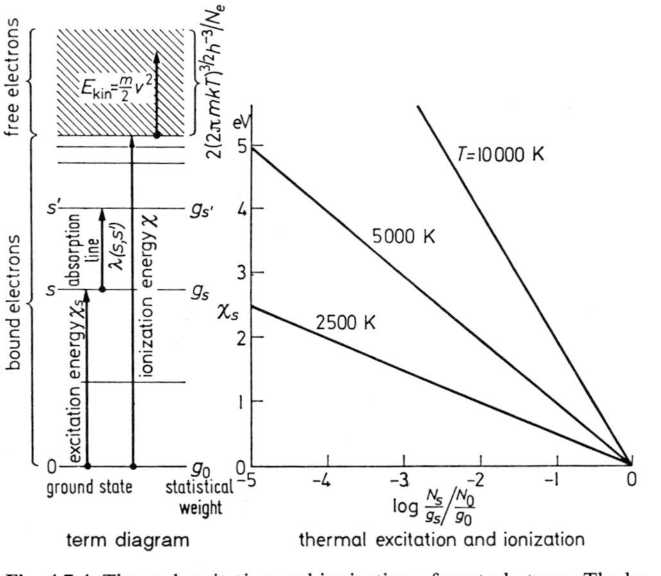

Fig. 4.7.4. Thermal excitation and ionization of neutral atoms. The basic concepts are explained on the (simplified) term scheme at the left. The diagram on the right shows the fraction of excited atoms, at various temperatures, referred to simple quantum states, i.e. with statistical weights of 1, as a function of the excitation energy χ_s in [eV] (*ordinate*). The ionization energy $\chi = 5.14$ eV corresponds to that of Na I

If all the quantum states are *nondegenerate*, then according to the fundamentals of statistical mechanics as developed by L. Boltzmann,

$$\frac{N_s}{N_0} = e^{-\chi_s/kT} , \qquad (4.7.5)$$

where k again means the Boltzmann constant.[9]

For example, if the energy level is g_s-fold degenerate, i.e. if it would split into g_s single levels on applying a suitable (magnetic) field, then we must ascribe to it a multiplicity or *statistical weight* g_s. Correspondingly, the ground state has a statistical weight of g_0. Then the *Boltzmann formula* holds generally:

$$\frac{N_s}{N_0} = \frac{g_s}{g_0} e^{-\chi_s/kT} ; \qquad (4.7.6)$$

its content is summarized graphically by the right-hand side of Fig. 4.7.4. If we wish to relate N_s not to the number of atoms in the ground state, N_0, but rather to the *total number* of all neutral atoms $N = \sum N_s$, then we find

$$\frac{N_s}{N} = \frac{g_s e^{-\chi_s/kT}}{\sum g_s e^{-\chi_s/kT}} . \qquad (4.7.7)$$

The quantity in the denominator is the important *partition function*:

$$Q = \sum g_s e^{-\chi_s/kT} . \qquad (4.7.8)$$

The *statistical weights* g_s are taken from the theory of atomic spectra: a level with the angular momentum quantum number J, for example, exhibits $2J+1$ different M_J in a magnetic field (Fig. 4.7.2) and therefore has the statistical weight:

$$g_J = 2J+1 . \qquad (4.7.9)$$

If we combine the levels of a multiplet term with the quantum numbers S and L, the term has the statistical weight

$$g_{S,L} = (2S+1)(2L+1) . \qquad (4.7.10)$$

The addition of the corresponding g_J's in Table 4.7.1, of course, leads to the same result.

The thermal excitation of the atoms into quantum states with higher and higher excitation energies χ_s as described by the Boltzmann formula (4.7.6) passes continuously into the region of positive-energy states with $E>0$. In this region, the atom receives the *ionization energy* χ (Fig. 4.7.4), which just suffices to remove an electron from the atom, as well as the kinetic energy $E = \frac{1}{2}mv^2$ with which the electron is ejected.

We now denote the number of (singly) ionized atoms by an additional index, i.e. N_1, and the number of neutral atoms by N_0 (for clarity), and correspondingly for higher ionization states:

Ionization State:	Neutral		Singly ionized		Doubly ionized		...	r-fold ionized
Free electrons per atom:	0		1		2		...	r
Ionization energy		χ_0		χ_1		$\chi_2 \cdots \chi_{r-1}$		
Spectra, e.g. iron, Fe:	Fe I		Fe II		Fe III		...	Fe$(r+1)$
Total atoms in the ionization state:	N_0		N_1		N_2		...	N_r
Atoms in the groundstate of the ion:	$N_{0,0}$		$N_{1,0}$		$N_{2,0}$		...	$N_{r,0}$
Atoms in a level $s, s' \ldots$:	$N_{0,s}$		$N_{1,s'}$		$N_{2,s''}$		...	$N_{r,s}$

We employ the corresponding notation for the statistical weights. We shall now calculate $N_{1,0}/N_{0,0}$, i.e. the ratio of the numbers of singly ionized to neutral atoms in their respective groundstates with the corresponding statistical weights $g_{1,0}$ and $g_{0,0}$.

Clearly, this problem can be reduced to the calculation of *the statistical weight of the ionized atom in its ground state plus its free electron*. The former has the statistical weight $g_{1,0}$; the latter by itself, corresponding to the two

[9] Formula (4.7.5) can be interpreted as a generalization of the *barometric pressure-altitude formula* (2.8.18), according to which the density profile in an isothermal atmosphere at temperature T is given as a function of the altitude h by

$$N(h)/N_0 = \exp\left(-\frac{mgh}{kT}\right) .$$

Here, m is the mass of the molecules and g the acceleration of gravity. mgh is thus the potential energy of a molecule at the altitude h above the surface; in (4.7.5), it corresponds to the excitation energy χ_s in the atom. However, while in classical statistics the potential energy may be continuously varied, in quantum statistics there are discrete states, with each nondegenerate state having the same statistical weight 1.

possible spin orientations in an applied field, has the statistical weight 2. In addition, we must include the statistical weight corresponding to the motion of the single free electron, i.e. the number of quantum cells h^3 in phase space.

In statistical mechanics, it can be shown that an electron of mass m occupies a volume $(2\pi m k T)^{3/2}$ in *momentum space*.[10] In real space, if we have N_e free electrons in a unit volume, they occupy a volume $1/N_e$ each. Thus we find the statistical weight for an

ionized atom in the groundstate +1 free electron $$g = g_{1,0} \cdot 2 \frac{(2\pi m k T)^{3/2}}{h^3 N_e} \; . \qquad (4.7.11)$$

If we insert this result into the Boltzmann formula (4.7.6), we immediately obtain the *Saha formula* with respect to the ground states of the atoms and ions

$$\frac{N_{1,0}}{N_{0,0}} N_e = \frac{g_{1,0}}{g_{0,0}} \cdot 2 \frac{(2\pi m k T)^{3/2}}{h^3} \mathrm{e}^{-\chi_0/kT} \; . \qquad (4.7.12)$$

For the total number of ionized or neutral atoms, we obtain using (4.7.7) the corresponding ionization formula

$$\frac{N_1}{N_0} N_e = \frac{Q_1}{Q_0} \cdot 2 \frac{(2\pi m k T)^{3/2}}{h^3} \mathrm{e}^{-\chi_0/kT} \; . \qquad (4.7.13)$$

Analogous expressions hold for the transition from the r-th to the $(r+1)$-th ionization state, where the $(r+1)$-th electron is ejected with the ionization energy χ_r, independently of other ionization processes:

$$\frac{N_{r+1}}{N_r} N_e = \frac{Q_{r+1}}{Q_r} \cdot 2 \frac{(2\pi m k T)^{3/2}}{h^3} \mathrm{e}^{-\chi_r/kT} \qquad (4.7.14)$$

etc. Instead of the number of electrons per unit volume, N_e, the *electron pressure* P_e can be just as well employed; it is the partial pressure of the free electrons

$$P_e = N_e k T \; . \qquad (4.7.15)$$

For numerical evaluations, it is often expedient to use, as a measure of the temperature, the quantity introduced by H.N. Russell

$$\Theta = \frac{5040}{T} \; ; \qquad (4.7.16)$$

one then obtains, after inserting the numerical constants, the Saha equation in logarithmic form

$$\log\left(\frac{N_{r+1}}{N_r} \cdot P_e\right) = -\chi_r \Theta + \frac{5}{2} \log T - 1.48 + \log \frac{2 Q_{r+1}}{Q_r} \; , \qquad (4.7.17)$$

where P_e is in [Pa = 10 dyn cm^{-2}], χ_r in [eV], and T in [K].

M.N. Saha originally derived his formula on the basis of thermodynamic calculations, employing the 3rd law of thermodynamics and the "chemical constant of electrons". The process of ionization of an atom A or the recombination of the ion A^+ with a free electron e^- is treated as a *chemical reaction*, whereby, in a state of chemical (thermodynamic) equilibrium, the reaction occurs with equal frequency in either direction:

$$A \leftrightharpoons A^+ + e^- \; . \qquad (4.7.18)$$

If we limit ourselves to sufficiently low pressures, the number of recombination processes (←) becomes proportional to the number of collisions between ions and electrons, i.e. $\propto N_1 N_e$. The number of radiation-induced ionization processes (→) will be proportional to the density of neutral atoms, N_0. The proportionality constants, which we have left undetermined here, depend only on the temperature T. Thus, one can understand the form of the ionization equation (4.7.13), and furthermore its generalization, as an application of Guldberg-Waage's *Law of Mass Action*,

$$\frac{N_1 N_e}{N_0} = \text{funct.}\,(T) \; . \qquad (4.7.19)$$

In Fig. 4.7.5, we have combined the formulas for thermal excitation and ionization for an electron pressure P_e = 10 Pa = 100 dyn cm^{-2} (which can be considered to be a rough average for the atmospheres of stars on the main sequence) and temperatures of 3000 to 50000 K, to give

[10] The a priori probability of a state with momentum $p = mv$ or kinetic energy

$$\frac{1}{2} m v^2 = \frac{p^2}{2m} \quad \text{is} \quad \exp\left(-\frac{p^2}{2m} \Big/ kT\right)$$

Integrated over momentum space, this gives

$$\int_0^\infty \exp\left(-\frac{p^2}{2m} \Big/ kT\right) \cdot 4\pi p^2 dp = (2\pi m k T)^{3/2}$$

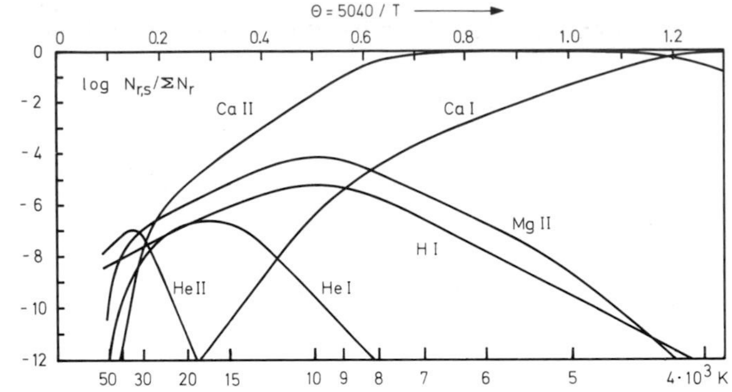

Fig. 4.7.5. Thermal ionization (4.7.14) and excitation (4.7.7) as functions of the temperature T or $\Theta = 5040/T$ for an electron pressure $P_e = 10$ Pa = 100 dyn cm^{-2} ($\simeq$average value for the atmospheres of main sequence stars). The temperature scale covers the whole region from the O stars (*left*) to the M stars (*right*). The Sun (G2) should be placed at about $T = 5800$ K. Our curves give an intuitive explanation of the interpretation of the sequence of spectral types given by M.N. Saha in 1920 (Sect. 4.5): for example, hydrogen (H I) is for the most part neutral up to $T = 10000$ K; the excitation of the 2nd quantum state, from which the Balmer lines in the visible spectral region are absorbed, increases with increasing T. Above 10000 K, hydrogen is rapidly ionized away. Thus, it can be understood that the hydrogen lines in the visible have their intensity maximum in the A0 stars with $T \simeq 10000$ K

Spectrum	Ionization energy χ_0 [eV]	State	Excitation energy $\chi_{r,s}$ [eV]
H I	13.60	$n = 2$	10.20
He I	24.59	$2\,^3P^o$	20.96
He II	54.42	$n = 3$	48.37
Mg II	15.04	$3\,^2D$	8.86
Ca I	6.11	$4\,^1S$	0.00
Ca II	11.87	$4\,^2S$	0.00

the fraction of H, He, Mg, and Ca atoms in several ionization and excitation states; the absorption lines of these states play a role in the MK classification of stellar spectra.

The maxima of the curves, which should correspond to the maximum intensity of the corresponding line(s), are due to the fact that within a particular ionization state, with increasing T at first the *excitation* increases. If T increases still further, this state will be "ionized away", so that the effective fraction of atoms again decreases. By assuming the temperatures of the maxima to be known (e.g., the maximum of the Balmer lines of hydrogen for the spectral class A0V is at $\simeq 10000$ K), R.H. Fowler and E.A. Milne in 1923 were able to estimate the mean electron pressure P_e in stellar atmospheres.

The Saha theory could furthermore qualitatively explain the increase of the intensity ratios of lines from singly ionized to neutral atoms on going from the main sequence stars to the giant stars as an increase in the ionization, i.e. of N_1/N_0, due to the lower pressure. On the other hand, the well-known differences between the spectra of sunspots and those of the normal solar atmosphere could be readily understood as being due to the lower temperature of the sunspots.

The ionization of *mixtures* of several elements is most simply calculated by considering the temperature T and the electron pressure P_e as independent parameters and then applying Saha's equation (4.7.13) to each element and its ionization states. The gas pressure P_g is easily calculated at the end as kT times the sum of *all* particles, including the electrons. Analogously, the *mean molecular mass* $\bar{\mu}$ can be written down. For example, completely ionized hydrogen has a molecular mass $\bar{\mu} = 0.5$, since on ionization, one proton and one electron are formed.

In Table 4.7.2, we have summarized the *composition of stellar matter* for reference in the theory of stellar atmospheres and of stellar interiors; this matter, of which as we shall see the Sun and most stars consist, is listed in order of decreasing atomic abundance. Table 4.7.3 gives the relation $P_e(P_g, T)$ for this mixture (the formation of hydrogen molecules at low temperatures is also taken into account). Under conditions characteristic of the atmospheres of main-sequence stars, $P_e \simeq 10$ Pa $\simeq$ 100 dyn cm^{-2}, at temperatures $T \simeq 10000$ K the stellar matter is nearly completely ionized and then $P_g/P_e \simeq 2$. At the solar temperature $T \simeq 6000$ K, essentially only the metals (Mg, Si, Fe) are ionized; corresponding to their abundances $\simeq 10^{-4}$, $P_g \simeq 10^4 P_e$. At still lower temperatures, the electrons are given up only by the most readily ionized group, Na, K, and Ca.

4.7.3 Kinetic Equations for Excitation and Ionization. Atomic Cross-Sections

The description of the excitation and ionization state of atoms according to the Boltzmann or Saha formulas (4.7.6, 12) is based on the validity of the assumption of thermodynamic equilibrium. Although in stellar interiors and for the most part also in stellar atmospheres (photospheres), the particle densities are sufficiently high that a thermodynamic equilibrium can be established through collision processes, this is not the case for the thinnest layers of the atmospheres, for gaseous nebulas,

Table 4.7.2. These elements (atomic number Z, atomic mass μ, and solar abundance relative to hydrogen = 100) make notable contributions to the electron pressure $P_e = N_e \cdot kT$ in stellar atmospheres ($P_e \simeq 10$ Pa $\simeq 100$ dyn cm^{-2}). In addition, the ionization energies and the statistical weights of the ground states for the first three ionization states are given. The three groups, separated according to the ionization potential of the neutral atom, χ_0, play a role in different temperature ranges

Z	Element	Atomic Mass μ	Abundance ε (H = 100)	Neutral atom		Singly ionized		Doubly ionized		At $P_e \simeq 10$ Pa important for
				χ_0 [eV]	g_0	χ_1 [eV]	g_1	χ_2 [eV]	g_2	
1	H Hydrogen	1.008	100	13.60	2	–	–	–	–	$T>5700$ K
2	He Helium	4.003	8.5	24.59	1	54.42	2	–	–	
12	Mg Magnesium	24.31	$2.6 \cdot 10^{-3}$	7.65	1	15.04	2	80.14	1	$6000>T>4500$ K
14	Si Silicon	28.09	$3.3 \cdot 10^{-3}$	8.15	9	16.35	6	33.49	1	
26	Fe Iron	55.85	$4.0 \cdot 10^{-3}$	7.87	25	16.16	30	30.65	25	
11	Na Sodium	23.00	$1.8 \cdot 10^{-4}$	5.14	2	47.29	1	71.64	6	$T<4700$ K
19	K Potassium	39.10	$8.9 \cdot 10^{-6}$	4.34	2	31.63	1	45.72	6	
20	Ca Calcium	40.08	$2.0 \cdot 10^{-4}$	6.11	1	11.87	2	50.91	1	

Table 4.7.3. The electron pressure P_e as a function of gas pressure P_g and temperature T for stellar matter of normal solar composition (Table 4.7.2). P_e and P_g are given in Pascal (1 Pa = 10 dyn cm^{-2})

T [K] \ P_g [Pa]	10^{-2}	10^{-1}	1	10^1	10^2	10^3	10^4	10^5
3000	$7.01 \cdot 10^{-7}$	$4.12 \cdot 10^{-6}$	$1.96 \cdot 10^{-5}$	$9.33 \cdot 10^{-5}$	$5.33 \cdot 10^{-4}$	$3.47 \cdot 10^{-3}$	$2.18 \cdot 10^{-2}$	$1.02 \cdot 10^{-1}$
4000	$2.36 \cdot 10^{-6}$	$1.26 \cdot 10^{-5}$	$1.00 \cdot 10^{-4}$	$8.69 \cdot 10^{-4}$	$5.98 \cdot 10^{-3}$	$3.09 \cdot 10^{-2}$	$1.38 \cdot 10^{-1}$	$6.78 \cdot 10^{-1}$
5000	$1.04 \cdot 10^{-4}$	$3.39 \cdot 10^{-4}$	$1.12 \cdot 10^{-3}$	$3.96 \cdot 10^{-3}$	$1.68 \cdot 10^{-2}$	$1.02 \cdot 10^{-1}$	$6.85 \cdot 10^{-1}$	3.64
6000	$1.49 \cdot 10^{-3}$	$5.38 \cdot 10^{-3}$	$1.78 \cdot 10^{-2}$	$5.77 \cdot 10^{-2}$	$1.88 \cdot 10^{-1}$	$6.31 \cdot 10^{-1}$	2.36	$1.09 \cdot 10^{1}$
8000	$4.77 \cdot 10^{-3}$	$4.58 \cdot 10^{-2}$	$3.54 \cdot 10^{-1}$	1.74	6.43	$2.14 \cdot 10^{1}$	$6.93 \cdot 10^{1}$	$2.24 \cdot 10^{2}$
10000	$4.82 \cdot 10^{-3}$	$4.80 \cdot 10^{-2}$	$4.77 \cdot 10^{-1}$	4.56	$3.46 \cdot 10^{1}$	$1.67 \cdot 10^{2}$	$6.13 \cdot 10^{2}$	$2.04 \cdot 10^{3}$
15000	$5.00 \cdot 10^{-3}$	$5.00 \cdot 10^{-2}$	$4.99 \cdot 10^{-1}$	4.96	$4.85 \cdot 10^{1}$	$4.74 \cdot 10^{2}$	$4.37 \cdot 10^{3}$	$2.94 \cdot 10^{4}$
20000	$5.00 \cdot 10^{-3}$	$5.00 \cdot 10^{-2}$	$5.00 \cdot 10^{-1}$	5.00	$5.00 \cdot 10^{1}$	$4.97 \cdot 10^{2}$	$4.86 \cdot 10^{3}$	$4.62 \cdot 10^{4}$
30000	$5.19 \cdot 10^{-3}$	$5.19 \cdot 10^{-2}$	$5.17 \cdot 10^{-1}$	5.08	$5.01 \cdot 10^{1}$	$5.00 \cdot 10^{2}$	$4.99 \cdot 10^{3}$	$4.96 \cdot 10^{4}$
40000	$5.19 \cdot 10^{-3}$	$5.19 \cdot 10^{-2}$	$5.19 \cdot 10^{-1}$	5.19	$5.18 \cdot 10^{1}$	$5.14 \cdot 10^{2}$	$5.04 \cdot 10^{3}$	$4.99 \cdot 10^{4}$

and especially for the interstellar medium. Here, we must return to the individual *elementary atomic processes.*

In general, we can assume that all the elementary processes occur so frequently that the plasma can be considered to be in a *stationary state.* Then the population density N_a for the energy state a of any particular particle species is constant on the average

$$\partial N_a/\partial t = 0 \ . \tag{4.7.20}$$

If we let P_{ab} [s^{-1}] be the probability or the rate per unit time with which a particle in state a is transformed into a (generally different) particle in state b, then the "reaction rate" is the number of such processes per unit volume and time, given by $N_a P_{ab}$. The *kinetic* or *statistical equilibrium* is then determined by the system of *rate equations*:

$$\frac{\partial N_a}{\partial t} = -N_a \sum_b P_{ab} + \sum_{b'} P_{b'a} = 0 \ , \tag{4.7.21}$$

which describes the balance for each energy state a going to all other states b and coming from all other states b'.

The atomic processes include on the one hand *radiative processes*, whose rate we denote by R_{ab}, and on the other, *collision processes* with a rate C_{ab}, so that we can set

$$P_{ab} = R_{ab} + C_{ab} \ . \tag{4.7.22}$$

In the following, we limit ourselves to excitation and ionization and their reverse processes for atoms and ions.

In the case of a *bound-bound transition* (Sect. 4.7.1) between states of energy E_m and E_n, a *spectral line* of frequency given by $h\nu = |E_m - E_n|$ is absorbed or emitted. According to A. Einstein, we can describe this interaction of radiation with matter in terms of the following transition processes:

a) *Spontaneous emission* of photons with the probability A_{mn}, associated with a transition from an excited state m to a lower-lying state n. The energy emitted per unit time is $h\nu A_{mn}$.

b) *Stimulated* or *induced emission*. An atom in an excited state m is caused by a photon of frequency ν to emit a photon of the same frequency and direction with a probability proportional to the intensity I_ν of the radiation field, $B_{mn}I_\nu$.

c) *Absorption* of a photon with the probability $B_{nm}I_\nu$ (the inverse process to stimulated emission), associated with the excitation of an atom from the state n into a higher state m. The total energy absorbed from the spectral line per unit time from a unit solid angle is $h\nu B_{nm}I_\nu/4\pi$.

The *rates* which correspond to these processes according to (4.7.22) are, when we simplify by taking into account only the angle-averaged intensity J_ν, given by $R_{nm} = B_{nm}J_\nu$ and $R_{mn} = A_{mn}+B_{mn}J_\nu$ (Fig. 4.7.6).

The following relations hold between the *Einstein coefficients* A_{mn}, B_{mn}, and B_{mn}:

$$A_{mn} = \frac{2h\nu^3}{c^2}B_{mn} \ ,$$

and

$$g_n B_{nm} = g_m B_{mn} \ , \tag{4.7.23}$$

where g_n and g_m are again the statistical weights of the levels n and m.

We have introduced another, phenomenological description of absorption in connection with the radiation transport equation, in the form of the absorption coefficient κ_ν or the *atomic absorption coefficient*, equal to the *absorption cross-section* $\kappa_{\nu,\mathrm{at}} = \kappa_\nu/N_\nu$ per particle in the lower state n (4.2.15). This absorption coefficient, which we initially treated *without* consideration of stimulated emission, will be denoted as $\tilde{\kappa}_\nu$. Its integral over the entire spectral line is:

$$\int\limits_{\mathrm{line}} \tilde{\kappa}_{\nu,\mathrm{at}}d\nu = \frac{1}{4\pi\varepsilon_0}\frac{\pi e^2}{mc}f = 2.65\cdot10^{-6}f\,[\mathrm{m^2\,s^{-1}}] \tag{4.7.24}$$

where e and m are the charge and mass of the electron and c the velocity of light.[11] $f \equiv f_{nm}$ is the (dimensionless) *oscillator strength*; it gives the "effective number

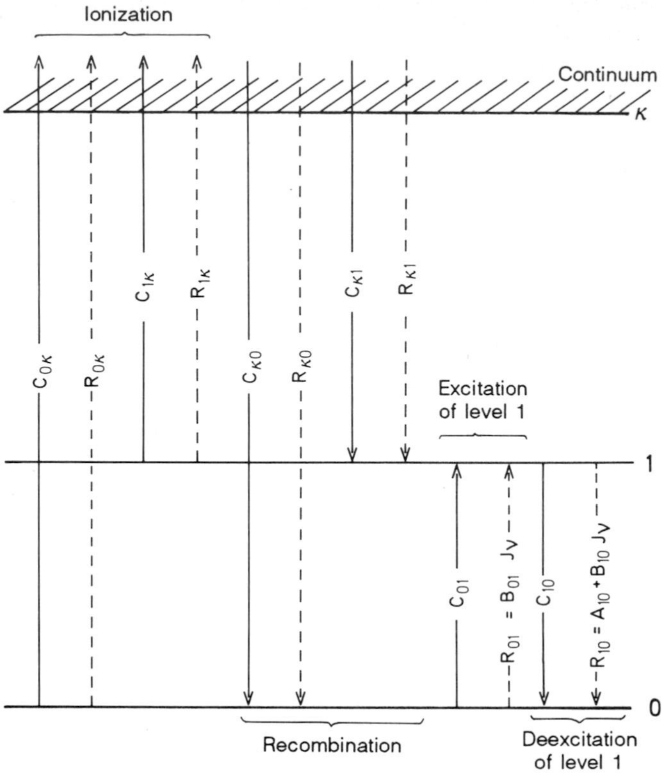

Fig. 4.7.6. Elementary processes in an atom which is schematically represented by two bound energy states 0 and 1 and the continuum κ (averaged over the Maxwell-Boltzmann distribution for free electrons). The dashed arrows indicate the radiative processes with their rates R_{ij}, and the fully-drawn arrows show impact processes with rates C_{ij}. A_{10}, B_{10}, and B_{01} are the Einstein coefficients for the line transition $0 \leftrightarrow 1$, and J_ν is the angle-averaged intensity in the corresponding line

of oscillators" and represents the quantum-mechanical extension of the expression already obtained in the framework of classical electron theory, which attempts to relate the spectral lines to harmonic electron-oscillators of the corresponding frequencies. The f value is related to the Einstein B coefficient by the equation

$$\frac{1}{4\pi\varepsilon_0}\frac{\pi e^2}{mc}f = \frac{h\nu}{4\pi}B_{nm} \ , \tag{4.7.25}$$

[11] In the Gaussian system, we find

$$\int\limits_{\mathrm{line}} \tilde{\kappa}_{\nu,\mathrm{at}}d\nu = \frac{\pi e^2}{mc}f$$

with $\pi e^2/mc = 2.65\cdot10^{-2}\,\mathrm{cm^2\,s^{-1}}$.

and can thus also be expressed in terms of the A coefficient using (4.7.23):

$$g_m A_{mn} = \frac{1}{4\pi\varepsilon_0}\frac{8\pi^2 e^2}{mc^3}\nu^2 g_n f$$
$$= \frac{6.670\cdot 10^{13}}{(\lambda\,[\mathrm{nm}])^2} g_n f \quad [\mathrm{s}^{-1}] \ . \qquad (4.7.26)$$

The frequency dependence of the spectral absorption coefficients is described by the *profile function* $\phi(\nu)$, normalized to one, so that

$$\tilde{\kappa}_{\nu,\mathrm{at}} = \frac{1}{4\pi\varepsilon_0}\frac{\pi e^2}{mc} f\phi(\nu)$$

and (4.7.27)

$$\tilde{\kappa}_\nu = \frac{1}{4\pi\varepsilon_0}\frac{\pi e^2}{mc} f\phi(\nu) N_n$$

with

$$\int_{\text{line}} \phi(\nu)\,d\nu = 1 \ .$$

We shall discuss the precise form of the profile function, which is resonant at $\nu = |E_m - E_n|/h$, in Sect. 4.9.2.

It is usual in the radiation transport equation (4.2.16) to combine the *stimulated emission* [12], proportional to the intensity I_ν, with the absorption $\tilde{\kappa}_\nu$ into the term $\kappa_\nu I_\nu$, i.e. to interpret it as a "negative absorption". Then, if we assume for simplicity the same profile function for emission and absorption, the *spectral absorption coefficient* becomes

$$\kappa_\nu = \frac{h\nu}{4\pi}(B_{nm}N_n - B_{mn}N_m)\phi(\nu) \ , \qquad (4.7.28)$$

or, with (4.7.23 and 25),

$$\kappa_\nu = \frac{h\nu}{4\pi} B_{nm} N_n \phi(\nu)\left(1 - \frac{N_m/g_m}{N_n/g_n}\right)$$
$$= \frac{1}{4\pi\varepsilon_0}\frac{\pi e^2}{mc} f N_n \phi(\nu)\left(1 - \frac{N_m/g_m}{N_n/g_n}\right) \ . \qquad (4.7.29)$$

The factor for stimulated emission in brackets is, for thermodynamic equilibrium using the Boltzmann formula (4.7.6), given by $[1 - \exp(-h\nu/kT)]$, and thus, especially in the Rayleigh-Jeans region ($h\nu/kT \ll 1$), leads to a strong reduction of the absorption by a factor $h\nu/kT$.

The contribution of a particular spectral line to the emission coefficient in the transport equation is, finally,

$$j_\nu = \frac{h\nu}{4\pi} A_{mn} N_m \phi(\nu) \ . \qquad (4.7.30)$$

A first-order feeling for the *absolute values* of the oscillator strengths can be obtained from the sum rule of W. Kuhn and W. Thomas: *from* a particular level n of an atom or ion with z electrons (as an approximation, z is normally restricted to the "valence electrons" which take part in the transitions considered), the absorption transitions $n \to m$ with oscillator strengths f_{nm} are possible; from lower-lying levels, the transitions $m' \to n$ with $f_{m'n}$ lead *into n*. Then

$$\sum_m f_{nm} - \sum_{m'} \frac{g_{m'}}{g_n} f_{m'n} = z \ . \qquad (4.7.31)$$

If essentially *one* strong transition leads from the ground-state term of an atom with z outer electrons, we can as an approximation set $f \simeq z$ for this multiplet. Thus, for example, we estimate for the two NaI D lines ($^2S - {}^2P^\circ$) *together* $f \simeq 1$. The exact value is $f = 0.98$; according to (4.7.26), the corresponding Einstein coefficient is $A = 0.6\cdot 10^9\,\mathrm{s}^{-1}$. The *relative* f values within a multiplet are given by quantum mechanical formulas of A. Sommerfeld and H. Hönl, H.N. Russell and others; e.g. the ratio gf for a doublet, such as the Na D lines $^2S_{1/2} - {}^2P^\circ_{1/2,3/2}$, is equal to 1 : 2; for a triplet, such as CaI $4p\ {}^3P^\circ_{2,1,0} - 5s\ {}^3S_1$ ($\lambda = 616$, 612, 610 nm; Fig. 4.7.3), it is equal to 5 : 3 : 1.

Precise transition probabilities can either be quantum-mechanically calculated (for simpler atoms or ions) or experimentally determined (emission from thermal sources such as arc discharges, absorption in a King oven, lifetime measurements using the beam-foil method ...). The voluminous data on transition probabilities and on the energy levels of atoms can be found in *compilations* ordered according to the atomic number of the elements, as well as in *bibliographies*; they are mainly collected and issued by the National Institute of Standards and Technology in Washington, D.C. Term schemes are given by Ch.E. Moore and P.W. Merrill, "Partial Grotrian Diagrams of Astrophysical Interest" (1968), and in

[12] In the case of lasers or masers, which also occur in cosmic sources (e.g., the OH maser, Sect. 5.3.4), the exact treatment of the stimulated radiation process is essential, but we will not go into details here.

several volumes by S. Bashkin and J. O. Stoner, "Atomic Energy-Level and Grotrian Diagrams" (from 1975).

Along with the bound-bound radiative processes, inelastic collision processes also lead to ("radiationless") transitions between the energy states of an atom. We limit ourselves here to *electronic collisions*, which are predominant at temperatures above a few 10^3 K; in cooler gases or in special cases, collisions by protons, hydrogen atoms, and hydrogen molecules, among others, would also have to be considered.

For the *excitation* of a level of energy E_m from a lower level E_n by electron impact, a minimal energy or velocity v_0 (relative to the atom) given by

$$\tfrac{1}{2} m v_0^2 = E_m - E_n$$

is required. The interaction cross section σ_{nm} for this process attains characteristic values of the order of $\pi a_0^2 = 8.80 \cdot 10^{-21}$ m^2 (a_0: first Bohr radius). In a given case, it depends on the relative velocity. Since an electron moves a distance v relative to the atom in a unit of time, the *collision rate* is obtained as an average over the velocity distribution:

$$C_{nm} = N_e \int_{v_0}^{\infty} \sigma_{nm}(v)\, v\, \Phi(v)\, dv = N_e \langle \sigma v \rangle \;, \tag{4.7.32}$$

where $N_e \Phi(v)\,dv$ is the number of electrons with a relative velocity between v and $v+dv$ per unit volume and N_e is the electron density. The integral of the distribution function $\Phi(v)$ over all velocities is equal to one.

In most astrophysical applications, a thermodynamic equilibrium in the electron system is established through collisions of the electrons among themselves; it corresponds to an *electron temperature* T_e, for which the *Maxwell-Boltzmann velocity distribution* holds:

$$\Phi_e(V)\,dV = \frac{4}{\sqrt{\pi}} \left(\frac{V}{V_0}\right)^2 \exp\left[-\left(\frac{V}{V_0}\right)^2\right] d\left(\frac{V}{V_0}\right) \tag{4.7.33}$$

with the most probable velocity

$$v_0 = \sqrt{\frac{2kT_e}{m}}$$

(m: mass of the electron). Because the electron's mass is small relative to that of the atoms, V is practically equal to the relative velocity v in (4.7.32) and $\Phi(v) = \Phi_e(V)$.

The rate of electronic collisions, C_{mn}, for the deexcitation $m \to n$ is related to C_{nm} through the "principle of detailed balance", which we cannot treat in detail here, by the equation:

$$g_m C_{mn} \equiv N_e \int_0^{\infty} g_m \sigma_{mn}(v)\, v\, \Phi(v)\, dv$$

$$= g_n C_{nm} \exp \frac{-(E_m - E_n)}{kT_e} \;. \tag{4.7.34}$$

The transition probability from m to n, taking into account collisions *and* radiative processes, is $A_{mn} + B_{mn} J_\nu + C_{mn}$.

The *ionization* of an atom or ion from a bound state n, whereby an electron makes a transition into a free state κ in the continuum, can occur on the one hand radiatively (bound-free transition, Sect. 4.7.1). Let $\alpha_{n\kappa}$ be the interaction cross section for this *photoionization*; then with the photon density N_ν (4.2.35) and the relative velocity c of the photons as seen by the atoms, the corresponding rate is

$$R_{n\kappa} = \int_{\nu_n}^{\infty} \alpha_{n\kappa}(\nu)\, c N_\nu\, d\nu = 4\pi \int_{\nu_n}^{\infty} \alpha_{n\kappa}(\nu) \frac{J_\nu}{h\nu}\, d\nu \;. \tag{4.7.35}$$

For hydrogen-like atoms having an effective nuclear charge Z, $\alpha_{n\kappa}$ decreases from the ionization edge $h\nu_n = E_{ion} - E_n$ (E_{ion}: ionization energy) to higher frequencies in proportion to ν^{-3}

$$\alpha_{n\kappa}(\nu) = \alpha_n \left(\frac{\nu}{\nu_n}\right)^{-3} \tag{4.7.36}$$

with $\alpha_n = 7.91 \cdot 10^{-22}\, n/Z^2$ [m^2] (n: principal quantum number).

On the other hand, an atom may be ionized by *electron impacts* with a rate:

$$C_{n\kappa} = N_e \int_{v_n}^{\infty} \sigma_{n\kappa}(v)\, v\, \Phi(v)\, dv \;, \tag{4.7.37}$$

where $\sigma_{n\kappa}$ is the cross section for impact ionization.

The cross sections or rates for the inverse processes of *recombination* can again be derived using the principle of detailed balance.

The interplay of the individual elementary processes is represented in Fig. 4.7.6 for a schematic atom, which consists of only two bound levels 0 and 1 and the continuum

κ. The index κ implies a summation over the Maxwell-Boltzmann distribution of free electron velocities. The *occupation numbers* of the atomic states are determined by three rate equations of the form (4.7.21) and by conservation of the overall particle density, $N = N_0 + N_1 + N_\kappa = \text{const}$. For example, the rate equation for the excited state 1 is given by:

$$N_1(R_{10}+C_{10}+R_{1\kappa}+C_{1\kappa}) = N_0(R_{01}+C_{01})+N_\kappa(R_{\kappa 1}+C_{\kappa 1}) \ . \quad (4.7.38)$$

Its mean lifetime with respect to transitions both into the ground state 0 and into the continuum κ is $(R_{10}+C_{10}+R_{1\kappa}+C_{1\kappa})^{-1}$.

To conclude our rather formal considerations in this section, we define the *source function* S_ν, i.e. the ratio of emission to absorption coefficients (4.2.21) for the spectral transition $0 \longleftrightarrow 1$ in a simple two-level atom, neglecting the continuum κ which is included in Fig. 4.7.6. According to (4.7.28 and 30), we have

$$S_\nu = \frac{A_{10}N_1}{B_{01}N_0 - B_{10}N_1} \ . \quad (4.7.39)$$

If we insert N_1/N_0 from the simplified-form rate equation (4.7.38),

$$N_1(A_{10}+B_{10}J_\nu+C_{10}) = N_0(B_{01}J_\nu+C_{01}) \ , \quad (4.7.40)$$

and then express C_{01} using (4.7.34) in terms of C_{10}, express B_{01} and B_{10} in terms of A_{10} using (4.7.23), and introduce the Kirchhoff-Planck function $B_\nu(T)$ (4.2.23), we obtain the source function in a straightforward form:

$$S_\nu = (1-\varepsilon)J_\nu + \varepsilon B_\nu(T) \ , \quad (4.7.41)$$

where the coefficient

$$\varepsilon = \frac{C_{10}(1-e^{-h\nu/kT})}{A_{10}+C_{10}(1-e^{-h\nu/kT})} \quad (4.7.42)$$

is a measure of the ratio of the collision rate for deexcitation of level 1 to the rate for spontaneous emission.

The system of kinetic equations for the occupation numbers developed in this section naturally contains the well-known relations for thermodynamic equilibrium as a limiting case; e.g. from (4.7.38), we must obtain the Boltzmann formula for N_1/N_0 (4.7.6), and the Saha formula (4.7.13) for $N_\kappa/(N_0+N_1)$. Using the representation (4.7.41) for the source function, the conditions for a transition to the *limiting case of thermodynamic equilibrium* can be readily demonstrated: we obtain Kirchhoff's theorem (4.2.30), i.e. $S_\nu = B_\nu(T)$, which must hold in thermodynamic equilibrium, either when collisional processes dominate radiative processes ($C_{10} \gg A_{10}$ or $\varepsilon \simeq 1$), or when the radiation field is that of a black body ($J_\nu = B_\nu$), or when both of these conditions are fulfilled.

4.8 The Physics of Stellar Atmospheres

With an eye to the quantitative explanation and evaluation of spectra from the Sun and the stars, we turn now to the physics of stellar atmospheres. The *atmosphere* by definition includes the layers of a star which send us radiation directly. In Sect. 4.8.1, we discuss the characteristic parameters of a stellar atmosphere and the basic equations for its structure. In Sect. 4.8.2, we consider the absorption coefficients in stellar atmospheres in more detail; they describe the interaction between the radiation field and matter. Finally, in Sect. 4.8.3 we treat the construction of a model atmosphere and show calculated spectral energy distributions for some stars. The exact treatment of the spectral lines is taken up in Sect. 4.9.

4.8.1 The Structure of Stellar Atmospheres

We limit ourselves here to *compact* atmospheres, whose thickness is small compared to the radius of the star, so that we can approximate the layer structure as a series of parallel planes and can assume the acceleration of gravity, cf. (2.8.14), to be constant.

A compact stellar atmosphere is characterized by the following *parameters*:

1) The *effective temperature* T_{eff}, which is defined in such a way that the *radiation energy flux*

$$F = \int_0^\infty F_\lambda d\lambda = \int_0^\infty F_\nu d\nu = \sigma T_{\text{eff}}^4 \quad (4.8.1)$$

is given by the Stefan-Boltzmann radiation law (4.4.20).

For the Sun, in (4.3.12) we obtained $T_{\text{eff},\odot} = 5780$ K directly from the solar constant, $F_\odot = 6.33 \cdot 10^7$ Wm^{-2}.

2) The *surface gravity* g of the star. For the Sun, we found in (4.3.6) $g_\odot = 274$ m s^{-2}; the gravitational ac-

celeration of the other main-sequence stars is, according to (4.6.7), not very different from this value.
3) The *chemical composition* of the atmosphere, i.e. the relative abundances of the elements.

In Table 4.7.2, we have already given some information on the latter quantities.

Possible additional parameters, such as rotation or oscillations of the star, stellar magnetic fields, etc., will not be considered.

One can readily reach the conclusion that with a given T_{eff}, g, and the chemical composition, the *structure*, i.e. the temperature and pressure distribution in a static stellar atmosphere, can be completely calculated. For this purpose, we need two equations (in the case of local thermodynamic equilibrium, see below): (a) an equation for *energy transport*, by radiation, convection, heat conduction, mechanical or magnetic energy, which determines the temperature distribution in the atmosphere; and (b) the *hydrostatic equation*, or, generally speaking, the basic equations of hydrodynamics or magnetohydrodynamics which determine the pressure distribution. The degree of ionization, the equation of state, and all the material constants of stellar atmospheres can in principle be calculated from the results of atomic physics, so long as we know the chemical composition. The theory of stellar atmospheres thus makes it possible, beginning with parameters 1, 2, and 3, to calculate a so-called *model atmosphere*, and furthermore to find out how certain *measurable quantities*, e.g. the intensity distribution in the continuum (the color indices), or the intensities of the Fraunhofer lines of certain elements in particular ionization and excitation states, depend on these parameters.

Once we have solved this problem in theoretical physics, we can (and this is the decisive point!) reverse the process and, in a series of successive approximations, find the answer to the question, "Which T_{eff}, g and abundance distribution of the chemical elements does the atmosphere of a particular star have, when its color indices (energy distribution in the continuum), intensities of various Fraunhofer lines, etc. have given *measured* values?" In this way, we arrive at a procedure for the *quantitative analysis* of the spectra of the Sun and stars.

The *temperature distribution* in a stellar atmosphere is determined, as mentioned above, by the type of energy transport. As was recognized in 1905 by K. Schwarzschild, the energy transport in most stellar atmospheres takes place predominantly through radiation; we thus speak of *radiation equilibrium*.

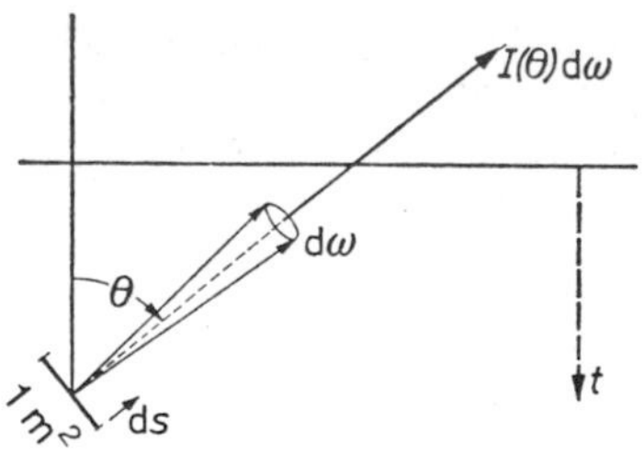

Fig. 4.8.1. Radiation equilibrium

In order to describe the *radiation field*, we imagine an element of unit surface area at a depth t (measured from an arbitrarily chosen zero level; see Fig. 4.8.1); its surface normal makes an angle θ with that of the stellar surface ($0 \leq \theta \leq \pi$). Then, within a solid angle element $d\omega$, the radiation energy $I_\nu(t,\theta)\,d\nu\,d\omega$ in the frequency interval ν to $\nu+d\nu$ passes through this surface element in the direction of its normal per unit time (Sect. 4.2). Along a path element $ds = -dt/\cos\theta$, the radiation intensity I_ν suffers an attenuation due to absorption, according to (4.2.14), of $-I_\nu(t,\theta)\kappa_\nu ds$ (where κ_ν in [m^{-1}] is the absorption coefficient per unit length at the frequency ν under consideration). On the other hand, I_ν is increased by emission, which we can calculate under the assumption of *local* thermal equilibrium (LTE), i.e. by applying Kirchhoff's theorem (4.2.30) to each individual volume element in the atmosphere, to be $+\kappa_\nu B_\nu(T)\,ds$, where $B_\nu(T)$ is the *Kirchhoff-Planck function* (4.2.23) for the local temperature T at the depth t and frequency ν. All together, (4.2.16) holds:

$$dI_\nu(t,\theta) = I_\nu(t,\theta)\kappa_\nu dt/\cos\theta - B_\nu(T(t))\kappa_\nu dt/\cos\theta \ . \tag{4.8.2}$$

In order to describe the different layers of the atmosphere, it is preferable to use the *optical depth* for radiation of frequency ν according to (4.2.19) instead of the geometric depth t:

$$\tau_\nu = \int_{-\infty}^{t} \kappa_\nu dt \quad \text{or} \quad d\tau_\nu = \kappa_\nu dt \ ; \tag{4.8.3}$$

it is calculated from the viewpoint of the observer ($t = -\infty$). We thus obtain from (4.8.2) the *radiation or transport equation* for radiation at frequency ν:

$$\cos\theta \frac{dI_\nu(t,\theta)}{d\tau_\nu} = I_\nu(t,\theta) - B_\nu(T(t)) \ . \tag{4.8.4}$$

If radiation equilibrium holds, i.e. if the overall energy transport takes place via radiation, we can apply energy conservation, which tells us that the *total radiation flux* must be independent of the depth t, i.e.

$$F=\int_{\theta=0}^{\pi}\int_{\nu=0}^{\infty} I_\nu(t,\theta)\cos\theta\, 2\pi\sin\theta\, d\theta\, d\nu=\sigma T_{\mathrm{eff}}^4 . \quad (4.8.5)$$

The boundary conditions of the system (4.8.4 and 5) are: (a) the incident radiation vanishes at the stellar surface, i.e. $I_\nu(0,\theta)=0$ for $0\leq\theta\leq\pi/2$, and (b) at great depths, the "enclosed" radiation field, which approximates that of a cavity radiator, cannot increase more rapidly than exponentially with increasing depth (we cannot treat this point in detail here). The solution of the system of equations now becomes relatively simple, if we first assume that the absorption coefficient is *in*dependent of the frequency ν; that is, we insert in (4.8.4) in place of κ_ν its harmonic mean value (with suitable weighting factors) averaged over all frequencies ν, i.e. the *Rosseland opacity coefficient* $\bar{\kappa}$ and the corresponding optical depth:

$$\bar{\tau}=\int_{-\infty}^{t}\bar{\kappa}\, dt . \quad (4.8.6)$$

For such a "grey atmosphere", according to E.A. Milne and others, the solution of (4.8.4, 5) yields to a good approximation the temperature distribution

$$T^4(\bar{\tau})=\tfrac{3}{4}T_{\mathrm{eff}}^4(\bar{\tau}+\tfrac{2}{3}) . \quad (4.8.7)$$

The effective temperature T_{eff} is thus obtained at an optical depth $\bar{\tau}=\frac{2}{3}$. At the stellar surface $\bar{\tau}=0$, T approaches a finite limiting temperature $T_0=T_{\mathrm{eff}}/\sqrt[4]{2}=0.84\ T_{\mathrm{eff}}$. It is frequently expedient to use an optical depth τ_0 instead of $\bar{\tau}$; it refers to the absorption coefficient κ_0 at a suitably chosen wavelength (e.g. at $\lambda=500$ nm). The relation between $\bar{\tau}$ and τ_0 (and correspondingly also between the optical depths for two different wavelengths) is readily found from $d\bar{\tau}=\bar{\kappa}\, dt$ and $d\tau_0=\kappa_0 dt$ to be

$$\frac{d\bar{\tau}}{d\tau_0}=\frac{\bar{\kappa}}{\kappa_0} \quad (4.8.8)$$

at a particular depth.

If the temperature distribution $T(\bar{\tau})$ or $T(\tau_0)$ in a stellar atmosphere is known, the calculation of the *pressure distribution* offers no difficulties. The increase of gas pressure P_g with depth t in a static atmosphere is determined by the *hydrostatic equation*

$$dP_g/dt=g\varrho , \quad (4.8.9)$$

where the density ϱ is related to P_g and T by the equation of state for an ideal gas; g again means the surface gravity. Dividing both sides by κ_0 and using $\kappa_0 dt=d\tau_0$ we obtain the relation

$$\frac{dP_g}{d\tau_0}=\frac{g\varrho}{\kappa_0} . \quad (4.8.10)$$

Since the right-hand side of this equation is known as a function of P_g and T, and since T is known as a function of τ_0 from the theory of radiation equilibrium, (4.8.10) can be (numerically) integrated without difficulty.

In hot stars, the radiation pressure must be taken into account in addition to the gas pressure. If currents are present in the stellar atmosphere, and their velocity v is no longer small compared to the local velocity of sound (in atomic hydrogen at 10000 K, for example, 12 km s^{-1}), then the dynamic pressure $\varrho v^2/2$ also plays a role. In sunspots and in the Ap stars with their strong magnetic fields, the magnetic forces must also be taken into consideration.

For a real analysis of stellar spectra, the "grey approximation" (4.8.7) is too imprecise; we need to take into account the frequency dependence of the absorption coefficient κ_ν, which is composed of the continuous absorption coefficient and of the very strongly frequency-dependent resonance line absorption (Sect. 4.8.2). In the radiation equilibrium, the total radiation flux F must then be equal at all depths t. The theory of the radiation equilibrium in such "non-grey" atmospheres is fraught with considerable mathematical difficulties, which can only be overcome by successive approximation methods.

For the calculation of stellar atmospheric models, besides the absorption coefficient κ_ν or $\bar{\kappa}$ as a function of temperature and pressure (Sect. 4.8.2), we need the relation between gas and electron pressure, $P_e(P_g, T)$ for various mixtures of elements (Table 4.7.3), and, in the equation of state, the mean molecular mass $\bar{\mu}$. For neutral (atomic) or fully ionized stellar matter having the solar elemental composition (Table 4.7.2), $\bar{\mu}$ is 1.26 or 0.60.

In the deep layers of the atmospheres of later spectral classes, energy transport by *convection* is predominant over radiative transport (Sect. 4.10.6). In this case, the condition of radiation equilibrium (4.8.5) must be extended to include the *total* energy flux from radiation plus convection, which must be independent of depth.

4.8.2 Absorption Coefficients in Stellar Atmospheres

We first discuss the *continuous* absorption coefficients, which are due to bound-free and free-free transitions (Sect. 4.7.1) of various atoms and ions.

In the case of *neutral hydrogen*, a series-limit continuum lies adjacent to the limits of the Lyman series at $\lambda = 91.2$ nm, the Balmer series at $\lambda = 364.7$ nm, and the Paschen series at $\lambda = 820.6$ nm; their absorption decreases as ν^{-3}, see (4.7.36). At long wavelengths, the series-limit continua overlap and pass over smoothly into the free-free continuum of H I.

As was noticed in 1938 by R. Wildt, in cooler stellar atmospheres, the bound-free and free-free transitions in the *negative hydrogen ion* H^-, which can be formed from a neutral hydrogen atom by electron attachment, play an important role. The H^- ion, with only one bound state (the ground state ^1S) has a binding energy or ionization energy of 0.75 eV. Corresponding to this small energy, the long-wavelength limit of the bound-free continuum lies in the infrared at 1.655 μm. The free-free absorption also increases (as in the H atom) at longer wavelengths.

The *atomic coefficients* for absorption from a particular energy level have been calculated quantum-mechanically for the H atom and the H^- ion with great precision. In order to obtain from them the absorption coefficient κ_ν of stellar matter of a given composition (in local thermodynamic equilibrium) as a function of the frequency ν, the temperature T, and the electron pressure P_e (or the gas pressure P_g), the ionization of the various elements and the excitation of their energy levels must first be calculated using the formulas of Boltzmann and Saha.

For the "normal" element mixture (Table 4.7.2), one first finds that in hot stars ($T \geq 7000$ K), the continuous absorption of H atoms, in cooler stars that of the H^- ions, is predominant. Although for example in the Sun's atmosphere (at roughly $\tau_0 = 0.1$, Fig. 4.8.2) the particle density of the H^- ion is only about 10^{-8} of that of H I, it nevertheless dominates the continuous absorption in the visible region, because the relative occupation density of the second excited H I level (12.1 eV), from which the "competing" absorption in the Paschen continuum originates, is much lower (about 10^{-11}), and the atomic cross sections are of comparable magnitude.

In addition to the absorption of H I and H^-, the following must be considered, in particular: bound-free and free-free absorption of He I and He II (in hot stars) and of the heavier elements C I, Mg I, and Si I (in cooler stars); also, in the coolest stars, the continuum of H_2^-. Along with these absorption processes, the *scattering* of light is important: in hot stars, Thomson scattering by free electrons, σ_T (3.4.6); and in cooler stars, Rayleigh scattering by neutral hydrogen atoms.

In Fig. 4.8.2 and 3, we give as examples the continuous absorption coefficients (from calculations by G. Bode, 1965) as functions of the wavelength for mean values of the state variables (Sect. 4.8.3) in the atmosphere of a cooler star: the Sun; and of a hotter star: τ Sco (B0V).

In calculating the temperature distribution, in addition to the continuous absorption, the *spectral line absorption* must also be considered. The optical spectra (Fig. 4.5.3) at first give the impression that absorption in the spectral lines is important only in cooler stars, e.g. the Sun, with their numerous Fraunhofer lines, but not in hotter stars. However, observations from satellites indicate that even in the case of the O and B stars, in the region of their

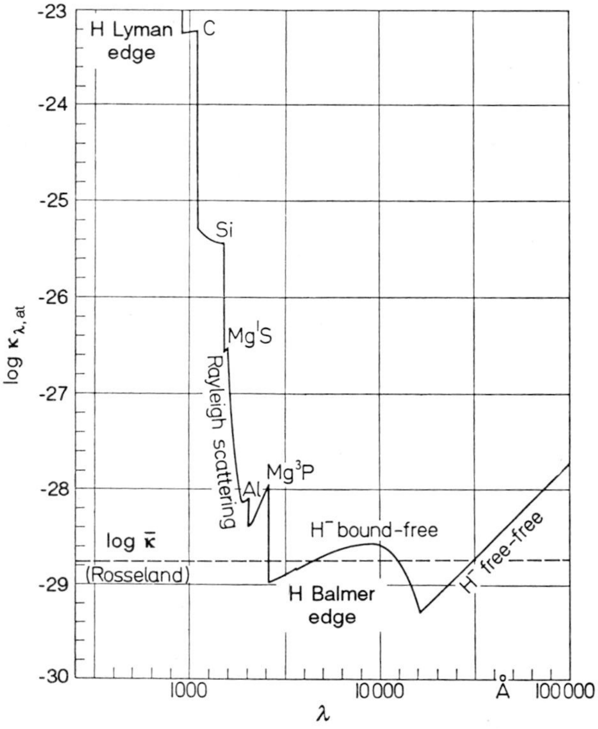

Fig. 4.8.2. The continuous absorption coefficient $\kappa_{\lambda,\,\mathrm{at}}$ per nucleus in [m^2] in the solar atmosphere (G2V) at $\tau_0 \simeq 0.1$ (τ_0 for $\lambda = 500$ nm), i.e. $T = 5040$ K, $P_e \simeq 0.32$ Pa, $P_g \simeq 5.8 \cdot 10^3$ Pa

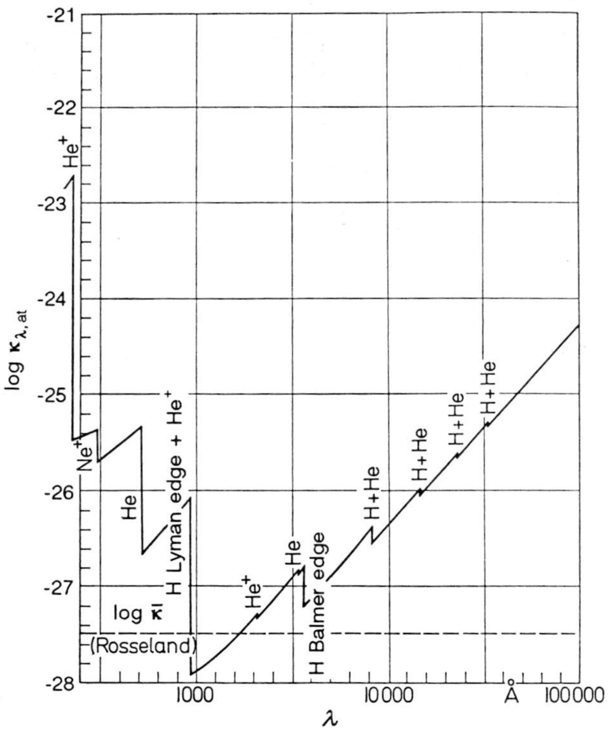

Fig. 4.8.3. The continuous absorption coefficient $\kappa_{\lambda,\,\mathrm{at}}$ per nucleus in [m^2] in the atmosphere of τ Scorpii (B0 V) at $\bar{\tau}_0 \simeq 0.1$, i.e. $T = 28\,300$ K, $P_e \simeq 320$ Pa, $P_g \simeq 640$ Pa

maximum energy fluxes (in the ultraviolet, $\lambda \simeq 200$ nm), the absorption lines are crowded together in a similar way as for the Sun in the blue spectral region.

For the calculation of a model atmosphere, in general a very large number ($\geq 10^6$) of lines must be taken into account. In this process, only a few strong lines can be calculated individually (and precisely); the majority of the medium-strong and weak lines are treated by *statistical* methods (e.g., sampling methods or distribution functions), in which great precision for the individual lines is not required. We shall discuss the theory of spectral line absorption coefficients in Sect. 4.9. Since the lines reduce the outflow of radiation from the depths, but increase its intensity in the higher layers, they tend to cause a steeper temperature decrease in the outer layers as compared with the temperature distribution calculated on the basis of continuous absorption only ("blanketing effect").

4.8.3 Model Atmospheres. The Spectral Energy Distribution

Having described the equations which govern the structure of a stellar atmosphere, the absorption coefficients, and other material functions, we present in Table 4.8.1 the *models* for atmospheres at various effective temperatures calculated according to the current state of the art (and with the assumption of local thermodynamic equilibrium). The details of the stellar atmosphere model calculations, which are carried out with the aid of large computers, go beyond the scope of this book. Table 4.8.1 contains the model of a cool star (the Sun, G2V), a star of medium temperature (α Lyr, A0V), and a hot main-sequence star (B0V), represented by τ Sco, which has been

Table 4.8.1. Models for stellar atmospheres by R. L. Kurucz (1979), for the solar element abundance distribution and various effective temperatures T_{eff} and surface gravities g. The line absorption is taken into account using opacity distribution functions; the optical depth τ_0 refers to κ_λ at $\lambda = 500$ nm, and $\bar{\tau}$ to the Rosseland mean $\bar{\kappa}$

	Sun G2 V $T_{\mathrm{eff}} = 5770$ K, $g = 274\ \mathrm{m\,s^{-2}}$				α Lyr A0 V $T_{\mathrm{eff}} = 9400$ K, $g = 89\ \mathrm{m\,s^{-2}}$				B0 V $T_{\mathrm{eff}} = 30\,000$ K, $g = 100\ \mathrm{m\,s^{-2}}$ *			
$\bar{\tau}$	τ_0	T [K]	P_g [Pa]	P_e [Pa]	τ_0	T [K]	P_g [Pa]	P_e [Pa]	τ_0	T [K]	P_g [Pa]	P_e [Pa]
10^{-3}	$1.1\cdot10^{-3}$	4485	$3.46\cdot10^2$	$2.84\cdot10^{-2}$	$0.6\cdot10^{-3}$	7140	6.52	$4.31\cdot10^{-1}$	$0.8\cdot10^{-3}$	19680	2.13	1.07
0.01	0.01	4710	$1.29\cdot10^3$	$1.03\cdot10^{-1}$	$0.5\cdot10^{-2}$	7510	$2.70\cdot10^1$	1.61	$0.9\cdot10^{-2}$	21450	$1.60\cdot10^1$	7.98
0.10	0.09	5070	$4.36\cdot10^3$	$3.78\cdot10^{-1}$	0.05	8150	$9.13\cdot10^1$	7.33	0.14	24880	$1.01\cdot10^2$	$5.03\cdot10^1$
0.22	0.19	5300	$6.51\cdot10^3$	$6.43\cdot10^{-1}$	0.11	8590	$1.22\cdot10^2$	$1.40\cdot10^1$	0.37	27030	$1.86\cdot10^2$	$9.31\cdot10^1$
0.47	0.40	5675	$9.55\cdot10^3$	1.34	0.24	9240	$1.53\cdot10^2$	$2.94\cdot10^1$	0.92	29840	$3.33\cdot10^2$	$1.66\cdot10^2$
1.0	0.84	6300	$1.29\cdot10^4$	4.77	0.53	10190	$1.79\cdot10^2$	$5.81\cdot10^1$	2.2	33490	$5.87\cdot10^2$	$2.95\cdot10^2$
2.2	1.8	7085	$1.52\cdot10^4$	$2.13\cdot10^1$	1.3	11560	$2.12\cdot10^2$	$9.21\cdot10^1$	5.5	38310	$1.04\cdot10^3$	$5.29\cdot10^2$
4.7	3.5	7675	$1.71\cdot10^4$	$5.86\cdot10^1$	3.6	13480	$2.99\cdot10^2$	$1.40\cdot10^2$	13.3	43940	$1.81\cdot10^3$	$9.43\cdot10^2$
10	7.1	8180	$1.89\cdot10^4$	$1.27\cdot10^2$	11.5	16000	$5.81\cdot10^2$	$2.77\cdot10^2$	37	51310	$3.60\cdot10^3$	$1.88\cdot10^3$

* Roughly equal to the parameter values for τ Sco (B0 V), corresponding to $T_{\mathrm{eff}} = 31\,500$ K and $g = 140\ \mathrm{m\,s^{-2}}$.

intensively investigated following the pioneering work by A. Unsöld (1941/44).

The radiation which is emitted at the surface of a star originates from various layers within its atmosphere; that from the deeper layers naturally undergoes more attenuation by absorption before emerging at the stellar surface than does radiation from further up. Radiation of frequency ν, emitted from a depth τ_ν at an angle θ relative to the surface normal of the atmosphere, is attenuated by the absorption factor $\exp(-\tau_\nu/\cos\theta)$ before leaving the atmosphere. We thus obtain from (4.8.2 or 4) the *radiation intensity* at the surface, $\tau_\nu = 0$:

$$I_\nu(0,\theta) = \int_0^\infty S_\nu(\tau_\nu) e^{-\tau_\nu/\cos\theta} d\tau_\nu/\cos\theta \ , \tag{4.8.11}$$

where the source function or yield S_ν, defined as the emission coefficient j_ν divided by the absorption coefficient κ_ν [see (4.2.21)], is in the case of local thermodynamic equilibrium equal to the Kirchhoff-Planck function at the local temperature, $B_\nu(T(\tau_\nu))$.

Since we can, on the one hand, calculate the *temperature* T as a function of the optical depth (e.g. τ_ν) using the theory of radiation equilibrium, and since on the other, the relation between the differently-defined optical depths $\bar{\tau}$, τ_ν, τ_0 ... can be established with (4.4.8) using the absorption coefficients κ_ν, the entire theory of *stellar spectra* is contained in (4.8.11), including (for the Sun) the *center-limb variation* ($\theta = 0$ corresponds to the $\odot$-center, $\theta = 90°$ to the edge).

Indeed, calculations (which we shall not describe in detail here), carried out on the basis of the model of the *solar atmosphere* in Table 4.8.1, yield the solar spectrum $I_\nu(0,0)$ [e.g., for the $\odot$-center (Fig. 4.3.4)], and the center-limb variation in the continuum (Fig. 4.3.3) with good accuracy.

In order to determine which layers of the atmosphere make the *principal contribution* to the emitted *intensity* at a frequency ν, we can carry out a greatly simplified calculation of $I_\nu(0,\theta)$ by expressing the source function $S_\nu(\tau_\nu)$ or $B_\nu(\tau_\nu)$ at a particular optical depth τ^*, which we shall initially leave undetermined, in the form of a series expansion:

$$S_\nu(\tau_\nu) = S_\nu(\tau^*) + (\tau_\nu - \tau^*)\left(\frac{dS_\nu}{d\tau_\nu}\right)_{\tau^*} + \dots \tag{4.8.12}$$

If we carry out the integration (4.8.11) using this approach, we readily obtain the result:

$$I_\nu(0,\theta) = S_\nu(\tau^*) + (\cos\theta - \tau^*)\left(\frac{dS_\nu}{d\tau_\nu}\right)_{\tau^*} + \dots \tag{4.8.13}$$

If we now set $\tau^* = \cos\theta$, the second term on the right vanishes, and we obtain the so-called *Eddington-Barbier approximation*:

$$I_\nu(0,\theta) \simeq S_\nu(\tau_\nu = \cos\theta) \ . \tag{4.8.14}$$

This means that on the surface of, e.g., the Sun, the emitted intensity corresponds to the source function or, in the sense of Planck's radiation law, to the temperature at the optical depth $\tau_\nu = \cos\theta$, measured perpendicular to the Sun's surface; i.e. $\tau_\nu/\cos\theta = 1$, measured along the line of sight from the observer. This is intuitively clear: we can say (with a grain of salt) that the radiation, measured along the line of sight, arises from the layer at optical depth equal to one.

For the *stars* we can now find the mean intensity of the stellar disk, that is (up to a factor π) the *radiation flux* at the surface. According to (4.2.7), we have:

$$F_\nu(0) = 2\pi \int_0^{\pi/2} I_\nu(0,\theta)\cos\theta\sin\theta d\theta \ . \tag{4.8.15}$$

Applying the *ansatz* of (4.8.13), we obtain the approximation

$$F_\nu(0) \simeq \pi S_\nu(\tau_\nu = \tfrac{2}{3}) \ . \tag{4.8.16}$$

With local thermodynamic equilibrium (4.2.27), the stellar radiation at a frequency ν thus corresponds to the local temperature at an optical depth $\tau_\nu = \frac{2}{3}$.

To illustrate the energy distribution of stellar radiation, we show in Figs. 4.4.2, 4 the radiation fluxes calculated for main-sequence stars with $T_{\rm eff} = 9400$ K (A0V) and 30000 K (B0V), using the models in Table 4.8.1. All similarity to the spectrum of a black-body has now vanished. Instead, the F_λ curve represents more or less a mirror image of the wavelength dependence of the absorption coefficients κ_λ or κ_ν (compare the wavelength dependence of the continuous absorption coefficient in Fig. 4.8.3). Since $d\tau_\lambda/d\bar{\tau} = \kappa_\lambda/\bar{\kappa}$, for *large* κ_λ the radiation comes from layers near the surface with *small* $\bar{\tau}$, these layers are relatively *cool*, and the radiation curve F_λ is decreased.

We can therefore understand that at each absorption edge of hydrogen ($\lambda = 91.2$ nm, Lyman edge; $\lambda = 364.7$ nm, Balmer edge, ...), F_λ shows a sudden decrease towards shorter wavelengths. (We shall discuss the energy flux in the Fraunhofer lines later, in Sect. 4.9.1). The hotter of the two stars emits its radiation flux mainly between the Balmer and the Lyman edges; in the case of the cooler star, the region between the Paschen and the Balmer edges contributes about the same order of magnitude to the emitted radiation as does the ultraviolet. It is clear that for hot stars, observations in the far ultraviolet from rockets and satellites are of decisive importance; measurements in the optical region, which for a long time formed the exclusive basis of our knowledge of these stars, must be extremely precise in order to make a reasonable contribution to the adjustment of theoretical calculations to the observations. This is particularly true of the *bolometric correction*, B.C. (4.4.19, 21), which is obtained by integration of F_λ over the whole spectrum and comparison with the radiation flux in the visual region; for example, for a B0V star, B.C. is $\simeq 3.2$ mag.

Finally, Fig. 4.8.4 clarifies the role of *spectral line absorption* in the far ultraviolet for hot stars: the energy distribution taking into account the Fraunhofer lines (they are "smeared out" over 2.5 nm wide wavelength regions) lies well below the "true" continuum. For the A stars

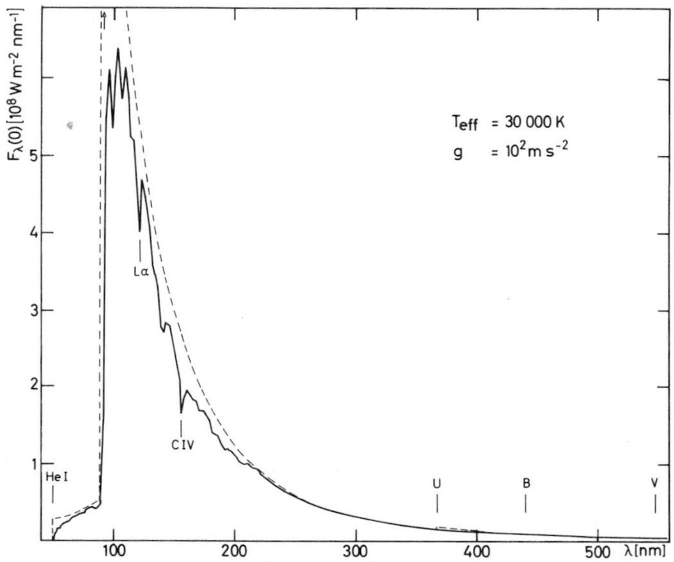

Fig. 4.8.4. The calculated spectral energy distribution $F_\lambda(0)$ of a main sequence star with an effective temperature $T_{eff} \simeq 30000$ K (from R. L. Kurucz, 1979). (———) Radiation flux including the Fraunhofer lines, which were calculated using opacity distribution functions with a spectral resolution of 2.5 nm. (– – –) Radiation flux in the continuum, maximum at $\lambda = 92$ nm (*arrow*) with the value $10.1 \cdot 10^8$ W m^{-2} nm^{-1}

(Fig. 4.4.2), the ultraviolet region is dominated by numerous lines from metallic elements, while in the optical region, the broad Balmer lines from hydrogen must be taken into account. In cooler stars, roughly below F5, the metal lines block out a major portion of the spectrum, especially in the blue and near ultraviolet regions (Fig. 4.3.4); finally, in the K and M stars, the absorption from molecular bands also comes into play (Table 4.5.1).

When greater precision is required, *deviations from local thermodynamic equilibrium* must be considered within the framework of the kinetic equations described in Sect. 4.7.3 for calculating the model atmospheres. These "non-LTE effects" become ever more important the hotter and brighter the star is, since then radiation processes become more important relative to collisional processes. For example, on the main sequence, the O and B stars show notable deviations from LTE, in particular for the occupation numbers of the lower energy states of hydrogen and helium.

Thus far, we have limited ourselves in this section and in Sect. 4.3 to a discussion of the *photospheres* of the stars and the Sun, where the absorption line spectrum is produced. We shall meet up with the thinner layers of the atmospheres lying above the photosphere (chromospheres, coronas, ...), for which local thermodynamic equilibrium is no longer a reasonable approximation and in which matter currents and magnetic fields play an important role in addition to the radiation field, in Sects. 4.10 and 4.11. First, however, we turn our attention to a more exact theory of the Fraunhofer lines, which is necessary for the quantitative analysis of the stellar spectra and for the determination of elemental abundances.

4.9 Theory of the Fraunhofer Lines. The Chemical Compositions of Stellar Atmospheres

In reference to observations of the solar spectrum, we have already mentioned the classic work of Rowland in Sect. 4.3.1; with modern diffraction gratings, resolutions up to $\lambda/\Delta\lambda \simeq 10^6$ can be attained. In the case of brighter stars, resolutions of $\lambda/\Delta\lambda \geq 10^4 \ldots 10^5$ can be reached in the optical region using, e.g., the coudé spectrographs at the large mirror telescopes.

For *photographically* recorded spectra, the microphotometer, automatically taking the sensitivity curve in-

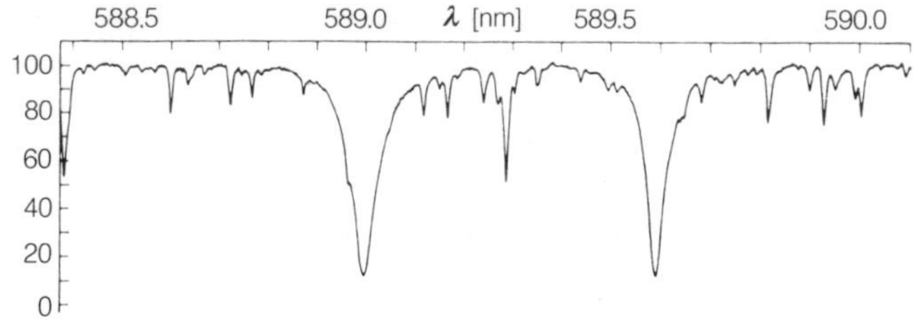

Fig. 4.9.1. Microphotometer curve; the Na D lines in the solar spectrum

to account using intensity markers, yields directly the intensity distribution in the spectrum (Fig. 4.9.1). In recent times, the direct recording of spectra using various types of *photoelectric* detectors (Sect. 3.2.3) has become increasingly important; here, the spectral intensity is usually recorded directly in digital form on magnetic tape or other data media, in a format appropriate for immediate further treatment and evaluation by a computer.

In spectral regions with relatively few Fraunhofer lines, it is not difficult to determine a "true" continuum, whose intensity is normalized to 1 or 100% and which is used as a reference level for measuring the intensity decrease in the lines, $R_\nu = 1 - I_\nu$. In this way, the intensity distribution in the line, I_ν or R_ν, is obtained as a function of ν or λ; it is called the *line profile*. Since the spectrograph itself reproduces even an infinitely sharp line with a finite width which is called the *instrumental profile*, it must be kept in mind that the width and structure of weak Fraunhofer lines always contains an instrumental contribution. It can be determined, for example, by measuring the profiles of sharp lines from a laboratory source. The wings of strong lines are only slightly influenced by the instrumental profile; the center portions, as well as weak lines, are, however, strongly affected. The absorbed energy is clearly independent of the instrumental distortion of the line profile. It is therefore frequently advantageous, as pointed out by M. Minnaert, to measure the *equivalent width* W_λ; this quantity is the width (in suitable units of length such as pm, Å, mÅ, ...) of a rectangular absorption line in the spectrum having the same area as that of the line profile (Fig. 4.9.2).

In spectral regions with a high density of lines, the determination of a continuum level and thus the calculation of an isolated spectral line is in general not possible. Here, a longer section of the spectrum must be calculated "in one piece", and then compared with the observed intensity distribution. This technique of *spectrum synthesis* however requires precise knowledge of atomic data for a large number of spectral line transitions. In the following, we shall limit ourselves to the discussion of individual lines, for which the neighboring continuum (background) can be determined.

From spectrophotometric measurements of the line and the adjacent continuum, we wish to extract the parameters mentioned above which characterize the stellar atmosphere, namely the effective temperature T_{eff}, the surface gravity g, and the abundance distribution of the elements.

The complexity of the problem suggests the use of a method employing successive approximations. We first ask ourselves: which equivalent widths or profiles of various lines would we expect theoretically for an atmosphere of *given* T_{eff}, g, and the chemical composition? Here, again, we use *model atmospheres*. We then consider how the observable quantities, i.e. along with the color indices and other quantities already discussed, in particular the W_λ of certain elements and ionization and excitation states, depend on the parameters chosen initially. These parameters are readjusted until the best possible agreement between the calculated spectrum of the model atmosphere and the observed spectrum of the star being studied is obtained. We discuss the details of the analysis of a spectrum in Sect. 4.9.3 and present the elemental abundances derived for the Sun and some chosen stars in Sect. 4.9.4. First, however, we shall have a look at the theory of Fraunhofer lines.

The *theory of Fraunhofer lines* (in the LTE) consists of two quite different parts, to wit:

1) the theory of *radiation transport* in the lines (Sect. 4.9.1), which relates their profiles and equivalent widths for a given model atmosphere to the line absorption coefficients κ_ν;

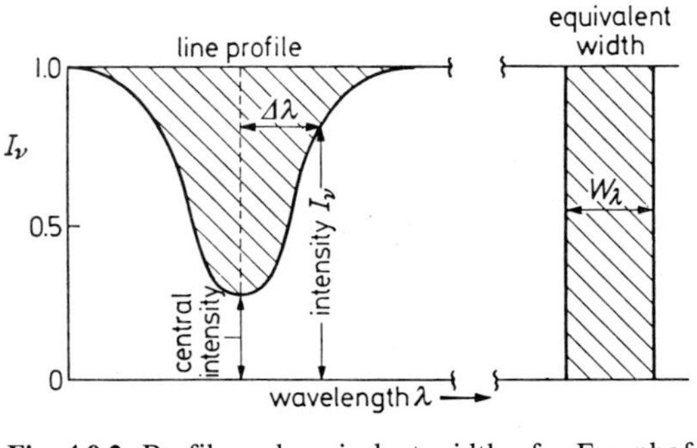

Fig. 4.9.2. Profile and equivalent width of a Fraunhofer line

2) the atomic theory of the *line absorption coefficients* κ_ν themselves (Sect. 4.9.2), which we have already treated in part in Sect. 4.7.3. According to (4.7.27), neglecting stimulated emission, we have:

$$\kappa_\nu = \frac{1}{4\pi\varepsilon_0}\frac{\pi e^2}{mc} f\phi(\nu)N_n \ . \qquad (4.9.1)$$

The identification and classification of the line under consideration in the stellar spectrum will be assumed to be known; along with numerous lists of spectral lines and monographs for individual stars, the multiplet tables of C. E. Moore are an indispensable aid in making this identification. Furthermore, we assume that the atomic constants f, the oscillator strengths, (or the Einstein coefficents A_{mn} for the transitions $n \leftrightarrow m$ which are proportional to them) are also known (Sect. 4.7.3). The number N_n of atoms in the lower state n is obtained as a function of the temperature T, the electron density N_e, or the electron pressure P_e from the conditions for ionization and excitation using the formulas given in Sect. 4.7. Here, we need only discuss the *profile function* $\phi(\nu)$, which is normalized to one.

4.9.1 Radiation Transport in the Fraunhofer Lines

In addition to the continuous absorption coefficient κ, which varies only slowly with the wavelength and can be considered constant in the neighborhood of a spectral line, we now consider the *line absorption coefficient* κ_ν, which decreases more or less rapidly to zero as a function of the distance $\Delta\nu$ or $\Delta\lambda$ from the center of the line (Fig. 4.9.3); at the line center, it exhibits a strong maximum. We thus distinguish between the following absorption coefficients, which each in a different manner depend on the state parameters T and P_e and the corresponding optical depths:

κ_ν
Line absorption coefficient:

$$\tau_\nu(t) = \int_{-\infty}^{t} \kappa_\nu(t')dt'$$

κ
Continuous absorption coefficient:

$$\tau(t) = \int_{-\infty}^{t} \kappa(t')dt'$$

$$\left.\right\} \quad x_\nu = \int_{-\infty}^{t} (\kappa_\nu + \kappa)dt' . \qquad (4.9.2)$$

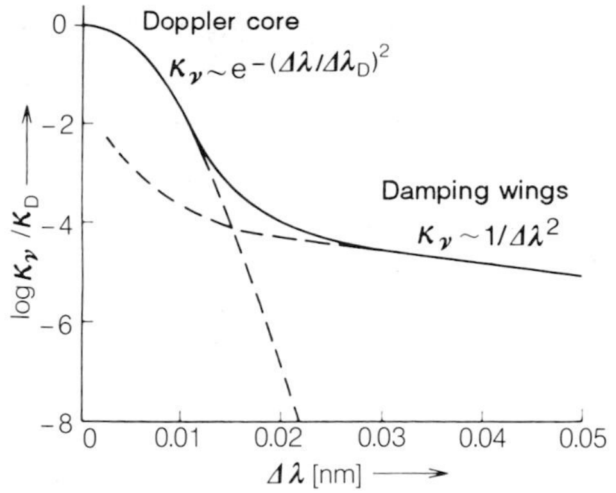

Fig. 4.9.3. Line absorption coefficient κ_ν (referred to the line center). Doppler core and damping wings of the Na D lines calculated according to (4.9.13) and (4.9.16) for $T = 5700$ K and purely radiative damping. Their superposition (———) gives the *Voigt profile*

We can now adopt all the approaches and calculations of Sect. 4.8 to compute the line profile, if we replace κ_ν there by $\kappa + \kappa_\nu$ or the optical depth τ_ν by x_ν.

We thus have for the *intensity* of the radiation leaving the surface of the atmosphere at the angle θ for the *line* (frequency ν):

$$I_\nu(0,\theta) = \int_0^\infty S_\nu(x_\nu)e^{-x_\nu/\cos\theta}dx_\nu/\cos\theta \qquad (4.9.3)$$

and for the neighboring *continuum* (no index):

$$I(0,\theta) = \int_0^\infty S(\tau)e^{-\tau/\cos\theta}d\tau/\cos\theta \ , \qquad (4.9.4)$$

where S_ν and S are again the source functions (4.2.21) giving the ratio of emission to absorption coefficients, and the relation between the two optical depths is given by

$$\frac{dx_\nu}{d\tau} = \frac{\kappa_\nu + \kappa}{\kappa} \quad \text{or} \quad x_\nu = \int \frac{\kappa_\nu + \kappa}{\kappa} d\tau \ . \qquad (4.9.5)$$

The intensity reduction in the line is then

$$r_\nu(0,\theta) = \frac{I(0,\theta) - I_\nu(0,\theta)}{I(0,\theta)} \qquad (4.9.6)$$

and its equivalent width, defined in Fig. 4.9.2, is

$$W_\lambda = \int r_\lambda(0,\theta)d\lambda \ . \qquad (4.9.7)$$

To get a rough feeling for these quantities, we can again employ the Eddington-Barbier approximation (4.8.14) and obtain:

$$r_\nu(0,\theta) \simeq \frac{S_\nu(\tau=\cos\theta)-S_\nu(x_\nu=\cos\theta)}{S_\nu(\tau=\cos\theta)} \ . \tag{4.9.8}$$

If we again make the LTE (*local thermodynamic equilibrium*) assumption for the radiation transport, which is a defensible approximation especially for atoms or ions with fairly complex term schemes in atmospheres which are not too hot, we have $S_\nu(\tau) = B_\nu(T(\tau_\nu))$.[13]. In the calculation of the Kirchhoff-Planck function $B_\nu(T)$, the small frequency range in the neighborhood of a line of course plays no role; the important point is that the radiation in the *line* comes from a higher layer, $x_\nu = \cos\theta$, with correspondingly lower temperature, than the adjacent *continuum*, which originates from $\tau = \cos\theta$. If the absorption coefficient at the center of the line is very much stronger than in the continuum ($\kappa_\nu \gg \kappa$), then the *central intensity* of the line simply corresponds to the Kirchhoff-Planck function at the surface temperature T_0 of the atmosphere. All of the sufficiently strong lines in a particular region of the spectrum then have the same, i.e. the largest possible, intensity reduction, $r_c(0,\theta)$.

For applications to *stars*, where the center-limb variation cannot be observed, we require the corresponding expressions for the radiation flux F_ν in the line and F in the adjacent continuum (4.8.15). The depression $R_\nu(0)$ in the line then becomes (assuming local thermodynamic equilibrium):

$$R_\nu(0) = \frac{F(0)-F_\nu(0)}{F(0)}$$

$$\simeq \frac{B_\nu[T(\tau=\frac{2}{3})]-B_\nu[T(x_\nu=\frac{2}{3})]}{B_\nu[T(\tau=\frac{2}{3})]} \tag{4.9.9}$$

and the corresponding equivalent width in the stellar spectrum is:

$$W_\lambda = \int R_\lambda(0)\,d\lambda \ . \tag{4.9.10}$$

For weak Fraunhofer lines and in the wings of stronger lines, where $\kappa_\nu \ll \kappa$, it follows from (4.9.5) within the Eddington-Barbier approximation that:

$$x_\nu \simeq \frac{2}{3}\left(1+\frac{\kappa_\nu}{\kappa}\right) \tag{4.9.11}$$

and for the intensity reduction in the line, using a series expansion of $B_\nu(x_\nu = \frac{2}{3})$ in (4.9.9):

$$R_\nu(0) \simeq \frac{2}{3}\frac{\kappa_\nu}{\kappa}\left(\frac{d\ln B_\nu}{d\tau}\right)_{\tau=2/3} \ , \quad \kappa_\nu \ll \kappa \ . \tag{4.9.12}$$

We see for one thing that in a stellar spectrum, the continuum as well as the wings of the Fraunhofer lines originate mainly in *those* layers of the atmosphere whose optical depth is equal to $\tau \simeq \frac{2}{3}$ for the neighboring continuum.[14] Secondly, the line intensities depend strongly on the temperature gradient in the atmosphere. We obtain *absorption* lines when the temperature *increases towards the stellar interior.* The central portions of the stronger lines, where $\kappa_\nu \gg \kappa$, originate in correspondingly higher layers, $\tau \simeq (2/3)[\kappa/(\kappa+\kappa_\nu)]$.

Employing the Eddington-Barbier approximation, we are able to get an overall picture of the origin of Fraunhofer lines; however, the correct calculation of the lines must proceed by direct integration of (4.9.6 or 9).

4.9.2 Broadening of the Spectral Lines

In order to actually apply the theory of radiation transport in the lines, we turn to the second point of our program, the calculation of the line absorption coefficients, in particular of the *profile function* $\phi(\nu)$, as a function of temperature T, the electron pressure P_e or the gas pressure P_g, and the position in the spectrum relative to the center of the line, $\Delta\nu$ in frequency units or $\Delta\lambda$ in wavelength units.

In classical optics, a wave train of finite length in time, having a characteristic length τ, corresponds (using a well-known theorem of Fourier analysis) to a spectral line whose absorption coefficient is given by a typical *damping profile* or *Lorentz lineshape*:

$$L(\nu) = \frac{\gamma}{(2\pi\Delta\nu)^2+(\gamma/2)^2} \ . \tag{4.9.13}$$

The integral $\int L(\nu)\,d\nu$ over the line (i.e. the line area) is normalized to one. The damping constant $\gamma\,[\mathrm{s}^{-1}] = 1/\tau$ is

[13] In general, the methods described in Sect. 4.7.3, which require considerably more effort, must be applied.

[14] An exception is the lines of neutral metals (Fe I, Ti I, ...) in the spectra of the Sun and similar cool stars. Here, the concentration of atoms decreases so rapidly with increasing depth in the atmosphere, due to the increasing ionization, that the "center of gravity" for the absorption producing the lines lies at $\tau \simeq 0.05$ to 0.1.

equal to the full width at half maximum (FWHM) of the absorption coefficient in units of circular frequency.

Depending on whether the time width of the radiation process is due to the atomic emission itself or to collision processes with other particles, we speak of *radiation* or of *collisional* damping. In quantum mechanics, we also obtain a line profile of the form (4.9.13), with

$$\gamma = \gamma_{\mathrm{rad}} + \gamma_{\mathrm{coll}} \, . \tag{4.9.14}$$

The *radiation damping constant* γ_{rad} is equal to the sum of the reciprocal lifetimes (decay constants) of the two energy levels n and m between which the transition takes place. In many cases, we can neglect stimulated emission; then we have

$$\gamma_{\mathrm{rad}} = \sum_{l<n} A_{nl} + \sum_{l<m} A_{ml} \, . \tag{4.9.15}$$

Since γ_{rad} has the order of magnitude 10^7 to $10^9\,\mathrm{s}^{-1}$ for allowed transitions, we expect this mechanism to contribute (e.g. at $\lambda = 400$ nm) about $10^{-6} - 10^{-4}$ nm to the FWHM of the absorption coefficient.

The *collisional damping constant* is given by $\gamma_{\mathrm{coll}} = 2$ times the number of effective collisions per s. The effective collisions, according to W. Lenz, V. Weisskopf and others, are those near-passages of a perturbing particle past the absorbing particle which cause a shift of the absorbed light wave by more than 1/10 of an oscillation period.

In cooler stars like the Sun, where the hydrogen is mainly neutral, collisional damping by neutral hydrogen atoms is predominant, the important mechanism being the van der Waals interaction between the emitting atom and the perturbing atom [interaction energy $\propto$ (distance)$^{-6}$]. In addition to van der Waals forces, other repulsive forces which have a stronger distance dependence may become important in collisions where the particles approach each other closely. The damping constant γ_{coll} is proportional to the gas pressure P_g, as long as it is determined by processes in which the emitting atom interacts with only one H atom at a time.

In the case of spectral lines which exhibit a large *quadratic Stark effect*, and in strongly ionized atmospheres, collisional damping by free electrons is occasionally the predominant process. The interaction then goes as the square of the field strength produced by the electron at the position of the absorbing particle, i.e. as (distance)$^{-4}$. The damping constant is in this case proportional to the electron pressure P_{e}.

With about 10^9 effective collisions per s, we expect values of the FWHM of the line absorption coefficients of the order of several 10^{-4} nm. It is not important whether the collisions occur mainly with hydrogen atoms, as in the Sun's atmosphere, or, as in hot stars, mainly with free electrons.

The lines of *hydrogen* and of singly ionized *helium* deserve particular attention. These atoms show especially large *linear Stark effects* in an electric field, and therefore exhibit linewidths in partially ionized gases which are due to the quasi-static Stark effect from statistically distributed electric fields generated by the ions, which move relatively slowly. Starting from the consideration that in a distance interval $r \ldots r+dr$ around a H atom, a perturbing ion may be present with a probability proportional to $4\pi r^2 dr$ and will generate a field $\propto 1/r^2$ (to which the line splitting is proportional), it can be shown that in the wings of the line, the absorption coefficient is proportional to $\Delta\lambda^{-5/2}$. This theory, originally developed by J. Holtsmark, has been refined to include nonadiabatic effects, collisional damping by electrons, and an improved calculation of the microfield in the plasma.

In addition to line broadening by radiative and collisional damping, there is broadening from the *Doppler effect* as a result of the thermal velocities of the particles and possibly also from turbulent flow. Corresponding to the Maxwell-Boltzmann velocity distribution of the atoms, the Doppler profile, again normalized to unit area, is given by

$$D(\nu) = \frac{1}{\sqrt{\pi}\,\Delta\nu_{\mathrm{D}}} \exp\left[-\left(\frac{\Delta\nu}{\Delta\nu_{\mathrm{D}}}\right)^2\right] \tag{4.9.16}$$

with a FWHM equal to $2\Delta\nu_{\mathrm{D}}\sqrt{\ln 2}$ (in frequency units). The Doppler width $\Delta\nu_{\mathrm{D}}$ or $\Delta\lambda_{\mathrm{D}}$ is determined by the most probable velocity $V_0 = (2kT/m)^{1/2}$ where m is the mass of the absorbing atom:

$$\frac{\Delta\nu_{\mathrm{D}}}{\nu_0} = \frac{\Delta\lambda_{\mathrm{D}}}{\lambda_0} = \frac{V_0}{c} \tag{4.9.17}$$

where ν_0 and λ_0 refer to the center of the line. The thermal velocity of, e.g., Fe atoms in the solar atmosphere at $T \simeq 5700$ K is 1.3 km s^{-1}, which for the Fe I line at $\lambda = 386$ nm leads to a Doppler width of $\Delta\lambda = 1.7 \cdot 10^{-3}$ nm. Turbulent flow in stellar atmospheres ("microturbulence") often produces similar velocities and makes a corresponding contribution to $\Delta\lambda_{\mathrm{D}}$.

Finally, we consider the *combined effect* of the Doppler effect and damping, which yields the profile function as a convolution integral:

$$\phi(\nu) = \int_{-\infty}^{+\infty} L(\nu-\nu')D(\nu')d\nu' \; . \qquad (4.9.18)$$

This is termed a *Voigt profile*; in it, the depth at the center instead of the area is normalized to one. Since the ratio a of half the damping constant, $\gamma/2$, to the Doppler width $\Delta\omega_D = 2\pi\Delta\nu_D$, also in circular-frequency units:

$$a = \frac{\gamma}{2\Delta\omega_D} \qquad (4.9.19)$$

is almost without exception <0.1 in stellar atmospheres, one might at first be tempted to assume that broadening due to damping could be neglected relative to Doppler broadening. This is however not correct, because the Doppler distribution (4.9.16) falls off exponentially away from the line center, while the damping distribution decreases only proportionally to $1/\Delta\lambda^2$. Each moving atom produces a damping distribution with a sharp center portion and broad wings, which is, as a whole, Doppler shifted depending on its velocity. Thus, the resulting profile function (Fig. 4.9.3) has a relatively sharply-defined Doppler core, flanked almost abruptly by the wing regions due to damping, which are $\propto 1/\Delta\lambda^2$.

All together, after some simple intermediate calculations, we obtain from (4.9.13, 16) the *absorption coefficient* κ_ν (4.9.1) in [$\mathrm{m^{-1}}$] at a distance $\Delta\lambda$ from the center of a line at λ_0, for the *Doppler core*:

$$\kappa_\nu = \sqrt{\pi}\frac{1}{4\pi\varepsilon_0}\frac{e^2}{mc^2}Nf\frac{\lambda_0^2}{\Delta\lambda_D}\exp\left[-\left(\frac{\Delta\lambda}{\Delta\lambda_D}\right)^2\right] \qquad (4.9.20\,\mathrm{a})$$

and for the *damping wings*:

$$\kappa_\nu = \frac{1}{4\pi}\frac{1}{4\pi\varepsilon_0}\frac{e^2}{mc^2}Nf\gamma\frac{\lambda_0^4}{c\Delta\lambda^2} \; .^{15} \qquad (4.9.20\,\mathrm{b})$$

The quantity $(4\pi\varepsilon_0)^{-1}\, e^2/mc^2 = 2.82\cdot 10^{-15}\; m$ is the so-called classical electron radius. The occupation number $N \equiv N_n\,[\mathrm{m^{-3}}]$ of the lower level is proportional to its statistical weight g, so that instead of f, usually the value of gf is quoted. The intensity of the absorption lines which originate in an optically thin layer is directly proportional to this value.

4.9.3 Curve of Growth. Quantitative Analysis of Stellar Spectra

We shall now investigate how the profile and the equivalent width W_λ of an absorption line increase as we increase the concentration of the absorbing atom, N, or its product with the oscillator strength f. The absorption coefficient κ_ν is given in the central part of the line by the Doppler effect, and in the wings by damping (Fig. 4.9.3), the damping constant being determined by radiative and collisional processes.

The relationship between the depression in the line, R_ν, and the line absorption coefficient κ_ν, using an *absorption tube* (without reemission) of length H and optical thickness $\tau_\nu = \kappa_\nu H$ in the laboratory, would be [from (4.2.18)]:

$$R_\nu = 1 - \mathrm{e}^{-\kappa_\nu H} \; . \qquad (4.9.21)$$

For a *stellar atmosphere*, it is given by (4.9.3 – 9) and can, in general, be calculated only by direct integration. When the requirements for accuracy are not too demanding, these tedious calculations can often be avoided by using the approximate interpolation formula

$$R_\nu = \left(\frac{1}{\kappa_\nu H} + \frac{1}{R_c}\right)^{-1} \; . \qquad (4.9.22)$$

Here, H is an effective height, or NH an *effective number* of absorbing atoms over a unit area on the star's surface (column density). For $\kappa_\nu H \ll 1$ (absorption in an optically thin layer), $R_\nu \simeq \kappa_\nu H$; for $\kappa_\nu H \gg 1$ (optically thick layer), R_ν tends towards the limiting value for very strong lines, R_c. H or NH can be calculated by comparing (4.9.22) with (4.9.3 – 9), cf. also (4.9.12) for $\kappa_\nu H \ll 1$. In a limited wavelength region, the effective layer thickness of an atmosphere may be treated as constant.

We shall immediately give the result of this calculation: in the lower part of Fig. 4.9.4, we have drawn the *line profiles* for various values of the quantity NHf. The depressions for the weak lines, $R_\nu \simeq \kappa_\nu H$, simply reflect the Doppler distribution of the absorption coefficient in the core of the line. With increasing NHf, the value at the line center approaches the maximum depth R_c, since here only radiation from the uppermost layers with the limiting temperature T_0 contributes. On the other hand,

[15] As we have already mentioned, (4.9.20) does not apply to the lines of H I and He II which are broadened by the linear Stark effect.

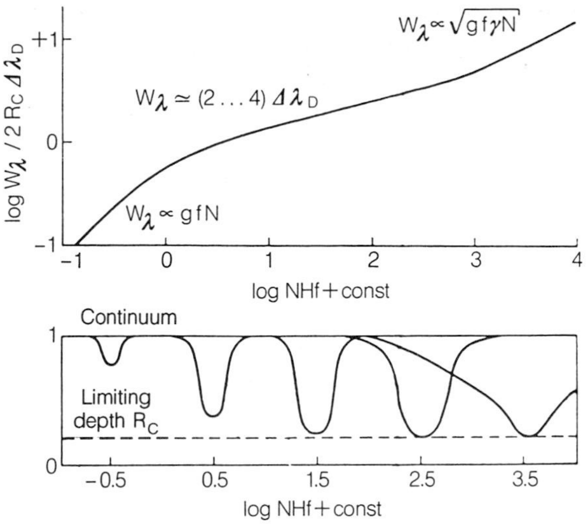

Fig. 4.9.4. Curve of growth (*above*). The equivalent width W_λ, referred to a rectangular profile of width equal to twice the Doppler broadening $\Delta\lambda_D$ and height given by the limiting depth R_c, is plotted as a function of the effective concentration of absorbing atoms, i.e. of $\log NHf+$const. The line profiles show how the curve of growth is generated

the line is initially not strongly broadened, since the absorption coefficient drops off steeply with increasing $\Delta\lambda$. This changes only with further increasing NHf, when the effective optical depth becomes important in the damping wings also. Since now $\kappa_\nu H \propto NHf\gamma/\Delta\lambda^2$, the line acquires broad "damping wings" and its width for a given reduction R_ν becomes $\propto \sqrt{NHf\gamma}$.

By integrating over the line profile, we readily obtain the *equivalent widths* W_λ of the lines. We have plotted them as a ratio to $2R_c \cdot \Delta\lambda_D$, i.e. relative to a rectangular absorption line whose depth corresponds to the maximum depression and whose width is equal to twice the Doppler width. We thus obtain the *curve of growth* (Fig. 4.9.4, upper part), which is important for the evaluation of stellar spectra, giving $\log(W_\lambda/2R_c\Delta\lambda_D)$ as a function of $\log NHf+$const.

As can be easily understood in light of our discussion of line profiles, for weak lines (on the left in the figure), W_λ increases proportionally to NHf; we are in the *linear* part of the curve of growth. This is followed by the *flat* or *Doppler region*, in which the equivalent width is equal to 2−4 times the Doppler width $\Delta\lambda_D$. For strong lines (right), corresponding to the increase in width of the profile, $W_\lambda \propto \sqrt{NHf\gamma}$; we have come to the *damping* or *square-root region* of the curve of growth. Here, in general, the damping constant γ is also of importance. In Fig. 4.9.4, we have used for the ratio of damping to Doppler width $\alpha = 0.03$, corresponding to an average value for the metal lines in the Sun's spectrum. For example, the strong D lines from Na I and the H- and K-lines from Ca II in the solar spectrum lie in the damping region of the curve of growth.

We now turn to our main goal: the *quantitative analysis of stellar spectra.* As an initial orientation, it is usual to begin with a simple approximation method, the so-called *coarse analysis,* in which calculations are carried out for a whole atmosphere using *constant average values* of the temperature T, the electron pressure P_e, the effective layer thickness H (over a large spectral region), etc. The universal curve of growth calculated with the interpolation formula (4.9.22) can then be immediately applied, in order to determine the column density NH of absorbing particles in the lower energy level of any given atom or ion, by comparison to the measured equivalent width W_λ of the corresponding Fraunhofer line. For this purpose, the f value and often the damping constant γ for the line must be known.

The *Doppler width* $\Delta\lambda_D$ or $\Delta\nu_D$ also enters the curve of growth (Fig. 4.9.4); it depends via the thermal motion V_0 on the temperature. Furthermore, turbulent flow fields often contribute the same order of magnitude as V_0 to the Doppler width. The corresponding *microturbulent velocity* ξ adds quadratically in (4.9.17) to the thermal velocity, so that $\Delta\lambda_D \propto (V_0^2+\xi^2)^{1/2}$. Since it is currently not possible to derive this "microturbulence" from the stellar parameters from first principles using a hydrodynamic theory, ξ must be determined in the analysis as an additional parameter along with T, P_e, NH, ... This is done using lines from the flat portion of the curve of growth, which, as we saw, depend sensitively on $\Delta\lambda_D$.

By comparison of NH for energy levels with various excitation energies and for various ionization states of the same element (e.g., Ca I and Ca II), it is finally possible using the formulas of Boltzmann and Saha (Sect. 4.7.2) to calculate the temperature T and the electron pressure P_e. Knowing them, however, we are led immediately from the numbers of atoms in particular energy levels to the total number of all particles of the element considered (independently of the ionization and excitation state) and thus to the *abundance distribution of the elements.* When the latter, as well as the degree of ionization of the different elements, is known with reasonable completeness, the step from the electron pressure P_e to the gas pressure P_g can be taken and the surface gravity can be calculated from the hydrostatic equation: the gas pressure P_g is just the weight, i.e. the mass times the gravitational acceleration g, of all particles above a unit area on the stellar surface.

Based on such a coarse analysis, one can carry out the more precise but much more tedious *fine analysis* of the stellar spectrum: a *model* of the stellar atmosphere under investigation as given in Sect. 4.8.3 is constructed using the most plausible values of T_{eff}, g, and the chemical composition. In this model, using the complete theory of radiation transport and of the continuous and line absorption coefficients (including their depth dependence), the profiles or equivalent widths W_λ of the lines are calculated, taking the microturbulent velocity initially as an undetermined parameter (see above). The results of the model calculations are then compared with spectral measurements for those elements which are represented in the spectrum in various states of ionization and excitation. The hydrogen lines also play an important role, since it is known that hydrogen is *the* most abundant element in nearly all stars; thus the element abundance is not relevant here. Furthermore, the energy distribution or color indices can be taken into account. It can be seen from our previous considerations which criteria are more strongly dependent on T_{eff}, g, ξ, or the abundance of a certain element, so the initial values of these quantities can now be improved systematically in a series of approximation steps. The beginner is usually shocked by the enormous number of lines in a spectrum. However, in carrying out an analysis, it is often found that the available observations are hardly sufficient to determine all parameters of interest!

4.9.4 Element Abundances in the Sun and in Other Stars

Having tried to illustrate briefly the methods of a quantitative analysis of stellar spectra, we now turn to the *results* of such an analysis:

For the *Sun* (spectral class G2V), we have incomparably better observational data than for any other star; furthermore, T_{eff} and g are known independently. Since the first quantitative analyses by C.H. Payne (1925), A. Unsöld (1928), and H.N. Russell (1929), the solar element abundances have been derived repeatedly using more precise observations and improved theoretical models. In the case of the Sun, we can determine the abundances even of rare elements, whose weak lines can be seen only in low-noise spectra having high dispersion. In addition to the Fraunhofer spectrum of the photosphere, the emission spectra of the solar chromosphere and the corona (and of the prominences) from the optical out to the X-ray regions (Sects. 4.10.2, 3) permits the determination of abundances of many elements, especially of the noble gases He, Ne, and Ar. They are not seen in the Fraunhofer absorption spectrum of the photosphere, due to its relatively low temperatures ($T \leq 8000$ K) and to their term structures (states of high excitation energy). To determine abundances of the more common elements, measurements of the solar wind (Sect. 4.10.7) and of solar cosmic rays (Sect. 5.3.8) have also been evaluated. The abundances obtained by different methods agree with each other well (for the elements where several methods can be applied).

In Table 4.9.1, the *relative abundances* N (ordered according to atomic number) in the Sun are set out; they are taken mainly from the analysis of the Fraunhofer spectrum. As usual, all abundances are given relative to that of hydrogen, H, and are normalized to

$$N(\text{H}) = 10^{12} \quad \text{or} \quad \log N(\text{H}) = 12.0 \ . \tag{4.9.23}$$

For example, a ratio $N(\text{C})/N(\text{H}) = 4.2 \cdot 10^{-4}$ corresponds to $\log N(\text{C}) = 8.6$. The imprecision of the solar abundance determination is about $|\Delta \log N| \leq 0.1$ for most elements and is mostly due to the uncertainties in the oscillator strengths f.

As an example, in Table 4.9.1 we also give the abundances for the *carbonaceous chondrites of type* C1 (Sect. 2.9.2), whose accuracy is $\Delta \log N \simeq \pm 0.05$ with a few exceptions. The agreement with the values for the Sun is quite good. These meteorites thus represent (except for the very volatile elements) nearly unchanged solar matter. The deviations for Li and in part for Be and B can be explained by the fact that these light elements are destroyed in the lower hydrogen convection zone (Sect. 4.10.6) by nuclear reactions with protons and their abundances at the solar surface have decreased over long periods of time due to the convective mixing of the outer layers of the Sun.

In Table 4.9.2, we list some selected *isotopic ratios* for the material in the Solar System; in most cases, they are again obtained with the greatest precision from the analysis of meteorite data.

Since the pioneering analysis of the B0V star τ Sco by A. Unsöld in 1941/44, precise element abundance analyses of numerous selected bright *stars*, representing the different spectral classes in the Hertzsprung-Russell

Table 4.9.1. *Element abundances* log N in the Solar System: *Sun* (⊙) from H. Holweger (1985); and *carbonaceous chondrites*, type C1, from E. Anders and M. Ebihara (1982). Normalized to hydrogen, log N(H) = 12.0. The solar and meteoritic abundance distributions are adjusted at silicon, log N(Si) = 7.6. The solar abundances are determined from photosphere measurements except for He, Ne, Ar (from the corona or prominences), and Tl (from sunspots). For the meteorites, C1 chondrites were used except for Be, B, Br, Rh, and I, for which other chondrites were taken. For Kr, Xe and Hg, the values are estimated from interpolations. For the radioactive elements Th and U, the *present* abundances are given; at the time of formation of the Solar System $4.5 \cdot 10^9$ years ago, their abundances were larger by an amount $\delta \log N = 0.2$ (Th) or 0.3 (U)

	⊙	C1		⊙	C1		⊙	C1		⊙	C1
1 H	12.0	–	22 Ti	5.1	5.0	44 Ru	1.8	1.9	66 Dy	1.1	1.2
2 He	*11.0*	–	23 V	4.1	4.1	45 Rh	1.1	*1.1*	67 Ho	0.3	0.6
3 Li	1.1	3.4	24 Cr	5.8	5.7	46 Pd	1.7	1.7	68 Er	0.9	1.0
4 Be	1.2	*1.5*	25 Mn	5.4	5.6	47 Ag	0.9	1.3	69 Tm	0.3	0.1
5 B	2.5	*3.0*	26 Fe	7.6	7.6	48 Cd	1.9	1.8	70 Yb	1.1	1.0
6 C	8.6	–	27 Co	4.9	5.0	49 In	1.7	0.9	71 Lu	0.8	0.2
7 N	8.0	–	28 Ni	6.2	6.3	50 Sn	1.9	2.2	72 Hf	0.9	0.8
8 O	8.9	–	29 Cu	4.2	4.3	51 Sb	1.0	1.1	73 Ta	–	0.0
9 F	4.6	4.5	30 Zn	4.6	4.7	52 Te	–	2.3	74 W	1.1	0.7
10 Ne	*7.6*	–	31 Ga	2.9	3.2	53 I	–	*1.6*	75 Re	–	0.3
11 Na	6.3	6.4	32 Ge	3.5	3.7	54 Xe	–	(2.2)	76 Os	1.4	1.5
12 Mg	7.5	7.6	33 As	–	2.4	55 Cs	–	1.2	77 Ir	1.4	1.4
13 Al	6.4	6.5	34 Se	–	3.4	56 Ba	2.1	2.2	78 Pt	1.8	1.7
14 Si	7.6	7.6	35 Br	–	*2.7*	57 La	1.1	1.3	79 Au	1.1	0.9
15 P	5.4	5.6	36 Kr	–	(3.3)	58 Ce	1.6	1.7	80 Hg	–	(1.3)
16 S	7.2	7.3	37 Rb	2.6	2.5	59 Pr	0.7	0.8	81 Tl	*0.9*	0.9
17 Cl	–	5.3	38 Sr	3.0	3.0	60 Nd	1.4	1.5	82 Pb	1.9	2.1
18 Ar	*6.7*	–	39 Y	2.2	2.3	62 Sm	0.8	1.0	83 Bi	–	0.8
19 K	5.1	5.2	40 Zr	2.6	2.6	63 Eu	0.5	0.6	90 Th	0.2	0.1
20 Ca	6.4	6.4	41 Nb	1.4	*1.5*	64 Gd	1.1	1.1	92 U	–	−0.4
21 Sc	3.1	3.1	42 Mo	1.9	2.0	65 Tb	0.2	0.4			

diagram, have been carried out on the basis of high-resolution spectra with the help of model atmospheres. We mention as examples the main-sequence stars 10 Lac (O9V), ι Her (B3V), α Lyr (A0V), β Vir (F8V); the giants and supergiants ζPer (B1Ib), α Cyg (A2Ia), ε Vir (G8III), βGem (K0III), α Boo (K2III); the white dwarf van Maanen 2 (DZ, previously DG); and the "metal deficient" stars HD 140283 (G0V), HD 122563 (K0III), and HD 161817 ($\simeq$A2).

The *accuracy* with which the abundance of an element in stars of different temperatures can be determined is naturally variable. For one thing, it depends on the quality of the spectrum (signal-noise ratio); for another, on the number of lines from the element seen in the spectrum, in which region of the growth curve they lie, and especially how accurately the f values are known. Furthermore, it must be remembered that the abundances can only be determined together with T_{eff} and g; for medium spectral classes, for example, a value of T_{eff} chosen too low would lead to abundances of metallic elements which are too small! In general, it can be said that, relative to hydrogen, helium and the lighter elements are more precisely determined in the hot stars, owing to their high ionization energies, while the more readily ionized metallic elements can be more precisely determined in cooler stars. The uncertainty of a careful abundance determination is at present of the order of $\Delta \log N = \pm 0.3$. In recent times, a number of metal abundances have been obtained even in hot stars by using lines in the far ultraviolet observed from satellites.

The quantitative analysis of stars whose temperature is not too different from that of the Sun is expediently performed *relative to the Sun*; thus the uncertainty in the absolute f values is avoided. They are currently one of the principal sources of error in absolute abundance deter-

Table 4.9.2. Some isotopic ratios in the Solar System

^{2}D/^{1}H	$2.0 \cdot 10^{-5}$	^{15}N/^{14}N	$3.7 \cdot 10^{-3}$
^{3}He/^{4}He	$1.4 \cdot 10^{-4}$	^{17}O/^{16}O	$3.8 \cdot 10^{-4}$
^{6}Li/^{7}Li	$8.1 \cdot 10^{-2}$	^{18}O/^{16}O	$2.0 \cdot 10^{-3}$
^{10}B/^{11}B	0.25	^{25}Mg/^{24}Mg	0.13
^{13}C/^{12}C	$1.1 \cdot 10^{-2}$	^{26}Mg/^{24}Mg	0.14

minations. For sunlike stars, a relative accuracy of $\Delta \log N \simeq \pm 0.1$ can be obtained in this way.

For stars, not only for sunlike stars, the abundances are referred to the *solar* or *normal abundances*. We denote the abundance of an element El by

$$\varepsilon(\mathrm{El}) = \left(\frac{N(\mathrm{El})}{N(\mathrm{H})}\right)_{\mathrm{star}} \bigg/ \left(\frac{N(\mathrm{El})}{N(\mathrm{H})}\right)_{\odot}$$

or

$$[\mathrm{El/H}] \equiv \log \varepsilon(\mathrm{El}) \ . \tag{4.9.24}$$

The *abundance distribution* of the elements is clearly related to the history of their formation and to their transmutation by *nuclear processes* in the course of *stellar evolution*. Without anticipating later discussions (Sects. 4.12, 5.4.4, and 5.8.5), we summarize the most important results of the quantitative analysis of the spectra of the Sun and stars:

The chemical composition of the *normal* stars (of the spiral arm population I and the disk population of the Milky Way; Sect. 5.5.4) is nearly constant; in particular, it agrees with that of solar matter. It thus seems justified to call this mixture of elements *cosmic matter*. We denote those stars by definition as "normal" which can be uniquely positioned in the MK classification. The feasibility of a two-dimensional classification in fact presumes that no further parameters besides T_{eff} and g are necessary, i.e. that these stars all have the same chemical compositions.

The *interstellar matter* of the H I and H II regions, in which young, hot stars such as 10 Lac or τSco are continually being formed, has nearly the same composition as these stars (Sects. 5.3.2, 5).

Many (but, as we shall see, by no means all) stars which are in advanced stages of development, such as the red giant εVir (G8III), have atmospheres or shells of unmodified cosmic matter.

The *oldest* stars such as the rapidly receding stars and members of globular clusters, which belong to the so-called stellar population II of the galactic halo (Sect. 5.5.4), are characterized by an *abundance deficiency of all heavier elements or "metals"* (from carbon on, i.e. for atomic numbers $Z \geq 6$), relative to hydrogen. We find reduction factors for the metals (M) in the whole range from 1 to at least 1000, or mean metal abundances $[\mathrm{M/H}] = \log \varepsilon(\mathrm{M})$ of 0 to about -3; for example, in the horizontal branch A star HD 161817, a high-velocity star with a radial velocity of -363 km s^{-1}, $[\mathrm{M/H}] \simeq -1.1$; in the G subdwarf HD 140283, $[\mathrm{M/H}] \simeq -2.3$; and in the K giant HD 122563, $[\mathrm{M/H}] \simeq -2.7$. The K giant CD$-38°$ 245 has, according to M.S. Bessell and J. Norris, an even more extreme deficiency of $[\mathrm{M/H}] \simeq -4.5$. (Conversely, there are also "metal-rich" stars, e.g. βVir, to be sure with only moderate metal enrichments, $[\mathrm{M/H}] \leq +0.4$.)

The *relative* abundance distribution of the metals among themselves in the metal-deficient stars (except for some of the most extreme cases) is *more or less the same as in the Sun* and in the population I stars. Significant deviations from this rule, which clearly exceed a factor of 2 or 3, are found for only a few elements, e.g. for C, N, O, and Al, as well as for the elements beyond the iron group which are most difficult to assess spectroscopically, such as Ba and Y.

The abundance distribution in a stellar atmosphere is determined in part by the composition of the *interstellar matter* at the time when the star was formed in it. Not yet mature (main-sequence) stars, both from the young and from the old stellar populations, still exhibit these original chemical compositions in their spectra, except for a few special cases. From the observed deficiency of the metals in the oldest stars, we can conclude that an enrichment in the heavy elements (>He) has occurred in the interstellar medium in the Milky Way since its formation (Sects. 5.4.4 and 5.8.5). Furthermore, the chemical composition of a stellar atmosphere can also change in the course of the star's evolution, e.g. when products of the *nuclear processes* which take place in its interior (Sect. 4.12.3) reach the surface through convection or flow processes, possibly aided by mass loss (stellar winds). The first clear evidence for nuclear processes was obtained with the discovery in 1952 of technetium lines in the spectra of some cool giant stars by P.W. Merrill. The most long-lived isotope of this element (^{98}Tc) has a half-life of only $4 \cdot 10^6$ years, which is much shorter than the time required for evolution of these stars. As we shall see in Sect. 5.4.5, abundance anomalies of He, C, N, O and heavy elements such as Y, Ba, and La are also indicators for nuclear processes during stellar evolution, accompanied by transport processes to the surface.

Not all abundance anomalies in stellar atmospheres originate, however, with nuclear processes. In particularly stable, unmixed stellar layers, a sedimentation of the elements can occur, the heaver elements diffusing out of the observable atmosphere into the layers below. Here, radiation pressure can oppose the sinking of certain atoms or ions as a sort of "selective buoyancy" when they have "favorable" absorption properties. Thus, the anomalies in the spectra of the so-called *Ap or Bp stars* and the *metallic line stars* (Sect. 4.11.2) can probably be explained by this kind of *diffusion processes*. The extreme metal deficiencies in the atmospheres of the *white dwarf stars*, which nearly all show either only hydrogen lines (class DA) or only helium lines (class DB), have also been explained by diffusion of the heavier elements into the layers lying below the atmosphere. The few white dwarfs whose spectra exhibit lines of C, Ca, Mg, Si, and Fe could have accreted these elements on their surfaces from interstellar matter (of normal chemical composition) on passing through dense clouds of it.

4.10 The Sun: The Chromosphere and the Corona. Flow Fields, Magnetic Fields, and Solar Activity

During total solar eclipses, the radiation from the Sun's disk (the photosphere) is blocked by the Moon, so that the outer layers of the solar atmosphere, which emit much less light in the optical region of the spectrum, become visible. They are called the *chromosphere* and, further out and extending far out into space, the *corona.*

A view of the higher layers of the solar atmosphere is also permitted by observations of the light of its *spectral lines*. As we saw in Eq. (4.8.14), this light arises at the optical depth for continuous *plus* line absorption:

$$x_\nu = \int (\kappa + \kappa_\nu) dt \simeq \cos\theta \ . \tag{4.10.1}$$

Thus, layers whose optical depth in the continuum is only $\tau \simeq 10^{-3}$ and less, and which therefore do not appear at all in ordinary photographs, can be observed separately. For this purpose, the following instruments are used:

1) The *spectroheliograph* (G.E. Hale and H. Deslandres, 1891), a large grating monochromator with which the image of the Sun in a precisely defined wavelength range of ≤ 0.01 nm within a Fraunhofer line can be photographed bit by bit. The most interesting images are obtained using the intense K line of Ca II ($\lambda = 393.3$ nm) or the Hα hydrogen line ($\lambda = 656.3$ nm). These calcium or hydrogen *spectroheliograms*, which originate in rather high layers of the solar atmosphere (chromosphere), are one of the most important tools for the investigation of solar activity;

2) The *Lyot polarization filter* (B. Lyot, 1933/38), which permits photography of the entire solar image in *one* short exposure, e.g. using the red Hα line of hydrogen, with a wavelength bandwidth (which is to some extent adjustable) of only about 0.05 to 0.2 nm. The "Hα filtergrams" thus attain a somewhat better image definition than the corresponding spectroheliograms. Lyot also constructed similar filters for the lines of the corona (Sect. 4.10.3).

The observation of the highest atmospheric layers at the rim of the Sun is made difficult by *scattered light* which comes to some extent from impurities in the Earth's atmosphere and in the optics, but in large part is due to diffraction of the light at the entrance slit of the instrument. *Instrumental* light scattering is avoided for the most part by the *Lyot coronagraph*, whose principle is illustrated in Fig. 4.10.1. Furthermore, total solar eclipses still provide an indispensible tool for research related to the outermost layers of the Sun.

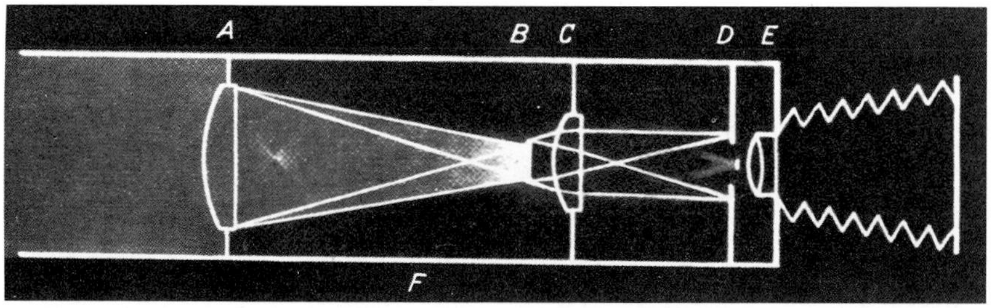

Fig. 4.10.1. A coronagraph as designed by B. Lyot (about 1930). The objective *A*, a simple plane-convex lens made of particularly homogeneous glass, projects an image of the Sun onto a circular plate at *B*; the plate extends 10″ to 20″ beyond the rim of the Sun's image and blocks the light coming from the solar disk. The field lens *C* forms an image of the objective aperture at *D*; here, a ring-shaped collimator removes the light diffracted from the edge of *A*. The objective *E* finally forms an image of the corona, prominences, etc. on the photographic plate or the entrance slit of a spectrograph

In recent years, observations made from rockets and especially from *space vehicles* have become increasingly important for the investigation of the upper solar atmosphere. Here, scattered light from the Earth's atmosphere is avoided, and also the ultraviolet and X-ray spectral regions are accessible, which would otherwise be absorbed by the air. In these wavelength regions, and in the radiofrequency region, the radiation from the chromosphere and the corona predominates over that from the photosphere. We mention the manned Skylab mission (1973/74) and the satellite observations by OSO-8 and the Solar Maximum Mission (launched in 1975 and 1980).

Before discussing the chromosphere (Sect. 4.10.2) and the corona (Sect. 4.10.3) in detail, we first turn in Sect. 4.10.1 to sunspots and the magnetic fields in the photosphere, which strongly influence the layers further up. Later, we consider the prominences (Sect. 4.10.4) and solar flares (Sect. 4.10.5). In Sect. 4.10.6, we attempt to gain a coherent overview within the framework of magnetohydrodynamics of all the phenomena of solar activity, which have their origin in the extended hydrogen convection zone below the photosphere. Finally, in Sect. 4.10.7, we deal with the plasma which streams out from the Sun into space, i.e. the solar wind.

4.10.1 Sunspots and the Activity Cycle. Magnetic Flux Tubes

The *sunspots*, discovered already by Galileo and his contemporaries, appear for the most part in two zones having the same heliographic north and south latitudes. A typical sunspot has roughly the following structure and dimensions:

	Diameter	Area in millionths of ⊙-hemispheres
Umbra (dark core)	18000 km	80
Penumbra (lighter rim)	37000 km	350

The reduced brightness in the spots is due to a reduced temperature. In the largest spots, the effective temperature decreases from 5780 K for the normal solar surface to 3700 K. As a result, the spectrum of a large sunspot is on the whole similar to that of a K star; we have already anticipated the explanation based on Saha's theory of ionization in Sect. 4.7.2.

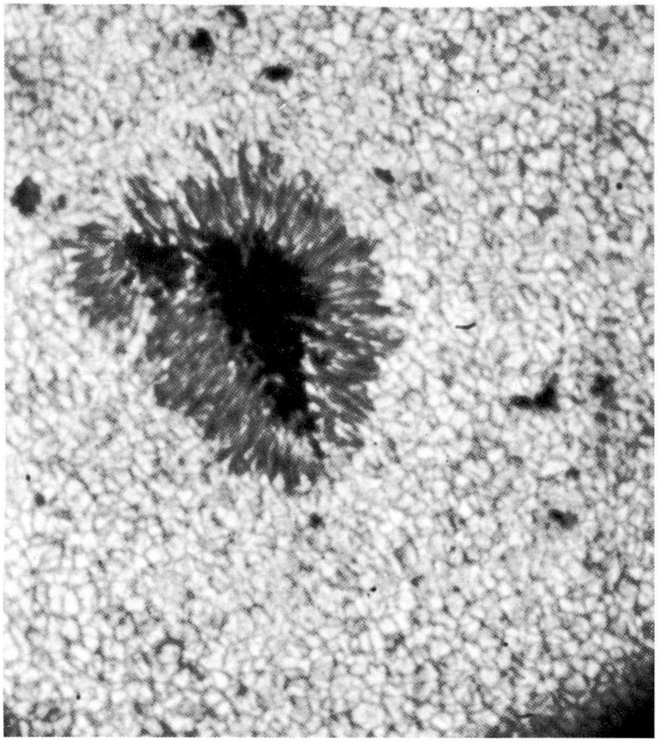

Fig. 4.10.2. Granulation and a sunspot. In the neighborhood of the spot, there are several dark "pores" with diameters of a few seconds of arc. The photograph was made with the 30 cm stratosphere telescope of M. Schwarzschild (1959) at an altitude of 24 km. The exposure time was 0.0015 s, the spectral range of sensitivity 547 ± 37 nm

Sunspots usually occur on the solar surface in groups. A *sunspot group* (Fig. 4.10.2) is surrounded by brighter *faculae*. Furthermore, there are so-called polar faculae, independent of sunspots. The brightness of the faculae is a few percent above that of the normal surface only at the limb of the Sun. Applying (4.8.14), we conclude that in the faculae, only the layers nearest the surface (about $\tau \leq 0.2$) are overheated by a few hundred degrees.

Using the sunspots and, at higher heliographic latitudes, the faculae, the *rotation* of the Sun can be observed. Heliographic latitude is measured from the equator. It is found that the Sun does not rotate as a rigid body, but rather that higher latitudes rotate more slowly than the equator:

Heliographic latitude:	0°	20°	40°	70°
Mean sidereal rotation:	14.5°	14.2°	13.5°	$\simeq 12°\ d^{-1}$
Sidereal period:	24.8	25.4	26.7	$\simeq 31$ d

Spectroscopic measurements of the Doppler effect at the Sun's perimeter (equatorial velocity about $2\,\mathrm{km\,s^{-1}}$) confirm this picture within the experimental accuracy. The synodic period (as seen from the Earth) is correspondingly longer; for the *sunspot zones*, a (rounded) value of 27 d is obtained; it determines the quasiperiodic behavior of many geophysical phenomena.

As was first shown by the pharmacist H. Schwabe about 1834, the abundance of sunspots varies with an average period of 11 years. All other phenomena of *solar activity*, to which we shall return later, follow this *sunspot cycle*; one therefore refers to the 11-year *solar activity cycle*. As a measure for the activity cycle, we use the

Relative (sunspot) number $R = k\cdot(10\times$the number of visible sunspot groups + total number of spots) (4.10.2)

which was introduced by R. Wolf in Zurich. Here, k is a constant which depends on the size of the telescope used. Another measure, in use at the Greenwich Observatory after R. Carrington, is the photographically determined area of the umbrae, the overall spots, and the faculae, either directly in projection (in units of 1 millionth of a solar disk) or corrected for foreshortening (in units of 1 millionth of a solar hemisphere).

The activity cycles are numbered consecutively; the maximum of the arbitrarily chosen 1st cycle was in 1761.5, and that of the 21st cycle was in 1980.0. There are occasionally longer "inactive" periods, most recently between 1645 and 1715, when only a few sunspots occur (Maunder minimum).

In each cycle, the area where new sunspots appear moves from high latitudes (±30 to 40°) at the maximum to low latitudes (±5°) at the minimum (Fig. 4.10.3). New spots or sunspot groups appear preferentially again and again in the same regions, called the *activity centers*.

The *evolution* in time of a (larger) sunspot group exhibits the following characteristic pattern: at first, two main sunspots form, each surrounded by several smaller spots. Their axis lies practically parallel to the equator. Gradually, the smaller spots disappear, and a "double spot" (bipolar group, see below) remains. The spot which trails behind with respect to the Sun's rotation becomes smaller and finally vanishes, while the leading sunspot remains visible for a long time.

The outer regions of a sunspot, the *penumbra*, show bright and dark, radially directed filaments (Fig. 4.10.2), which at a very high spatial resolution appear as a series of bright elements (granules) strung together on a darker background. Observations of the Doppler effect in the spectral lines near the edge of the solar disk show that the

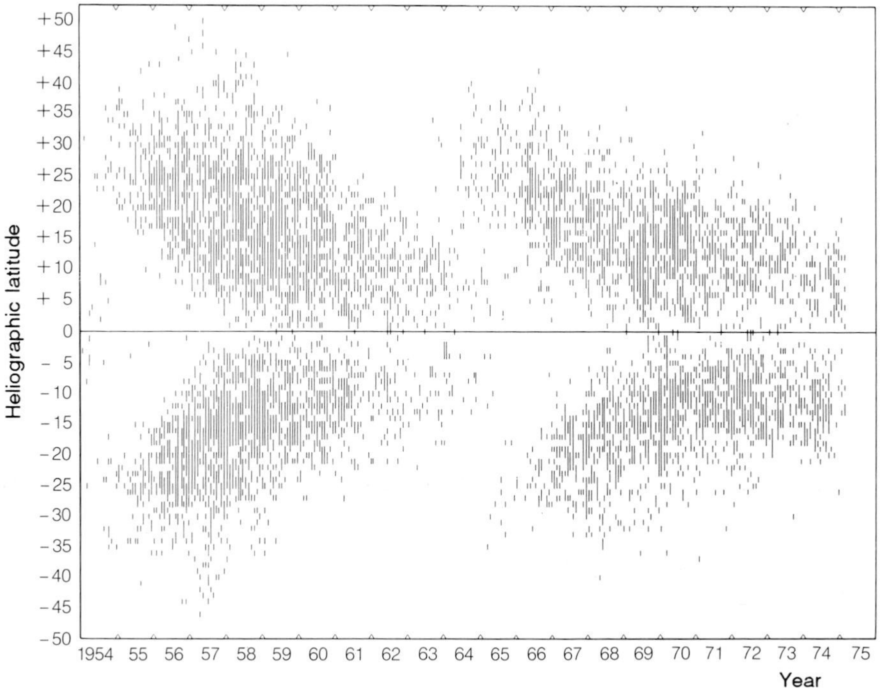

Fig. 4.10.3. A butterfly diagram (Spörer's law) showing the movement of sunspot zones in an activity cycle, from observations made at the Mt. Wilson Observatory. Each vertical dash represents a sunspot group, which is observed at the corresponding heliographic latitude within a synodic rotational period of about 27 d. (R. Howard, 1977). (Copyright © 1977 by D. Reidel Publishing Company, Dordrecht, Holland; reproduced by permission)

matter in the dark zones is moving radially outwards at about 6 $km\ s^{-1}$, i.e. parallel to the solar surface, while the brighter granules are moving more slowly inwards.

Structures like the radial filaments of the penumbra, and also the vortex-shaped flow fields which are observed on Hα spectroheliograms in the neighborhood of sunspots, the filamentary composition of the prominences (Fig. 4.10.8), and the polar bundles and rays of the corona at minimum (Fig. 4.10.6), all suggested some time ago that in solar physics, hydrodynamics alone is insufficient; rather, in addition, *magnetic fields* must play an important role. Thus G.E. Hale in 1908 searched for the *Zeeman effect*, i.e. a magnetic splitting, for example of the Fe I spectral line at $\lambda = 617.3$ nm, using polarization optics which alternately remove the right-hand and the left-hand circularly polarized outer Zeeman components as observed parallel to the magnetic field. He discovered magnetic fields which attain roughly 0.4 T or 4000 G in the largest sunspots. It could also be shown that the two spots (or halves) of a *bipolar sunspot group* always represent a north and a south pole, like the poles of a horseshoe magnet. Long series of observations have since revealed that the leading sunspot (relative to the Sun's rotation) in such a group always has the opposite sign in the northern or in the southern hemisphere, and that this sign alternates from cycle to cycle, so that the true period of a sunspot cycle is $2 \cdot 11$ years (Hale cycle).

The magnetic fields are a more sensitive indicator of disturbances in the activity centers than are the visible sunspots themselves, since they are frequently observable before the appearance of a spot or after its disappearance. Within the umbra of a sunspot, the magnetic field lines are perpendicular to the Sun's surface, i.e. they form a *flux tube*. The *magnetic flux*, the integral of the magnetic flux density (or "induction") B over the area through which field lines pass, lies in a range between about $5 \cdot 10^{12}$ to $3 \cdot 10^{14}$ T m^2 or $5 \cdot 10^{20}$ to $3 \cdot 10^{22}$ G cm^2. The *energy density* of the magnetic field, which gives also the lateral magnetic *pressure* of the flux tubes, is $B^2/2\mu_0$ ($\mu_0 = 4\pi \cdot 10^{-7}$ Vs A^{-1} m^{-1} = the permeability constant of vacuum).[16] Taking $B = 0.1$ T (1 kG) as a typical value for a large sunspot, we obtain a pressure of about 10^4 Pa, which is comparable to the pressure of the surrounding photosphere (Table 4.8.1). In order to maintain pressure equilibrium between the sunspot and the photosphere, the gas pressure within the spot must be lower than in the gas surrounding the spot. In the upper regions of the solar atmosphere, the magnetic field lines fan out as a result of the rapidly decreasing gas pressure.

An explanation of the lower temperature in the spots using purely thermodynamic approaches has remained unsuccessful; we must assume that in the deeper regions of the sunspots, the convective energy flow (Sect. 4.10.6) is strongly impeded. Clearly, this energy flow is not simply deflected to reach the surface at another point as an area of higher temperature. Instead, it is stored for long periods of time (ca. 10^5 years) below the photosphere in the convection zone. This idea has been suggested by recent precise measurements of the solar luminosity (Sect. 4.3.2), which is found to vary by 0.1 to 0.3%. The variation corresponds precisely to the dark areas of the sunspots present at the time of the measurement.

In 1952, H.W. and H.D. Babcock, using a considerably more sensitive apparatus, succeeded in recording Zeeman effects on the Sun corresponding to 1 to $2 \cdot 10^{-4}$ T (1 to 2 G) (with a moderate spatial resolution of several seconds of arc); the spectral line splitting is only a small fraction of the linewidth. Among other things, it was found that a weak, extended magnetic field of the order of a few 10^{-4} T exists at high heliographic latitudes (above 55°), having a total flux of $3 \cdot 10^{14}$ T m^2, and its polarity reverses with the 11-year period. In this polarity-reversal process, both hemispheres may exhibit the same polarity for up to several months. The field is composed of a number of magnetic regions, which are the remains of the fields in the activity centers and which drift towards the poles.

Within the experimental precision of about 10^{-4} T, the Sun has *no* general magnetic field. Instead, it has become clear since the 1970's as a result of spatially highly-resolved observations that the magnetic fields are concentrated in *thin flux tubes* at the limit of resolution (diameter $\lesssim 300$ km), each having a magnetic flux of about $3 \cdot 10^9$ T$\cdot m^2$ and a flux density of 0.1 to 0.2 T, or 1 to 2 kG. They thus attain nearly the field strength found in large sunspots. These flux tubes, which are embedded in the photosphere and are directed perpendicular to the solar surface, are spread over the entire Sun, but are more numerous at the rims of flow cells of the supergranulation (Sect. 4.3.3), and naturally also in the activity centers. All together, they take up about 1% of the Sun's surface area. The regions between the flux tubes are probably free of magnetic fields. Above the photosphere, the flux tubes broaden out and merge to some extent.

---

[16] The energy density in the Gaussian system of units is $B^2/8\pi$.

We shall defer the question of the origin of the solar magnetic fields and first turn to the description of the higher atmospheric layers.

4.10.2 The Chromosphere

If we observe the *limb of the Sun*, at first in the optical continuum, its brightness decreases rapidly on going outwards, as soon as the optical thickness along the direction of observation becomes <1. The corresponding optical thickness for a strong Fraunhofer line however remains ≥ 1 for hundreds or thousands of kilometers further outwards. This means that the Fraunhofer spectrum with its absorption lines becomes an *emission* spectrum in the highest layers of the solar atmosphere, the *chromosphere*. This was first noticed by J. Janssen and N. Lockyer during the total solar eclipse of 1868: when the Moon covers the Sun out to its outer edge, the emission spectrum of the chromosphere "flashes" for a few seconds, producing the *flash spectrum* (Fig. 4.10.4). Spectrographic observations with moving-picture cameras (time resolution about 1/20 s) yield excellent information about the structure of the solar chromosphere, when combined with the relative motion of the Moon with respect to the Sun, calculated from the ephemerides. The scale height (corresponding to a decrease in density of a factor e) is larger than expected for an isothermal atmosphere at the limiting surface temperature of $T_0 \simeq 4000$ K, and it even increases on going outwards.

While the lower chromosphere up to about 1000 km above the rim of the Sun is nearly homogeneously layered, like the photosphere, the upper chromosphere shows strong spatial and temporal density fluctuations. In Ca II spectroheliograms and in the light of other chromospheric spectral lines, bright and dark graininess can be seen. The bright grains are arranged into the *chromospheric network* and are contiguous with the rims of the supergranulation cells (Sect. 4.3.3).

Observed from the limb of the Sun, that is seen from the side, the higher layers of the chromosphere e.g. in

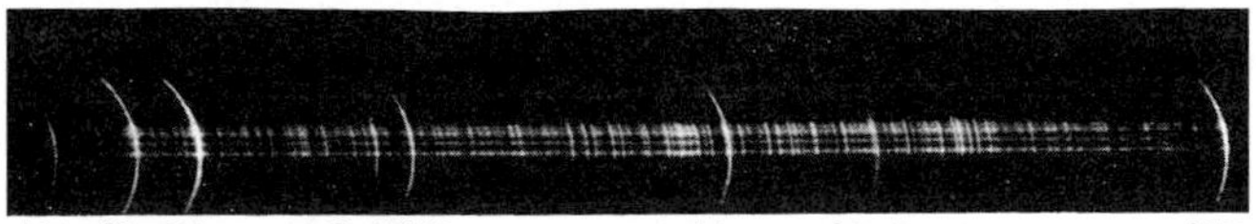

Fig. 4.10.4. A flash spectrum = emission-line spectrum from the solar chromosphere. Taken by J. Houtgast during the total solar eclipse in Khartoum in 1952 with an objective-prism camera

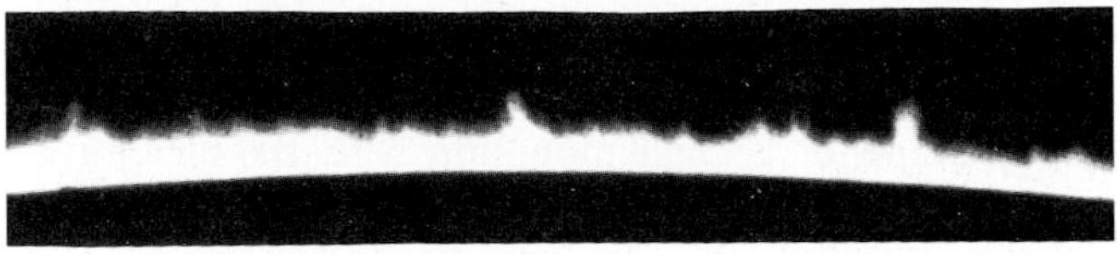

Fig. 4.10.5. Spicules at the limb of the Sun's disk (which itself is blocked out) photographed in Hα light. The altitude of these structures, which were first described by A. Secchi, attains up to 10000 km above the rim of the Sun. Their thickness is about 900 km; they move with velocities of around 25 km s^{-1} up- (and more rarely down-)wards; their lifetimes are about 5 min. The direction of the spicules follows local magnetic fields

Hα light look like a "burning prairie" with small "flamelets", the so-called *spicules*, which move with velocities of about 25 km s^{-1} upwards or downwards (Fig. 4.10.5). The spicules are not distributed uniformly, but rather are also concentrated at the rims of the flow cells of the supergranulation. They probably represent wave motions in thin magnetic flux tubes.

In the *active regions* near sunspots, we find in spectroheliograms extended bright emission regions called the *chromospheric plages*, which lie above the photospheric faculae (Figs. 4.10.10, 11). Here we can also observe an intensification of the magnetic fields, which are in the range of 10^{-2} T, averaged over large areas.

In the chromosphere, the *temperature* increases up to about 20000 K; in the corona, which lies further up, the temperatures are considerably higher, about 10^6 K, as we shall see in the next section. Since the temperature must somehow increase within a thin transition layer (≤ 15000 km) up to the higher value in the corona, it does not seem surprising that lines of greater ionization and excitation energies are already observed with strong intensities in the upper chromosphere: among others the Balmer lines of hydrogen (excitation energy 10.2 eV) and the lines of neutral and ionized helium such as the He I D_3 $\lambda = 587.6$ nm (20.9 eV) and He II $\lambda = 468.6$ nm (48.2 eV) lines. Incidentally, the D_3 line of the "Sun element" was first observed in the solar spectrum by J. Janssen in 1868; helium was first isolated from minerals on the Earth by W. Ramsay only in 1895.

In the *ultraviolet*, the strongest line by far is the Lα H I line at $\lambda = 121.6$ nm. Additional strong chromospheric lines are the resonance lines of He I at $\lambda = 58.4$ nm and of He II at $\lambda = 30.4$ nm, as well as Mg II at $\lambda = 279.5/280.2$ nm. The solar spectrum below $\lambda \simeq 160$ nm is dominated by numerous emission lines from the chromosphere.

4.10.3 The Corona

During the totality of a solar eclipse or using the Lyot coronagraph on a high mountain with the clearest air possible, the solar corona can be observed out to several solar radii (Fig. 4.10.6). Its form (flattening, radial structure, etc.) and brightness are functions of the 11 year cycle. Spectroscopic analysis distinguishes the following phenomena, which we shall in part attempt to explain immediately:

The *inner corona* ($r \simeq 1$ to 3 solar radii) exhibits a completely *continuous* spectrum in the visible region; its energy distribution corresponds to that of normal sunlight (K corona). The light is partially linearly polarized. We attribute this to Thomson scattering of photospheric light by the free electrons of the completely ionized gas (plasma) in the corona. The Fraunhofer lines are therefore completely smeared out by the Doppler effect, corresponding to the high electron velocities.

Fig. 4.10.6. The solar corona, near the sunspot minimum. Taken during the solar eclipse in Khartoum in 1952 by G. van Biesbroeck. The corona at minimum exhibits extended "rays" in the region of the sunspot zones; above the polar regions, there are finer "polar plumes". The corona at maximum has a more rounded shape

From the brightness distribution of the K corona in white light, the *average* (i.e. neglecting inhomogeneities) electron density N_e as a function of the distance r from the Sun's center can be calculated, since the distribution of the luminosity and the Thomson scattering coefficient per free electron, $\sigma_T = 6.65 \cdot 10^{-29}$ m^2, are known. For the (round) corona at maximum, we find:

$r = 1.03$	1.5	2.0	3.0	Solar radii
$N_e \simeq 3 \cdot 10^{14}$	$2 \cdot 10^{13}$	$3 \cdot 10^{12}$	$3 \cdot 10^{11}$	Electrons [m^{-3}]

These values should, however, be taken only as a rough indication of the radial density distribution, due to the extremely inhomogeneous structure of the corona.

In the *outer corona*, which builds up rapidly on going outwards from the K corona, scattered photospheric light with *unmodified Fraunhofer lines* can be observed. Following W. Grotrian, who termed this component the *F corona* (Fraunhofer corona), it was pointed out in 1946/47 by C.W. Allen and H.C. van de Hulst that the light was due to Tyndall scattering, i.e. mainly forward scattering by small particles (somewhat larger than the wavelength of the light); they are so far from the Sun that they are not heated sufficiently to be vaporized. Measurements of the distributions of brightness and polarization in the outer corona have shown that the F corona is simply the innermost portion of the zodiacal light (Sect. 2.9.3). The F or dust corona, and the zodiacal light, thus do not belong to the Sun at all and are influenced by it only to a relatively small degree.

In the *inner* corona, the emission of the so-called *corona lines* was also observed in the optical region; their identification however remained one of the great riddles of astrophysics, until in 1941, B. Edlén succeeded in explaining them as arising from forbidden transitions of metastable levels of the ground states of *highly ionized* atoms. The strongest and most important are (χ: ionization energy):

	λ [nm]	Identification	χ [eV]
Red corona line	637.4	[Fe X] $3s^2 3p^5\ ^2P_{3/2} - P_{1/2}$	235
Green corona line	530.2	[Fe XIV] $3s^2 3p\ ^2P_{1/2} - ^2P_{3/2}$	355
Yellow corona line	569.4	[Ca XV] $2s^2 2p^2\ ^3P_0 - ^3P_1$	820

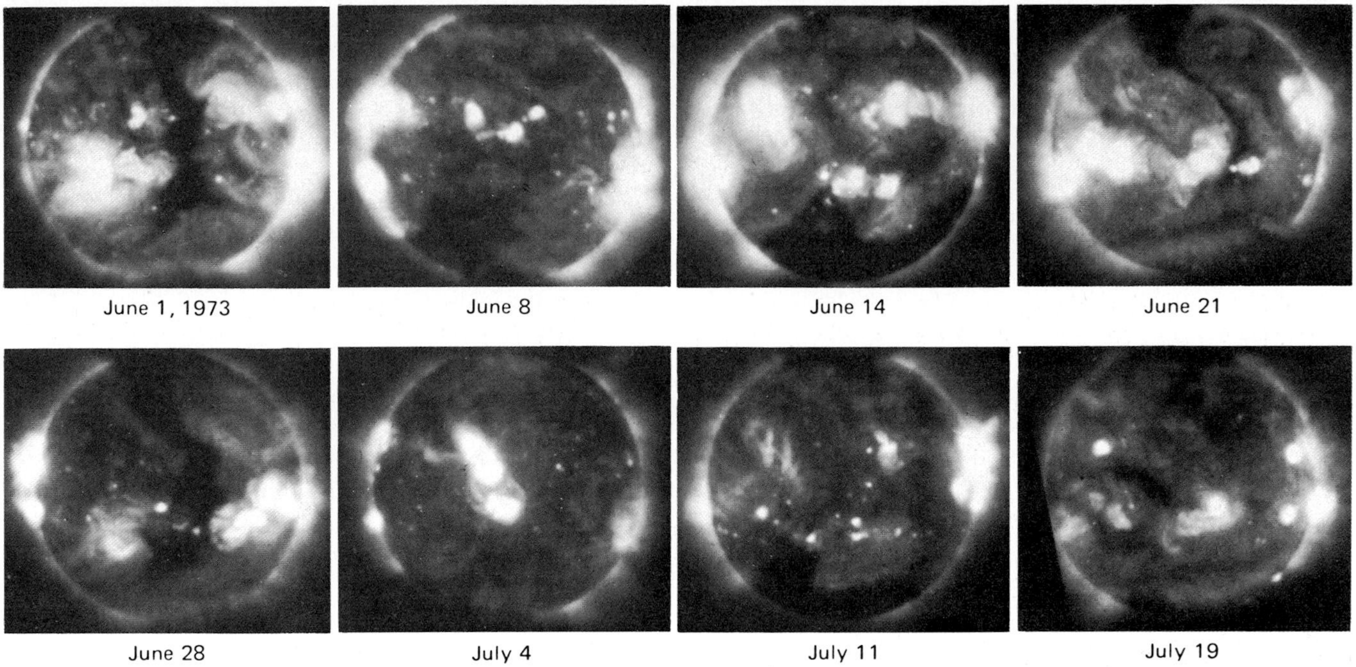

Fig. 4.10.7. Images of the solar corona in the soft X-ray region at intervals of 7 days over nearly two rotational periods. Photographs made with the X-ray telescope of American Science and Engineering, Inc. on Skylab, 1973 (From J. A. Eddy, 1979)

Additional, similar lines of the elements Ar, K, Ca, V, Cr, Mn, Fe, and Co have been identified with certainty. The ionization energies of several hundred eV indicate clearly that the electron temperature is of the order of a million degrees.

The following facts also support corona temperatures of several 10^6 K:

a) the density distribution $N_e(r)$, together with the hydrostatic equation (2.8.16) or the scale height (2.8.17);
b) the linewidths of the corona lines, which are due to the thermal Doppler effect (4.9.16);
c) the spectra in the far ultraviolet and X-ray regions, which in the main reflect the emission spectrum of the inner corona, with numerous allowed and forbidden lines from *high* ionization states of the more abundant elements; and
d) the free-free and free-bound continua of the coronal plasma in the X-ray region.

The *ionization* and *excitation* in the corona *cannot* be calculated under the assumption of local thermodynamic equilibrium (Saha and Boltzmann formulas), since in the extremely tenuous plasma, there is no radiation field corresponding to 10^6 K. Instead, the ionization, recombination, and excitation processes must be considered individually (Sect. 4.7.3). From detailed calculations, temperatures of (1 to 5)$\cdot 10^6$ K are found; within about this range, the temperature in the corona varies with time and place. The *abundances* of the elements relative to hydrogen, determined from the line intensities compared with the electron-scattering continuum of the K corona, agree well with the values determined using the absorption lines of the photosphere (Table 4.9.1).

Photographs of the Sun in the *soft X-ray region* using a Wolter telescope and filters to limit the wavelength range (Fig. 4.10.7) clearly show the structure of the corona and its changes with time. Relative to the thermal emission of the corona at several 10^6 K, the continuum of the photosphere at about 6000 K, which is prodominant in the optical region, is not at all visible in the X-ray region. The X-ray emission of the corona is extremely variable and is non-uniformly distributed. While spatially extended structures can be followed through several rotational periods, other smaller emission regions change form in days or even hours. Since the magnetic energy density is larger than the thermal energy density in the corona, magnetic structures dominate its density distribution and flow properties (Sect. 4.10.6). In X-ray photographs, we can see the magnetic field lines more or less directly, so to speak.

Pictures having higher resolution show that nearly all the X-ray emissions consist of a number of loop- or arch-shaped structures, which occur both in quiet regions and in active regions. These *coronal loops* represent *closed* magnetic flux tubes. Outside the active regions, we find long, extended arches, which reach up to several tenths of a solar radius above the limb of the Sun, with plasma densities of about $2 \cdot 10^{14}$ to 10^{15} m^{-3} and temperatures in the range of 1.5 to $2 \cdot 10^6$ K. Active regions are often connected together by long arches (≥ 700000 km), in which somewhat higher temperatures (2 to $3 \cdot 10^6$ K) are found.

Above sunspot groups or activity centers, both in the optical and in the ultraviolet, as well as in the light of the corona lines and in the X-ray region, we can observe the hot and more dense *coronal condensations*, whose area is for the most part the same as that of the chromospheric facular regions or the plages. In a dense ($\leq 10^{16}$ m^{-3}) core region with a lifetime of several days, temperatures $\geq 3 \cdot 10^6$ K are attained. At a high spatial resolution it becomes clear that the corona above the active regions consists of bundles of small coronal arches (of 10^4 to 10^5 km length), tightly crowded together and having temperatures of $\leq 2.5 \cdot 10^6$ K. Above them, we find roughly radially-directed *coronal rays* in a brush or fan-shaped arrangement, whose densities are about 3 to 10 times higher than those of the surrounding medium and which reach out as far as 10 solar radii into space (Sect. 4.10.6).

The coronal rays above the active regions, as well as the polar brushes of the corona at minimum, represent long, for the most part *open* magnetic field lines. The extended *coronal holes*, which appear dark in X-ray pictures and occur preferentially at higher heliographic latitudes, exhibit open magnetic field structures. In them, relatively cool matter ($T \leq 10^6$ K) streams outwards with velocities up to 800 km s^{-1} (solar wind, see Sect. 4.10.7).

Finally, the X-ray images also show small *bright spots* (diameter ≤ 20000 km), which are distributed over the whole Sun and have lifetimes of only several hours up to a few days. They are formed (at a rate of about 1500 per day) when a small flux tube loop rises up to the surface of the Sun from the depths and becomes recognizable on photospheric magnetograms as a small bipolar region. The magnetic flux in all the bright spots is greater than that in the active centers at sunspot minimum; at maximum, it is still almost 50% of the total flux which emerges from the solar surface.

The *thermal radiofrequency radiation* of the Sun can be explained in terms of free-free radiation from the chromosphere (millimeter and centimeter waves) and the corona (meter waves). In the radiofrequency range, the free-free absorption coefficient of the plasma at frequency ν is proportional to $N_e^2 T^{-3/2} \nu^{-2}$. With increasing frequency, one can thus "look" deeper and deeper into the solar atmosphere. For example, the radio spectrum of the quiet Sun in the decimeter-wave range gives practically an image of the temperature and pressure distribution in the transition layer between the chromosphere and the corona.

The fluctuations in the intensity of the thermal radiation in the course of days and months allow us to decompose them into the everpresent radiation from the quiet Sun and the slowly varying radiation which comes from the active regions (S component in the range $1.5\,\mathrm{cm} \leq \lambda \leq 70\,\mathrm{cm}$). At the rim of the Sun and in the denser parts of the corona, an optical thickness >1 is reached in some places, so that we can measure the blackbody radiation directly. Its temperature, which corresponds to the electron temperature, is several 10^6 K.

4.10.4 Prominences

At the limb of the Sun during eclipses or using the coronagraph, the prominences can be seen as bright "long, extended clouds in the corona". In front of the Sun's disk, e.g. in Hα light, they appear as thin, dark filaments (Figs. 4.10.8 and 4.10.11).

The spectra of the prominences are for the most part similar to those of the upper chromosphere (spicules). Along with weak lines from neutral metals (e.g., Na I and Ca I), they exhibit strong Ca II H and K lines and the He I D_3 line ($\lambda = 587.6$ nm) in the optical region. The ultraviolet region is dominated by numerous emission lines. We see here a dense, cool plasma with about 10^{17} particles m^{-3} and excitation temperatures in the range from 5000 to 12000 K, which is embedded in the less dense, hotter corona.

The *quiescent prominences* (filaments) retain their form with minor changes (flow fields of the order of 10 km s^{-1}) often for weeks at a time. They are characterized by a peculiar threadlike structure (Fig. 4.10.8). The transition from the cool threads to the surrounding hot corona at about 10^6 K takes place in an extremely thin "skin" only 10 to 100 km thick, as can be seen in images made with the ultraviolet light from highly excited or

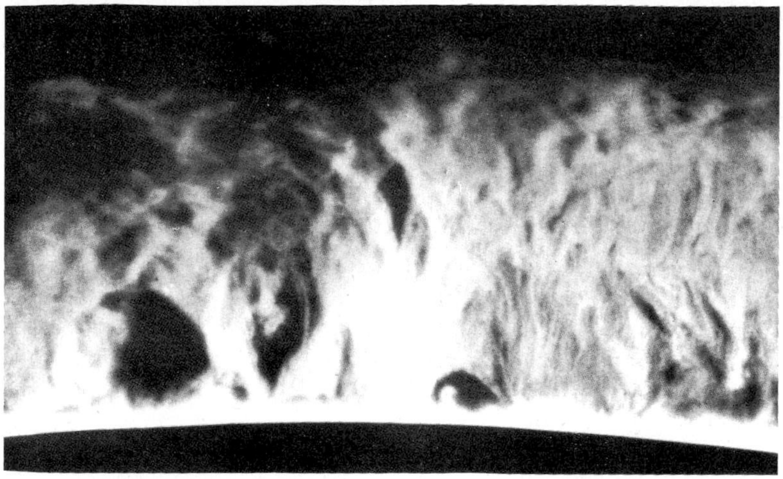

Fig. 4.10.8. The quiescent prominences (filaments) have, on the whole, the form of a thin sheet standing upright on the Sun's surface on several "feet": their thickness is about 7000 km (4000 to 15000 km), height about 45000 km (15000 to 120000 km), and length around 200000 km (up to 10^6 km). This detail photograph from the Sacramento Peak Observatory (in Hα light) shows threadlike structures in which matter flows upwards or downwards with velocities on the order of 10 to 20 km s^{-1}. The shape of the prominences is clearly determined in part by solar magnetic fields

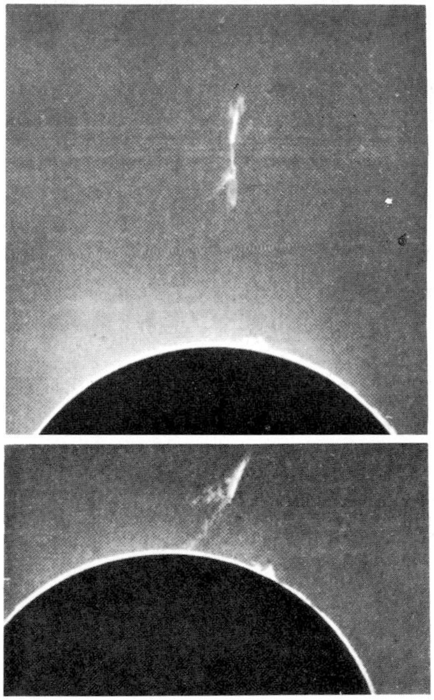

▲
Fig. 4.10.9. An eruptive or ascending prominence. The time interval between the two photographs was 71 min. Maximum altitude: 900000 km above the rim of the Sun; maximum velocity: 229 km s^{-1}

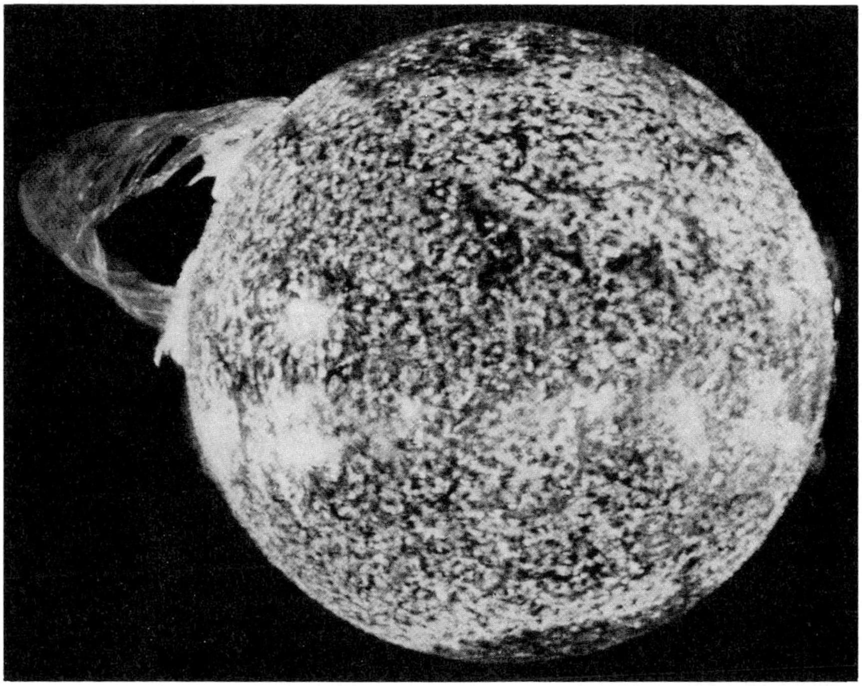

▲
Fig. 4.10.10. An enormous eruptive prominence photographed on December 19th, 1973 in the light of the He II resonance line at $\lambda = 30.4$ nm, using the extreme-ultraviolet spectrograph of the U.S. Naval Research Laboratory on Skylab (from J.A. Eddy, 1979)

highly ionized states. Magnetic fields of $\leq 10^{-3}$ T (≤ 10 G) have been measured in the filaments.

Sometimes, the prominences or parts of them are accelerated more or less suddenly to velocities of 100 km s^{-1} or occasionally up to 600 km s^{-1}, without any previous external indications. These *eruptive* or *ascending prominences* (Figs. 4.10.9, 10) can then escape into interplanetary space. In other cases, a disturbed prominence "rains" back onto the solar surface, its material streaming down along arch-shaped paths (following the magnetic lines of force).

Above sunspot groups or activity centers, we observe manifold forms of *active prominences* as jets, sprays, or arches with lifetimes of only a few minutes to hours; some of them are accelerated to above the escape velocity from the Sun. The temperatures and magnetic fields in the active prominences are considerably higher than in the static prominences.

The activation of static prominences and the occurrence of active prominences are often connected with solar eruptions or flares (Sect. 4.10.5). Along with eruptive prominences, expanding clouds of matter or arches with radially-directed velocities up to 1200 km s^{-1} can be observed in the outer corona (coronal transients). The corona above an active region can thus be "blown away" within about an hour.

4.10.5 Solar Eruptions or Flares

The impressive phenomenon of *solar eruptions* (not to be confused with the eruptive prominences!) or *flares* can be most readily observed as a "brightening" in the Hα line; they are, however, accompanied by an intensification of the radiation in the whole electromagnetic spectrum as well as the production of energetic particles.

When an active region is observed in the Hα or the CaIIK line (Fig. 4.10.11), the chromospheric facular areas or plages between the sunspots and in their neighborhood show structures of the order of several thousand km across with smaller irregular motions and variations in brightness. Suddenly, many of the brighter structures merge together; then, in a major eruption, a region of (2 to 3)$\cdot 10^{-3}$ of the solar hemisphere, corresponding to a whole sunspot group, flares up in the Hα, CaIIK, and other lines. The Hα emission in the middle attains an intensity up to about 3 times that of the normal continuum and a width of several tenths of a nanometer. The lifetime of the flares varies from the order of one second in the case of the tiny "microflares" of about 1″ diameter, up to several hours in the case of major flares. The "importance" of the eruptions is classified according to the area of the Hα emission at the maximum as S (subflare, $<10^{-4}$ hemispheres), 1, 2, 3, and 4 ($>10^{-3}$ hemispheres).

The increase of intensity in Hα is accompanied by increased emission in the whole spectrum, which, except for the hard X-ray and the radiofrequency regions, is of *thermal* origin. The flaring up of the upper chromosphere in Hα and other lines is however only a secondary phenomenon. In the far ultraviolet and especially in the soft X-ray regions (≥ 10 eV), an increase in the radiation intensity can be observed about 10 min before the beginning of the Hα flare; it indicates heating of the lower corona. In the case of large flares, at the beginning of the intensity rise in Hα a burst occurs in the microwave and hard X-ray regions, lasting ≤ 5 min. This non-thermal radiation is produced by the interaction of electrons which are accelerated to high energies during the flare (see below) with the solar plasma. In this phase, in the soft X-ray

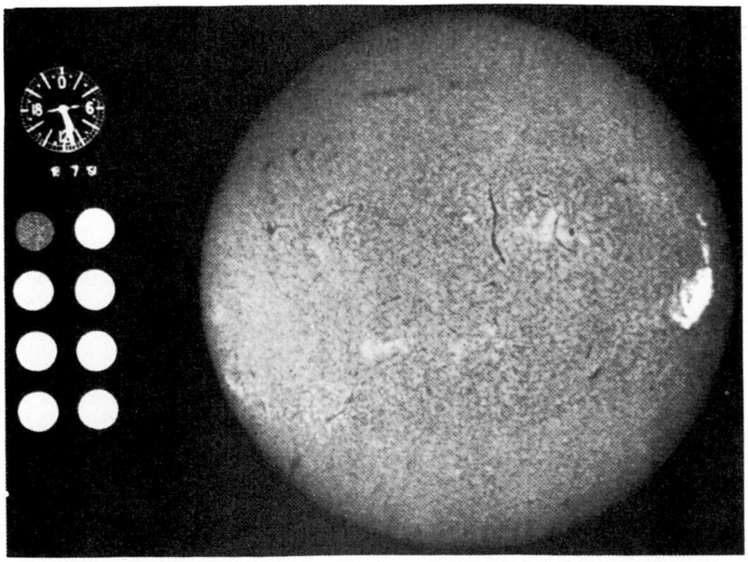

Fig. 4.10.11. A large solar eruption (type 4 flare), taken on July 18th, 1961 (*right*); it was accompanied by strong cosmic-ray particle emissions. This is an Hα solar observation photo taken at the Cape Observatory. Somewhat above the midpoint of the solar disk, an extended filament can be seen, i.e. a prominence (compare Fig. 4.10.8) in absorption. (*Left*): Photometric intensity calibrations and a time marker

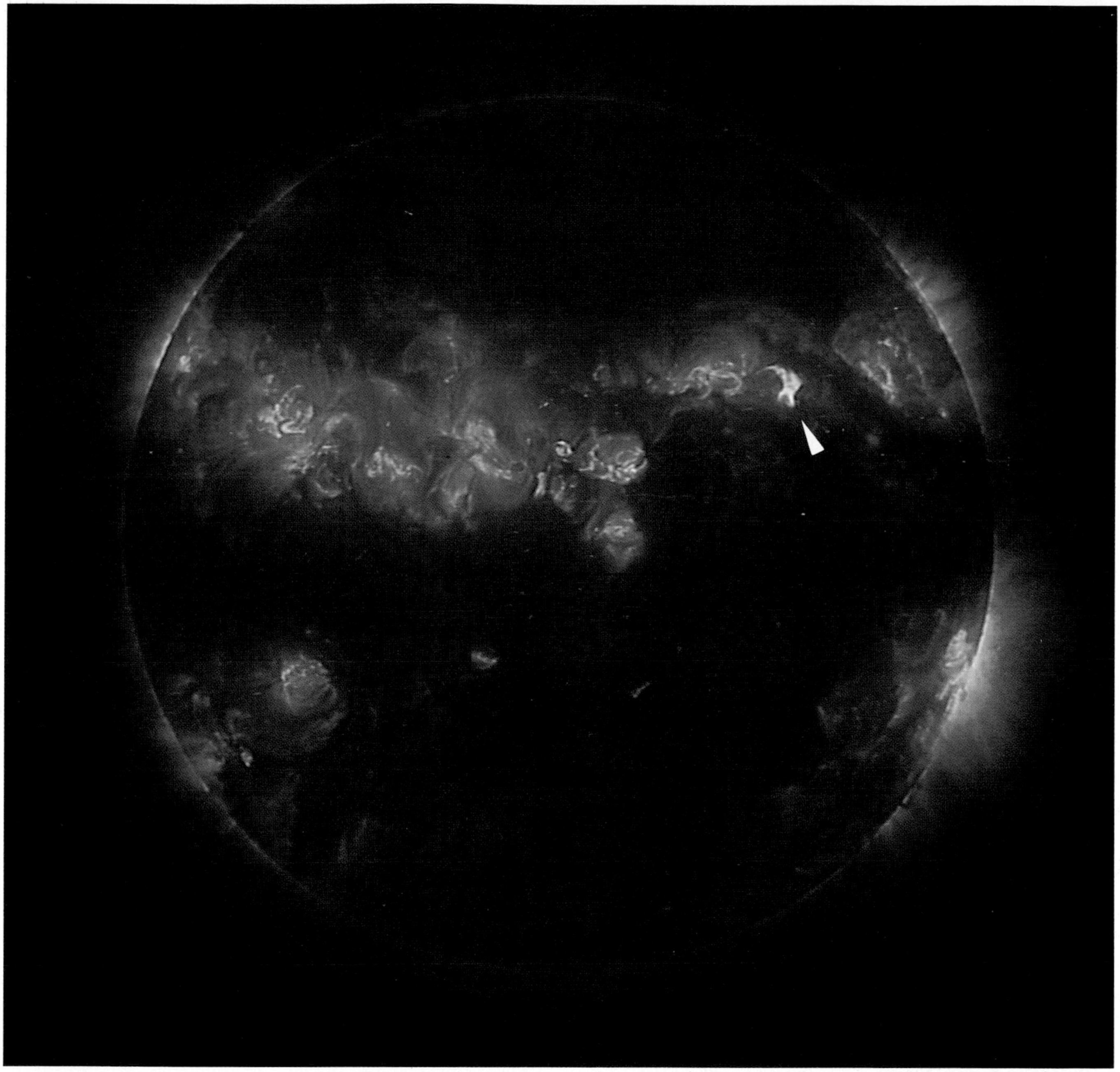

Fig. 4.10.12. A high-resolution X-ray image of the solar corona with active regions and a flare (indicated by the arrow). The angular resolution is 0.75″, and the temperature of the brightest areas is about $3 \cdot 10^6$ K. This image was made by L. Golub et al. (1989) from a rocket with the 0.25 m X-ray telescope NIXT (*N*ormal *I*ncidence *X*-ray *T*elescope) in a narrow wavelength range around $\lambda = 6.35$ nm; this range includes the emission lines of Fe XVI at $\lambda = 6.37/6.29$ nm and of Mg X at $\lambda = 6.33/6.32$ nm. In contrast to an image-forming X-ray telescope of the Wolter type, with grazing incidence (Fig. 3.5.2), here the image is produced at *normal* incidence as with an ordinary optical mirror. Reflection of the X-rays is made possible by vapor-deposition of alternating thin layers of cobalt and carbon, so that constructive interference results for the wavelength 6.35 nm. NIXT was developed by L. Golub and coworkers at the Smithsonian Astrophysical Observatory, Cambridge, together with the IBM – Thomas J. Watson Research Center, Yorktown Heights, New York. (Photo courtesy of SAO and IBM Corp.)

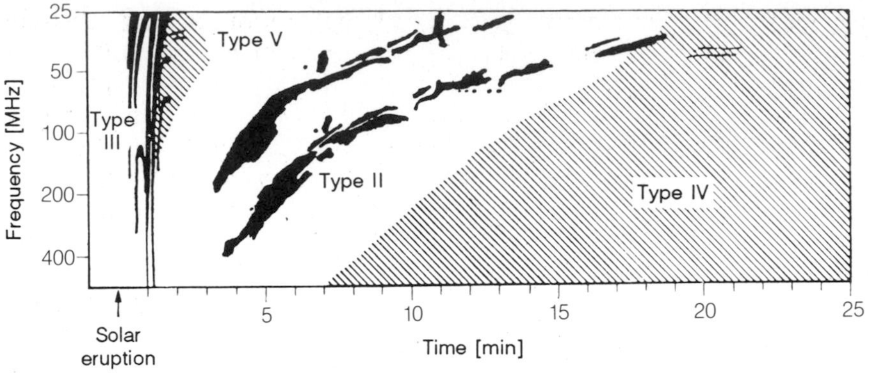

Fig. 4.10.13. The dynamic (i.e. time-dependent) radio spectrum in the meter wave region (schematic), from J. P. Wild: the time evolution of bursts of different types following a major solar eruption. Lower/higher frequencies are emitted in general by higher/lower layers of the corona. Type II and type III bursts are often accompanied by 2:1 harmonics

region a bright, dense, and hot core (diameter ≤ 4000 km) with temperatures up to $3 \cdot 10^7$ K can be observed; it is often at the highest point of a corona arch (Fig. 4.10.12).

The emission of *corpuscular radiation* over a large energy range accompanies major flares. Plasma clouds or magnetohydrodynamic shock waves, which reach the Earth about one day later and thus have a velocity of about 2000 km s^{-1}, cause magnetic storms (sudden commencement) and auroras. Major flares also make a solar contribution to cosmic rays, as found by S. E. Forbush and A. Ehmert, with particle energies of 10^8 to 10^{10} eV. Their chemical composition corresponds closely to that of the usual solar matter.

The interaction of the energetic particles with atomic *nuclei* in the solar matter also causes the emission of low-energy *gamma radiation.* Aside from a continuum, several strong emission *lines* are observed: nuclear transitions in the abundant nuclei ^{12}C and ^{16}O at 4.43 and 6.14 MeV, and, with a time delay, lines at 0.511 and 2.23 MeV. The 0.511 MeV line is produced through annihilation of positrons, which are generated by the flare, and electrons; the 2.23 MeV line comes from the formation of deuterium, ^{2}D, through reaction of neutrons which are also produced in the flare with protons (hydrogen nuclei) in the solar atmosphere.

Accompanying the flares and also the less spectacular phenomena of solar activity is a *non-thermal* contribution to the *radiofrequency radiation* of the Sun.

Its analysis with the aid of a radiofrequency spectrometer, registering the intensity in a large frequency range as a function of time, has been carried out by J. P. Wild and coworkers in order to distinguish between several different types of "bursts" or intense radiofrequency emissions, initially in the range below 400 MHz (Fig. 4.10.13). In order to understand this analysis, we must first remark that electromagnetic waves of frequencies below the critical or *plasma frequency* can*not* be emitted from a plasma of electron density N_e:

$$\nu_0 = \sqrt{\frac{1}{4\pi\varepsilon_0}\frac{e^2}{\pi m}N_e}$$

or

$$\nu_0\,[\mathrm{MHz}] = 9.0\sqrt{N_e\,[\mathrm{m}^{-3}]}\ , \qquad (4.10.3)$$

since then its index of refraction is <0. Thus, radiofrequency radiation of a particular frequency in the corona can only have had its origin above a certain layer.

The *type II and type III bursts* (Fig. 4.10.13) exhibit a slow or a rapid shift of their frequency bands to lower frequencies, respectively. From this it can be concluded that the agent which produces them sweeps through the corona with velocities of the order of 1000 km s^{-1} for type II and up to 40% of the velocity of light for type III. Simultaneously with the type III bursts, in the *microwave region* above $\nu = 1.5$ GHz ($\lambda \leq 20$ cm), pulses of 1 to 5 min length (with a smoother time envelope) are detected. The *type IV* events emit a continuum over a long period of time, covering a broad frequency band; this is due to synchrotron radiation from the fast electrons.

In 1967, J. P. Wild and coworkers at the Culgoora Observatory in Australia succeeded in observing the motion of various types of burst sources at 80 MHz ($\lambda = 3.75$ m) on and outside the solar disk, directly on the display monitor of the *radioheliograph.* Using 96 parabolic antennas each having a 13 m diameter, suitably connected together and mounted in a circle of diameter 3 km, the Sun and the solar corona could be probed over a field of view of 2° diameter with 2′ to 3′ resolution. One image per second of each polarization (right and left circular polarization or two directions of linear polarization) could be recorded. The observations verify the results from the radio spectra. We unfortunately cannot discuss here the complicated plasma-physical treatment of the different types of bursts.

The total *energy* released in a flare ranges from 10^{22} J in the case of a subflare up to $3 \cdot 10^{25}$ J in a type 4 flare; it is divided about equally among electromagnetic radiation, kinetic energy of the ejected plasma, and high-energy particles. Although the complex processes which are responsible for solar flares are not understood in detail, we can assume that in the end, they involve *magnetic energy* (energy density $B^2/2\mu_0$), which is transformed into thermal energy that is then radiated, and also causes the acceleration of particles. In order to produce the energy of $3 \cdot 10^{25}$ J in a major flare, for example in a volume with a diameter of about 30000 km, a magnetic field of $5 \cdot 10^{-2}$ T (500 G) would have to be "annihilated", perhaps by an instability which leads to a regrouping of field lines (magnetic reconnection). Measurements of the solar magnetic fields support this idea: flares occur preferentially in activity regions with complex field configurations and steep field gradients, and there mostly in regions where the polarity of the magnetic field changes its sign.

4.10.6 Magnetohydrodynamics in the Solar Atmosphere. The Hydrogen Convection Zone

We must now consider how to *explain* the connection between sunspots, prominences, corona, flares, etc. within the framework of the 2×11 year cycle of solar activity. Clearly, the picture of a static atmosphere in radiation equilibrium is no longer sufficient here; instead, everything is in motion and in a state of flux, and magnetic fields play a central role. The high electrical conductivity of the ionized matter leads to a strong interaction of the flow fields with magnetic fields. In order to describe these phenomena, the basic equations of electrodynamics and hydrodynamics must be combined into those of *magnetohydrodynamics*.

We first however ask the question: what "thermodynamic machine" produces the mechanical energy which these flow fields continually require, whereby (according to the 2nd law of thermodynamics) thermal energy must be transported from regions of higher to those of lower temperature? This function is fulfilled by the *hydrogen convection zone* (A. Unsöld, 1931): from the deeper photospheric layers, i.e. from an optical depth of $\tau_0 \simeq 2$ (at $\lambda = 500$ nm), with a gas pressure $P_g \simeq 1.5 \cdot 10^4$ Pa and a temperature $T \simeq 7000$ K, down to $P_g \simeq 10^{11}$ Pa and $T \simeq 10^6$ K, the solar atmosphere is *convectively unstable*. The thickness of this layer, about 150000 km, corresponds to $\simeq 0.2$ solar radii. Above the layer, hydrogen (the most abundant element!) is practically neutral, within the layer it is partially ionized, and below the layer, it is completely ionized. Now the following happens: when a volume element with partially ionized gas rises, the hydrogen begins to recombine, and for each recombination process 13.6 eV (corresponding to 16 kT at 10000 K) is added to the thermal energy ($\frac{3}{2} kT$ per particle). This so strongly reduces the cooling by adiabatic expansion that the effective ratio of specific heats, c_p/c_v, approaches one and the rising gas volume becomes *warmer* than its new surroundings which are in radiation equilibrium. The volume element thus rises further. A sinking volume element undergoes exactly the reverse process. This effect is amplified further by the opposite influence of ionization on the radiation temperature gradient, and we obtain a zone with *convective flow*. At $P_g \geq 1.5 \cdot 10^4$ Pa, convection takes over practically the whole energy transport; radiation energy transport becomes unimportant in the lower layers of the convection zone.

While the thermodynamics of the hydrogen convection zone are relatively simple and straightforward, its *hydrodynamics* are among the most difficult problems in the study of flow phenomena. Models are usually calculated using the rough *mixing length* approach of W. Schmidt and L. Prandtl: a gas volume of dimension l is supposed to move through a sort of mean free path with the same length l and then suddenly to give up its temperature excess, its momentum, etc. to the surroundings by mixing; this is obviously a very rough schematic description of the highly complex convectional flow process. More detailed calculations of the processes in the convection zone are still in the early stages.

The *solar granulation* (Sect. 4.3.3) can now be attributed to the particularly strong instability of the surface layer of the convection zone, having a thickness of only a few hundred km. The granula are of the order of the thickness of this layer, but also not much larger than the equivalent height of the atmosphere at this point. The larger network of *supergranulation* (Sect. 4.3.3) consists of cells with a diameter of up to about 40000 km. It therefore appears attractive to attribute it to flow processes which include a major portion of the depth of the convection zone.

In the solar *spectrum*, the flow velocities which we have discussed above, ranging from about 0.2 to 3 $\mathrm{km\,s^{-1}}$, contribute a Doppler shift of the Fraunhofer lines and cause them to have a sawtooth-like shape and a broadening on the average. As long as the moving volume elements are not optically dense, the Doppler effects due to their convective motion simply add to those due to thermal motion: this gives the so-called *microturbulence*, which increases the purely thermal linewidths $\Delta\lambda_D$ (Sect. 4.9.2) by a factor of the order of 1.2 to 2. The Doppler effect due to the somewhat higher velocities in the spicules of the upper chromosphere has an analogous effect on the Doppler cores of the stronger Fraunhofer lines.

Instead of spending too much time on details, we turn to a more exciting question: Why, in going away from the Sun from the chromosphere to the corona, i.e. within the region of the *transition layer* which is only $\lesssim 15000$ km thick, does the temperature *increase* from ca. 4000 K to $\gtrsim 10^6$ K? From the Second Law of thermodynamics, this heating of the highest layers of the solar atmosphere, opposing the "natural" (i.e. entropy-increasing) temperature gradient from within the Sun to the outside, can only arise through *mechanical* energy or other "ordered" energy forms which have a large negative entropy.

Indeed, following M. Schwarzschild and L. Biermann, we can show that in upper layers of a partially convective atmosphere, mechanical energy transport becomes increasingly important relative to radiative energy transport. In the turbulent flow fields of the photosphere, *acoustic waves* (with periods in the range from 30 to 60 s) are produced; this was first investigated in detail by I. Proudman and M.J. Lighthill. Furthermore, due to interactions with the magnetic fields present in the plasma of the solar atmosphere, *magnetohydrodynamic waves* are also generated; they couple the oscillations of the plasma to correspondingly oscillating magnetic fields. All of these waves penetrate into the higher and thinner layers of the solar atmosphere and are concentrated there to *shock waves.*[17] Their energy is relatively rapidly dissipated, i.e. converted back to heat, and since the ability of the solar matter to radiate heat away becomes poor at low densities and high temperatures, the temperature T increases until the atmosphere finds a new mode of energy transport. This is due, as noted by H. Alfvén, to the fact that at sufficiently high temperatures and with a large temperature gradient inwards (!), the *thermal conductivity* due to the free electrons in the plasma (analogous to the high thermal conductivities of metals) is large enough to carry the energy back *inwards*, where it is finally radiated away.

While the physical processes which lead to the steep temperature increase in the corona are by no means understood in detail, we can estimate the required (minimal) *mechanical energy flux density* from observations of the total radiation flux from the chromosphere and the corona (using model calculations). If we neglect the energy losses of the acoustic and magnetohydrodynamic waves in the convection zone and the photosphere, we find that for the chromosphere, about 5 $\mathrm{kW\,m^{-2}}$, and for the corona, about 0.4 $\mathrm{kW\,m^{-2}}$ are required; i.e. only 10^{-4} or 10^{-5} of the overall energy flux of the Sun (4.3.9) need be tapped from the convection zone.

We now turn to the theory of the flow of conducting matter together with magnetic fields, i.e. to *magnetohydrodynamics* (MHD); it will provide us with the basic physics needed to understand the phenomena of solar activity and many other astrophysical processes. We must of course leave the complex mathematical apparatus by the wayside. Initially, however, an intuitive understanding of the physical fundamentals should be more important, in any case.

From the basic work of H. Alfvén, T.G. Cowling, and others, we take the following ideas:

Practically all cosmic plasmas have a very high electrical *conductivity* σ (reciprocal of the electrical resistivity). For fully ionized hydrogen gas, we find:

$$\sigma \simeq 10^{-3}\,T^{3/2} \quad [\Omega^{-1}\,\mathrm{m}^{-1}] \tag{4.10.4}$$

(T = temperature in K); e.g., at corona temperatures of 10^6 K, σ is about $10^6\,\Omega^{-1}\,\mathrm{m}^{-1} = 10^6\,\mathrm{V}^{-1}\,\mathrm{A\cdot m}^{-1}$, thus only an order of magnitude less than for pure metallic copper ($6\cdot 10^7\,\Omega^{-1}\,\mathrm{m}^{-1}$). For a conductor *at rest*, the current density j (current per cross-sectional area) depends on the electric field strength $\boldsymbol{E}$ through *Ohm's Law*:

$$j = \sigma \boldsymbol{E}\ . \tag{4.10.5}$$

If in addition a magnetic field of flux density $\boldsymbol{B}$ is present in the conducting medium, a change in its strength will

[17] The energy density of the acoustic waves is of the order of the density $\varrho\cdot$(velocity amplitude $v)^2$. If the wave travels at the velocity of sound c into a less dense medium, then v^2 increases correspondingly. When v approaches c, Mach's number $M = v/c \to 1$, and shock waves are formed.

induce currents which are only slowly damped out. The *decay time* τ can be derived from Maxwell's equations; here, we shall content ourselves with an estimate of the order of magnitude: if the volume of conducting plasma considered has a characteristic dimension x, then from the law of induction for the change in magnetic field $\partial B/\partial t \simeq B/\tau$, the electric field strength and induced current density are given by $E = j/\sigma \simeq xB/\tau$. (The induced voltage $V = Ex$ is equal to the change in the magnetic flux Bx^2/τ.) On the other hand, j itself produces a magnetic field given by $H = B/\mu_0 \simeq jx$ (Ampère-turns per meter!). Eliminating B/j from these two equations, we find the characteristic decay time[18] to be:

$$\tau \simeq \mu_0 \sigma x^2 \, , \qquad (4.10.6)$$

where $\mu_0 = 4\pi \cdot 10^{-7}\ \mathrm{V \cdot s \cdot A^{-1}\, m^{-1}}$ is again the permeability constant of vacuum. The form of the dependence, τ proportional to x^2, shows that the propagation and decay of a magnetic field in a conductor at rest has the character of a diffusive process.

If we now permit *motions* of the conducting medium, the result will depend upon whether the magnetic field $\boldsymbol{B}$ can move more quickly by diffusion, independently of the motions of the matter, or is *"frozen into the matter"*. Magnetohydrodynamic waves or disturbances propagate with the Alfvén velocity:

$$v_A = \frac{B}{\sqrt{\mu_0 \varrho}} \qquad (4.10.7)$$

(ϱ = matter density). When the associated propagation time x/v_A is shorter than the diffusion time (4.10.6), i.e. when *Alfvén's condition*:

$$\frac{\sqrt{\mu_0}\, \sigma B x}{\sqrt{\varrho}} > 1 \qquad (4.10.8)$$

is fulfilled, and this is frequently the case in cosmic plasmas, the magnetic field remains frozen into the plasma matter. The matter can then move essentially only *parallel* to the magnetic lines of force, like glass beads on a string. Since the *magnetic pressure* (Maxwellian tension) is roughly given by $B^2/2\mu_0$, the dynamic pressure $\varrho v^2/2$ will frequently be of the same order (in magnetohydrodynamic flows which are not too peculiar).

Applying the results of MHD to solar physics, we find it basically understandable that the distribution of *radiation emission* in the higher layers of the solar atmosphere represents in the final analysis an *image of the paths of the magnetic field lines*: not only matter flows, but also the energy transport by thermal conduction and magnetodynamic waves follow the field lines or the flux tubes and are strongly hindered in directions perpendicular to them. For example, a coronal arch is "filled up" with matter, so that its density is higher than that of the surrounding material; it can be heated by waves propagating along the curved flux tubes. Within an "unperturbed" arch, we should expect for one thing a hydrostatic equilibrium, and for another, thermal equilibrium among heating, thermal radiation, and thermal conduction. The magnetic field prevents the "overly dense" matter from spreading out laterally. The precise physical behavior of matter in the solar atmosphere, with its widely varying densities, its heating, dynamics, and stability, of course requires a more detailed investigation.

We can also understand with the help of MHD that the *quiescent prominences* can often remain suspended for weeks in the much hotter corona which surrounds them. We have already seen that only cooler matter can radiate energy away effectively. That is, cool matter remains cool, hot remains hot, even when a certain energy transport takes place. The pressure equalization between prominences and the corona in a horizontal direction requires that $p \propto \varrho T$ is about equal in each. Thus the density of the prominences must be about 300 times larger than in the corona. The fact that the static prominences do not fall down can now be explained, according to R. Kippenhahn and A. Schlüter (1957), by their lying on a "pillow" of magnetic lines of force (which cannot be penetrated by matter), like rain water in the depressions of a plastic tarpaulin covering a haystack. We must still explain why individual prominence-nodes do not fall down along the lines of their magnetic guide fields according to Newton's laws, but rather float down much more slowly and often with velocities which remain constant over long periods of time. This behavior, which is quite reminiscent of the clouds in our Earth's atmosphere, is clearly based on the fact that the corona, as a result of the high velocities and thus the long mean free paths of its electrons, has an enormous *viscosity*. The fact that the prominences do not fall down into the photosphere at the first opportunity, but rather float like real clouds, is due (in both cases) to the predominance of

---

[18] In alternating current technology, the decay time is given by the ratio of the self-inductance L to the resistance R. The latter is proportional to $1/\sigma$, while L is given by $L \simeq B/j$.

viscous forces over the forces related to pressure and inertia; we have "creeping flow" with small Reynolds' numbers. We cannot take up here the dynamics of the perturbed prominences.

Finally, MHD also gives us the physical fundamentals for understanding the *sunspots* and the *solar activity cycle*:

As early as 1946, T. G. Cowling pointed out that the magnetic field of a sunspot would decay in a solar atmosphere *at rest* only over a time of the order of 1000 years by "diffusion", owing to the very high electrical conductivity of the plasma (4.10.6). In fact, however, sunspots have lifetimes of only several days up to a few months. Only in 1969 was it noted by M. Steenbeck and F. Krause that the turbulence in the hydrogen convection zone contributes strongly to vortex formation in the magnetic fields and in the electrical current and matter velocity fields which are associated with them; in the solar atmosphere, this produces a reduction of the electrical conductivity by a factor of 10^4, so that now (4.10.6) yields the right order of magnitude for the lifetimes of the sunspots.

The formation of a *bipolar sunspot group* (others can be traced back to this case) was proposed by V. Bjerknes (1926) to proceed as follows: in the Sun, there are always toroidal "tubes" of magnetic field lines (i.e. parallel to the lines of latitude). Since in these tubes the pressure is partly of magnetic origin, the gas pressure and the density are lower than in the surrounding medium. The tubes are therefore pushed up towards the surface ("magnetic buoyancy") and are "cut open" there. The two open ends of such a field tube form a bipolar sunspot group.

As is shown by Spörer's law of sunspot zone motion (Fig. 4.10.3), and Hale's law for the magnetic polarity of the sunspot groups, the entire *activity cycle* is based upon a reversal of the overall flow- and magnetic fields in the interior of the Sun (mainly in the lower portion of the hydrogen convection zone) with a period of $2 \cdot 11$ years, and their stability (on the average) in the intervening time. The theory of this solar *dynamo*, whose driving force is the hydrogen convection zone together with the Sun's rotation, has become clearer piece by piece, beginning with the work of H.W. Babcock (1961), R.B. Leighton (1969), and M. Steenbeck and F. Krause (1969). The initially assumed toroidal magnetic field must, in the appropriate phase of the cycle, give rise to a meridional magnetic field, and *vice versa*. This is made possible, according to Steenbeck and Krause, by the turbulence; it generates a current parallel to the field lines (in a manner which we cannot explain in detail here), giving rise to a meridional field component, i.e. perpendicular to the original field lines. In other words, for our dynamo, not only induction due to the *differential rotation* of the Sun, but also that due to the statistically distributed *turbulent flow* is essential. The dynamo theory must finally take into account the recently-observed fact that the magnetic flux of the photosphere is almost completely concentrated into thin *flux tubes* at the rims of the supergranulation convection cells.

The extremely complex, complete calculation of the problem based on the approaches mentioned above would seem to be fundamentally able to yield a quantitatively correct explanation of the phenomena of solar activity.

4.10.7 The Solar Wind

As we already mentioned in Sect. 2.9.1, it was suggested by L. Biermann in 1951 that the plasma tails of comets are blown away from the Sun not by radiation pressure, but rather by an everpresent corpuscular solar radiation. Measurements from satellites and space vehicles later showed that this plasma, which has roughly the composition of solar matter, streams away from the Sun with velocities – in the neighborhood of the Earth's orbit – of about 470 km s^{-1} (with variations between about 300 and 700 km s^{-1}). Its density corresponds to about $9 \cdot 10^6$ protons and electrons per m^3, also with large variations. The (more or less) statistical part of the particle velocities corresponds to a temperature of about 10^5 K. Associated with the plasma are magnetic fields of the order of $6 \cdot 10^{-9}$ T or $6 \cdot 10^{-5}$ G. In 1959, E.N. Parker named this phenomenon the *solar wind* and suggested that it could be explained on the basis of *hydrodynamics* or of magnetohydrodynamics. If we calculate the pressure distribution $p(r)$ of the corona (schematically assumed to be isothermal as a first approximation) at a large distance r from the Sun, we find that the finite limiting value of the pressure there, for corona temperatures $< 500\,000$ K, lies *below* that of interstellar matter (Sect. 5.3). In this case, the interstellar matter would stream into the Sun. However, for $T > 500\,000$ K, as is true of the real corona, matter streams continually outwards: this is just the solar wind. Its *velocity* v as a function of the distance from the center of the Sun is determined by the Bernouilli equation, i.e. hydrodynamic energy conservation, together with the equation of continuity and the equation of state of the gas, $p = \varrho k T/\mu m_u$ (2.8.15), where ϱ is again the density, k Boltzmann's constant, μ the mean molecular mass, and m_u the atomic mass constant. For a fully ionized plasma of 90% H and 10% He, $\mu = 0.65$.

In the case of a spherically-symmetric gas stream, the *equation of continuity* is given by the condition that the same amount of matter passes through every spherical shell $4\pi r^2$ per unit time, i.e.

$$4\pi r^2 \varrho(r) v(r) = \text{const} . \qquad (4.10.9)$$

This statement must be drastically modified if the matter is guided by a magnetic field. Then, a particular matter current must move along a particular magnetic force tube, as long as the pressure of the matter on its "walls" does not exceed the magnetic pressure. We cannot pursue the solution of this extremely difficult problem of magnetohydrodynamics any further here.

Rather, we first ask what conclusions we can draw from *Bernoulli's equation*. It states that along a streamline, i.e. a line in the velocity field $\boldsymbol{v}$, the sum of the kinetic energy $v^2/2$ (all energies calculated per unit mass), the potential energy or potential $\Phi(r)$, and the pressure energy $\int dp/\varrho$ for a stationary flow $(\partial/\partial t = 0)$ is constant. The gravitational potential of the Sun (mass $\mathcal{M}$, gravitational constant G) is given from (2.6.34) by $\Phi = -G\mathcal{M}/r$; we thus obtain:

$$\frac{v^2}{2} + \int \frac{dp}{\varrho} - \frac{G\mathcal{M}}{r} = \text{const} , \qquad (4.10.10)$$

where the (pressure) integral is calculated up to the distance being considered. For example, in order to calculate the velocity of the solar wind $v_\oplus$ near the Earth's orbit $r \simeq r_\oplus$, we assume that it originates at the base of the corona, at $r \simeq r_\odot$, with $v \simeq 0$. The potential difference through which it passes, $G\mathcal{M}\,[(1/r_\oplus)-(1/r_\odot)]$, recalculated in terms of kinetic energy $v^2/2$, corresponds to the *escape velocity* of the Sun (Sect. 2.7), $v_E = 620$ km s^{-1}. Applying Bernoulli's equation (4.10.10) to the acceleration of the solar wind between $r_\odot$ and $r_\oplus$, we obtain

$$v_\oplus^2 = -2\int \frac{dp}{\varrho} - v_E^2 . \qquad (4.10.11)$$

The numerical value of the integral depends decisively on the temperature distribution $T(r)$ between the Sun and the Earth. If we rely on the observation that T decreases relatively little from $T_\odot \simeq 10^6$ K in the corona to $T_\oplus \simeq 10^5$ K near the Earth's orbit, we can as an approximation use the constant mean temperature $\bar{T} \simeq 10^6$ K and then readily carry out the integration $(dp = d\varrho \cdot k\bar{T}/\mu m_u)$:

$$v_\oplus^2 = \frac{2k\bar{T}}{\mu m_u} \ln(\varrho_\odot/\varrho_\oplus) - v_E^2 . \qquad (4.10.12)$$

$\sqrt{2k\bar{T}/\mu m_u}$ corresponds to the thermal velocity of the particles (165 km s^{-1}) or, within about 10%, the velocity of sound in the corona. With the numerical values indicated, we obtain $v_\oplus \simeq 400$ km s^{-1}.

It is important to realize that the fact that our calculation leads to a result which is in agreement with the measurements is a result of our *implicit* assumption, in the chosen numerical value for $\bar{T}$, that the outer corona or the solar wind is *heated* by some process or other (dissipation of wave energy?) and has only very small radiation losses. If we had for example calculated the case of an *adiabatic* outflow from the corona, we would have obtained as the maximum value of the integral in (4.10.11) the enthalpy per unit mass of the coronal matter, $c_p T_\odot$, where c_p is the specific heat at constant pressure. As can be readily verified, the material streaming out adiabatically would cool rapidly and remain "stuck" near the Sun. Our hydrodynamic theory of the solar wind indeed gives the correct order of magnitude of the observed result, but it is to some extent unsatisfying since the mean free path is of the same order as the characteristic lengths in the model. Therefore, a calculation based on the kinetic theory of gases, in which the magnetic field would necessarily be taken into account ("collision-free plasma"), would actually be required.

When the solar wind streams out into interplanetary space, it takes magnetic field lines along with it. We first consider the motion of particles in the neighborhood of the Sun's equatorial plane, where we introduce polar coordinates r and ϕ. A particle with velocity v arrives in the time t at a distance $r = vt$ from the Sun. The Sun has, in the meantime, rotated through an angle $\phi = \omega t$, where the angular velocity $\omega = 2\pi/(\text{sidereal rotational period})$. The particles which are emitted from a particular point on the Sun thus lie along an Archimedian spiral at a particular time,

$$\phi = \frac{\omega r}{v} , \qquad (4.10.13)$$

which traverses a radius vector everywhere at the same angle, which is determined by

$$\tan \alpha = r \frac{d\phi}{dr} = \frac{\omega r}{v} . \qquad (4.10.14)$$

With $v = 470 \text{ km s}^{-1}$, we obtain $\alpha = 45°$, in good agreement with observations.

The magnetic field lines follow this spiral. Since they must always be closed curves, they form loops whose beginnings or ends lie in regions of opposite magnetic polarity on the Sun. Such regions fill up a considerable fraction of the solar surface area, so that enormous field loops are continually pulled out into interplanetary space by the solar wind. Using observations from the satellite IMP-1, J. M. Wilcox and N. F. Ness in 1965 first detected in the neighborhood of the Earth's orbit the *sector structure* of the interplanetary magnetic field (Fig. 4.10.14), which corresponds to the field distribution on the Sun. The number (usually two or four) and the distribution of the sectors is variable and corresponds to the arrangement of the field configuration on the Sun.

It is essential for an understanding of the variations in the sector structure to have a knowledge of the *three dimensional* distribution of the solar or interplanetary

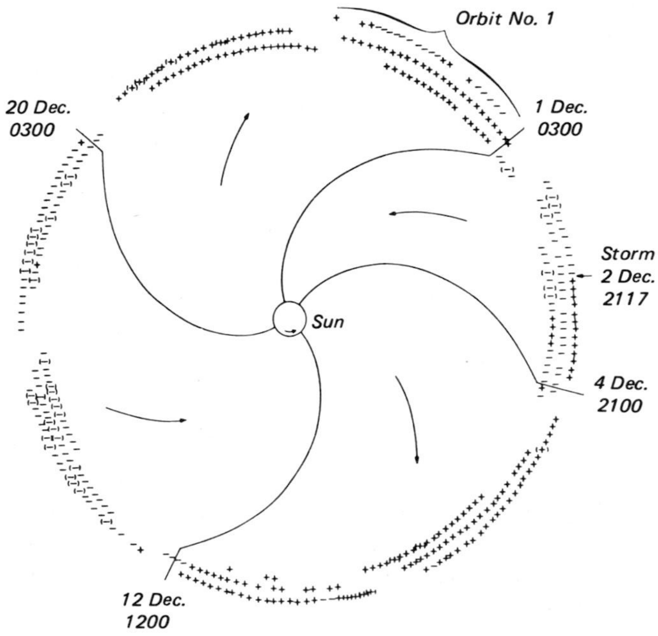

Fig. 4.10.14. The sector structure of the interplanetary magnetic field. Observations by IMP-1, from J. M. Wilcox and N. F. Ness (1965). The + and − signs indicate magnetic fields which are directed outwards or inwards, respectively. The portion of the field near the Sun is extrapolated schematically

magnetic field over a large region of space. Based on the measurements made by the space probe Pioneer 11 on its way to Saturn, which showed no signs of a sector structure at 16° above the plane of the ecliptic, we can construct the following picture:

Near the Sun's equator, there are many active regions with closed field lines; however, at higher heliographic latitudes, we find *open* field lines, along which the solar wind streams outwards (Sect. 4.10.3). These field lines, which emerge from or enter the Sun's surface roughly perpendicular to it, are deflected towards the equatorial plane at a few solar radii away from the surface and thereafter run essentially parallel to it. Since the magnetic field (except for localized regions) exhibits opposite magnetic polarity in the northern and southern solar hemispheres, field lines of opposite direction run close to each other in the equatorial plane. They are separated by a thin "*neutral sheet*" (in which electric currents flow, owing to the field reversal). This layer is, however, not precisely in the plane of symmetry, due to asymmetries in the field distribution, activity regions, etc.; rather, it is "*deformed*" up to about ±15° of latitude. The boundaries of the sectors, which for example are observed in the neighborhood of the Earth, result finally from the intersection of the plane of the ecliptic, which is inclined by 7°15′ relative to the equatorial plane of the Sun, with this rotating, deformed neutral layer.

The region filled by the solar wind, the *heliosphere*, ends where the dynamic pressure of the wind becomes equal to the pressure of the interstellar medium. The transition zone (heliopause) lies at about 50 to 100 AU distance from the Sun.

4.11 Variable Stars. Flow Fields, Magnetic Fields, and Activity in Stars

The first observations of variable stars at the turn of the 16th to the 17th centuries represented at the time a weighty argument against the Aristotelian dogma of the immutability of the heavens. Tycho Brahe's and Kepler's discoveries of the *supernovae* of 1572 and 1604 have continued to contribute to knowledge of these mysterious objects even in our day and have permitted the radioastronomical identification of their remains. Fabricius' discovery of the Mira Ceti should also be mentioned in this connection.

Variable stars are denoted by capital letters, R, S, T, ... Z, and the genitive case of the name of the constellation; these are followed by RR, RS, ... ZZ, AA, ... AZ, BB, ... QZ. After using all of these 334 combinations,[19] further variables are denoted by V335, V336, ..., followed by the name of their constellation (Lat. genitive), etc. (V: variable star).

It should be clear from the outset that the investigation of variable stars promises us much deeper insights into stellar structure and evolution than that of the static stars, which are "eternally the same". On the other hand, the observation and theory of variable stars presents much greater difficulties. A warning against easy, *ad hoc* hypotheses is not inappropriate here.

It is impossible within the scope of this book for us to describe the numerous classes of variable stars (usually named for a prototype) with any degree of completeness. We shall leave aside the eclipsing binaries, which have already been discussed, and consider a few interesting and important types of *physically variable stars* which we group according to common physical aspects of their descriptions. We first treat the pulsating variables (Sect. 4.11.1) and the magnetic variables with their variable spectra (Sect. 4.11.2). After having in Sect. 4.10 been introduced to the outer layers of the *solar* atmosphere and solar activity, we now discuss in Sects. 4.11.3 and 4.11.4 the indications for activity, chromospheres, coronas, etc. in *stars* of different classes. The cataclysmic variables, to which the novae belong (Sect. 4.11.5) and the very diverse stellar X-ray sources (Sect. 4.11.6) represent quite different groups of variable stars. The phenomena observed in these two groups are due to matter flows within close binary systems, which are "pulled in" by a compact star component (a white dwarf or neutron star). Finally, in Sect. 4.11.7, we turn to the spectacular phenomenon of the burst of brightness observed in a *supernova*, which is accompanied by the casting off of a shell of stellar matter (supernova remnant). These powerful stellar explosions often leave a neutron star behind, which can be observed as a pulsar.

In discussing all types of variable stars, it will become clear that we must consider them as particular stages in stellar *evolution*. However, we shall not develop this important idea further until Sects. 4.12 and 5.4.

4.11.1 Pulsating Stars. R Coronae Borealis Stars

The following groups of variable stars, among others, belong to the *pulsating variables*; they are, for the most part, giant stars, but there are also pulsating stars along the main sequence stars and among the white dwarfs.

RR Lyrae Stars or *Cluster Variables*. These are stars with regular changes in brightness having periods of about 0.2 to 1.2 days, brightness amplitudes of the order of 1 mag (ranging from about 0.4 to 2 mag), spectral types A and F, and masses from 0.5 to 0.6 $\mathcal{M}_\odot$. They are found in the halo and core of the Milky Way galaxy and are important in the globular clusters.

δ Cephei Stars (Classical Cepheids). Stars of high luminosity (class Ia to II) which also exhibit very regular brightness changes having periods from 1 to 50 d and about the same brightness amplitudes as the cluster variables (0.1 to 2 mag), but belonging to later spectral types (F5 – K5) and having masses between about 5 to 15 $\mathcal{M}_\odot$. They occur in the spiral arms of the Galaxy.

W Virginis Stars with very similar properties to the δ Cep stars, but weaker in absolute magnitude (by 1 to 2 mag) and with low masses (0.4 to 0.6 $\mathcal{M}_\odot$). They are found in the halo and core regions of the Milky Way.

Dwarf Cepheids and *δ Scuti Stars*. Short-period variables near the main sequence of spectral types A and F (masses 1 to 2 $\mathcal{M}_\odot$), with periods between 0.03 to 0.2 d, and brightness amplitudes of 0.3 to 0.8 mag in the case of the dwarf cepheids and, as a rule, ≤0.1 mag in the δ Sct variables.

ZZ Ceti Stars. White dwarfs of spectral type DA with very short periods in the range from 3 to 20 minutes and small amplitudes between 0.01 and 0.3 mag.

Mira Variables or *Long-Period Variables* are all giant stars of the late spectral types (M, C, and S), usually with emission lines; Mira Ceti = *o* Cet (M7 IIIe) is a member of this group. The light curves are not so stable as in the case of the cepheids; they have periods of from 80 d up to more than 500 d, and large brightness amplitudes of more than 2.5 to 8 mag in the visible region. Their masses are of the order of 1 $\mathcal{M}_\odot$, and their radii range from 100 to 1000 $R_\odot$. Mira variables occur both in the young and in the old stellar populations of the Milky Way.

---

[19] J is not used.

RV Tauri Stars. Bright giants and supergiants with spectral types F to K having alternately deep and shallow minima in their light curves, periods between about 30 and 150 d, and amplitudes up to 3 mag.

Semiregular variables. Giants to supergiants of medium or late spectral types (≥F) with quasi-periods in the range of 30 to over 1000 d.

The first indication of the physical nature of the groups of variable stars described here was given by their *radial velocity curves*. These are closely connected to their light curves. Initially, it was attempted to attribute the very regular velocity fluctuations of e.g. the classical cepheids to a binary star motion. Integration over the velocity yields (without further hypotheses) the dimensions of the "orbit", since

$$\int_{t_1}^{t_2} \frac{dx}{dt}\,dt = x_2 - x_1 \qquad (4.11.1)$$

(x: coordinate along the line of observation).

It soon became apparent, however, that the star would have no room in this orbit alongside the postulated companion star. Thus, in 1914, H. Shapley returned to the possibility of a *radial pulsation* of stars, which had been discussed in the 1880's by A. Ritter as a purely theoretical problem. The pulsation theory of the cepheids (and related variables) was then developed further in 1917 by A. S. Eddington. This, in turn, gave the impulse for his pioneering work on the internal structure of stars (Sect. 4.12).

An important aspect of the theory of pulsating stars, which is difficult in its details, is already revealed by a simple estimate: we take the pulsation to be a kind of *acoustic wave* in the star. Its velocity is $c_s = \sqrt{\gamma p/\varrho}$, where $\gamma = c_p/c_v$ is the specific heat ratio, p the pressure, and ϱ the density of the stellar matter. The average pressure $\bar{p}$ in stellar interiors is of the order of magnitude of the gravitational force on a column of matter with a unit cross-sectional area extending from the stellar surface to the interior of a star of mass $\mathcal{M}$ (2.8.7), i.e.:

$$\bar{p} \simeq \underbrace{\bar{\varrho}}_{\text{mean density}} \cdot \underbrace{R}_{\text{radius}} \cdot \underbrace{\frac{G\mathcal{M}}{R^2}}_{\text{acceleration}} . \qquad (4.11.2)$$

The period of oscillation P now has the order of magnitude:

$$P \simeq \frac{R}{c_s} \simeq R\left(\gamma \frac{G\mathcal{M}}{R}\right)^{-1/2} . \qquad (4.11.3)$$

Due to $\mathcal{M} = (4\pi/3)R^3\bar{\varrho}$, we find from this the important relation between the period P and the mean density $\bar{\varrho}$ of the star:

$$P \simeq \frac{1}{\sqrt{G\bar{\varrho}}} , \qquad (4.11.4)$$

which has been convincingly confirmed on the basis of a large amount of observational data.

A second test of the pulsation theory was suggested by W. Baade: the brightness of a star is proportional to the area of its "disk", πR^2, times the radiation flux F_λ at its surface. The temporal variation of the stellar radius R can be obtained directly by integrating the radial velocity curve as in (4.11.1). On the other hand, the radiation flux F_λ can be determined independently from the theory of stellar atmospheres using the color indices or other spectroscopic criteria. (We shall apply this method later in Sect. 5.5.1 to the determination of the distances of galaxies).

In Fig. 4.11.1, some of the quantities of state for δ Cep stars, with their time variation, are summarized. The expected proportionality of the measured magnitudes to $R^2 F_\lambda$ is indeed well fulfilled.

Miss H. Leavitt at the Harvard Observatory in 1912 discovered a relation between the periods P and, initially, the apparent magnitudes m_V of the many hundred cepheid variables in the Magellanic Clouds. Since all these stars have the same distance modulus, she had actually discovered a *period-luminosity relation*. H. Shapley determined the zero point of the scale of *absolute magnitudes* using the modest amount of observational data of proper motions then available (Sect. 5.2.2). It thus became possible to determine the *distance* to every cosmic object in which cepheids could be found (the problems of interstellar absorption were not yet known). For example, H. Shapley in 1918, using observational data of S. Bailey (1895), was able to determine the distances to many cluster variable stars (RR Lyr stars) for the first time and thus to fix the boundaries of the galactic system in the modern sense. Then in 1924, E. Hubble, using the same methods, but employing classical cepheids (with longer periods), determined the distances to some of the neighboring spiral nebulas and showed definitively that they are *galaxies* of a similar scale to our Milky Way. We shall report this "penetration of deep space" in Sects.

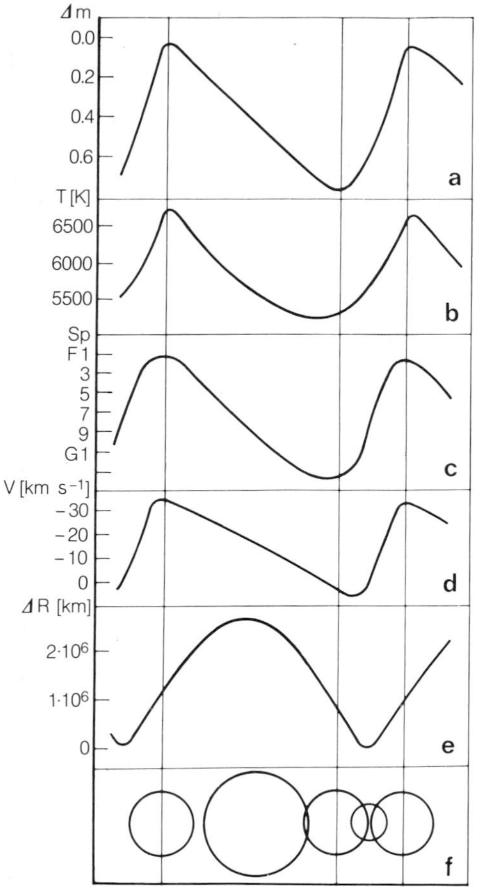

Fig. 4.11.1 a–f. Periodic variations of δ Cephei. From above, **(a)** the brightness (light curve in [mag]), **(b)** color temperature, **(c)** spectral type, **(d)** radial velocity, **(e)** change in radius, $\Delta R = R - R_{\min}$, and **(f)** the stellar disk are shown

5.5 and 5.9. Here, however, we should discuss an important correction to the fundamentals of the cepheid method, discovered in 1950 by W. Baade. He was able to show that the zero point of the period-luminosity relation is different for different types of cepheids. In particular, the classical cepheids of population I (cf. Sect. 5.1 for a discussion of the stellar populations) are 1 to 2 magnitudes brighter than the W Virginis stars of population II with the same period.

The *amplitude of the light variation*, measured e.g. in visual magnitudes m_v, increases systematically on going to cooler stars. This is essentially based on Planck's radiation law. If we write m_v in Wien's approximation analogously to (4.4.10), we find

$$m_v = \frac{1.56 \cdot 10^7}{\lambda_v\,[\mathrm{nm}]\; T\,[\mathrm{K}]} + \mathrm{const}_v \; . \qquad (4.11.5)$$

A particular temperature variation ΔT thus corresponds to a brightness amplitude given by

$$\Delta m_v = -\frac{1.56 \cdot 10^7}{\lambda_v T^2}\,\Delta T \; , \qquad (4.11.6)$$

which in cooler stars increases proportionally to $1/T^2$.

An interesting theoretical problem is the *conservation of pulsation*: what "valve" guarantees that the stellar oscillations, like the pistons of a thermal engine, are always pushed with the right phase? The production of thermal energy by nuclear processes near the center of the star is practically not influenced by the pulsation. Rather, the critical factor is the temperature and pressure dependence of the opacity, which regulates the flow of radiation energy and thus determines the temperature of a particular layer. This "κ mechanism" is found to be particularly effective, together with the change in the adiabatic temperature gradient, in the region of the second ionization of helium. Similar (very difficult) calculations can make it clear theoretically which combination of quantities of state, i.e. which regions of the Hertzsprung-Russell diagram, permit the existence of pulsation. Thus, for example, one can understand that the δ Sct stars and the dwarf cepheids as well as the RR Lyr, W Vir, and δ Cep stars are all found close to an *instability strip* which stretches from an effective temperature of about 8000 K near the main sequence "upwards and to the right" to 5000 K at the cepheids.

The cooler pulsating variable stars with longer periods of light variation, such as the RV Tauri variables and long-period variables, have an increasingly irregular light variation. The theory of the inner structure of the stars (Sect. 4.12) shows that in cooler stars, the hydrogen convection zone becomes stronger and stronger. It is therefore tempting to attribute the observed semi-regular light variation to a coupling of the pulsation with the turbulent flow processes of convection.

The shells of the *Mira variables* expand with velocities of the order of 10 km s^{-1}, as can be determined from measurements of the radial velocities of absorption and emission lines. The stars lose a corresponding amount of mass at a rate of about 10^{-8} to 10^{-6} $\mathcal{M}_\odot$ yr^{-1}. From some Mira variables, lines in the radiofrequency region can also be observed, in particular the *maser emission of the OH radical* at $\lambda = 18$ cm (Sect. 5.3.4), where the 1612

MHz component is strongest (so-called type II maser). The characteristic double structure of this component also indicates an expanding motion of the gas shell.

Radio surveys at $\lambda = 18$ cm, searching for OH sources, together with infrared observations, have in recent times led to the discovery of optically invisible *OH/IR stars*. These cool stars, which are hidden behind a thick shell of dust, exhibit brightness variations in the infrared corresponding to the type of the Mira variables; the periods extend up to 2000 d. The period-luminosity relation of the OH/IR stars indicates that we can take them to represent a continuation of the Mira variables. Their mass loss rates of about 10^{-5} to 10^{-4} $\mathcal{M}_\odot$ yr^{-1} are higher than those of the Mira stars.

In the region of the red giant stars, there is quite a different kind of slow variable stars, the *R Coronae Borealis stars*. Their brightness decreases suddenly by several magnitudes from a constant normal value, and then slowly returns to the normal value. Spectral analysis of these relatively cool stars shows that their atmospheres have low hydrogen contents, but high contents of carbon (and probably helium). One might suspect that the R CrB stars at times emit clouds of colloidal carbon, i.e. a sort of soot cloud, which darken the star for a while.

4.11.2 Magnetic or Spectrum Variables. Ap Stars and Metallic-Line Stars

In the region of the main sequence, various types of stars are found which do *not* fit into the two-dimensional MK classification. They are all characterized by peculiarities in their spectra. The hotter Ap stars are for the most part variable, while the cooler metallic-line stars are not. Whether these two groups have anything in common remains an open question.

The (cooler) Ap stars (peculiar A stars) or *spectrum variables* exhibit anomalous intensities and a periodic variation in the intensities of certain spectral lines, with different spectral lines showing different behaviors. The prototype is α^2 *Canum Venaticorum* with a period of 5.5 d, in which the lines of Eu II and Cr II change with opposite phase, while e.g. the Si II and Mg II lines remain nearly constant. Along with the spectral changes, there are usually brightness variations of around 0.1 mag. H.W. Babcock was able to show by Zeeman effect measurements that these stars have magnetic fields of 0.1 to 1 T (10^3 to 10^4 G), which exhibit periodic changes in intensity and often even in sign. According to A. Deutsch, at least the major part of the observations can be explained by the hypothesis that these stars possess enormous magnetic *spots*, in which, depending on their polarity, one or another group of spectral lines is intensified. The variations observed are attributed to the *rotation* of the stars.

In the color-magnitude diagram, the Ap stars lie on or near the main sequence roughly in the region -0.20 mag $\leq$ (B-V) $\leq +0.20$ mag corresponding to effective temperatures between about 18000 and 8000 K. A large number of the Ap stars are thus in fact Bp stars, which were originally classified as A stars due to their characteristic weak He I lines. In general, however, the classification as Ap stars has been retained for all the stars of this group.

Among the late B and A stars, there are at least 10 to 15% Ap(Bp) stars. These are subdivided into the following groups, according to their most noticeable spectral anomalies: (a) at the cooler end of the region are the *Eu-Cr-Sr stars*, which border towards higher effective temperatures ($T_{\rm eff} \geq 12000$ K) on the *Si stars*; among the latter is e.g. α^2 CVn. (b) At temperatures similar to those of the Si stars are the *Hg-Mn stars*, which however, in contrast to the magnetic Eu-Cr-Sr and Si stars, show no observable magnetic fields and are not spectrum variables. (c) The *helium-weak stars*, with less noticeable anomalies, represent the continuation of both groups towards the highest effective temperatures up to about ≤ 20000 K.

In some Ap stars, unusual anomalies are observed: for example, a subgroup of the helium-weak stars exhibits the lines of ^{3}He with comparable intensities to those of ^{4}He, but somewhat shifted in wavelength. In some of the Eu-Cr-Sr stars, lines of otherwise rare elements such as Os I and II and Pt II, and perhaps also U II, occur with unexpectedly strong intensities. In the spectrum of HR 465, the radioactive promethium isotope ^{145}Pm, with a half-life of 17.7 years, has been identified. This can be regarded as an indication that the anomalous element abundances could in part be a result of neutron irradiation. On the other hand, most of the abundance anomalies of the Ap stars, and of the metallic-line stars (see below), can hardly be understood as resulting from nuclear processes. F. Praderie, E. Schatzman, and G. Michaud (1967/70) have therefore suggested an explanation on the basis of *diffusion processes*, by which the elements can be separated by the interplay of selectively-acting radiation pressure as a "buoyancy force", and sedimentation resulting from the gravitational acceleration, in layers of these stars near to their surfaces. The details, which are by no means understood, probably depend sensitively on the strength of turbulent or convective mixing and of magnetic fields.

Along the main sequence going to cooler temperatures, the non-variable *metallic-line stars* (Am stars) border on the Ap stars, with a considerable region of overlap. They are mostly members of binary star systems, and their rotational velocities are lower (≤ 100 km s^{-1}) than those of the normal A stars. Strong magnetic fields are not observed. About 10% of the (brighter) A stars are metallic-line stars. They are classified according to their hydrogen lines, for example as A0 and F1. On the basis of their hydrogen types, these stars lie on the main sequence. However, the lines of calcium (especially H and K) and/or scandium are too weak, while the metal lines of the iron group and the heavier elements are too strong. Precise analyses yield effective temperatures between about 7000 and 10000 K and show, by comparison of lines from various ionization and excitation stages, that the *abundances* of the elements are indeed *anomalous*, rather than the observed line intensities being due to some sort of anomalous excitation processes (deviations from LTE).

Both the Ap and the Am stars occur on the main sequence of relatively young (only 10^6 to 10^7 years old) star clusters and associations. This also indicates that the anomalies in these stars must have been produced quickly and in the immediate neighborhood of the main sequence.

4.11.3 Activity, Chromospheres, and Coronas of Cool Stars

In Sect. 4.10, we found that the Sun can be considered to be a variable star, with its $2 \cdot 11$ year magnetic activity cycle. The manifold phenomena of solar activity are, in the final analysis, based on the interaction of the *rotation* of the Sun with the flow- and magnetic fields of the *hydrogen convection zone*; this interaction gives rise to a dynamo process which maintains the activity cycle. The temperature increase towards the outer layers of the solar atmosphere, the chromosphere and the corona, is also due to the convection zone, in which acoustic and magnetohydrodynamic waves are generated that then give rise to heating of the upper layers. On the other hand, the theory of the inner structure of stars shows that all cooler stars, later than about F0 ($T_{\rm eff} \leq 6500$ K), must have more or less extensive hydrogen convection zones near their surfaces. It is thus tempting to search for indications of chromospheres and coronas and for solar-like activity phenomena in other stars from this region of the Hertzsprung-Russell diagram.

On the Sun, the active regions or their plages are characterized by Ca II H and K emission lines, sometimes superposed on a finer absorption structure. Exactly the same spectral features are observed in many main sequence and giant stars of types G through M. This *chromospheric emission* or *stellar activity*, discovered by K. Schwarzschild and G. Eberhard in 1913 and later thoroughly investigated by O.C. Wilson, often exhibits variations with time which are an indication for the rotation or an activity cycle in the star.

Then in 1957, a completely unexpected phenomenon was observed by O.C. Wilson and M.K.V. Bappu: the *width* of the Ca II emission lines was found (independently of their intensity) to be a function of the *absolute magnitude* of the stars. This provides an excellent method for the determination of the spectroscopic parallax of cooler stars, which is valid over 15 magnitudes or 6 powers of ten in the luminosity. Just why the turbulent velocity in the chromospheres of the cooler main sequence and giant stars is related to their other atmospheric parameters in the observed manner remains for the most part an unanswered question.

In several cool dwarf stars, G.E. Kron (1952) observed small periodic brightness variations, which he attributed to "starspots", analogous to sunspots. Following B.V. Kukarkin and others, we refer to this type of rotating stars as *BY Draconis variables*. They include main sequence stars of the later spectral types (with emission lines) and quasiperiodic brightness variations (≤ 0.6 mag). The periods lie in the range from several tenths of a day up to a few days; the amplitudes may change in the course of some years.

Many cool dwarf stars, usually of type M with hydrogen emission lines, also show *flares* at irregular intervals; these differ from those observed on the Sun only in their sometimes greater brightness. The optical brightness of the star increases within 3 to 100 s by amounts of 6 to 7 mag; the decrease back to normal magnitude is considerably slower. Smaller fluctuations in brightness merge into nearly continuous variations. The spectra of these *flare stars*, or, as they are named for their prototype, the *UV Ceti stars*, show a superposed continuum in the ultraviolet during a flare outburst, so that the brightness in this region may increase up to 10 mag. They also exhibit strong emission lines from Ca II, H I, He I, and even He II as in solar eruptions. However, while on the Sun the continuous spectrum of the flare reaches at most a few percent relative to the brightness of the photosphere, this ratio reverses in cool stars (≤ 4000 K), since the radiation flux from their photospheres is weaker by several magnitudes. Accompanying an optical flare, B. Lovell et al. discovered in 1963 the corresponding *radio flares* in the meter wavelength range. Likewise, in the *X-ray region*, outbursts of flare stars amounting to 10^{23} to 10^{24} W were observed from the Einstein satellite; they showed a great similarity to solar flares.

Since about 1970 it has been possible, through observations from satellites in the ultraviolet and X-ray regions, to investigate *chromospheres*, *coronas*, and *stellar winds* on a large scale in stars of different types. In many stars of the later spectral types, we find, as in the Sun, ultraviolet emission lines such as H I Lα, $\lambda = 121.6$; O I, $\lambda = 130.4$; C I, $\lambda = 155.7/156.1$; Si II, $\lambda = 180.8/181.7$; and Mg II (h and k), $\lambda = 279.6/280.3$ nm, which are of *chromospheric* origin. These occur together with lines of higher ionization states such as Si IV, $\lambda = 139.4/140.3$; C III, $\lambda = 97.7/117.5$; C IV, $\lambda = 154.8/155.1$; or N V, $\lambda = 123.9/124.3$ nm, which indicate temperatures above $3\cdot 10^4$ K and are formed in the *transition layer* to a corona or in a *cool corona* (Fig. 4.11.2). In contrast, in the cooler, more luminous stars such as α Ori (M2 Iab), the lines from highly excited states are lacking, so that their chromospheres at temperatures $\leq 3\cdot 10^4$ K probably merge directly, with no hot transition zone, into cool, massive *stellar winds* having relatively low velocities of around 10 to 100 km s^{-1}.

Based on the *blue shifted* circumstellar absorption components of strong lines such as the Ca II H and K lines in the spectra of cool, luminous giants, A. J. Deutsch recognized in 1956 that the expansion velocities of the extended shells are greater than the escape velocities; thus, these stars give off matter into the interstellar medium. The *rate of mass loss* $\dot{\mathcal{M}}$ can now be derived not only from the optical circumstellar lines, but also among other possibilities from the infrared radiation of the dust in the shell as well as from lines of OH, H_2O, or CO in the radiofrequency region.

According to D. Reimers (1975), the order of magnitude of the mass loss from cool giants and supergiants is given by

$$\dot{\mathcal{M}} \simeq 4\cdot 10^{-13}\frac{L/L_\odot}{(g/g_\odot)(R/R_\odot)}\,[\mathcal{M}_\odot\,\mathrm{yr}^{-1}]\;, \tag{4.11.7}$$

which relates it to the fundamental stellar parameters luminosity L, radius R, and the surface gravity, $g = G\mathcal{M}/R^2$. The rates of mass loss for the coolest supergiants attain 10^{-7} to 10^{-5} $\mathcal{M}_\odot$ yr^{-1}; for the OH/IR stars, they go as high as 10^{-4} $\mathcal{M}_\odot$ yr^{-1}. On the other hand, a weak stellar wind comparable to the solar wind ($\dot{\mathcal{M}} \sim 10^{-14}$ $\mathcal{M}_\odot$ yr^{-1}), as expected for cool main sequence stars, lies below the current detection limit.

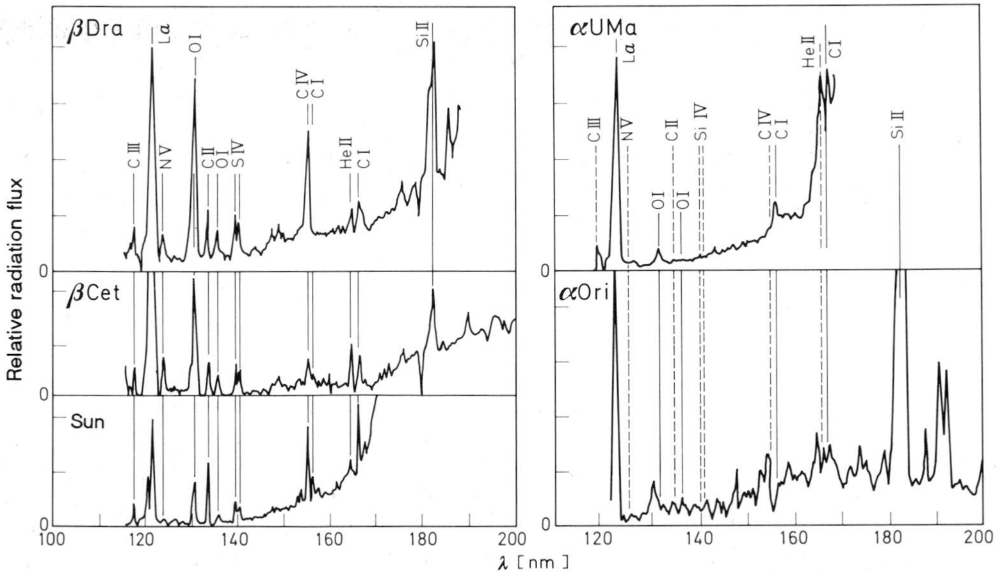

Fig. 4.11.2. Ultraviolet spectra of some cool stars (observations with the International Ultraviolet Explorer by J.L. Linsky and B.M. Haisch, 1979); β Dra (G2 II), β Cet (K1 III), α UMa (K0 II-III), and α Ori (M2 Iab), and, for comparison, the quiet Sun. The positions of solar spectral lines which are not present in the stellar spectra are indicated by dashed vertical lines

The (generally variable) stellar winds are expected to be due, as in the Sun, to the transport of energy by acoustic and magnetohydrodynamic waves from the hydrogen convection zone. In the case of the highest mass loss rates, acceleration by radiation pressure from the star probably also plays a role; it acts on the circumstellar dust particles with their large absorption cross sections.

In the X-ray region (0.2 – 3 keV), corresponding to its corona temperature of several 10^6 K, the Sun emits at a rate of $L_X \sim 5 \cdot 10^{19}$ W ($= 5 \cdot 10^{26}$ erg s^{-1}) (quiet corona) up to $2 \cdot 10^{22}$ W ($= 2 \cdot 10^{29}$ erg s^{-1}) (many active regions). Only with the sensitivity of the Einstein satellite and later of EXOSAT did it become possible to detect *stellar coronas* of a similar kind up to distances of about 100 pc by means of their X-ray emissions. The X-ray luminosity of the cool *main sequence stars* of spectral types G to M is in the range 10^{19} to 10^{21} W, and is thus quite comparable to that of the Sun at an "average" activity level; on the whole, it varies only slightly with spectral type. While the X-ray emission of a G dwarf makes up only about 10^{-7} of the overall radiation, the relative amount in the case of M dwarfs is noticeably higher, 10^{-3} to 10^{-2}, due to their much smaller total luminosities. The X-ray luminosity depends strongly on the *rotational velocity* $v_{\rm rot}$ of the star; the approximate relation $L_X \propto v_{\rm rot}^2$ holds independently of L.

While for some giant stars (earlier than K2), luminosities L_X similar to those of the dwarf stars are observed, it is noticeable that in the case of the cool giants and supergiants, *no* X-ray emissions have been discovered, although the detection limit corresponds to a value of L_X/L well below that of the quiet Sun ("coronal holes").

Many variables of the type *RS Canum Venaticorum*, detached binary stars with late supergiants and dwarfs as components, exhibit variable, strong emissions of about $4 \cdot 10^{24}$ W in the soft X-ray region, an additional sign of chromospheric activity and starspots. Their activity is probably also based esentially on the rapid rotation of the stars, which is synchronized with their orbital motions by tidal interactions.

From the work of O. C. Wilson, R. P. Kraft, A. Skumanich and others it has become clear since the 1960's that there is a connection between the *chromospheric activity* or Ca II emission with the *rotation* of the stars, and, through comparisons of the stars in stellar clusters of differing ages, with their *ages* also. With increasing stellar ages (on the average), both the rotational velocities and the activities of stars decrease. This is true not only of chromospheric emission, but probably also for all the other indicators of stellar activity, such as flares and X-ray emission. In the course of the evolution of a cool star, angular momentum is transferred to the interstellar medium by the outward streaming of plasma with its magnetic fields, similar to the solar wind, so that the star's rotation is slowed down. This in turn causes the dynamo process which maintains the stellar activity cycle and produces the star's magnetic field to become less effective, and the activity phenomena, which depend essentially on the magnetic field, thus decrease.

Strong activity or variability, which shows similarities with those of the Sun and other cool dwarf stars, are also found in *very young stars* in the spectral types F to M, which have not yet evolved onto the main sequence and lie above it in the Hertzsprung-Russell diagram (Sect. 5.4.7). The *T Tauri stars* or the *RW Aurigae stars*, named for their prototypes, are characterized spectroscopically by strong emission lines, especially from Ca II H and K, Hα, and other Balmer lines of hydrogen. Their luminosities "flicker" in an irregular manner, of the order of 1 mag within a few days; many are strong X-ray sources ($\leq 5 \cdot 10^{24}$ W). V. A. Ambartsumian, G. Haro, and others have shown that these variables are to be found in the sky particularly in the neighborhood of the dark clouds and young star clusters: in Orion, where a star cluster is being formed in the region of the well-known nebula; in Taurus with the Pleiades, etc. These are stars which were formed from interstellar matter relatively recently ($\leq 4 \cdot 10^8$ years ago). Along with them, a large number of flare or UV Cet stars are found.

In rare cases, strong brightness outbursts can occur in T Tau stars, which are accompanied by ejection of matter (FU Orionis phenomenon). For example, in 1969, the luminosity of V1057 Cyg increased by 6 mag in about 300 d, and then dropped back very slowly. Shortly after the outbreak, the spectrum of V1057 Cyg was similar to that of an A supergiant.

4.11.4 Coronas, Stellar Winds, and Variability of Hot Stars

Many, probably in fact all *supergiants* show irregular brightness variations of the order of 0.1 to 0.2 mag as well as variations in the radial velocities of their Fraunhofer lines up to several km s^{-1} with typical time scales of days to months; for example, α Cyg (A2 Ia) has been known since 1896 to be a spectrum variable. In the more luminous supergiants, emission lines are also observed

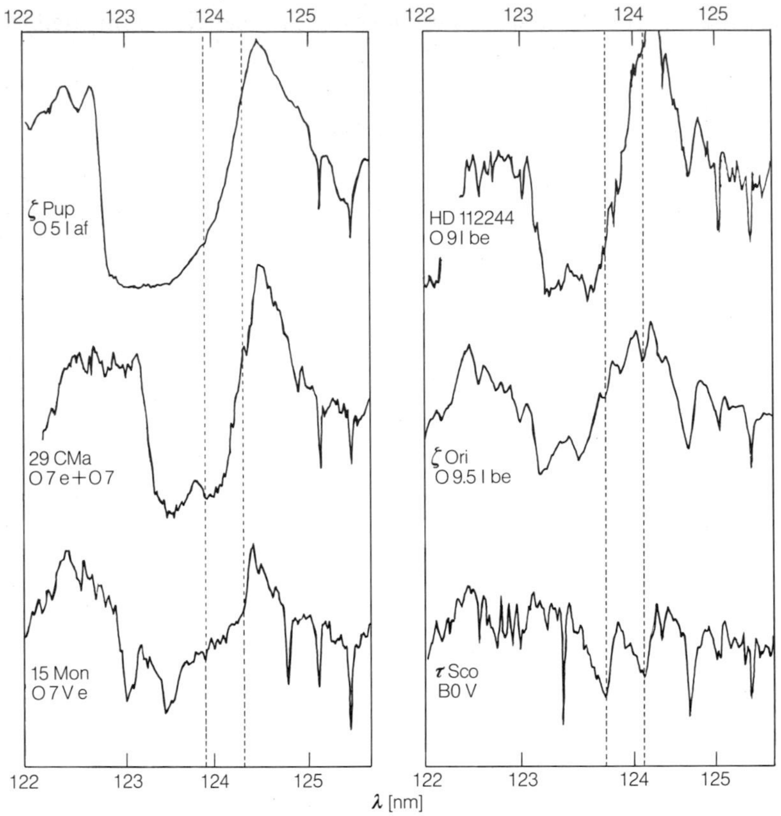

Fig. 4.11.3. Stellar winds and mass losses in OB stars: as an example, the P Cyg profile of the resonance doublet of N V at $\lambda = 123.88/124.28$ nm is shown (the wavelengths at rest are indicated by dashed lines). Observations made with the Copernicus Satellite by T. P. Snow and D. C. Morton (1976). (With the kind permission of the International Astronomical Union)

and are also variable, such as Hα and other Balmer lines, He II, $\lambda = 468.6$ nm, or (in the so-called Of stars) N III lines around $\lambda = 463$ nm.

The rare supergiants which have the brightest absolute magnitudes, with M_V in the range of -7.5 to -9.5 mag and spectral types between B and F (hypergiants), the *Luminous Blue Variables* (LBV), include the Hubble-Sandage variables and the S Dor variables (in the Large Magellanic Cloud), as well as η Car (Fig. 5.3.10) and P Cyg, and exhibit a wide range of brightness variations, from rapid, small variations (of a few 0.01 mag) on a time scale of hours, up to slow changes of the order of 1 mag in years to decades and occasional stronger bursts of brightness. Their spectra are rich in emission lines, which frequently show a so-called *P Cygni profile*: a broad emission with an absorption component shifted to the short-wavelength side (Fig. 4.11.3), which can be attributed to *ejected matter*. The velocities reach several 100 km s^{-1}, and the mass loss caused by this stellar wind is considerable ($\geq 10^{-5} \mathscr{M}_\odot$ yr^{-1}).

Although in the optical region, only the particularly dense, massive stellar winds from the most luminous stars can be recognized on the basis of their emission lines or P Cygni profiles, the far *ultraviolet* offers the advantage that here the *resonance lines* of many ions of abundant elements are to be found; therefore, relatively low densities of matter can be detected. The first ultraviolet observations with rockets by D.C. Morton and coworkers (1967) already indicated strong P Cygni profiles from some brighter stars. Using more sensitive spectrographs, in particular those of the Copernicus and the IUE satellites, it became clear that *stellar winds* or *mass losses* occur in all OB stars with luminosities $\geq 10^4 L_\odot$.

The mass-loss rate $\dot{\mathscr{M}}$ due to a stationary wind is given by the equation of continuity (4.10.9):

$$\dot{\mathscr{M}} = 4\pi r^2 \varrho(r) v(r) = \text{const} \ , \qquad (4.11.8)$$

where $\varrho(r)$ is the density and $v(r)$ the wind velocity at a distance r from the center of the star. The Doppler shift of the observed absorption components of a P Cyg profile gives us information about the maximum velocity v_∞, while the equivalent width W_λ, which in the optically thin case is proportional to the column density

$\int_R^\infty \varrho(r)dr$ (R: star's radius), essentially determines the density. The rate of mass loss is thus proportional to the product $R v_\infty W_\lambda$.

In the region of the OB stars, $\dot{\mathcal{M}}$ increases for the most part with increasing luminosity; empirically, the relation $\dot{\mathcal{M}} \propto L^{1.7}$ is found. On the one hand, the supergiants with their well-developed P Cyg profiles show mass losses up to several $10^{-5}\ \mathcal{M}_\odot\ \mathrm{yr}^{-1}$ at wind velocities up to $3500\ \mathrm{km\ s}^{-1}$. The mass losses of the Wolf-Rayet stars (Sects. 4.5 and 5.4.5), $4 \cdot 10^{-5}\ \mathcal{M}_\odot\ \mathrm{yr}^{-1}$, surpass even those of the OB stars of *comparable* luminosity by about a factor of 10. Mass losses of this magnitude during the "dwell time" in the supergiant phase make up a considerable portion of the original mass of a star and must be taken into account in considering *stellar evolution* (Sects. 5.4.4, 5). On the other hand, in the early B main sequence stars, such as τ Sco (B0 V), $\dot{\mathcal{M}}$ is a few $10^{-8}\ \mathcal{M}_\odot\ \mathrm{yr}^{-1}$ or less and can only be recognized through the *asymmetric* absorption lines with an extended wing on the short wavelength side (Fig. 4.11.3).

The extended, expanding shells of the supergiants can also be detected by their free-free radiation in the radiofrequency and infrared ranges. The first of these stars to be observed in the radio region was P Cyg, by H. J. Wendker *et al.* (1973) at 5 and 11 GHz.

The occurrence of higher ionized states such as C IV, N V, and O VI in the ultraviolet spectra of the stellar winds indicates temperatures from 10^5 to 10^6 K. The discovery by the Einstein satellite in 1979 that OB stars belong among the strongest *X-ray sources*, excepting close binary systems (Sect. 4.11.6), likewise indicates temperatures of 10^6 K. The X-ray emission of the *coronas* of the O and B stars is to a large extent independent of the spectral type and luminosity and is in the range $L_X \simeq 10^{-7} L$, i.e. in the case of O supergiants from 10^{25} to $4 \cdot 10^{26}$ W.

The existence of coronas in hot stars is surprising, since here, in contrast to the cooler stars, *no* extended hydrogen convection zones which could serve as a source of heating for a corona by mechanical transport are present. Although the acceleration of the corona is probably due to the high *radiation pressure* of the hot stars, the physical processes which lead to the coronal temperatures, high degrees of ionization, and X-ray emission are still for the most part unexplained. At any rate, a "mechanical power" of $\frac{1}{2}\dot{\mathcal{M}} v_\infty^2$ must be generated by the star to drive the stellar wind, which e.g. in the case of an O supergiant like ζ Pup with $\dot{\mathcal{M}} = 6 \cdot 10^6\ \mathcal{M}_\odot\ \mathrm{yr}^{-1}$ and $v_\infty = 2700\ \mathrm{km\ s}^{-1}$ is equal to nearly $10^4\ L_\odot$, i.e. about 1% of the star's luminosity (L_X is negligible compared to $\frac{1}{2}\dot{\mathcal{M}} v_\infty^2$).

4.11.5 Cataclysmic Variables: Novae and Dwarf Novae

The group of the *eruptive variables* offers quite different perspectives; they are characterized by single or multiple, sudden increases in brightness. They include the novae, with an increase in brightness of about 7 to 20 mag within a few days; the dwarf novae, with weaker outbursts (2 to 6 mag) and a more or less regular rhythm; and some nova-like variables. These all belong to close, semi-detached binary star systems, in which matter from a cool main sequence star (secondary component) that fills up its Roche lobe (Sect. 4.6.5) continually overflows onto a somewhat more massive white dwarf (primary component, with about $1\ \mathcal{M}_\odot$). They are therefore called *cataclysmic variables* or binaries (from the Greek κατακλυσμός, flood). The X-ray binary stars, which we shall discuss in the next section, are related to them. In contrast, the supernovae, considering the strength of their outbursts (≥ 20 mag) and their causes, represent a completely different phenomenon.

A *nova outburst* occurs roughly as follows: the initial state, the *prenova*, is a hot object with an absolute magnitude $M_V \simeq +5$ mag, a white dwarf surrounded by an accretion disk (see below). Within at most 2–3 days, its brightness increases to a maximum at about $M_V \sim -8$ mag ($L \simeq 10^5\ L_\odot$), i.e. by about 5 powers of ten. (We shall not consider the differences between fast and slow novae here). The spectrum during this process is similar to that of a supergiant such as α Cyg A2 Ia. The increase in brightness is thus not a result of a temperature increase, but, as is confirmed by the radial velocities, is related to an enormous expansion of the star. After the passage through the maximum brightness (during the decrease, the brightness sometimes exhibits fluctuations like those of cepheids), broad emission lines are observed, whose Doppler effects show that the nova is now ejecting a shell of gas with velocities of the order of $2000\ \mathrm{km\ s}^{-1}$ and masses from 10^{-5} to $10^{-4}\ \mathcal{M}_\odot$. In several cases, e.g. in the case of Nova Aquilae 1918 (V 603 Aql), this shell and its expansion could be observed for more than a decade in direct photographic images.

In some (slow) novae, an increase in the infrared brightness can be observed about 40 to 150 d after the initial outburst: in the expanding shell, condensing small dust particles absorb the radiation of the nova and reemit it in the infrared.

In the course of many years, the nova returns to its initial brightness and to its original state. This can be most clearly seen in the *recurrent novae* such as T Pyx, which has suffered several nova-like outbursts ($\Delta m \simeq 7$ mag) at intervals of ca. 10 yr and shows a nearly constant magnitude in between. Based on theoretical models (see below), it is assumed that the outbursts of "common" novae also repeat themselves, to be sure at longer intervals of perhaps 10^3 to 10^6 yr. All together, in our galaxy there are about 100 nova outbursts per year; the rate of recurrent nova outbursts is probably 10 to 100 times greater.

In the X-ray region, *transients* are observed; their intensity remains above the detection limit for only a limited time. Some of them can be regarded as *X-ray novae* on the basis of their light curves: a rapid increase in brightness (to $L_X \leq 10^5 L_\odot$) is followed by a basically exponential decrease which is observable over several months, interrupted by irregular flares which last for a few days and are sometimes observable also in the optical region.

The *U Geminorum* and *Z Camelopardalis variables*, again named for their prototypes, are likewise hot, blue stars below the main sequence. These *dwarf novae* exhibit less violent outbursts at irregular intervals of about 10 d up to several months. In the Z Cam stars, the sequence of the outbursts is occasionally interrupted by a period of relative stability at an intermediate level of brightness.

The *energy* which is released by a nova is of the order of 10^{38} J; its estimation, taking the bolometric correction into account, is relatively imprecise. This corresponds to the thermal energy content of a thin layer of matter, at for example $5 \cdot 10^6$ K and having only 1/1000 the mass of the Sun. All indications are that novae represent a sort of "skin disease" of stars. More precise observations, especially of several favorable cases of eclipsing binaries (Sect. 4.6.2), yield the following picture (Fig. 4.11.4): in the cataclysmic binary star systems whose periods lie in the range of about 1.3 to 15 h, the matter ejected from the secondary component, due to its angular momentum, does not fall directly into the white dwarf, but instead forms a rapidly rotating *accretion disk* around the star. In this disk, the matter loses angular momentum through friction and thus gradually moves inwards to the surface of the white dwarf. At the point where the gas flow meets the disk, a *hot spot* is formed. The main contribution to the luminosity comes not from the white dwarf, but, depending on the particular case, from the disk or the hot spot.

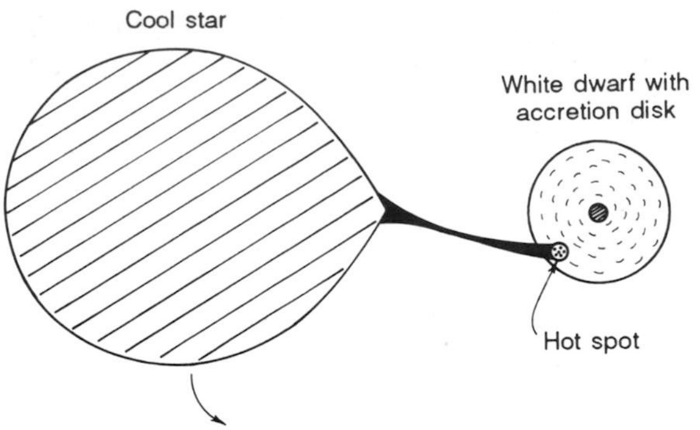

Fig. 4.11.4. A model of a cataclysmic binary star system. The geometric relationships correspond to those of the dwarf nova Z Cam. (Reproduced, with permission, from the Annual Review of Astronomy and Astrophysics, Vol. **14**. © 1976 by Annual Reviews, Inc.)

The outburst of *dwarf novae*, in which the luminosity of the accretion disk sharply increases and no matter is ejected, is probably due to *instabilities* of the disk. The energy source is in the final analysis the gravitational energy released by the material accreted at an average rate of the order of $10^{-10} \mathscr{M}_\odot \mathrm{yr}^{-1}$. In the case of true *novae*, this hydrogen-rich matter overflows and collects until nuclear hydrogen fusion (CNO burning, Sect. 4.12.3) is initiated explosively on the surface of the white dwarf, leading to a burst of radiation intensity and to the ejection of a shell of matter.

An interesting group of cataclysmic variables are the *AM Herculis stars*, which are notable on the one hand for the high (variable) degree of circular and linear polarization of their radiation in the optical and ultraviolet regions, and on the other for their strong emissions in the soft X-ray region. The cause is probably a *strong magnetic field* of the white dwarf, which prevents the formation of an accretion disk and, instead, concentrates the gas flow onto a spot around one of the magnetic poles, so that there, in a stationary shockwave front, gravitational energy is released, primarily in the form of polarized X-rays and ultraviolet radiation.

4.11.6 X-Ray Pulsars, X- and Gamma-Ray Bursters: Accretion onto Neutron Stars

As a result of the progress in X-ray astronomy, a large number of "point sources" in our Milky Way galaxy have been discovered by satellite observations in recent years; they radiate *mainly in the X-ray region* and are characterized by a strong *variability*. Intensity fluctuations and bursts of radiation are observed, in which the X-ray emission increases by factors of 10 to 100; they show an impressive variety of behaviors, including irregular flickering on a time scale of a few milliseconds, series of short bursts lasting several seconds, nova-like outbursts followed by a slow decrease of the intensity back to the original value, X-ray flares which occur at irregular intervals and last for several days, and regular or irregular pulses on a time scale of seconds. Some sources exhibit more or less constant emission, in which often several different periods can be discerned, with active and quiet phases succeeding each other on a time scale of weeks or months.

In spite of this multiplicity of phenomena, it is possible to describe practically all of these variable galactic X-ray sources within the framework of a single theory, that of *accretion* of matter in *close binary star systems*. They thus differ from the cataclysmic variables (Sect. 4.11.5), where X-rays can, to be sure, also be observed, but the main contributions to the luminosity are in the visible and ultraviolet spectral regions; by contrast, in the *X-ray binary systems*, the luminosity L_X in the X-ray region is in the range 10^{29} to 10^{32} W (10^{36} to 10^{39} erg s^{-1}), and the matter is not accreted by a white dwarf star, but rather by a *neutron star* or possibly by a black hole (Sect. 4.12.9). This theoretical model is in many cases supported by optical identification of one of the binary system components and by the occurrence of X-ray pulses of short periods, which can only be explained by the rotation of a neutron star ("lighthouse effect", Sect. 4.11.7).

The *pulsing X-ray source* or *X-ray pulsar* Her X-1 could be identified with the *eclipsing binary system* HZ Her discovered by C. Hoffmeister in 1936. Its X-radiation shows first of all pulses with a period of 1.24 s, which are attributed to the rotation of a neutron star. This period is modulated by the Doppler effect due to the orbital motion (Sect. 4.6.6) with a period of 1.7 d. The latter period is seen both as an eclipsing of the X-ray source and as an intensification of the visible light, which occurs whenever the side of the optical component facing us is irradiated by the X-ray source. An additional period of 35 d is probably related to the precession of the axis of the neutron star. Her X-1 can also "switch off" its X-ray emissions over periods of several months, as a result of occultation by the accretion disk surrounding the neutron star. The mass of the latter is estimated to be about 1.3 $\mathcal{M}_\odot$, while that of the other component is about 2.2 $\mathcal{M}_\odot$.

Her X-1 = HZ Her is the prototype of the *low mass* (semi-detached) X-ray binary stars, in which the secondary component, a main sequence star with $\mathcal{M} \leq 2.5\,\mathcal{M}_\odot$, fills its Roche lobe and allows matter to overflow to its companion neutron star. In addition, we also find X-ray pulsars in (detached) binary star systems with a *more massive* component ($\geq 15\,\mathcal{M}_\odot$), usually an OB supergiant, which can be observed spectroscopically with relative ease in the optical region. In these systems, the neutron star accretes matter from the strong *stellar wind* of the supergiant. Examples are Vel X-1 = HD 77581 (B0.5 Ib) and SMC X-1 = Sk 160 (B0 I).

The pulsation periods, which reflect the rotation of the neutron star, range from 0.07 up to several 100 s. In some cases, a secular *decrease* of the period is observed, interrupted by rapid fluctuations; this acceleration of the rotation is probably due to angular momentum transfer by the accreting matter. The orbital periods are in the range of several days.

Determination of the masses of X-ray binary star systems (Sect. 4.6.6) yields a relatively narrow range of 1.2 to 1.6 $\mathcal{M}_\odot$ for the neutron star. An exception is the large mass (3...10 $\mathcal{M}_\odot$) of the compact component of Cyg X-1 = HDE 226868, which presumably lies above the limiting mass for neutron stars (Sect. 4.12.8); it may thus be an example of a black hole.

The *energy source* for the X-ray emissions is the gravitational potential energy released by the accreting gas. If matter of mass m falls from a large distance to a distance R from the center of a (spherical) mass $\mathcal{M}$, according to (2.6.34), an energy equal to $mG\mathcal{M}/R$ is released. Acceleration by the strong gravitational field of a *compact object* (white dwarf, neutron star, or black hole) and then braking and heating near its surface generates X-rays very effectively, with a luminosity of the order of:

$$L_X \simeq \dot{m}\,\frac{G\mathcal{M}}{R}\,, \tag{4.11.9}$$

where $\dot{m}$ is the *rate* of accretion of mass. For example, a relatively modest gas flow of $\dot{m} \simeq 10^{-8}\,\mathcal{M}_\odot\,\mathrm{yr}^{-1}$ in the gravitational field of a neutron star ($\mathcal{M} \simeq 1\,\mathcal{M}_\odot$, $R \simeq 10$ km) is sufficient to explain the value 10^{31} W observed from the stronger X-ray pulsars. This rate can readily be produced either by mass flow through the Roche surface or by stellar winds.

The details of the dynamics of the accreting gas in a close binary system with a neutron star which possesses a strong magnetic field (Sect. 4.11.7) are quite complex. The dipole-like magnetic field makes the formation of an accretion disk more difficult; it diverts and concentrates

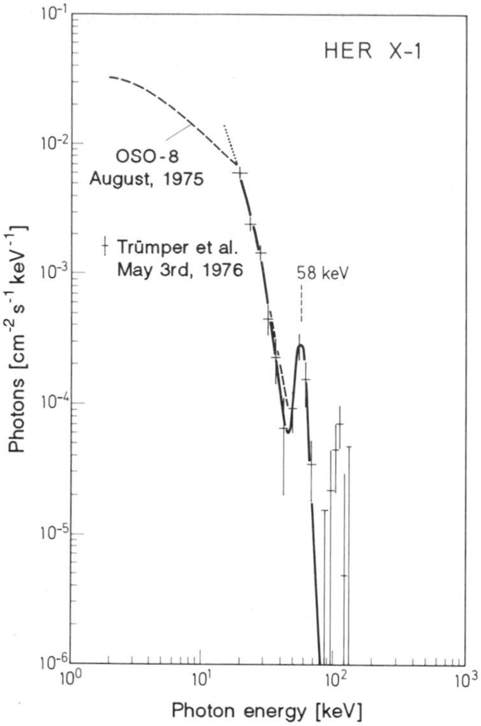

Fig. 4.11.5. The cyclotron emission line at 58 keV in the pulsed X-ray spectrum of Her X-1; from observations by J. Trümper et al. (as well as observations of Her X-1 from the OSO-8 satellite)

the inflowing gas towards the polar regions and has a considerable effect on the X-ray pulses.

The line at $\simeq 58$ keV which was discovered in 1976 by J. Trümper and coworkers in the spectrum of Her X-1 (Fig. 4.11.5) allows a direct estimate of the magnetic flux density B in the neighborhood of the surface of the neutron star. The line, in which about $2 \cdot 10^{28}$ W are emitted, is too strong to be explained as an atomic or nuclear transition of a heavy element. Instead, it must be regarded as *cyclotron emission*, which is produced by the spiral motion of nonrelativistic electrons in magnetic fields at multiples of the cyclotron frequency $2\pi\nu_c = eB/m$ (5.3.30). For relativistic electrons, by the way, the cyclotron emissions become continuous *synchrotron radiation* (Sect. 5.6.1), which is important for the explanation of nonthermal radio emissions. In the extremely strong magnetic fields of pulsars or neutron stars, the quantization of the gyrational motion into discrete energy states (Landau levels), which are separated by the energy $h\nu_c$, must be taken into account. For Her X-1, the observed energy of 58 keV corresponds to an estimated magnetic field of about 4 to $5 \cdot 10^8$ T (4 to $5 \cdot 10^{12}$ G).

An unusual binary star system is SS 433 = V 1343 Aql (SS refers to the list of stars with Hα emission by C.B. Stephenson and N. Sanduleak, 1977); it shows irregular brightness fluctuations in the optical, X-ray, and radio spectral regions. The Doppler shifts of the strong, broad emission lines are notable: they are unusually large for stars, especially those of the Balmer and He I lines. Each line consists of three components: a "normal" component with a small radial velocity amplitude and a period of 13.1 d, which reflects the orbital motion of the binary system; and two strongly shifted components, each with a period of 164 d, one with red shifts up to +50000 km s^{-1}, the other with blue shifts up to −30000 km s^{-1}! This binary system probably consists of an O star and a neutron star (or a black hole) with an accretion disk which makes the main contribution to the emitted radiation. Two collimated beams are emitted in opposite directions roughly perpendicular to the orbital plane with nearly relativistic velocities ($\simeq$80000 km s^{-1}); they show a precessional motion with a period of 164 d. This kinematic model reproduces the observational data from SS 433 well, but the physical processes are not yet understood in detail. SS 433, with its two highly energetic jets, represents a rare stellar "miniature version" of a phenomenon which we shall meet later in many galaxies (Sect. 5.6), with much larger dimensions.

The *X-ray bursters* are characterized by series of very short, intense, and non-periodic bursts of radiation. The X-ray emission in one burst rises within about 1 s to roughly 10^{31} to 10^{32} W and decays again in a few seconds or minutes (Fig. 4.11.6). During the active phases, the bursts occur in long series, often following each other at nearly regular intervals of hours to days; many bursters show inactive phases which last for weeks or months. In a few cases, bursts in the visible region, delayed by several seconds, are also observed.

The X-ray bursters are concentrated towards the center of our galaxy, and many of them have been identified with sources in the central regions of *globular star clusters*. They therefore probably represent *old* objects ($\geq 10^9$ yr). We thus expect that the magnetic field of the neutron star has for the most part decayed away, so that the gas stream from the cool companion star can distribute itself into an accretion disk or over the whole surface of the neutron star and is not concentrated onto the polar regions as in the X-ray pulsars. This model is supported by the fact that the spectrum and intensity of

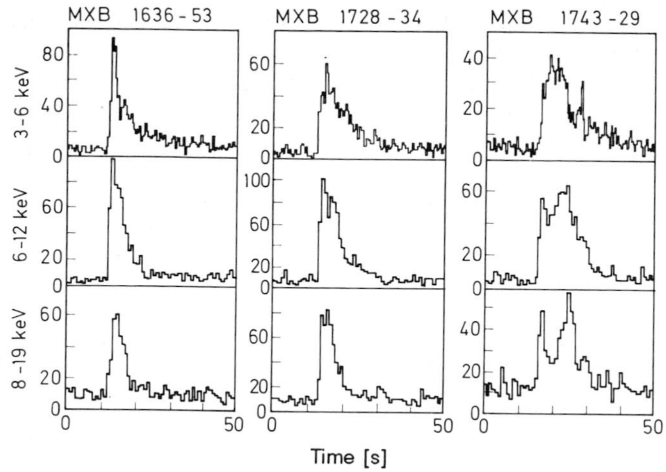

Fig. 4.11.6. The time dependence of the radiation emissions from three X-ray bursters in different energy ranges; observations from the SAS-3 satellite. The intensity is expressed as the number of counts in 0.4 s (*upper region*) or in 0.8 s (*lower region*)

the X-ray bursts are close to those calculated for a black-body source with a surface area of about (10 km)^2 at a temperature of about $3 \cdot 10^7$ K. The bursts themselves probably result from an unstable, pulsed thermonuclear helium fusion (Sect. 4.12.3) on the surface of the neutron star, occurring after the hydrogen-rich accreted matter has first been converted to helium by hydrogen fusion.

The unusually large fraction of X-ray bursters in globular clusters, in comparison with the other stars in the galaxy, indicates that the conditions for formation of close binary systems are particularly favorable in the dense centers of the clusters. An interesting case is the burster 4U 1820−30 (U = UHURU survey), which lies at the center of the cluster NGC 6624: a modulation of its X-ray emissions with a period of 11.4 min was discovered in 1984/85 in EXOSAT observations; it has been interpreted as the *orbital period* of a system consisting of a neutron star and a very low-mass (0.055 $\mathcal{M}_\odot$) white dwarf. This is the shortest known period of a binary star system.

The *"rapid burster"* (MXB 1730-335) in a globular cluster at a few degrees from the galactic center, which was unknown before the discovery of the X-ray source, is also notable. During its active periods, which last for a few weeks, it produces several 1000 bursts per day (Fig. 4.11.7).

In recent years, it has been possible in certain cases also to observe *gamma radiation* from X-ray binary stars: Her X-1 emits pulsed gamma radiation above 10^{12} eV, which again occurs with the pulse period known from the X-ray and optical observations, 1.24 d. The system Cyg X-3 = V 1521 Cyg was found to be a strongly variable pulsed gamma source, in which a modulation at the orbital period of 4.8 h can be discerned. Extremely high-energy gamma quanta of 10^{15} to 10^{16} eV from Cyg X-3 were discovered in air-shower experiments, which indicated that the luminosity in this energy range alone is equal to at least $10^3\ L_\odot$. The neutron star in this binary system is probably a rapidly rotating, active pulsar, through which protons are accelerated to extreme relativistic energies of 10^{17} to 10^{18} eV and then produce the gamma quanta by their interactions with the matter in the system (accretion flows and disk?). Thus, Cyg X-3

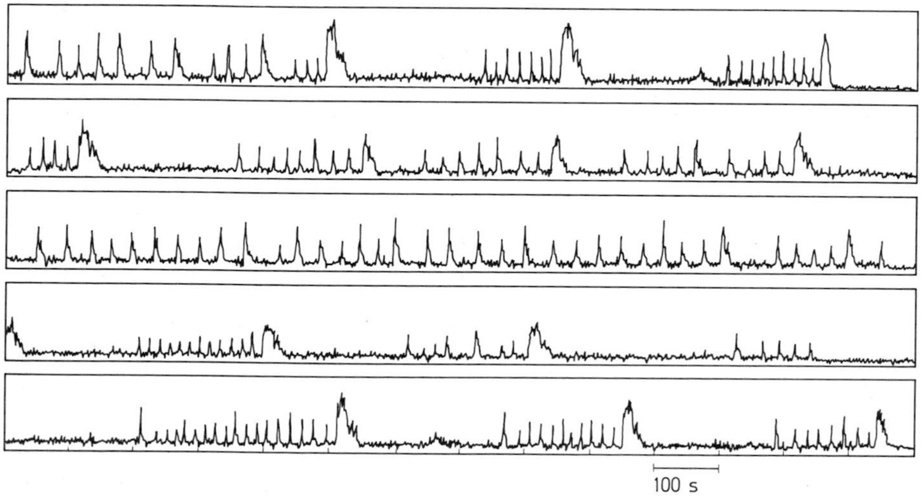

Fig. 4.11.7. The time dependence of radiation emissions from the "rapid X-ray burster" MXB 1730-335. Portions of observations made from the SAS-3 satellite in March, 1976

should be a strong source of energetic cosmic radiation (Sect. 5.3.8).

In the 1970's, the Vela satellites[20] discovered short, intense radiation flashes in the gamma radiation region. Since then, several hundred *gamma bursters* have been observed from satellites and space probes, mainly in the energy range from 0.1 to 100 MeV with a nearly isotropic distribution on the celestial sphere. The length of the bursts ranges from a few hundredths to several thousands of seconds, with the longer-lasting bursts usually showing a complex structure. The origin of the gamma bursters is still an unsolved riddle, since they cannot be identified with sources in the optical, radio, or X-ray regions, and their distances are unknown. It is assumed that they (like the X-ray bursters) result from thermonuclear explosions on the surfaces of (isolated?) magnetic neutron stars. As a rule, the gamma bursts from a particular object do not seem to repeat themselves.

A particularly intense gamma burst was recorded by eleven space probes on March 5, 1979. Its short rise time (≤ 0.25 ms), together with the large distances of the space probes from one another, made an exact position determination by "time delay triangulation" possible: this gamma source, GBS 0526-66, concides (coincidentally?) with the supernova remnant N49 in our neighboring galaxy, the Large Magellanic Cloud. Repeated gamma bursts, and also flashes of optical radiation, have been observed from this source.

4.11.7 Supernovae, Pulsars, and Supernova Remnants

Among the eruptive variables, the *supernovae* represent cosmic explosions of a much greater magnitude than the novae, as was recognized by W. Baade and F. Zwicky in 1934. A supernova at its maximum can attain the luminosity of an entire galaxy (to which it belongs). Two types of supernovae may be distinguished:

a) The *type I supernovae* reach a mean absolute magnitude in the blue (corrected for interstellar absorption) of $M_B = -19.7$ mag[21] at maximum. Their *light curves* (Fig. 4.11.8) are very similar: in the first 20 to 30 d following the maximum, the brightness decreases by 2 to 3 magnitudes; thereafter, it decreases roughly exponentially with time, having a half-life of 40 to 70 d. The *spectra* are rich in lines and seem to be dominated by metals (Fe II?); hydrogen lines are lacking completely or are very weak.

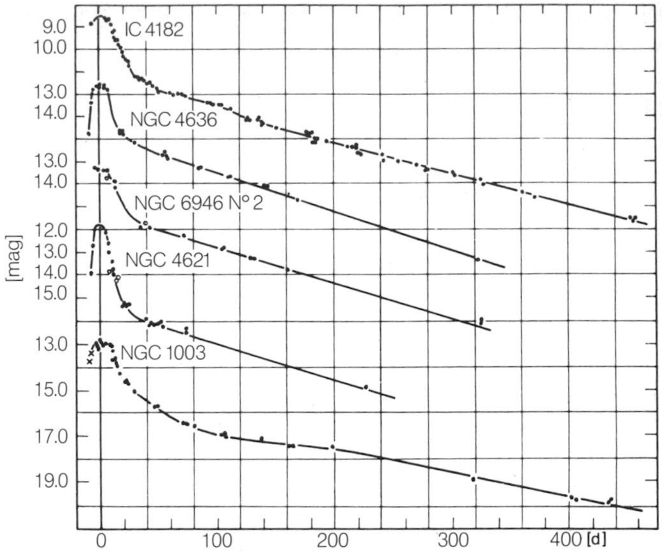

Fig. 4.11.8. Photographic light curves of Type I supernovae in different galaxies

b) The *type II supernovae* attain on the average "only" $M_B \simeq -18.0$ mag (with a scatter of about 1 mag). This, after all, corresponds to about 10^4 times the luminosity of an ordinary nova! The decrease of brightness with time after the maximum is initially more rapid, then slower than for type I. The light curves furthermore show greater individual differences. The spectrum of a type II supernova and its time evolution are amazingly like those of ordinary novae. Along with metal lines, there are strong Balmer lines, so that supernovae of type II on the whole seem to have a "normal" composition.

The basic difference between the two types of supernovae thus seems to be determined by their different chemical compositions.

The emission lines of the supernova spectra are often accompanied by a short-wavelength absorption component (P Cyg profile, Sect. 4.11.4), whose Doppler shifts indicate *ejection velocities* of up to $2 \cdot 10^4$ km s^{-1}. The intensity distribution in the continuum, which is particularly difficult to determine for supernovae I owing to the large number of spectral lines, indicates that the temperature of the emitting layers decreases from $\geq 10^4$ K at

[20] US satellites intended to detect the explosions of nuclear weapons.

[21] This magnitude applies for distances based on a value of the Hubble constant $H_0 = 50$ km s^{-1} Mpc^{-1} (Sect. 5.9.1). For different values of H_0, $M_B = -9.7 + 5 \log (H_0/50)$.

the maximum to about 6000 K. These temperatures, which are not very high, correspond to photospheric radii of the order of $10^4\,R_\odot$ using the relation $L = 4\pi R^2 \sigma T_{\mathrm{eff}}$ (4.4.26), due to the extremely high luminosities of supernovae.

In our *Milky Way galaxy*, the following supernovae have been identified with certainty in historic times: the very bright supernova of 1006 A.D. in Lupus reached the brightness of the half-full Moon and is mentioned in Asiatic, European, and Arabian sources. The supernova of 1054 A.D. in Taurus, which was the source of the Crab Nebula (see below), was identified by J. G. Bolton as the radio source Tau A, and is described in Chinese and Japanese annals; but, remarkably, it is not mentioned in European sources, although its brightness at maximum was greater than that of Venus. We have already mentioned *Tycho Brahe's* supernova of 1572 and *Kepler's* supernova of 1604, in Ophiuchus. Finally, the strong radio source Cas A and the nebula at the same position, which is expanding with a velocity of 7400 km s^{-1}, have been clearly identified as the remains of a supernova which must have exploded around 1650-1700, but was not observed.

All together, about 700 supernovae (as of 1991) have been discovered in *other galaxies*; the rate of discovery increased sharply after F. Zwicky began systematic observations at Mt. Palomar in 1934. Based on extensive statistics, G.A. Tammann (1981) estimates that in a galaxy like ours, *one* supernova occurs on the average every 25 yr, with an uncertainty in this rate of a factor of 2. Due to the interstellar extinction in the disk of our galaxy, we can, to be sure, observe only a tenth of these explosions.

A first indication of the origin of supernovae is provided by their occurrence in galaxies of different types (Sects. 5.5.2, 4). While supernovae I occur in all types of galaxies, supernovae II are not observed in elliptical galaxies, but only in spiral and irregular galaxies, where they are mainly concentrated in the spiral arms. It therefore can be expected that supernovae of type II are to be attributed to young, massive population I stars. The origin of supernovae I is, by contrast, largely unexplained (Sect. 5.4.5).

The total *energy release E* from the explosion of a supernova can be estimated on the one hand by integration of the *light curve* with an approximate bolometric correction, in favorable cases also using observations in the ultraviolet and infrared, to obtain the total emitted radiation; on the other hand, the kinetic energy of the ejected gas shell (0.1 to 10 $\mathcal{M}_\odot$ at about 10^4 km s^{-1}) can be taken into account. The order of magnitude is $E \simeq 10^{44}$ J or 10^{51} erg. In comparison, one can readily calculate that the energy released e.g. by the fusion of 1 $\mathcal{M}_\odot$ of hydrogen to helium is $1.3 \cdot 10^{45}$ J (Sect. 4.12.3). Furthermore, the collapse of a star, e.g. of $\mathcal{M} = 1\,\mathcal{M}_\odot$, to a *neutron star* of $R \simeq 10$ km radius with a density comparable to that of nuclear matter (10^{17} kg m^{-3}) can, as was recognized already in 1934(!) by Baade and Zwicky, release about 10^{46} J of gravitational energy (Sect. 4.12.4), and thus could readily provide the required energy for a supernova. The question of what actually causes the explosion of a supernova will be treated further in Sect. 5.4.5, together with the topic of instabilities in stellar evolution.

The bright type II *supernova SN 1987 A* has particular significance; it was observed on February 23rd, 1987, in our neighboring galaxy, the Large Magellanic Cloud (Fig. 4.11.9), which is only about 50 kpc away, and reached its maximum (apparent) visual magnitude $m_V = 2.9$ mag after about three months. After 120 d, it entered the phase of exponential decrease in light intensity. SN 1987 A is the first supernova which has been visible to the unaided eye since Kepler's SN 1604; its absolute magnitude at maximum, $M_V = -15.5$ mag, however remained well below the average value for supernovae of type II. With SN 1987 A, for the first time in the case of a supernova, the *presupernova* could be clearly identified; it is a blue B3 supergiant with an absolute magnitude of $M_V = -6.6$ mag.

Due to its large apparent magnitude and the flat light curve at maximum, SN 1987 A was observed early on in a wide wavelength range with high spectral resolution. Furthermore, it was possible to detect its *neutrino radiation*, for the first time for any object excepting the Sun (see also Sect. 4.12.6); the first neutrino signals were in fact recorded several hours *before* the initial optical observation of the supernova was registered. In spite of the uncertainties in the analysis of the neutrino burst, the observation of neutrinos with a total energy of 10^{45} to 10^{46} J and an average particle energy of the order of 10 MeV supports our basic theoretical understanding of supernova explosions of type II as the collapse of a massive star, in which gravitational energy is released almost exclusively in the form of neutrinos, with only about 1% as electromagnetic radiation and kinetic energy of the ejected gas shell (Sect. 5.4.5). Finally, the measurement of the time during which neutrinos were emitted offers the possibility in principle (by means of a time-

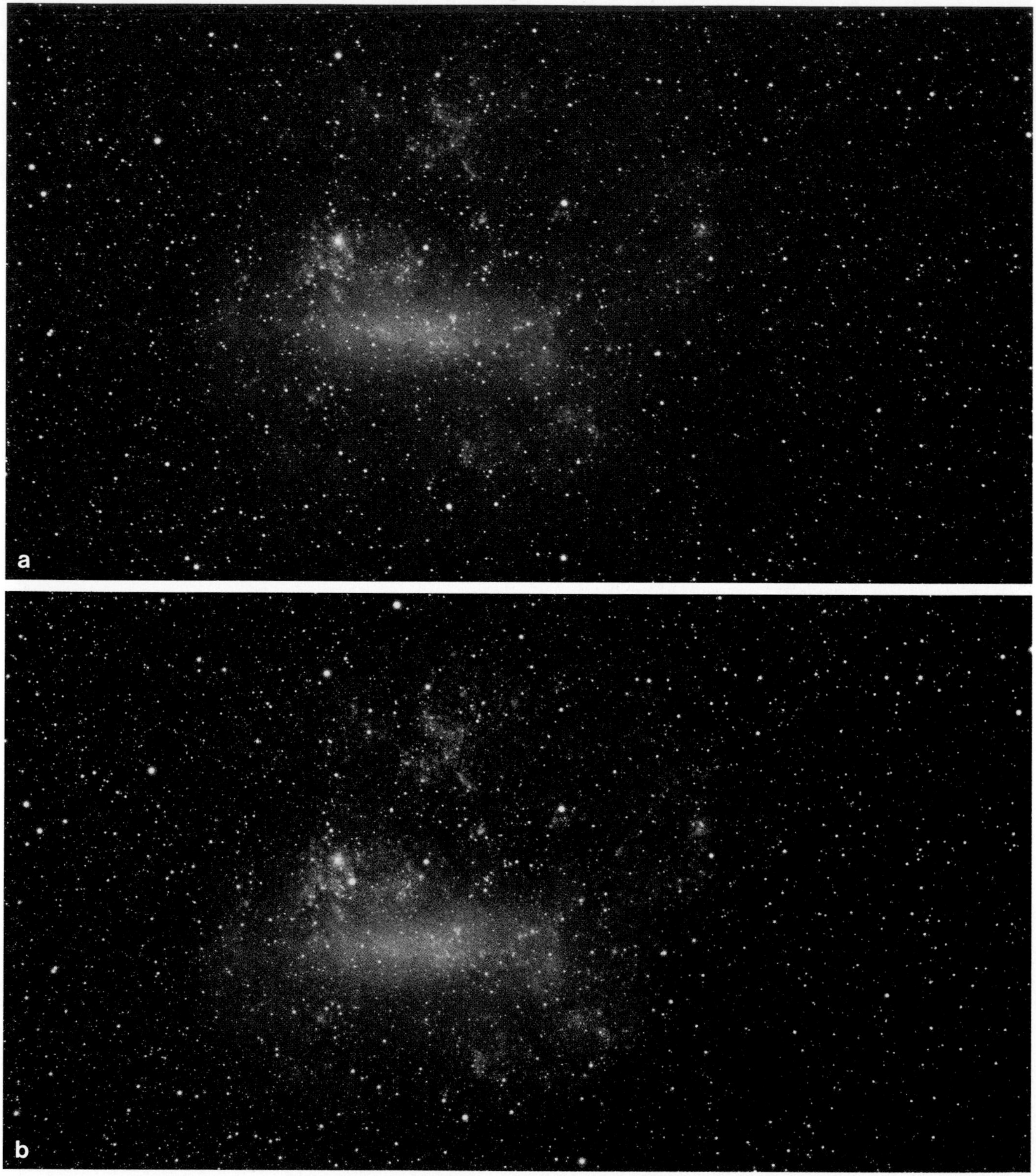
a
b

of-flight analysis) of determining the rest mass of the neutrino, a fundamental quantity both for elementary particle theory and for cosmology (Sect. 5.9.5). The current models for a supernova explosion indicate an upper limit for the mass of the (electron) neutrino in the range of 10 to 30 $\mathrm{eV\,c^{-2}}$.

In the last part of this section, we shall concern ourselves with the situation *following* a supernova explosion, i.e. with the observation of the *remaining stars* and the fate of the ejected shells, the *supernova remnants*, SNR.

Although the structure of *neutron stars* was already investigated theoretically in the 1930's (Sect. 4.12.8), the first one was discovered only in 1967 in a completely unexpected manner by A. Hewish and his coworkers. At the Cambridge radiotelescope, they noticed signals in the meter wavelength range which repeated themselves very regularly with a period of 1.337 s. Observations of the apparent motion in the sky showed that these were not due to terrestrial disturbances, but rather to a cosmic object. Today, about 500 of these *pulsars* or *radio pulsars* are known. They are denoted by PSR and their (approximate) right ascension and declination; e.g. the first pulsar discovered at $\alpha \simeq 19$ h 19 min and $\delta \simeq 21°47'$ is called PSR 1919+21. Their *periods* lie mostly in the range between 0.3 and 2 s; the shortest known period is $1.56 \cdot 10^{-3}$ s, the longest 4.31 s (as of 1990). The signals, whose length is about 3% of the period, all arrive somewhat later at lower frequencies than at higher ones, because the propagation velocity of electromagnetic waves in the interstellar plasma decreases at lower frequencies. The delay is proportional to the number of electrons in a column (of unit cross-sectional area) reaching from the observer to the pulsar, the so-called dispersion measure:

$$DM = \int N_e dl \ . \tag{4.11.10}$$

If, as a rough average, a value $N_e \simeq 3 \cdot 10^4 \ \mathrm{m^{-3}}$ is used, the *distances* to the pulsars can be estimated.

It represented a great step forward when in 1968/69 the pulsar PSR 0531+21, with the shortest known period at the time, $P = 0.0332$ s, was identified optically with the *central star of the Crab Nebula* (Fig. 4.11.10). It was then soon possible to demonstrate that this star emits similar "pulses" in the optical and even in the X-ray and gamma-ray regions (Fig. 4.11.11), using observations through a chopper (rotating collimator) adjusted to the period *P*. The pulse shape is about the same from the gamma region out to wavelengths of about 1 m; at still longer wavelengths, the pulses are broadened due to rapid local fluctuations in the interstellar electron density, the "interstellar scintillation". The high luminosity in the gamma region is surprising; it is about 10^5 to 10^6 times greater than in the radiofrequency region. Besides the Crab pulsar, the *Vela pulsar*, PSR 0833−45, with $P = 0.089$ s, emits pulses extending from the radio to the gamma regions. In the few cases in which a pulsar can be observed within a supernova remnant, a precise distance

Fig. 4.11.10. The Crab Nebula, M1 = NGC 1952: a photograph in red light made with the 5 m Hale telescope by W. Baade. In 1949, J. G. Bolton first succeeded in optically identifying a radio source, and showed that Tau A is the Crab Nebula. The inner, homogeneous part emits continuous synchrotron radiation, while the outer parts, the "legs of the crab", emit a nebula spectrum, particularly the red hydrogen line Hα. The Crab Nebula was formed in 1054 A. D. by a supernova explosion. The remnant star is the pulsar PSR 0531+21, a neutron star which "ticks" with a period $P = 33.2$ ms over the whole range from the gamma-ray spectral region to the meter-wave region

Fig. 4.11.9 a, b. The Large Magellanic Cloud and the Supernova 1987 A: a color photograph made by C. Madsen, ESO; **(a)** a few hours before the outburst on February 23rd, 1987; and **(b)** 2 days after the outburst. The H II regions in the galaxy appear reddish due to their strong Hα emissions. SN 1987 A can be clearly recognized to the southwest of the giant H II complex 30 Doradus (at the left). The distance to the Large Magellanic Cloud is 50 kpc. (With the kind permission of the European Southern Observatory)

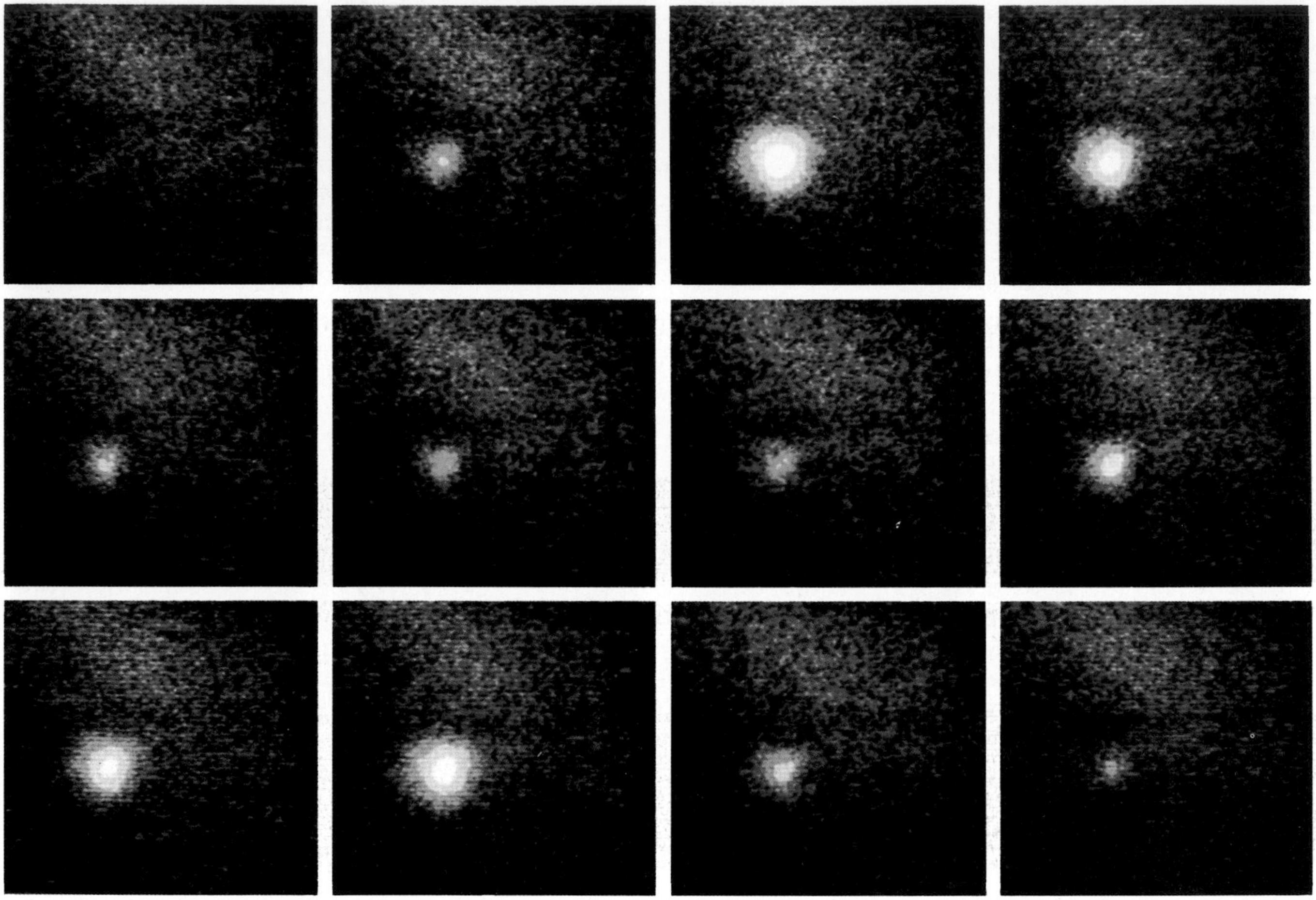

Fig. 4.11.11. X-ray emissions of the Crab Pulsar and the Crab Nebula in the range from 0.1 – 4.5 keV; observations from the Einstein Satellite with an angular resolution of about 4″. The pulsar period of 33.2 ms is time-resolved in 16 images, each comprising about 2′ · 2′; 12 of these are shown here (3rd picture: main pulse; 10th picture: secondary pulse). The large-area X-ray emissions of the Crab Nebula are shifted to the NW relative to the pulsar. From F. R. Harnden and F. D. Seward, Astrophys. J. **283**, 279 (1984). (Reproduced by permission of The University of Chicago Press, The American Astronomical Society, and the authors)

determination becomes possible; e.g. the Crab pulsar is at a distance of 2000 pc, and the Vela pulsar at 500 pc.

The extraordinarily short periods and the regularity of the pulses limits possible explanations of pulsars to *neutron stars rotating with the period P*, emitting a beam with an opening angle of about 20° like a lighthouse, over the enormous frequency range from meter waves to gamma rays. The requirement that the surface of the star cannot exceed the velocity of light already implies a considerable restriction on its possible radius.

The mechanism of emission in pulsars is not yet clear in all its details, but it can be considered that a rapidly rotating neutron star with a strong *magnetic field* ejects plasma outwards; the plasma initially rotates with the star, reaching the velocity of light c on a cylindrical surface of radius $cP/2\pi$ – in the case of the Crab pulsar, for example, at 1580 km. The conditions for emission of a *beam* of radiation are then favorable. Why the Crab pulsar emits a weaker secondary pulse between the main pulses, while other pulsars do not, as well as the finer details of the pulse structure, are not yet understood satisfactorily.

The radius of a neutron star is of the order of 10 km (Sect. 4.12.8). If a star like the Sun initially rotates with a period of 25 d and collapses to form a neutron star, then this period will become 1 millisecond(!), assuming that angular momentum is completely conserved. (The surface of the star will still be far from attaining the velocity of light). If the star initially has a reasonably ordered magnetic field of about $5 \cdot 10^{-4}$ T, its lines of force will be compressed corresponding approximately to the cross-section of the star, and the neutron star will obtain an enormous magnetic field of about 10^6 T. Further-reaching considerations, based on models for the emission of radiation, require even higher fields in the range of 10^7 to 10^9 T at the surface of the rotating neutron stars.

The extraordinary precision of the observations of pulsars quickly made it possible to determine that the period P of all pulsars, neglecting occasional small "jumps", is *increasing*, i.e. the rotation is slowing down. As a readily-understood measure of the slowing down of their "ticking", we define the time $\tau = P/\dot{P}$, after which the period would double for a constant rate $\dot{P} = dP/dt$. The Crab pulsar is changing the most rapidly, with $\tau = 2500$ yr; for most pulsars, times τ in the range of 10^6 to 10^7 yr are found. This indicates a *lifetime* or an *age* of the pulsars of the same order of magnitude.

With a few exceptions (Sect. 4.6.6), the radio pulsars are *individual stars*. In several cases, their proper motions could be determined, and indicate *high spatial velocities* (≥ 100 km s^{-1}). According to A. Blaauw, these could be the result of the supernova explosion of a rapidly rotating binary star system, in which the main component was scattered out into space.

In contrast to the radio pulsars, the related X-ray pulsars (see Sect. 4.11.6) are all neutron stars in close binary systems, whose binding to their companion stars was evidently not destroyed by the supernova explosion (Sect. 5.4.6).

The first "*millisecond pulsar*", PSR 1937+21, was discovered by D. C. Backer *et al.* in 1982 and has a period of $P = 1.558$ ms; it at first seemed to be an extremely young object because of its very rapid rotation. However, the quite small rate of change of its period and the lack of a supernova remnant argue for an *old* pulsar, which probably obtained its high angular momentum during an earlier phase by matter accretion in a close (X-ray) binary system. This picture is also supported by the discovery (in 1987) of another millisecond pulsar, PSR 1821−24, with $P = 3.05$ ms, in a globular cluster (M 28) where the formation of binary stars should be favored. Today, these two pulsars are single objects. On the other hand, several of the pulsars later discovered in globular clusters also belong to binary systems. The group of old pulsars, whose age is comparable to that of the Milky Way, includes those pulsars which belong to at least one of the three types: millisecond pulsars (with periods ranging from a few ms to several 100 ms); binary-system pulsars; or members of globular clusters. All together (as of 1990) we know of about 20 such objects.

Among the *supernova remnants*, the Crab Nebula had earlier already played a decisive role in the development of astrophysics. In 1942, W. Baade and R. Minkowski investigated carefully this highly interesting object, M1 = NGC 1952 with $m_{pg} = 9.0$ mag (Fig. 4.11.10). It consists of an inner, nearly amorphous region of dimensions $3.2' \cdot 5.9'$, with a continuous spectrum and a gas shell, whose bizarre filaments (the "legs of the crab") radiate mainly in Hα. As was known from Asiatic annals, the Crab Nebula is the remnant of a supernova explosion in the year 1054 A.D. Comparison of the radial expansion velocities of about 1000 to 1500 km s^{-1} with the corresponding, measured outward proper motions confirmed the date of the explosion and, in particular, the distance of the Crab Nebula, about 2 kpc.

Following I.S. Shklovsky's daring suggestion in 1953, we can assume the continuous radiation of the Crab Nebula, which ranges from the radiofrequency to the X-ray regions, to be *synchrotron radiation* (Sect. 5.6.1), produced by energetic electrons in a magnetic field of the order of 10^{-8} T or 10^{-4} G. The following arguments support this idea: if the radiation were thermal, the electron temperature T_e would have to be larger than the highest radio radiation temperatures, i.e. $\geq 10^9$ K, although T_e remains $\leq 10^4$ K in all other gas nebulas. Furthermore, the amorphous core exhibits only a continuum in the optical wavelength region, and no lines. Finally, a clearcut argument in favor of synchrotron radiation was provided by the discovery of the required polarization of the continuum radiation (Fig. 5.6.1). It was measured in the optical spectral region by Dombrowsky and Vashakidze in 1953/54, then with better resolution and improved precision by Oort, Walraven, and Baade. Later, in spite of difficulties caused by the Faraday effect, it proved possible to carry out measurements in the cm and dm wavelength ranges. A thorough consideration of the lifetimes of relativistic electrons in the Crab Nebula led Oort to the opinion that they are still being continuously generated today, nearly a thousand years after the explosion.

The *X-ray emission* of the Crab Nebula also originates, as shown by spatially resolved observations making use of lunar occultations, in an extended source about 1′

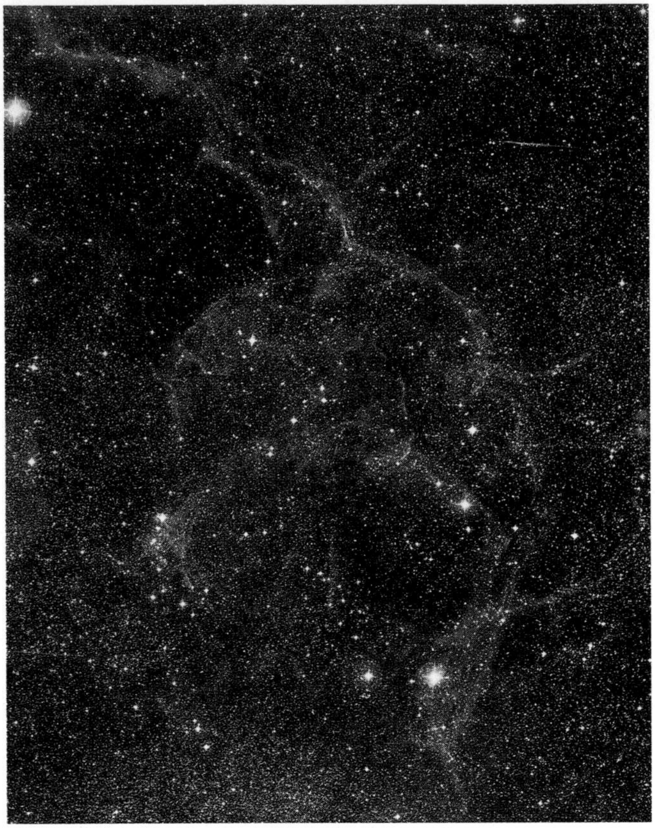

Fig. 4.11.12. The Vela supernova remnant (a section of angular size $2° \cdot 2.5°$). This is a color photograph taken by H.-E. Schuster with the 1 m Schmidt telescope at the ESO. The clearly recognizable filamentary structure is due to the interaction of the supernova shell, ejected in the explosion which occurred 12000 yr ago, with the surrounding interstellar medium. Collisional heating is greatest at the front edges of the filament arcs, where spectral lines from ionized oxygen produce a blue glow. The inner, cooler regions appear reddish due to the $H\alpha$ neutral hydrogen emissions. The remnant star from the supernova explosion is observed as the Vela pulsar PSR 0833-45, with a period of 89 ms from the radiofrequency to the gamma-ray regions. (With the kind permission of the European Southern Observatory)

across. The high angular and time resolution of the Einstein satellite clearly shows the pulsar and the extended source which is located northwest of it (Fig. 4.11.11). The average contribution of the pulsar in the X-ray region is only about 4%; during about 1/8 of the pulse period, its radiation intensity is below the detection limit of the satellite. This lack of its own thermal radiation yields an upper limit for the temperature of the neutron star of $2.5 \cdot 10^6$ K, assuming a radius of 10 km.

Soon after the discovery of the Crab pulsar, it became clear that its rotation is in fact the source of the Crab Nebula's energy, and that as byproducts of the rotation and its braking by the surrounding plasma with its magnetic field, enormous numbers of superthermal particles would be generated. The production of synchrotron radiation in the radiofrequency range requires electrons of $\simeq 10^9$ eV, and in the X-ray range, of $\simeq 10^{14}$ eV. Together with these electrons, the corresponding positively-charged nuclei, i.e. cosmic ray particles (Sect. 5.3.8) would certainly also be accelerated. Theoretical considerations make it seem not impossible that in this way, particle energies up to the highest yet observed, about 10^{20} eV, may have been produced.

A comparison of the energy emitted per unit time in the form of superthermal particles with the rotational energy of the Crab pulsar confirms our earlier estimate of its lifetime.

The Vela remnant and the remnants of the supernovae of 1572 (Tycho Brahe) and 1604 (Kepler) could also be observed by optical *and* radioastronomical methods (Fig. 4.11.12, 13). Cassiopeia A, the strongest radio source in the northern sky, coincides with a visible nebula of 4 pc diameter, which is expanding at a velocity of 7400 km s^{-1}. Cas A can be regarded as the remnant of a supernova which exploded between 1650 and 1700. Furthermore, $H\alpha$ photographs show a large number of ring-shaped or circular nebulas, which also emit in the radiofrequency region, e.g. the well-known loop in Cygnus (Veil Nebula). Finally, the long-enigmatic "radio spur", which stretches across the sky from the neighborhood of the galactic center to the galactic north pole (and likewise several other similar formations) seems to be part of a gigantic ring. All of these radio sources are no doubt related to the remnants of old supernovae.

The *radio spectra* of the supernova remnants, of which about 150 are known in our galaxy, are characterized by *nonthermal* radiation, whose intensity decreases with frequency following a power law (Sect. 5.6.1). They can thus be distinguished from the spectra of other nebulas, such as the luminous gas nebulas (H II regions). The younger remnants can be classified according to their *radio images* into two types, the shell or ring-shaped remnants like the remnant of Tycho Brahe's supernova (Fig. 4.11.13), and the "filled" remnants, whose prototype is the Crab Nebula. The latter are presumably filled with highly energetic particles as a result of the "activity" of a rapidly rotating pulsar, and these particles can emit synchrotron radiation when they move in magnetic fields.

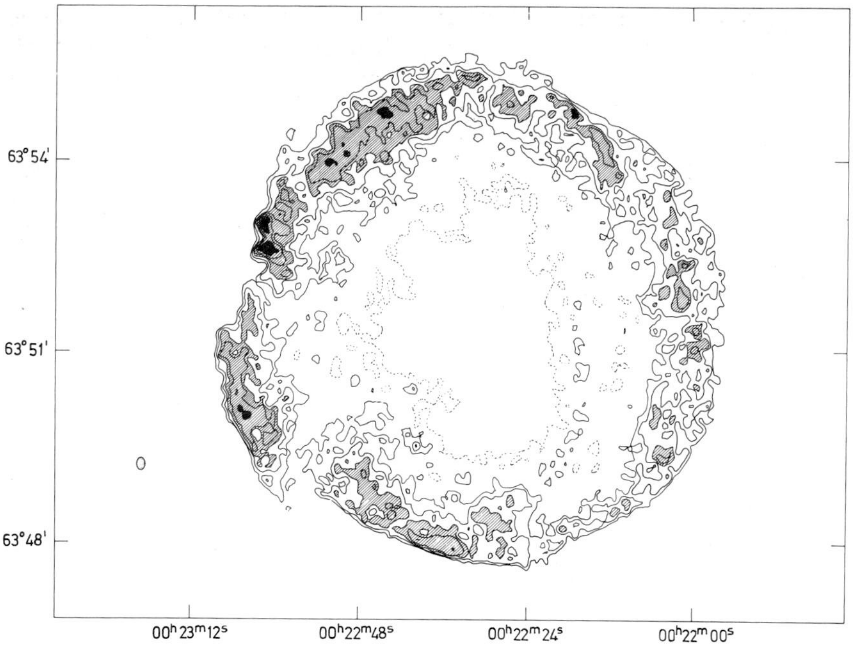

Fig. 4.11.13. Radio contour lines at $\lambda = 6$ cm (5 GHz) of the remnant 3C 10 of Tycho Brahe's supernova from the year 1572. Observations made with the Westerbork Synthesis Telescope at an angular resolution of 7″ · 8″ by R. M. Duin and R. G. Strom (1975)

Most supernova remnants have been detected as *X-ray sources*. With the exception of the Crab Nebula, the X-ray spectra of the supernova remnants observed up to now can be explained in terms of *thermal emission* from a very hot plasma source. Theoretically, we would expect essentially a continuum at temperatures $\geq 10^7$ K, with only a few lines from highly ionized, abundant elements such as Fe superposed on it; at low temperatures, numerous emission lines dominate the continuum. The (spatially unresolved) X-ray spectrum of the strong radio and X-ray source Cas A fits well into this picture, if we assume the existence of two components at different temperatures, one at $\simeq 4.5 \cdot 10^7$ K, which emits the continuum in the range ≥ 5 keV and the Fe XXIV + XXV line at 6.7 keV, the other at $\simeq 7 \cdot 10^6$ K, represented by many lines in the soft X-ray region, which are unresolved except for those of Si XIII (1.9 keV) and S XV (2.45 keV) (Fig. 4.11.14).

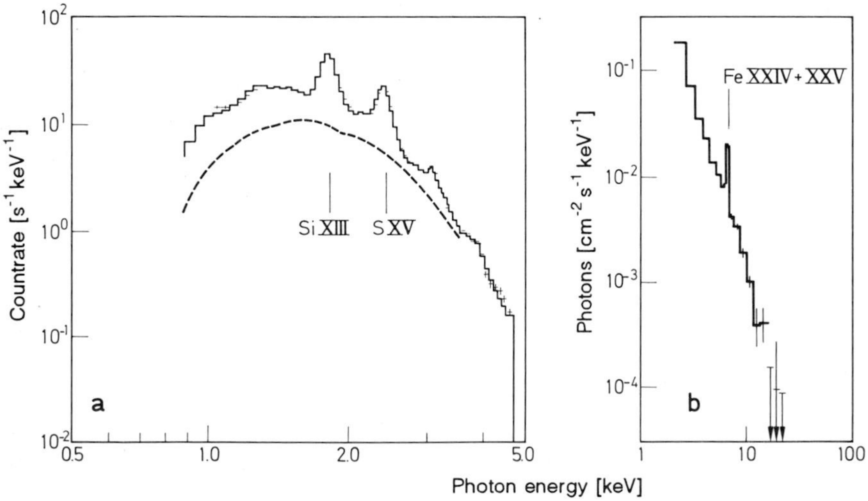

Fig. 4.11.14 a, b. Spectral lines in the X-ray spectrum of the supernova remnant Cas A. **(a)** Observations with the solid-state spectrometer on the Einstein Satellite (HEAO-2) by R. H. Becker et al. (1979). The main contribution to the soft X-ray emissions is from unresolved lines; the dashed curve gives an estimate of the continuum. **(b)** Proportional counter observations from the OSO-8 satellite by S. H. Pravdo et al. (1976)

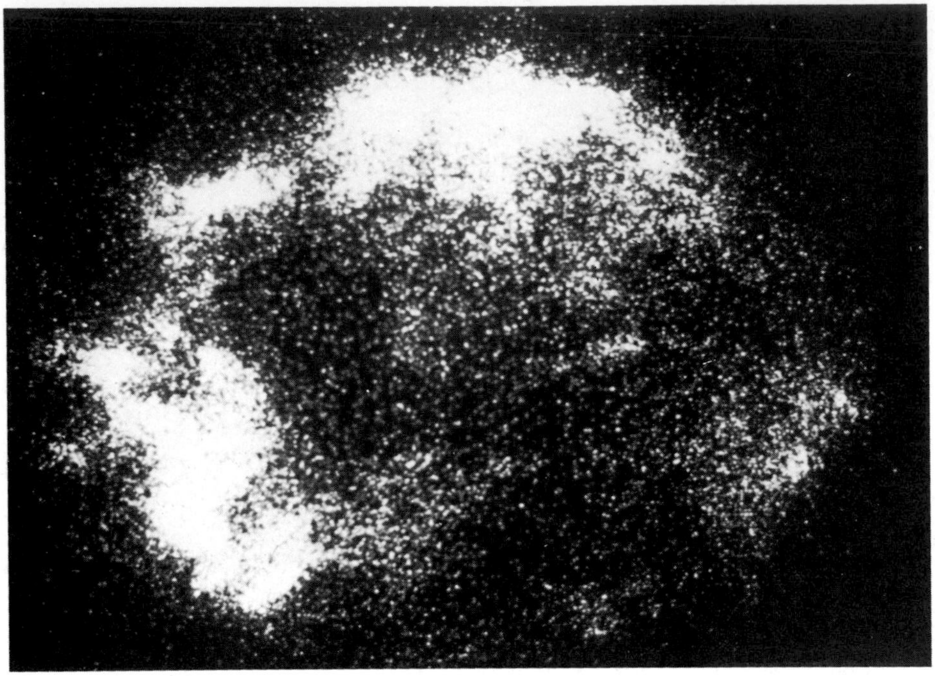

Fig. 4.11.15. X-ray image of the supernova remnant Cassiopeia A, made from the Einstein Satellite with an angular resolution of about 4″, corresponding to about 0.05 pc at the distance of 2.8 kpc to Cas A. The diameter of Cas A is about 5′ or 4 pc. (Photograph from Stephen S. Murray, Center for Astrophysics, Cambridge, MA)

The interaction of the matter ejected by the supernova with the interstellar gas gives the remnant a very complex spatial and kinetic structure. In optical photographs, numerous arches, threads, knots, and "flakes", with in part widely differing velocities, can be seen. The hot plasma in the supernova remnant also presents a similarly complicated picture, as the high-resolution X-ray photographs from the Einstein satellite show, for example in the case of Cas A (Fig. 4.11.15).

The Crab Nebula has a unique position among supernova remnants. For one thing, we know the time of the supernova explosion in this case; for another, we can observe the pulsar (neutron star) in it, and can investigate the ejected gas shell and the interaction between the pulsar and its surroundings in detail. Furthermore, we can expect essential information about the development of the star remaining after a supernova, and the formation and chemical composition of the gas shell or supernova remnant, from observations of the nearby supernova SN 1987A in the Large Magellanic Cloud, continuing up to several years after the explosion.

In general, however, some important questions remain open: does a star remain after every supernova explosion, does every neutron star become active as a pulsar, and how do the remaining star and the supernova remnant depend on the type of supernova? We cannot answer these questions in a satisfactory way at present. The question of which stars develop into supernovae in the course of their evolution, and what conditions are required for such an instability, will be dealt with in Sect. 5.4.5. First, we must take up the basic theory of stellar structure.

4.12 Stellar Structure and Energy Production in Stars

H. N. Russell clearly realized the significance of his diagram for investigating stellar evolution already in 1913. But its deeper understanding, and thus a theory of stellar evolution founded on observational facts, only became possible in connection with the study of the *inner structure of stars*. The older works of J.H. Lane (1870), A. Ritter (1878–89), R. Emden (the "Spheres of Gas" appeared in 1907) and others could only be based on classical thermodynamics. A. S. Eddington succeeded in combining these approaches with the theory of radiation equilibrium and with Bohr's theory of atomic structure, which had in the meantime been formulated; his book "The Internal Constitution of Stars" (1926) gave the starting signal for the whole development of modern astrophysics.

Starting from the knowledge that the *Sun* has essentially not changed its luminosity since the formation of the Earth $4.5 \cdot 10^9$ years ago, J. Perrin and A.S. Eddington recognized in 1919/20 that the mechanical or radioactive energy sources which had been considered up to that point could not have been sufficient to supply the Sun's radiative energy. They made the assumption that *nuclear energy* is released in the interior of the Sun and the stars, by transmutation, or, as it is termed, "burning" of hydrogen into helium. The rapid developments in nuclear physics in the 1930's then made it possible in 1938 for H. Bethe and others to find out which *nuclear reactions* would be possible at temperatures of 10^6 to 10^8 K in solar matter and other element mixtures, and to calculate their consequences. Experimental investigations of the reaction cross-sections at low proton energies, especially those of W. A. Fowler, made essential contributions to our understanding of nuclear energy release in stars.

Although we can understand the fundamental ideas of Eddington's theory with a very modest mathematical basis, the solution of the basic equations, together with the complicated material equations (for energy release, etc.) and the equation of state of stellar matter, requires a considerable numerical effort. The application of increasingly powerful computers has given the field of stellar structure and evolution a strong impulse since the mid-1950's.

We shall first limit ourselves here to a discussion of the basic equations and the associated material equations for the *structure* of a star, with applications to the Sun and to the final states of a white dwarf and a neutron star. The formation of the stars and the time variation of their structures, i.e. their evolution up to their final stages, can be treated reasonably only in Sect. 5.4, after we have learned about the properties of star clusters.

In Sect. 4.12.1, we discuss the hydrostatic equation together with the equation of state for stellar matter. This is followed in Sect. 4.12.2 by a treatment of the transport of energy by radiation and convection, which is essential for the determination of the temperature profile within a star. The release of energy can take place on the one hand by thermonuclear processes, of which we give an overview in Sect. 4.12.3; on the other, it can occur through the conversion of gravitational binding energy. The latter, together with the important question of the stability of a star, will be treated in Sect. 4.12.4. In Sect. 4.12.5, we summarize the system of basic equations and derive some general statements about stellar structure.

As a first application, we concern ourselves with the structure of the Sun and its predicted neutrino flux in comparison with observations (Sect. 4.12.6). The final stages of stellar evolution are characterized by extremely high densities of matter: in the interior of white dwarf stars (Sect. 4.12.7), the equation of state of an ideal gas is no longer valid; here, the degeneracy of the electrons in the sense of Fermi-Dirac statistics must be taken into account. Finally, we meet up with still higher densities in the neutron stars (Sect. 4.12.8); they correspond to those of the matter within atomic nuclei.

To calculate the precise structure of a neutron star with its enormous concentration of mass, Newton's theory of gravitation is no longer sufficient. We must therefore replace it here by *Einstein's general theory of relativity*. It is thus appropriate, also in view of later applications to the activity in galactic cores and especially to cosmology, to give at this point an overview of the starting assumptions and most important results of this theory (Sect. 4.12.9).

4.12.1 Hydrostatic Equilibrium and the Equation of State of Stellar Matter

We first consider a star as a sphere with the mass $\mathscr{M}$ and radius R in hydrostatic equilibrium: on each volume element with density $\varrho(r)$ at a distance r from the star's center, the gravitational force of the mass $\mathscr{M}(r)$ which lies within r acts, and produces a radial variation in the pressure P within the volume element. The *hydrostatic equation*, which we have already derived in connection with planetary structures in Sect. 2.8.3, is then:

$$\frac{dP}{dr} = -\varrho(r)\frac{G\mathscr{M}(r)}{r^2} , \qquad (4.12.1)$$

where $\mathscr{M}(r)$ is given by

$$\frac{d\mathscr{M}(r)}{dr} = 4\pi r^2 \varrho(r) . \qquad (4.12.2)$$

In most stars, P is practically equal to the gas pressure P_g; only in very hot and massive stars is it necessary to include the radiation pressure P_r explicitly, i.e. $P = P_g + P_r$. Using (4.12.1), we can readily estimate the pressure, for example at the center of the Sun, $P_{c\odot}$, by setting $\varrho = \text{const}$ (2.8.7). We thus obtain $P_{c\odot} \simeq 1.3 \cdot 10^{14}$ Pa; exact model calculations yield a central pressure which is about a factor of 100 larger ($2.5 \cdot 10^{16}$ Pa).

The relation (at each point) between the pressure P, the density ϱ, and the temperature T as the third state variable is contained in the *equation of state*. Following Eddington, we begin with the equation of state of an *ideal gas*:

$$P_g = \varrho \frac{kT}{\mu m_u} , \quad (4.12.3)$$

where $k = 1.38 \cdot 10^{-23}$ J K^{-1} is Boltzmann's constant, $m_u = 1.66 \cdot 10^{-27}$ kg the atomic mass constant ($\simeq$ the proton's mass), and μ the average molecular mass. Equation (4.12.3) can be used so long as the interactions of neighboring particles are sufficiently small relative to their thermal (kinetic) energies. On Earth, we are accustomed to the fact that this no longer applies, i.e. that condensation begins, roughly at densities $\varrho \gtrsim 500$ to 1000 kg m^{-3}. In stars, this limit is shifted up to much higher values by *ionization*. In particular, the most abundant elements H and He are completely ionized at relatively shallow depths in all stars.[22]

The *mean molecular mass* μ is, for complete ionization, equal to the atomic mass divided by the total number of particles, i.e. nuclei plus electrons. We thus obtain for

Hydrogen	Helium	Heavy Elements
$\mu = \frac{1}{2}$	$\frac{4}{3}$	$\simeq 2$

From these values, μ can easily be calculated for any mixture of elements.

The average temperature in the interior of the Sun can be estimated from (4.12.3) with $\varrho = \bar{\varrho}_\odot$, $\mu = \mu_H = 0.5$, and $\bar{P}_g \simeq P_{c\odot}/2$ (see above) to be $\bar{T}_\odot \simeq 6 \cdot 10^6$ K.

4.12.2 The Temperature Distribution and Energy Transport in Stellar Interiors

To calculate the temperature distribution $T(r)$ in the interior of a star, we must first investigate the mechanisms of energy transport. Poor energy transport leads to a steep temperature gradient, good energy transport yields a flat temperature gradient. (Whoever doubts this can test it by holding with bare fingers first a glass rod and then a nail in a flame!).

We first consider the energy transport via *radiation*, i.e. radiation equilibrium (cf. Sect. 4.8.1, also for nomenclature); we immediately integrate all the radiation quantities such as I_ν, F_ν, and B_ν over the *total* frequency spectrum. These integrals are denoted as I, F, and B. Furthermore, we again introduce *Rosseland's opacity* κ (4.8.6) as an average over all frequencies, here using *mass* absorption coefficients, as is usual in the theory of stellar structure. The *radiation transport equation* (4.8.2 or 4) then takes on the form:

$$\cos\theta \frac{dI}{\kappa \varrho \, dr} = -I + B . \quad (4.12.4)$$

From the radiation intensity I, we again calculate the radiation flux F (4.8.5) by multiplying with $\cos\theta$ and integrating over all directions. For the total radiation we thus obtain:

$$F = -\int_0^\pi \frac{dI}{\kappa \varrho \, dr} \cos^2\theta \cdot 2\pi \sin\theta \, d\theta . \quad (4.12.5)$$

(Since the Kirchhoff-Planck function B is isotropic, the integral of $B \cos\theta$ over all directions vanishes.)

In a *stellar interior*, the radiation field I is nearly isotropic, so that we can extract it from the integral and carry out the integration over θ on the right side of (4.12.5); this gives

$$\int_0^\pi \cos^2\theta \cdot 2\pi \sin\theta \, d\theta = \frac{4\pi}{3} . \quad (4.12.6)$$

Furthermore, we can write down I in this case as a function of temperature using the Stefan-Boltzmann law. It is

$$I = \frac{\sigma}{\pi} T^4 = \frac{ac}{4\pi} T^4 \quad (4.12.7)$$

and thus

$$F = -\frac{c}{3\kappa\varrho} \frac{d}{dr}(aT^4) , \quad (4.12.8)$$

i.e. the radiation flux is proportional to the gradient of the energy density $u = aT^4$ (4.2.39) and to the "conductivity" $(\kappa\varrho)^{-1}$ of the matter with respect to photons.

[22] In the range of higher densities (and lower temperatures), at $T = 10^7$ K, above $\varrho \simeq 10^6$ kg m^{-3}, a Fermi-Dirac degeneracy of the electrons arises, so that (4.12.3) is not applicable. The equation of state of these degenerate electron components, which are important in the later phases of stellar evolution, will be discussed in Sect. 4.12.7.

The total radiation energy which passes outwards per unit time through a spherical surface of radius r is then given by

$$L(r) = 4\pi r^2 F = -\frac{16\pi a c r^2}{3} \frac{T^3}{\kappa \varrho} \frac{dT}{dr} . \tag{4.12.9}$$

At the surface of the star, $r = R$, $L(r)$ becomes the directly measurable luminosity $L = L(R)$. At temperatures near $\simeq 6 \cdot 10^6$ K, the maximum in Planck's radiation function $B_\lambda(T)$ lies at $\lambda_{max} = 0.5$ nm (4.2.26), i.e. in the X-ray region. The *absorption coefficient* is determined here by bound-free and free-free transitions of the atomic states which are not yet completely "ionized away". In an element mixture such as we find in the Sun and in population I stars, these are higher ionization states of the more abundant, heavier elements such as O, Ne, ... In the "metal poor" stars of population II, the contributions of H and He must also be taken into account. Detailed tables or computer programs for the absorption coefficients in stellar interiors as functions of ϱ and T for various element mixtures are collected and published by the Los Alamos National Laboratory, USA.

Besides energy transport by radiation, a hydrogen and helium convection zone can occur in stellar interiors, as we saw in the example of the Sun; the conditions for its occurrence were discussed in Sect. 4.10.6. In addition, further convection zones can be formed in connection with nuclear energy release. In all such convection zones, the *energy transport by convection* (rising of hotter and sinking of cooler matter) is strongly predominant over transport by radiation, aside from the boundary regions of the zone. The relation between the temperature T and the pressure P is then given by the adiabatic equation (2.8.20):

$$T \propto P^{1-1/\gamma} , \tag{4.12.10}$$

which we have already used in treating planetary atmospheres. $\gamma = c_p/c_v$ is the ratio of specific heats at constant pressure and at constant volume. By logarithmic differentiation with respect to r, we obtain (2.8.21), the *temperature gradient* in a convection zone:

$$\frac{dT}{dr} = \left(1 - \frac{1}{\gamma}\right) \frac{T}{P} \frac{dP}{dr} . \tag{4.12.11}$$

The methods for calculating convective energy transport are still very unsatisfactory in quantitative terms.

4.12.3 Energy Release from Nuclear Reactions in Stellar Interiors

In *hydrogen fusion*, the overall reaction consists of the combination of four hydrogen nuclei to form one helium nucleus, $4\,{}^1\mathrm{H} \rightarrow {}^4\mathrm{He}$. The energy released, ΔE, can be readily calculated using the mass-energy relation discovered by A. Einstein in 1905,

$$\Delta E = \Delta m c^2 , \tag{4.12.12}$$

from the mass difference Δm between the products and the reactants in the above nuclear reaction. The mass of 4 hydrogen atoms is $4 \cdot 1.007825$ atomic mass units ($1\,m_u = 1.6605 \cdot 10^{-27}$ kg $= 931.49$ MeV/c^2); that of the helium atom is $4.0026\,m_u$[23]. The difference corresponds to:

$$\begin{aligned} 0.0287\,m_u &= 4.766 \cdot 10^{-29}\,\mathrm{kg} \\ &= 4.28 \cdot 10^{-12}\,\mathrm{J} = 26.73\,\mathrm{MeV} . \end{aligned} \tag{4.12.13}$$

If we imagine that e.g. one solar mass of pure hydrogen were to be converted into helium, then an energy of $1.28 \cdot 10^{45}$ J would be released; this would supply the radiant energy of the Sun at the current level for a period of $1.05 \cdot 10^{11}$ years.

The continued formation of elements heavier than helium from lighter elements allows further energy releases in accord with (4.12.12), up to ^{56}Fe, at which point the nuclear binding energy is maximal, at 8.4 MeV per nucleon. The main portion of this energy gain is already obtained in the fusion of hydrogen to helium, with 6.7 MeV per nucleon.

In order to calculate the rate of nuclear energy release in stars, we must consider the individual nuclear reactions in detail. In the following, we summarize the most important reactions, employing the usual notation:

$$x(a,b)y$$

reactant nucleus x (reacts with a, ejects b) product nucleus y.

Here, we use the symbols

p: proton
α: He^{2+} particles

[23] Actually, we require here the mass of the atomic *nuclei*. However, since the number of electrons remains unchanged in the reaction, we can use the atomic masses, whose experimental values are more precisely known, risking only very small errors due to differences in the electronic binding energies.

n: neutron
e^-: electron
e^+: positron
γ: quantum of radiation
ν: electron neutrino.

A nucleus (nuclide) is denoted by adding its mass number A to the upper left of the chemical symbol X and, where needed, its atomic number (number of protons) Z to the lower left: ${}^AX$ or ${}^A_ZX$. For example, we write the two stable helium isotopes as ^{3}He and ^{4}He or ${}^3_2$He and ${}^4_2$He. A nucleus in an excited state is denoted by an asterisk (*).

A positron which is released in a nuclear reaction immediately forms two γ quanta by annihilating with a thermal electron. The *neutrinos* ν, owing to their extremely small interaction cross sections, escape from even the interiors of stars (excepting those in the very dense final stages of stellar evolution). The "usable" energy Q released is thus reduced by the amount carried away by the neutrinos from that calculated with the mass defect (4.12.12).

The *rate of energy production* ε, i.e. the energy released per unit time and mass,

$$\varepsilon = Q\frac{\varkappa}{\varrho} \ , \tag{4.12.14}$$

depends essentially on Q and the rate at which the nuclear reactions occur. $\varkappa$ is the number of reactions $x(a,b)y$ in a unit volume per unit time, cf. (4.7.32)

$$\varkappa = N_x N_a \langle \sigma v \rangle \ , \tag{4.12.15}$$

where N_x and N_a are the particle densities of the reaction partners x and a, v is their relative velocity, and σ the reaction cross-section for the fusion reaction. The average $\langle \sigma v \rangle$ is obtained by integration over the distribution function for the relative velocities. The Maxwell-Boltzmann distribution (4.7.33) also holds for the *relative* velocities in the thermal motion of the nuclei if the particle mass in the expression for the most probable velocity V_0 is replaced by the reduced mass $m_x m_a/(m_x+m_a)$. Due to the short range of the nuclear forces, the positively-charged, mutually repulsive reactant nuclei must approach each other to within a distance of the order of a nuclear diameter, $r_0 \simeq 10^{-15}\,A^{1/3}$ [m], for a reaction to take place; i.e., they must overcome a "Coulomb barrier" of height

$$B \simeq \frac{1}{4\pi\varepsilon_0}\frac{Z_x Z_a e^2}{r_0} \ . \tag{4.12.16}$$

B is in the range of several MeV, while the mean thermal energy, e.g. at a temperature of 10^8 K, is only about 0.01 MeV. The Coulomb barrier can thus only be "tunneled" through, with a probability which decreases exponentially with B, as predicted by quantum mechanics. This leads to a very strong temperature dependence of the *thermonuclear reactions*; one can speak of an "ignition temperature" for the initiation of energy production, particularly in the case of nuclei with higher proton numbers.

We now consider the reactions of *hydrogen burning*, i.e. the combination of four protons to a helium nucleus, in detail. A first possibility is the *proton-proton reaction* or *pp-chain* (4.12.17).

The slowest reaction, which represents the rate-limiting step for the whole chain, is ^{1}H(p, $e^+\nu$) ^{2}D. The neutrinos which are released in this step carry away on the average 0.50 MeV of energy, so that 26.23 MeV of energy is available to the star from the main PP I chain. The branches PP II and PP III become more significant with increasing temperature.

A second possibility for the conversion of hydrogen into helium is offered by a reaction *cycle* investigated in 1938 by H. Bethe and C. F. von Weizsäcker, in which the elements C, N, and O are involved and essentially determine the reaction rate, but are quantitatively "returned" at the end. Energy release by the *CNO triple cycle* (Fig. 4.12.1) is mainly due to the principal or CN cycle (24.97 MeV), whose slowest reaction is ^{14}N(p, γ) ^{15}O. The two subcycles take place when a γ-quantum instead of an α-particle is emitted following the proton capture by ^{15}N; they occur e.g. in the Sun about 1000 times less frequently than the principal cycle.

$$
\begin{array}{lllll}
 & \nearrow {}^3\mathrm{He}({}^3\mathrm{He},2\mathrm{p})\,{}^4\mathrm{He} & & \mathrm{PP\,I} & 26.23\ \mathrm{MeV} \\
{}^1\mathrm{H}(\mathrm{p},e^+\nu)\,{}^2\mathrm{D}(\mathrm{p},\gamma)\,{}^3\mathrm{He} & & \nearrow {}^7\mathrm{Be}(e^-,\nu)\,{}^7\mathrm{Li}(\mathrm{p},\alpha)\,{}^4\mathrm{He} & \mathrm{PP\,II} & 25.67\ \mathrm{MeV} \\
 & \searrow {}^3\mathrm{He}(\alpha,\gamma)\,{}^7\mathrm{Be} & & & \\
 & & \searrow {}^7\mathrm{Be}(\mathrm{p},\gamma)\,{}^8\mathrm{B}(e^+\nu)\,{}^8\mathrm{Be}(2\alpha) & \mathrm{PP\,III} & 19.28\ \mathrm{MeV}\ .
\end{array}
\tag{4.12.17}
$$

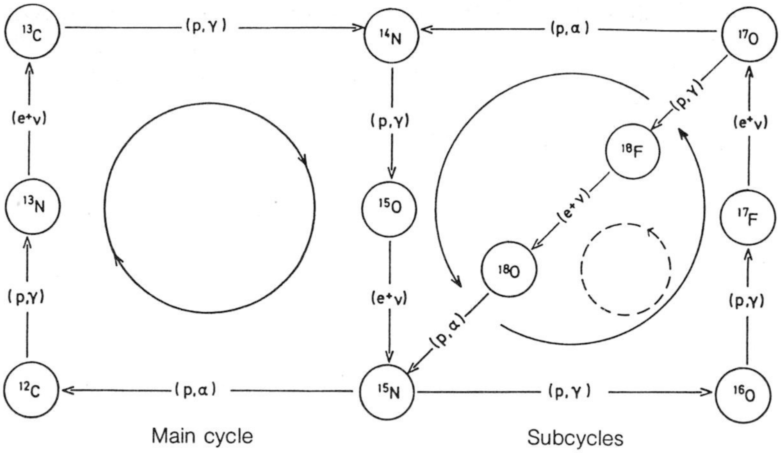

Fig. 4.12.1. The reactions of the CNO triple cycle

In the stationary state, the *abundance ratios* of the isotopes involved are determined by the *rates* of the individual reactions as in a "radioactive equilibrium", although in each *single* cycle, the C, N, and O nuclei, as "catalysts", are not consumed. At temperatures from about 10^7 to 10^8 K, most of the matter originally present as C, N, and O is converted by the CNO cycle into ^{14}N. Starting from the normal element mixture (Tables 4.9.1 and 2), the nitrogen abundance increases by about a factor of 10. The abundance ratio of the two carbon isotopes ^{12}C/^{13}C, which is about 90 in terrestrial and solar matter and between 40 and 90 in interstellar matter, settles down to a much smaller value, $\simeq 4$. This is an important indicator for (possibly earlier) hydrogen burning by the CNO cycle.

In Fig. 4.12.2, the mean *energy release* ε of hydrogen burning for an element mixture as found in the Sun and in population I stars is plotted as a function of temperature over the range 5 to $50 \cdot 10^6$ K. Due to the higher electric charges of the reactant nuclei, the CNO cycle is initiated only at higher temperatures than the pp chain. Cool main sequence stars with central temperatures of up to $1.8 \cdot 10^7$ K, in particular also the Sun with $T_{c\odot} \simeq 1.5 \cdot 10^7$ K, therefore extract their energy mainly from the pp chain; hotter stars in the upper portion of the main sequence derive theirs mainly from the CNO cycle. For the old stars of population II, the contribution of the CNO cycle relative to the pp chain is proportional to their abundances of C, N, and O relative to H.

If a major portion of the hydrogen is used up in the interior of a star, and its temperature increases to above 10^8 K (as a result of contraction), then the burning of helium is initiated, as noted by E. J. Öpik and E. E. Salpeter in 1951/52, leading to ^{12}C by the *3α process*: $3\alpha \rightarrow {}^{12}$C.

This process begins with the slightly endothermic fusion of two He nuclei:

$$^4\text{He} + {}^4\text{He} + 95\ \text{keV} \rightarrow {}^8\text{Be} + \gamma\ . \qquad (4.12.18)$$

The ^{8}Be is present in thermal equilibrium (^{4}He + ^{4}He $\leftrightarrow$ ^{8}Be) at a very small concentration which is then further decreased by the reaction

$$^8\text{Be} + {}^4\text{He} \leftrightarrow {}^{12}\text{C}^* \rightarrow {}^{12}\text{C} + \gamma \qquad (4.12.19)$$

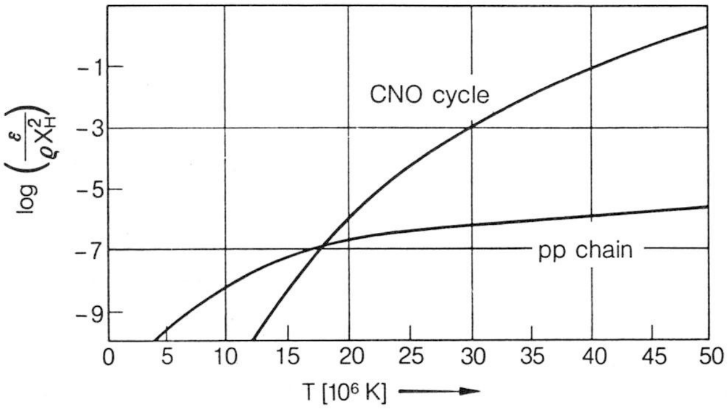

Fig. 4.12.2. Temperature dependence of the energy release ε by hydrogen fusion in equilibrium for the element mixture of the Sun and Population I stars. ε in [W kg^{-1} = 10^4 erg s^{-1} g^{-1}], ϱ in [kg m^{-3} = 10^{-3} g cm^{-3}], X_H = relative mass abundance of hydrogen. $\varepsilon/\varrho X_H^2$ depends only on T in the pp chain, but for the CNO cycle, it is also proportional to the abundance $X_{C,N}/X_H$. At the center of the Sun, X_H is currently 0.36 (originally 0.73), $\varrho = 1.6 \cdot 10^5$ kg m^{-3}, $T = 1.5 \cdot 10^7$ K, and thus $\varepsilon \simeq 1.8 \cdot 10^{-3}$ W kg^{-1}

with an energy release of 7.28 MeV per ^{12}C nucleus. The decay of the excited ^{12}C* nucleus back to ^{4}He and ^{8}Be with 7.65 MeV energy release is more than 1000 times more rapid than the transition to its ground state. Only 2.4 MeV are released per helium nucleus by its fusion, i.e. about 10% of the energy which was obtained from its formation from hydrogen. Starting with ^{12}C, the following nuclei (with A a multiple of four) can be formed by further (α,γ) reactions

$$^{12}\mathrm{C}(\alpha,\gamma)^{16}\mathrm{O}(\alpha,\gamma)^{20}\mathrm{Ne}(\alpha,\gamma)^{24}\mathrm{Mg}(\alpha,\gamma)^{28}\mathrm{Si}\ , \tag{4.12.20}$$

whereby the last two reactions hardly contribute to the energy release, due to the slow reaction rate of $^{16}\mathrm{O}(\alpha,\gamma)^{20}\mathrm{Ne}$.

At somewhat higher temperatures than required for the 3α process, ^{14}N, the main product of the CNO cycle, is consumed by the reaction series

$$^{14}\mathrm{N}(\alpha,\gamma)^{18}\mathrm{F}(\mathrm{e}^+\nu)^{18}\mathrm{O}(\alpha,\gamma)^{22}\mathrm{Ne}\ . \tag{4.12.21}$$

The reaction

$$^{22}\mathrm{Ne}(\alpha,\mathrm{n})^{25}\mathrm{Mg} \tag{4.12.22}$$

can continue this series; it provides *free neutrons* for the *synthesis of heavy elements* of $A \geq 60$ (through the so-called *s*-process in red giant stars, see Sect. 5.4.5). This synthesis could in no case occur by means of charged-particle reactions, due to the strong Coulomb fields of the heavy nuclei.

After the cessation of helium fusion and the α reactions which follow it, at temperatures of 6 to $7\cdot 10^8$ K, in the course of the evolution of a star *carbon burning* can take place; its most important reactions:

$$\begin{aligned} &^{12}\mathrm{C}+{}^{12}\mathrm{C}\rightarrow{}^{23}\mathrm{Na}+\mathrm{p}\ , \quad {}^{23}\mathrm{Na}(\mathrm{p},\alpha)^{20}\mathrm{Ne}\ , \\ &^{12}\mathrm{C}+{}^{12}\mathrm{C}\rightarrow{}^{20}\mathrm{Ne}+\alpha\ , \end{aligned} \tag{4.12.23}$$

lead to ^{20}Ne and release 2.3 MeV of energy per ^{12}C nucleus. Finally, above 1.5 to $2\cdot 10^9$ K, the energies of thermal photons γ are sufficiently high to destroy ^{20}Ne by *photodisintegration* and, by further reactions with the helium nuclei formed in the disintegration, to convert it to ^{24}Mg and ^{28}Si:

$$\begin{aligned} &^{20}\mathrm{Ne}(\gamma,\alpha)^{16}\mathrm{O}\ , \\ &^{16}\mathrm{O}(\alpha,\gamma)^{20}\mathrm{Ne}(\alpha,\gamma)^{24}\mathrm{Mg}(\alpha,\gamma)^{28}\mathrm{Si}\ . \end{aligned} \tag{4.12.24}$$

This *neon burning* is followed at somewhat higher temperatures by oxygen burning ($^{16}\mathrm{O}+{}^{16}\mathrm{O}$) and silicon burning, a quasi-equilibrium between photodisintegration and α capture, which leads to the formation of heavier elements up to the nuclei of the iron group.

In the carbon burning and the later burning phases, the temperatures and densities in stellar interiors are so high that *neutrinos* are produced in large numbers at the expense of the thermal energy of the stellar matter, via the weak interaction between electrons, positrons, and photons; they leave the star practically without hindrance. The energy loss from these neutrinos is of the same order of magnitude as the energy release from thermonuclear processes (in contrast, the neutrinos formed by β-decay of the nuclear reaction products represent only a moderate energy loss). Formally, the neutrinos can be included in the reactions as a negative energy release.

The application of nuclear energy release in the theory of the internal structure of stars is (fundamentally) quite simple: if $\varepsilon(\varrho, T)$ is the energy released per unit of time and mass by *all* the nuclear reactions which occur at the corresponding temperatures T and densities ϱ, then in a spherical shell $r\ldots(r+dr)$ in unit time, the energy $\varrho\varepsilon 4\pi r^2 dr$ is released and the energy flux $L(r)$ increases according to the equation:

$$\frac{dL(r)}{dr} = 4\pi r^2 \varrho\varepsilon\ . \tag{4.12.25}$$

4.12.4 Gravitational Energy and Thermal Energy. The Stability of Stars

When the internal temperature of a star (which has a particular chemical composition) becomes too low to permit thermonuclear processes to continue, the only energy source in the intermediate period until nuclear reactions can again be ignited is the *gravitational energy* E_G, i.e. energy of contraction.

It seems appropriate to mention here that as early as 1846, soon after his discovery of the law of energy conservation, J. R. Mayer raised the question of the origin of the radiation energy emitted by the Sun. He considered the possibility that a mass m of meteorites which fall into the Sun would release their energy as heat according to the relation (2.6.34)

$$mG\mathcal{M}/R \tag{4.12.26}$$

(where G is again the gravitational constant, $\mathcal{M}$ and R the mass and radius of the Sun). Since in fact the mass of meteorites falling into the Sun is rather small, H. von Helmholtz pointed out in 1854, and Lord Kelvin in 1861, that the *contraction of the Sun itself* would be a more effective source of gravitational energy.

As one can readily see from (4.12.26), the energy which is released by the formation of a sphere of gas of radius R from matter which was originally far away, or also by the strong contraction of a star to the radius R, is given by

$$E_G \simeq \frac{G\mathcal{M}^2}{R} . \qquad (4.12.27)$$

For the Sun, for example, this is $\simeq 3.8 \cdot 10^{41}$ J; the Sun's thermal energy content is of the same order of magnitude. The energy E_G could supply the Sun's radiant power at the current level of $L_\odot = 3.85 \cdot 10^{26}$ W for only $E_G/L \simeq 3 \cdot 10^7$ yr.

The close connection between the gravitational energy of a star and its thermal energy becomes clear from the *virial theorem* (2.6.25), which states that in a system of point masses with mutual gravitational attraction, on the average over time the kinetic energy E_{kin} and the potential energy E_{pot} are related by

$$\bar{E}_{kin} = -\tfrac{1}{2}\bar{E}_{pot} . \qquad (4.12.28)$$

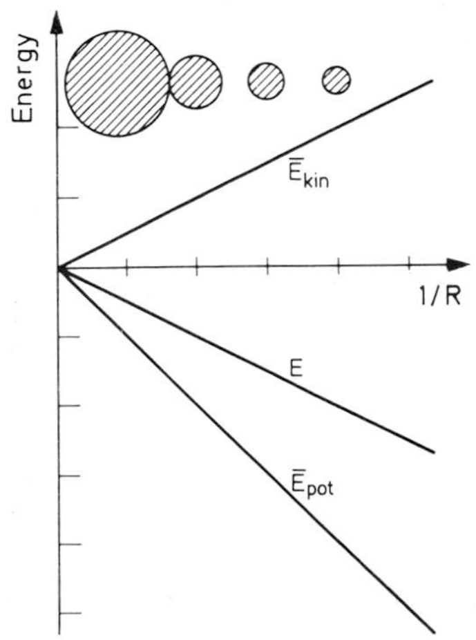

Fig. 4.12.3. The virial theorem for a system of point masses under the influence of mutual gravitational attraction, $\bar{E}_{kin} = -\frac{1}{2}\bar{E}_{pot}$. A contracting star loses potential energy; half of the amount lost is, however, added to its reserves of kinetic energy. $2\bar{E}_{kin} < -\bar{E}_{pot}$ leads to gravitational instability, according to J. Jeans (Sect. 5.8.1)

This relation is presented more clearly in Fig. 4.12.3. Since the total energy E is given by $E = E_{kin} + E_{pot}$, we also have $\bar{E}_{kin} = -E$.

We apply the virial theorem initially to a star (in hydrostatic equilibrium) which we consider for simplicity to be a homogeneous sphere (mass $\mathcal{M}$, radius R, density ϱ). If we further assume it to consist of a monatomic gas of atomic mass μ, then the only kinetic energy present is the *thermal energy* E_T. It is on the average $3kT/2$ per particle, i.e. $3kT/2\mu m_u$ per unit of mass (m_u = atomic mass unit), and for the whole star

$$E_T = \frac{3}{2}\frac{kT}{\mu m_u}\mathcal{M} . \qquad (4.12.29)$$

The potential energy or the *gravitational energy* of a homogeneous sphere is calculated as follows: if we add a shell of mass $d\mathcal{M}(r)$ (as an intermediate stage, so to speak) to a sphere of radius r and mass $\mathcal{M}(r) = 4\pi\varrho r^3/3$, the gain in potential energy is $dE_G = -G\mathcal{M}(r)\,d\mathcal{M}(r)/r$. With $r = [3\mathcal{M}(r)/4\pi\varrho]^{1/3}$, we immediately obtain:

$$E_G = -G\int_0^R \frac{\mathcal{M}(r)\,d\mathcal{M}(r)}{r} = -\left(\frac{3}{4\pi\varrho}\right)^{-1/3} G\frac{3}{5}\mathcal{M}^{5/3}$$
$$= -\frac{3}{5}\frac{G\mathcal{M}^2}{R} . \qquad (4.12.30)$$

From the *virial theorem*

$$2E_T + E_G = 0 \qquad (4.12.31)$$

we can now derive directly the fact that the temperature in the stellar interior must be 10^6 to 10^7 K, using the known data for main sequence stars, and as long as radiation pressure plays no essential role. (We had already estimated this temperature in Sect. 4.12.1 by a different route.)

However, a star's energy does *not* remain constant, due to the energy which is radiated away; instead, it decreases in a time interval δt according to $\delta E = L\,\delta t$ (luminosity L). If no nuclear energy sources are available to the star, it can replace this energy loss only through contraction; from (4.12.27), in a time δt, contraction provides the quantity $\delta E_G = G\mathcal{M}^2\,\delta(1/R)$. If this contraction takes place sufficiently slowly, so that *hydrostatic equilibrium* can be maintained in the star, then from (4.12.31) we find

$$2\delta E_T + \delta E_G = 0$$

and

$$\delta E_{\mathrm{T}} + \delta E_{\mathrm{G}} = L\,\delta t \ . \tag{4.12.32}$$

This result shows that one-half of the gravitational energy is required to increase the thermal energy, and the other half to supply the radiation energy (Fig. 4.12.3). When energy is *extracted* (by radiation), the star thus reacts by *increasing* its temperature, i.e. it represents a system with "negative specific heat". The star is stabilized by the initiation of nuclear energy release as its temperature rises;[24] the contraction is halted and the luminosity is then maintained from nuclear energy sources.

The time interval during which a star can supply its radiation energy at the expense of gravitational energy E_{G}, the *Helmholtz-Kelvin time*, is of the order of

$$t_{\mathrm{HK}} \simeq \left|\frac{E_{\mathrm{G}}}{L}\right| \simeq \frac{G\mathcal{M}^2}{RL} \simeq \frac{E_{\mathrm{T}}}{L} \ . \tag{4.12.33}$$

From the virial theorem, it is comparable to the thermal relaxation time. In general, t_{HK} is much shorter than the nuclear evolution time t_{E} (5.4.3).

For those evolutionary stages of a star in which gravitational energy plays a role due to configurational changes, we must thus extend (4.12.25) to include the *rate of release of gravitational energy*. There is then no longer an equilibrium in a mass-containing shell from $r \ldots (r+dr)$ simply between the nuclear energy release rate $4\pi r^2 \varrho \varepsilon\, dr$ and the net radiative energy flux $dL(r)$. Instead, the time rate of change of the inner structure in each mass shell gives an additional (positive or negative) contribution, which, following the 1st Law of thermodynamics, arises from the sum of the change in thermal energy per unit mass, $\frac{3}{2}kT/\mu m_{\mathrm{u}}$, and the work performed by pressure forces, $-P\,dv$ (v = specific volume, $v = 1/\varrho$), so that:

$$4\pi r^2 \varrho \varepsilon - \frac{dL(r)}{dr} = 4\pi r^2 \varrho \left(\frac{3}{2}\frac{k}{\mu m_{\mathrm{u}}}\frac{dT}{dt} - \frac{P}{\varrho^2}\frac{d\varrho}{dt}\right) \tag{4.12.34}$$

holds.

Finally, we ask the question, under which circumstances is a star able to maintain its hydrostatic equilibrium (4.12.1) in the face of disturbances or changes in its structure? Here, the *compressibility* of its matter, i.e. the variation of the pressure P with density ϱ, plays a decisive role. If we take P proportional to ϱ^{Γ}, then the gravitational force (per unit volume) averaged over the star varies (due to $\varrho \propto \mathcal{M}/R^3$) with the star's radius R according to:

$$\langle F_{\mathrm{G}}\rangle \propto \varrho \frac{G\mathcal{M}}{R^2} \propto R^{-5} \ , \tag{4.12.35}$$

while the pressure gradient is given by

$$\left\langle \frac{dP}{dr} \right\rangle \propto \frac{P}{R} \propto \frac{\varrho^{\Gamma}}{R} \propto R^{-3\Gamma-1} \ , \tag{4.12.36}$$

and therefore the ratio of the two terms varies with R as

$$\left\langle \frac{dP}{dr} \right\rangle \Big/ \langle F_{\mathrm{G}}\rangle \propto R^{-3(\Gamma-4/3)} \ . \tag{4.12.37}$$

If $\Gamma > 4/3$, then for a *contraction*, $\langle dP/dr\rangle$ increases more rapidly than $\langle F_{\mathrm{G}}\rangle$, i.e. the star will finally be able to find a new configuration in hydrostatic equilibrium. On the other hand, for $\Gamma \le \frac{4}{3}$ the gravitational force increases more and more if the radius is reduced, and the star will not find a stable configuration. We therefore obtain as a *criterion for the stability* of hydrostatic equilibrium that the relation:

$$\Gamma > \tfrac{4}{3} \tag{4.12.38}$$

must be fulfilled in the interior of stars for the major portion of the stellar mass. This condition is obeyed e.g. by a star consisting of a monatomic ideal gas, since then the adiabatic equation (4.12.10) has the form $P \propto \varrho^{\Gamma}$ with $\Gamma \equiv \gamma = \frac{5}{3}$ (specific heat ratio).

If the stability criterion is violated, for example by the excitation of internal degrees of freedom or by a "phase transition" (Sect. 5.4.5), then a stellar *collapse* results, practically in free fall (as long as Γ remains $\le \frac{4}{3}$). The characteristic time for this collapse is the *free fall time* t_{ff}, which we estimate by extending (4.12.1) to include an acceleration term $\varrho\, d^2r/dt^2 \sim \varrho R/t_{\mathrm{ff}}^2$; it is of comparable magnitude to the gravitational force $\varrho G\mathcal{M}/R^2$ or the pressure gradient P/R. With $\varrho \propto \mathcal{M}/R^3$, we find for the dynamic time scale:

$$t_{\mathrm{ff}} \simeq \sqrt{\frac{1}{G\varrho}} \ . \tag{4.12.39}$$

For example, t_{ff} for the Sun ($\bar{\varrho}_{\odot} \simeq 1400\ \mathrm{kg\,m^{-3}}$) is of the order of 1 h, i.e. much, much shorter than the Helmholtz-Kelvin time. Conversely, we can conclude from the stability of the Sun over thousands of millions

---

[24] The prerequisite for this is that the pressure increase with increasing temperature, i.e. that the equation of state (4.12.3) be valid.

of years that its hydrostatic equilibrium (4.12.1) must be maintained with an extremely high precision. Large-scale pressure disturbances are balanced within t_{ff}. As can be seen from the virial theorem (4.12.31), the free fall time corresponds to the characteristic propagation time of an acoustic wave through the star with the velocity of sound $c_s \simeq (kT/\mu m_u)^{1/2}$, or also to the fundamental period of oscillation of the star (Sect. 4.11.1).

4.12.5 The Fundamental Equations for the Internal Structure of Stars, and Their General Results

In order to give a better overview, we summarize the four basic equations of the theory of stellar interiors from the preceding sections. Their numerical solution is carried out exclusively using fast electronic computers. In addition, we need the *equation of state* and the equations which relate the *materials constants* ε, κ, and γ to two of the state variables P, T, or ϱ. All of these relationships depend essentially on the *chemical composition* of the stellar matter – this is important for what follows!

The mass within r (4.12.2):

$$\frac{d\mathscr{M}(r)}{dr} = 4\pi r^2 \varrho \ ,$$

hydrostatic equilibrium (4.12.1):

$$\frac{dP}{dr} = -\varrho \frac{G\mathscr{M}(r)}{r^2} \ ,$$

nuclear energy production (4.12.25):

$$\frac{dL(r)}{dr} = 4\pi r^2 \varrho \varepsilon \ , \tag{4.12.40}$$

energy transport by radiation (4.12.9):

$$\frac{dT}{dr} = -\frac{3\kappa\varrho}{4acT^3}\frac{L(r)}{4\pi r^2} \ ,$$

and by convection (4.12.11):

$$\frac{dT}{dr} = \left(1 - \frac{1}{\gamma}\right)\frac{T}{P}\frac{dP}{dr} \ .$$

Finally, our problem is completely determined by the *boundary conditions*:

a) At the center $r = 0$ of a star, we must of course have $\mathscr{M}(0) = 0$ and $L(0) = 0$.

b) At the surface, the equations for the stellar interior must make a smooth transition to those discussed previously for the stellar atmosphere. As long as we are interested *only* in the interior of the star, we shall frequently be able to apply the above equations out to $T \to 0$ or $T \to T_{eff}$ for $r = R$. Furthermore, $\mathscr{M}(R)$ is naturally $\mathscr{M}$, the total mass, and $L(R) = L$, the luminosity of the star.

We can readily gain an intuitive understanding of some *general results* from the theory of stellar structure; of course, they can also be derived by formal calculation.

We imagine the total mass $\mathscr{M}$ to be provided with given nuclear energy sources. We then suppose in a thought experiment that this object, initially not at all defined in terms of its spatial structure, consolidates itself into a star of mass $\mathscr{M}$ and luminosity L. It will adjust itself to a particular radius R, assuming that a stable configuration is possible. Since on the other hand, R and L are related to the effective temperature T_{eff} by $L = 4\pi R^2 \sigma T_{eff}^4$, i.e. luminosity = surface area × total radiation flux (4.4.26), the effective temperature T_{eff} is determined. There must then exist for stars of similar structure and composition (we should not forget that factor!), so-called homologous stars, a unique relationship between the *mass* $\mathscr{M}$, *the luminosity* L, *and the radius* R or the *effective temperature* T_{eff}:

$$\phi(\mathscr{M}, L, T_{eff}) = 0 \ . \tag{4.12.41}$$

Such a relation was discovered in 1924 by A.S. Eddington. According to his calculations, the dependence of the function on T_{eff} was so weak that he referred simply to a *mass-luminosity relation*. Its agreement with observations initially appeared to be rather good; later, several exceptions were found, which however in the light of the general theory are by no means unexpected.

If we also take into account the fact that theory relates the nuclear energy release ε to the state variables (e.g. T and ϱ) for a given element mixture, there is an additional relation between the three quantities $\mathscr{M}$, L, and T_{eff}. For stationary stars of similar structure, an equation of the form

$$\Phi(L, T_{eff}) = 0 \tag{4.12.42}$$

thus holds.

These stars must lie on a particular *line* in the Hertzsprung-Russell diagram or in the color-magnitude diagram. This result is sometimes called the *Russell-Vogt theorem*. We shall indeed meet up with such a line in the so-called *zero age main sequence*. On the other hand, the very existence of the red giants and supergiants tells us that at least one other parameter enters the picture. We shall see that it is the *chemical composition* of the stars, which changes along with their *ages* as a result of nuclear reactions. The interior structure of the stars therefore changes and our above considerations do not necessarily remain applicable.

Conversely, in addition to the mass of a star, its chemical composition must be known before we can calculate its structure from the fundamental equations. The composition is however in general known only for very young stars, which have not yet undergone any essential changes in their element mixtures by nuclear reactions since their formation from the interstellar medium. In practice, we must therefore follow the evolution of a star "from the beginning", i.e. starting from a homogeneous composition, in order to learn about its structure in the later stages of evolution.

A direct comparison of the theory of stellar structure and evolution with observations is, to be sure, not possible, due to the extremely long times required for the evolution as a rule; however, investigations of *star clusters* represent an important test for model calculations. In a star cluster, all members were formed nearly *at the same time* and with the *same chemical composition*; on the other hand, in a "snapshot view", they exhibit a wide range of different stages of evolution, as a result of their differing masses. We shall therefore treat the results of the theory of stellar evolution, whose basic ideas originally grew out of the study of star clusters, in Sect. 5.4, along with the color-magnitude diagrams of star clusters. In this chapter, we first restrict ourselves to a discussion of the internal structure of the Sun, and of the possibilities for testing this model of our nearest-neighboring star through observations of its emitted neutrino radiation. At the conclusion of the chapter, we provide some extensions of the basic physical considerations to the region of extremely high densities, which we shall need in dealing with the later evolutionary stages and the final phases of stellar evolution.

4.12.6 The Internal Structure of the Sun. Solar Neutrinos

In order to calculate a model for our Sun in its present state from the system of fundamental equations (4.12.40),

Table 4.12.1. Internal structure of the Sun. Age: $4.5\cdot10^9$ yr. Original element mixture $X:Y:Z = 0.73:0.25:0.015$

$r/R_\odot$	$\mathcal{M}(r)/\mathcal{M}_\odot$	$L(r)/L_\odot$	P [Pa]	ϱ [kg m^{-3}]	T [K]
0.0	0.00	0.00	$2.5\cdot10^{16}$	$1.6\cdot10^{5}$	$1.5\cdot10^{7}$
0.1	0.08	0.45	$1.4\cdot10^{16}$	$9.2\cdot10^{4}$	$1.3\cdot10^{7}$
0.15	0.20	0.79	$8.4\cdot10^{15}$	$5.8\cdot10^{4}$	$1.1\cdot10^{7}$
0.2	0.35	0.94	$4.5\cdot10^{15}$	$3.6\cdot10^{4}$	$9.3\cdot10^{6}$
0.3	0.62	1.00	$1.1\cdot10^{15}$	$1.2\cdot10^{4}$	$6.6\cdot10^{6}$
0.4	0.80	1.00	$2.6\cdot10^{14}$	$3.8\cdot10^{3}$	$5.0\cdot10^{6}$
0.5	0.89	1.00	$7.0\cdot10^{13}$	$1.4\cdot10^{3}$	$3.9\cdot10^{6}$
0.8	0.99	1.00	$1.9\cdot10^{12}$	$9.0\cdot10^{1}$	$1.7\cdot10^{6}$
1.0	1.00	1.00	(10^4)[a]	$(3\cdot10^{-4})$[a]	$(6\cdot10^{3})$[a]

[a] Typical values for the photosphere

we begin with a star of $1\,\mathcal{M}_\odot$ with a *homogeneous* element mixture: H:He:heavier elements (relative mass abundances) of

$$X:Y:Z = 0.73:0.25:0.015 \ , \qquad (4.12.43)$$

which corresponds to the atmosphere of the Sun and the population I stars as well as to the interstellar medium. Following the beginning of nuclear energy release by hydrogen fusion,[25] we construct a series of models representing the time evolution in such a way that after $4.5\cdot10^9$ yr, the age of the Sun and the Solar System (2.8.24), we find the presently observed luminosity $L_\odot$ and effective temperature $T_{\mathrm{eff},\odot}$ at one solar radius, $R_\odot$. We thus obtain the *model for the internal structure of the Sun* as given in Table 4.12.1.

Half of the mass is found to be within about $0.25\,R_\odot$. In most of the interior, energy transport occurs via radiation and only outside $r \geq 0.8\,R_\odot$ (out to the lowest layers of the photosphere), i.e. for only about 1% of the overall mass, is convection important. From hydrogen burning during $4.5\cdot10^9$ yr, about half of the hydrogen originally present at the center of the Sun has been converted to helium; here, we have $X:Y:Z = 0.36:0.63:0.015$. Outside about $0.2\,R_\odot$ we still find the original chemical composition. The energy release in the central regions is mainly due to the pp chain (4.12.17); at $r = 0.2\,R_\odot$, $L(r)$ has already reached 95% of the luminosity $L_\odot$.

Several percent of the energy released in the Sun's interior by nuclear reactions escape as *neutrino radiation* di-

[25] The preceding, relatively short evolutionary stage after formation from the interstellar medium [of the order of the Helmholtz-Kelvin time $t_{\mathrm{HK}} \simeq 3\cdot10^7$ yr (4.12.33)] has only a slight influence on the present structure of the Sun.

rectly into space. The extraordinarily small interaction cross-sections for neutrinos with all types of matter allows them to pass through a star, indeed even through the whole universe, practically without any collisions. *Neutrino astronomy* can thus give us direct information about the energy-releasing core of the *Sun*, while the flux of neutrino radiation from distant stars is far below the present limit of detection. An exception was the explosion of the supernova SN 1987A in the "neighboring" Large Magellanic Cloud (Sect. 4.11.7).

The *detection* of neutrinos, which is very difficult owing to their weak interactions with matter, can be carried out via their nuclear reaction with ^{37}Cl, which makes up about 24% of naturally-occurring chlorine:

$$^{37}\mathrm{Cl}(\nu, e^-)^{37}\mathrm{Ar} \; . \tag{4.12.44}$$

The ^{37}Ar which is produced in this reaction decays with a half-life of 35.0 days back to ^{37}Cl; the Auger electrons emitted as a result of this decay are detected and counted. Using this radiochemical method, R. Davis Jr. has carried out measurements intended to detect *solar neutrinos* since 1964. In order to avoid disturbances due to cosmic radiation as much as possible, the apparatus is located 1.5 km deep in the Homestake gold mine in South Dakota. The "receiver" is a tank containing about 610 tons of tetrachloroethylene, C_2Cl_4, which is normally used as a cleaning fluid. After each 2 to 3 half-lives, the noble gas ^{37}Ar is washed out with helium and its decay electrons are counted. The currently accepted result (average of the observations from 1970–1985) is

$$(2.1 \pm 0.3) \cdot 10^{-36} \text{ neutrino processes s}^{-1} \text{ per } ^{37}\text{Cl atom} \tag{4.12.45}$$

or (2.1 ± 0.3) SNU (solar neutrino units), corresponding to a production rate of about 0.4 ^{37}Ar atoms per day.

According to *theory*, i.e. according to the standard model of the Sun (Table 4.12.1), the *neutrino spectrum* shown in Fig. 4.12.4 is to be expected; in it, the neutrinos from the reaction $^1\mathrm{H}(p, e^+ \nu)^2\mathrm{D}$ of the pp chain (4.12.17) are strongly predominant, with a continuous energy distribution below 0.42 MeV. However, since the ^{37}Cl apparatus registers only neutrino energies above 0.81 MeV, the threshold energy for the reaction (4.12.44), the experiment of Davis essentially measures only the continuous neutrino emission from the β^+ decay of ^{8}Be in the PP III chain, which plays only a subordinate role in energy release by hydrogen fusion. The calculated neutrino production corrected for the detection sensitivity of the apparatus is, however, *3 times larger* than the measured value (4.12.45). The reason for this discrepancy is not yet understood. The neutrino flux from the decay of ^{8}B depends very sensitively on the temperature in the core region of the Sun, so that it is presumed among other explanations that the theory of convection and mixing in the interior of the Sun is still insufficiently precise. Another possibility is that the *electron* neutrinos produced in the Sun are in part converted into μ neutrinos (neutrettos) and τ neutrinos (Sect. 5.9.5) on their way from the Sun, and these would *not* react according to (4.12.44). Such *neutrino oscillations*, predicted by elementary particle theory, would take place mostly within the Sun via interactions with the electrons there.

It is expected that new light will be shed on the "neutrino problem" by experiments begun in 1989, in which the detection of neutrinos is accomplished through the capture reaction $^{71}\mathrm{Ga}(\nu, e^-)^{71}\mathrm{Ge}$, whose threshold energy is low enough (0.23 MeV) that the main contribution to the measured neutrino flux comes from the fusion reaction $^1\mathrm{H}(p, e^+ \nu)^2\mathrm{D}$. The ^{71}Ge produced in the detector decays back to ^{71}Ga with a half-life of 11.4 d; the accompanying Auger electrons are counted, in a similar

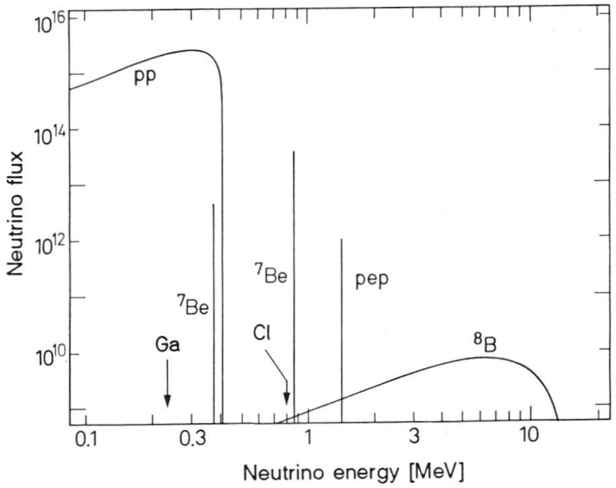

Fig. 4.12.4. The theoretical neutrino spectrum of the Sun. The neutrino flux at the location of the Earth from the reactions of the pp chain (4.12.17) and the "pep" reaction, $p + e^- + p \rightarrow {}^2\mathrm{D} + \nu + 1.44$ MeV, is shown. The contributions from ^{13}N and ^{15}O from the CNO cycle are not included here. The continuous neutrino radiation is given in [neutrinos $m^{-2}\,s^{-1}\,MeV^{-1}$], the monoenergetic radiation in [neutrinos $m^{-2}\,s^{-1}$]. The thresholds for detection by ^{37}Cl and ^{71}Ga reactions are indicated by arrows. From T. Kirsten (1983) (with the kind permission of the VCH Verlagsgesellschaft, Weinheim)

manner to the chlorine detector. The GALLEX experiment, carried out by a mainly European cooperation under the Gran Sasso in Italy, employs 30 tons of gallium in the form of $GaCl_3$ solution. The Soviet-American gallium experiment SAGE, in a mine near Baksan in the Soviet Caucasus, will initially use 30 tons of liquid metallic gallium (60 tons in the final phase).

For the detection of *high-energy* neutrinos (≥ 8 MeV), elastic scattering from electrons, $\nu + e^- \rightarrow \nu + e^-$, can be used; it gives information about the energy and the *incident direction* of the neutrinos. For example, the detector in the Kamioka mine in Japan consists of a tank containing 3000 tons of water, which is surrounded by large-area photomultiplier tubes to register the Cherenkov radiation from the electrons (see also Sect. 3.4.1). This type of detector is also employed to search for a possible decay of the proton, a fundamental experiment for both elementary particle physics and for cosmology (Sect. 5.9.5). Neutrino radiation from the supernova 1987A in the Large Magellanic Cloud could be observed with the Kamiokande detector; since then, it has also been used to detect solar neutrinos.

4.12.7 The Equation of State of Degenerate Matter. The Structure of White Dwarfs

The average densities of white dwarf stars (Sect. 4.5), whose radii are comparable to the that of the Earth, while their masses are of the order of the Sun's mass, are in the range 10^8 to $10^9\,\mathrm{kg\,m^{-3}}$; they are thus about 10^6 times greater than the mean density of the Sun. At such densities, the equation of state for an ideal gas (4.12.3) is no longer applicable. As R.H. Fowler recognized in 1926, the matter in these stars (more precisely, the electron component of the matter) is degenerate in the sense of *Fermi-Dirac statistics*, with the exception of a quite thin atmospheric layer. This means the following: in classical or Maxwell-Boltzmann statistics, the velocity distribution of electrons or the equations of state of gases, for example, are derived by determining the distribution of the particle states in *phase space* according to the rules of probability theory. A 6-dimensional volume element $\Delta\Omega$ in phase space consists of the usual volume element $\Delta V = \Delta x \Delta y \Delta z$ in position space, multiplied by a volume element in momentum space, $\Delta p_x \Delta p_y \Delta p_z$. The applicability of this method is however limited for an electron gas by the *Pauli principle*, which requires that *one* quantum state or *one* quantum cell of phase-space volume $\Delta\Omega = h^3$ (h: Planck's constant) can be occupied by at most *one* electron of each spin direction, i.e. all together by 2 electrons. This fact is taken into account by Fermi-Dirac statistics. In the interior of a white dwarf star, the matter is so strongly compressed due to the enormous pressures and relatively low temperatures that all phase-space cells h^3 are completely occupied up to a particular limiting energy E_0 or the corresponding momentum p_0, i.e. each one contains two electrons. This state is referred to as *complete* degeneracy of the Fermi gas.

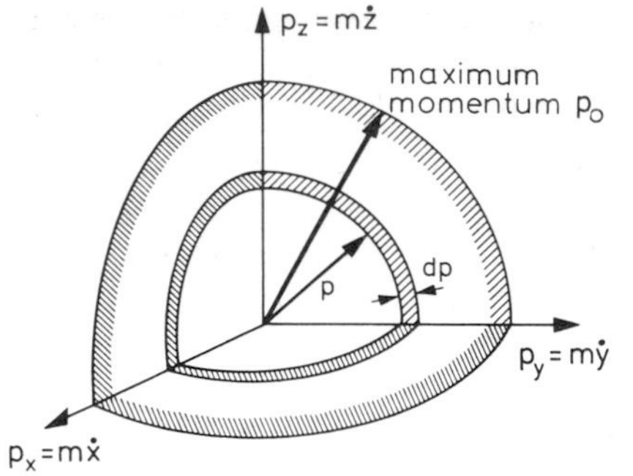

Fig. 4.12.5. A degenerate electron gas (Fermi-Dirac statistics). In momentum space, of which only one octant is shown here, the electrons fill uniformly a sphere of radius p_0, the so-called Fermi sphere

The *equation of state* of the degenerate electron gas is readily derived (the associated proton gas becomes degenerate only at much higher densities; its pressure can be neglected):

In a volume V, with a particle density N, there are a total of $V \cdot N$ electrons. In momentum space, $p_x p_y p_z$ (Fig. 4.12.5), these electrons fill a homogeneous sphere up to a maximum limiting momentum p_0, corresponding to the maximum energy (the Fermi energy) E_0. In phase space, we thus have a volume $V(4/3)\pi p_0^3$, and with 2 electrons per phase-space cell of volume h^3, we obtain the relation:

$$NV = \frac{2}{h^3} V \frac{4\pi}{3} p_0^3 \quad \text{or} \quad p_0 = \left(\frac{3h^3}{8\pi}\right)^{1/3} N^{1/3} \ . \tag{4.12.46}$$

For a *nonrelativistic* electron gas, i.e. for particle velocities $v \ll c$ or energies $E \ll mc^2$ (m = rest mass of the electron), the relation $E = p^2/2m$ holds between the kinetic energy $E = mv^2/2$ and the momentum $p = mv$; thus the Fermi energy is given by $E_0 = p_0^2/2m$. The pressure P is, from the well-known relation of kinetic gas theory, as for an ideal gas,

$$P = \tfrac{2}{3} N\bar{E} \ , \tag{4.12.47}$$

where $\bar{E}$ is the mean energy per electron. The relationship between $\bar{E}$ and E_0 can be calculated by referring to Fig. 4.12.5. We find namely:

$$\bar{E} = \int_0^{p_0} E 4\pi p^2 dp \Big/ \int_0^{p_0} 4\pi p^2 dp$$

$$= \frac{3}{5} \frac{p_0^2}{2m} = \frac{3}{5} E_0 \qquad (4.12.48)$$

and we obtain the equation of state of a completely degenerate electron gas:

$$P = \frac{1}{5m} \left(\frac{3h^3}{8\pi} \right)^{2/3} N^{5/3} \ . \qquad (4.12.49)$$

The temperature does not enter this equation at all; this is a characteristic of complete degeneracy. It can readily be verified that the pressure of the equally dense *proton* gas is much smaller than that of the degenerate electron gas.

The relation between N and the mass density ϱ is most simply expressed in terms of the mass μ_e (in atomic mass units) which corresponds to *one* electron. We then have:

$$\varrho = \mu_e m_H N \ , \qquad (4.12.50)$$

which shows that for completely ionized matter, in the case of hydrogen, $\mu_e = 1$, and for helium and heavier elements, $\mu_e \simeq 2$.

An estimate of the density of a white dwarf, $\varrho = 10^9 \text{ kg m}^{-3}$ (with $\mu_e = 2$), can clarify the situation: according to (4.12.50), we then have $N \simeq 3 \cdot 10^{35}$ electrons per m^3, and, from (4.12.49), a pressure of the degenerate gas equal to $P \simeq 3 \cdot 10^{21}$ Pa, i.e. about 10^5 times higher than in the center of the Sun. An ideal gas at this pressure would have to be at a temperature of about 10^9 K.

The equation of state $P \propto \varrho^{5/3}$ can be used together with our earlier estimate of the pressure in the interior of a star with mass $\mathcal{M}$ and radius R from (4.12.1), i.e. $P \propto \varrho G \mathcal{M}/R$, and the trivial relation $\varrho \propto \mathcal{M}/R^3$, to obtain an approximate[26] *mass-radius relation* for white dwarf stars, namely:

$$R \propto \mathcal{M}^{-1/3} \ . \qquad (4.12.51)$$

That is, with increasing mass, the radius of the star decreases.

We can now also write down the luminosity $L = 4\pi R^2 \sigma T_{\text{eff}}^4$. White dwarf stars of mass $\mathcal{M}$ should thus be ordered according to their mass in the (theoretical) color-magnitude diagram on lines:

$$L \propto \mathcal{M}^{-2/3} T_{\text{eff}}^4 \ . \qquad (4.12.52)$$

Comparison with observations yields the initially surprising result that nearly *all* white dwarfs have masses between 0.5 and $0.6 \mathcal{M}_\odot$.

All of our estimates however still require an extension and, mostly in their finer details, some corrections. With increasing density ϱ, the energy of the electrons at first increases $\propto \varrho^{2/3}$, until at $E \geq mc^2$ (i.e. in more massive and therefore more dense stars), the *relativistic mass variation* of the electrons becomes noticeable. In the case of *fully relativistic* degeneracy ($\bar{E} \gg mc^2$), $E = pc$ and $P = \frac{1}{3} N \cdot \bar{E}$. A repetition of our above calculation, taking special relativity into account, yields the equation of state:

$$P = \frac{c}{4} \left(\frac{3h^3}{8\pi} \right)^{1/3} N^{4/3} \ , \qquad (4.12.53)$$

that is, we now have $P \propto \varrho^{4/3}$, corresponding to a weaker compressibility of the stellar matter.

As was recognized by S. Chandrasekhar in 1931, the radius of a white dwarf star tends with increasing relativistic degeneracy to the limiting value $R \to 0$ even at a finite *limiting mass*:

$$\mathcal{M}_{Ch} \simeq \frac{5.8}{\mu_e^2} [\mathcal{M}_\odot] \ . \qquad (4.12.54)$$

Since $\mu_e \simeq 2$ as a result of the evolutionary history of the white dwarfs, these stars are stable or able to exist only for $\mathcal{M} \leq 1.44 \ \mathcal{M}_\odot$. The appearance of Chandrasekhar's limiting mass is caused by the violation of the stability condition for hydrostatic equilibrium (4.12.38) as a result of the pressure dependence $P \propto \varrho^{4/3}$ of the relativistically degenerate electron gas.

The formation and evolution of white dwarfs will be discussed further in Sect. 5.4.5. We now ask one more question: what happens to stars, or to the central portions of stars, whose mass at the end of their nuclear evolution is greater than Chandrasekhar's limiting mass of about $1.4 \mathcal{M}_\odot$? The only way for them to form a stable star is through a *collapse* to much higher densities. The

[26] Here, we are neglecting the fact that the outer part of the star is not degenerate and that the interior is relativistically degenerate.

electrons and protons will finally be pressed together so strongly that, in a reversal of β-decay, they fuse to form neutrons and thus give rise to a neutron star.

4.12.8 Neutron Stars

In the year of the discovery of the neutron, 1932, L. Landau already discussed the possibility of the existence of stable neutron stars. W. Baade and F. Zwicky predicted in 1934 that neutron stars are formed in supernova explosions (Sect. 4.11.7), in which the energy for the explosion is provided by the gravitational energy released (Sect. 4.12.4) in the collapse to higher densities. The first models for stars consisting of degenerate neutron matter were calculated by J.R. Oppenheimer and G.M. Volkoff in 1939; neutron stars were finally discovered in 1967 as *pulsars* (Sect. 4.11.7) by A. Hewish and his coworkers.

Neutron matter has great similarities to the *nuclear matter* of heavy atoms, since the interaction forces between protons and neutrons are equally strong (aside from the electrostatic Coulomb force, which is unimportant here). On this basis, we first make an elementary estimate, by considering the neutron star as a giant atomic nucleus, so to speak:

A nucleus of e.g. ^{56}Fe has a mass $\mathcal{M}_{Fe} = 56 \cdot 1.67 \cdot 10^{-27} = 9.4 \cdot 10^{-26}$ kg and a radius of about $5.6 \cdot 10^{-15}$ m, and thus a mean density of $\varrho \simeq 10^{17}$ kg m^{-3}. If a star of, for example, $1\,\mathcal{M}_\odot$ is compressed to have a density equal to that of nuclear matter, it must shrink its radius to about $5.6 \cdot 10^{-15}\,(\mathcal{M}_\odot/\mathcal{M}_{Fe})^{1/3}$ m or about 16 km! The comparison between atomic nuclei and neutron stars should, however, not be taken too literally, since the latter cannot have an electric charge. While roughly equal numbers of protons and neutrons are present in an atomic nucleus, neutrons are strongly predominant in the star, where each proton has been "squeezed together" with an electron to form a neutron: $p^+ + e^- \to n + \nu$. A free neutron decays with a half-life of 617 s ($n \to p^+ + e^- + \bar{\nu}$), but in the star, this decay is prevented by the high electron density. If the Fermi energy E_0 of the electrons namely exceeds 0.78 MeV, the equivalent of the mass difference between neutron and proton (1.29 MeV) minus that of the electron's rest mass (0.51 MeV), then "the decay electrons would find no more room in phase space". This value of $E_0 = pc$ is attained according to (4.12.53), and using $P = \frac{1}{3} NE$, at N about $3 \cdot 10^{36}$ electrons m^{-3}, or a mass density of $1 \cdot 10^{10}$ kg m^{-3}, i.e. somewhat above the density characteristic of white dwarfs. (Above 0.51 MeV, the electrons are relativistically degenerate.)

The transition from the low densities at the surfaces of neutron stars to a neutron fluid with $\gtrsim 10^{17}$ kg m^{-3} in their interiors poses a number of very interesting problems for the theoretician. For a neutron star with a radius $R \simeq 16$ km, the density initially increases in the *outer crust* from about 10^7 kg m^{-3} to about $4 \cdot 10^{14}$ kg m^{-3} at 1 km depth. Here, the matter, similar to the interiors of white dwarfs, consists of a degenerate electron gas and atomic nuclei, which form a crystal lattice. While in the outer layers ^{56}Fe nuclei are most abundant, further inside we find, with increasing density, more and more neutron-rich nuclei. Above $4 \cdot 10^{14}$ kg m^{-3}, in the *inner crust*, the nuclei begin gradually to dissociate and free neutrons are present. Finally, above about $2 \cdot 10^{17}$ kg m^{-3}, for $R \lesssim 11$ km, the strongly incompressible *neutron fluid* is reached; it still contains a small percentage of protons and electrons. The central density of the neutron star is in the range $\gtrsim 4 \cdot 10^{17}$ kg m^{-3}. In this region, the equation of state of matter is only imprecisely known; hyperons, pions, or quarks may be present.

This neutron matter exhibits interesting physical properties: the neutrons can to some extent interact pairwise with each other and form a *superfluid* constituent, while the proton constituent is *superconducting*. We cannot go into the details here of the influence of the superfluidity on the rotation of neutron stars in connection with their extremely strong magnetic fields.

Following its origins in a supernova explosion (Sect. 5.4.5), the interior of a neutron star is initially very hot ($\simeq 10^7$ K), and cools in the course of several thousand years to a few 10^6 K. During this phase, there is a possibility of observing the thermal X-radiation in favorable cases.

The *gravitational energy* $G\mathcal{M}^2/R$ (4.12.30) of a neutron star of mass $= 1\,\mathcal{M}_\odot$ and radius $R = 16$ km is about $2 \cdot 10^{46}$ J, i.e. about one-tenth of the energy equivalent of its rest mass, $\mathcal{M}c^2$. In such strong gravitational fields, Newton's theory of gravity is no longer sufficient for the calculation of its structure; instead, Einstein's general relativity theory (see following section) must be applied. The deviations from Newton's theory are of the order of $G\mathcal{M}^2/R)/\mathcal{M}c^2 = G\mathcal{M}/Rc^2 \simeq 0.1$.

In particular, the (gravitational) mass $\mathcal{M}$ which an external observer would derive e.g. from the motion of a neutron star in a binary star system, is of the order of 10% *smaller* than the mass which is required to form the star from an originally extended configuration. The difference (mass defect) is for the most part due to the negative gravitational binding energy.

As was first shown by J.R. Oppenheimer and G.M. Volkoff in 1939, a *limiting mass* exists for neutron stars, just as for the white dwarfs. Owing to the uncertainties in the equation of state of material at densities above those of nuclear matter, the limiting mass is not very accurately known; it probably lies near $1.8\mathcal{M}_\odot$. If this mass is exceeded, a stable final configuration is no longer possible.

4.12.9 Strong Gravitational Fields: The General Theory of Relativity

If massive bodies or particles move at velocities near to that of light, Newtonian mechanics and gravitational theory (Sect. 2.6) can no longer be applied. We first make a rough estimate of the strength which the gravitational field or the gravitational potential $G\mathcal{M}/R$ of a sphere of mass $\mathcal{M}$ and radius R must have, in order that the escape velocity v from its field region becomes of the order of the velocity of light, c. If we naively set $v = c$ in the relation $v^2 = 2G\mathcal{M}/R$ (2.6.42), we obtain:

$$\frac{2G\mathcal{M}}{Rc^2} = 1 \ . \qquad (4.12.55)$$

Thus, if the gravitational potential $G\mathcal{M}/R$ becomes comparable to c^2 or the gravitational energy $G\mathcal{M}^2/R$ becomes comparable to the relativistic rest energy $\mathcal{M}c^2$, then a *relativistic theory* of gravitation must replace the Newtonian theory: this is the *General Theory of Relativity* formulated by A. Einstein in 1916.

While on the surface of the *Sun*, $2G\mathcal{M}_\odot/R_\odot c^2 \simeq 4\cdot 10^{-6}$, so that we can expect only very small deviations from Newtonian theory in the Solar System and for ordinary stars, we have learned that *neutron stars* (Sect. 4.12.8) represent a mass concentration $G\mathcal{M}/Rc^2$ of the order of 0.1. In the calculation of their structure, especially of their limiting mass, the effects of general relativity are important. The strong gravitational fields in the sense of (4.12.55), characteristic of *relativistic astrophysics*, and the highly energetic particles and radiation fields which occur in connection with them, will be found again for one thing in galactic centers (Sect. 5.6). For another, cosmology, which treats the structures and evolution of the universe as a whole (Sect. 5.9), is based on general relativity as the theory of gravitation.

This section is intended merely to give an impression of the structure of general relativity theory; it cannot take the place of a proper introductory study of relativity theory.

The precursor of general relativity was the *Theory of Special Relativity* developed by Einstein in 1905. This starts from the result of the Michelson-Morley experiment by requiring that the propagation of a light wave in different coordinate systems, which may be in (nonaccelerated) translational motion relative to one another, has the same form, i.e. it obeys the same equation. If the light moves along a path element $dr = \sqrt{dx^2+dy^2+dz^2}$ in a time element dt with the vacuum light velocity c, then

$$dx^2+dy^2+dz^2-c^2dt^2 = 0 \ . \qquad (4.12.56)$$

This quantity can be considered as a line element ds^2 in a four-dimensional space with three spacelike coordinates x, y, and z and the fourth coordinate ct [the light path; or, following H. Poincaré and H. Minkowski, still more "intuitively", $\mathrm{i}ct$ (with $\mathrm{i} = \sqrt{-1}$)]. More generally, we can require that in the transformation from one Cartesian system to another one, which is in a state of nonaccelerated motion relative to the first, the four-dimensional line element

$$ds^2 = dx^2+dy^2+dz^2-c^2dt^2$$

or

$$ds^2 = dx^2+dy^2+dz^2+d(\mathrm{i}ct)^2 \qquad (4.12.57)$$

must be conserved. Such a transformation can clearly not be confined to the spacelike coordinates $x, y, z \to x', y', z'$, but rather must also include the time, $t \to t'$. This *Lorentz transformation* is, as can be easily read from (4.12.57), nothing other than a rotation in a four-dimensional space, x, y, z, $\mathrm{i}ct$, in which by definition the element of length ds is *invariant*. The essential advantage of special relativity as compared to Newtonian mechanics is based upon the fact that it considers the particularly important role of the vacuum velocity of light c from the beginning. Correspondingly, its consistent application leads to the recognition of the fact that no motion of matter and no signal of any kind can be faster than the velocity $c = 3\cdot 10^8$ m s^{-1}. Considering physics as a whole, special relativity demands the invariance (independence) of all natural laws with respect to Lorentz transformations or (physically speaking) translational motions.

Should it not be possible to formulate the natural laws in such a way that they are invariant with respect to even *arbitrary* transformations? In 1916, Einstein had the ingenious idea of combining this appealing requirement with the theory of gravitation in his *Theory of General Relativity*. The identity of *gravitational* and *inertial* mass, independent of the type of matter, was so to speak a "miracle" in the framework of Newtonian mechanics. Newton himself, then Bessel and later Eötvös had verified it experimentally with great precision; but what did it mean? Einstein raised the experimental observation that in a freely falling coordinate system (elevator), the gravitational force (mg) appears to be compensated by the inertial force ($m\ddot{z}$) to the status of a basic postulate. This means that gravitational force and inertial force are, in the end, *the same*. These forces can be transformed away by local transformations in a four-dimensional Cartesian coordinate system with the "Euclidian" Minkowskian metric (4.12.57). Conversely, as is found, the coefficients g_{ik} of the Riemannian metric

$$ds^2 = \sum_{i,k} g_{ik} dx^i dx^k \qquad (4.12.58)$$

of an arbitrary coordinate system, and its spatial connectivity, determine the gravitational and inertial fields acting there.

This connection is formulated in *Einstein's field equations*

$$G_{ik} = -\frac{8\pi G}{c^4} T_{ik} , \qquad (4.12.59\text{a})$$

in which $G = 6.67 \cdot 10^{-11}\,\mathrm{m^3\,s^{-2}\,kg^{-1}}$ is the usual Newtonian gravitational constant, G_{ik} is the Einstein tensor, and T_{ik} is the energy-momentum tensor ($i = 1, 2, 3, 4$). G_{ik} contains the geometrical properties of the four-dimensional space and depends only on the metric g_{ik} and its first and second derivatives ($\partial g_{ik}/\partial x^i, \partial^2 g_{ik}/\partial x^l \partial x^m$), while T_{ik} contains the energy and momentum and thus affects the properties of matter, the electromagnetic field, etc. Since the tensors G_{ik} and T_{ik} are symmetric ($G_{ik} = G_{ki}$, $T_{ik} = T_{ki}$), (4.12.59a) represents a system of 10 coupled, nonlinear differential equations.

For applications in cosmology (Sect. 5.9.2), Einstein later added another term to the field equations, containing the *cosmological constant* Λ, which becomes important only at "cosmological distances":

$$G_{ik} + \Lambda g_{ik} = -\frac{8\pi G}{c^4} T_{ik} . \qquad (4.12.59\text{b})$$

The field equations (4.12.59a and b) of course contain the Newtonian theory of gravitation as a limiting case for weak gravitational fields.

The motion of a point mass under the influence of gravitation alone follows *geodesic* (i.e. shortest) *curves*, in a generalization of the Galilean law of inertia. The propagation of light is along the so-called *null geodesic*

$$ds^2 = 0 . \qquad (4.12.60)$$

In the simple spherically symmetric case, K. Schwarzschild showed in 1916 that the gravitational field of a mass $\mathcal{M}$ in external space ($T_{ik} = 0$), for example the Sun, can be represented using the metric:

$$ds^2 = \frac{dr^2}{1-(R_s/r)} + r^2(d\theta^2 + \sin^2\theta\, d\phi^2) - [1-(R_s/r)]\,c^2 dt^2 , \qquad (4.12.61)$$

where r, θ, and ϕ are the spatial spherical coordinates and t is the time. The theory of planetary motions can thus be based on this metric. The constant of integration

$$R_s = 2G\mathcal{M}/c^2 \qquad (4.12.62)$$

is called the *Schwarzschild radius* or *gravitational radius* of the mass $\mathcal{M}$; for $\mathcal{M} = 1\,\mathcal{M}_\odot$, for example, $R_s = 2.9$ km. Its value determines the deviations from the Euclidian metric of empty space [cf. (4.12.55)].

In 1963, R. P. Kerr developed a metric which takes into account not only the mass $\mathcal{M}$, but also its *angular momentum* (and electric charge).

Within the Solar System, the predictions of general relativity differ only slightly from those of Newtonian mechanics and gravitational theory. The observations which can be used to distinguish between the two require great metrological precision. From an experimental viewpoint, the situation is roughly as follows:

1) In his classical experiment to test the *equality of gravitational and inertial mass*, which makes use of the gravitational force of the Earth and the centrifugal force of its rotation, R. v. Eötvös attained a precision of 10^{-9} as early as 1922. V. B. Braginsky and V. N. Rudenko improved this to 10^{-12} by 1970.

2) *Deflection of Light*: the gravitational field of the Sun should deflect the light from a star which passes by the Sun (during a total eclipse), at a distance of R solar radii from its midpoint, by $1.75/R$ seconds of arc. The extremely difficult observations yielded values for the constant with considerable variation in the range 1.4″ to 2.7″.

Completely new possibilities have opened up since 1969 with the construction of *very-long-baseline interferometers* in the centimeter-wave region; they have an angular measurement precision of better than 0.001″. Fortunately, the two intense quasars 3C273 and 3C279 are completely or partially occulted by the Sun every year, so that measurements of their distance from the Sun's midpoint can be used to determine the Einsteinian light deflection. The theoretical value has been verified to within the limits of experimental error of less than ±1%.

3) *Delay of Radar Signals*: I. Shapiro in 1964 discovered that, according to general relativity, a radar signal which passes near the Sun should be delayed by a time of the order of $2 \cdot 10^{-4}$ s. Reflections of such radar signals could initially be obtained from Mercury and Venus (passive) and from the space probes Mariner 6 and 7 (active; i.e. the arrival of the signal triggers a transmitter). The predictions of theory were verified to within a few percent. Later, an agreement of ±0.1% was obtained using transmitters on Mars placed there by the Viking probes.

4) *Red Shift*: a light quantum $h\nu$, which passes through a gravitational potential difference, e.g. from the Sun to the Earth, $G\mathcal{M}_\odot/R_\odot$, should exhibit a red shift relative to a laboratory source; it is given by:

$$-\Delta(h\nu) = \frac{G\mathcal{M}_\odot}{R_\odot}\frac{h\nu}{c^2} , \tag{4.12.63}$$

which corresponds formally to a Doppler effect of $c\Delta\nu/\nu = 0.64 \text{ km s}^{-1}$. The measurements showed agreement to within about 1%, but the separation of the relativistic red shift from the Doppler effect due to flow motions in the solar atmosphere is very uncertain. A considerably higher precision is possible using the extremely sharp γ-ray lines of the *Mössbauer effect* in the gravitational field of the Earth. With the "recoil-free" γ line of ^{57}Fe over a height of 22.6 m, R.V. Pound and G.A. Rebka in 1960 were able to verify the predicted frequency shift of only $\Delta\nu/\nu = 2.5 \cdot 10^{-15}$ within ±10%. In 1965, R.V. Pound and J.L. Snider attained an accuracy of ±1% using an improved apparatus.

5) *The Perihelion Precession of the Planets*: the small rotation of the perihelion of Mercury calculated by Einstein agrees well with the old calculations of Leverrier. Later, the electronic data analysis of a large mass of observational material by G.H. Clemence and R.L. Duncombe (about 1956) yielded the following values for the rotation of the perihelion of the three innermost planets per 100 yr:

	Mercury	Venus	Earth
Observed:	43.11″±0.45″	8.4″±4.8″	5.0″±1.2″
Calculated:	43.03″	8.6″	3.8″

(4.12.64)

The determination of the rotation of the perihelion of Mercury using radar observations of its orbit by I. Shapiro et al. (1972) also shows good agreement with the theoretical value.

Since the rotation of the perihelion is a second-order effect, while the deflection of light, the delay of radar signals, and the red shift are all first-order effects, we may speak of an altogether excellent verification of the theory of general relativity. The newer tests have eliminated practically all competing theories of gravitation.

The Schwarzschild metric (4.12.61), besides the very small effects described above, predicts also a much more spectacular possibility, whose significance will be treated together with stellar and galactic evolution: the existence of *black holes*.

If, namely, the radius of our mass $\mathcal{M}$ becomes smaller than the gravitational radius R_s (4.12.62), the coefficients of the spacelike and temporal elements dr^2 and $-c^2dt^2$ in (4.12.61) reverse their signs within the sphere $r \leq R_s$. The result, which we cannot derive in detail here, is that neither matter nor radiation quanta, i.e. no signals at all, can escape from the region $r \leq R_s$ to the outside world ($r > R_s$). A black hole, more precisely a Schwarzschild-type black hole, is observable *only* through its gravitational field.

This result, as shown by J.D. Bekenstein, S.W. Hawking and others in 1972/75, must be modified when gravitational theory is combined with thermodynamics and quantum mechanics: a black hole of mass $\mathcal{M}$ has an associated entropy or temperature

$$T = \frac{hc^3}{16\pi^2 kG\mathcal{M}} \simeq 10^{-7}\frac{\mathcal{M}_\odot}{\mathcal{M}} \text{ [K]} . \tag{4.12.65}$$

Corresponding to this temperature, which is very low for stellar or larger masses, a black hole "evaporates" exceedingly slowly through emission of particles (with a thermal energy spectrum) owing to quantum-mechanical effects.

If we write (4.12.62) in the form $G\mathcal{M}^2/R_s = \frac{1}{2}\mathcal{M}c^2$, it can be seen that the gravitational collapse of a mass $\mathcal{M}$ to a black hole or a related configuration offers the only possibility for releasing a major portion of its relativistic rest energy $\mathcal{M}c^2$. Even with the highest-yield nuclear process, $4\,^1H \rightarrow {}^4He$, according to (4.12.17) only 0.7% of this energy would be available.

To conclude this section, we discuss briefly the interesting phenomenon of *gravitation waves*. Soon after his discovery of the gravitational equations, A. Einstein derived from them the result that a system of moving masses emits gravitation waves, which propagate with the velocity *c*. If a (polarized) gravitation wave interacts with matter, the latter will be compressed in a direction perpendicular to the wave propagation vector (at a particular time), and expanded in the direction perpendicular to both; after a half-oscillation, the reverse deformation occurs. The oscillation pattern for the "other" polarization is found by performing a 45° rotation about the propagation vector.

Since about 1958, J. Weber has carried out intensive efforts to *detect gravitation waves* from outer space. As detectors, he used massive (≤ 3 tons) aluminum cylinders, which are suspended horizontally and protected as much as possible from (terrestrial) vibrations and other disturbances. The deformation is detected by piezoelectric crystals mounted on a circumference of the cylinder, and amplified electronically. In this manner, deformations of the order of 10^{-16} m at the resonance frequency of the cylinder can be measured. The reality of the short pulses detected by Weber in coincidence from two detectors placed about 1000 km apart, and attributed to gravitation waves, has been questioned, since they could not be verified in independent series of observations. The detection of gravitation waves from space will probably be attained only by future, considerably more sensitive detectors.

An *indirect detection method* for the emission of gravitation waves is obtained from the observation of the "binary pulsar" PSR 1913+16, which belongs to an unusual, very close binary star system (Sect. 4.6.6). The observed small decrease in the orbital period corresponds to a decrease in the energy of the system which agrees well with the theoretical value for the gravitation-wave emission of the moving masses according to the theory of general relativity.

5. Stellar Systems: The Milky Way and Other Galaxies. Cosmogony and Cosmology

In the second and third decades of this century, a new development in astronomy took place which is hardly less important than the nearly simultaneous formulations of general relativity and quantum mechanics: the universe as a whole, the cosmos with its spatial structure and its temporal evolution, became the object of exact scientific investigation. In this chapter, we continue the direction begun in Sect. 4.1, giving in Sect. 5.1 an overview of the development of the most important ideas of this field in historical perspective. This short summary is intended to make it easier for the reader to understand the detailed treatments in the later sections, which are ordered according to their objective interrelations rather than historically.

We begin in Sect. 5.2 with our Milky Way galaxy, describing its structure and dynamics, and continue the discussion of interstellar matter (including cosmic radiation) and its distribution in the Milky Way in Sect. 5.3. Then, in Sect. 5.4, we turn to the simplest stellar systems (after binary and multiple stars), the star clusters. Following a description of their color-magnitude diagrams, we discuss stellar evolution in connection with star clusters, building on the results of Sect. 4.12.

We then leave our own galaxy and treat the properties of the numerous other galaxies: Sect. 5.5 is devoted to the "normal" galaxies and infrared galaxies, and Sect. 5.6 to the radio galaxies, the Seyfert galaxies, quasars, and the phenomena of activity in galactic centers. The fundamentals of the genesis and the dynamic and chemical evolution of galaxies are presented in Sect. 5.8. Before that, in Sect. 5.7, we consider the fact that galaxies do not occur alone, but rather form groups and clusters; these clusters of galaxies are, in turn, ordered into superclusters.

Finally, in Sect. 5.9, we take up cosmology: the structure and evolution of the universe as a whole.

5.1 The Advance into the Universe

A Historical Introduction to the Astronomy of the 20th Century

Around the turn of the century, H. v. Seeliger (1849–1924), J. Kapteyn (1851–1922), and others attempted (following the "star gauging" of W. and J. Herschel) to apply statistical methods to investigate the *structure of the Milky Way galaxy*. Although they did not attain their goal, the enormous amount of work expended has proved to be extremely valuable in other respects.

The decisive step forward was made in 1918 with H. Shapley's method of photometric distance determinations using cepheid variables (cluster variables). The *period-luminosity relation*, i.e. the relationship between the period of their brightness variations and their absolute magnitudes, made it possible to measure the distance to any cosmic object in which cepheids could be found.

More precise investigations of the basic assumptions of this method revealed two difficulties: (1) the lack of a correction for interstellar absorption, and (2) the question of the applicability of a single period-luminosity relation to all types of cepheids. These later made it necessary to apply significant corrections: in 1930, R. J. Trumpler discovered the general *interstellar absorption and reddening*, and in 1952, W. Baade recognized that the period-luminosity relationships of the classical cepheids and of the W Virginis stars, i.e. the pulsation variables of stellar populations I and II (see below), differ by 1 to 2 magnitudes. In the following paragraphs, we shall consistently include these corrections where relevant in all numerical values quoted, and thus deviate from the purely historical standpoint.

The distances to the *globular clusters* determined by H. Shapley made it clear that they form a somewhat flattened system whose center lies about 10 kpc or 30000 light years from the Earth, in the direction of Sagittarius.

From this beginning, our present view of the Milky Way galaxy rapidly developed: the main portion of its stars form a flat disk about 30 kpc in diameter, containing the spiral arms. The outer portions of its center can be viewed as bright swarms of stars in Scorpio and Sagittarius; the galactic center itself is hidden from us by thick, dark interstellar clouds. Only after the inception of radio astronomy did it become accessible to direct observation. Our own Solar System is in the outer portion of the disk, about 9 kpc from its center. The disk is surrounded by a considerably less flattened halo, which contains the globular clusters and certain types of individual stars.

As early as in 1926/27, B. Lindblad and J.H. Oort were able to elucidate the *kinematics and dynamics of the Milky Way* to a considerable extent. The stars in the disk circle the galactic center under the influence of the gravitational force of the large mass which is concentrated there. In particular, the Sun completes its circular orbit with a radius of $\simeq 9$ kpc at a velocity of $\simeq 220\,\mathrm{km\,s^{-1}}$ in $\simeq 250$ million years. We initially detect only the *differential rotation*, however: away from the galactic center, the stars move somewhat more slowly (like the planets around the Sun), and towards the center, they move somewhat more rapidly than we do. From this motion, an estimate of the mass can be obtained; after applying various corrections, the mass of the whole system is found to be about $2 \cdot 10^{11}$ solar masses.

While the stars of the galactic disk move on circular orbits, the globular clusters and the stars of the halo describe extended elliptical orbits about the galactic center: their velocities relative to the Sun are therefore of the order of $100-300\,\mathrm{km\,s^{-1}}$. This is J.H. Oort's explanation of the socalled "high-velocity stars".

In addition to the stars, the *interstellar matter* in the Milky Way system plays an important role, although it makes up only a small percentage of the mass. Precise distance measurements became possible only after the interstellar absorption and reddening of starlight by cosmic dust were quantitatively described by R.J. Trumpler in 1930. Somewhat earlier, in 1926, A.S. Eddington had clarified the physics of interstellar gas and the interstellar absorption lines, while H. Zanstra and I.S. Bowen (1927/28) had dealt with the light emission from the *galactic and planetary nebulas*. The surprising discovery of the polarization of starlight by W.A. Hiltner and J.G. Hall (1949) finally made it clear that there is a galactic magnetic field in the disk of $\lesssim 10^{-9}$ Tesla.

In 1924, E. Hubble, using the 2.5 m telescope of the Mt. Wilson Observatory, succeeded after considerable refinement of photographic techniques in resolving the outer regions of the *Andromeda nebula* (and other spiral nebulas which are not too distant) to a considerable extent into individual stars, and in finding there (classical) cepheids, novae, bright blue O and B stars, etc. These made it possible to determine by photometric methods the distance, which was found to be about 700 kpc or 2 million light years. It was thus settled, after long and tedious controversies, that the Andromeda nebula and our own Milky Way system are basically similar cosmic objects. Later investigations by W. Baade have verified this down to the finest details. We can therefore combine results on the Andromeda nebula (M 31 = NGC 224) and on the Milky Way to a large extent into a *single* picture; some observations are best performed "from outside", others "from inside". Following Hubble's investigations, it has become customary to call the "relatives" of our Milky Way *galaxies*, and to reserve the term "nebula" for gas or dust masses *within* the galaxies.

In 1929, E. Hubble made a second discovery of great import: the spectra of the galaxies show a *red shift* proportional to their distances. We interpret this as due to a uniform *expansion* of the cosmos; one can readily imagine that dwellers in other galaxies would observe exactly the same effect as we have. If the spreading of the galaxies is extrapolated (somewhat schematically) backwards, it is found that the whole cosmos would have been concentrated closely together at a time around $\tau_0 \simeq 10^{10}$ yr ago. We call τ_0 the Hubble time; it gives a first indication of the *age* of the universe. What might have occurred further back lies outside the grasp of our research, and in any case, the universe at the time $-\tau_0$ must "have been very different" from its present state. The forerunners of Hubble's discovery, V.M. Slipher, C. Wirtz et al., as well as M. Humason's collaboration at the 2.5 m telescope, can only be named briefly here. These observations made the cosmos as a whole the object of exact science for the first time. Theoretical approaches to *cosmology* had been developed within the framework of general relativity theory, which initially aimed at an explanation of gravitation and inertial forces, as early as 1916 by A. Einstein, then W. de Sitter, A. Friedmann, G. Lemaître, and others. Conversely, for the investigation of distant galaxies, it is of fundamental importance that their distances, and thus their absolute magnitudes, their true dimensions, etc. can be determined from the red shift of their spectral lines.

The recognition that the age of the universe is of the order of 10^{10} yr, not very much more than the age of the Earth, $4.5 \cdot 10^9$ yr, created a strong impulse for the study of the evolution of stars and stellar systems.

After H. Bethe and C.F. von Weizsäcker in 1938 described the thermonuclear reactions which allow hydrogen to undergo fusion to helium in the interiors of main sequence stars, Bethe and then A. Unsöld in 1944 pointed out that the nuclear energy sources would have lasted for a time of the order of the age of the universe only in the cooler main sequence stars. For the hotter stars, shorter lifetimes were obtained, going down to only about 10^6 yr in the case of the O and B stars, with their high luminosities. In such a short time, the stars cannot have moved far from the places where they were formed; they must therefore have arisen at nearly the same locations where we observe them today. In reply to numerous speculative hypotheses, W. Baade has pointed out that the close spatial connection of the blue OB stars with dark clouds, e.g. in the Andromeda galaxy, indicates that they were formed from interstellar matter.

The investigations by A.R. Sandage, H.C. Arp, H.L. Johnson and others, likewise mostly based on suggestions by W. Baade, of the color-magnitude diagrams (CMD) of the globular clusters and the galactic star clusters, have led even further. Together with the theory of stellar structure, they reveal the following picture: a star which has been formed from interstellar matter at first passes through a relatively short contraction phase. On the main sequence, hydrogen fusion begins; the star remains there until it has burned about 10% of its hydrogen. Then it moves to the right in the CMD (M. Schönberg and S. Chandrasekhar, 1942) and becomes a red giant star. The place where the main sequence deviates to the right in the CMD of a star cluster (Fig. 5.4.2) indicates which stars have used up about 10% of their hydrogen since the formation of the cluster. The time of evolution necessary for this process likewise indicates the age of the cluster. Clusters with blue OB stars such as h and χ Persei are thus very young, while in older star clusters, the main sequence remains only below G 0. The theoretical studies of the *evolution of stars* which began with F. Hoyle and M. Schwarzschild in 1955 represent a fundamental continuation of A.S. Eddington's theory of stellar structure; however, the earlier assumption of a continual mixing of the matter in stellar interiors had to be abandoned under the pressure of observational results. The newer interpretation indicates that a burned-out helium zone is formed at the center of a star. The resulting forced transfer of the nuclear burning to regions further out causes the star to expand and become a red giant. The study of numerous color-magnitude diagrams led to the fundamental result that all globular clusters have the *same* age; recent determinations yield a value of 14 to $18 \cdot 10^9$ yr. In the case of open star clusters, by contrast, there are young and old objects. While the youngest are hardly a million years old, the oldest galactic cluster, NGC 188, has an age of 6 to $8 \cdot 10^9$ yr, and is thus still younger than the globular clusters.

The thermonuclear reaction stages which follow hydrogen burning, i.e. helium burning, carbon burning etc., take place at increasingly higher temperatures and densities and on increasingly shorter time scales. These *later evolutionary stages* are characterized by large mass reductions due to stellar winds and the ejection of gas shells; considerable energy losses through neutrino emission; and a complex interaction between hydrodynamic and nuclear processes, so that model calculations can be carried out only by using the fastest computers. Knowledge of the nuclear reactions which are important for energy release was improved greatly by W.A. Fowler's excellent determinations of nuclear reaction cross-sections. Although many detailed questions remain open, e.g. in connection with the explosion of supernovae, we at present understand the essentials of stellar evolution even for the final evolutionary stages: stars with less than about 8 solar masses end, in part after major mass reductions, as white dwarfs; more massive stars (≥ 10 solar masses) as neutron stars or black holes. We have already described in Chap. 4 their structures in the final configurations of their evolution.

Turning from stellar evolution, we now consider the distribution of various types of stars in our Milky Way and in other galaxies. It was pointed out in 1944 by W. Baade that different portions of the Milky Way differ not only in their dynamics, but also in their color-magnitude diagrams. Thus, the concept of *stellar populations* was developed. It soon became apparent that these populations differ essentially in their ages and in their abundances of heavy elements relative to hydrogen (i.e. elements heavier than helium; one usually refers for short to metal abundances). Leaving aside fine distinctions and transitional groups, we define:

1) Halo Population II: globular clusters, the metal-poor subdwarfs, etc. having the same color-magnitude diagrams. They follow elongated galactic orbits and form a structure which is slightly flattened, but with a higher concentration towards the center. Their metal abundance corresponds to about 10^{-3} to 1/5 of the "normal" values. Their age, about $1.7 \cdot 10^{10}$ yr, may be roughly equated with that of the Milky Way itself.

2) The Disk Population: most of the stars in our neighborhood belong to this group; they form a strongly flattened structure with an increasing concentration towards the galactic center. These stars follow nearly circular orbits and have "normal metal abundances", such as e.g. those in the Sun.

3) The Spiral-Arm Population I: it is characterized by young, blue stars of high luminosities. Within the disk, the interstellar matter is concentrated in the spiral arms; from it, associations and clusters of young stars are formed.

The *galactic classification scheme* developed in 1926 by E. Hubble, based on the shapes of galaxies, later proved to be basically a classification according to the predominance of population II or population I characteristics: the elliptical galaxies have only a small amount of interstellar matter left, and therefore contain in the main old stars, similar to the halo population or the disk population in the Milky Way. At the other end of the scale, the Sc and Irr I galaxies, which have quite strongly differentiated shapes, contain much gas and dust, well-developed spiral arms or other structures, and bright blue O and B stars. It was later recognized that the luminosities and masses of similar-appearing galaxies may in fact differ by several orders of magnitude. Therefore, we refer to giant and dwarf galaxies.

In recent times, it has been found that the rotational velocities of stars and gas in the outer portions of the Milky Way (and many other spiral galaxies) do not decrease on going outwards (as would be expected for large distances from the main concentration of mass), but instead are nearly constant. This means that an important, probably even the major portion of the mass of the galaxy is "invisible" and forms a roughly spherical system of about 50 to 100 kpc diameter. The nature of the matter in this *dark halo* is still a complete riddle.

An entirely new era in the investigation of our Milky Way and the more distant galaxies was ushered in by *radio astronomy*, beginning with K. G. Jansky's discovery of the meter-wave radiation from the Milky Way in 1932.

The thermal free-free radiation from plasmas at $\simeq 10^4$ K in the interstellar gas, in H II regions, planetary nebulas etc. was soon detected. In 1951, the first observations were made of the 21 cm line of atomic hydrogen, predicted in 1944 by H. C. van de Hulst; its Doppler effect gave completely new insights into the structure and dynamics of interstellar hydrogen and thus into those of whole galaxies. Additional lines in the mm to dm ranges could be attributed to transitions between states of very high quantum number in hydrogen and helium atoms and to diatomic and polyatomic molecules, which are in some cases surprisingly complex. The 2.6 mm line of the abundant *CO molecule*, discovered in 1970, has particular significance: it has been used to investigate the distribution of cold (≥ 10 K) constituents of interstellar matter which are concentrated in massive *molecular clouds* in the Milky Way. These concentrations, which are mainly composed of hydrogen molecules, have been found to be the real locations of *stellar genesis*; only after the formation of highly luminous O and B stars do they become visible in the optical wavelength region as glowing nebulas (H II regions).

The radio continuum originally observed by Jansky is nonthermal *synchrotron radiation*. It was postulated in 1950 by H. Alfvén and N. Herlofson, and soon thereafter given strong support by I. S. Shklovsky, V. L. Ginzburg, and others, that this radiation is produced when high-energy electrons move on spiral orbits around the force lines of cosmic magnetic fields.

Along with the elucidation of the mechanisms which can lead to the emission of radiofrequency radiation, the experimental development of better and better angular resolution and more precise radio position determinations was of equal importance. We have already described the construction of continually improved and larger radiotelescopes and of radiointerferometers up to intercontinental very-long-baseline interferometers in Sect. 3.3; here, we add some historic dates in radioastronomical research: in 1946, J. S. Hey employed the fluctuations of radio emissions (later recognized as scintillation originating in the ionosphere) to find the first cosmic *radio source*, Cygnus A; in 1949, J. G. Bolton, G. J. Stanley, and O. B. Slee identified the radio source Taurus A with the Crab nebula. Only in 1952, using the meanwhile much improved accuracy of radio position determinations, could W. Baade and R. Minkowski demonstrate that the radio source Cassiopeia A is, like Taurus A, the remnant of an earlier supernova. Cygnus A, in contrast, could be identified with a peculiar galaxy having emission lines from highly excited states: the first *radio galaxy*. This opened the door to the rapid development of the field of extragalactic radioastronomy, which is unparalleled in the whole history of astrophysics. In 1962/63, M. Schmidt at the Mt. Palomar Observatory was able to recognize the *quasars*, whose optical images are hardly distinguishable from those of stars, as very distant galaxies with extremely high optical and radio luminosities. Their red shifts, which are much larger than those known previously, opened unsuspected perspectives for cosmology. It was further found that in the *centers* of the quasars, radio galaxies, Seyfert galaxies, and, to a lesser extent, even in normal galaxies like our own, a characteristic "*activity*" occurs in the form of nonthermal optical and radiofrequency emissions and emission lines with nonthermal excitation. Often, two narrow beams of relativistic particles or plasma (jets) are found to emerge from the galactic center in opposite directions along the axis of rotation of the galaxy and to extend out over enormous distances of several 100 kpc; their synchrotron radiation covers the spectral regions from the radio to the X-ray range. Structural changes in the central emission regions, which apparently correspond to velocities exceeding that of light, can be explained in terms of relativistic effects in jets which are moving with near-light velocity close to the direction towards the observer. The *compact cores* of galaxies thus contain energy sources of unimaginable strength, whose physical nature (release of gravitational energy in the neighborhood of an extremely massive object?) is still to a large extent unknown.

An important complement to the radioastronomy of synchrotron radiation, which arises from relativistic electrons, has been provided since the 1960's by *X-ray astronomy* (H. Friedman, R. Giacconi, B. Rossi and others) and by *gamma ray astronomy*, which developed a few years later. In both adjacent energy regions, observations must necessarily be carried out from rockets and space vehicles, or at least from stratospheric balloons, as in the ultraviolet region.

Parallel to the rapid progress of astronomy in these "high energy" spectral ranges, in recent years a turbulent development in *infrared astronomy* has begun; it is also based for the most part on observations from outside the Earth's atmosphere. In 1983/84, a sky survey was carried out by the first infrared satellite, IRAS. The opening up of new spectral regions observable only from space has been accompanied since about 1970 by an enormous extension of the range of *earthbound optical astronomy*, to weaker and weaker light sources; this has been made possible in particular by the development of new types of highly sensitive photodetectors.

These improvements in astronomical observation techniques have led to a great increase in our knowledge and to quite new ideas about the structure and the evolution of the Milky Way and other galaxies. The observations in the ultraviolet and X-ray regions have demonstrated the existence within the Milky Way galaxy of *hot interstellar gas* (10^4 to 10^6 K), which extends in part far out into the halo. On the other hand, the galactic gamma radiation is characteristic of locations where high-energy cosmic particle radiation interacts with the dense, cool portions of interstellar matter. Optical, infrared, and radioastronomical measurements have shown surprisingly energetic *mass flows* in the neighborhood of very *young stars*, which are directed conically in a "bipolar" manner outwards and are also frequently narrowed down to form a pair of jets. The optical emissions of the central star are in general absorbed by a dust-filled *accretion disk*.

Infrared astronomy has yielded important contributions in recent times to the problem of the formation of planetary systems: indications of cool, weakly luminous companions in binary star systems with masses corresponding to those of large planets, and of rings of dust particles or disks around some type A stars.

From our own Milky Way galaxy, we now turn to the current ideas about *galactic evolution*, the formation of the chemical *elements* and their abundance distributions, and to *cosmology* as a whole. A first venture into these areas was made by G. Lemaître and G. Gamow in 1939, with their proposal that the expansion of the universe began with a primordial explosion ("Big Bang"), during whose initial stages the "cosmic" (i.e. roughly the solar) *abundance distribution of the elements* already became established. The discovery of metal-poor stars and especially the difficulty of explaining the continued formation of the elements above mass number $A = 5$ at first seemed to discredit this theory. Only after the detection of the cosmic *3 K radiation* in the microwave region by A. A. Penzias and R. W. Wilson (1965), which has been explained as a remnant of the Big Bang, did it return in a more modest form: in an initial "Big Fireball", essentially only hydrogen and helium atoms are produced, with an abundance ratio $\simeq 10:1$.

Dealing with the formation of the galaxy and the abundance distribution of the elements (nucleosynthesis), the theory developed by E. M. and G. R. Burbidge, W. Fowler, and F. Hoyle in 1957 begins with a nearly spherical pregalaxy composed of hydrogen plus 10% helium and possibly also traces of heavier elements. Then, the first halo stars are formed; they produce heavy elements, decay (by supernova explosions), and thus enrich the interstellar matter in heavier elements. From this matter, a new generation of metal-richer stars is formed, etc.

Before the inventory of heavy elements has reached that in current "normal" stars, the galactic disk with its spiral arms is formed by collapse of the remaining halo material, which has not yet condensed into stars, after a period of only about 10^8 to 10^9 years.

The role of the galactic centers and their activity in the formation and evolution of galaxies, which has been pointed out since 1958 by V. A. Ambartsumian, is still to a large extent unclear. In future theories of the *activity of galactic centers*, high-energy astronomy in particular should have a decisive word, since the galactic centers all the way out to the quasars have proved to be strong sources of nonthermal radiation, emitted in large part in the *X-ray* and *gamma-ray regions*. As we have previously mentioned, the source of energy for these and other activity phenomena is still undetermined.

Among the strongest extragalactic sources in the X-ray region are the *clusters of galaxies*. Their radiation, discovered by the Einstein satellite in 1979, originates in extremely hot gases ($\simeq 10^7 - 10^8$ K) in the central areas of the dense regular clusters; these were "lost" by the galaxies in the course of collisions with each other. The fact that *interactions* between the galaxies within groups and clusters influence their development, and thereby the morphological type of galaxies long after their formation, has been made clear in recent times in particular through infrared observations: near collisions definitely cause a strong increase in *star formation*. The radiation of young, hot stars is absorbed by dense dust clouds around them, heating the clouds and thus converting them into infrared emitters. In the course of the survey by the IRAS satellite, many *infrared galaxies* were discovered; their luminosities are confined almost exclusively to the infrared.

Optical spectroscopy of the faintest and most distant galaxies with new, sensitive detectors, beginning about 1976, was able to confirm the organization of the clusters of galaxies into *superclusters*, which had earlier been postulated on the basis of their distribution on the celestial sphere. The unique large-scale *structure of the cosmos*, consisting of connected filaments and disks of the superclusters surrounding large empty regions, is likely to be a key to theories of the formation of the galaxies and galaxy clusters.

At the conclusion of our historical introduction, we return to the *evolution of the cosmos as a whole*. While cosmology was dominated in the first half of this century by models of the universe based on Einstein's *general theory of relativity* and Hubble's discovery of the *expanding universe*, the observation of the 3 K radiation by Penzias and Wilson in 1965 made it possible to establish *physical* models of cosmic evolution. Progress in *elementary-particle physics* and the fundamental interactions since the 1960's has allowed a physical description of the earliest stages of the universe, with their extremely high energies and temperatures, and has initiated a fascinating development in modern cosmology, which is at present by no means completed.

5.2 The Structure and Dynamics of the Milky Way Galaxy

We can obtain an initial overview of the structure of our Milky Way by means of "star gauging" (Sect. 5.2.1). This method is, however, severely limited by interstellar absorption. We shall take up the subject of the interstellar medium in Sect. 5.3; here, we discuss the distribution and motions of the stars and star clusters in the Milky Way galaxy. Sect. 5.2.2 treats stellar kinematics. In Sect. 5.2.3, we then consider the star clusters: the open or galactic clusters, and the globular clusters. With their help, distances in the whole galaxy can be determined, so that we can obtain an initial overview of its structure. After introducing the galactic coordinate system appropriate to the investigation of the Milky Way galaxy in Sect. 5.2.4, we turn in Sect. 5.2.5 to its rotation. It was noticed as early as 1926/27 by B. Lindblad and J. H. Oort that the stars in the neighborhood of the Sun undergo a differential rotation around the center of the galaxy; later, radioastronomical observations of the 21 cm line of neutral hydrogen were able to be used to characterize the rotational velocities throughout the Milky Way. Following a brief discussion of stellar orbits and of the density of matter in the immediate neighborhood of the Sun (Sect. 5.2.6), we conclude this subchapter in Sect. 5.2.7 with a treatment of the distribution of mass in the Milky Way galaxy.

5.2.1 Star Gauging

W. Herschel (1738–1822) was the first to attempt to penetrate the secrets of galactic structure with his *star gauging*, in which he counted the number of stars which he could see in different directions down to a particular limiting magnitude.

What should we expect if space were completely transparent and uniformly filled with stars? For stars of a given absolute magnitude, the apparent magnitude decreases as a function of r proportional to $1/r^2$ [cf. (4.4.15, 16)]. The stars brighter than m thus fill a sphere with $\log r = 0.2\,\mathrm{m} + \mathrm{const}$. Their *number* $N(m)$ is $\propto r^3$; we thus find:

$$\log N(m) = 0.6m + \text{const} \ . \tag{5.2.1}$$

In Fig. 5.2.1, we compare the number of stars per square degree according to F. H. Seares (1928) for the galactic plane and for those in the direction towards the galactic pole (i.e. $b = 0°$ and $90°$) with what is expected from (5.2.1). The much slower increase of $N(m)$ observed at fainter magnitudes can have only two causes: (1) a decrease in the density of stars at larger distances, or (2) interstellar absorption; or both. H. v. Seeliger and J. Kapteyn took the scatter of the absolute magnitudes of the stars into account and showed how their numbers $N(m)$ can be correctly represented by a superposition (mathematically speaking, by a convolution) of (1) the *density function* $D(r)$ = the number of stars per pc^3 at the distance r and a given direction with (2) the *luminosity function* $\Phi(M)$ = the number of stars per pc^3 in the interval of absolute magnitudes between $M - \frac{1}{2}$ to $M + \frac{1}{2}$, possibly also taking into account (3) an *interstellar absorption* of $\gamma(r)$ magnitudes per pc. Even taking the luminosity function $\Phi(M)$ to be everywhere constant and determining it by using stars of known parallax in a region of 5 or 10 pc, one could not separate the functions $D(r)$ and $\gamma(r)$. We can therefore discard the results of older stellar statistics. The concepts just introduced remain important, as does the great sampling survey of the entire sky (magnitudes, color indices, spectral types) in the Kapteyn fields.

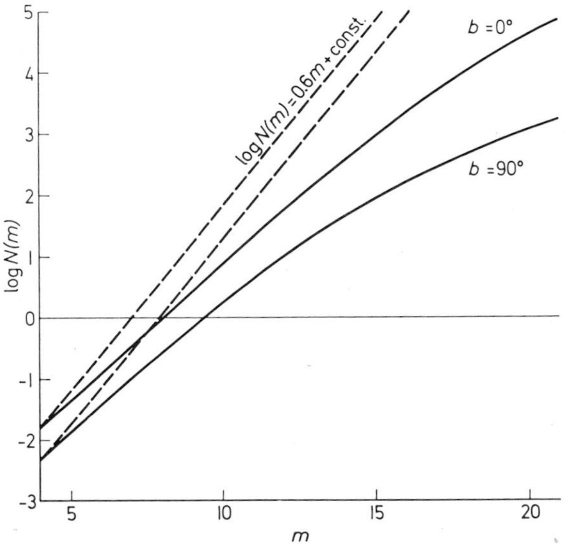

Fig. 5.2.1. Numbers of stars $N(m)$, i.e. the number of stars brighter than m per square degree, from the star counts by F. H. Seares (1928), at the galactic equator ($b = 0°$) and at the galactic pole ($b = 90°$) (*full curves*). Calculated curves (*dashed*): $\log N(m) = 0.6\,m + \mathrm{const.}$ for constant stellar density, without galactic absorption. (The constant was adjusted to the observational data for $m = 4$ mag)

5.2.2 Spatial Velocities of the Stars. Stream Parallaxes

The study of the motions of stars initially proved to be more fruitful than did stellar statistics. We have already mentioned the spectroscopic determination of the *radial velocities* V_r from the *Doppler effect* (H. C. Vogel, 1888; W. W. Campbell and others):

$$V_r = c\frac{\Delta\lambda}{\lambda_0} \ , \tag{5.2.2}$$

which gives us V_r in absolute units, e.g. in $[\mathrm{km\,s^{-1}}]$ (2.5.2). The sign is defined in such a way that positive (negative) radial velocities correspond to a red shift (violet shift), i.e. moving away from the Sun (approaching the Sun). The interfering effects of the Earth's orbital motion and rotation can be immediately eliminated by using the *Sun* as reference point.

We must also consider the *proper motions* μ (abbreviated PM) of the stars on the celestial sphere, discovered much earlier by E. Halley (1718). They are usually quoted in seconds of arc per year. Relative proper motions (referred to more distant stars with smaller PM) are measured by comparing two photographs ("blinking"), taken if possible using the same instrument at a time interval of 10 to 50 years. The reduction to *absolute* proper motions presumes that absolute positions on the meridian circles of several stars have been determined in different epochs. Following C. D. Shane, distant galaxies and, more recently, quasars are employed as extragalactic reference points.

The *tangential component* V_t $[\mathrm{km\,s^{-1}}]$ of the stellar velocity is related to the proper motion μ as follows:

If p is the *parallax* of the star in seconds of arc, then μ/p is equal to V_t in astronomical units per year. The latter is equal to $1/2\pi$ times the orbital velocity of the Earth, or 4.74 km s^{-1}. We thus find:

$$V_t\,[\mathrm{km\,s^{-1}}] = 4.74\frac{\mu\,[''\mathrm{yr}^{-1}]}{p\,['']} , \qquad (5.2.3)$$

and the *spatial velocity* v of the star is given by

$$v = \sqrt{V_r^2 + V_t^2} . \qquad (5.2.4)$$

The angle θ which the star's motion makes with a line of sight from the observer is determined by the relations:

$$V_r = v\cos\theta \quad \text{and} \quad V_t = v\sin\theta . \qquad (5.2.5)$$

Since, for example, a star of 6th magnitude has on the average a parallax of 0.012″, but a proper motion of 0.06″ yr^{-1}, we can delve out ca. 100 times further into interstellar space by studying the proper motions (over a period of 20 years) than with parallax measurements.

Let us first make the simplifying assumption that the stars are at rest and that only the Sun moves relative to them with a velocity $v_\odot$ towards the *apex*; then we expect the distribution of radial velocities V_r and tangential velocities V_t or proper motions μ shown in Fig. 5.2.2 as a function of the angular distance χ of the star from the apex.

If the motions of the stars are distributed randomly in space, we can clearly still use this model, by *averaging* over many stars. The first apex determination was made in 1783 by W. Herschel, using only a few proper motions.

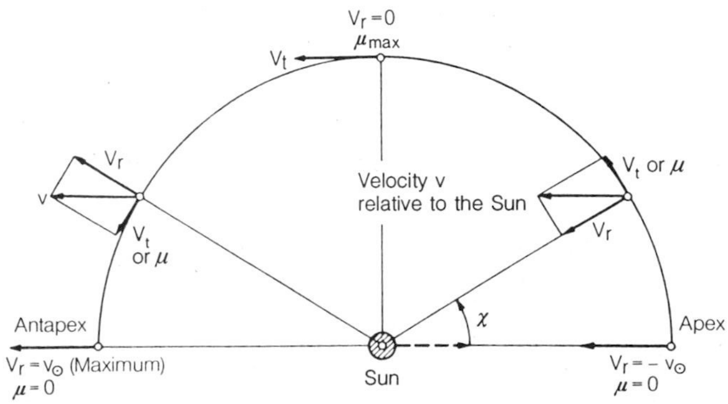

Fig. 5.2.2. Motion of the Sun, relative to the surrounding stars, with velocity $v_\odot$ towards the apex. The observed parallactic motions of the stars are the reflection of this solar velocity. The figure explains the dependence of the radial velocity V_r and the tangential velocity V_t (or the proper motions μ) of the stars on their angular distance χ from the apex

Later, it was found that, at a higher precision, the motion of the Sun depends upon *which* stars are used to determine it; this was the first indication of the systematic effect of the stellar motions. The socalled standard solar motion which is generally used to reduce stellar motions to the *locality of the Sun* (Local Standard of Rest) is obtained from the proper motions and the radial velocities of the stars in our neighborhood as:

$$\begin{aligned}&\text{Solar motion:}\quad v_\odot = 20\ \mathrm{km\,s^{-1}}\\ &\text{to the apex:}\quad \alpha = 18\,\mathrm{h}\ ,\ \delta = +30^\circ .\end{aligned} \qquad (5.2.6)$$

Knowledge of the *solar motion* can then be used to separate the statistical part of the proper motions, the *peculiar motions*, from the reflection of the solar motion (the parallactic motion) within the measured *proper motions* of a carefully chosen group of stars. This parallactic motion clearly depends on the mean parallax $\bar{p}$ of the star group: the part of the tangential velocity due to the solar motion is, from (5.2.5), given by $V_t = v_\odot \sin\chi$, where χ is again the angular distance of the star group[1] from the apex. We can now apply (5.2.3), if we restrict ourselves in calculating the average over the proper motions μ to their components in the direction of the apex. The *mean* or *secular parallax* of our star group is thus given by:

$$\bar{p} = \frac{4.74\bar{\mu}}{v_\odot\sin\chi} . \qquad (5.2.7)$$

The hypothesis of randomly distributed peculiar motions must, however, be applied with great caution. It was namely discovered in 1908 by L. Boss that, for example in a large group of stars in Taurus, which are gathered around the galactic star cluster of the *Hyades*, the proper motion vectors on the sphere are directed towards a *convergence point* at $\alpha = 93°$, $\delta = +7°$. The stars of this stellar stream or *moving cluster* thus clearly move together in space like a school of fish, their parallel motions tending towards the convergence point. Let the velocity of the cluster (relative to the Sun) be v_C. If we now know the proper motion μ of a star in the cluster (Fig. 5.2.3) and its radial velocity V_r relative to the Sun, and if θ is the angle in the sky from the star to the convergence point, then we can apply the considerations of (5.2.3) and (5.2.5) and obtain:

[1] We assume here that our star group was chosen to lie within a relatively small area of the sky, so that *one* mean value of χ suffices.

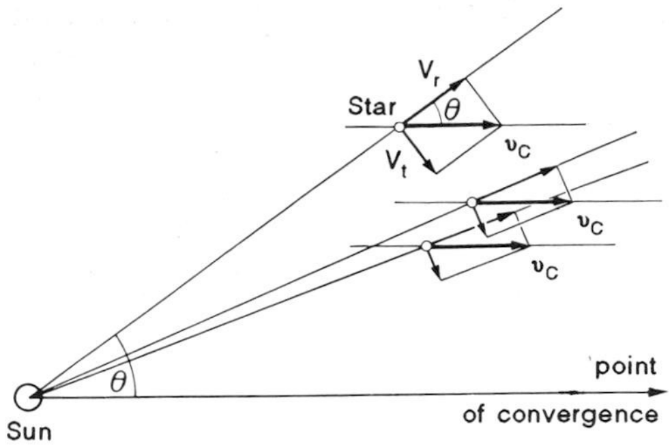

Fig. 5.2.3. A moving cluster or star stream

$$V_r = v_C \cos\theta \quad \text{and} \quad V_t = v_C \sin\theta = 4.74\frac{\mu}{p} \ ,$$

from which we immediately find the parallax of the star:

$$p = \frac{4.74\mu}{V_r \tan\theta} \ . \tag{5.2.8}$$

The Taurus cluster, for example, is 45 pc away from us, its velocity is 31 km s^{-1}, and the major portion of its stars are within a region of about 10 pc diameter. This method of *stream parallaxes* is superior to the trigonometric parallax method in terms of the range of distances to which it can be applied and often in its precision.

The common small proper motions of the stars in galactic clusters (the Pleiades, Praesepe, etc) to some extent allow the determination of their parallaxes; however, they are especially important for testing whether or not particular stars are members of the cluster.

5.2.3 Star Clusters and the Structure of the Milky Way

All of our knowledge about the structure and size of the Milky Way galaxy is based essentially on the method of *spectrophotometric distance measurement*. From the $1/r^2$ law of photometry, we see a star of absolute magnitude M and parallax p or distance $r = 1/p$ with the apparent magnitude m, so that the distance modulus (4.4.17) is given by:

$$m - M = 5 \log r\,[\mathrm{pc}] - 5 + A\,[\mathrm{mag}] \ . \tag{5.2.9}$$

The interstellar extinction A [mag] will be treated in Sect. 5.3.1; we shall, however, take it into account in all quoted numerical values of distances, etc.

In the final analysis, we must always rely on the absolute magnitudes of certain objects which are known from trigonometric and secular parallaxes, stream parallaxes, etc. Because of their fundamental importance, we have first treated the methods of geometric distance determination in considerable detail.

In order to gain an insight into the structure of our Milky Way, it is reasonable to begin with groups of stars whose structures can be readily recognized, in part by the unaided eye, in part in suitable photographs: the *globular clusters*, in which the stars appear to be gathered together like bees in a swarm; the less "concentrated" *galactic* or *open clusters*; and *stellar associations*. The classical position catalogues, containing also all types of "nebulas", which we now classify separately into galaxies on the one hand and into planetary and galactic gas nebulas on the other, are the Messier Catalogue (M) of 1784 and Dreyer's New General Catalogue (NGC) of 1890 with continuations in the Index Catalogue (IC) of 1895 and 1910.

1) *Globular clusters.* The two brightest globular clusters, ω Centauri and 47 Tucanae, are in the southern sky. In the northern sky, M13 = NGC 6205 in Hercules can be seen with the unaided eye; its brightest stars have

Fig. 5.2.4. The globular cluster M 13 = NGC 6205 in the constellation Hercules. Distance 6.4 kpc

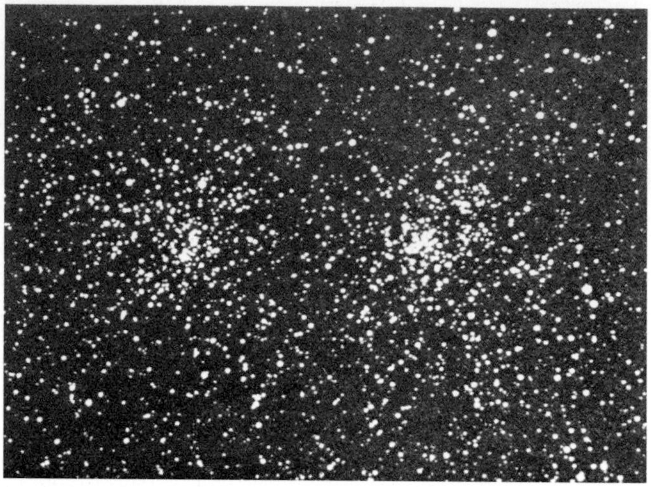

Fig. 5.2.5. The binary galactic cluster h and χ Persei

magnitudes of about 13.5. Photographs made with large telescopes (Fig. 5.2.4) show more than 50000 stars in the brighter globular clusters; in the center, the stars can no longer be separated from one another. The globular clusters show a strong concentration in the direction of Scorpio-Sagittarius in the sky.

2) The *galactic* or *open clusters* (Fig. 5.2.5) are found in the sky along the whole bright band of the Milky Way. Some have a relatively large number of stars, of the order of several hundreds (but still not nearly as many as the globular clusters); others have few stars, only some dozens. The concentration of stars towards their centers, i.e. the compactness of the clusters, is also very diverse. The best-known are the Pleiades and the Hyades in Taurus and the double cluster h and χ Persei.

3) The OB *associations* are relatively loose groups of bright O and B stars, which often surround a galactic cluster, such as for example the ζ Per association around the double cluster h and χ Per mentioned above. On the other hand, the *T associations* are corresponding groupings of T Tauri or RW Aurigae variables and other stars close to the lower part of the main sequence. The importance to cosmogony of the OB and T assocations as extremely young formations was pointed out in 1947 by V.A. Ambartsumian.

About 130 globular clusters, 1000 galactic clusters, and 100 associations are known. Along with the classical work of H. Shapley, "Star Clusters" (1930), we should in particular mention the "Catalogue of Clusters and Associations" by G. Alter, J. Ruprecht, and V. Vanýsek (Budapest 1970).

We therewith close this brief overview and turn again to the cardinal problem of *distance determinations.*

The decisive turning point towards modern astronomy was taken by H. Shapley in 1917, when he determined the distances to numerous *globular clusters* by means of the cluster variables contained in them (RR Lyr stars); the essential difficulty lay in finding out the absolute magnitudes of these variables, or in calibrating the period-luminosity relation (Sect. 4.11.1). For those

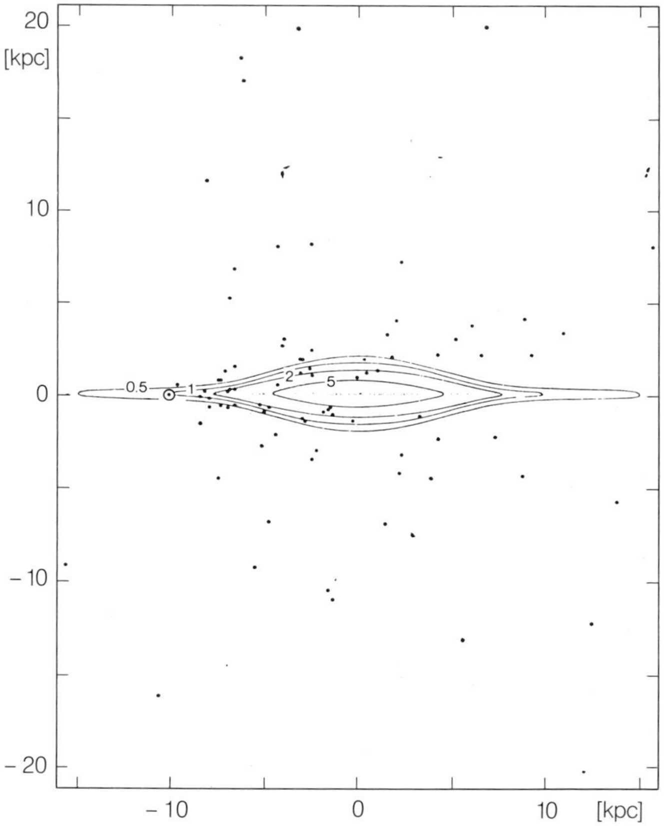

Fig. 5.2.6. The Milky Way system. The spatial distribution (halo) of the globular clusters is projected onto a plane perpendicular to the galactic plane and passing through the Sun; the contours show surfaces of constant mass density (relative to the vicinity of the Sun). In the galactic plane, the thin layer of interstellar matter and the extreme Population I is indicated by dots (after J.H. Oort, 1965). The disk of the galaxy and the halo of globular clusters are, as confirmed by newer dynamic models, embedded in a considerably larger system with a radius of about 60 to 100 kpc, the invisible "*dark*" *outer halo* or galactic corona, which contains a major part of the mass of the Milky Way system; its composition is currently *unknown* (Sect. 5.5.3)

globular clusters which contained no variable stars, Shapley used the brightest stars of the cluster as a secondary criterion. The five brightest stars, which are possibly foreground stars, are discarded; the absolute magnitudes of the next brightest, up to perhaps the 30th, have proven to be well defined. Furthermore, the overall brightness of the cluster, or its angular diameter, can also be used (with certain precautions) as criteria. From his observational data of (at that time) 69 clusters, Shapley was able to draw the conclusion that they form a system which is hardly flattened in the plane of the Milky Way, the *halo* (Fig. 5.2.6); its center lies about 9 kpc from us (the value quoted by Shapley was 13 kpc) in Sagittarius. This work laid the framework for further investigations of the Milky Way.

We can draw *color-magnitude diagrams* for the *galactic clusters* and the associations, for example with (B−V) color indices as abcissa and the apparent (visual) magnitude m_V as ordinate. If we may assume that the *main sequence* is the same for all systems, as well as for the socalled field stars in our neighborhood which are not apparently a part of any system, then the vertical distance of the main sequence in the (B−V, m_V) diagram of the cluster and in the (B−V, M_V) diagram of our immediate neighborhood yields directly the distance modulus $m_V - M_V$ and thus the distance to the cluster.[2] (We shall return to the finer details of the CM diagrams and the influence of interstellar extinction in Sect. 5.4.1.) The fundamental work of R. Trumpler should be mentioned; in Fig. 5.2.7 we show the results for the distribution of the young galactic star clusters and the OB associations based on newer observations. Their arrangement into long, stretched-out regions is obvious at first glance; these are the neighboring portions of the galactic *spiral arms*. Like the OB stars, the H II regions and the absolutely brightest cepheids (with periods ≥ 11 d) are concentrated along the spiral arms.

Observations in the visible spectral range within the galactic disk are limited to distances ≤ 4 kpc even for very bright objects, due to interstellar extinction. On the other hand, in the near infrared ($\lambda \simeq 2\,\mu$m), star clusters and individual K and M supergiants can be distinguished even near the center of the galaxy.

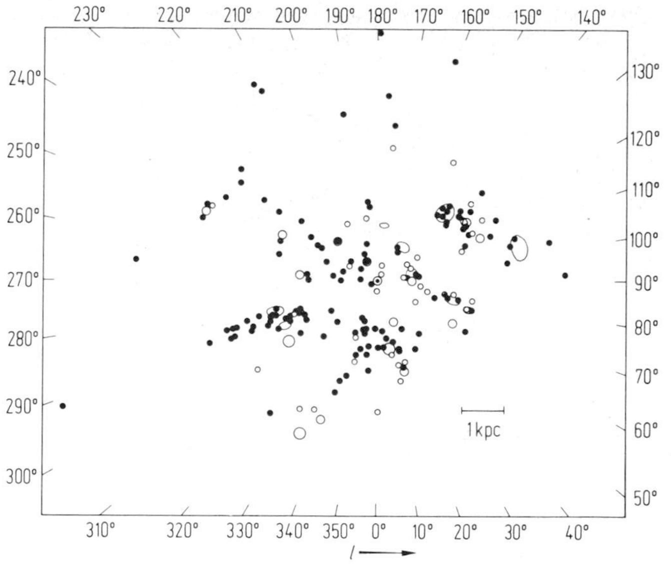

Fig. 5.2.7. The distribution of young galactic clusters (●) (which contain early spectral types, O to B3 stars) and the OB associations (○) in the plane of the Milky Way, after R.M. Humphreys (1979). The Sun ⊙ is at the origin of the coordinate system, and the galactic longitude $l = 0°$ points towards the center of the galaxy. The objects of extreme Population I are distributed along the spiral arms (from the center outwards: Sagittarius arm, local or Orion-Cygnus arm, Perseus arm)

5.2.4 Galactic Coordinates and Velocity Components

It is appropriate at this point to introduce a system of galactic coordinates which we can use to describe the Milky Way galaxy; these are the *galactic longitude* l in the plane of the galaxy, and the *galactic latitude* b in the perpendicular direction, which is positive towards the north and negative towards the south. In 1958, a system of galactic coordinates (l, b) was introduced,[3] which had been improved in particular through the use of radioastronomical data; they are defined by the equatorial coordinates (1950.0) of the galactic *north pole*:

$$\alpha = 12\,\mathrm{h}\ 49\,\mathrm{min}\ , \quad \delta = +27.40°\ . \tag{5.2.10}$$

Galactic longitude is measured starting from the galactic *center*:

$$\alpha = 17\,\mathrm{h}\ 42.4\,\mathrm{min}\ , \quad \delta = -28.92°\ ; \tag{5.2.11}$$

[2] In principle, the same method can also be used to determine the distances to the *globular clusters*. However, they are in the main further away from us than the galactic clusters, so that it is difficult to obtain precise magnitudes for their main sequence stars; furthermore, the influence of their different metal abundances must be taken into account (Sect. 5.4.2).

[3] During a transition period, these "new" coordinates were denoted as $(l^{\mathrm{II}}, b^{\mathrm{II}})$ to distinguish them from the older system.

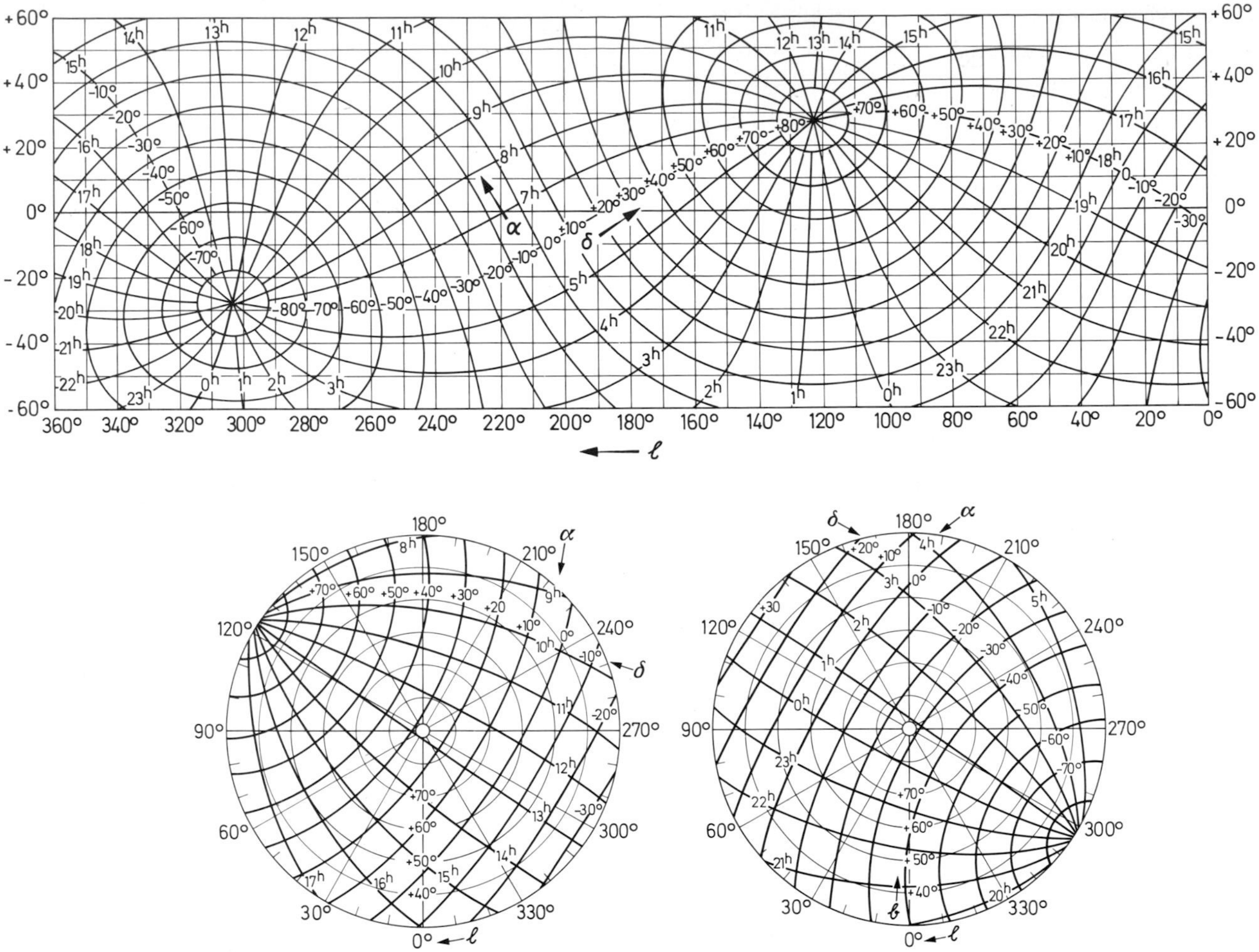

Fig. 5.2.8. Charts for converting galactic coordinates l, b (plotted above as abcissa and ordinate, respectively; below, around the circumference or as the radius of the circles, respectively) into right ascension α and declination δ, and *vice versa*, for Epoch 1950. *Above*: the galactic equatorial zone. *Lower left*: the galactic north pole; *lower right*: the galactic south pole. After G. Westerhout

the plane of the galaxy is inclined to the celestial equator by an angle of 62.6°.

Tables for converting equatorial into galactic coordinates and *vice versa* were published by the Lund Observatory in 1961. In Fig. 5.2.8, we reproduce the maps drawn by G. Westerhout for converting (l, b) to (α, δ) for the epoch 1950.

The galactic components of the *spatial velocity* of stars are defined by (positive sign):

U = radial component, away from the galactic center,
V = component in the direction of the rotation of the galaxy ($l = 90°$)
W = component perpendicular to the galactic plane, towards the galactic north pole. (5.2.12)

Here, we must remember to note whether the *motion of the Sun* has been subtracted or not.

The *standard solar motion* relative to the average motions of neighboring stars (5.2.6) is:

$$U_\odot = -10\ \mathrm{km\,s^{-1}}\ , \quad V_\odot = +15\ \mathrm{km\,s^{-1}}\ ,$$
$$W_\odot = +8\ \mathrm{km\,s^{-1}} \qquad (5.2.13)$$

towards the apex, $l \simeq 56°$ and $b \simeq +23°$.

5.2.5 Galactic Rotation

We now turn to the kinematics and dynamics of the Milky Way galaxy, as developed by B. Lindblad and J. H. Oort in 1926/27 in their theory of the *differential rotation* of the galaxy.

We first assume that all motions (Fig. 5.2.9) take place on planar *circular orbits* around the galactic center (Z).

Let the angular velocity ω of a star P at the galactic longitude l as a function of its distance R from the center be $\omega = \omega(R)$, and thus $\omega R = V$ its linear velocity on its galactic circular orbit. For the Sun, let $R = R_0$, $\omega(R_0) = \omega_0$ and $\omega_0 R_0 = V_0$. Precisely stated, we always relate the motion here and in the following sections to the *neighborhood of the Sun*, by subtracting the solar motion (5.2.6) from all observed coordinates.

We now decompose the velocity vector $V - V_0$ of the star P *relative* to the Sun into its components in the direction ⊙ P and perpendicular to this direction, in order to obtain the radial velocity:

$$V_r = V \sin\alpha - V_0 \sin l \qquad (5.2.14)$$

and the proper motion μ or the tangential velocity:

$$V_t = V\cos\alpha - V_0 \cos l = \mu r \ . \qquad (5.2.15)$$

Here, r is the distance of the star from the Sun. The auxiliary angle α can be eliminated by using $|ZQ| = R\sin\alpha$ and $|PQ| = R\cos\alpha$, which we can read off from the triangle ⊙ZQ (see Fig. 5.2.9):

$$\begin{aligned} R\sin\alpha &= R_0 \sin l \ , \\ R\cos\alpha + r &= R_0 \cos l \ . \end{aligned} \qquad (5.2.16)$$

We obtain finally

$$V_r = R_0(\omega - \omega_0)\sin l \ , \qquad (5.2.17)$$

$$V_t = R_0(\omega - \omega_0)\cos l - \omega r \ . \qquad (5.2.18)$$

These equations are valid for stars or interstellar gas on circular orbits at *arbitrary* distances r from the Sun.

If we now consider our *immediate neighborhood*, $r \ll R_0$, in the Milky Way galaxy, then we can use the series expansion:

$$\omega - \omega_0 \simeq \left(\frac{d\omega}{dR}\right)_0 (R - R_0) \simeq -\left(\frac{d\omega}{dR}\right)_0 r\cos l \qquad (5.2.19)$$

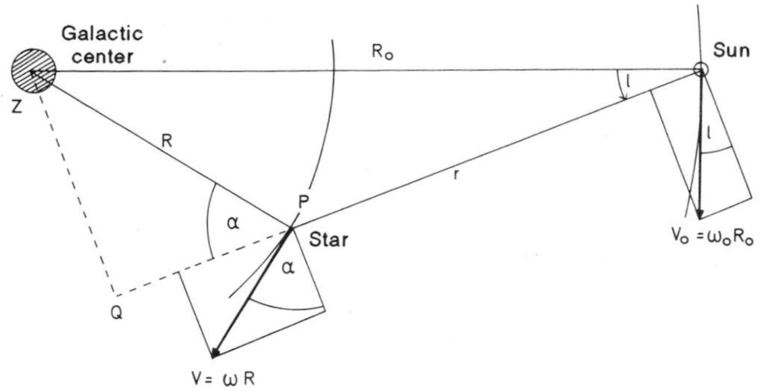

Fig. 5.2.9. Galactic rotation

(the index 0 always means $R = R_0$). We now introduce *Oort's constants* of the differential galactic rotation:

$$\begin{aligned} A &= -\frac{R_0}{2}\left(\frac{d\omega}{dR}\right)_0 = \frac{1}{2}\left[\frac{V_0}{R_0} - \left(\frac{dV}{dR}\right)_0\right] \\ B &= -\frac{R_0}{2}\left(\frac{d\omega}{dR}\right)_0 - \omega_0 = -\frac{1}{2}\left[\frac{V_0}{R_0} + \left(\frac{dV}{dR}\right)_0\right] \end{aligned} \qquad (5.2.20)$$

or

$$A + B = -\left(\frac{dV}{dR}\right)_0 \quad \text{and} \quad A - B = \frac{V_0}{R_0} = \omega_0 \ . \qquad (5.2.21)$$

We have written these constants using the orbital velocity $V(R)$, i.e.

$$\omega = \frac{V(R)}{R} \ , \quad \frac{d\omega}{dR} = \frac{1}{R}\left(\frac{dV}{dR} - \frac{V}{R}\right) \ . \qquad (5.2.22)$$

With Oort's constants A and B, Eq. (5.2.19), and the relation $2\cos^2 l = 1 + \cos 2l$, the radial and tangential velocities for our neighborhood as functions of the galactic longitude l finally take on the simple form:

$$V_r = A r \sin 2l \ , \qquad (5.2.23)$$

and

$$V_t = A r\cos 2l + B r \qquad (5.2.24)$$

(compare Fig. 5.2.10). After averaging out the peculiar motions, observations verify this "double wave" ($\sin 2l$!) of the two velocity components very well. While the amplitudes of V_r and V_t increase proportionally to the distance r, the amplitude of the proper motion $\mu = V_t/r$ is

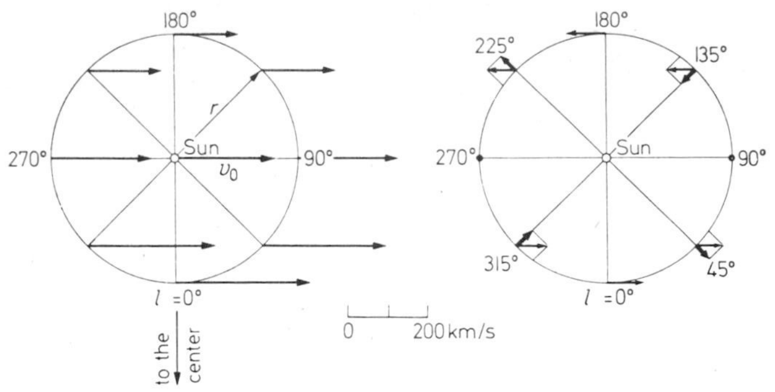

a b

Fig. 5.2.10 a,b. Differential galactic rotation. (**a**) Absolute velocities of the stars at a distance r from the Sun. In our drawing, $r = 3$ kpc. The length of the velocity vectors corresponds to the distance traveled by the stars in 10 million years. (**b**) Velocities of the same stars relative to the Sun and their radial components (heavy arrows), illustrating the double-wave dependence of the radial velocities from (5.2.23)

independent of r. The numerical values of Oort's constants are found to be:

$$A = 14\,\mathrm{km\,s^{-1}\cdot kpc^{-1}} \quad \text{and} \quad B = -12\,\mathrm{km\,s^{-1}\cdot kpc^{-1}} \,. \tag{5.2.25}$$

The distance of the Sun from the galactic center is as a first approximation taken to be that to the center of the system of globular clusters. However, the period-luminosity relation can also be applied directly to the RR Lyr stars in the regions of the Milky Way which are not too strongly covered by dark clouds of cosmic matter. With an uncertainty of about 15%, the result is

$$R_0 = 8.5\,\mathrm{kpc} \,. \tag{5.2.26}$$

Thus, with (5.2.21), we obtain the orbital velocity of the Sun,

$$V_0 = 220\,\mathrm{km\,s^{-1}} \tag{5.2.27}$$

or $\omega_0 = 26\,\mathrm{km\,s^{-1}\cdot kpc^{-1}}$, corresponding to a rotational period of $2.4 \cdot 10^8$ yr. Since the beginning of the middle age of the Earth in the Triassic (Table 2.8.2), we have therefore made one trip around the center of the galaxy.

The numerical values for R_0 and V_0 (and for Oort's constants) are based on the evaluation of a large amount of observational data, which we cannot discuss further here. They were adopted in 1985 by the International Astronomical Union and replace the somewhat higher values recommended by the IAU in 1963, $R_0 = 10$ kpc and $V_0 = 250\,\mathrm{km\,s^{-1}}$.

In the immediate *neighborhood of the Sun*, according to (5.2.20, 21) and using the values of Oort's constants (5.2.25), both the angular velocity and the linear orbital velocity decrease with increasing R; for example, $V(R)$ changes by $-2\,\mathrm{km\,s^{-1}\cdot kpc^{-1}}$.

Outside the range of optical observations in the plane of the galaxy, we must depend on *radioastronomical* measurements of the 21 cm line of neutral hydrogen for deriving the *rotation curve*. In contrast to stars and star clusters, for interstellar gas the distance cannot be determined directly. The method which is applied in this case will be discussed in Sect. 5.3.3; in Fig. 5.2.11, we simply summarize the results of the 21 cm observations.

5.2.6 Galactic Orbits of the Stars. Local Mass Density

How well does our previous assumption of circular galactic orbits hold up? If we allow noticeable *eccentricities e* of the stellar orbits, we find that relative velocities of $100\,\mathrm{km\,s^{-1}}$ and more would occur in the immediate

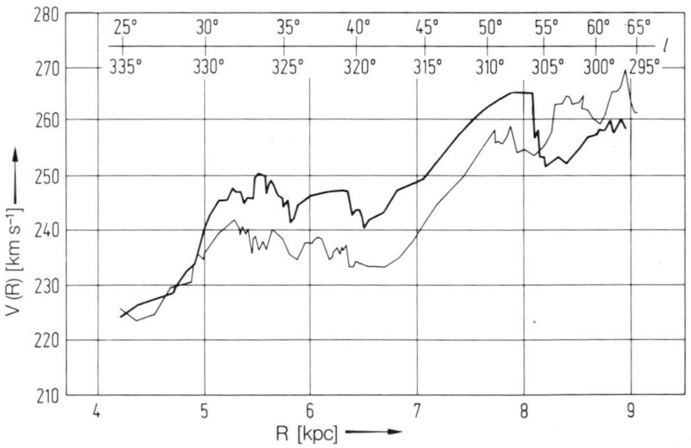

Fig. 5.2.11. Rotation curve $V(R)$ of the Milky Way, from observations of the 21 cm line of neutral hydrogen for $4\,\mathrm{kpc} \le R \le 10\,\mathrm{kpc}$, after F. J. Kerr (1964). The observations from the Northern Hemisphere ($l < 90°$, *heavy curve*) and from the Southern Hemisphere ($l > 270°$, *light curve*) deviate systematically from each other (large-scale asymmetry of the galaxy or slight expansion?). The irregularities in both curves ($\le 10\,\mathrm{km\,s^{-1}}$) are probably related to the spiral structure of the Milky Way. Within $R < 4$ kpc, there are strong deviations of the motion of H I from circular orbits. The numerical values in this figure are based on older values of R_0 and V_0 which are somewhat too high; compare (5.2.26, 27)

neighborhood of the Sun. We can thus understand the phenomenon of *high-velocity stars*, as J. H. Oort remarked in 1928.

If we initially limit our considerations to stellar orbits *in the galactic plane*, then their orbits are determined by their galactic velocity components U and V or by the analogous velocity components relative to the Sun's vicinity,

$$U' = U \quad \text{and} \quad V' = V - 220\,\mathrm{km\,s^{-1}} \ . \tag{5.2.28}$$

The position coordinates of stars which are accessible to precise observation can namely be set equal to those of the Sun with sufficient accuracy. In a diagram with the coordinates U' and V', and for a given galactic force or potential field, we can therefore draw for example curves of constant eccentricity e, curves of constant apogalactic distance R_1, etc. Calculations of this type were first carried out by F. Bottlinger in 1932 for a $(1/R^2)$-force field; Fig. 5.2.12 shows a corresponding *Bottlinger diagram* for a force field which is adjusted to better fit the actual Milky Way.[4] The velocity vectors of the high-velocity stars indicate that these stars move on orbits of large eccentricity e around the galactic center, in some cases in the direct sense, in some cases with a retrograde motion. The "normal" stars in our vicinity, in contrast, have small values of U' and V', i.e. they all move (like the Sun) in the direct sense on nearly circular orbits.

The motions of the stars in our neighborhood *perpendicular* to the plane of the Milky Way can be understood to a large degree, according to J. H. Oort (1932, 1960), by considering the distribution of mass density ϱ in our region of the galactic disk to be *planar*. We thus take into account here only the dependence on the distance z of the mass density from the galactic plane. We can then consider the W components (5.2.12) of the stellar velocity vectors independently of their motions (U and V) parallel to the galactic plane. The stars undergo oscillations through the galactic plane and perpendicular to it with periods of about 10^8 yr. The distribution of the stellar density perpendicular to the plane is connected with the gravitational field of the galactic disk on the one hand, and with the velocity distribution of the W components on the other, in an analogous manner to the relation between the density distribution of molecules in an atmosphere and the gravitational field as well as the Maxwell-Boltzmann velocity distribution or the temperature. In the case of the galaxy, however, the gravitational field is itself directly related to the matter density ϱ by

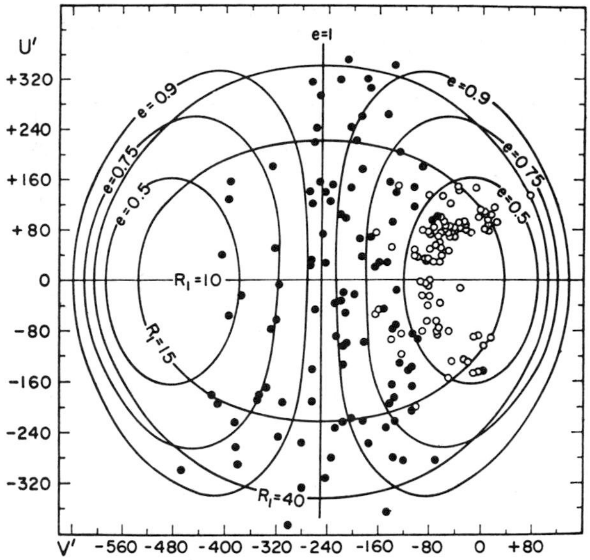

Fig. 5.2.12. A Bottlinger diagram for stars with spatial velocities $> 100\,\mathrm{km\,s^{-1}}$. The galactic velocity components U' (to the anticenter) and V' (in the direction of the rotation) are plotted relative to the neighborhood of the Sun; the axes correspond to the absolute velocity components U and V. Velocities are in $[\mathrm{km\,s^{-1}}]$. The eccentricity e of the orbit and its apogalactic distance R_1 in [kpc] can be read off the two families of curves. ● Stars with an ultraviolet excess, $\delta(\mathrm{U-B}) > +0.15$ mag (Sect. 5.4.3), i.e. metal-poor stars of Halo Population II; these are all high-velocity stars with large spatial velocities. ○ Stars with $\delta(\mathrm{U-B}) < 0.15$ mag; these stars make up the transition region from Halo Population II to the disk population, i.e. to stars with more nearly circular orbits. (After O. J. Eggen)

Newton's law of gravitation (or the Poisson equation). Therefore, Oort (1960) was able to estimate the overall matter density in the galactic plane near the Sun; he obtained:

$$\varrho = 1.0 \cdot 10^{-20}\,\mathrm{kg\,m^{-3}} = 0.15\,\mathcal{M}_\odot\,\mathrm{pc^{-3}} \ . \tag{5.2.29}$$

More recent determinations yield $0.19\,\mathcal{M}_\odot\,\mathrm{pc^{-3}}$.

We compare this matter density with the overall density of the *stars* which have been observed in our immediate neighborhood (within 20 pc), which is about $0.05\,\mathcal{M}_\odot\,\mathrm{pc^{-3}}$, and the contribution of the *interstellar matter* which is around $0.04\,\mathcal{M}_\odot\,\mathrm{pc^{-3}}$. Owing to the uncertainties both in the number of faint M dwarfs and white dwarfs, which make up the main contribution to

[4] In a field which deviates from the $1/R^2$ law, the stellar orbits are in general not closed curves and e is not an orbital element in the strict sense, as it is in the case of planetary orbits.

the stellar mass density, and in the local density determined from gravitation, we cannot at present say just how much *dark matter* contributes; its contribution can be estimated *only* by Oort's method. (The contribution from unknown dark matter in the galactic halo, Sects. 5.2.7 and 5.5.3, is about 0.01 $\mathscr{M}_\odot\,\mathrm{pc}^{-3}$; it is thus unimportant for the mass balance in the vicinity of the Sun.)

Analysis of the *W*-velocity distribution again exhibits clearly the two types of stars mentioned above: the disk stars with $|\bar{W}| \simeq 12\,\mathrm{km\,s^{-1}}$, and the high-velocity stars with considerably larger values. While the disk stars indeed describe nearly circular, planar orbits, the high-velocity stars mostly move on strongly eccentric orbits which are also *inclined* with respect to the galactic plane.

The stellar dynamics also lead us again to the fundamental concept of *stellar populations* introduced in 1944 by W. Baade. We prefer, however, to postpone their discussion to Sect. 5.5.4, where we consider the star populations of different types of galaxies; here, we first deal with the large-scale distribution of mass densities in our galaxy, which we can obtain from the observed rotation curve.

5.2.7 The Mass Distribution in the Milky Way Galaxy

If the entire mass $\mathscr{M}$ which determines the circular orbit of the Sun were concentrated in the center of the galaxy, then the following relation would hold, as for the motions of the planets:

$$V_0^2 = \frac{G\mathscr{M}}{R_0}\,. \tag{5.2.30}$$

From this relation, as a first approximation, one obtains for the mass of the Milky Way $\mathscr{M} \simeq 2\cdot 10^{41}\,\mathrm{kg} \simeq 10^{11}\,\mathscr{M}_\odot$. From Kepler's 3rd law (2.6.38), for a "point mass" $\mathscr{M}$, the rotational velocity $V(R)$ would be proportional to $R^{-1/2}$, or the angular velocity would be $\omega(R) \propto R^{-3/2}$. A glance at the rotational curve for our galaxy (Figs. 5.2.11 and 5.5.10) shows that the assumption of a gravitational potential $\propto 1/R$ gives rather poor agreement with the observed curve.

The *mass distribution* or the *density distribution* can be obtained from models which are constructed in such a way that their gravitational potentials yield the observed rotation curve. Here, simplifying assumptions such as rotational symmetry and neglect of the spiral structure are made. For a spherically-symmetric mass distribution, $V(R)$ depends only on the mass $\mathscr{M}(R)$ which is located *within* R,

$$V^2(R) = \frac{G\mathscr{M}(R)}{R}\,. \tag{5.2.31}$$

Our estimate above would thus correspond to the mass within a sphere having a radius equal to that of the Sun's orbit, $R_0 = 8.5$ kpc. In the case of a homogeneous ellipsoid, $V(R)$ is determined only by the mass within its surface (which is an equipotential surface).

The often-used *mass model* of the Milky Way galaxy due to M. Schmidt (1965), which in spite of its simplicity reproduces the kinematic observations sufficiently well, is based on an inhomogeneous, strongly flattened ellipsoid of rotation (eccentricity $e = 0.999$, or flattening ratio $c/a = \sqrt{1-e^2} = 0.05$) having a mass $1.7\cdot 10^{11}\,\mathscr{M}_\odot$, and a small central point mass of $7\cdot 10^{9}\,\mathscr{M}_\odot$. The overall mass of the Milky Way is thus $1.8\cdot 10^{11}\,\mathscr{M}_\odot$ in this model; roughly half of this mass lies within the distance R_0 from the Sun to the galactic center.

More precise models take into account *several* components in the mass distribution of the Milky Way, which can be distinguished from each other on the basis of their dynamics and star populations (see also Fig. 5.2.6): the *disk component*, with a mass density which decreases exponentially on going outwards; the *spheroidal component*, which is hardly flattened; and the *nucleus of the Milky Way* ($R \leq 0.1$ kpc). The spheroidal component includes the halo with a radius of about 20 kpc, containing the globular clusters and high-velocity stars; it merges on going towards the galactic center into the *central region* ("central lens" or "central bulge") with half axes of about $2.5\cdot 1.5$ kpc. The central region of our galaxy is hardly accessible to observations in the optical spectral region, but it can be readily observed in the infrared at $\lambda \simeq 2.2\,\mu\mathrm{m}$.

The galactic nucleus will be treated in Sect. 5.6.3, in connection with activity in galactic centers.

The rotation curve in the Milky Way is more difficult to observe for $R > R_0$ than within the Sun's orbit. Recent measurements in the visible of H II regions and galactic star clusters and in the radiofrequency range (H I and CO lines, recombination lines in H II regions) show that $V(R)$ is essentially flat out to $R \simeq 30$ kpc, perhaps even increasing slightly to about $300\,\mathrm{km\,s^{-1}}$. The Keplerian situation has thus not been attained out to this distance, and therefore the total mass of the system does not lie within this radius. Based on these observations and on the analysis of the motion of the Sun within the local group of galaxies (Sect. 5.5.1), as well as on theoretical considerations concerning the stability of rotating flat disks, the following picture of our galaxy has crystallized out since about 1975:

The long-known "visible" subsystems, the central region and the (inner) halo of globular clusters, are probably embedded in an enormous, spheroidal *outer halo* (or galactic corona) of 60 to 100 kpc radius, which consists of "dark matter" of still unknown composition, and which makes up the main portion of the overall mass of the system, about 3 to 10 times as much as the visible portions.

We shall return to the question of this halo, with its dark populations, in connection with the discussion of rotation curves of other galaxies in Sect. 5.5.3.

5.3 Interstellar Matter. Cosmic Radiation

The matter which is finely distributed between the stars of the Milky Way at first came to the attention of astronomers in the form of *dark clouds*, which weaken and redden the light of those stars which are behind them, due to absorption and scattering. But it was only in 1930 that R. J. Trumpler was able to show that even outside the recognizable dark clouds, *interstellar* extinction and reddening are by no means negligible in the photometric determination of *distances* throughout the Milky Way Galaxy. Already in 1922, E. Hubble had recognized that the galactic (diffuse) *reflection nebulas* (like the one which surrounds the Pleiades, for example) are due to scattering of the light from relatively cool stars by cosmic dust clouds, while in the galactic (diffuse) *emission nebulas*, interstellar gas is excited by the radiation from hot stars and therefore emits line spectra. Following his observations, the investigation of the *interstellar gas* quickly gained momentum in the years 1926/27. The "stationary" Ca II lines had already been discovered in 1904 by J. Hartmann; they occur in the spectra of binary stars but do not show Doppler shifts corresponding to the orbital motion. Only in 1926 was an explanation developed, theoretically by A. S. Eddington, and based on observations by O. Struve, J. S. Plaskett, and others: the interstellar Ca II, Na I, . . . lines are produced in a gas layer which is partially ionized by the stellar radiation. This gas layer fills the entire disk of the Milky Way and takes part in its (differential) rotation. On the other hand, in 1927, I. S. Bowen succeeded in making the long-sought identification of the "nebulium lines" in the spectra of gaseous nebulas, finding that they are due to *forbidden transitions* in the spectra of [O II], [O III], [N II], . . .; and H. Zanstra developed the theory of nebular luminescence. Only about ten years later was it recognized that in the interstellar gas, as in stellar atmospheres, hydrogen is the strongly predominant constituent. O. Struve and his coworkers discovered with the aid of their nebula spectrograph, which had great light-gathering power, that many O and B stars or groups of such stars are surrounded by well-defined regions which fluoresce in the red hydrogen recombination line, Hα. Here, the interstellar hydrogen must thus be *ionized*. The theory of these H II regions was formulated in 1938 by B. Strömgren.

Neutral hydrogen (one speaks of H I regions) at first seemed not to be directly observable, until in 1944, H. C. van de Hulst calculated that the transition between the two hyperfine levels of its ground state must lead to a *radiofrequency* emission of measurable intensity at $\lambda = 21$ cm. This line was first observed in 1951, almost simultaneously at the Harvard Institute, in Leiden, and in Sydney, and led to completely new insights into the structure and dynamics of interstellar hydrogen and thus of the galaxies. Thanks to progress in amplifier technology in the mm to dm range, numerous other lines in the radiofrequency region have been detected, for example transitions between energy states with very large quantum numbers in hydrogen and helium atoms, lines of the OH radical at $\lambda = 18$ cm with unusual intensities produced by the maser amplification principle, and the rotational transition of the abundant CO molecule at $\lambda = 2.6$ mm.

The surprising discovery of the first *polyatomic* molecule, NH_3, in interstellar space by C. H. Townes and coworkers in 1968 has been followed by the detection up to the present of over sixty species of di- and polyatomic molecules.

Progress in *radio* and *infrared* astronomy has led us to recognize that large, dense clouds of molecules are, in fact, the locations where stars are formed. The luminous nebulas, very noticeable in the optical region, appear in the border areas of these molecular clouds in connection with the formation of OB stars.

At the other end of the spectrum, observations from satellites in the *ultraviolet* and *X-ray* regions show the existence of very hot interstellar gases (10^4 to 10^6 K). The most abundant interstellar molecule, H_2, could be observed with the aid of its bands at $\lambda \simeq 100$ nm. Finally, using the tools of *gamma-ray* astronomy, the interaction

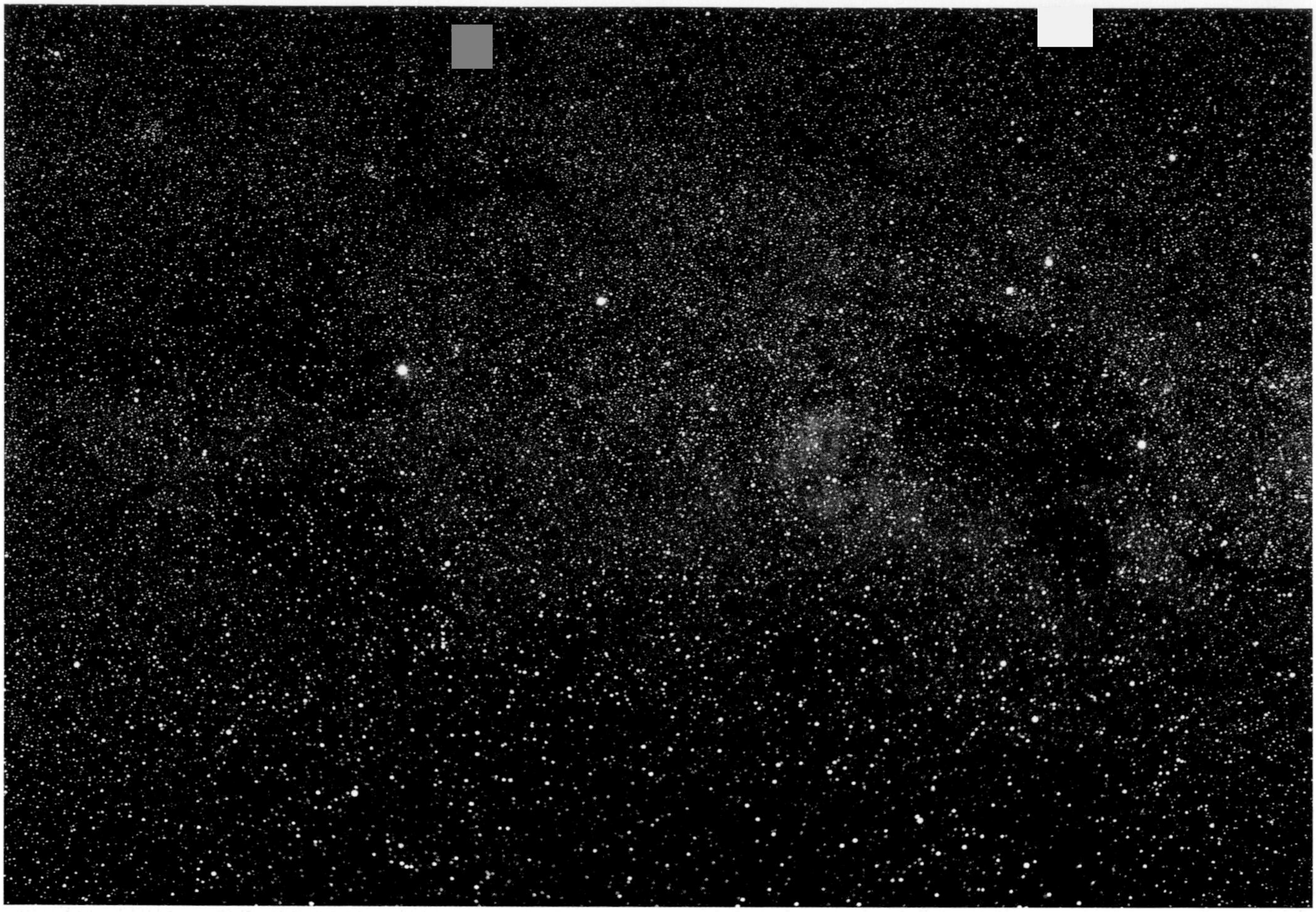

Fig. 5.3.1. The southern Milky Way in the region of the constellations Centaurus and Crux; this picture was taken by C. Madsen, European Southern Observatory (ESO). At the far right is the Southern Cross, and to its left the "Coal Sack", a dark cloud which is near to us, with an extension of 5° times 8° at a distance of 170 pc. The bright star to the right near the Coal Sack is α Cru; the two bright stars in the left half of the picture are α Cen (*left*) and β Cen. The galactic equator is a horizontal line near the center of the picture, passing somewhat to the north of α Cen and α Cru, and through the center of the Coal Sack. (With the kind permission of the European Southern Observatory)

of cosmic radiation with the interstellar gas in the Milky Way has been investigated. This *cosmic radiation* itself consists for the most part of highly energetic protons (and some heavier atomic nuclei); it was discovered in 1912 by V. Hess as a result of its ionizing effect in the Earth's upper atmosphere.

In the following sections, we discuss in sequence the various components of the interstellar matter in our galaxy: the dust (Sect. 5.3.1) makes itself apparent for one thing through the extinction and reddening of starlight, and for another by diffuse spectral bands and its own thermal radiation in the infrared. Neutral atomic gas can be observed by means of many absorption lines of the more abundant elements, which are mostly in the ultraviolet (Sect. 5.3.2), as well as, throughout the entire Milky Way, by the 21 cm line of neutral hydrogen (Sect. 5.3.3). The cool, dense molecular clouds are characterized by a large number of interstellar molecular lines, mostly observable with radioastronomical methods (Sect. 5.3.4); the strongest, at $\lambda = 2.6$ mm, arises from the abundant carbon monoxide molecule. The luminous gas nebulas or H II regions (Sect. 5.3.5) contain gas at about 10^4 K, which is ionized by the ultraviolet radiation of the OB stars. Still hotter components of about $5 \cdot 10^4$ to 10^6 K (Sect. 5.3.6) are detected using the lines of highly ionized atoms and their emissions in the soft X-ray region. Following a short overview of the interstellar magnetic field (Sect. 5.3.7), we turn to the highly energetic components or processes in the interstellar medium, the cosmic radiation (Sect. 5.3.8) and the gamma radiation from the Milky Way (Sect. 5.3.9). In Sect. 5.3.10, we finally give a concluding summary of interstellar matter.

5.3.1 Interstellar Dust

Even with the unaided eye, we can see *dark clouds* against the background of the bright starfields, especially in the southern Milky Way, such as the well-known "Coal Sack" in the Southern Cross (Crux), the dark cloud in Ophiuchus, etc. E.E. Barnard, F. Ross, M. Wolf and others have made the most beautiful photographs with relatively small cameras of high light-gathering power (Fig. 5.3.1); they show a strong concentration of dark clouds towards the plane of the Milky Way. The well-known "division of the Milky Way" is clearly caused by a long, extended dark cloud. Pictures of distant galaxies give an even clearer image of the relationship between these dark nebulas and the *spiral arms*. M. Wolf was the first to estimate the distances to some dark nebulas using the diagram which bears his name: the number of stars in the magnitude range from $m-\frac{1}{2}$ to $m+\frac{1}{2}$ per square degree, $A(m)$, is counted in the region of the dark cloud and in one or more comparison fields. If all stars had the same absolute magnitude M, a dark cloud in the distance range r_1 to r_2, or with the reduced (i.e., "absorption-free") distance moduli m_1-M to m_2-M (4.4.15) would produce an extinction of Δm magnitudes and reduce the number of stars $A(m)$ in a manner readily seen from the schematic drawing in Fig. 5.3.2. Because of the scatter in the absolute magnitudes which is in reality present, the precision of the method is not very great, but it is sufficient to show that many of the noticeable dark clouds are no more than a few hundred parsecs from us. Probably even the large complexes in Taurus and Ophiuchus are connected together past the direction of the Sun.

Bright, diffuse nebulas with continuous spectra, the *reflection nebulas*, such as for example the one surrounding the Pleiades, occur where a dust cloud is illuminated by bright stars having temperatures below 30000 K. Often, the transition from a dark to a bright nebula can be seen directly in photographs.

Both in our Milky Way galaxy and in more distant galaxies, the form of the dark clouds gives the impression

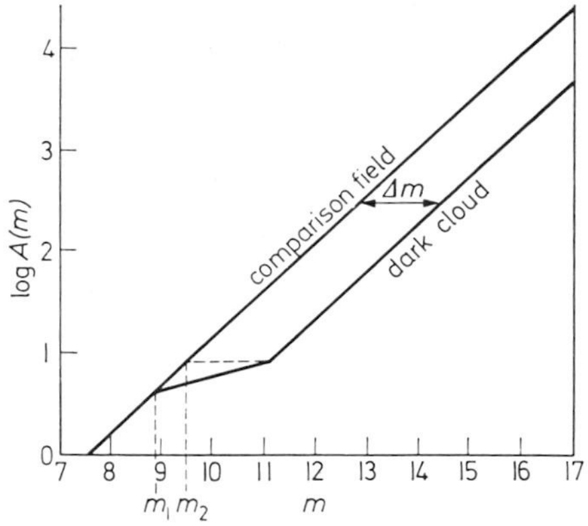

Fig. 5.3.2. A Wolf diagram for determining the distance to galactic dark clouds. The number $A(m)$ of stars per square degree in the magnitude interval $m-\frac{1}{2}$ to $m+\frac{1}{2}$ is plotted as a function of m. The front and back limits of the cloud correspond to the mean stellar magnitudes m_1 and m_2; their extinction corresponds to Δm magnitudes

that structures of only a few parsecs in cross-section are streched out to a length of a hundred and more parsecs.

Although the extinction of starlight in the extended and often not sharply bounded regions of the dark nebulas can readily be detected, it was only in 1930 that the idea of a *general interstellar extinction* (and reddening) became accepted; it plays a decisive role in the photometric determination of larger distances.

If a star of absolute magnitude M is located at a distance r [pc], without interstellar extinction its apparent magnitude m would be given by (4.4.15) in terms of the *true distance modulus*, as we now more precisely term it:

$$(m-M)_0 = 5\log r\,[\mathrm{pc}] - 5 \quad [\mathrm{mag}] \;, \qquad (5.3.1)$$

which is thus simply a measure of its distance. If, however, the starlight is subject to an extinction of γ [mag pc^{-1}] on its way from the star, so that overall $A = \gamma r$ [mag], then we obtain, as the difference between the actually measured apparent magnitude and the absolute magnitude, the *apparent* distance modulus (4.4.17) or (5.2.9):

$$m-M = 5\log r\,[\mathrm{pc}] - 5 + \gamma r \quad [\mathrm{mag}] \;. \qquad (5.3.2)$$

In Fig. 5.3.3, we have plotted the relation between $m-M$ and r, on the one hand, and between $(m-M)_0$ and r on the other, for $\gamma = 0$ (no extinction) and for varying degrees of extinction γ. At extinctions of 1 to

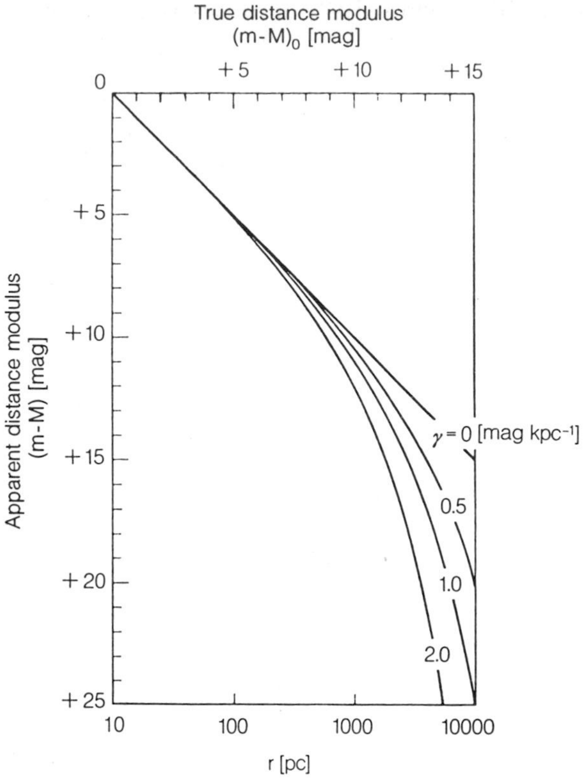

Fig. 5.3.3. The relationship between the apparent distance modulus $m-M$ and the distance r [pc] of stars without interstellar extinction ($\gamma = 0$), and with interstellar extinctions, taken to be uniform, of $\gamma = 0.5$, 1, and 2 mag kpc^{-1}

2 mag kpc^{-1}, our view out to distances of more than a few thousand parsecs is evidently practically cut off.

The contribution γ to the mean interstellar extinction in the plane of the Milky Way was first estimated quantitatively in 1930 by R. Trumpler using a comparison of the angular diameters and the magnitudes of open star clusters of similar structure as functions of their distances. He was able to find an elementary relation between geometric and photometric distance determinations. Trumpler's discovery that the extinction is always accompanied by a *reddening* of the starlight is equally important.

On the average, one can expect a (visual) extinction *within* the plane of the Milky Way but outside the directly recognizable dark clouds of $\gamma \simeq 0.3$ mag kpc^{-1}; if we do not exclude dark clouds, this value rises to $1-2$ mag kpc^{-1}. Concerning the distribution of matter *perpendicular* to the galactic plane, or the dependence of γ on galactic latitude b, some information was obtained through E. Hubble's discovery of the "*zone of avoidance*" (1934) in his investigations of the distribution in the sky of galaxies brighter than a particular limiting magnitude. Their number per square degree is nearly constant towards the galactic poles. Below 30° to 40° galactic latitude, it decreases faster and faster towards the galactic equator, so that in the neighborhood of the latter, a nearly *galaxy-free zone* is found. From this, Hubble concluded that the absorbing matter in the Milky Way forms a flat disk, at whose center we are located, so that extragalactic objects experience a visual extinction $\simeq 0.2 \operatorname{cosec} b$ [mag]. Observations of stars in our galactic neighborhood then showed further that the (full) half-absorption length of the absorbing layer in our vicinity is $\simeq 300$ pc, corresponding roughly (as was later discovered) to that of the hydrogen.

The interstellar reddening is described in the framework of spectrophotometry by socalled *color excesses*:

$$E_{X-Y} = (X-Y)-(X-Y)_0 \ , \tag{5.3.3}$$

which indicate the increase of a color index X–Y (4.4.6) relative to its extinction-free value (X and Y are the measured magnitudes in two wavelength regions for some color index system).

Spectrophotometric measurements have shown that in the optical region, the dependence of the interstellar extinction $A(\lambda)$ on the wavelength λ is given to a good approximation by a proportionality[5] to $1/\lambda$. From this relation, as well as from the photometry of objects of known color, we obtain as an *average* relation between, for example, the decrease of magnitude in the visual, A_V, and the color excess E_{B-V}

$$A_V = (3.1 \pm 0.1) E_{B-V} \ . \tag{5.3.4}$$

The dependence of A_λ on wavelength over a large range of λ is shown in Fig. 5.3.4. Although the interstellar extinction is not strong in the infrared (and in the radiofrequency region), it increases through the optical and on going into the far ultraviolet region. We shall deal with the noticeable broad maximum near 220 nm later.

[5] Using this dependence of the interstellar extinction, *reddening independent* indices can be empirically defined, as for example for the UBV system by S. van den Bergh,

$$Q = (U-B) - 0.72(B-V) \ ,$$

and with a corresponding index Δm_1 for Strömgren's narrow-band photometry.

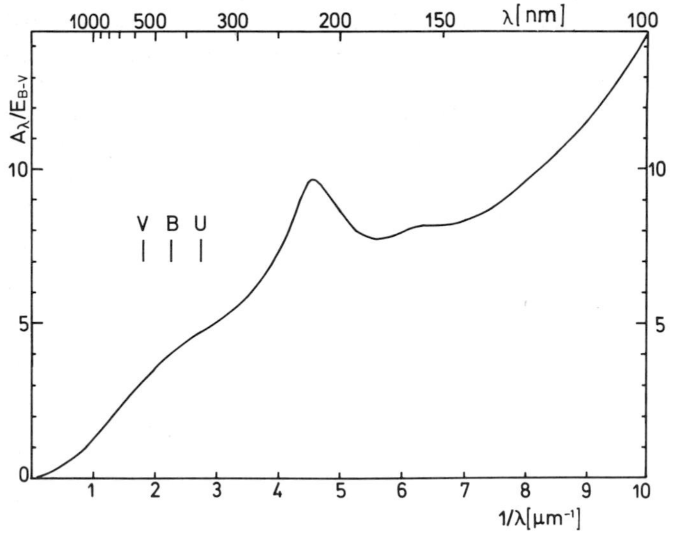

Fig. 5.3.4. The curve of average interstellar extinction, A_λ, after B.D. Savage and J.S. Mathis (1979). The normalization in the visual region is given by $A_V = 3.1\,E_{B-V}$, see (5.3.4)

The distribution of interstellar dust in the Milky Way galaxy is so nonuniform (Fig. 5.3.1) that it is best to determine *directly* the extinction A_V, e.g., for a particular star cluster. This can be accomplished by measuring a color index [usually (B−V)] for stars (e.g., bright B stars) for which the extinction-free color indices are known from examples in our immediate neighborhood, and thereby determining the extinction from the color excess.

Since the color excesses, e.g. for the color indices U−B and B−V, are proportional to one another ($\propto \lambda_{\text{eff}}^{-1}$), the interstellar reddening displaces a star in the *two-color diagram* (Fig. 4.5.5) along a straight *interstellar reddening line*, whose direction was already shown in that figure. If, for example, we know of a star that it belongs to the main sequence, we can extrapolate from the measured color indices U−B and B−V along an interstellar-reddening line of the slope indicated towards the "main-sequence line" and read off the two color excesses and the unshifted color indices. From the color excess E_{B-V}, according to (5.3.4) we immediately obtain the magnitude of the (visual) interstellar extinction. This technique, which can of course be modified in several ways, is one of the most important methods of stellar astronomy.

In 1949, W.A. Hiltner and J.S. Hall made the startling observation that the light of distant stars is partially *linearly polarized* and that the degree of polarization increases roughly proportionally to the interstellar reddening E_{B-V} or the interstellar extinction A_V. The electric vector of the light waves (perpendicular to the conventional plane of polarization) oscillates preferentially parallel to the galactic plane.

If we denote the intensity of the light oscillating parallel to the plane of polarization by $I_\parallel$ and that oscillating perpendicular to it by $I_\perp$, then the *degree of polarization* is defined as:

$$P = \frac{I_\parallel - I_\perp}{I_\parallel + I_\perp} . \tag{5.3.5}$$

The polarization is often quoted in terms of magnitudes, Δm_P:

$$\Delta m_p = 2.5 \log \frac{I_\parallel}{I_\perp}$$

or (5.3.6)

$$\Delta m_p = 2.17P \quad \text{for} \quad P \ll 1 .$$

The largest values of the degree of polarization P are of the order of a few percent; as a function of wavelength, it shows a flat maximum at around 550 nm. The interstellar polarization (in the visual) is correlated with the interstellar reddening E_{B-V} and the extinction A_V. We find:

$$\Delta m_P \lesssim 0.065 A_V , \tag{5.3.7}$$

where, on the average, $\Delta m_P \simeq 0.03\,A_V$. The interstellar polarization indicates that the particles which cause the extinction and reddening are *anisotropic*, i.e. that they are needle-shaped or disk-shaped and partially oriented. The orientation of the particles has been attributed by L. Davis and J.L. Greenstein to a galactic magnetic field of at least a few 10^{-10} T (10^{-6} G). In this field, the particles precess, so that the axes of their largest moments of inertia stay parallel to the magnetic lines of force, while the remaining components of their motion are damped.

Figure 5.3.5 shows the preferred directions of oscillation of the electric vector, and the degree of polarization, for light from about 7000 stars. The strongest polarization is apparently observed under otherwise similar conditions when the lines of force are perpendicular to the line of sight.

In addition to the known interstellar lines, notable for their sharpness, in the spectra of stars, P.W. Merrill in 1934 discovered several broad *interstellar absorption bands*. The strongest lies at $\lambda = 443$ nm, with a full width

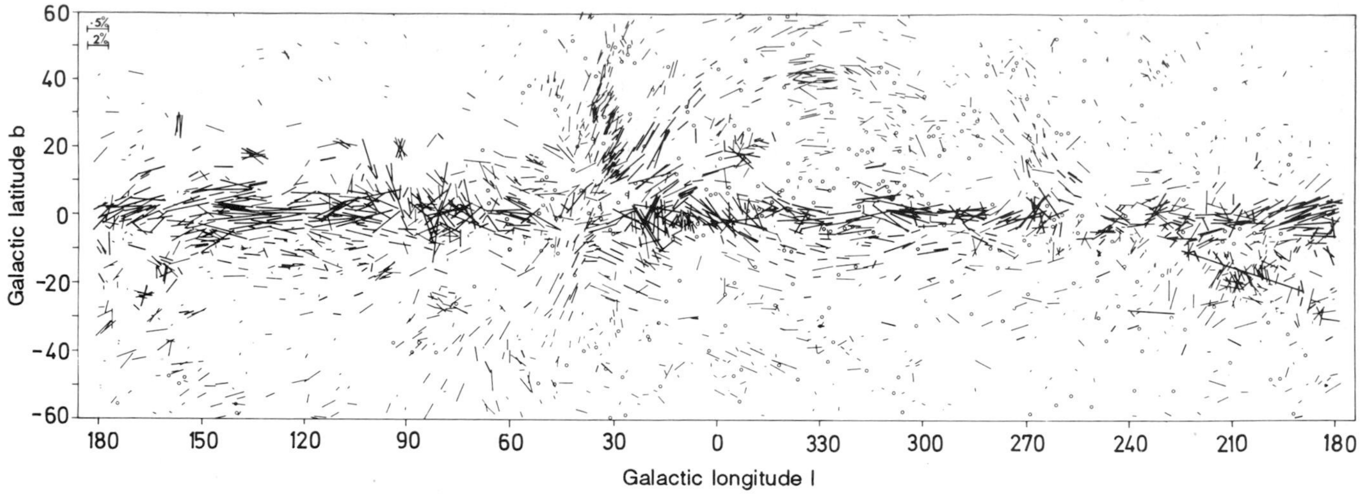

Fig. 5.3.5. Interstellar polarization, from D.S. Mathewson and V.L. Ford (1970), plotted in galactic coordinates. The lines, each of which represents a star at its midpoint, denote the direction of the electric field vector of the optical polarization, and their length gives the degree of polarization P; small circles denote stars with $P<0.08\%$. The scales for the degree of polarization (*upper left corner*) are to be understood as follows: the upper scale for stars with $P<0.6\%$ (*light lines*), the lower scale for stars with $P\geq 0.6\%$ (*heavy lines*). Roughly speaking, this picture can be regarded as the analog of the familiar experiments with iron filings sprinkled onto a sheet of paper above a magnet

at half-maximum of 3 nm. All together, in the optical region there are about 40 such bands, whose identification presents a tedious problem. Owing to their broadness, they can be due only to solid particles or large molecules. A recent suggestion seems promising, according to which these absorption bands as well as some diffuse infrared emission bands are to be attributed to (positively charged) *polycyclic aromatic hydrocarbons* with perhaps 10 to 100 carbon atoms, such as for example coronene, $C_{24}H_{12}$. These relatively stable molecules consist of several benzene rings in a plane.

Coronene

In the *ultraviolet*, we observe a strong, broad interstellar *absorption band* at $\lambda = 220$ nm with a full width at half-maximum of about 40 nm (Fig. 5.3.4). Based on model calculations and laboratory experiments, this band can probably be assigned to small *graphite particles*.

In the *infrared*, where the interstellar extinction is in general weaker, a series of interstellar bands can be observed in absorption and to some extent in emission. The strongest are a band at $\lambda = 9.7$ µm, which occurs together with a weaker structure at 18 µm, and a band at 3.1 µm, whose strength is not correlated with that of the 9.7 µm band. The extinction at 9.7 and 18 µm is probably caused by vibrations of the SiO_4 tetrahedra which occur as functional groups in silicates such as olivine, $(Mg, Fe)_2SiO_4$. Laboratory experiments show that, since there are no structures within the bands, the absorption must be due to *amorphous silicate particles* rather than crystalline silicates. The source of the 3.1 µm band is probably water or ammonia *ice*.

Finally, we wish to obtain a picture of the *composition* and *structure* of the interstellar dust which will explain the observations listed above. The *theory of scattering and absorption* of light by colloidal particles, formulated by G. Mie, H.C. van de Hulst, and others, indicates the following results:

Large particles ("sand"), with radii $a \gg \lambda$, absorb and scatter independently of the wavelength λ, in a manner corresponding approximately to their geometrical cross-section πa^2; very small particles, of $a \ll \lambda$, show cross-sections proportional to λ^{-4} (Rayleigh scattering). The dust particles which give rise to interstellar extinction and reddening must therefore have *radii* of the order of the wavelength of visible to ultraviolet light, i.e. $a \simeq 0.3\ \mu m$. Assuming a density of $3000\ kg\ m^{-3}$, their average *mass* is then about $3 \cdot 10^{-16}$ kg.

The average *particle density* of the interstellar dust in the Milky Way can now be readily estimated from the observed extinction, e.g. in the visual. Its order of magnitude of $1\ mag\ kpc^{-1}$ corresponds over a distance of $L \simeq 1$ kpc to an optical thickness of $\tau_V \simeq 1$ (Sect. 4.4.2). On the other hand, from (4.2.19) we have:

$$\tau_V = Q_{e,V} \pi a^2 N_d L \ , \qquad (5.3.8)$$

where N_d is the particle density of the dust particles. The extinction coefficient per particle (4.2.15) has been replaced here by the *extinction factor* $Q_{e,V}$ expressed in units of the geometric cross-section:

$$k_{\nu,d} = Q_{e,\nu} \pi a^2 \ . \qquad (5.3.9)$$

For particles with $a = 0.3\ \mu m$, $Q_{e,V}$ in the visual is $\simeq 1$, so that we obtain a mean dust particle density of $N_d \simeq 10^{-7}\ m^{-3} \simeq 10^{-13}\ cm^{-3}$, or several $10^{-23}\ kg\ m^{-3}$ in terms of mass density. The mass due to dust is thus only about 1% of the total mass of interstellar matter; only about one dust particle occurs in interstellar space for every 10^{11} hydrogen atoms (Sect. 5.3.3). In spite of this low particle density, in the optical region the extinction due to dust particles is dominant, owing to their enormous interaction cross-sections ($\pi a^2 \simeq 3 \cdot 10^{-13}\ m^2$) compared to atomic or molecular scattering cross-sections.

According to the *Mie Theory*, the frequency dependence of the extinction factor can be calculated for various materials (different indices of refraction of dielectric or conducting materials) and for simple geometric forms (spheres, long needles, etc.). It is composed of an absorption part and a scattering part:

$$Q_e = Q_{abs} + Q_{sca} \ . \qquad (5.3.10)$$

However, these idealized assumptions do not lead to a unique interpretation of the interstellar extinction curve, in particular since it must be taken into account that the dust particles have a broad distribution of "radii" a. The interstellar polarization indicates that the dust particles are anisotropic and partially oriented; the absorption bands in the ultraviolet and infrared show that the dust consists of a mixture of various kinds of particles (graphite, silicates, ices).

The interstellar dust is subjected to the radiation field of the stars in the Milky Way and is "heated" by it to a *temperature* T_d; this temperature corresponds to an equilibrium between irradiation and re-radiation. Analogously to our considerations of the global thermal balance of a planet (2.8.2), we obtain for a (spherical) dust particle of radius a, which is struck by a radiation flux $f_\nu(r)$ at a distance r from a star,

$$\int_0^\infty \pi a^2 Q_{abs,\nu} f_\nu(r) d\nu = \int_0^\infty 4\pi a^2 Q_{abs,\nu} B_\nu(T_d) d\nu \ , \qquad (5.3.11)$$

where the re-radiation is assumed to follow Kirchhoff's law (4.2.30). [$B_\nu(T_d)$ is the Kirchhoff-Planck function for a dust temperature T_d]. Only the absorption part of the extinction factor (5.3.10) enters here; the fraction Q_{sca}/Q_e ("albedo") of the incident radiation is reflected. Since the absorption of stellar radiation is predominantly in the ultraviolet, but re-radiation predominates in the infrared, we carry out suitable averaging over the corresponding spectral regions for simplicity:

$$\bar{Q}_{abs,UV} \bar{f}_{UV}(r) = 4 \bar{Q}_{abs,IR} \sigma T_d^4 \ . \qquad (5.3.12)$$

Here, σ is the Stefan-Boltzmann radiation constant (4.2.28).

Depending on the size a and on the composition of the dust particles, we find from (5.3.12) for dust in the vicinity of hot stars, e.g. in H II regions (Sect. 5.3.5), temperatures T_d of up to several 100 K. According to Wien's displacement law (4.2.26), these dust particles radiate mainly at wavelengths $\lambda \lesssim 30\ \mu m$. This infrared *thermal self-radiation* of the "warm" dust is observed from the H II regions as an excess over the free-free emission of the ionized gas (Fig. 5.3.12).

Surprisingly, the IRAS infrared satellite discovered a radiation flux from the nearby A0V standard star *Vega* (α Lyr), in the 60 μm and 100 μm bands, which is the of order of 10 times as strong as expected from the theoretical atmospheric models; these however reproduce the optical and ultraviolet spectra well (Table 4.8.1 and Fig. 4.4.2). From (5.3.11) and using the observed diameter of the infrared source of $\simeq 45''$, this excess in the thermal emission can be attributed to a *circumstellar dust disk* of about 100 AU diameter with a temperature of only 85 K. The dust particles have radii of at least 10 μm and are thus considerably larger than those in the general interstellar dust (see above). The mass of the halo is of the order of the mass of our planetary system, so that here, and in the cases of some other stars observed by IRAS, we might be dealing with a "protoplanetary system".

Dust which is subjected only to the *general* stellar ultraviolet radiation field of the Milky Way attains temperatures between 15 and 50 K and thus emits only in the long-wavelength portion of the infrared spectrum at $\lambda \geq 100\,\mu m$, a region which was also investigated by IRAS.

The observation of this radiation from our galaxy and from other galaxies gives us information on the *large-scale distribution* of dust. In the Milky Way, the dust is clearly concentrated within the galactic disk; on the other hand, IRAS observed filamentary regions of emission in the far infrared from cool dust even at higher galactic latitudes, the socalled "*galactic cirrus*".

Independently of their material, dust particles evaporate at temperatures above about 1500 to 1800 K. *Dust formation* can take place through condensation ("smoke") of the gas in the cooler outer atmospheres of red giant stars, in the expanding shells of novae, and in connection with the formation of stars and planets. The relative importance of the different processes for the overall balance of dust formation and removal in the Milky Way is still unclear.

5.3.2 Interstellar Absorption Lines

The discovery of the interstellar calcium lines by J. Hartmann in 1904 gave the first indication at all of the existence of interstellar atoms or ions. It was recognized that the H and K lines from the binary star δ Ori show no Doppler shifts corresponding to its orbital motion; they were therefore initially called "stationary lines". In the course of time, interstellar lines of the following atoms, ions, molecules, and molecule-ions were discovered in the *optical* region:

Na I; K I; Ca I, II; Ti II; Fe I; CH; CH^+; CN .

By far the most intense are the H and K lines of Ca II at $\lambda = 393.3/396.8$ nm and the D lines of Na I at $\lambda = 589.0/589.6$ nm.

Using the Copernicus satellite, a group of researchers from Princeton in 1972/73 discovered interstellar lines in the *ultraviolet* from numerous, in part multiply-ionized atoms and molecules. In the spectral region from 95 to 300 nm, nearly 400 lines are known. By far the strongest line is the Lα line from H I at $\lambda = 121.6$ nm (Fig. 5.3.6).

Additional strong lines are due to:

H I; C I, II; N I, II; O I; Mg I, II; Al II, III;
Si II, III; P II; S II, III; Ar I; Mn II; Fe II; Zn II .

A major portion of the ultraviolet interstellar lines belongs to the Lyman and Werner bands of the molecules H_2 and HD around $\lambda \simeq 100$ nm; furthermore, lines from CO are observed.

In all the interstellar lines, the corresponding transitions originate from the ground state term, i.e. they are *resonance lines*. The *ionization* of interstellar matter so far from thermodynamic equilibrium takes place at the low pressures present only through photo processes starting from the ground state; thus, for example, in neutral calcium, with an ionization energy of 6.1 eV, it requires radiation of $\lambda \leq 204$ nm. At a distance r from a star of radius R, its radiation acts with a *dilution factor*:

$$W = \frac{\pi R^2}{4\pi r^2} = \frac{1}{4}\left(\frac{R}{r}\right)^2 \quad . \tag{5.3.13}$$

Recombination, on the other hand, takes place only by means of two-particle collisions between ions and electrons. The degree of ionization reaches a dynamic equilibrium value for which the rates of ionization and recombination are equal (Sect. 4.7.3).

If we now compare the interstellar lines, for example from Ca II and Ca I or Na I, we find, surprisingly, that their intensity ratios are not too different from those found for an F star. How can we understand this? In the interstellar gas, as well as in the stellar atmosphere, the abovementioned processes must be in equilibrium. However, in interstellar matter, the electron density is, on the one hand, about 10^{15} times smaller than in the stellar atmosphere (10^5 compared to $10^{20}\,m^{-3}$), and the recombination rate is correspondingly smaller. On the other hand, the radiation field in interstellar space is diluted relative to that in the stellar atmosphere by about the same factor, $W \simeq 10^{-15}$, if we assume as orders of magnitude $R \simeq 2R_\odot$ and $r \simeq 1$ pc, so that the ionization rate is also reduced by this factor. Thus, in interstellar space, both processes occur with rates about 10^{15} times slower than in the stellar atmosphere, but the degree of ionization remains about the same!

The unexpected observations of ultraviolet absorption lines of *higher* ions such as C IV, N V, and O VI, which indicate very much hotter gas in the interstellar medium, will be discussed in Sect. 5.3.6.

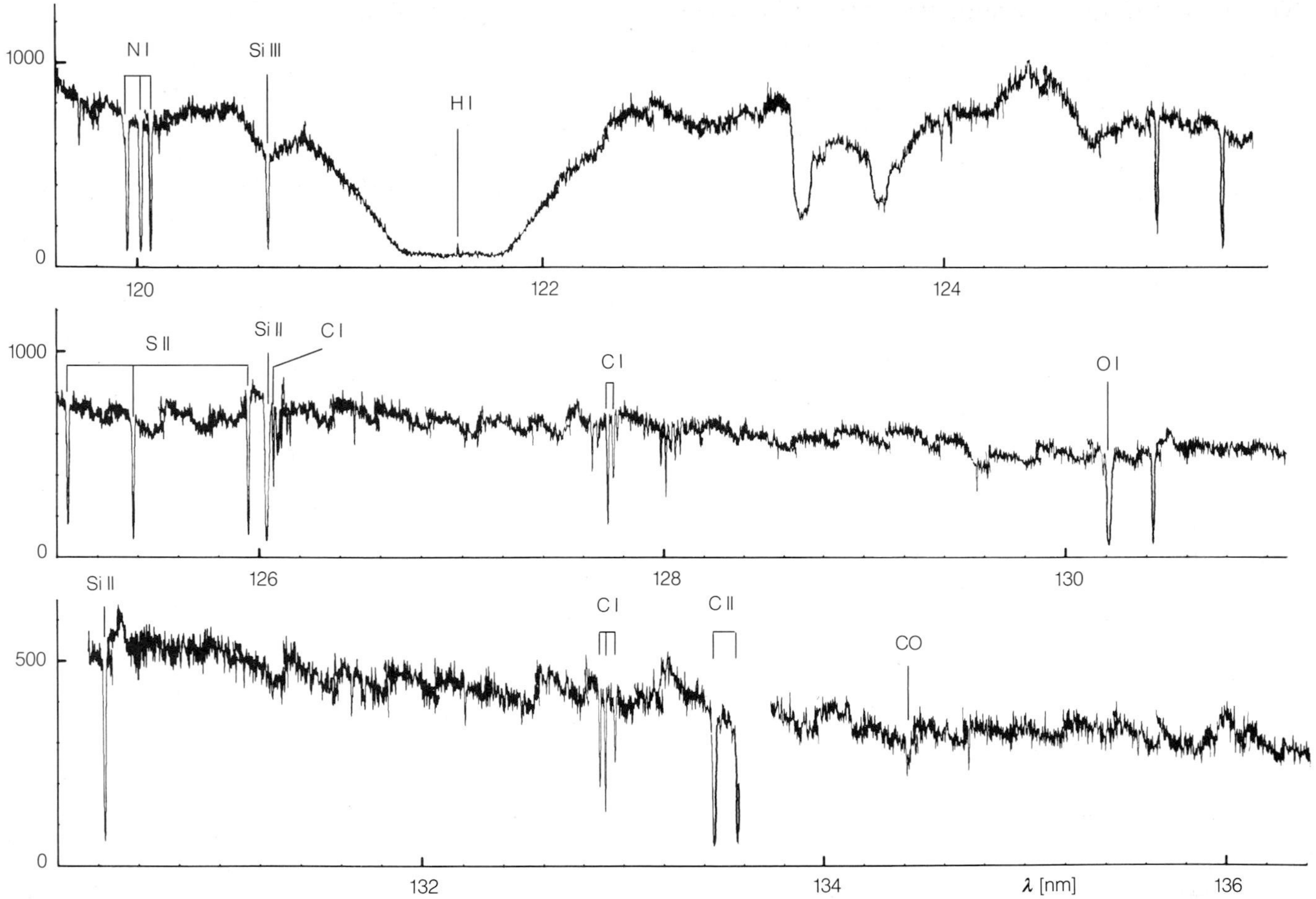

Fig. 5.3.6. Interstellar absorption lines in the ultraviolet spectrum (119.7 – 135.7 nm) of the O9 V star ζ Oph, from observations with the Copernicus satellite. The intensities are not corrected for sensitivity variations of the spectrometer, etc. The strongest interstellar line is the broad, saturated Lα line of H I at $\lambda = 121.6$ nm (in the middle of which the weak emission of the geocorona can be seen). Some of the strongest interstellar lines, which can be distinguished from stellar lines by their sharpness, are identified by the absorbing ion. (From D.C. Morton, Astrophys. J. **197**, 85 (1975); reprinted courtesy of D.C. Morton and *The Astrophysical Journal*, published by the University of Chicago Press; © 1975, The American Astronomical Society)

The equivalent widths of the interstellar lines increase with increasing *distance* of the star in whose spectrum they are observed, in a fairly regular manner. Therefore, they can, conversely, be used to estimate the star's distance. Observations with high spectral resolution (corresponding to 0.5 to 1 km s^{-1}) show that, for example, interstellar calcium lines usually consist of *several components*, which can be attributed to several regions of concentration of interstellar matter, or interstellar *clouds*, through which the light has passed. On the average, about 5 to 10 such clouds occur within a distance of 1 kpc. The radial velocities of the stronger components correspond to those of the spiral arms of the Milky Way.

Taking into account the component structure as well as local differences in the ionization and excitation, the *chemical composition* of the interstellar gas can be derived from the measured equivalent widths of the absorption lines. It is found that along lines of observation with little interstellar extinction or reddening ($E_{B-V} \lesssim 0.05$ mag), i.e. when the light has not passed through any large concentration of interstellar matter, the abundance distribution of the chemical elements in the gas is roughly the same as that on the Sun. On the other hand, in the interstellar clouds, many elements show *underabundances* of varying degrees compared to the solar composition: for example, Ca and Al are up to 1000 times and Fe up to 100 times less abundant than on the Sun, while C, N, and O are only slightly underabundant and S has practically its normal abundance. Clearly, many elements are not to be found in the gas phase in the cooler, more dense areas; instead, they are bound up in the dust components.

From the intensity of the H I and H_2 lines in the ultraviolet (as well as from radioastronomical observations), we can derive the fact that in our vicinity in the Milky Way, about half of the interstellar *hydrogen* is present in the atomic and half in the molecular state. The optical and ultraviolet spectral regions can give no information about the global properties of the interstellar matter in the Milky Way, owing to the strong extinction originating with the dust. This task must be taken up by observations in the infrared and in particular in the radiofrequency regions.

5.3.3 The Neutral Hydrogen Line at 21 cm

The 21 cm line, which is of great importance for the investigation of the large-scale structure and dynamics of the interstellar gas in the Milky Way galaxy, corresponds to a transition within the $1s^2S_{1/2}$ ground state of H I, between its two hyperfine levels with total spin $F=1$ (nuclear and electronic spins parallel) and $F=0$ (nuclear and electronic spins antiparallel). The small energy difference of $6\cdot 10^{-6}$ eV gives a line in the radiofrequency range at

$$\lambda_0 = 21.1\ \text{cm} \quad \text{or} \quad \nu_0 = 1420.4\ \text{MHz} .$$

This transition is "forbidden" (magnetic dipole radiation) and has an extremely small transition probability (Einstein coefficient), $A = 2.87\cdot 10^{-15}\ \text{s}^{-1}$ or, from (4.7.26), an oscillator strength of $f = 5.8\cdot 10^{-12}$. The average lifetime of the upper level with respect to emission of a 21 cm photon is $A^{-1} = 1.1\cdot 10^7$ yr. Under the conditions in the interstellar medium, an equilibrium distribution of hydrogen atoms in the two hyperfine levels 0 and 1 is established by (electron-exchange) collisions of the atoms among themselves, with an average time between collisions of "only" 400 yr. From the Boltzmann formula (4.7.6), the ratio of the populations N is given by the statistical weights $g = 2F+1$, since the exponential function is practically equal to one because of the smallness of $h\nu_0$. We thus have $N_1/N_0 \simeq 3$, or, since the overall density of H atoms is $N_\mathrm{H} \simeq N_1 + N_0$, $N_1 = \frac{3}{4} N_\mathrm{H}$ and $N_0 = \frac{1}{4} N_\mathrm{H}$.

The intensity of the emitted radiation for emission in an *optically thin* layer of thickness l (optical thickness $\tau_\nu = \kappa_\nu l \ll 1$), in which we here assume a constant temperature, pressure, etc. for simplicity, is given by (4.2.32):

$$I_\nu = \kappa_\nu \frac{2\nu^2 kT}{c^2} l . \tag{5.3.14}$$

For the Kirchhoff-Planck function, owing to $h\nu/kT \ll 1$ we can use the Rayleigh-Jeans approximation (4.2.24b). T is the temperature which corresponds to the thermal velocity distribution of the H atoms. The *absorption coefficient* of the 21 cm line is found from (4.7.29) using $N_0 = \frac{1}{4} N_\mathrm{H}$ to be:

$$\kappa_\nu = \frac{1}{4\pi\varepsilon_0} \frac{\pi e^2}{mc} f \frac{h\nu_0}{kT} \phi(\nu) \frac{1}{4} N_\mathrm{H} , \tag{5.3.15}$$

where, in the Rayleigh-Jeans approximation, *stimulated emission* has to be taken into account by the factor $[1-\exp(-h\nu_0/kT)] \simeq h\nu_0/kT \ll 1$.

The case of optically *thick* layers occurs only in a very few places in the Milky Way (see below). These can be recognized by the fact that the intensity takes on a constant maximum value over a wide range of frequencies:

$$I_\nu = B_\nu(T) = \frac{2\nu^2 kT}{c^2} , \tag{5.3.16}$$

from which the temperature T of the interstellar hydrogen can be immediately calculated to be $T \simeq 125$ K. The precise value is fortunately unimportant for the emission from the optically thin layers, since T then drops out of (5.3.14) due to (5.3.15).

The *frequency dependence* of the absorption coefficient or the profile function $\phi(\nu)$ is determined entirely by the *Doppler effect* due to the motion of the interstellar hydrogen. If there are $N(V_\mathrm{r})dV_\mathrm{r}$ H atoms per unit volume in the radial velocity interval V_r to $V_\mathrm{r}+dV_\mathrm{r}$, then we find:

$$\frac{N(V_\mathrm{r})dV_\mathrm{r}}{N_\mathrm{H}} = \phi(\nu)d\nu = \phi(\nu)\frac{\nu_0}{c} dV_\mathrm{r} . \tag{5.3.17}$$

Measurement of the line profiles thus initially yields the number of H atoms along the direction of observation in a column of cross-sectional area 1 m², whose radial velocities V_r or frequencies ν lie in the given interval:

$$V_\mathrm{r} \text{ to } V_\mathrm{r}+dV_\mathrm{r} \quad \text{or} \quad \nu_0 - \frac{V_\mathrm{r}}{c}\nu_0 \text{ to } \nu_0 - \frac{V_\mathrm{r}+dV_\mathrm{r}}{c}\nu_0 . \tag{5.3.18}$$

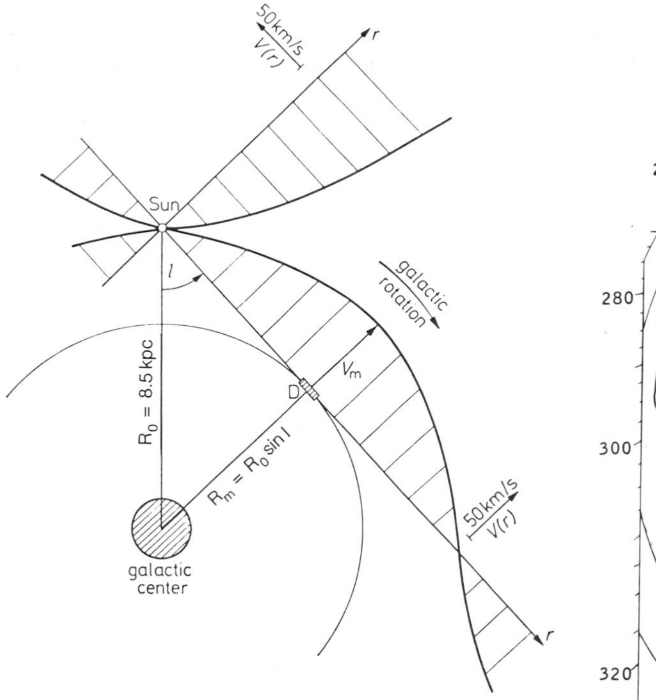

Fig. 5.3.7. The differential galactic rotation of interstellar hydrogen. The radial velocity V_r of the interstellar hydrogen relative to the vicinity of the Sun is plotted along a line of sight of galactic longitude l as a function of the distance r. It has a maximum V_m at D, where the line of sight is tangential to a circular orbit. Compare this plot with the approximation for $r \ll R_0$ in Fig. 5.2.10

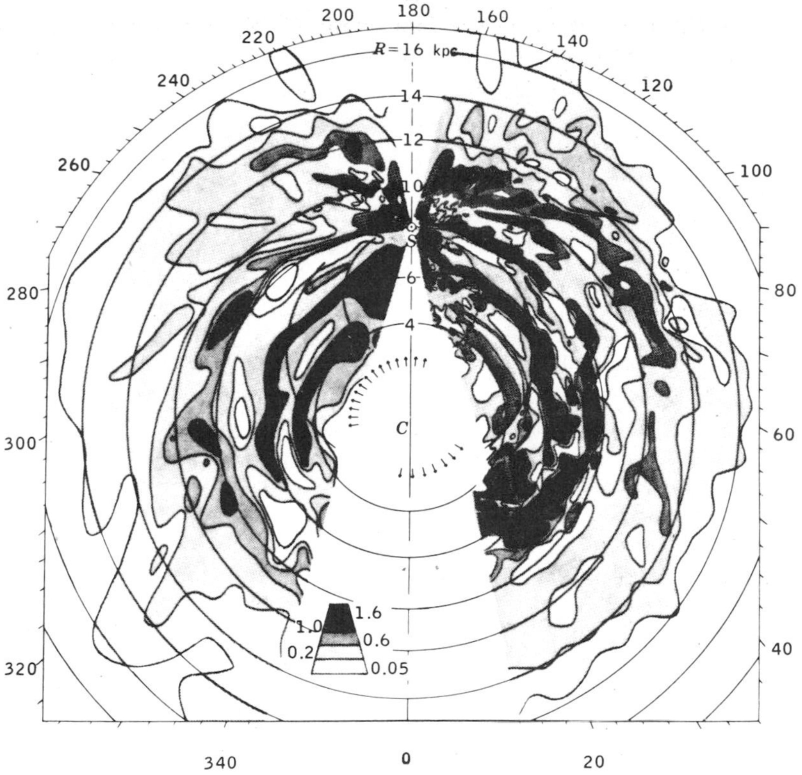

Fig. 5.3.8. The distribution of neutral hydrogen in the galactic plane (maximum densities projected onto the plane). The density scale is in number of atoms per cm^3. *Outer scale*: galactic longitude l

The *velocity distribution* of the interstellar hydrogen is composed of two parts: on the one hand, there are statistical velocities, a kind of turbulence, whose distribution function is similar to a Maxwell-Boltzmann distribution. Their mean value was known from the interstellar Ca II lines to be about 6 km s^{-1}. More important, however, is the contribution from the *galactic differential rotation* (Sect. 5.2.5).

Before we consider this question further, we can estimate the *optical thickness* for the 21 cm radiation in our galaxy. We assume a Gaussian distribution (4.9.16) for the statistical velocities, with a full width at half-maximum of ΔV_r; its value at maximum is then $N(V_0)/N_H \simeq 1/\Delta V_r$ and the optical thickness at the center of the 21 cm line (index 0) is found from (5.3.15 and 17) to be:

$$\tau_{v_0} \simeq \frac{1}{4\pi\varepsilon_0}\frac{\pi e^2}{mc} f \frac{h v_0}{kT}\frac{c}{v_0 \Delta V_r}\frac{1}{4} N_H l$$

or

$$\tau_{v_0} \simeq 1.69 \cdot 10^{-3} \frac{N_H[\mathrm{m}^{-3}]\, l[\mathrm{kpc}]}{T[\mathrm{K}]\, \Delta V_r[\mathrm{km\,s}^{-1}]} . \qquad (5.3.19)$$

For typical values of $N_H = 10^6\,\mathrm{m}^{-3}$ or $1\,\mathrm{cm}^{-3}$ (Fig. 5.3.8), $T = 100$ K, and $\Delta V_r = 10$ km s^{-1}, a distance of 1 kpc would already be optically thick for the 21 cm line. It is only through the "stretching out" of the velocity distribution by the differential rotation, to the order of $\Delta V_r \simeq 100$ km s^{-1}, that it becomes possible for us to observe the *whole* of the Milky Way with the 21 cm line.

The differential rotation in the Milky Way has as its first effect in our neighborhood, according to (5.2.23), a linear increase of the radial velocity V_r (taking the motion of the Sun into account) with increasing distance r. The *directional dependence* of V_r shows the characteristic double wave form, $\propto \sin 2l$. For larger distances from the Sun, we must use the exact relation (5.2.17). It can be seen from Fig. 5.3.7 that V_r attains its maximum value along an observation direction in the direction of the galactic longitude l ($|l|<90°$):

$$V_m = R_0[\omega(R_m)-\omega_0]\sin l\ , \qquad (5.3.20)$$

at that point where the line of sight touches the galactic circular orbit of "minimal" radius $R_m = R_0 \sin l$, i.e. at the point D. The line profile at V_m therefore shows a steep decrease towards larger radial velocities. By combining the V_m for various galactic longitudes l, one can find the *rotational velocity* $V(R)$ as a function of the distance R from the center of the galaxy (Fig. 5.2.11).

Conversely, from the dependence $V(R)$, the distribution of radial velocities V_r along any line of observation can be calculated; and thus from the *measured line profiles* I_ν or $I(V_r)$, which generally show a complex structure composed of several components, the density distribution of hydrogen can be found. The uncertainty for $|l|<90°$ as to whether a particular V_r belongs to the corresponding point in front of or behind D can often be resolved by making use of the consideration that the more distant object usually has a smaller extension in a direction perpendicular to the galactic plane, i.e. in b. Furthermore, the self-consistency of the picture which emerges from the evaluation of the 21 cm measurements must be constantly kept in mind and tested.

The distribution of neutral hydrogen in the galactic plane derived from the observations of Dutch and Australian radioastronomers is shown in Fig. 5.3.8. Its increased abundance in *spiral arms* is apparent; however, the large-scale structure of the spiral arms, which we can readily recognize in other galaxies as "outside observers" (Figs. 5.5.1 and 5.5.13), is not immediately clear here. This picture of the distribution of interstellar hydrogen is essentially based upon the assumption that the gas is moving around the galactic center on *circular orbits* (distance determination!). If major departures from this assumption occur, the galactic distribution can*not* be determined from the 21 cm line profiles alone. Outside $R \geq 4$ kpc in the galactic plane, no systematic differences of more than about 10 km s^{-1} are found relative to the circular-orbit velocities; however, in the inner part of the Milky Way, there are surprisingly strong deviations from this picture. There, we see the 21 cm line against the strong continuous radio emission from the galactic center in absorption, with a radial velocity $V_r = -53$ km s^{-1}. A detailed investigation of the dependence of the line profiles on galactic longitude indicates a spiral arm at $R \simeq 3.7$ kpc, which takes part in the rotation of the galaxy, but is in addition *expanding* away from the center with a velocity of about 50 km s^{-1}. There is a counterpart to this socalled "3 kpc arm" behind the galactic center, with $V_r = +82$ km s^{-1} (and a further arm segment with $V_r = +135$ km s^{-1}). We shall meet up with still further unusual gas flows in the central region of our galaxy in Sect. 5.6.3.

Observations of the 21 cm emission *perpendicular* to the galactic plane show that the neutral hydrogen on the average forms a *flat disk*. The distance between the surfaces where the density has decreased to half its value in the galactic plane is about 240 pc within the range $4 \leq R \leq 10$ kpc. Further outwards, the disk becomes increasingly thick [due to tidal forces caused by the Magellanic Clouds (Sect. 5.5.2)?] and is deflected systematically from the galactic center plane (cf. also Fig. 5.5.14).

The mean *particle density* of H I in the galactic plane between 4 and 14 kpc distance from the center is about $4 \cdot 10^5$ m^{-3} or 0.4 cm^{-3}; it decreases on going further inwards or outwards (Fig. 5.3.9). The total mass of the interstellar H I in the Milky Way is about $2.5 \cdot 10^9\ \mathcal{M}_\odot$. Sky surveys of the 21 cm line with high angular and velocity resolutions ($\delta\phi \simeq 10'$, $\delta V_r \simeq 1$ km s^{-1}) show that the spiral-arm fragments are composed of numerous regions of concentration or *diffuse* H I *clouds* which can have a wide range of sizes. Characteristic values for an H I cloud are a diameter of roughly 5 pc, a mean density of $2 \cdot 10^7$ m^{-3}, and an average temperature of about 80 K, corresponding to a mass of about 30 $\mathcal{M}_\odot$.

The concentration of hydrogen in the diffuse H I clouds is correlated with the concentration of interstellar dust clouds, as long as the extinction of the latter is not too strong. In the denser regions with higher extinctions due to dust, i.e. in the actual dark clouds ($E_{B-V} \gtrsim$ 0.3 mag), the hydrogen is present mainly in molecular form and can thus not be observed by means of the 21 cm line.

In the line profiles of the 21 cm line, often a very *broad component* can be distinguished as a background to the contributions from individual H I clouds. This emission originates from "*warm*" gas at temperatures around 6000 K and low densities of a few 10^{-5} m^{-3}, in which the hydrogen is partially (10 to 20%) ionized. The diffuse H I clouds are presumably embedded in this socalled "inter-cloud gas", which occupies a large part of the volume of the Milky Way.

5.3.4 Interstellar Molecular Lines. Molecular Clouds

In the diffuse H I clouds, where hydrogen occurs mainly in atomic form at densities $\leq 10^8\,m^{-3}$, we find only a few *simple* molecules such as CH, CH^+, and CN, which have long been known from their optical absorption lines; the hydroxyl radical OH, discovered in 1963 at $\lambda = 18$ cm; and the molecules H_2, HD, and CO, which can be observed by means of their absorption lines in the ultraviolet (Sect. 5.3.2). Following the discovery in 1968/69 through observations of their radiofrequency lines of the first *polyatomic* interstellar molecules, ammonia (NH_3), water (H_2O), and formaldehyde (H_2CO), continuing up to the present time over 60 different molecules have been observed, including some surprisingly complex ones, in the more dense and cooler regions of the Milky Way. In these *molecular clouds*, hydrogen is present mainly in molecular form.

The *hydrogen molecule* H_2, which is by far the most abundant interstellar molecule, has no spectral lines in either the radiofrequency or the infrared regions which could be used to determine the large-scale distribution of cool, molecular hydrogen gas in the Milky Way. The Lyman and Werner bands in the ultraviolet can give information only from our immediate galactic neighborhood, owing to the strong interstellar extinction at these wavelengths; the rotational-vibrational spectra and the quadrupolar transitions in the infrared are sensitive to considerably hotter H_2 gas, which is present only near regions where star formation takes place.

The galactic distribution of H_2 can, however, be determined *indirectly* by observations of the second-most abundant molecule, stable *carbon monoxide*, CO; its abundance relative to the hydrogen molecule ($CO/H_2 \simeq 10^{-4}$ in particle numbers) varies only slightly from cloud to cloud, in contrast to those of other molecules. The transition from the first excited rotational level ($J = 1$) to the ground state ($J = 0$) occurs for the most abundant isotope, $^{12}C^{16}O$, at

$$\lambda_0 = 2.60\,\text{mm} \quad \text{or} \quad \nu_0 = 115.27\,\text{GHz} \quad (^{12}CO) .$$

Since the emission in this line is optically thick in many directions, the weaker line corresponding to a transition in the molecule $^{13}C^{16}O$ containing the isotope ^{13}C ($^{12}C/^{13}C \simeq 40 \ldots 80$) at

$$\lambda_0 = 2.72\,\text{mm} \quad \text{or} \quad \nu_0 = 110.20\,\text{GHz} \quad (^{13}CO)$$

may also be used.

We cannot give a complete treatment of the fundamentals of *molecular spectroscopy* here; instead, we must limit ourselves to a few basic remarks. As in the case of atoms (Sect. 4.7.1), the energy states in molecules are essentially determined by the electrons, corresponding to their quantum numbers. Here, however, the electric field of the nuclei is, in the simplest case of a *diatomic* molecule AB, a dipole field. The transitions between two different electronic states fall, as a rule, in the optical or the ultraviolet spectral ranges. Each electronic state is split as a result of the oscillations of the atoms within the molecule into *vibrational levels* (quantum number v), and these are in turn split into a series of *rotational levels* due to the rotation of the molecule (quantum number J). Transitions between (pure) vibrational or rotational levels are typically in the infrared or the radiofrequency regions. For the cool molecular clouds, transitions between the lowest rotational levels are especially important. In the simplified case of a *rigid rotator* ("dumbbell"), in which the two nuclei of masses m_A and m_B, spaced apart at a fixed distance a, rotate about their common center of gravity on an axis perpendicular to the line joining them (Fig. 2.6.4), we can define a moment of inertia (2.6.44) using the relation (2.6.36):

$$I = \frac{m_A m_B}{m_A + m_B} a^2 . \tag{5.3.21}$$

The *rotational energy*[6] is then:

$$E_{rot} = \frac{h^2}{8\pi^2 I} J(J+1) = hcBJ(J+1) , \tag{5.3.22}$$

where $J = 0, 1, 2 \ldots$ is the rotational quantum number and B is the rotational energy constant (usually quoted in units of cm^{-1}). The selection rule for electric dipole radiation is $\Delta J = 0, \pm 1$ (except $0 \leftrightarrow 0$).

The appearance of the *reduced mass* $(m_A m_B)/(m_A + m_B)$ in I and in E_{rot} demonstrates that molecular spectra allow the observation of different *isotopes*; for example, there is, from (5.3.22), an energy or frequency shift of a factor of 1.046 between the lines of $^{12}C^{16}O$ and $^{13}C^{16}O$ corresponding to the reduced masses of 6.86 and 7.17, respectively (cf. above).

[6] According to classical mechanics, the rotational energy would be $E_{rot} = I\omega^2/2$ (ω = angular velocity) or, introducing the angular momentum $S = I\omega$ (2.6.43), $E_{rot} = S^2/2I$. In the quantum-mechanical expression, S occurs in units of $h/2\pi$, or more precisely, $S = (h/2\pi)\sqrt{J(J+1)}$.

Before we turn to the variety of interstellar molecules observed, we first describe the *spatial distribution of molecular hydrogen* in the Milky Way, as found by observing the CO lines (Fig. 5.3.9). In contrast to atomic hydrogen, the H_2 molecules are concentrated in a flat, broad ring between about 4 and 8 kpc distance from the center of the galaxy; its thickness in the direction perpendicular to the galactic plane is only about 100 pc. The total mass of H_2 in the Milky Way is about $2 \cdot 10^9 \mathscr{M}_\odot$ and is thus comparable to that of neutral atomic hydrogen. Like the interstellar dust and atomic hydrogen, molecular hydrogen occurs in non-uniform concentrations, as clouds. The "*molecular clouds*" or "*CO clouds*" occur in a great variety of sizes (roughly 1 to 200 pc), densities (about 10^9 to $10^{12}\,m^{-3}$), and masses (from 10 to $10^6 \mathscr{M}_\odot$); they are correlated with dark clouds (Sect. 5.3.1). Their extinction in the visible region, A_V, ranges from about 1 mag to over 25 mag. The *giant molecular clouds* with masses $\geq 10^5 \mathscr{M}_\odot$ are, along with the globular star clusters, the most massive objects in the galaxy; their number is estimated to be around 4000. Near the center of the galaxy, we find particularly massive giant molecular clouds (Sect. 5.6.3). Sgr B 2 is notable among all the molecular clouds for the large variety of molecules that it has been observed to contain.

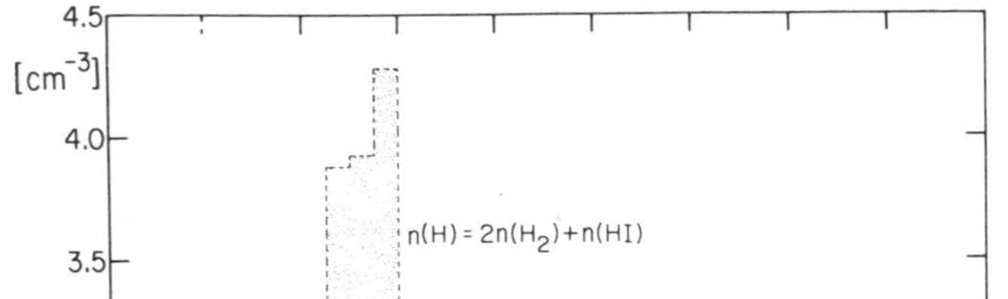

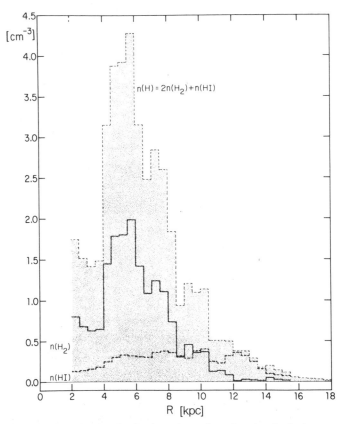

Fig. 5.3.9. The density distribution of atomic and molecular hydrogen in the plane of the Milky Way as a function of the distance to the center of the galaxy, after M. A. Gordon and W. B. Burton (1976). The distribution of H_2 is derived from observations of the $\lambda = 2.6$ mm line of the CO molecule

The *temperatures* of the molecular clouds are in the range 10 to 30 K. Often, dense *compact condensates* with diameters of the order of 1 pc are observed in the far infrared ($\lambda \geq 20\,\mu m$) in the larger molecular clouds; in them, and in their immediate vicinities, the temperatures are often considerably higher (10^2 to 10^3 K). These condensates are *protostars* which are surrounded by a thick shell of dust and can only be recognized by their radiation in the infrared. We shall return in Sect. 5.4.7 to the subject of star formation, which takes place in the most dense parts of the molecular clouds.

We now have a look at the great variety of *interstellar molecules* which can be observed by molecular spectroscopic methods in the radiofrequency and infrared ranges. Favorable observation conditions, with a large number of molecular species, are offered for example by the giant molecular cloud Sgr B 2 in the region of the galactic center, the nearby molecular cloud Ori MC 1 (Orion *M*olecular *C*loud) in Orion (Fig. 5.4.13), the dark cloud Tau MC 1 in the Taurus complex, or the cold gas shell around the carbon star IRC +10° 216.

Nearly all interstellar molecules are composed of the abundant elements H, C, N, O, S, and Si. Along with the simple molecules and radicals already mentioned, such as H_2CO, CH, and OH, the following polyatomic molecules are relatively abundant and thus observable also in other galaxies: ammonia, NH_3; water, H_2O; hydrogen cyanide, $HC{\equiv}N$ and its isomer, hydrogen isocyanide, $H{=}N{=}C$; formaldehyde (methanal), $H_2C{=}O$; and the oxomethylium ion, HCO^+.

Further "inorganic" molecules which are observed are mainly oxides and sulfides such as NO, NS, SiO, SiS, SO, SO_2, and O_3, as well as H_2S, HNO, N_2H^+, and SiH_4. Because of carbon's ability to form complicated chain compounds, the "organic" molecules also exhibit great variety in interstellar space. We find (in addition to those already named) simple molecules, ions, and radicals such as C_2, CS, CSO, HCO, C_2H, C_3H, C_4H, C_3N; HCS^+, HCO^+, and also HOC^+. We also observe among others the hydrocarbons methane, CH_4, ethyne (acetylene), $HC{\equiv}CH$, and propyne, $CH_3{-}C{\equiv}CH$; the alcohols methanol, CH_3OH, methane thiol, CH_3SH, and ethanol, CH_3CH_2OH; formic acid, $HC\overset{O}{\underset{OH}{\diagdown}}$ and its methyl ester (methyl formate) $HC\overset{O}{\underset{O-CH_3}{\diagdown}}$; dimethyl ether, CH_3-O-CH_3; and ethanal (acetaldehyde), CH_3-CHO. In addition, numerous nitrogen-containing compounds occur, such as methylamine, CH_3-NH_2, methanimine, $CH_2=NH$, cyanamide, $NH_2-C\equiv N$, formamide, $HC\overset{O}{\underset{NH_2}{\diagdown}}$, isocyanic acid, $O=C=NH$, and isothiocyanic acid, $S=C=NH$, as well as several nitriles. The series of long, chain-like polyinenitriles (cyanopolyines) which are observed is astounding: they range from hydrogen cyanide, $HC\equiv N$, up to undecapentainenitrile,

$$HC\equiv C-C\equiv C-C\equiv C-C\equiv C-C\equiv C-C\equiv N \ ,$$

and some of them have yet to be synthesized under terrestrial conditions. Further "unusual" molecules include tricarbonmonoxide, $\overset{\ominus}{}:C\equiv C-C\equiv O:^{\oplus}$; 1,3-pentadiyne, $CH_3-C\equiv C-C\equiv CH$; the deuterated ethinyl radical, $\cdot C\equiv C-D$; and common salt, $NaCl$.

In contrast to terrestrial chemistry, at first glance under the conditions of the interstellar medium, organic ring compounds and branched chains would seem to be less favored than the linear chain molecules. However, recently (1984/85), radiofrequency lines from the ring molecules SiCC and cyclopropenylidene, C_3H_2, have been discovered.

$$\begin{array}{cc} Si & \ddot{C} \\ \diagup \diagdown & \diagup \diagdown \\ C\equiv C & HC=CH \end{array} \ .$$

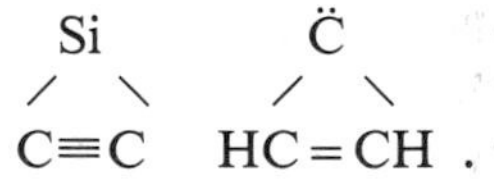

Finally, polycyclic aromatic hydrocarbons have been suggested to explain the diffuse interstellar bands (Sect. 5.3.1).

Even in the most dense molecular clouds, the densities remain so small ($\leq 10^{12} \, m^{-3}$) and the temperatures so low that *molecule formation* takes place far from thermodynamic equilibrium. Dust, which nearly always accompanies the gas, shields it from ultraviolet radiation and prevents the destruction (photodissociation) of the molecules once they are formed. On the other hand, the more energetic protons of cosmic radiation can also lead to partial ionization in molecular clouds. The (positive) ions which are produced in this manner can react efficiently with H_2 and other molecules and lead to the formation of still more complex molecular species. For these usually exothermic *ion-molecule reactions* in the gas phase (E. Herbst, W. Klemperer, 1973), the H_3^+ ion formed in the reaction $H_2^+ + H_2 \rightarrow H_3^+ + H$ seems to play a key role; it produces the observed ions HCO^+ and N_2H^+ by reacting with CO and with N_2.

In contrast, H_2 and the complex chain molecules are probably formed through *catalytic reactions* on the interstellar *dust particles* (D. J. Hollenbach and E.E. Salpeter, 1971). Gas atoms or molecules which strike a dust particle can remain adsorbed and can find reaction partners by diffusion on its surface. Under favorable conditions, the resulting molecule may then desorb from the surface.

In order to derive *element abundances* quantitatively from the observed intensities of the molecular lines, the individual processes of excitation, dissociation, ionization, charge exchange, etc., must be considered separately under the conditions in the interstellar gas, which deviate strongly from thermodynamic equilibrium conditions; this is analogous to the case of atoms (Sect. 4.7.3). However, the number and variety of relevant processes for molecules is considerably greater than for atoms.

If several isotopes are present in a particular molecule, the excitations of its energy levels in general depend only weakly on the difference in isotopic masses. We can thus determine *relative isotopic abundances* rather precisely by observing the deexcitation transitions. Using the molecules CO, HCO^+, H_2CO, HCN, CS, and SiO, it is possible to gain information about the less abundant isotopes ^2D, ^{13}C, ^{15}N, $^{17,18}O$, $^{33,34}S$, and $^{29,30}Si$. The isotopic ratios in interstellar space, in turn, can help elucidate the nuclear reactions which take place in the course of stellar evolution (see Sect. 5.4.4).

Quite unusually high intensities are found for the lines of the *OH radical* at $\lambda = 18$ cm; expressed formally in terms of brightness temperatures (5.6.1), they correspond to 10^{12} to 10^{15} K! The lines are exceedingly sharp and are emitted by very compact sources (≤ 10 AU), as is shown by angular measurements using very-long-baseline interferometry; in many cases, circular polarization is also observed. These extreme non-equilibrium properties can only be explained by the *maser amplification mechanism*[7]: the upper levels corresponding to the transitions

---

[7] Maser = *m*icrowave *a*mplification by *s*timulated *e*mission of *r*adiation.

are strongly overpopulated by a socalled pump process (which is not understood in detail), so that *stimulated emission* (Sect. 4.7.3), produced coherently by a radiation field of the same frequency, predominates over spontaneous emission. This "natural *OH maser*" functions in the microwave range in a manner similar to an optical laser, which produces concentrated, intense beams of radiation in the optical or infrared regions.

The OH masers are classified into two types, depending on their relative line intensities. Type I usually occurs in groups in the neighborhood of strong infrared sources, i.e. in regions where star formation is taking place (Sect. 5.4.7); the other type (II) is found, in contrast, to be correlated with late phases of stellar evolution (supergiants, Mira variables; see Sect. 4.11.1). Maser lines from other molecules are also observed, e.g. from H_2O at $\lambda = 1.35$ cm, and, less strongly, from SiO, H_2CO, and others.

5.3.5 Luminous Gas Nebulas. H II Regions

In the neighborhoods of bright O and B stars, the interstellar gas, and in particular the hydrogen, becomes ionized and is excited to luminosity; we then observe a diffuse nebula or an H II region. This happens, as was recognized by H. Zanstra in 1927, in the following way: when a neutral hydrogen atom absorbs the light from a star in the Lyman region at $\lambda \leq 91.1$ nm, it will be ionized (cf. Sect. 4.7.1). The resulting photoelectron will later be recaptured by a positive ion (proton). This recombination leads only rarely directly to the ground state; usually, there are cascade transitions via several energy levels, accompanied by the emission of small light quanta, $h\nu$. As an estimate, we can say that for each Lyman quantum absorbed, with $h\nu > 13.6$ eV or $\lambda < 91.1$ nm, among other quanta, about *one* Hα quantum will be emitted. If it can further be assumed that the nebula absorbs practically all the Lyman radiation from the star, then following H. Zanstra we can use the Hα radiation of the nebula to estimate the *Lyman radiation* of the star. Comparing it with the optical radiation, we can estimate the *temperature* of the star using the theory of stellar atmospheres. For the O and B stars, values are obtained which lie roughly within the range of temperatures determined spectroscopically.

The same process can be considered from another point of view: the number of *recombination processes* in a unit volume is proportional to the number of electrons, N_e, *times* the number of recombining ions in the unit volume. Since, however, each H atom releases *one* electron on ionization, the number of ions is also $\propto N_e$. The Hα luminosity at a particular place in the nebula will therefore be proportional to its *emission measure* as introduced by B. Strömgren:

$$EM = \int N_e^2 dr \ . \qquad (5.3.23)$$

The integration is to be carried out along the line of observation; r is usually measured in [pc], and the electron density N_e in [cm^{-3}]. For *diffuse nebulas*, EM is of the order of 10^3 cm^{-6} pc and higher. Estimating N_e by assuming the longitudinal and transverse dimensions of the nebula (which are in the range of about 1 to 100 pc) to be equal, we find that the electron density in the H II regions is of the same order of magnitude as the density of neutral atoms in the H I clouds, i.e. 10^7 to 10^8 m^{-3}. In the large diffuse nebulas, such as for example the *Orion Nebula*, values of $N_e \simeq 5 \cdot 10^9$ m^{-3} are attained. Within many nebulas, compact H II regions with considerably higher densities can be observed at radio wavelengths (see below).

The optical and ultraviolet *spectra* of gaseous nebulas (Figs. 5.3.10, 11) are produced under conditions which are far from those of thermal equilibrium, so that a treatment of the individual elementary processes is necessary to describe them (Sect. 4.7.3). The effect of radiation from the star which excites them is reduced by a *dilution factor*, $W \simeq 10^{-16}$ to 10^{-14} (5.3.13). Therefore, in their atoms and ions, only the *ground states* are occupied to any extent, as well as long-lived *metastable states* in lower terms with very small transition probabilities.

In particular, the following processes contribute to the excitation of nebular luminosity:

1) *Hydrogen* and *helium*, the two most abundant elements, are, as we have seen, ionized by the diluted

Fig 5.3.10. The central portion of the η Carinae nebula NGC 3372. This color picture is a composite image (made using blue and red filters and narrow-band filters in the region of the green [O III] line) by S. Laustsen and J. Surdej with the 3.6 m telescope at the ESO. The whole nebula has a diameter of about 3° or 130 pc at a distance of 2.5 kpc; the section shown has angular dimensions of 10′·13′. The bright star in the center of the nebula (to the left of the center of the picture) is η Car, one of the absolutely brightest stars of the "Luminous Blue Variable" type. η Car, which itself is surrounded by a small nebula, exhibits strong brightness fluctuations and a rapid mass loss due to its stellar wind. Although η Car today has a magnitude of about 6, in 1843 it was the second-brightest star in the sky, with a magnitude of -0.8. (With the kind permission of the European Southern Observatory)

Fig. 5.3.11. The Orion Nebula (M 42 = NGC 1976), a diffuse or galactic nebula, and its emission spectrum in the blue. N_2 and N_1 are the "nebulium lines" of [O III] at $\lambda = 495.89$ and 500.68 nm

stellar radiation ($\lambda \leq 91.1$ or ≤ 50.4 nm, resp.). Recombination of the ions and electrons takes place into all possible quantum states. From these initial recombination states, the electrons drop back to the ground state, for the most part via cascade transitions. We thus obtain the entire spectra of H, He I, and possibly He II and some other ions as *recombination lines.*

2) In a few special cases, certain transitions are excited by *fluorescence* (I. S. Bowen). For example, in the recombination of He^+, its resonance line at $\lambda = 30.38$ nm is emitted; it can, by coincidence, be absorbed by the ground state of O^{2+}, exciting its $3d\ ^3P_2$ state. This, in turn, emits a number of O III lines, which can be observed in the near ultraviolet.

3) In 1927, I. S. Bowen made a discovery which created a certain sensation among astronomers: he found that the strong lines at $\lambda = 495.89$ and 500.68 nm, observed in all nebular spectra, and which had long been attributed to a mysterious element called "nebulium" (Fig. 5.3.11), could be interpreted as *forbidden lines* in the O III spectrum, originating with a low-lying *metastable* level in the ground state term of the ion. While the "normal" allowed lines have transition probabilities of the order of $A \simeq 10^8\ s^{-1}$ (electric dipole radiation, Sect. 4.7.3), these are e.g. for the [O III] nebula lines[8] only 0.007 and 0.021 s^{-1}, respectively. They represent examples of magnetic dipole radiation; in other cases, electric quadrupole emissions from ions are observed. The *excitation* of the metastable levels takes place through *electronic collisions* with photoelectrons which are produced by the photoionization of H and He.

Now how does it happen that in nebulas, forbidden and allowed transitions occur with comparable intensities? We consider a line at the frequency ν_0 corresponding to a transition from an excited state 1 into the ground state 0 (Fig. 4.7.6). Under the conditions which apply in a nebula, the excitation takes place only by electron impact (rate C_{01}), while the depopulation of level 1 can occur both through collisions (rate C_{10}), *and* through spontaneous emission (rate A_{10}). From (4.7.40), the population ratio is then given by:

$$\frac{N_0}{N_1} = \frac{C_{01}}{A_{10}+C_{10}} . \tag{5.3.24}$$

Since C_{01} and C_{10} are both proportional to the electron density N_e, for a given A_{10} and with *sufficiently low densities*, $C_{10} \ll A_{10}$ and therefore $N_1/N_0 = C_{01}/A_{10}$ ($\ll 1$). This occurs, e.g. for the [O III] lines, corresponding to their value of $A_{10} \geq 0.01\ s^{-1}$, already when $N_e \ll 10^{11}\ m^{-3}$ or $\ll 10^5\ cm^{-3}$. The *line intensity*, which in the optically thin case is given by the integral over the emission coefficients (4.2.12), then becomes:

$$\int j_\nu d\nu = \frac{h\nu_0}{4\pi} A_{10} N_1 \simeq \frac{h\nu_0}{4\pi} C_{01} N_0 . \tag{5.3.25}$$

It is therefore *independent* of the transition probability A_{10} and is proportional to the electron density times the density of the atom in question, $N_0 + N_1 \simeq N_0$. For very small N_e, the electronic collisions are so rare that even with the small probability of a forbidden transition (and quite certainly for allowed transitions!), the excited state is nearly always depopulated by spontaneous emission.

Now that we have understood how in nebulas the forbidden and allowed lines can have similar intensities, if

[8] Forbidden transitions are denoted by setting the symbol for the spectrum in square brackets.

other conditions are the same, e.g. the same densities N_0 or similar element abundances, it remains to be explained how the [O III] lines can be even as strong as the allowed Hβ line from the much more abundant element hydrogen (Fig. 5.3.11). The difference in abundances is compensated here by the fact that electron impact excitation is much more effective for O^{2+} than is excitation by recombination of H.

The electrons of course lose energy in the course of excitation collisions; it is then radiated away from the nebula, mainly in the forbidden lines. The *electron temperature* T_e is therefore lower than would be expected from the temperatures of the stars, about 7000 to 10000 K.

The forbidden transitions between fine structure levels within the ground state term are, in the more abundant ions, mostly in the *infrared*; for example, the lines [O III] $\lambda = 88.3$ and 51.8 μm, [Ne II] $\lambda = 12.8$ μm, [S III] $\lambda = 18.7$ and 33.5 μm, and [C II] $\lambda = 157$ μm are observed from H II regions. In contrast to the optical lines, these lines experience no attenuation due to extinction by interstellar dust and can thus be observed from the entire Milky Way.

With respect to the physics of their luminosity, but *not* with respect to their place in the cosmos, the diffuse gaseous nebulas are related to the *planetary nebulas*, socalled because of their appearance (Figs. 5.3.12, 5.3.13 and 3.2.14). Zanstra's method gives temperatures for their central stars (Sects. 5.4.4 and 5) from 30000 up to 150000 K. The luminous shells, whose apparent size is several minutes of arc in the case of nearby objects, have radii of the order of 10^4 AU, electron densities of about 10^9 to 10^{10} m^{-3}, and electron temperatures of 10^4 K. The observed splitting of the spectral lines indicates that the shells are expanding: the front side is approaching us, and the back side is moving away at about 20 km s^{-1}.

Other luminous nebulas include the *supernova remnants* (Sect. 4.11.7), which are not, however, excited by radiation from a central star, but rather by the high temperatures in shockwave fronts. They can be distinguished from the H II regions in the optical spectral region in particular by their noticeably larger ratios for the line intensities of [S II] $\lambda = 671.6/673.1$ nm to the neighboring Hα lines.

After this excursion, let us return to the H II regions. In the *radiofrequency range*, we can observe their *thermal* radiation as a *free-free continuum*. This radiation is produced (Fig. 4.7.1) when an electron is not actually captured by a proton in a near collision, but instead is only deflected. The intensity of the free-free radiation is, like that of the recombination lines, proportional to the emission measure EM (5.3.23). Observations in the optical spectral range (e.g. of the Hα emissions) are severely limited, especially near the plane of the galaxy, by interstellar extinction; but radio waves, like infrared radiation, pass through the interstellar dark clouds practically without hindrance and thus allow the observation of H II regions in the entire Milky Way.

The free-free radiation can be recognized by the fact that its intensity I_ν for emission in an optically *thin* layer

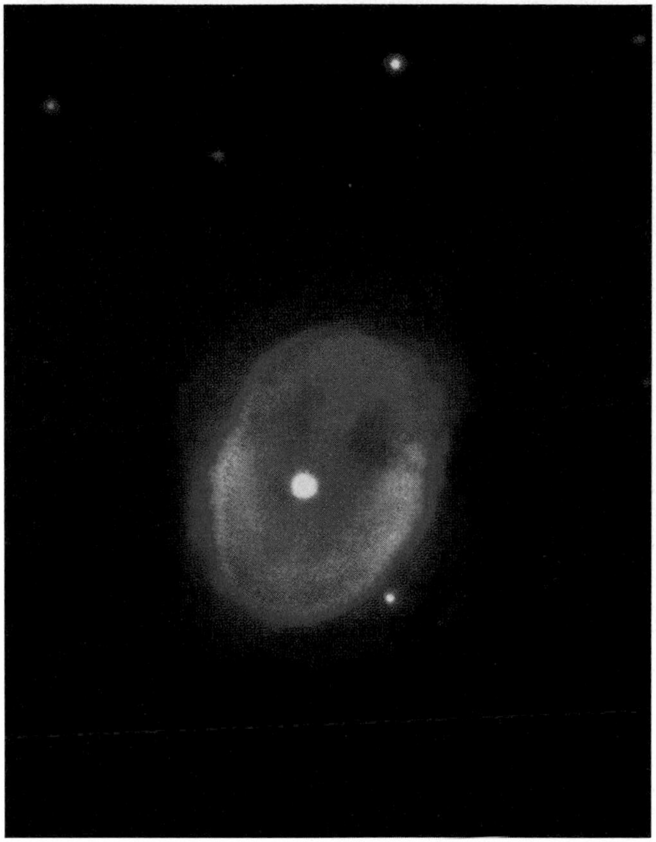

Fig. 5.3.12. The planetary nebula NGC 3132 in Vela; a color image made by S. Laustsen using the 3.6 m telescope at the ESO. The angular diameter of the nebula, about 50″, corresponds to 0.24 pc at its distance of around 1 kpc. The inner, hotter portion appears bluish due to the emission lines of ionized oxygen, and the outer portion appears reddish from the Hα emission of neutral hydrogen. The nebula is excited to luminosity by an extremely hot central star ($T_{\mathrm{eff}} \simeq 150000$ K), which is very faint in the optical region and cannot be seen in this picture. It is only 1.65″ from the bright star (of spectral type A) in the middle of the nebula, which itself is too cool to produce significant excitation. (With the kind permission of the European Southern Observatory)

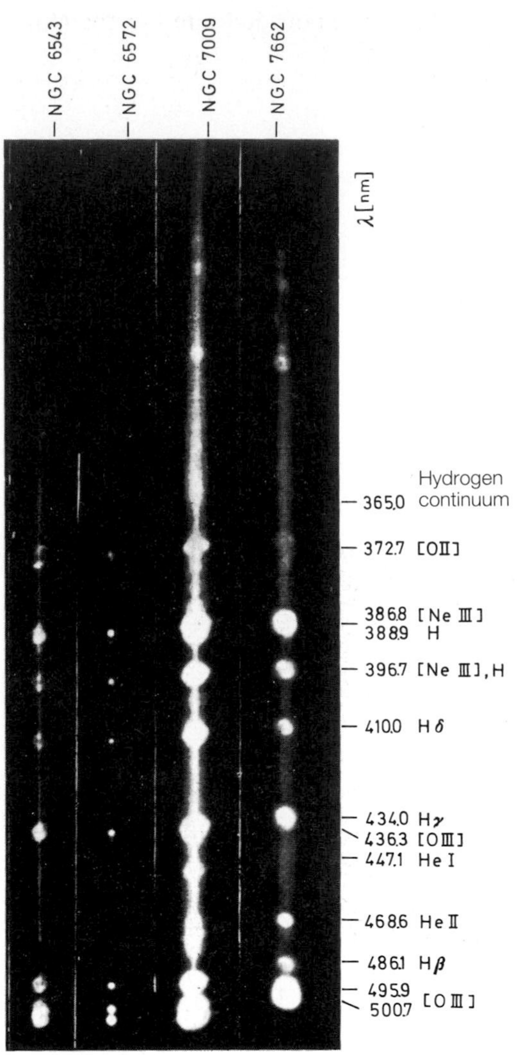

Fig. 5.3.13. Spectra of planetary nebulas, taken by a spectrograph without an entrance slit. This allows the distribution of emission of different lines in the nebular shell to be seen. NGC 6543 also clearly shows the continuous spectrum of the central star

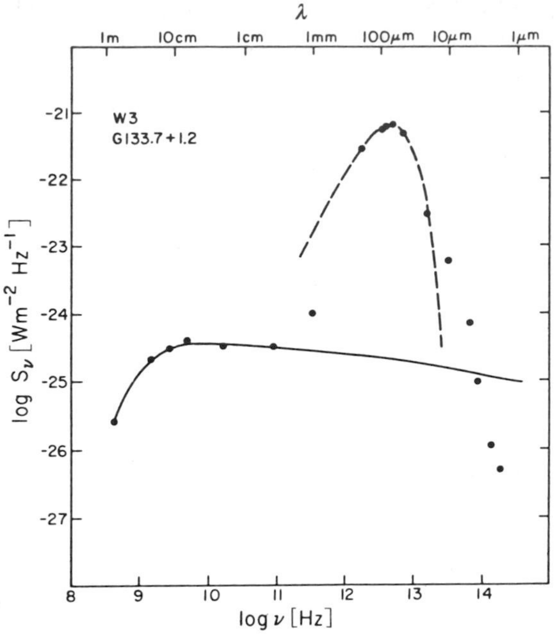

Fig. 5.3.14. Radio and infrared spectra of the H II region W 3 = IC 1795 = G 133.7 + 1.2 (W denotes the number in the sky survey by G. Westerhout, 1958; G = galactic coordinates, $l = 133.7°$ and $b = +1.2°$). The observations (●) of the (spatially integrated) H II region show the thermal free-free radiation of the *ionized* gas (——— theoretical model), which is optically thick for $\lambda \geq 20$ cm, as well as the thermal emission from "warm" *dust* for $\lambda \leq 1$ mm (– – – black body radiation for $T = 70$ K). With a higher angular resolution, several compact H II regions or infrared sources can be seen in W 3; they have qualitatively similar spectra. (Courtesy of the *Publications* of the Astronomical Society of the Pacific)

is practically independent of the frequency ν (Fig. 5.3.14). According to theory, the absorption coefficient κ_ν or the optical thickness τ_ν are proportional to $N_e^2 \nu^{-2}$, so that from (4.2.32), the frequency dependence in the Rayleigh-Jeans limit (4.2.24b) cancels out. At lower frequencies (or higher electron densities), τ_ν becomes $\gg 1$ and the intensity depends on the frequency, $I_\nu = B_\nu(T) \propto \nu^2$. In the infrared, thermal radiation from dust in the H II regions which is heated to about 100 K predominates over the free-free continuum.

In addition to the free-free radiation, in the region of mm to dm waves, *recombination lines* from neutral H, He, and C have been discovered; they correspond to transitions between neighboring energy levels with quantum numbers $n \simeq 100$ to 200. Such excited atoms (Rydberg atoms) are "expanded" to a radius $\simeq a_0 n^2$ (Sect. 4.7.1), i.e. to the order of 1 μm! By making use of the Doppler effect of these lines in the radiofrequency range, it is possible to investigate the kinematics of ionized hydrogen in the Milky Way and the velocity fields in the H II regions. Comparison of the intensities of corresponding lines furthermore allows the determination of the *abundance ratio* of He/H. Suitable lines for this purpose are e.g. the line pairs at $\lambda \simeq 6$ cm: H 109α (5.009 GHz), He 109α (5.011 GHz), and H 137β (5.005 GHz), He 137β (5.007 GHz).[9]

[9] A transition from $n+1 \rightarrow n$ is denoted by $n\alpha$, a transition from $n+2 \rightarrow n$ by $n\beta$, etc.

The result is of the order of He/H $\simeq 0.1$, in reasonable agreement with the analyses of OB stars, which were formed from the interstellar gas only 10^6 to 10^8 years ago. The He/H ratio increases on going from the outer reaches of the Milky Way towards the center, from about 0.07 to 0.11 (Sect. 5.5.4).

Radioastronomical sky surveys have revealed a wide range of sizes and electron densities for the H II regions. At the one extreme, we find the *giant H II regions*, whose radio emissions are many times greater than those of the Orion Nebula. About 80 of these are known in the Milky Way galaxy; we shall treat the giant H II regions near the galactic center in Sect. 5.6.3. At the other extreme, with high angular resolution we can detect *compact H II regions* with diameters of about 0.05 to 1 pc or less, densities $N_e \geq 10^9\,\mathrm{m}^{-3}$, and emission measures EM $\geq 10^7\,\mathrm{cm}^{-6}\,\mathrm{pc}$, which are embedded in extended, less dense regions of ionized hydrogen and are generally not detectable in the optical spectral range. The more dense of these regions often coincide with sources of molecular line emissions (Sect. 5.3.4) and indicate areas where star formation is occurring (Sect. 5.4.7).

The close connection between the more dense molecular clouds, as regions of stellar genesis, and the H II regions, which are ionized and excited by the ultraviolet radiation from the young OB stars, is also expressed in the similarity of the *large-scale density distributions* of ionized and of molecular hydrogen in the Milky Way (Fig. 5.3.9). Ionized hydrogen, like molecular hydrogen, exhibits a concentration in a flat ring between about 4 and 8 kpc distance from the galactic center, as well as in the central region of the Milky Way. The total mass of H^+ in the Milky Way is only about $2\cdot 10^8\,\mathcal{M}_\odot$.

5.3.6 Hot Interstellar Gas

A major modification of our picture of the interstellar medium was effected by the discovery of a *hot* gas component by satellite observations in the ultraviolet and X-ray regions; its temperature is considerably higher than that of the partially ionized hydrogen gas which can be observed by means of the 21 cm line, or the ionized gas in the H II regions.

The occurrence of interstellar absorption lines from *multiply ionized* atoms (Sect. 5.3.2) can be explained only by correspondingly high temperatures, due to the high ionization energies required. Thus, the lines of C II and S III indicate temperatures around $5\cdot 10^4$ K, and the lines of C IV $\lambda = 154.8/155.1$ nm, Si IV $\lambda = 139.4/140.3$ nm, and N V $\lambda = 123.8/124.2$ nm indicate temperatures $\lesssim 10^5$ K. This hot matter presumably occurs in "cloudlike structures" which are moving at relatively high velocities of $\gtrsim 20$ to $100\,\mathrm{km\,s^{-1}}$ relative to the colder gas. Finally, observations of the O VI resonance doublet at $\lambda = 103.2/103.8$ nm show a component at 2 to $8\cdot 10^5$ K (with radial velocities $\lesssim 2000\,\mathrm{km\,s^{-1}}$).

The spectral region *below* the Lyman edge at 91.1 nm has not yet been investigated in detail, since here, the strong absorption in the Lyman continuum of neutral hydrogen limits the observations to our immediate vicinity in the Milky Way. However, as a result of the cloudlike distribution of interstellar matter, objects out to a distance of about 100 pc can be observed in certain directions. In the *X-ray range* at λ below about 10 nm, the interstellar gas again becomes transparent. Although the view is restricted in the soft X-ray region by the absorption due to helium and the K and L absorption edges of the more abundant elements, below about 1 nm wavelength the interstellar medium is almost completely transparent. Observations of a thermal radiation in the soft X-ray range (about 2–12 nm), whose intensity is concentrated in the plane of the Milky Way, indicate the existence of a very hot, tenuous interstellar gas at $\gtrsim 10^6$ K and with electron densities $\lesssim 10^4\,\mathrm{m}^{-3}$, which fills $\lesssim 50\%$ of interstellar space.

The high temperatures and velocities of the hot gas require a considerable energy input to the interstellar medium. Supernova explosions and the stellar winds from OB stars are probably the sources of this energy.

Observations of "interstellar" lines in the ultraviolet spectra of bright stars in the Magellanic Clouds by the IUE satellites showed, surprisingly, several line components with radial velocities which lie *between* those of our local galactic vicinity and those of the two neighboring galaxies. We can conclude the existence of a gaseous *galactic halo*, whose nonuniformly distributed matter stretches out to several kpc from the plane of the Milky Way. Since the corresponding absorption lines are observed from Fe II, S II, and Si II, as well as from C IV and S IV, the halo must contain regions at different temperatures (up to about 10^5 K).

5.3.7 Interstellar Magnetic Fields

Keeping in mind our later considerations of the production, motion, and storage of cosmic-ray particles and of synchrotron electrons in the Milky Way and other galaxies, we deal briefly here with the methods for measuring interstellar magnetic fields.

The *polarization* of starlight by dust particles which are partially oriented through magnetic fields was already mentioned in Sect. 5.3.1. Another possibility for determining the interstellar magnetic fields is based on the *Zeeman effect* of the 21 cm line.

We shall also see that the *synchrotron radiation* of many radio sources is *polarized*. On the one hand, polarization measurements give information about the magnetic fields at the place where the radiation was generated. However, the polarization is rotated when the radiation passes through an interstellar plasma in the presence of a magnetic field; this is none other than the well-known *Faraday effect*. The theory of this effect shows that the rotation ψ of the plane of polarization is proportional to λ^2 times the wavelength-independent *rotation measure RM*:

$$\psi = \lambda^2 RM \ , \tag{5.3.26}$$

where

$$RM = 8.1 \cdot 10^3 \int_0^L N_e B_\parallel \, ds \quad [\mathrm{rad\,m^{-2}}] \ . \tag{5.3.27}$$

Here, we give ψ in units of arc [rad], the wavelength λ in [m], the electron density N_e in [$\mathrm{m^{-3}}$], the longitudinal component of the magnetic induction, $B_\parallel$, in [T], and the distance L (or ds) to the radio source in [pc]. From measurements at several wavelengths, (5.3.26) can be used to determine the original position of the plane of polarization and the rotation measure. The largest observed values of RM for extragalactic sources lie in the range 300 rad $\mathrm{m^{-2}}$. If the electron density "in between" is known, then (5.3.27) yields a correspondingly averaged value of $B_\parallel$.

To summarize, the following methods are available for the investigation of interstellar magnetic fields:

Method	Contributing Medium	Information on
a) Polarization of starlight	Dust	$B_\perp$
b) Zeeman effect of the 21 cm line	Neutral hydrogen	$B_\parallel$
c) Synchrotron radiation	Relativistic electrons	$B_\perp$
d) Faraday rotation	Thermal electrons	$B_\parallel$

Corresponding to the different spatial distributions of the contributing media, methods (a–d) yield different average values of the fields, which are not readily comparable. It is therefore not surprising that attempts to construct a model of the galactic magnetic field have not yet converged to a final result. From all the methods, the order of magnitude of the general galactic magnetic field is about $2 \cdot 10^{-10}$ T or $2 \cdot 10^{-6}$ G, with large-scale variations of the same order of magnitude superposed over regions of about 50 pc in size. In the more dense H I clouds and H II regions, we find fields about 10 times stronger; in molecular clouds, the fields are up to 100 times stronger. Measurements of the Zeeman splitting of the OH maser lines at $\lambda = 18$ cm (Sect. 5.3.4) indicate fields of up to 10^{-7} T in spatially very limited regions.

5.3.8 Cosmic Radiation

In 1912, during a balloon flight, V. Hess discovered that the electrical conductivity of the atmosphere increases with increasing altitude, and explained this as being due to the ionizing effect of energetic *cosmic radiation* from outer space. The nature of these *cosmic rays*, as they are now called, the fact that they consist mainly of *positively-charged particles*, was established after about 1930, first on the basis of their dependence on the Earth's magnetic field and then by direct observation. Their penetrating power (into deep mine shafts) shows that we are dealing with particle energies which by far exceed those encountered in everyday phenomena.

We consider initially the *primary* cosmic rays, which have *not* yet interacted with the matter in the Earth's atmosphere and can be investigated from balloons and rockets; or, before they are influenced even by the geomagnetic field, from satellites and space probes. We shall return later to the additional *solar* component of the primary cosmic rays.

We first deal with the *energy distribution* of *galactic* cosmic rays for the *nucleon component*, consisting of protons, α-particles (He^{2+}), and heavier nuclei (i.e. completely ionized atoms). At low energies, this distribution can be determined on the basis of deflection of the particles by magnetic and electric fields. At higher energies ($\geq 10^{14}$ eV), the ionization energy is measured by observing *major air showers* in the Earth's atmosphere. The results of such measurements over the energy range $10^8 < E < 10^{20}$ eV are summarized in Fig. 5.3.15. Above 10^{10} eV, the influence of the interplanetary magnetic field is negligible. From this energy onwards up to the

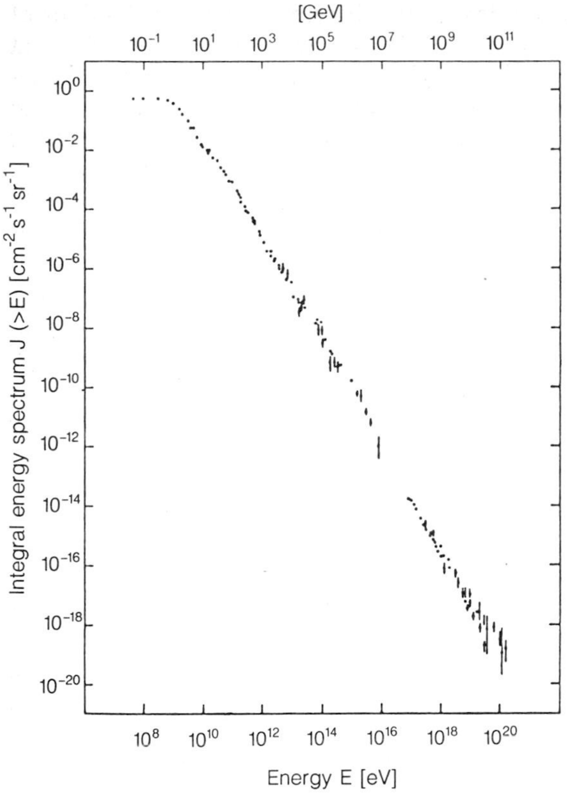

Fig. 5.3.15. The integral energy spectrum of galactic cosmic radiation. The number $J(>E)$ of particles with kinetic energies $>E$ which pass through an area of 1 cm^2 per second and unit solid angle is plotted against the energy E in [eV] or [GeV] (1 GeV = 10^9 eV)

highest measured energies of $\geq 10^{20}$ eV, the energy spectrum follows a power law to a good approximation. At about 10^{15} eV, it exhibits a slight break and becomes steeper; above 10^{19} eV, the measurements become increasingly uncertain due to the rarity of such strong air shower events. The particle flux $J(>E)$ of particles having kinetic energies $>E$ can be represented in sections by the interpolation formula[10]:

$$J(>E) = J_0 \left(\frac{E}{E_0}\right)^{-\beta} . \qquad (5.3.28)$$

If $J(>E)$ and J_0 are given in units of $[\mathrm{cm}^{-2}\,\mathrm{s}^{-1}\,\mathrm{sr}^{-1}]$ and E_0 in $[10^{12}\,\mathrm{eV}]$, then between 10^{11} and 10^{14} eV, $J_0 = 3.0 \cdot 10^{-6}$ and $\beta \simeq 1.5$; while between 10^{16} and 10^{19} eV, $J_0 = 5.7 \cdot 10^{-4}$ and $\beta \simeq 2.0$.

We cannot here go into the details of the complex processes which occur when primary cosmic rays penetrate and interact with the Earth's atmosphere. However, we must consider more precisely the deflection of these charged particles by magnetic fields, namely the geomagnetic field, the interplanetary magnetic field from the Sun, and the interstellar magnetic field in the Milky Way.

In a (homogeneous) magnetic field of flux density $\boldsymbol{B}$, a particle of charge e, rest mass m_0, and velocity v moves, in a plane perpendicular to $\boldsymbol{B}$, on a circular orbit with the cyclotron frequency $\omega_c = 2\pi\nu_c$ and radius $r_c = v/\omega_c$. The relativistic mass of the (moving) particle is $m = \gamma m_0 = m_0/\sqrt{1-(v/c)^2}$, its momentum is $p = mv$, and its energy is $E = mc^2 = \gamma m_0 c^2$ (where γ is the Lorentz factor). For the motion perpendicular to $\boldsymbol{B}$, the centrifugal force mv^2/r_c equals the Lorentz force evB, so that the *cyclotron radius* is related to the momentum by:

$$p = eBr_c \qquad (5.3.29)$$

and the *cyclotron frequency* is given by:

$$\omega_c = \frac{eB}{m} = \frac{1}{\gamma}\frac{eB}{m_0} \qquad (5.3.30)$$

where B is measured in [Tesla][11]. The socalled *magnetic rigidity* Br_c is thus directly related to the momentum p of the particle. For *relativistic* particles ($v \simeq c$, $E \gg$ rest energy $m_0 c^2$; i.e. for electrons, $E \gg 0.51$ MeV, or for protons, $E \gg 938$ MeV), we have $E \simeq cp$ and therefore

$$r_c \simeq \frac{E}{ceB} . \qquad (5.3.31)$$

If we express E in units of [GeV = 10^9 eV] and r_c in [pc], then we have:

$$r_c[\mathrm{pc}] \simeq 1.08 \cdot 10^{-16} \frac{E[\mathrm{GeV}]}{B[\mathrm{T}]} . \qquad (5.3.32)$$

[10] The integral spectrum $J(>E) \propto E^{-\beta}$ corresponds to a power law also for the flux $J(E)$ or the density $N(E)$ of particles in the range E to $E+dE$, with $J(E)$, $N(E) \propto E^{-p}$, where $p = \beta + 1$ [cf. also (5.6.7)].

[11] In the Gaussian system of units, the Lorentz force is evB/c, and B is expressed in Gauss, so that, instead of (5.3.30), we have $\omega_c = eB/mc$, etc. 1 G corresponds to 10^{-4} T. The cyclotron frequency ω_c is twice as large as the Larmor frequency ω_L, which occurs in the description of the Zeeman effect.

The general shape of the orbits of charged particles in a magnetic field results from the superposition of this circular motion with the component of translational motion parallel to the magnetic force lines, which is not affected by the field. The particles thus move on *helical orbits* around the lines of magnetic force.

The *geomagnetic field* has the effect that relatively low-energy charged particles can reach the Earth's surface only in limited zones around the geomagnetic poles. This *latitude effect* becomes negligible, as can be shown by setting r_c = Earth's radius and $B = 10^{-4}$ T, at energies above about 100 GeV.

When the cosmic rays enter the Solar System, they interact with the solar wind and with the *interplanetary magnetic field*, of order 10^{-8} T, which is swept out along with its plasma. It is therefore not surprising that the intensity of cosmic radiation within the Solar System shows a dependence on the 27-day cycle of the synodic solar rotation and on the 11-year cycle of solar activity.

In the *galactic magnetic field* of $2 \cdot 10^{-10}$ T, the *cyclotron radius* r_c for singly-charged particles of energy E is given by:

$$\begin{array}{llllll} E = 1 & 10^3 & 10^6 & 10^9 & [\mathrm{GeV}] & \\ r_c = 5 \cdot 10^{-7} & 5 \cdot 10^{-4} & 0.5 & 500 & [\mathrm{pc}] \ . & \\ \quad (0.1\ \mathrm{AU}) & & & & & \end{array} \tag{5.3.33}$$

Owing to the deflection of the particles by the galactic magnetic field (and, for lower energies, also by the interplanetary and terrestrial fields), their directional distribution is isotropic within the accuracy of the measurements. This underlines the importance of the methods of radioastronomy and of gamma-ray astronomy (Sect. 5.3.9) for investigating their origins.

We have been able to observe the production of cosmic-ray particles in at least one case from "close up", so to speak, since S.E. Forbush was able to demonstrate in 1946 that the *Sun* emits particles having up to several GeV during major eruptions (flares). We cannot describe these and further investigations by A. Ehmert, J.A. Simpson, P. Meyer, and others in more detail here; they give information in particular on the propagation of solar cosmic rays in the interplanetary plasma and magnetic field. The physical mechanisms responsible for the acceleration of charged particles in the magnetic plasma of the solar atmosphere or corona up to energies of 10^9 to 10^{10} eV are still not understood. Whether the principal mechanism is an induction effect, as in a betatron, or whether the particles are trapped between two magnetic "mirrors" (shock-wave fronts?) and thereby accelerated (E. Fermi) is not certain at present. Empirically, however, it is clear and should be remembered that the acceleration of particles to high energies in the Sun as well as in the giant radio sources (Sect. 5.6) always occurs in a highly turbulent plasma with a magnetic field.

We now return to the discussion of the *galactic cosmic radiation* and consider first its nucleon component. Its nuclear charge Z can be found from the tracks the particles leave in nuclear emulsion plates, by using suitable arrangements of counters, or from their tracks in solids,

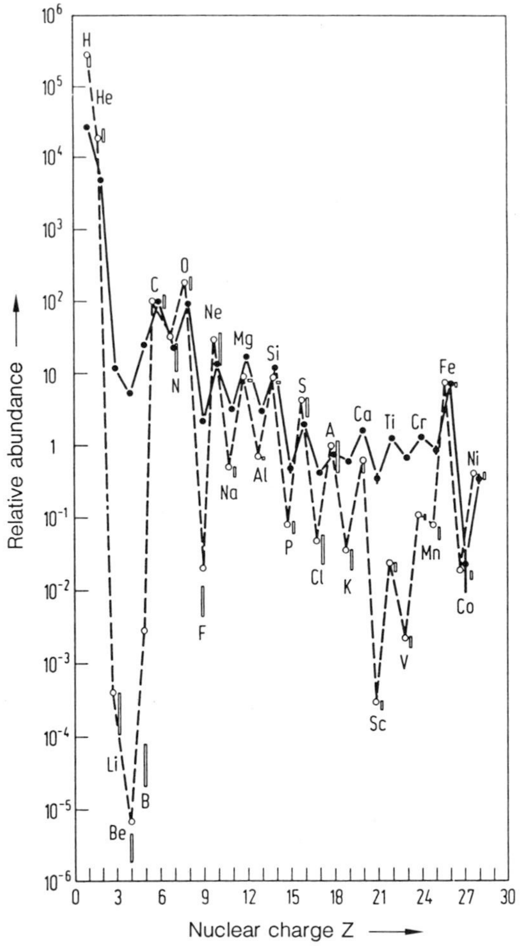

Fig. 5.3.16. Abundance distribution of nuclei in the cosmic radiation (in the energy range from about 0.2 to 0.3 GeV per nucleon) from observations (●) with the IMP-8 satellite by the group at the University of Chicago (P. Meyer, 1981). For comparison, the relative element abundances in the Solar System (○) and in the stars and nebulas in the vicinity of the Sun (□) are indicated. The abundance distributions are set equal for C ($Z = 6$), and are plotted against the atomic number

and thus its *abundance distribution* can be determined (Fig. 5.3.16). This is, on the one hand, relatively similar to the solar abundance distribution (Table 4.9.1); but on the other hand, it exhibits some noticeable differences: the light elements Li, Be, and B, which are exceedingly rare in stars, have nearly the same abundances in cosmic radiation as the following heavier elements. Also, the abundance minimum from Sc to Mn is "filled in" in the cosmic radiation. These overabundances are caused by *spallation*, i.e. as a result of the splitting of heavier nuclei, especially C and Fe, by collisions with protons and α particles in the interstellar medium; these are highly energetic in the rest system of a cosmic-ray particle. The interaction cross-sections for these spallation reactions can be measured with large particle accelerators up to energies of about 10^3 GeV; they are of the order of the geometrical nuclear cross-sections. From the ratio of the numbers of the lightest and heaviest nuclei in the cosmic radiation, it can thus be calculated that the particles have passed through an amount of matter corresponding to about $50\,\mathrm{kg\,m^{-2}}$ or $5\,\mathrm{g\,cm^{-2}}$. This number should however be regarded as an upper limit, since nuclei have been detected up to the end of the Periodic Table. Such "large" nuclei and furthermore all the less energetic nuclei can have passed through only a much smaller amount of matter. On the basis of suitable assumptions about the distribution of the layers of interstellar matter through which the cosmic rays have passed, and using measured or calculated interaction cross-sections for the production and annihilation of the energetic nuclei, M.M. Shapiro and others (1972) were the first to draw some conclusions about the chemical composition of cosmic rays at their points of origin. The *abundance distribution* of the elements in cosmic radiation at their *point of origin* is, within plausible error limits, the same as the abundance distribution in the Sun and other normal stars (Table 4.9.1), perhaps with the exception of the Ne isotopes and "ultraheavy" nuclei of $Z>26$ (overabundance in the Pt region). Before we follow up this important conclusion, we first consider the energy density of cosmic rays.

We start by comparing the *energy densities u* in our galactic vicinity for the most important constituents (5.3.34).

It would seem to be ruled out even by the 2nd Law of thermodynamics that in the Milky Way, roughly the same amount of energy would be generated in the form of extremely nonthermal cosmic radiation as in the form of thermal radiation. Indeed, the orbits of the charged cosmic-ray particles are "wound up" by the galactic

	$u\,[\mathrm{J\,m^{-3}}]$	$u\,[\mathrm{eV\,cm^{-3}}]$
Cosmic rays	$\simeq 1\cdot 10^{-13}$	$\simeq 0.7$
"Thermal radiation", i.e. total starlight	$5\cdot 10^{-14}$	0.3
Kinetic energy $\varrho v^2/2$ of interstellar matter ($\simeq 10^6$ proton masses $\mathrm{m^{-3}}$ at $\simeq 7\,\mathrm{km\,s^{-1}}$)	$\simeq 4\cdot 10^{-14}$	$\simeq 0.2$
Galactic magnetic field $B^2/2\mu_0$ (with $B\simeq 2\cdot 10^{-10}\,\mathrm{T}$)	$\simeq 2\cdot 10^{-14}$	$\simeq 0.1$

(5.3.34)

magnetic field $\boldsymbol{B}$; i.e. they move perpendicular to $\boldsymbol{B}$ on cyclotron orbits whose midpoints carry out a translational motion parallel to $\boldsymbol{B}$. The particles are thus stored in the Milky Way on these helical orbits. According to (5.3.33), we would expect that this mechanism would be effective up to energies of about 10^8 GeV. From the amount of matter the cosmic rays have passed through, about $50\,\mathrm{kg\,m^{-2}}$, we can estimate their path length, using an average density of $\simeq 2\cdot 10^{-21}\,\mathrm{kg\,m^{-3}}$ for the interstellar matter, and find $\simeq 800$ kpc or $2\cdot 10^6$ light years, corresponding to a lifetime of $\simeq 2\cdot 10^6$ yr. Compared to particles which are not deflected, e.g. photons, which have paths in the galactic disk of the order of $\simeq 300$ pc, this means an average path lengthening by a factor of $\simeq 2\cdot 10^3$. The fact that the storage of particles of different Z is about the same shows that the history of a cosmic-ray particle is in general not determined by nuclear collisions, but instead by its eventual escape from the galaxy.

The *energy densities* given in (5.3.34) correspond in each case to about the same *pressure* ($\mathrm{J\,m^{-3}} = \mathrm{N\,m^{-2}}$). The order-of-magnitude equality of the magnetic pressure and the turbulence pressure in interstellar matter seems plausible in the light of magnetohydrodynamics. The approximate equality of the pressure of the cosmic rays to the magnetic and turbulence pressures can be understood by considering that cosmic radiation collects in the Milky Way, "frozen in" by the magnetic field, until its pressure is sufficient to allow it to escape into intergalactic space, probably together with a certain amount of interstellar matter and its associated magnetic field.

From the abundances of certain isotopes, which are produced in *meteorites* by spallation processes, we know that the intensity of cosmic rays has remained practically constant over a period of time of at least 10^8 yr.

Since the average dwell time of a cosmic-ray particle in the Milky Way is only $\simeq 2\cdot 10^6$ yr, but the cosmic-ray intensity has remained rather constant over $\leq 10^8$ yr, it must be "regenerated" continually. Its sources are to be searched for in highly turbulent plasmas with magnetic fields, whose energy densities $B^2/2\mu_0$ adjust themselves to correspond roughly to the kinetic energy density $\varrho v^2/2$. These locations correspond precisely to the strongly nonthermal radio emission regions (A. Unsöld, 1949).

This connection is emphasized by the *electron component* of cosmic radiation discovered in 1961 by P. Meyer and others. Its energy distribution (in the range from about 1 to 300 GeV) obeys a power law, $N(E) \propto E^{-p}$, with $p \simeq 2.5$. As will be shown in Sect. 5.6.1, relativistic electrons produce *synchrotron radiation* as a result of their motion in magnetic fields: nonthermal radiation with a spectrum of the form $I_\nu \propto \nu^{-\alpha}$. The two exponents p and α are related by the equation $p = 2\alpha + 1$ (5.6.2, 8). The exponent $p \simeq 2.5$ of the cosmic electron component would thus lead to a radio spectrum with $\alpha \simeq 0.75$, in sufficiently good agreement with the values found for the galactic synchrotron radiation ($\alpha \simeq 0.4 \ldots 0.9$).

It was uncertain for a long time whether the electron component of cosmic radiation was a secondary product from the interaction of protons with the interstellar medium, through the decay chain $\pi^\pm$ meson→muon $\mu^\pm$ (penetrating component)→electron; or whether it is produced by *direct* acceleration of electrons *together* with the nucleons of the plasma. In the former case, somewhat more positrons, e^+, would be produced than electrons, e^-; while in the latter case, the electrons would strongly predominate. The measured ratio $e^-/e^+ \simeq 10$ indicates that the electron component of the cosmic radiation in the Milky Way is produced by the same sources as the nucleon component.

Which celestial objects are these sources? The abundance distribution of the elements in the sources of cosmic radiation (see above) indicates that its matter was initially formed in the interiors of stars by the nuclear processes which take place there. We have already seen that the Sun produces cosmic rays and synchrotron radiation. But all together, the solar flares and the flare stars make only a tiny contribution to the galactic cosmic radiation. I. S. Shklovsky, V. L. Ginzburg and others have therefore pointed out the importance of *supernovae*. In particular, their (possible) remnants, the *pulsars*, can accelerate considerable amounts of matter to high energies (Sect. 4.11.7). Furthermore, the extremely hard gamma radiation from the X-ray binary star system Cyg X-3 indicates that here, we are probably observing a source of highly energetic particles (from about 10^{17} to 10^{18} eV) (Sect. 4.11.6). The contribution of all these objects to cosmic rays, about which only rather uncertain estimates can be made, attains the correct order of magnitude. To what extent the Fermi mechanism, in which magnetic fields in the interstellar medium move together in shock-wave fronts, also plays a role in the acceleration to high energies is still uncertain.

A further source of synchrotron electrons and cosmic rays has been found, in particular by radioastronomical observations, in the *galactic centers*, that of our Milky Way as well as in the more distant galaxies out to the quasars (Sects. 5.6.3, 5.6.6). This might offer the solution to the riddle of the *most energetic particles* of the cosmic radiation, with energies up to $\simeq 10^{20}$ eV. Since the cyclotron radius at 10^{18} eV is already comparable to the thickness of the galactic disk, and at 10^{20} eV, it corresponds to the dimensions of the whole galactic system, such particles can no longer be stored. They are therefore reduced in number relative to the softer particles by a factor of $\simeq 2 \cdot 10^3$ in the Milky Way system. This suggests that in the energy spectrum of cosmic rays, the contributions of distant galaxies (galactic centers) and quasars become increasingly important at the highest energies.

Cosmic rays with energies below 10^{18} to 10^{19} eV are, in contrast, of *galactic* origin. The contribution from the center of our galaxy probably plays no significant role here, since observations of other, similar galaxies show no correlation between the activities of their centers and the strength of nonthermal radiofrequency radiation from their disks.

Significant contributions to this area are made by *gamma-ray astronomy*, which is closely connected with the observation of cosmic radiation (see the following section); it has the advantage, compared to the latter, that it can give us information about the *direction of origin* of the cosmic rays.

5.3.9 Galactic Gamma Radiation

In the gamma-ray region, i.e. for photon energies above several 100 keV, astronomical observations must be made outside the Earth's atmosphere. Only extremely energetic γ quanta ($\geq 10^{12}$ eV) are observable from the ground by detecting the air showers which they produce in the atmosphere, like those of the energetic cosmic-ray particles.

Since the beginning of the 1970's, large-scale observations in the gamma-ray region from satellites, space vehicles, and stratospheric balloons have become possible due to the development of sufficiently sensitive gamma-ray telescopes. The sky surveys from the satellites SAS-2 and COS-B (launched in 1972 and 1975) are particularly worthy of mention; they achieved an energy resolution of about 50% above 100 MeV with an angular resolution of a few degrees.

Before we discuss the results of gamma-ray astronomical observations, let us attempt to gain an overview of the most important processes leading to the *production* of cosmic gamma radiation. We first note that a thermal origin for these most energetic photons is hardly to be considered, since temperatures of more than several 10^9 K ($kT \geq 0.5$ MeV) would be necessary; such temperatures do not occur, with the exception of the very early stages of the cosmos and some late phases of stellar evolution (supernova explosions). Instead, we must consider a *nonthermal* origin for the gamma radiation. Relativistic electrons (with energies above $mc^2 = 0.511$ MeV) and energetic protons, which become relativistic above 938 MeV, play an important role, since γ quanta can be efficiently produced in the course of their interactions with matter and with fields. Particularly significant for the production of galactic gamma radiation are the *cosmic rays* with their highly energetic protons and electrons.

One possibility is that gamma radiation originates as *bremsstrahlung* from the interaction of energetic electrons with the Coulomb fields of charged particles. In this process, photons are produced at energies of the order of the electron energies E (on the average with $0.5\,E$) with an interaction cross-section which decreases roughly in inverse proportion to the energy of the bremsstrahlung photon. If the energy spectrum of the electrons falls off according to a power law, $\propto E^{-p}$, then the resulting gamma spectrum also follows a power law, $\propto E_\gamma^{-p}$.

Inelastic collisions of cosmic-ray protons with protons of the interstellar matter produce π^0 *mesons* (above a threshhold energy of about 300 MeV) with a cross-section of roughly $10^{-30}\,\mathrm{m}^2$. After about 10^{-16} s, a neutral pion decays into two γ quanta, each with an energy in the center of mass system of $m(\pi^0)c^2/2 = 67.5$ MeV. This component of the cosmic gamma radiation exhibits a broad maximum centered around 67.5 MeV, due to the velocity distribution of the relativistic pions.

Relativistic electrons can also give up a considerable portion of their energy E to low-energy photons such as those of starlight in the Milky Way or the 3 K black-body background radiation (Sect. 5.9.4), and thereby transform them into X-ray or gamma-ray quanta, by *inverse Compton scattering*. Photons of mean energy $\bar{E}_\gamma$ obtain in this process, on the average, an energy

$$\bar{E}_{\gamma'} \simeq \frac{4}{3}\bar{E}_\gamma \left(\frac{E}{mc^2}\right)^2 , \tag{5.3.35}$$

so that, for example, the transformation of photons of the 3 K radiation ($\bar{E}_\gamma \simeq 6\cdot 10^{-4}$ eV) into γ quanta of 10 MeV energy requires electrons of $E \simeq 60$ GeV. In the center of mass system of the electron, inverse Compton scattering can be considered simply as Thomson scattering, with an interaction cross-section of $\sigma_\mathrm{T} = 6.65\cdot 10^{-29}\,\mathrm{m}^2$, since the photon energy "seen" by the electron, $E_\gamma E/mc^2$, is small compared to its rest energy mc^2 over a wide range.

Finally, synchrotron radiation (Sect. 5.6.1), which results from the motion of relativistic electrons in magnetic fields, can also extend into the gamma-ray region. From (5.6.5), the electron energy and the strength of the magnetic field must be sufficiently high to permit this; e.g. in a field of about $2\cdot 10^{-10}$ T or $2\cdot 10^{-6}$ G, which is typical of the interstellar medium, extremely energetic electrons of $E \simeq 10^{16}$ eV would be required for the emission of gamma radiation at 10 MeV. This process thus probably plays no significant role in the production of galactic gamma radiation; on the other hand, it is important in pulsars or neutron stars, with their much stronger magnetic fields of the order of 10^6 to 10^8 T (Sect. 4.11.7).

Aside from the processes mentioned above, in which continuous gamma radiation is produced, we also find *spectral lines* of nuclear origin in the low-energy gamma-ray region (≤ 10 MeV). They correspond to transitions from excited nuclear states into lower-energy states, which take place in nuclides in the course of radioactive decay or following excitation of the nucleus by energetic particles.

When *positrons* are produced in nuclear interactions, we can expect to observe their *annihilation radiation* at an energy of $mc^2 = 0.511$ MeV. Positron annihilation, $e^+ + e^- \rightarrow \gamma + \gamma$, is the inverse process of pair production, which is important in the detection of energetic gamma rays (Sect. 3.4.1).

We now turn to the *observations* in the gamma-ray region.

The gamma radiation of some X-ray binary stars such as Cyg X-3, as well as the intense, short radiation pulses

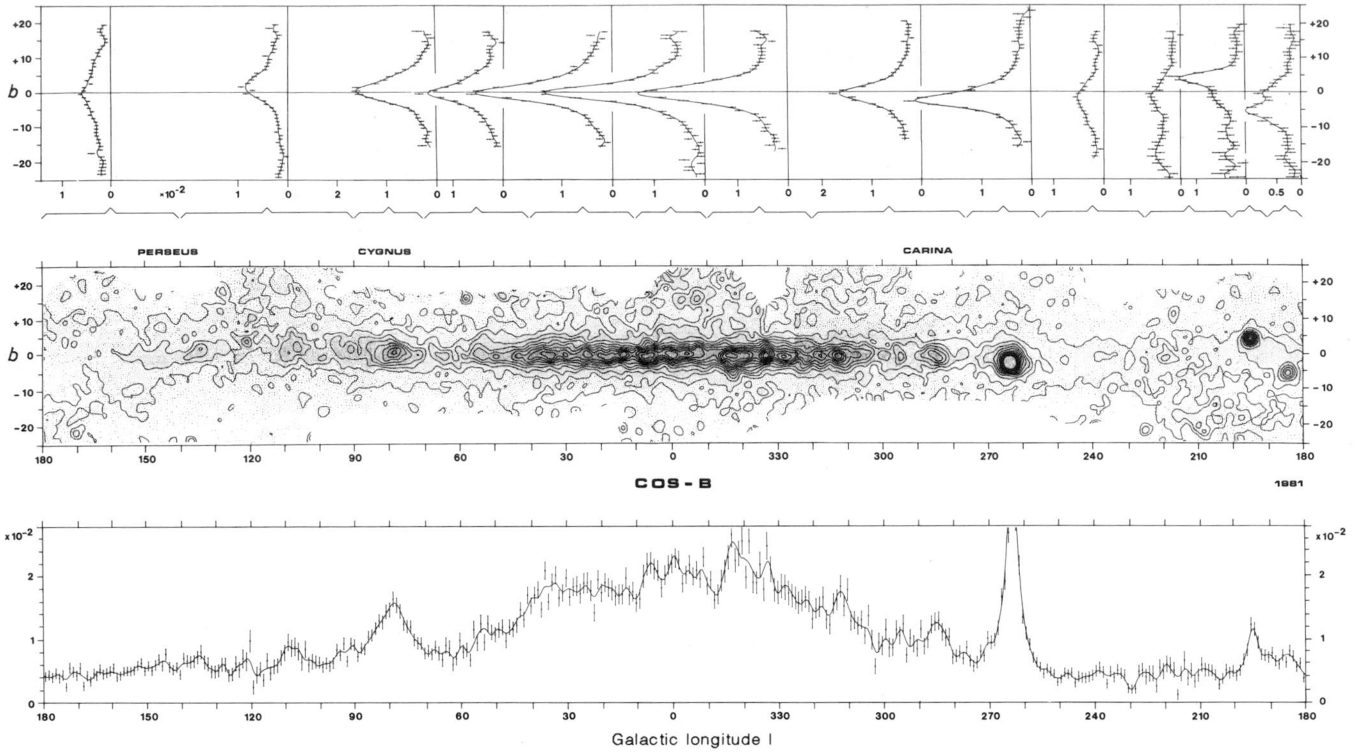

Fig. 5.3.17. A picture of the Milky Way in the gamma-ray region (70-5000 MeV), from observations with the gamma-ray telescope on the COS-B satellite by H. A. Mayer-Haßelwander et al. (1982). The count-rates are given; they are proportional to the intensity and are normalized to the maximum sensitivity for incidence parallel to the detector axis due to the detection efficiency, which is different for each image element of the map. The error bars give the measured points and the smooth curve shows a fit to the data. *Center*: contour lines of constant gamma-ray emission intensity in galactic coordinates (l, b); the separation of the contours corresponds to $3 \cdot 10^{-3}$ counts sr^{-1}. *Upper part*: Dependence of the intensity on the galactic latitude b, averaged over the range of galactic longitudes indicated. *Lower part*: The intensity along the galactic equator $b = 0°$, averaged over $|b| \leq 5°$. The strongest gamma source at $l \simeq 264°$ is the Vela pulsar PPSR 0833−45

from certain sources (γ bursters) were already discussed in Sect. 4.11.6. In the case of the Sun, we have seen that spectral lines in the gamma-ray region are emitted in connection with solar flares (Sect. 4.10.5).

Given the sensitivity of today's gamma-ray telescopes, it is still difficult to detect *gamma-ray lines* from galactic distances. A point of particular interest would be the nuclear spectroscopy of lines which occur in connection with the *nucleosynthesis* of the elements in supernova explosions, for example the observation of the gamma lines which accompany the decay $^{56}Ni \rightarrow {}^{56}Co \rightarrow {}^{56}Fe$ (Sect. 5.4.5). The first gamma-ray line from a radioactive product of nucleosynthesis in the Milky Way to be detected was the 1.809 MeV line which follows the β decay of ^{26}Al (half-life $7.4 \cdot 10^5$ yr) to an excited nuclear state, $^{26}Mg^*$. However, the observed intensity of this line seems to be too great to be explained by supernovae alone.

In the *survey charts* in the gamma-ray region, particularly at energies ≥ 100 MeV, the emissions of the *Milky Way* predominate as a narrow band, only a few degrees wide, around the galactic equator. The intensity exhibits several maxima which seem to be correlated with spiral arm structures; within $|l| \leq 40°$ from the galactic center, we find especially strong emissions (Fig. 5.3.17). Initially, about 30 "point-like" galactic gamma-ray sources stand out above the background; owing to the still limited angular resolution, a definite identification with known objects is in general not possible. Only two of the strongest point sources can be attributed without doubt to *pulsars*, the Vela and the Crab pulsar (Sect. 4.11.7), on the basis of their characteristic variability. Some other point sources are probably associated with dense molecular clouds. Two strong gamma-ray sources can be identified with extragalactic objects, the Seyfert

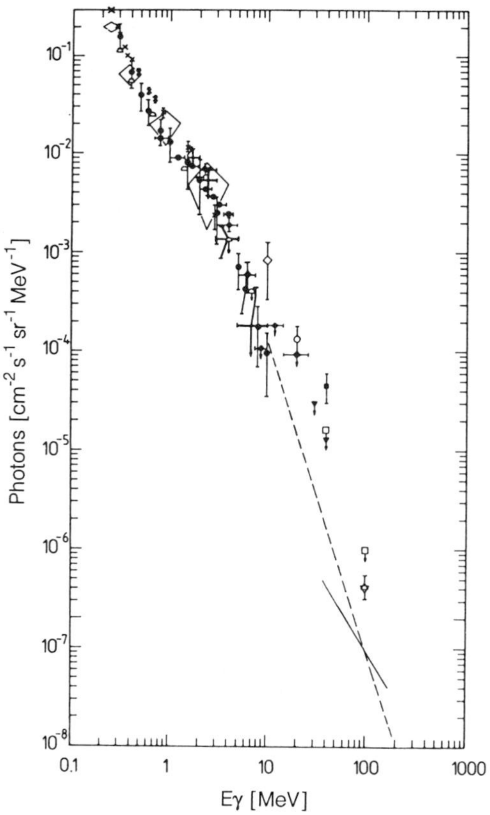

Fig. 5.3.18. The energy spectrum of diffuse gamma radiation, from various authors. The dashed line at high energies gives the *extragalactic* contribution, which is obtained by subtraction of the *galactic* contribution (*fully drawn curve*) from the observed spectrum

galaxy NGC 4151 and the Quasar 3 C 273 (Sects. 5.6.4 and 5.6.5). The strongest source which has not yet been identified is 2 CG 195+4 = Geminga at $l = 195°$ and $b = 4°$.

In addition to the point sources, a genuinely *diffuse component* contributes to the galactic gamma radiation; it has its origin in the interaction of cosmic rays with the interstellar matter. Its intensity distribution and spectrum (Fig. 5.3.18) indicate that presumably two processes are important in the production of this diffuse gamma radiation: the bremsstrahlung from relativistic electrons in the fields of interstellar nuclei, and the decay of neutral pions formed in collisions between cosmic-ray protons and the protons of the interstellar medium. In contrast, the inverse Compton effect appears to be of little importance in the production of the gamma radiation of the galactic disk, but it may contribute more strongly to the radiation from higher galactic latitudes (and also from the more compact extragalactic sources).

In the direction of the *galactic center*, a time-variable gamma radiation is observed at 0.511 MeV, which is interpreted as annihilation radiation from positrons and electrons (Sect. 5.6.3).

Finally, in the gamma-ray region, a nearly isotropic *extragalactic* component is observed, whose energy spectrum (Fig. 5.3.18) above 10 MeV is steeper than that of the galactic component. Since gamma radiation is practically not absorbed, it contains information from "cosmological" distances, i.e. from a very early phase of the cosmos, corresponding approximately to a redshift of $z \simeq 100$ (Sect. 5.9).

5.3.10 The Interstellar Medium: A Summary and Overview

In the preceding sections, we have dealt with the various constituents of the interstellar medium in our galaxy, from dust particles to gas at a wide range of temperatures to the energetic cosmic radiation. The gas component includes on the one hand very hot ($T \simeq 10^6$ K), tenuous gas, which was discovered by observations in the ultraviolet and X-ray regions, and extends down to the dense, cool ($T \simeq 10$ K) molecular clouds on the other hand; they have been investigated especially in the infrared and mm-wave regions with continually improving sensitivity and angular resolution. The large amount of newer observational data has led to major changes in our picture of the interstellar medium, and has shown that our understanding of important processes, such as for example star formation, the energy input to the interstellar medium, or the formation of interstellar dust, is still far from complete.

Nevertheless, we shall attempt to put together a provisional picture of the structure and importance of the interstellar medium in the Milky Way galaxy, leaving out however the central portion ($R \leq 4$ kpc): gas (at temperatures below about 10^4 K) and dust are concentrated in a flat disk, whose thickness between $R \simeq 4$ to 10 kpc is about 200 pc. In this disk, the gas and dust are concentrated in spiral arms; denser regions (clouds) are distributed along the arms, and are themselves divided into even finer structures. The lines of force of the interstellar magnetic field are in general frozen into the electrically conducting plasma along the spiral arms. In the densest clouds, hydrogen is present in molecular form.

The hot ($T \geq 10^4$ K) gas component of the interstellar medium takes up about half of the volume in the disk and extends out to considerably greater distances from the galactic plane than does the cooler gas and the dust.

The gas pressure $P = NkT$ is of the same order of magnitude both for the cool, diffuse H I clouds and for the partially ionized "warm" gas and the hot gas components (10^{-14} to 10^{-13} Pa or $\mathrm{J\,m^{-3}}$), so that a *pressure equilibrium* holds to a good approximation between these components in the disk. From (5.3.34), the corresponding energy densities are also comparable to those of the interstellar magnetic field and of the cosmic radiation. On the other hand, the equilibrium of the different components of the interstellar medium and their energy balance are influenced sensitively by *dynamic processes.* An important contribution to these is made by the supernova outbursts, with their expanding gas shells and X-ray emissions, but also stellar winds and ultraviolet radiation from hot stars, as well as the large-scale spiral density perturbations in the Milky Way (Sect. 5.8.2), play a role. The shock-wave fronts which pass through the interstellar gas and magnetic fields are probably responsible for some part of the acceleration of cosmic-ray particles. The details of the heating and dynamics of the interstellar medium are still poorly understood.

The OB associations of young, blue stars and the younger galactic star clusters are for the most part embedded in diffuse nebulas. These remarkable nebulas (H II regions), which are ionized by the stellar ultraviolet radiation and excited to luminosity, are mainly found in connection with dark, dense molecular clouds or groups of molecular clouds, which we can detect only through their radiofrequency and infrared radiations. They represent the locations of *star formation* or stellar genesis (Sect. 5.4.7). The very young stars are still surrounded by dense dust shells and appear as infrared sources. Only at a later point in their stellar evolution do the more massive of them appear as OB stars, surrounded by H II regions. This picture of the formation of stars from condensations of the interstellar matter, and its association with molecular clouds, infrared objects, and blue stars, will be taken up again in Sect. 5.4.7; then, in Sects. 5.5–5.7, it will be given a further basis and extended in relation to the investigation of distant galaxies.

5.4 Color-Magnitude Diagrams of Star Clusters. Stellar Evolution

Our present concepts of the formation, evolution, and final fate of stars were derived from investigations of the color-magnitude diagrams of star clusters. These clusters represent groups of stars which are all at the same distance from us. From the magnitudes and colors, which can be determined photometrically with high precision, subtraction of the common distance modulus and a uniform correction for interstellar absorption and reddening leads to the (generally used) values of the true

absolute magnitudes, $M_{\mathrm{V},0}$ and
color indices $(\mathrm{B-V})_0$.

As we have seen, the Hertzsprung-Russell diagram is fundamentally equivalent to the color-magnitude diagram. However, the color indices can still be measured even for extremely faint stars, for which classifiable spectra can no longer be obtained.

For completeness, we again summarize briefly the most important properties of the two types of star clusters: the distances to the *galactic* or *open* clusters are determined either by combining the proper motions μ and the radial velocities V, by the method of spectroscopic parallaxes, or by photometric comparison of the stars in the lower part of the main sequence to corresponding stars of a standard cluster (usually the Hyades) or to stars in our neighborhood. Galactic clusters (Fig. 5.2.5) contain a few dozen up to several hundred stars, and their diameters are of the order of 1.5 to 20 pc; they are found (Fig. 5.2.7) entirely in or in the vicinity of the spiral arms of the galaxy and, like the other objects of stellar Population I, they move in nearly circular orbits about the galactic center. Around 1000 galactic clusters are known; taking into account those which are more distant or are covered by dark clouds, their overall number is estimated to be about 20000. Closely related to the galactic clusters are the open moving clusters and the OB and T associations, which were also already mentioned in Sect. 5.2.

Open star clusters, whose stellar density is not much higher than that of their surroundings, will decompose, simply for kinematic reasons, after about one galactic rotation ($\simeq 2.5 \cdot 10^8$ yr). Even more compact clusters are gradually pulled apart by the gravitational fields of passing gas and star clouds. Nevertheless, more detailed calculations predict, e.g., for the Pleiades, a still fairly compact cluster, a lifetime of $\simeq 10^9$ yr.

In our Milky Way galaxy, about 150 *globular clusters* are known. We have already treated in detail the photometric determination of their distances making use of cluster variables or RR Lyrae stars. A typical globular cluster (Fig. 5.2.4) contains several hundred thousand stars within a region of $\simeq 40$ pc diameter, so that the average stellar density is roughly ten times larger than in galactic clusters. The density increases so strongly towards the center of the cluster that the night sky would be very bright there! The absolute visual magnitudes of the globular clusters are in the neighborhood of -7.3 mag. The total mass of a globular cluster can be estimated from the scatter of the radial velocities of the stars, using well-known approaches from the kinetic theory of gases, and is found to be several $10^5\,\mathcal{M}_\odot$. Since the globular clusters move around the galactic center on long, extended elliptical orbits, they pass through the galactic disk about every 10^8 yr. Due to their compact structures, the "jerk" which results from this passage generally has no serious effects.

Following these initial remarks, we turn to the *color-magnitude diagrams*, first those of the galactic star clusters (Sect. 5.4.1) and then of the globular clusters (Sect. 5.4.2). In Sect. 5.4.3, we discuss the influence of metal abundances on the two-color diagrams of star clusters.

We then make use of our knowledge of the structures and energy release in stars (from Sect. 4.12) to give a compact description of *stellar evolution*. In Sect. 5.4.4, we consider the first phases of nuclear burning, hydrogen and helium burning, and compare the theoretical positions and evolutionary paths of stars of different masses with the color-magnitude diagrams of star clusters. Sect. 5.4.5 is devoted to the late and final phases of stellar evolution and, in connection with them, the formation of the chemical elements in stars. Following a brief discussion of the fundamentals of stellar evolution in close binary systems in Sect. 5.4.6, we turn in Sect. 5.4.7 to stellar genesis, treating the early evolutionary phases of stars before they begin hydrogen burning, and to the extensive observations of regions of stellar formation, especially in the radiofrequency and infrared spectral ranges. Finally, in Sect. 5.4.8, we introduce the luminosity function or mass function for stellar genesis, and discuss the rates of star formation.

5.4.1 The Color-Magnitude Diagrams of Galactic Star Clusters

We can only mention briefly the pioneering work of R. Trumpler in the 1930's on the Hertzsprung-Russell diagrams of galactic star clusters. In the following section, we turn immediately to the more recent results, which were obtained photometrically or at least using photoelectric brightness scales by H. L. Johnson, W. W. Morgan, A. R. Sandage, O. J. Eggen, M. Walker, and others. With regard to a definitive decision as to whether this or that particular star belongs to a neighboring cluster, we must still rely on proper motions and possibly on radial velocity determinations: progress is naturally not so rapid in this area.

In Figs. 5.4.1a and b, we first show the raw observational data for Praesepe and for NGC 188. Figure 5.4.2 then gives a summary of the color-magnitude diagrams (M_V vs. B$-$V) of the galactic clusters NGC 2362, h$+\chi$ Persei, the Pleiades, M11, NGC 7789, the Hyades, NGC 3680, M 67, and NGC 188. The lower parts of the main sequence (up to about the "Sun", G 2) can be readily brought onto one curve. Further up, in contrast, the main sequence bends over to the right: earlier, in h$+\chi$ Persei already at the O and B stars, or later, in M 11 and Praesepe about at the A stars. Close to the absolute magnitude at this bend, the socalled "*knee*", some red giant stars are found to the right, at larger (positive) values of B$-$V. In NGC 188, M 67, ..., the transition from the main sequence to the red-giant branch is continuous. We discuss further details in connection with the theory of stellar evolution.

Finally, we quote the modern value of the true visual distance modulus for the "standard star cluster", the *Hyades*: $(m-M)_0 = 3.25$ mag, corresponding to a distance of 45 pc. The absolute luminosities of many star clusters are based on this value.

5.4.2 The Color-Magnitude Diagrams of the Globular Clusters

The structure of the color-magnitude diagrams of the globular clusters remained unclear until in 1952, under the leadership of W. Baade, a group of young astronomers at the Mt. Wilson and Mt. Palomar observatories, including A. R. Sandage, H. C. Arp, and W. A. Baum, took up the task of determining their *main sequence*, for favorable objects in the range from 19 to 21 mag. Only as a result of this work has it been possible to make comparisons with the stars in our neighborhood and with the galactic star clusters. Figure 5.4.3 shows, as an example, the color-magnitude diagram of Messier 92. The main sequence extends from the faintest observable stars with $V = 22$ mag up to the "knee" at $V = 18.4$ mag,

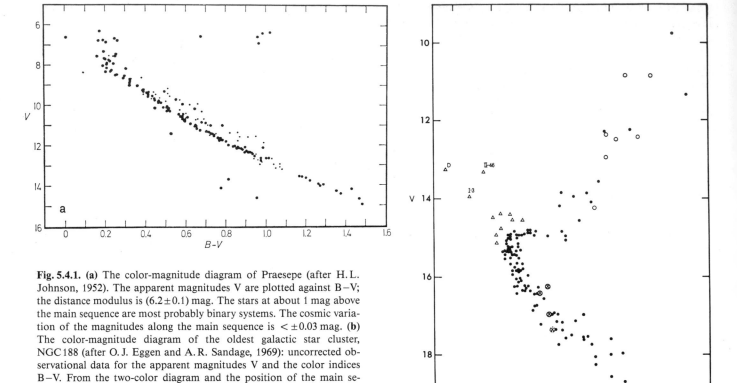

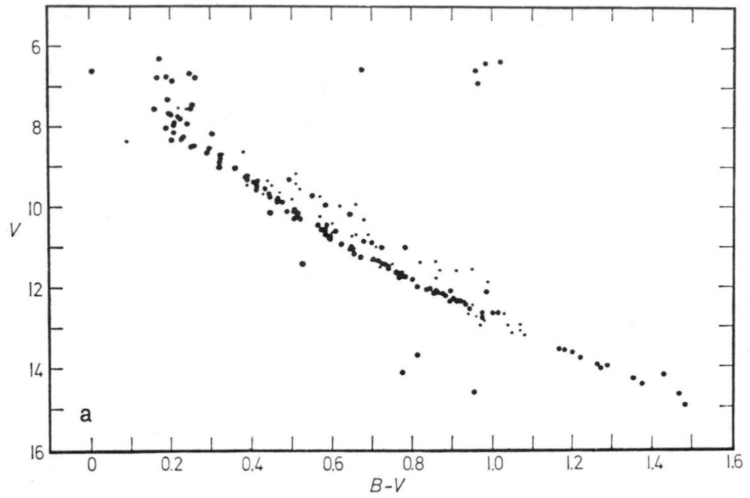

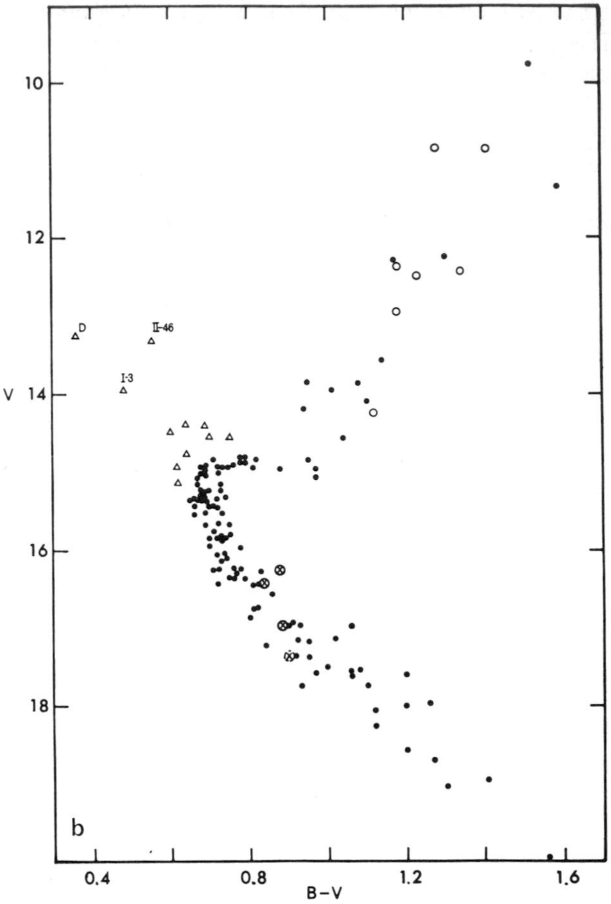

Fig. 5.4.1. (a) The color-magnitude diagram of Praesepe (after H. L. Johnson, 1952). The apparent magnitudes V are plotted against B−V; the distance modulus is (6.2 ± 0.1) mag. The stars at about 1 mag above the main sequence are most probably binary systems. The cosmic variation of the magnitudes along the main sequence is $< \pm 0.03$ mag. **(b)** The color-magnitude diagram of the oldest galactic star cluster, NGC 188 (after O. J. Eggen and A. R. Sandage, 1969): uncorrected observational data for the apparent magnitudes V and the color indices B−V. From the two-color diagram and the position of the main sequence, the interstellar reddening, extinction, and the true distance modulus, $(m-M)_0 = 10.85$ mag, can be computed. ● and ○ represent newer and older data, respectively; △ are possible "blue stragglers" which were formed by the evolution of close binary star systems. ⊗ are four eclipsing binary systems (each plotted as one star)

and is continued upwards by the *subgiants* B and then the *red giants* A. From the tip of this sequence, the *asymptotic* sequence C leads down and to the left, separate from the sequences of the giants. The *horizontal branch* D continues to the left after a well-defined gap, in which the *cluster variables* (not shown in Fig. 5.4.3) are located. The color-magnitude diagrams of other globular clusters also indicate that all the stars in this region are variables. (The red part of the horizontal branch, to the right of the region of variables, is lacking in M 92 and other metal-poor clusters.)

In Fig. 5.4.4, we show a combined diagram for several globular clusters which have been studied with particular precision in recent years by A. Sandage and coworkers. Making use of a careful determination of the distance modulus and the interstellar reddening, the data have been reduced to absolute magnitudes $M_{V,0}$ and true color indices, $(B-V)_0$. The shape of the main sequence curve and the subgiant and giant sequences is similar for all of the globular clusters; but, as we shall see, depending on their metal abundances, they differ markedly in their positions on the color-magnitude diagram.

A comparison of the diagrams for galactic clusters and for globular clusters (Figs. 5.4.2 and 5.4.4) will be discussed together with many other questions in connection with the theoretical approaches to stellar evolution.

5.4.3 Metal Abundance and Two-Color Diagrams

Spectroscopic analyses (Sect. 4.9.4) show that in the field stars of the halo population, the *abundances of all the heavier elements* (of nuclear charge $Z \geq 6$; usually referred to as "*metals*") relative to hydrogen are reduced in comparison to the Sun by factors up to 10^3, although the abundance *ratios* of the heavier elements among themselves are more or less the *same* as in the Sun. Deviations of more than a factor of 2 or 3 with respect to the relative mixture in the Sun are found only for a few elements such as C, N, O, and Al, Ba, and Y. *Helium* (Z = 2) is in general *not* reduced like the metals; both in Population I and in Population II stars, we find essentially the original atomic ratio of helium to hydrogen of about 1/10, which resulted from the Big Bang (Sect. 5.9.4).

For many purposes it is sufficient to employ the *metal abundance* ε of the halo stars and the closely related globular clusters, obtained by relating the abundances of the heavier elements M to that of hydrogen and comparing with the solar value:

$$\varepsilon = (\mathrm{M/H})_{\mathrm{star}}/(\mathrm{M/H})_{\odot}$$

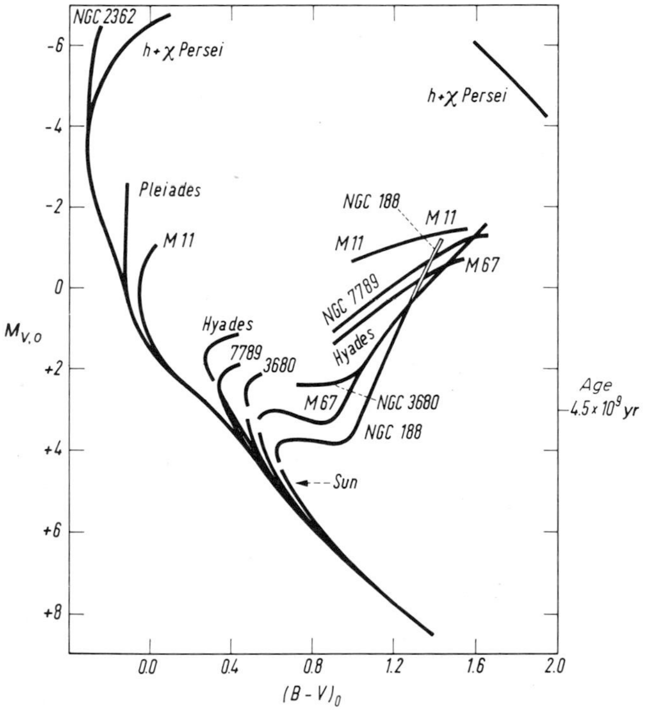

Fig. 5.4.2. The color-magnitude diagrams of galactic star clusters (after A. R. Sandage and O. J. Eggen, 1969; A. R. Sandage, 1957). The absolute magnitudes $M_{\mathrm{V,0}}$ corrected for interstellar extinction and reddening are plotted against the color indices $(\mathrm{B-V})_0$. The bending of the main sequence to the right ("knee") reveals the age of the star cluster. While the youngest clusters, NGC 2362 and h + χ Persei, are only a few million years old, the oldest cluster, NGC 188, has an age of 5 to $6\cdot10^9$ yr. The position of the knee for $4.5\cdot10^9$ yr, corresponding to the age of the Sun, is indicated on the right. The Sun itself is still on the (nearly) "unevolved" main sequence

or

$$\log\varepsilon = [\mathrm{M/H}] \ . \tag{5.4.1}$$

Since detailed spectroscopic analyses can, for practical reasons, be carried out only for relatively few, usually brighter objects, it is important to be able to determine the metal abundances of the stars at least globally by referring to the *two-color diagram.* In the cooler metal-poor halo stars, in comparison to normal metal-rich stars, the metal lines, which are crowded increasingly closely together towards the short-wavelength end of the spectrum, are so much weaker that the color index U−B is rather strongly shifted to smaller values, while the index B−V is only slightly shifted. Because of the large number of iron lines in the optical spectra of cooler stars, the photometrically derived metal abundance is practically equal to that of *iron*, $[\mathrm{M/H}] \simeq [\mathrm{Fe/H}]$.

In the two-color diagram (Fig. 4.5.5), the curve derived from normal stars is shifted upwards in the region of $\mathrm{B-V} \geq 0.35$ mag by a maximum (i.e. for extremely metal-poor stars) of $\delta(\mathrm{U-B}) \simeq 0.25$ mag; cf. the two-color diagram of the extremely metal-poor globular cluster M 92 in Fig. 5.4.3. This of course has to be taken into consideration in determining the interstellar reddening from the two-color diagram.

The *UV excess* $\delta(\mathrm{U-B})$, which is more precisely defined as the difference in U−B compared to main-sequence stars of the Hyades with the same B−V values, also gives a measure of the metal abundances of fainter stars. Still more precise results are obtained from narrow-band photometry by the method of B. Strömgren, which is however limited to somewhat brighter stars.

Applying the theory of stellar atmospheres, a calibration of the dependence of $\delta(\mathrm{U-B})$ and other corresponding indices on the metal abundance can be carried out for various effective temperatures and surface gravities of the stars.

5.4.4 Nuclear Evolution of the Stars: Interpretation of the Color-Magnitude Diagrams of Star Clusters

In the theory of stellar structure, we have seen that for *homologous* stars, i.e. for stars having similar structures and chemical compositions, a relation of the form $\Phi(L, T_{\mathrm{eff}}) = 0$ holds (Sect. 4.12.5). When central *hydrogen burning* begins, the young, still chemically homogeneous stars are all on one curve in the (L, T_{eff}) or the color-magnitude diagram, with stars of greater mass

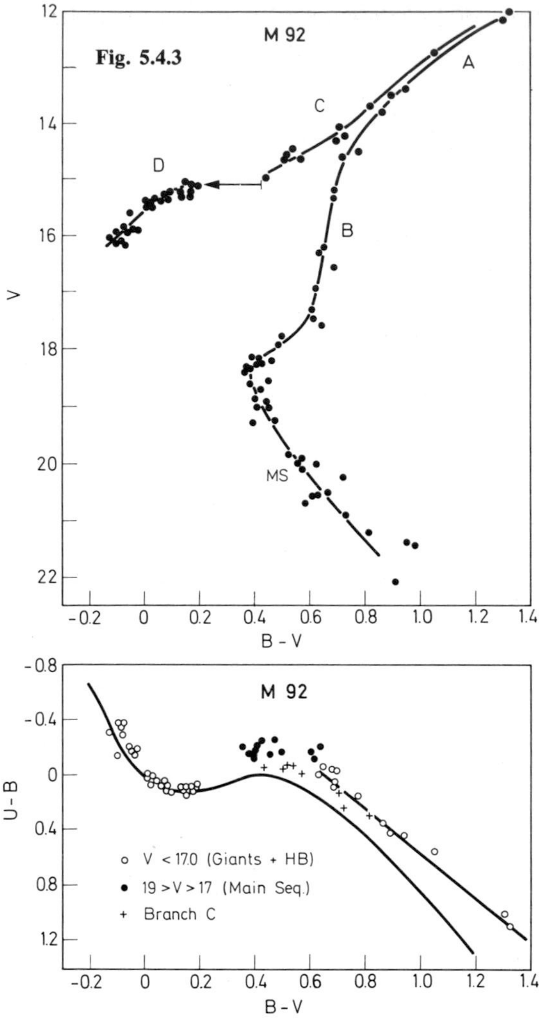

Fig. 5.4.3. The color-magnitude diagram of the globular cluster M 92, after A.R. Sandage (1970). The apparent magnitudes V are plotted against the color indices B−V without the (very small) corrections for interstellar extinction and reddening. A: giant branch; B: subgiant branch; C: asymptotic giant branch; D: blue horizontal branch; MS: main sequence. The cluster variables occur in the gap in the horizontal branch at V = 15 mag. The apparent visual distance modulus is $m-M = 14.4$ mag. The two-color diagram is also shown below. The heavy curve indicates the two-color diagram of the Hyades; the curve at the right shows its increase for the extremely metal-poor stars

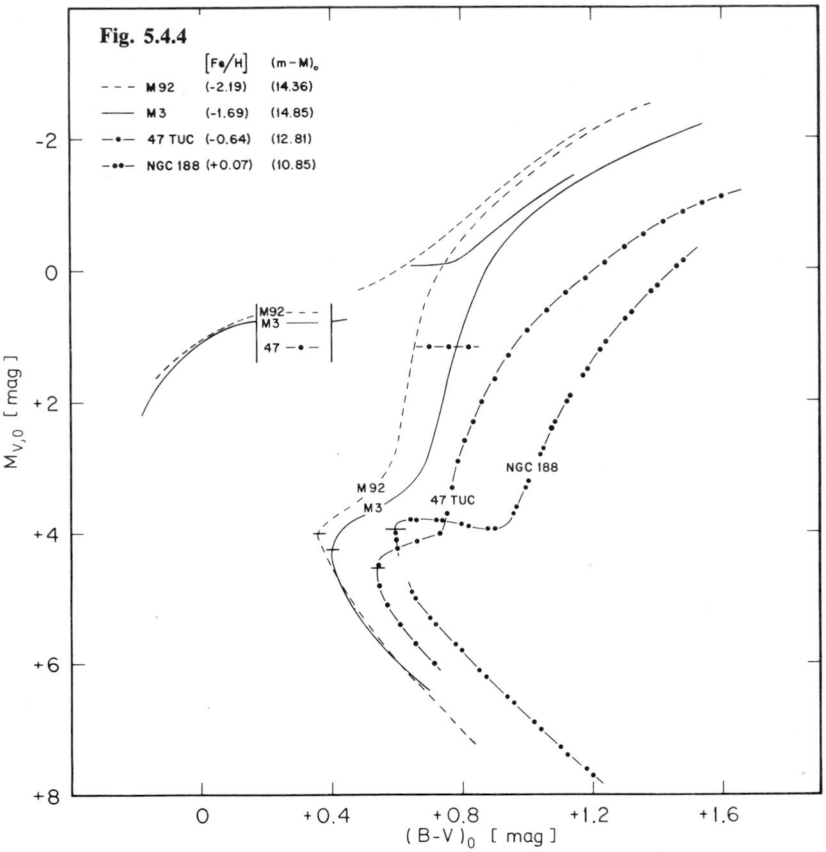

Fig. 5.4.4. Color-magnitude diagrams for the three globular clusters M 92, M 3, and 47 Tuc, and the oldest galactic cluster, NGC 188. The absolute magnitudes $M_{V,0}$ corrected for interstellar extinction are plotted against the true color indices $(B-V)_0$; also, the metal abundance $[Fe/H] \simeq \log \varepsilon$ and the true distance modulus, $(m-M)_0$, are given for each object. Although the globular clusters are considerably older (about $16 \cdot 10^9$ yr) than NCG 188 (about $5 \cdot 10^9$ yr), their giant branches lie to the left of the giant branch of the metal-richer galactic cluster; this results from the different metal abundances. From A. R. Sandage, Astrophys. J. **252**, 574 (1982); reprinted courtesy of A.R. Sandage and *The Astrophysical Journal*, published by the University of Chicago Press; © 1982, The American Astronomical Society

having higher luminosities and those of lesser mass having lower luminosities. This curve is termed the *initial main sequence* or also (not quite correctly) the *Zero Age Main Sequence* (ZAMS); from it, nuclear evolution takes place towards the upper right in the diagram (see below).

Empirically, the initial main sequence (for Population I stars) is defined by the envelope of all the color-magnitude diagrams of the galactic star clusters in Fig. 5.4.2, out to values of M_V corresponding to rather bright stars. In the upper part, it lies a bit below the main

Table 5.4.1. Initial main sequence (Zero Age Main Sequence) of Population I, after T. Schmidt-Kaler (1982). (B−V) and M_V in [mag]

B−V	M_V	B−V	M_V	B−V	M_V
−0.30	−3.3	+0.30	+2.8	+1.00	+ 6.7
−0.20	−1.1	+0.40	+3.4	+1.20	+ 7.5
−0.10	+0.6	+0.50	+4.1	+1.40	+ 8.8
0.00	+1.5	+0.60	+4.7	+1.60	+12.0
+0.10	+1.9	+0.70	+5.2	+1.80	+14.2
+0.20	+2.4	+0.80	+5.8	+2.00	(+16.7)

sequence of luminosity class V (Fig. 4.5.4), and merges with the main sequence for the G stars (Table 5.4.1).

The position of the zero age main sequence in the theoretical color-magnitude diagram (Fig. 5.4.5) is determined by the chemical composition of the stars. The curve for Population I stars with a normal (solar) element mixture is given by a mass ratio of hydrogen, X, to helium, Y, to heavier elements, Z, of[12]:

$$X:Y:Z = 0.73:0.25:0.02 \ . \tag{5.4.2}$$

(This corresponds to an *atomic* abundance of helium given by He/H = 0.09.)

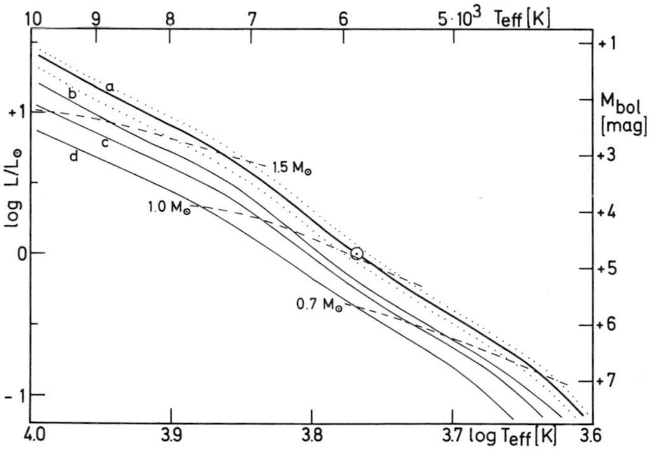

Fig. 5.4.5. The initial main sequence in the theoretical color-magnitude diagram for various chemical compositions, after P.M. Hejlesen (1980). a: $(X, Y, Z) = (0.70, 0.27, 0.03)$, characteristic curve for Population I; the adjacent dotted curves differ by $\Delta Z = \pm 0.01$. b: (0.70, 0.29, 0.01); c: (0.70, 0.296, 0.004); d: (0.70, 0.30, 0.0004), corresponding to $\log \varepsilon = -1.9$. This is typical of extremely metal-poor halo stars. Curves of constant mass for 1.5, 1.0, and 0.7 $\mathscr{M}_\odot$ are also drawn. The helium-rich mixture (0.60, 0.38, 0.02) yields an initial main sequence which is practically identical to the metal-poor sequence b

The zero age main sequence of the *metal-poor* halo population is *below* that of Population I, for the same helium abundance. The metals make the main contribution to the absorption coefficient κ at the cooler end of the sequence, and thus the luminosity L increases for a given mass $\mathscr{M}$ when the metal abundance Z decreases (and thus κ also decreases), following the mass-luminosity relation $L \propto \mathscr{M}^3/\kappa$ (4.12.41). Since, however, from the equations governing stellar structure (whose details we omit here), the effective temperature T_{eff} also increases with decreasing κ, the overall result is a metal-poor main sequence which lies below the normal main sequence but approximately parallel to it. The mass scales along the sequences are relatively displaced, so that for example, a Population I star with 1 $\mathscr{M}_\odot$ has the same effective temperature as an extremely metal-poor halo star having about 0.7 $\mathscr{M}_\odot$. A change in the *helium* abundance also leads to a parallel shift of the initial main sequence, in the opposite sense to that produced by a change in metal abundance (Fig. 5.4.5).

From the theory of stellar structure, we can thus understand the different main sequences in the color-magnitude diagrams of the galactic star clusters and the globular clusters with their different metal abundances (Figs. 5.4.2 and 5.4.4). For a quantitative calibration of the sequences, along with a possible variation of the helium abundance, the dependence of the color indices on metal content (Sect. 5.4.3) must be taken into account in going from a theoretical color-magnitude diagram to an (M_V, B−V) diagram.

The metal-poor main sequence stars of the halo population are also referred to as *subdwarfs* (sd). This has the following historical background: originally, the metal-poor stars were classified as too "early" in comparison to normal stars of the *same* effective temperature T_{eff}, since their Fraunhofer lines are weaker than those of normal stars with the same T_{eff} and g and since, on the other hand, in going along the main sequence to earlier spectral types, the metal lines become weaker. For example, HD 140 283 was originally classified as A2, then as sd F5, although its $T_{eff} \simeq 5900$ K corresponds to about G0 V. In this way, such stars lay *below* the main sequence in the Hertzsprung-Russell diagram and were therefore termed "subdwarfs". If, on the other hand, T_{eff} and the absolute bolometric magnitudes or the luminosities are used as

[12] Often, somewhat different mixing ratios are assumed for Population I, such as e.g. 0.74:0.24:0.02 or 0.70:0.27:0.03.

coordinates in the diagram, then the *early* subdwarfs ($\simeq$ F) lie nearly *on* the usual main sequence. It was later recognized, especially from precise distance determinations, that in contrast, the *cooler* subdwarfs ($>$G) also lie about 1 mag *below* the normal main sequence in an (L, T_{eff}) diagram, in agreement with the theory of stellar structure, i.e. they are *genuine* subdwarfs. The position of the early subdwarfs near to the normal sequence is interpreted as an evolutionary effect (see below): they have already evolved somewhat to the right and upwards from their original metal-poor initial main sequence and have thus arrived "coincidentally" at the sequence of Population I stars.

The stars remain in the immediate vicinity of the main sequence until a considerable portion of their hydrogen has been consumed. In order to estimate this dwell time, we first consider the *energy balance of the Sun*. Its *central temperature* (calculated from theory) of $T_c = 1.5 \cdot 10^7$ K has clearly adjusted itself (Table 4.12.1) in such a way that the pp process has taken over the energy production. Now the solar mass, $\mathcal{M}_\odot = 1.98 \cdot 10^{30}$ kg, consists of about 70% hydrogen, if we treat it in a preliminary assumption as homogeneous. The complete conversion of this hydrogen into helium by the pp process would yield $8.8 \cdot 10^{44}$ J of energy. At its current luminosity, $L_\odot = 3.85 \cdot 10^{26}$ W, the Sun would consume of the order of *10% of its hydrogen*, which should cause a noticeable change in its properties, after $7.3 \cdot 10^9$ yr. Thus, in the time since the Earth has had a solid crust, the Sun can hardly have changed.

How does the energy balance of the other *main sequence stars* look? Their central temperatures T_c increase from low values at the cool end of the sequence to about $3.5 \cdot 10^7$ K for the B0 stars, etc. Somewhat above the Sun, the CNO cycle (Fig. 4.12.2) therefore takes over energy production, without noticeably changing its efficiency. From the known values of the masses $\mathcal{M}/\mathcal{M}_\odot$ and the luminosities $L/L_\odot$ (i.e. the energy production) of the main sequence Population I stars, we can now readily calculate the time in which they would consume 10% of their hydrogen; we call it their *evolution time* t_E for short (Table 5.4.2):

$$t_E\,[\text{yr}] = 7.3 \cdot 10^9 \, \frac{\mathcal{M}/\mathcal{M}_\odot}{L/L_\odot} \quad . \qquad (5.4.3)$$

Since they were formed, which, as we state here in advance, cannot have been more than about $2 \cdot 10^{10}$ yr ago, the *main sequence stars* below about G0 have thus consumed only a small fraction of their hydrogen. On the other hand, the hot stars of the early spectral types burn their hydrogen so quickly that they can only have "come into being" relatively recently, at times of the order of t_E. The age of the B and O stars is in fact considerably less than the rotation period of the Milky Way in our

Table 5.4.2. The main sequence stars and their evolution times

Spectral type	Effective temperature T_{eff} [K]	Mass $\mathcal{M}/\mathcal{M}_\odot$	Luminosity $L/L_\odot$	Evolution time t_E [yr]
O5 V	44500	60	$7.9 \cdot 10^5$	$5.5 \cdot 10^5$
B0 V	30000	18	$5.2 \cdot 10^4$	$2.4 \cdot 10^6$
B5 V	15400	6	$8.3 \cdot 10^2$	$5.2 \cdot 10^7$
A0 V	9500	3	$5.4 \cdot 10^1$	$3.9 \cdot 10^8$
F0 V	7200	1.5	6.5	$1.8 \cdot 10^9$
G0 V	6050	1.1	1.5	$5.1 \cdot 10^9$
K0 V	5250	0.8	$4.2 \cdot 10^{-1}$	$1.4 \cdot 10^{10}$
M0 V	3850	0.5	$7.7 \cdot 10^{-2}$	$4.8 \cdot 10^{10}$
M5 V	3250	0.2	$1.1 \cdot 10^{-2}$	$1.4 \cdot 10^{11}$

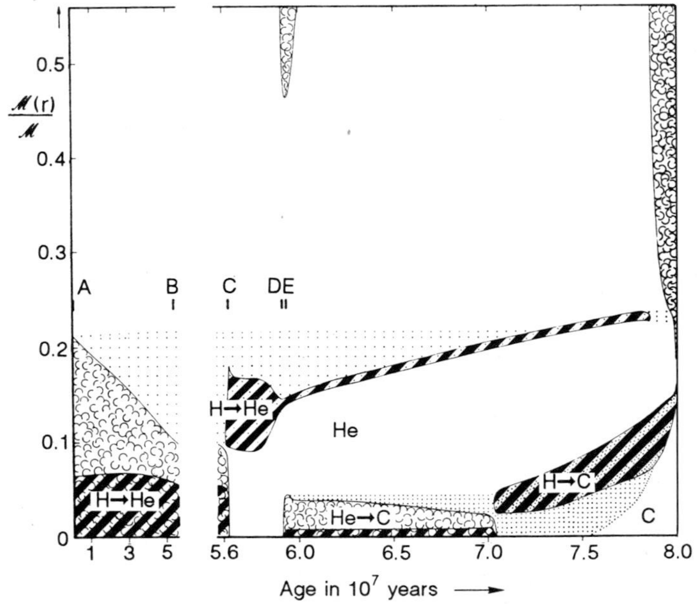

Fig. 5.4.6. Variations with time in the interior of a (Population I) star of $5\,\mathcal{M}_\odot$. The abcissa gives the age, reckoned in 10^7 yr since the star left the main sequence. The letters A to E denote the correspondence to the evolutionary paths shown in Fig. 5.4.7a. The ordinate is $\mathcal{M}(r)/\mathcal{M}$, the fraction of the mass within r. The stippled regions correspond to convection zones, and the barred regions are zones of nuclear energy release; the dotted regions are those in which the H or He content decreases on going inwards. (After R. Kippenhahn, H. C. Thomas, and A. Weigert, 1965)

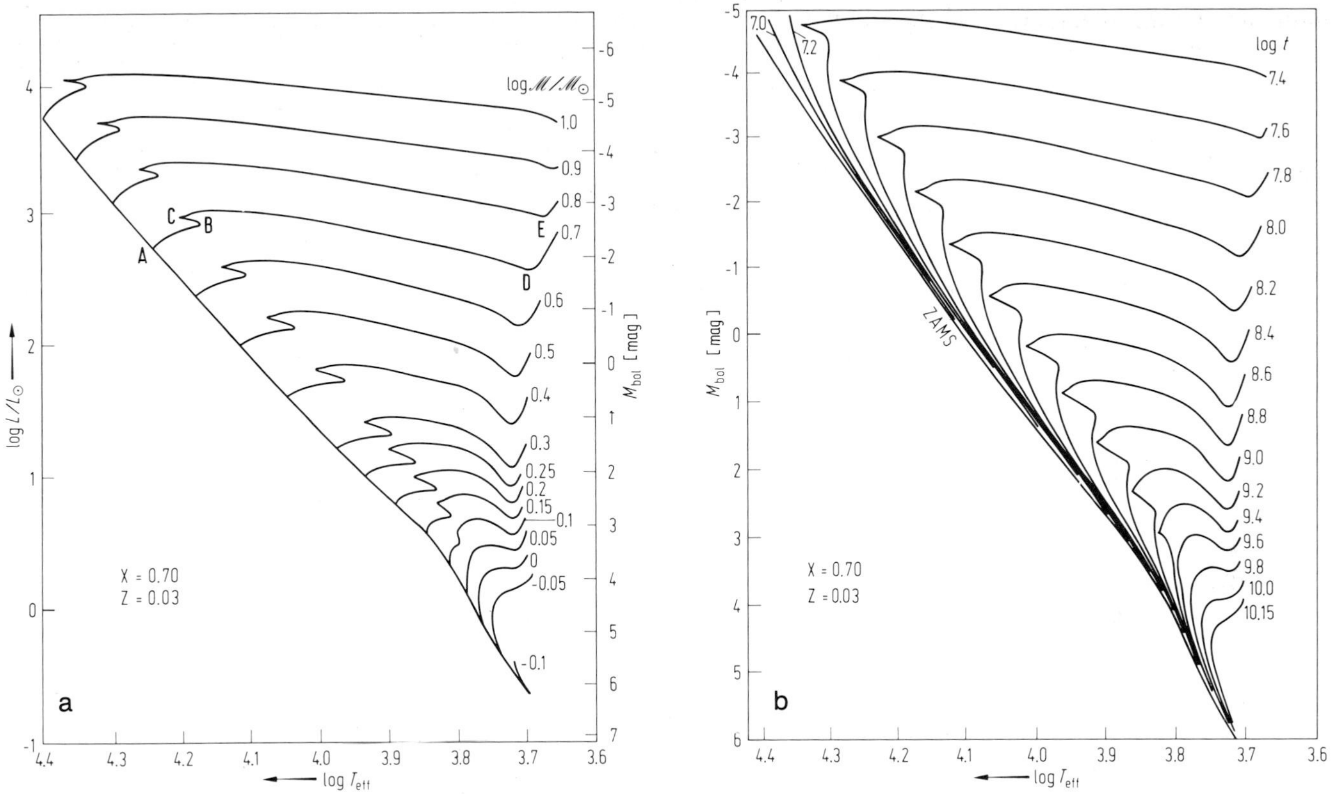

Fig. 5.4.7 a,b. Theoretical color-magnitude diagrams for stars of Population I. **(a)** Evolutionary paths for different masses $\mathcal{M}/\mathcal{M}_\odot$. The letters A to E refer to the changes in the inner structure of a star of $5\,\mathcal{M}_\odot$ as given in Fig. 5.4.6. **(b)** Isochrones for the evolutionary paths given in (a), with the ages t in [yr] as indicated. The lower envelope is the initial main sequence (ZAMS = Zero Age Main Sequence)

vicinity ($2.4\cdot 10^8$ yr); such stars must therefore have been formed near where we see them today.

The course of evolution of the stars *away from the main sequence* depends essentially on whether the transformed matter in the stellar interior mixes with the remaining matter, or whether it remains where it was formed in the core or the particular convection zone, if one is present. F. Hoyle and M. Schwarzschild were the first to show, in 1955, that only the latter possibility leads to an acceptable theory of stellar evolution and can also be made dynamically plausible.

What happens in detail, we shall first describe using the example of a star of $5\,\mathcal{M}_\odot$ (Fig. 5.4.6). It begins its evolution as a completely homogeneous B5 V Population I star with the chemical composition given in (5.4.2), an effective temperature $T_{\text{eff}} = 17\,500$ K, a radius of $2.6\,R_\odot$, and thus an absolute bolometric magnitude of $M_{\text{bol}} = -2.2$ mag. In its center, the temperature is $T_c = 2.6\cdot 10^7$ K and the pressure is $P_c = 5.5\cdot 10^{15}$ Pa. At the very center, there is a *hydrogen burning zone*, the nuclear fusion reactor, so to speak, where hydrogen is burned to helium by the CNO cycle. Adjacent to this hydrogen burning zone is a convection zone, within which the reaction products are thoroughly mixed. This phase of evolution (A→B→C, Fig. 5.4.7a) lasts about $6\cdot 10^7$ yr, corresponding roughly to our estimated evolution time t_E (Table 5.4.2). When the core is burned out, a hydrogen burning zone in the shape of a shell around the core is formed for a short time (C→D→E: $0.3\cdot 10^7$ yr). At E, a *helium burning zone* comes into being, initially in the core; in it, at central temperatures of $T_c \simeq 1.3$ to $1.8\cdot 10^8$ K, the 3α process (4.12.18 and 4.12.19) takes over energy production. When this He core is also burned out, a shell-shaped helium burning zone then also forms around it; but the thin hydrogen burning shell, which is continually moving outwards, still makes a considerable contribution to energy production after E. Finally, the star moves quickly into the stages of red giant and supergiant, where it loses part of its mass by the stellar wind which increases with the luminosity (4.11.7). The final, late stages of stellar evolution (from carbon fusion onwards) will be discussed in the next section.

The evolutionary paths of the massive, luminous stars in the Hertzsprung-Russell diagram depend sensitively on the rate of *mass loss*, $\dot{\mathcal{M}}$, which at present can only be taken into account in theoretical calculations as an empirical, rather uncertain parameter. Stars with $\geq 10\,\mathcal{M}_\odot$ on the initial main sequence lose on the order of magnitude of $\frac{1}{4}$ to $\frac{1}{3}$ of their mass before reaching the region of red supergiants. Their essentially "horizontal" evolution away from the main sequence at a roughly constant luminosity (Fig. 5.4.7a) takes place at smaller luminosities for increasing $\dot{\mathcal{M}}$. In the case of extreme mass losses (those which persist for long times, with $\geq 10^{-5}\,\mathcal{M}_\odot\,\mathrm{yr}^{-1}$), the evolutionary track bends downwards towards the main sequence, so that the red giant stage is not reached. (In the hypothetical limiting case of $\dot{\mathcal{M}} \to \infty$, the evolutionary path would coincide with the main sequence.)

All stars with masses $\geq 2.5\,\mathcal{M}_\odot$ have an evolution during the stages of hydrogen and helium burning which is qualitatively similar to what we have described for $\mathcal{M} = 5\,\mathcal{M}_\odot$. In particular, helium burning takes place "*hydrostatically*", i.e. the star remains continuously in hydrostatic equilibrium (4.12.1). This follows essentially from the equation of state (4.12.3) of an *ideal gas*, according to which a change in the temperature T is accompanied by a change in pressure. If, for example, T increases, and with it the rate of nuclear energy release, then the accompanying increase of pressure causes an expansion and cooling, which leads to a decrease in the energy production, thus stabilizing the star.

In contrast, for stars of mass $\leq 2.5\,\mathcal{M}_\odot$, helium burning takes place *explosively*. These stars attain such high densities in their interiors in the course of their evolution "straight up" on the Hertzsprung-Russell diagram, on the socalled first giant branch (Figs. 5.4.2 and 5.4.3), that a *Fermi-Dirac degeneracy* of the electron gas becomes established before helium burning can start at $T \simeq 8 \cdot 10^7$ K. The pressure of this Fermi gas, unlike that of an ideal gas, does *not* depend on the temperature T (4.12.49), so that when the energy release increases, no expansion and cooling can occur. On the contrary, a *helium flash* takes place: a strong temperature rise within a very short time, of the order of the free fall time (4.12.39). This temperature rise stops only when T becomes so high that the electron degeneracy is again lifted. The resulting explosive shock wave in the stellar interior is damped by the massive shell outside the helium zone, as is shown by tedious detailed calculations, so that the star "survives" the central helium flash at the tip of the giant branch (at roughly $2 \cdot 10^3\,L_\odot$). Afterwards, it finds a new equilibrium configuration with central, hydrostatic helium burning and a hydrogen burning shell. Both during its evolution on the giant branch and in the helium flash, the star loses a considerable portion of its mass; e.g. a star of originally $1\,\mathcal{M}_\odot$ loses in the range of 0.1 to $0.5\,\mathcal{M}_\odot$.

Before we consider the further evolution of stars of $\leq 2.5\,\mathcal{M}_\odot$, we turn to a comparison of the theoretical results with observations. In Fig. 5.4.7a, the initial *evolutionary paths* of Population I stars of varying masses are represented in a *theoretical color-magnitude diagram*, where the luminosity L or the absolute bolometric magnitude M_{bol} is plotted against the effective temperature T_{eff}; for $5\,\mathcal{M}_\odot$, the evolutionary phases (A to E) are taken from Fig. 5.4.6.

The most important features of nuclear evolution can be summarized as follows:

Young stars which are still homogeneous are found along the zero age main sequence on the color-magnitude diagram; its position in the diagram depends on their chemical composition. Their interiors contain hydrogen burning zones in which energy release is maintained by the CNO cycle for higher internal temperatures (larger masses) and by the proton-proton chain for lower internal temperatures (smaller masses). These stars remain near the main sequence until they have consumed about 10% of their hydrogen, i.e. during a time interval of the order of t_{E}. Thereafter, their evolution proceeds on a very short time scale, first to the right and upwards into the region of red giants, where helium burning is initiated. Using the evolution times which correspond to the evolutionary paths in Fig. 5.4.7a, we can draw the curves in the color-magnitude diagram which a group of stars that all started on the zero age main sequence at $t = 0$ will have reached after a time t. Such calculated *isochrones* (Fig. 5.4.7b) then allow us to interpret the color-magnitude diagrams of the galactic star clusters (Fig. 5.4.2) as an *age sequence* and thus to refine our earlier age estimates (t_{E}).

The star cluster h and χ Persei, with its extremely bright blue supergiants, is a very *young* cluster. Its deviation from the main sequence, the socalled knee at $M_{\mathrm{V}} = -6$ mag, indicates an age of a few million years. The several red supergiants to the right of the upper end of the main sequence are separated from it by the *Hertz-*

sprung gap, which has long been known empirically; it stretches down, becoming narrower, to about the F0 III stars. This can be explained by the fact that, for example, for a star of 5 $\mathcal{M}_\odot$ in Fig. 5.4.7a, the path C→D requires only $3 \cdot 10^6$ yr, as opposed to $2 \cdot 10^7$ yr for the following red giant stage, or $6 \cdot 10^7$ yr for the hydrogen burning phase on the main sequence.

The color-magnitude diagrams of e.g. the Pleiades, ... Praesepe, ... down to NGC 188, whose main sequences bend over to the giant branch at increasingly lower points in the diagram, indicate increasing ages along the same series. There are without doubt no diagrams which bend over much below that of NGC 188; i.e. for the galactic clusters, there is a maximum age, which is about 5 to $6 \cdot 10^9$ yr.[13]

The well-known color-magnitude diagram of the field stars in our immediate vicinity can best be interpreted as that of a mixture of stars from the remnants of numerous associations and star clusters which have drifted apart in the course of time. The calculated evolution times make it quite understandable that the main sequence connects closely with the zero age main sequence. The concentration of yellow and red giant stars in the giant branch was attributed by A. R. Sandage to the fact that in this region, the evolutionary lines of the more massive and brighter stars merge from left to right, and those of the less massive and fainter stars from the lower part of the main sequence to the upper right, in the shape of a funnel (Fig. 5.4.7a). The giant stars in our neighborhood can thus not be treated as a homogeneous group, particularly with regard to their masses.

We now turn to the color-magnitude diagrams of the *globular clusters* (Figs. 5.4.3, 5.4.4), which indeed formed the starting point for the modern theory of stellar evolution. They are to a large extent similar to the diagrams of older galactic clusters, with the distinction that their giant branches are steeper. According to calculations based on the theory of stellar evolution, this is a result of the fact that the globular clusters, as members of the extreme halo population, have very *low* metal abundances (of the order of magnitude $\varepsilon \simeq 1/10$ to $1/100$), as is verified by the spectra of their red giants. This has a considerable influence on their opacity and energy production. The bending over of the color-magnitude diagrams (at $M_V \simeq 4$ mag, similar to that of NGC 188) indicates, according to the most recent model calculations, that all the globular clusters have an *age of about 14 to 18·10⁹ yr*. The scatter in the numerical values, amounting to a few times 10^9 yr, which is mainly due to uncertainties in the theory, is probably larger than the true differences in the ages of different globular clusters.[14]

In the color-magnitude diagrams of the globular clusters, a distinction is made between the red giant branch (upper right) and the *asymptotic giant branch* at somewhat lower effective temperatures (Fig. 5.4.3), and finally the *horizontal branch* out to B stars of absolute magnitude $M_V \simeq +2$ mag. Embedded in the horizontal branch is the "gap" containing the pulsating cluster variables or *RR Lyrae stars*. The end of the evolutionary path at the tip of the first giant branch is attributed to the sudden ignition of central helium burning ("helium flash"). After the helium flash, and after a considerable loss of mass in the red giant stage, the metal-poor stars of Population II move onto the horizontal branch. According to the theory of stellar structure, they now produce their energy through central helium burning and burning of hydrogen in a concentric shell, whereby their masses amount to about 0.5 $\mathcal{M}_\odot$ at the blue end and 0.9 $\mathcal{M}_\odot$ at the red end of the horizontal branch. In the case of Population I stars, with normal metal abundances, the horizontal branch corresponds to a not-very-noticeable clump of stars in the color-magnitude diagram near the giant branch at about $10^2 L_\odot$.

Now, how does the evolution of a star proceed after it has burned up its helium core? Initially, a concentric helium burning shell is formed: the star departs from the horizontal branch (or the corresponding stage for Population I stars) and moves a second time towards higher luminosities (insofar as its mass is $\geq 0.6 \mathcal{M}_\odot$), into the region of the red giants and supergiants on the *asymptotic giant branch* (Fig. 5.4.3a). This phase is characterized by energy production in a *double-shell source*, an inner helium burning zone surrounding the core, which consists of ^{12}C and ^{16}O and a degenerate electron gas, and an outer hydrogen burning shell. The energy release in the helium zone is thermally unstable and takes place

[13] The reduction of the age of NGC 188 as compared to the earlier value of 8 to $10 \cdot 10^9$ yr (A. R. Sandage and O. J. Eggen, 1969) is mainly due to a revision of the distance modulus of the *Hyades*, which form the standard for the color-magnitude diagrams of all the other star clusters.

[14] Earlier age determinations yielded similar ages of around $10 \cdot 10^9$ yr for the globular clusters and the oldest open cluster, NGC 188. In contrast, newer calculations, taking a particularly detailed account of the effect of the different chemical compositions on the conversion of (B−V) and M_V into the physical parameters T_{eff} and L, lead to distinctly greater ages for the globular clusters, although their bend-over points in the color-magnitude diagram are near to that of NGC 188.

as a series of pulses with a typical period of about 1000 yr. The evolution of the more massive stars ($\geq 2.5\,\mathcal{M}_\odot$) also merges, as we saw above (Fig. 5.4.6), into a stage with double-shell burning. The asymptotic giant branch probably ends with the casting off of a *planetary nebula* having a few tenths of a solar mass (Sect. 5.3.5); the masses of the remaining *central stars* of the planetary nebulas lie in a relatively narrow range around $\simeq 0.6\,\mathcal{M}_\odot$.

The evolutionary stages which follow that of helium burning are passed through in very much shorter times and can hardly be followed on the color-magnitude diagram. Before we turn to these stages, we mention briefly the evolution of stars at the *lower end of the mass spectrum*. If the original mass is $\leq 0.4\,\mathcal{M}_\odot$, the star never attains the ignition temperature necessary for helium burning. If its mass is below $\simeq 0.1\,\mathcal{M}_\odot$, then it cannot even reach the stage of hydrogen burning, which is a defining characteristic of stars. These "substellar masses" or "*brown dwarf stars*" can be considered to be giant planets, related in their structures to Jupiter.

5.4.5 Stellar Evolution: Late and Final Stages. Nucleosynthesis in Stars

The *mass loss* in the region of red giants and supergiants is of decisive importance for the later evolutionary stages of stars with an original mass of $\leq 8\,\mathcal{M}_\odot$. If the loss is so great that the stellar remnant has a mass lower than Chandrasekhar's limit of $1.4\,\mathcal{M}_\odot$ (4.12.54), then the star finishes its evolution, after its nuclear energy sources (hydrogen and helium burning shells) have been exhausted, as a stable configuration in the form of a *white dwarf star*, containing ^{12}C and ^{16}O. In such stars, the pressure of the *degenerate* electron gas holds the gravitational forces in balance (Sect. 4.12.7). The observation of white dwarfs in galactic star clusters whose bend-over point from the main sequence in the color-magnitude diagram still corresponds to a mass of $8\,\mathcal{M}_\odot$ demonstrates that, indeed, a large fraction of stars of $\mathcal{M} \leq 8\,\mathcal{M}_\odot$ lose sufficient mass through stellar winds and the casting off of planetary nebulas that they become white dwarfs. On the asymptotic giant branch, these white dwarfs are already "hidden" within the innermost $10^{-2} R_\odot$ of the red giants, which themselves have radii of about $10^2 R_\odot$; the dwarf takes the form of a dense (10^8 to 10^9 kg m^{-3}) core of $\mathcal{M} \geq 0.6\,\mathcal{M}_\odot$.

In Fig. 5.4.8, the evolutionary path of a star of $1\,\mathcal{M}_\odot$ ending at the white dwarf stage is shown on the Hertzsprung-Russell diagram. Corresponding to a mass loss as given by (4.11.7), the final mass of $0.6\,\mathcal{M}_\odot$ is reached shortly after the end of the time spent on the asymptotic giant branch, some 10^6 yr. The remaining evolution occurs rapidly, in 10^4 to 10^5 yr, on a line going horizontally to the left towards higher effective temperatures into the region of the central stars of planetary nebulas. At T_{eff} between $3 \cdot 10^4$ and 10^5 K, the planetary nebula previously cast off becomes ionized and is excited to luminosity. Finally, in the interior of the star, nuclear energy release ceases to be possible; the evolutionary path bends down and merges into the sequence of white dwarfs. In some 10^9 yr, the star cools gradually to effective temperatures below 4000 K, the lowest values which have been observed in white dwarfs.

Those stars which have originally $\leq 8\,\mathcal{M}_\odot$ and whose mass *exceeds* the limiting mass for white dwarfs near the end of their evolutionary phase on the asymptotic giant branch finally attain such high temperatures in their helium cores with degenerate electron gas that, in spite of large energy losses due to the emission of neutrinos

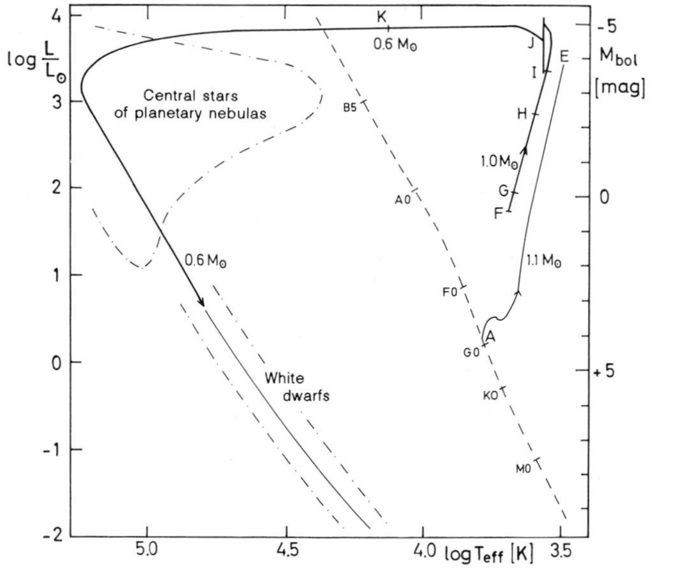

Fig. 5.4.8. The evolutionary paths in the Hertzsprung-Russell diagram of Population I stars having 1.0 and $1.1\,\mathcal{M}_\odot$, from central hydrogen fusion (A) to the helium flash (E), without taking mass losses into account. After A. V. Sweigart and P. G. Gross (1978). The ejection of a mass of $0.1\,\mathcal{M}_\odot$ during the helium flash was assumed. The further evolution of the star of $1.0\,\mathcal{M}_\odot$ was calculated taking the mass loss according to (4.11.7) into account, after D. Schönberner (1979). G→J: the asymptotic giant branch; only one of the thermal pulses (helium flashes) which occur after I is drawn in, at J. The mass loss becomes important at H and leads to a final mass of $0.6\,\mathcal{M}_\odot$, which is reached at K

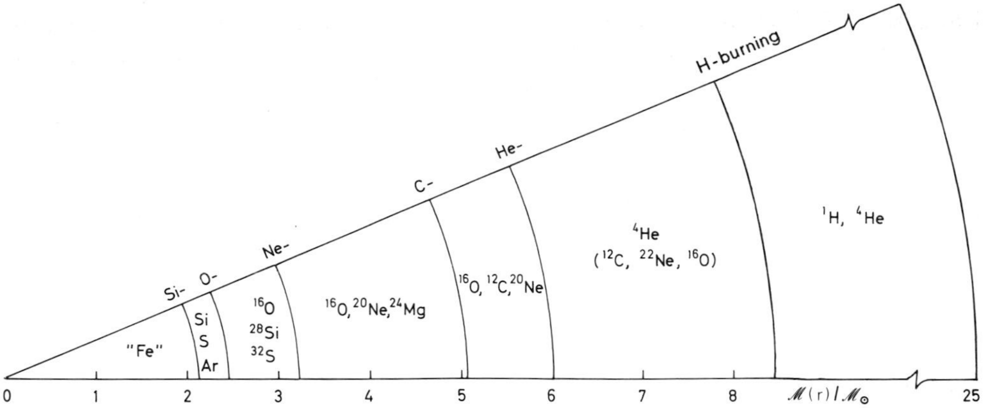

Fig. 5.4.9. The "onion-layer structure" of an evolved Population I star of $\mathcal{M} = 25\,\mathcal{M}_\odot$ after the conclusion of silicon burning. The zones of maximum energy release and the most abundant product nuclides within each burning shell are indicated (after J. R. Wilson et al., 1985). Outside of $\mathcal{M}(r) = 8.43\,\mathcal{M}_\odot$ or $r = 0.55\,R_\odot$, the star still has its original H- and He-rich composition. The neutronized Fe core contains $\mathcal{M}(r) = 2.1\,\mathcal{M}_\odot$, corresponding to $r = 4.2 \cdot 10^{-3}\,R_\odot = 2900$ km; at the center, the temperature is $T_c = 8.2 \cdot 10^9$ K and the density is $\varrho_c = 2.2 \cdot 10^{12}$ kg m^{-3}. In the Hertzsprung-Russell diagram, this star is in the region of red supergiants (effective temperature 4400 K, luminosity $3 \cdot 10^5\,L_\odot$, and radius around $10^3\,R_\odot$)

(Sect. 4.12.3), *carbon burning* $^{12}C + {}^{12}C$ (4.12.23) is ignited. This process proceeds *explosively*, similar to helium burning in degenerate matter. Complex hydrodynamic model calculations for the energy balance and expansion of the combustion front, which are not yet completely satisfactory in their details, indicate that the star explodes completely, without leaving a remnant. This explosive carbon burning is possibly the cause of some of the observed supernova outbursts (Sect. 4.11.7).

We may now ask the question, "How does the stellar evolution proceed following helium burning in the more *massive stars* ($\geq 8\,\mathcal{M}_\odot$)?" Here, at temperatures of 5 to $8 \cdot 10^8$ K, the reaction $^{12}C + {}^{12}C$ is initiated in *nondegenerate* matter, so that the stability of the star is maintained during the phase of *hydrostatic carbon burning* which lasts only the order of 100 yr. After the formation of a concentric-shell burning zone, a core region consisting of ^{16}O, ^{20}Ne, and ^{24}Mg is left at the center of the star. In stars of $\geq 10\,\mathcal{M}_\odot$, a rapid sequence occurs, with increasing temperatures and densities in the core, consisting of neon burning (lasting on the order of 1 yr), oxygen burning (several months), and silicon burning (1 day) (see Sect. 4.12.3). The star forms an "*onion-shell structure*" (Fig. 5.4.9), with an "*iron*" *core* of mass in the range 1.3 to $2.5\,\mathcal{M}_\odot$. Depending on the physical conditions, different nuclides are predominant in this core, e.g. ^{56}Fe or the neutron-poorer radioactive ^{56}Ni, which decays finally to ^{56}Fe. The synthesis of the nuclides in the iron group marks the maximum of the nuclear binding energy and thus the end of nuclear energy release in the star.

The nuclear evolution of stars in the mass range from about 8 to $10\,\mathcal{M}_\odot$, in which the densities in the (^{16}O, ^{20}Ne, ^{24}Mg) core of $\leq 1.4\,\mathcal{M}_\odot$ can be so high (2 to $3 \cdot 10^{13}$ kg m^{-3}) that electron degeneracy is produced and the following burning processes are no longer hydrostatic, is exceedingly complex; we cannot treat it further here.

After the star no longer has any more nuclear energy sources at its disposal, the central region contracts further; this is accompanied by an increase in temperature, until a "*phase transition*" takes place, whose nature depends on the range of (ϱ, T). In this transition, the compressibility of the stellar matter becomes so high that the stability condition $\Gamma > \frac{4}{3}$ (4.12.38) is violated and a *collapse* occurs within the free fall time (4.12.39).

For $\mathcal{M} \geq 100\,\mathcal{M}_\odot$, the instability is caused by the production of *electron-positron pairs*,

$$\gamma + \gamma \rightarrow e^+ + e^- \quad , \tag{5.4.4}$$

as soon as the temperature exceeds a few 10^9 K, i.e. when $kT \gtrsim mc^2$ ($mc^2 = 0.511$ MeV = rest energy of the electron and positron).

In the mass range 10 to $100\,\mathcal{M}_\odot$, above about $5 \cdot 10^9$ to 10^{10} K, the *photodisintegration* of nuclides by energetic thermal γ quanta triggers the collapse: in particular,

$$\gamma + {}^{56}Fe \rightarrow 13\,{}^4He + 4n \quad ,$$

followed by (5.4.5)

$$\gamma + {}^4\mathrm{He} \rightarrow 2\mathrm{p} + 2\mathrm{n} \ .$$

At the beginning of the collapse, for example in a star of $25\mathscr{M}_\odot$, the temperature in the core is about $8\cdot 10^9$ K and the density about $4\cdot 10^{12}\ \mathrm{kg\,m^{-3}}$; the corresponding free fall time is about 0.1 s.

At smaller stellar masses ($\lesssim 10\mathscr{M}_\odot$), the collapse of the degenerate electron-gas core region is caused by "*neutronization*" of the stellar matter; this occurs when the electron density, and with it the Fermi energy of the electrons, has become so high that the threshold for *electron capture* by the more abundant nuclei is exceeded. The maximum mass $\mathscr{M}_{\mathrm{Ch}}$ which can be held in equilibrium by the degeneracy pressure of the electrons is, from (4.12.54), proportional to μ_e^{-2}, where μ_e is the "atomic mass" referred to an electron (4.12.50). Therefore, $\mathscr{M}_{\mathrm{Ch}}$ is reduced from originally $1.4\mathscr{M}_\odot$ to about $0.8\mathscr{M}_\odot$ for the neutron-enriched matter.

Independently of the cause of the instability, the collapse of the inner region can come to a standstill only when about half of its mass has reached a density of $\gtrsim 2\cdot 10^{17}\ \mathrm{kg\,m^{-3}}$. At this density, which corresponds to that in the interior of normal atomic nuclei, matter consists mainly of neutrons and becomes practically incompressible. In the interior of a massive star ($\gtrsim 10\mathscr{M}_\odot$), a *neutron star* (Sect. 4.12.8) is thus formed within a short time ($\lesssim 1$ s); its surface then brakes the fall of the remaining material from the star.

The neutron star oscillates like an elastic sphere, and initially springs back somewhat; this creates a shock wave which passes out through the matter which is still falling inwards. Behind the shock wave front, the motion of this matter is reversed, but at the initially high temperatures, the atomic nuclei dissociate into free protons and neutrons and thereby strongly damp the shock wave. Its original kinetic energy of several 10^{44} J is dissipated after it has passed through only a few 100 km or about $0.5\mathscr{M}_\odot$ of stellar material. The subsequent fate of the star depends sensitively on its density structure and on energy transport in the layers which are outside the central region: if the shock wave has to pass through only a relatively small amount of matter, or if it can acquire sufficient energy, e.g. by absorption of high energy neutrinos from the inner regions, then it can reach the surface of the star and lead to the *casting off of a shell*. The neutron star remains as a remnant. If, on the other hand, the shock wave comes to a standstill within the star, more and more matter will be collected within the standing wavefront until finally the limiting mass for a neutron star, about $1.8\mathscr{M}_\odot$, is exceeded. Then a stable configuration no longer exists; the matter will collapse into a *black hole* (Sect. 4.12.9).

In the first case, we can identify the theoretical model of this final phase in the evolution of massive stars with the phenomenon of *supernova explosions* (of type II), in which an energy of about 10^{44} J is observed in the form of electromagnetic radiation (light curve) and kinetic energy of the expanding shell (Sect. 4.11.7). This amount of energy represents only about 1% of the total gravitational binding energy of about 10^{46} J which is released during the formation of a (proto) neutron star (corresponding to the compression of a mass of around $1\mathscr{M}_\odot$ to a radius of about 10 km). The major portion of this energy goes into the production of high-energy *neutrinos* (mainly during the formation of neutron-rich matter) and into an increase in the internal energy, and thus is finally lost to the star. In the case of the relatively near, bright supernova SN 1987A in the Large Magellanic Cloud, supernova neutrinos, with a total energy of 10^{45} to 10^{46} J, were detected for the first time; they were emitted several hours before the optical luminosity increase (Sect. 4.11.7).

As we have seen, the shock wave which moves outwards at first has a kinetic energy of only a few times 10^{44} J, so that only very precise numerical calculations can show whether our theoretical picture of stellar evolution is, in fact, able to explain the observed light curves and the expansion of the gas shell in supernovae. These hydrodynamic calculations require knowledge of the equation of state of stellar matter at high densities, of a complex chain of nuclear reactions, and in particular, of the cross-sections for energy and momentum transfer from neutrinos to the stellar matter, as well as of the internal structure of the pre-supernova. In spite of their extremely small cross-sections (of the order of magnitude of $10^{-47}\ \mathrm{m}^2$ at an energy of 1 MeV), at densities of $\gtrsim 4\cdot 10^{14}\ \mathrm{kg\,m^{-3}}$, the neutrinos are almost completely absorbed in the central region and are thus "stored" for a short time, before they can escape from the star. Even in the outer layers, the absorption of a small fraction of the enormous number of neutrinos may be sufficient to overcome the damping of the outgoing shock wave.

Our concept: that the evolution of a more massive star after the collapse of its interior can lead to a supernova explosion (of type II), is given strong support by, for one thing, the observation of neutrinos from the supernova

SN 1987A. For another, the complicated model calculations which have thus far been carried out fundamentally confirm it. However, at present, a number of important questions remain unanswered. For example, it is still not known with certainty how much stellar mass above $8 \ldots 10 \mathcal{M}_\odot$ is required for a supernova explosion, whether a neutron star always remains as a remnant after the explosion, whether neutron stars can also be formed without a supernova explosion, how much mass is required for the evolution to end with a black hole, etc.

In summary, we give a greatly simplified schematic representation of the possible final evolutionary stages of a star:

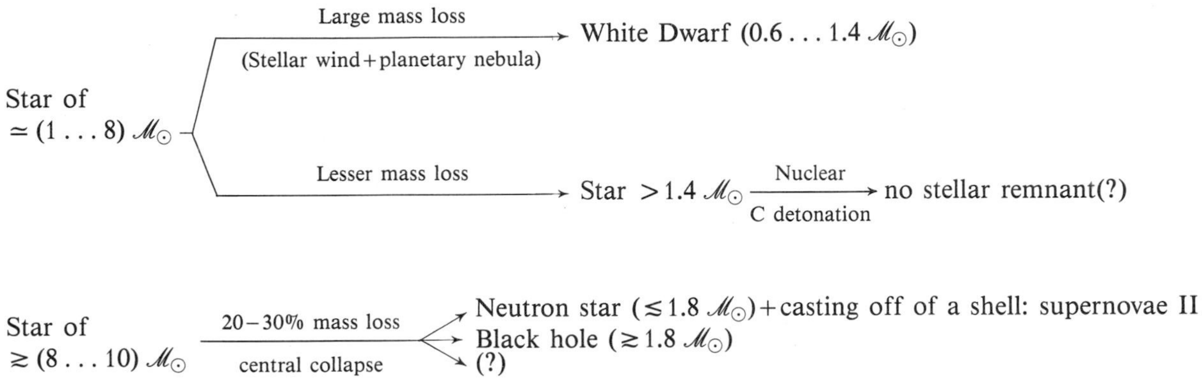

Large mass loss
White Dwarf (0.6 . . . 1.4 $\mathcal{M}_\odot$)
(Stellar wind + planetary nebula)
Star of
$\simeq (1 \ldots 8) \mathcal{M}_\odot$
Lesser mass loss
Star $>1.4 \mathcal{M}_\odot$
Nuclear
C detonation
no stellar remnant(?)
Star of
$\gtrsim (8 \ldots 10) \mathcal{M}_\odot$
20 – 30% mass loss
central collapse
Neutron star ($\lesssim 1.8 \mathcal{M}_\odot$) + casting off of a shell: supernovae II
Black hole ($\gtrsim 1.8 \mathcal{M}_\odot$)
(?)

While our theoretical picture for supernovae of type II seems to be well established, at least in its basic features, the origin of *supernova explosions of type I* is still not certain. On the one hand, the spectra indicate that the star has already lost its hydrogen shell before the explosion; and on the other, the shape of the light curves may best be explained by the explosion of a compact star. Thus, *one* model of a supernova I starts with a *white dwarf star*, which consists mostly of ^{12}C and ^{16}O and, as was shown above, can be formed by the evolution of a star of $\leq 8 \mathcal{M}_\odot$ as a result of major mass losses. If the white dwarf is in a *close binary system*, it can collect matter from its companion star (Sect. 5.4.6). If the accretion rate is sufficient, the mass of its (^{12}C, ^{16}O) core increases through continual hydrogen and helium burning until it exceeds Chandrasekhar's limit of $1.4 \mathcal{M}_\odot$. The contraction which results ignites carbon burning in the core, with its degenerate electron gas, i.e. an explosion occurs. The star is torn apart, without leaving a remnant star. The nuclear reactions in the expanding combustion front convert about $0.7 \mathcal{M}_\odot$ into nuclides of the iron group, mainly into ^{56}Ni. This *carbon deflagration model* can explain in principle the similar shapes of the light curves from supernovae of type I and their exponential brightness decreases. The principal energy source for the emissions is the radioactive decay of about $0.5 \mathcal{M}_\odot$ of ^{56}Ni, with a half-life of 6.1 d, to ^{56}Co; the latter nuclide further decays with a half-life of 78.3 d into stable ^{56}Fe. This model is supported by the fact that in the optical spectrum of e.g. SN 1981B, absorption structures were observed which could be attributed to lines from Co II.

To conclude this section, we consider the nuclear reactions in the course of stellar evolution from the point of view of *nucleosynthesis* or, as it is often called, the *formation of the elements in the stars*.

In the first thermonuclear fusion phase, that of hydrogen burning, H is initially converted into ^{4}He, and, in the CNO cycle, C, N, and O are converted mainly into ^{14}N, while the isotopic ratio of $^{12}C/^{13}C \simeq 4$ is produced (Sect. 4.12.3). For a portion of this matter, ^{4}He is later burned in the star, yielding mainly ^{12}C. Mixing processes, together with mass losses, as well as matter exchange between the components in close binary systems, can move the products of nuclear reactions from the stellar interior to the surface and thus make them "visible" in the spectra. The fact that this really does occur in some cases, even in the early evolutionary phases, is shown by the *anomalous spectra* of particular groups of stars, which exhibit enhanced intensities for the lines of product nuclides from hydrogen or helium burning.

Thus we find among the hotter stars evidence for the CNO cycle in the *helium stars*, whose spectra are dominated by strong helium lines, and to a lesser extent in many OB stars with minor *anomalies of the C and N lines*. In the *Wolf-Rayet stars* (Sect. 4.5), with effective temperatures around 50000 K and masses of about $10\mathcal{M}_\odot$, an extremely high mass loss has apparently laid bare the layers of helium- and nitrogen-rich matter from CNO burning (type WN) or even the carbon-rich matter from helium burning (type WC).

A great variety of anomalous spectra are observed in the cool *red giants*, which, depending on their evolutionary phase, can be either on the first giant branch or on the asymptotic giant branch. For example, low $^{12}C/^{13}C$ ratios, as predicted to result from CNO burning, are derived from the molecular bands of otherwise normal G and K giants. The spectra of the *carbon stars* (spectral types C or R and N, CH stars, etc.; cf. Sect. 4.5), with their strong bands from CN, CH, and C_2, indicate that in their atmospheres, in contrast to the normal stars (G, K, M), the ratio C/O is >1, so that here, matter which has been modified by helium burning is present on the star's surface.

Particularly noteworthy are the lines of elements such as Sr, Y, Zr, and Ba which occur with enhanced intensities in the spectra of many types of red giants (C and S stars, CH stars, barium stars), as well as the lines of Tc, which has no stable isotope, in some S stars (Sect. 4.5). These elements must have been produced in the stars in a few 10^6 yr during the red giant stage. Thermonuclear fusion reactions cannot be the source of these elements, due to the high Coulomb barriers for their production; instead, they must be due to *neutron capture* by the nuclides of the iron group, in the so-called *s process* (*s*low), where the reaction times are slow compared to the competing β-decays (Sect. 5.8.5).

Model calculations for stellar evolution show that on the *asymptotic giant branch*, as a result of the thermally unstable, pulsed helium-shell burning (Sect. 5.4.4), for one thing, sufficient neutrons are released, and for another, convection zones are formed, which can mix the "s-process elements" up to the surface. The most important source of these neutrons is probably the reaction $^{22}Ne(\alpha, n)\,^{25}Mg$ (4.12.22), which is initiated at the base of the helium-rich zone at sufficiently high temperatures according to (4.12.21) by reaction of ^4He with the ^{14}N which is produced by the CNO cycle. Another possibility for neutron production is the series

$$^{12}C(p,\gamma)\,^{13}N(e^+\,\nu)\,^{13}C(\alpha,n)\,^{16}O \; , \qquad (5.4.6)$$

which can occur when fresh, hydrogen-rich matter is mixed downwards in the star and comes into contact with the ^{12}C resulting from helium burning. Although the complex events in this phase of stellar evolution are not yet understood in detail, the asymptotic giant branch can still be considered with some certainty to be the place where the s-process elements are produced.

In the notable case of the F supergiant FG Sge, we can, so to speak, watch the mixing upwards to the surface of nuclear products: its spectrum shows changes within a few years. In particular, in the last 10 yr a decrease in the intensity of the lines of the iron group elements has been observed, while the lines of s-process elements such as Zr II, Ba II, and some of the rare earths are very strong. In the neighborhood of the star, there is a large (protoplanetary) nebula which was cast off about 6000 yr ago.

Stars with $\mathcal{M} \geq (8\ldots10)\,\mathcal{M}_\odot$, which pass through the various phases of nuclear fusion from carbon to silicon burning on a very short time scale and finally explode as *supernovae II*, play a decisive role in the nucleosynthesis of the elements between oxygen and the iron group. On the one hand, the major portion of the Fe-Ni core of the "onion-shell structure" (Fig. 5.4.9) remains trapped in the newly-formed neutron star. On the other hand, at the high temperatures and densities behind the expanding shock wave front, nuclear reactions take place which change the nuclear composition of those layers which then are cast off by the star and mix with the surrounding interstellar medium. For example, model calculations for a star of $25\,\mathcal{M}_\odot$ show that the re-ignition of oxygen and silicon burning increases the abundances of, in particular, the elements O, Si, S, Ca, and Fe by an order of magnitude in comparison to the solar mixture, while retaining about the same *relative* abundance ratios as in solar matter. (The products of these "explosive" nuclear processes behind the shock wave front differ from those of the corresponding "hydrostatic" fusion reaction phases, for which sufficient time is available for equilibrium to be established.) The quantitative results of calculations of nuclear synthesis in supernova explosions naturally also reflect the large uncertainties which we have already met in the models for the supernovae themselves. In a supernova explosion, presumably also the *neutron-rich isotopes* of the heavy elements following the iron group can be formed by neutron capture on time scales which are short compared to the competing β-decays (r-process; see Sect. 5.8.5).

As we have seen, the evolution of stars is inseparably bound up with the question of the formation of the elements. In particular, supernovae release matter to their surroundings which contains heavy nuclides formed by a series of nuclear reactions. From this matter, new generations of stars condense, with a chemical composition which is enriched in the heavy elements. Up to this point, we have discussed only the individual contributions of stars of different masses to nuclear synthesis; the interactions of all the stars, through stellar genesis and mass loss, with the interstellar matter in our Milky Way system and in other galaxies will be treated in Sect. 5.8.5.

5.4.6 The Evolution of Close Binary Star Systems

Nuclear evolution in *close* binary systems can follow quite a different course than in the individual stars which we have considered up to now; in such systems, the components can influence each other through their gravitational fields, through stellar winds, and in particular through an exchange of gas flows. Following the fundamental observations of O. Struve in the 1940's and 50's, R. Kippenhahn and A. Weigert, B. Paczyński, and M. Plavec and their coworkers began in the mid-1960's to apply theory and to develop complex computer programs to describe stellar evolution in close binary systems.

We have already met up with the common gravitational potential of a binary star system, with its *Roche surface* and the Lagrange points (Fig. 4.6.4); we have also identified important groups of stars, such as the cataclysmic variables (novae, ...) and the X-ray pulsars, as close binary systems (Sects. 4.11.5 and 6). In this section, we describe the basics of the evolution of close systems, building on the results for the evolution of individual stars. What happens in detail depends on the two original masses $\mathcal{M}_1$ and $\mathcal{M}_2$ and on the distance a between them, as well as on the losses of mass and angular momentum which the system sustains. We can give only an outline here of the great variety of possible binary star configurations.

The originally more massive component (star 1) is the *primary component*. This term is retained even when the mass ratio is reversed in the course of the evolution of the system. Corresponding to the evolution times t_E for the different stellar masses (Table 5.4.2), component 1 first evolves into a giant star, increasing its radius towards the end of the hydrogen burning phase, during hydrogen-shell burning, and then during helium burning (Fig. 5.4.7a). If it grows larger than the Roche lobe during one of these phases, then matter will flow through the inner Lagrange point L_1 (Fig. 4.6.4a) onto component 2. In this way, a binary system can be formed in which the further-evolved component has the smaller mass and no longer fits on the mass-luminosity curve (Fig. 4.6.2).

The change in separation a of the two components as a result of mass exchange can be given for the case, which often occurs in fact, that the system loses no mass and that the total orbital angular momentum is conserved:

$$L = a_1^2 \mathcal{M}_1 \omega + a_2^2 \mathcal{M}_2 \omega = \text{const} \ . \tag{5.4.7}$$

Here, ω is the circular frequency of the orbital motion (angular velocity, $\omega = 2\pi/\text{orbital period } T$) and a_i is the distance of component i from the common center of gravity. Inserting the definition of the center of gravity (2.6.36) and Kepler's 3rd law, $\omega^2 a^3 = G(\mathcal{M}_1 + \mathcal{M}_2)$ (2.6.38) into (5.4.7), we find that the separation is proportional to the following function of the mass ratio $q = \mathcal{M}_1/\mathcal{M}_2$:

$$a \propto \frac{(1+q)^4}{q^2} \tag{5.4.8}$$

(Fig. 5.4.10). The two stars have their minimum separa-

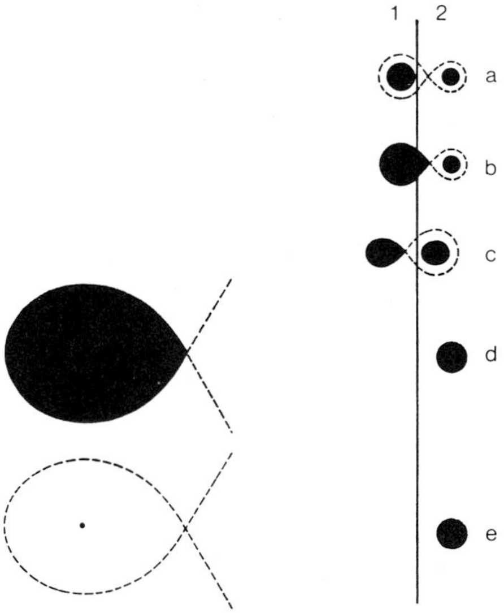

Fig. 5.4.10. The schematic evolution of a close binary star system with an initial mass ratio of $q = \mathcal{M}_1/\mathcal{M}_2 = 2$ (time a); star 1 = primary component. The axis of rotation around the center of gravity of the system is drawn in as a vertical line, and sections through the Roche surfaces are shown as dashed curves; the volumes of the two components are indicated by the filled areas. Time b: beginning of mass loss from star 1 as it fills its Roche lobe. Time c: $q = 1/2$. Time d: the end of mass loss at $q = 1/10$; the last contact of star 1 with its Roche surface. The primary component contracts into a compact star, for example a white dwarf (at time e)

tion when q equals 1. The radius of the two Roche surfaces depends upon q, and is also directly proportional to the separation a.

While a few binary star systems, the contact systems or W UMa stars (Fig. 4.6.4b), are so close that the Roche surfaces of the two stars touch each other already during the main sequence stage, in general, a strong interaction of the two components begins only when the more massive star has evolved away from the main sequence. Even before it has filled its Roche lobe, it can heat up its cooler companion and, among other things, can initiate surface activity on the companion. The strong X-ray and radiofrequency emissions of the RS CVn variables (Sect. 4.11.3) are probably produced in this way.

Once star 1 has reached its Roche lobe, there is at first a phase with a strong mass flow through the inner Lagrange point within a relatively short time, of the order of the Helmholtz-Kelvin time, until the masses of the two components have become equal. Thereafter follows a phase with a slower gas flow, as a rule on a nuclear time scale. The observations of eclipsing binaries (Sect. 4.6.2) show that, for example, β Lyr is in the first phase and Algol in the second. The secondary component is often surrounded by a dense accretion disk during the rapid phase. Observations of a common gas shell around both components indicate that star 2 can sometimes not accommodate the gas flow, and the overall system then loses mass and angular momentum.

During the further evolution of the primary component, its radius decreases again due to loss of the hydrogen-rich outer layers or to the ignition of helium burning (Sect. 5.4.4). The star draws back from its Roche surface and the mass flow comes to an end. We thus obtain a relatively widely-separated binary star system (Fig. 5.4.10), with a main sequence star as the more massive secondary component and a white dwarf or, as long as the supernova explosion does not tear the system apart, a neutron star as primary component.

On the way to this configuration, anomalous element abundances can appear on the surface of one or both components as a result of hydrogen and helium burning in combination with strong mass exchange (Sect. 5.4.5). This phase can be attributed to e.g. the helium stars, the OB stars with CNO anomalies, the Wolf-Rayet stars (at least those which belong to binary systems) and the barium stars.

Owing to the strong dependence of the evolution time t_E along the main sequence on the mass of the star, the primary star usually ends its evolution before the *secondary star* has left the main sequence and evolved into a giant star. Then, the mass transfer is replayed with reversed roles. The compact primary component can, as a result of its deep gravitational potential, at first effectively collect the stellar wind from the more massive star 2 and then its gas flow through the Lagrange point. Accretion onto a neutron star as primary component yields the various forms of X-ray binary stars (Sect. 4.11.6); continued accretion can lead to the formation of a black hole. If, on the other hand, the matter flows over onto a white dwarf star, the result, in connection with nuclear fusion processes, is the great variety of cataclysmic variables (Sect. 4.11.5) such as novae, recurring novae, etc.

Finally, the evolution of the secondary component also ends with the formation of a white dwarf plus planetary nebula, a neutron star, or a black hole. In the first case, the result is a relatively widely-spaced binary system consisting of two white dwarfs or of a white dwarf and a neutron star (primary component). On the other hand, if the more massive star 2 explodes as a supernova of type II, the binding of the binary system is destroyed and each component flies away as a "runaway star" with a velocity of the order of 100 km s^{-1}. This explains the high spatial velocities observed for many radio pulsars (Sect. 4.11.7). If star 2 had lost a great deal of mass, however, a weaker supernova explosion is possible, leaving a bound system consisting of two neutron stars. The binary pulsar PSR 1913+16 (Sect. 4.6.6) was perhaps formed in this way. The remaining possible final configurations, pairs consisting of one neutron star and one black hole, or of two black holes, would be rather difficult to observe.

5.4.7 The Initial Phases of Stellar Evolution. Star Formation

After our excursion into the evolution of close binary star systems, we return to the starting point of our considerations and ask the questions, "How do stars evolve onto the standard main sequence?", and then, "How and from what are the stars formed?".

The close spatial conjunction of the young O and B stars of high absolute magnitudes with gas and dust clouds in the spiral arms of our own galaxy, and in others such as the Andromeda galaxy, leads us to assume that quite generally, stars are formed from, or in, cosmic clouds of diffuse matter. The only energy source which is initially at a star's disposal, i.e. until the ignition of some sort of thermonuclear reaction, is *gravitational energy*, $E_G \simeq G\mathcal{M}^2/R$ (4.12.27); it is released during the *contrac-*

tion of the stellar mass $\mathcal{M}$ from originally widely-distributed matter to a radius R.

The characteristic evolution time for the phase of gravitational contraction is the *Helmholtz-Kelvin time* $t_{HK} \simeq G\mathcal{M}^2/RL$ (4.12.33), whereby a hydrostatic equilibrium (4.12.1) can be established in the (whole) star, as long as the dynamic time scale or *free fall time:*

$$t_{ff} \simeq \frac{1}{\sqrt{G\bar{\varrho}}} \ll t_{HK} \qquad (5.4.9)$$

(4.12.39) is much shorter than the Helmholtz-Kelvin time. In the neighborhood of the main sequence, this condition is fulfilled very well; for example, for $\mathcal{M} = 1\,\mathcal{M}_\odot$, a luminosity $L = 1L_\odot$, and a mean density $\bar{\varrho} = 1400\,\mathrm{kg\,m^{-3}}$, it follows that $t_{ff} \simeq 3\cdot 10^3\,\mathrm{s} \simeq 10^{-4}\,\mathrm{yr}$ and $t_{HK} \simeq 2\cdot 10^7\,\mathrm{yr}$. However, with increasing radius, t_{ff} increases proportionally to $R^{3/2}$, while t_{HK} decreases as R^{-3} (keeping the effective temperature constant for simplicity) $[L = 4\pi R^2 \sigma T_{eff} \propto R^2$, (4.4.26)]. Beginning at $R \geq 300\,R_\odot$, we thus find $t_{ff} \geq t_{HK}$. During the evolution of *protostars*, with their much greater radii, *dynamic* calculations are therefore necessary; these were first carried out by R. B. Larson in 1969. Here, the condition for hydrostatic equilibrium must be extended by inclusion of an acceleration term, $\varrho\, d^2r/dt^2$. Only when the main sequence has been nearly reached can *global hydrostatic equilibrium* be established in the *pre-main-sequence stars.*

We begin with these pre-main-sequence stars and ask, "Which path does a star follow on the color-magnitude diagram, when it has been formed by *contraction* from widely-distributed matter and is presumed to remain in hydrostatic equilibrium?" Giant stars with effective temperatures below 3000 to 4000 K have essentially *convective* structures (Sect. 4.12.2); their interiors are occupied for the most part by extensive hydrogen convection zones. C. Hayashi showed in 1961 that *no* hydrostatic equilibrium can be established in these stars. On the (theoretical) color-magnitude diagram there thus exists, for each mass, a line (the *Hayashi line*) which is nearly vertical, i.e. at constant effective temperature, and which separates the "forbidden zone" of the unstable stars (to the right) from the region of the stable stars (left).

A pre-main-sequence star in hydrostatic equilibrium forms an extended convective envelope as soon as its hydrogen is partially ionized. It moves downwards on the *color-magnitude diagram* (Fig. 5.4.11) from high luminosities, with a slightly increasing temperature, remaining just to the left of the Hayashi line, until (after $\simeq 10^7$ yr for $1\,\mathcal{M}_\odot$) it reaches the nearly horizontal line representing radiation equilibrium (calculated earlier by L. G. Henyey and coworkers). On this line, it continues to fill its energy requirements through release of gravitational energy. Ignition of the well-known thermonuclear reactions takes place only shortly before it reaches the initial main sequence.

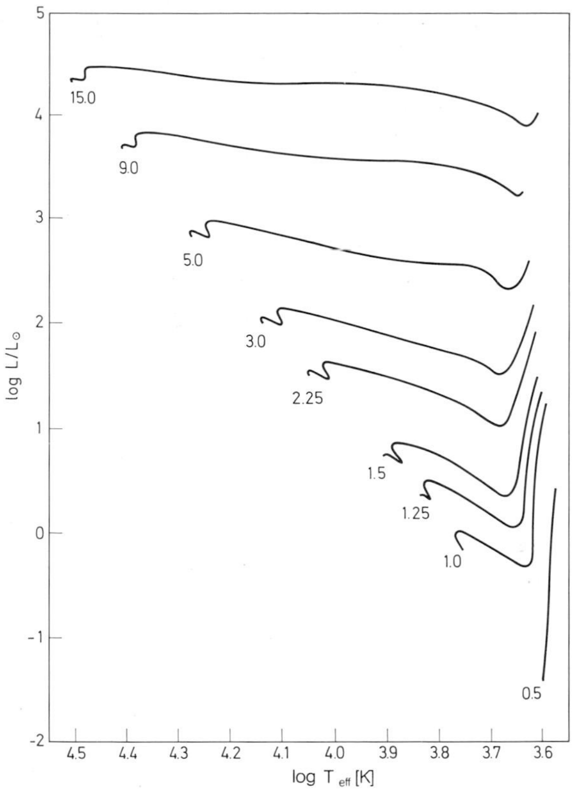

Fig. 5.4.11. The evolution of hydrostatic pre-main-sequence stars of various masses (0.5 to $15\,\mathcal{M}_\odot$) after I. Iben (1965). Stars of $\geq 1\,\mathcal{M}_\odot$ require less than several 10^7 yr to reach the main sequence

It is notable that stars can also not cross over the Hayashi line "in the opposite direction". Their evolutionary lines in the region of giant stars therefore point steeply upwards in Figs. 5.4.7 and 8.

Towards the upper part of the main sequence, the contraction time t_{HK} becomes shorter and shorter, according to (4.12.33); for a B0 star, it is only about 10^5 yr. At the other end, stars of smaller mass are formed more slowly; an M star of $0.5\,\mathcal{M}_\odot$ requires about $1.5\cdot 10^8$ yr.

In young galactic star clusters, which can be recognized by their bright blue stars, M. Walker indeed found stars of middle and later spectral types to the right of the lower part of the main sequence, whose T Tauri variability, Hα emission lines, and in part rapid rotation etc. identify them as young stars. Figure 5.4.12 shows, as an example, the color-magnitude diagram of the cluster NGC 2264, after M. Walker; calculating on the basis of the brightest star, this cluster has an age of only $3 \cdot 10^6$ yr. In agreement with theory, stars below spectral type A0 ($\leq 3\,\mathcal{M}_\odot$) have not yet reached the standard main sequence.

Contracting masses below a certain mass limit never reach the temperature necessary for ignition of nuclear processes. S.S. Kumar has shown (1963) that this mass limit lies somewhat below $0.1\,\mathcal{M}_\odot$. Smaller masses form cool, completely degenerate objects, which are called *brown dwarfs* (or also black dwarfs).

We now ask how a star, following its formation from condensation of the interstellar medium, can arrive at the (hydrostatic) stage of a pre-main-sequence star. We have already estimated that dynamic time scales (free fall times) are the determining factors in these earliest evolutionary phases.

Stability calculations (Sect. 5.8.1) indicate that only relatively large masses of about 10^2 to $10^5\,\mathcal{M}_\odot$, which are much greater than observed stellar masses, can condense from the interstellar medium. Indeed, on the basis of observations, V.A. Ambartsumian in 1947 had already arrived at the important conclusion that the stars (at least for the most part) are formed over periods of time of the order of 10^7 yr in *groups*, with overall masses of $10^3\,\mathcal{M}_\odot$. These are the socalled *OB associations*, with bright blue stars, and the *T associations*, with cool stars, especially T Tauri variables of low absolute magnitudes (frequently, both kinds occur together). An OB association, such as those in Orion or Monoceros, with enormous masses of gas ionized by the short wavelength radiation of the O and B stars embedded in it, appears to the observer in the optical range at first as an H II region, dominated by emission of strong Hα radiation. The stars of an association move away from its center with velocities of the order of $10\,\mathrm{km\,s^{-1}}$; associations are thus *not* stable. The *expansion age* estimated by extrapolating the motions of the stars backwards in time agrees in general with the evolutionary age of the brightest stars.

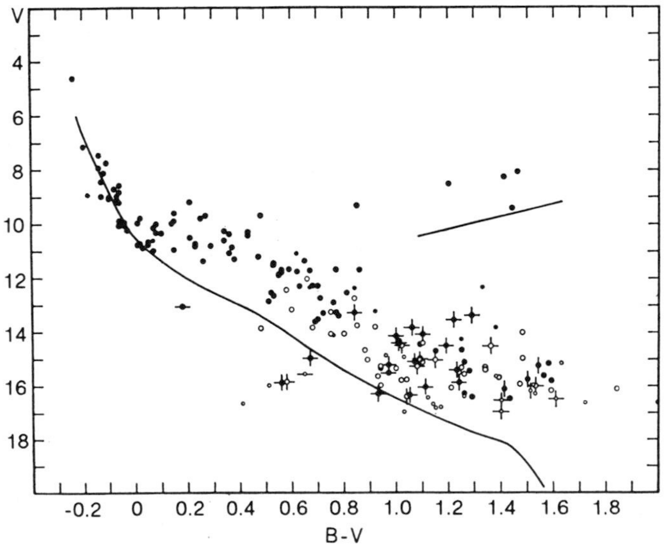

Fig. 5.4.12. The color-magnitude diagram of the very young galactic star cluster NGC 2264, after M. Walker (1956). ● Photoelectric measurements; ○ photographic measurements; | variable stars; – stars with Hα emission. The apparent magnitudes V are plotted against B−V, corrected for uniform interstellar reddening of the cluster. The curves indicate the standard main sequence and the giant branch. The apparent distance modulus is 9.7 mag, and the distance is 800 pc

As a result of progress in radioastronomy in the millimeter-wave region, and in infrared astronomy, it has become clear since the 1960's that the gas nebulas which emit strongly in the visible, with their OB stars, are only a part of the gigantic *molecular cloud complexes* with masses above $10^5\,\mathcal{M}_\odot$. These make themselves apparent especially through line emissions from molecules such as CO, NH_3, H_2CO, and others (Sect. 5.3.4). The most dense and coolest parts of these molecular clouds can be regarded as the actual locations of *stellar genesis*. The major area of stellar genesis which is closest to us, in *Orion*, at a distance of about 500 pc, is particularly well suited for detailed investigations. Figure 5.4.13 clearly shows the large-scale correlation of the density distribution of CO molecules with both the dark cloud complexes around Lynds 1640/41 and 1603, and with the association Ori OB 1, which includes the stars in Orion's belt and sword, as well as with the H II regions which are excited by their OB stars. The center of the CO distribution, the molecular cloud Ori MC 1 coincides with the Orion Nebula, M 42 = NGC 1976 (Fig. 5.3.11), which is excited by the four trapeze stars θ^1 Ori A, B, C, and D. More precise observations show that the molecular cloud as seen from our position lies directly behind the Orion Nebula. In it, several "point-like" infrared sources can be seen, among them the *Becklin-Neugebauer Object* and the *Kleinmann-Low Object*, named for their discoverers; the luminosities of these objects are several $10^3\,L_\odot$, and their surface temperatures are a few 100 K. These infrared sources form a cluster of very young *protostars* (see below), with altogether about $10^5\,L_\odot$, which is however

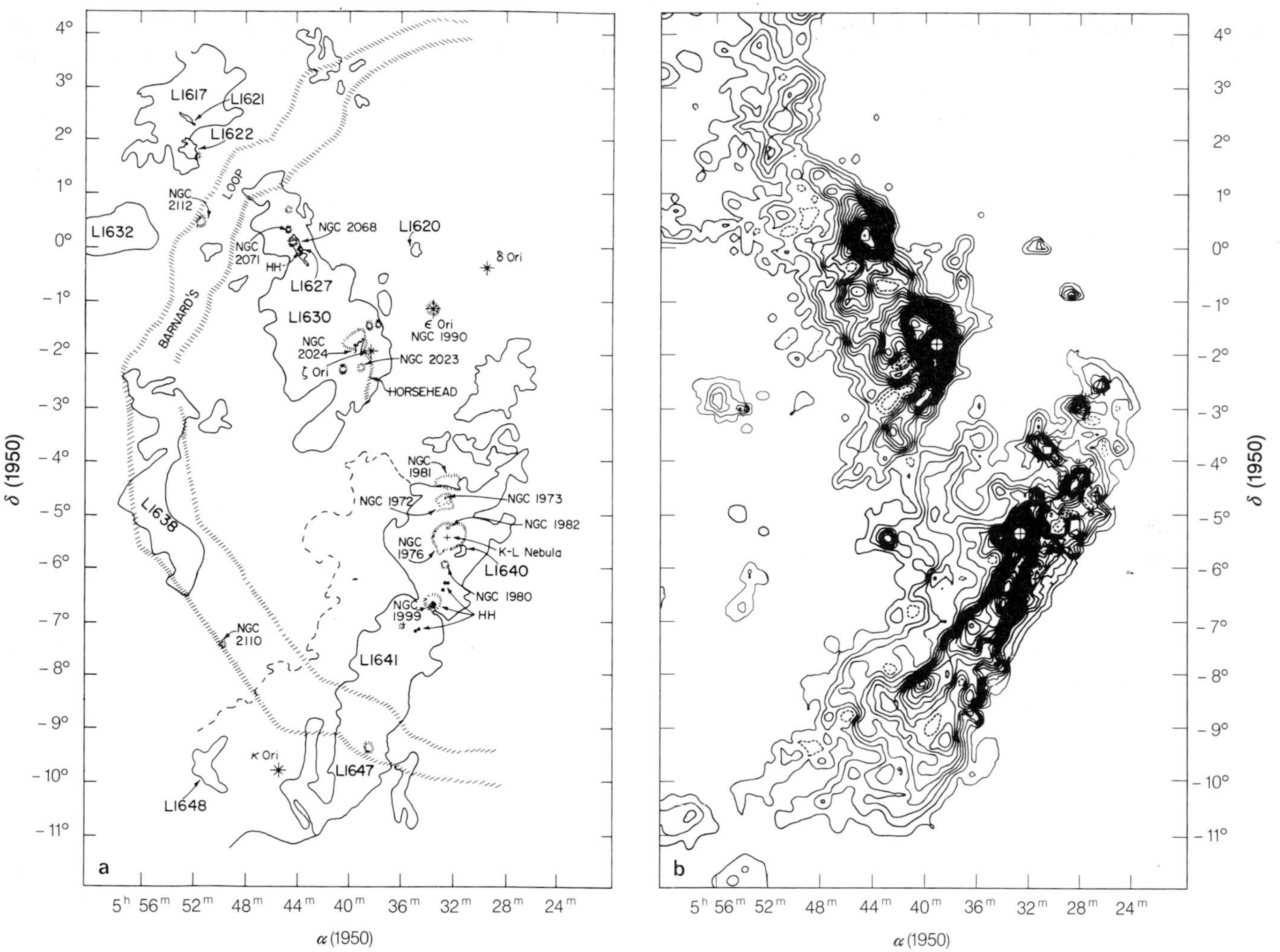

Fig. 5.4.13 a,b. A region of star formation in Orion. **(a)** A sketch of the most noticeable structures in the optical region. *Dashed*: the boundaries of the emission and reflection nebulas (NGC 1976 = M 42 = Orion Nebula); *full curves*: boundaries of the dark dust clouds with the notation of B. T. Lynds (L); HH denotes Harbig-Haro objects; δ, ε, and ζ Ori are stars in Orion's belt. The position of the protostar cluster around the Becklin-Neugebauer and Kleinmann-Low objects is indicated as "K–L Nebula". (From M.L. Kutner et al., Astrophys. J. **215**, 521 (1977); reprinted courtesy of M.L. Kutner and *The Astrophysical Journal*, published by the University of Chicago Press; © 1977 The American Astronomical Society). **(b)** Intensity contours of the integrated line emissions from the rotational transition $J = 1 \to 0$ at $\lambda = 2.6$ mm in the CO molecule. From the Goddard-Columbia Sky Survey with a 1.2 m telescope having a beam width of 8′. (P. Thaddeus, Ann. New York Academy of Sciences **395**, 9 (1982); with the kind permission of the publisher)

not observable in the optical region due to dense dust clouds.

Closely related to the formation of young stars are a variety of interesting phenomena: we thus find *emission lines of* H_2 from excited vibrational states, which originate in thin layers at about 2000 K and are probably excited in shock wave fronts. Furthermore, we observe point sources of intense microwave radiation with extreme excitation conditions, i.e. *masers* based on transitions in OH as well as in H_2O and SiO (Sect. 5.3.4). They evidently originate in regions of quite dense gas, whose proper motions of order $10\ \mathrm{km\,s^{-1}}$ indicate an expansion away from the infrared sources. Also the *Herbig-Haro Objects*, glowing small spots of nebulosity, are found in large numbers in the vicinity of young stars (Fig. 5.4.13). Their emission spectra, in particular the high in-

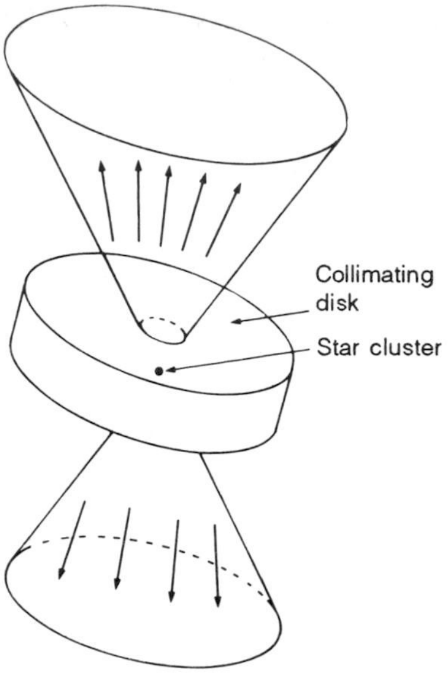

Fig. 5.4.14. Schematic drawing of the bipolar protostellar wind in the Orion Nebula, which originates in the star cluster around the Becklin-Neugebauer object and is collimated by a thick, dust-filled disk of gas. The axis of the disk is inclined in such a way that the flow in the upper cone (NW part) has a component directed towards the observer, giving a blue shift of the spectral lines. From B. Zuckerman, Nature **309**, 403 (1984); reprinted by permission from Nature, **309**, 403. Copyright © 1984 Macmillan Magazines Ltd

tensities of the lines [S II] $\lambda = 671.6/673.1$ nm, indicate excitation by shock wave fronts. Finally, detailed investigations of line profiles, both in the optical and in the radiofrequency and infrared regions, lead to surprisingly high velocities of up to a few 100 km s^{-1} for the gas and the HH objects in the neighborhood of Ori MC 1. All these observations together can be explained in terms of a massive "*protostellar wind*", which originates at the infrared sources around the Becklin-Neugebauer Object and streams away in two oppositely-directed cones. The collimation of this gas flow is probably effected by a thick, dust-filled gas disk in the vicinity of the protostars, which also prevents observations in the optical region (Fig. 5.4.14). Such *bipolar nebulas* with *outflowing* matter are observed in many cases near young stars also in other parts of the sky.

But after taking this view into the regions where stars are formed, let us return to the question of stellar genesis itself, and of the initial evolution of newly-formed stars! Concerning the process of *fragmentation*, i.e. how the molecular clouds, which have condensed from the interstellar medium and which have masses from 10^2 to $10^4\,\mathscr{M}_\odot$, are broken up into the observed stellar masses in the range of 0.1 to $10^2\,\mathscr{M}_\odot$, we know only very little. From the observational side, there is still a gap of an order of magnitude between the smallest sources which can be resolved in the radio and infrared regions and the initial diameters predicted by theory for the protostars. Therefore, following R. B. Larson, we begin the evolutionary calculations by assuming a homogeneous sphere of molecular hydrogen and dust to have a *stellar* mass, and choose its density and temperature to keep it just gravitationally unstable according to the *Jeans criterion* (5.8.4). In this process, we orient our considerations to the physical properties of the denser parts of cool molecular clouds (Sect. 5.3.4). For $T = 10$ K and $\varrho = 10^{-16}$ kg m^{-3}, or $N \simeq 10^{11}$ m^{-3}, masses $\geq 1\,\mathscr{M}_\odot$ are unstable. The radius of a sphere of $1\,\mathscr{M}_\odot$ with $\varrho = 10^{-16}$ kg m^{-3} is $R = 1.7 \cdot 10^{15}$ m $\simeq 10^4$ AU $\simeq 0.5$ pc, and the corresponding free fall time (4.12.39) is $t_{ff} \simeq 4 \cdot 10^5$ yr.

Already within a time of the order of t_{ff} after the beginning of the collapse, within the originally homogeneous protostellar cloud of mass $\mathscr{M}$, a *core region* of $\leq 0.01\,\mathscr{M}$ has formed which is many orders of magnitude denser. The collapse of the core is halted when it becomes optically thick as a result of the efficient absorption by dust particles of the radiation which it emits, mainly in the infrared. The impact of matter which continues to fall from the outer regions of the cloud onto the surface of the core produces a shock wave front, which, together with the slow gravitational contraction of the core, causes a continual rise in its temperature. When $T \geq 2000$ K, the H_2 dissociates and the effective adiabatic index Γ drops below $\frac{4}{3}$. According to (4.12.38), in the innermost part a (second) collapse then takes place, lasting only a few months or years, owing to the high densities; it stops when the density has reached 1 to 10 kg m^{-3} and the temperature is about 10^4 K. This leads to the formation of a small core of mass $\leq 10^{-3}\,\mathscr{M}$. After some time, when the remaining material from the first core has fallen onto the second, the *protostar* has the following characteristics: onto a dense *core* of mass $\mathscr{M}_c$ and radius R_c in hydrostatic equilibrium, the matter of the extended, much less dense *gas shell* drops practically in *free fall*; it is braked at the surface of the core by a *shock wave front*. The kinetic energy released in this process, $\dot{\mathscr{M}} G \mathscr{M}_c / R_c$ ($\dot{\mathscr{M}}$ = rate of mass fall onto the core), supplies nearly the total *luminosity* of the protostar. In the course of time, $\mathscr{M}_c$ increases continually, while $\dot{\mathscr{M}}$ decreases; for example, for $1\,\mathscr{M}_\odot$, after about 10^5 yr, $\mathscr{M}_c \simeq 0.6\,\mathscr{M}$ and $\dot{\mathscr{M}}$ has dropped from an initial value of some $10^{-3}\ \mathcal{M}_\odot\ \mathrm{yr}^{-1}$ to a few $10^{-6}\mathcal{M}_\odot\ \mathrm{yr}^{-1}$.

In the *less massive stars* ($\mathcal{M} \leq 3\,\mathcal{M}_\odot$), the entire shell falls within a relatively short time onto the core; e.g. for $1\ \mathcal{M}_\odot$, after 10^6 yr practically the whole mass of the protostar is contained in the core. With decreasing $\dot{\mathcal{M}}$, the contribution of the kinetic energy to the luminosity also decreases. When it becomes negligible compared to the contraction energy $G\mathcal{M}_c^2/R_c$ of the core, the protostar has become a pre-main-sequence star, and its evolutionary path on the Hertzsprung-Russell diagram merges into a "hydrostatic curve" (Figs. 5.4.11 and 15).

In the early evolutionary stages, in which the protostars are still unobservable in the optical region due to their dust-filled shells, there is hardly a chance of observing them in the infrared either, because of the short time scales. These young stars can first be seen, in part as infrared objects and especially in the visible range, as *T Tauri stars* in the transition stage between the dynamic and the hydrostatic phases (Sect. 4.11.3). Some of them, the YY Ori stars, evidence the falling matter directly in their spectra through absorption components shifted towards long wavelengths (inverse P Cygni profiles).

In the case of the *more massive stars* ($\mathcal{M} \geq 3\,\mathcal{M}_\odot$), the Helmholtz-Kelvin time for the contraction of the hydrostatic core is shorter than the free-fall time of the shell, in contrast to the less massive stars. Therefore, in

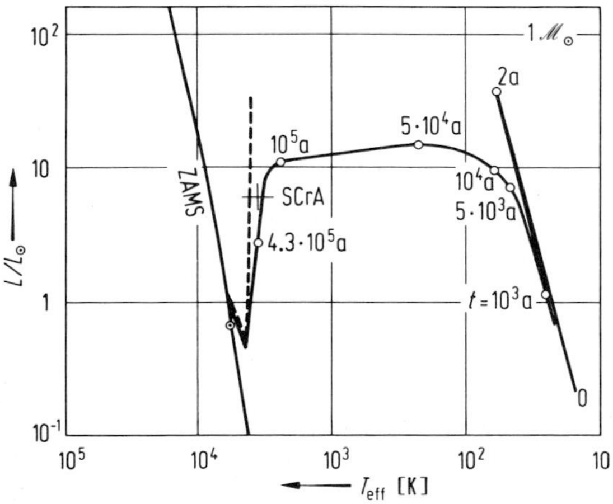

Fig. 5.4.15. The evolution of a protostar of $1\,\mathcal{M}_\odot$ in the Hertzsprung-Russell diagram, with times given in years since the formation of the hydrostatic core. Shortly before the initial main sequence (ZAMS), the evolutionary path merges into a Hayashi curve (*dashed line*) for fully convective stars in hydrostatic equilibrium. The position of the T Tauri star S CrA is indicated by a cross

these protostars, *hydrogen burning* begins already while a considerable portion of the shell is still falling onto the core; they thus never pass through the hydrostatic phase of a pre-main-sequence star (Fig. 5.4.11). Hydrogen burning causes a strong increase in the luminosity of the core; the resulting radiation pressure, which acts in particular on the dust particles in the shell, slows the falling matter until finally its motion is reversed. The star casts off its shell, and with it, a major portion of its original mass; for example, a protostar of originally $60\,\mathcal{M}_\odot$ becomes a main sequence star of only $17\,\mathcal{M}_\odot$.

Young stars ($\geq 3\mathcal{M}_\odot$) are thus observable before they enter the main sequence as relatively cool infrared sources of very high luminosities, whose spectra are typified by their dense dust shells (cocoons). Characteristic temperatures (of the dust) are in the range 100 to 600 K, and the luminosities are 10^3 to $10^6 L_\odot$. An example of a massive protostar is the *Becklin-Neugebauer Object* in the Orion association, which radiates mainly in the region from 3 to 10 µm and shows strong absorption structures at 3.1 µm (ice) and at 9.7 µm (silicates).

Along with the mass, the initial *angular momentum* J_0 of a collapsing cloud is decisive for its evolution. For small values of J_0, the spherically symmetrical models described above represent good approximations, but numerical calculations for large J_0 show that instead of a dense core, a rotating *ring-shaped* density distribution is at at first formed, when the angular momentum is retained in that part of the cloud. Presumably, the ring later breaks up into several fragments, so that in this way, binary stars and multiple systems can be formed. However, if the *angular momentum* is *transferred* from the inner part of the protostellar cloud to the outer parts, e.g. by viscosity in turbulent flow patterns, then a starlike core is formed, surrounded by a *rotating disk*. This kind of model can describe the fundamental aspects of the formation of our planetary system (Sect. 6.1).

An indirect indication of disks in the immediate neighborhood of protostars or young stars is given by the *outward flows of gas* with *bipolar characteristics* observed in many cases (Figs. 5.4.14 and 16). Here, a disk probably hinders the radial flow of gas outwards from a central source within its plane, so that a focussing effect towards the rotation axis results. The central object is eclipsed in the optical region by the dust absorption of the disk and can be observed only in the infrared.

As we have seen, the *early stages of stellar evolution* are characterized on the one hand by a concentration of originally widely-distributed matter, especially by a *fall-*

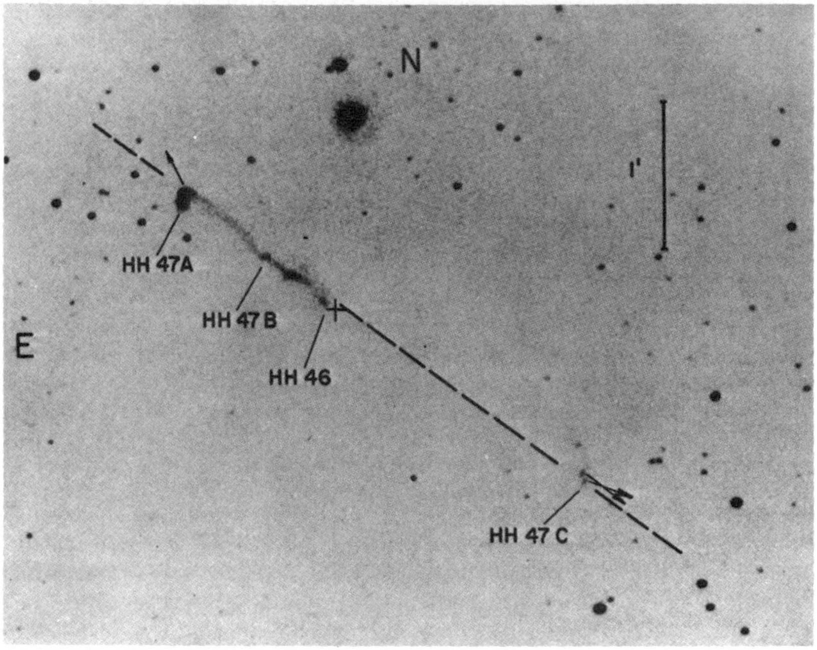

Fig. 5.4.16. The bipolar mass outflow from an infrared object of $12 L_{\odot}$: Herbig-Haro Objects HH 46/47 and a jet in the globule ESO 210-6A. The image was made in the light of the forbidden [S II] transitions at $\lambda = 671.6$ and 673.1 nm by J.A. Graham and J.H. Elias (1983). The distance is 400 pc, and 1′ corresponds to 0.12 pc. The central infrared source (+) is obscured in the visual region by dust absorption of $A_V \simeq 20$ mag. Measurements of the proper motions of the HH objects were made by R.D. Schwartz et al. (1984); the arrows show the distance covered in 500 yr

ing inwards of its mass. On the other hand, recent observations show ever more clearly that in addition, a strong mass *outflow* often accompanies the early phases of star development. We thus find energetic protostellar or stellar winds with velocities of several 100 km s^{-1} not only in massive, bright objects, but also in less massive stars with luminosities only a few times that of the Sun. The wide wings of the emission lines in the millimeter range from molecules such as CO indicate that also in the vicinity of these stars, the gas flows outwards, perhaps influenced by the stellar wind. In the neighborhoods of T Tauri stars, as well as of their "antecedents", infrared objects with similar luminosities and masses, there are often *Herbig-Haro Objects*, whose proper motions of the order of 100 km s^{-1} are directed away from the central star. In some cases, such as for example in the spherical dust cloud or globule ESO 210-6A, with about 0.4 pc diameter and 25 $\mathcal{M}_{\odot}$ (Fig. 5.4.16), strongly focussed gas flows or *jets* of up to 0.1 – 0.2 pc length have been discovered; they move from the central star towards the HH objects. The HH objects, which have emission line spectra excited by collisions, are probably "nodes" of particularly high energy dissipation in the gas which is flowing out at several hundred km s^{-1}. Frequently, an HH object, such as HH 47A, is found at the end of a jet; it probably results from the interaction of the supersonic jet flow (which has Mach numbers of the order of 10) with the matter of the surrounding molecular cloud.

A theoretical explanation of *how* the T Tauri stars and related infrared objects with luminosities of only 1 to $100 L_{\odot}$ can produce these energetic, strongly collimated bipolar gas flows has yet to be offered.

5.4.8 Stellar Statistics and Evolution. The Rates of Formation of Stars.

In the 1920's, J. Kapteyn, P.J. van Rhijn and others carried out a statistical analysis of stars of known parallax and obtained the *luminosity function* for stars in our vicinity. We shall limit ourselves here to the stars on the main sequence in early stages of their evolution, and define the luminosity function $\Phi(M_V)$ as the number of main sequence stars with absolute magnitudes M_V per pc^3 and per luminosity interval $\Delta M_V = 1$ mag (sometimes 0.5 mag instead). In the vicinity of the Sun, $\Phi(M_V)$ decreases rapidly from $\simeq 3.5$ mag towards brighter magnitudes (Fig. 5.4.17). This was explained by E.E. Salpeter (1955) with the hypothesis that stars fainter than 3.5 mag have collected since the formation of the galaxy roughly $T_0 \simeq 2 \cdot 10^{10}$ yr ago, without essential changes, while the brighter stars depart from the main sequence after about one evolution time t_E (Table 5.4.2), reckoned

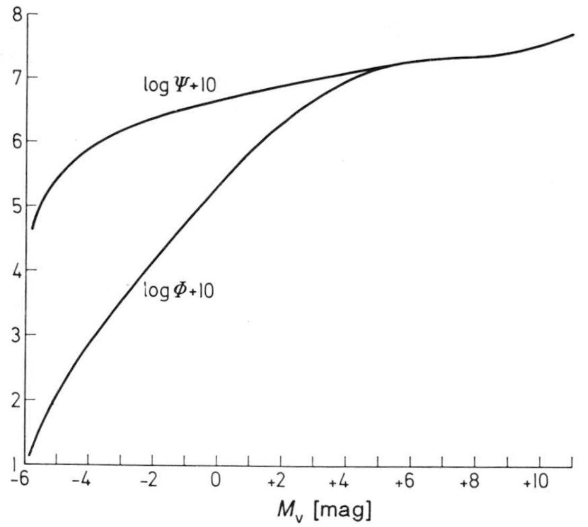

Fig. 5.4.17. The luminosity function $\Phi(M_V)$ and the initial luminosity function $\Psi(M_V)$ of the main sequence stars in the vicinity of the Sun. Φ and Ψ give the number of stars per pc^3 in the magnitude range $M_V - \frac{1}{4}$ to $M_V + \frac{1}{4}$ which are present or have been formed during the existence of the Milky Way

from the time of their formation, and finally, after a time which is likewise $\ll T_0$, become white dwarfs, neutron stars, etc. Thus, we can readily calculate the *initial luminosity function* $\Psi(M_V)$, which tells us how many stars of magnitude M_V in the interval ΔM_V have been formed in the Milky Way per pc^3 during the time of its existence (the external conditions are assumed to have remained constant). For the brighter stars, we find:

$$\Psi(M_V) = \Phi(M_V)\frac{T_0}{t_E(M_v)} \; ; \qquad (5.4.10)$$

for the fainter stars, $\Psi(M_V)$ merges continuously into $\Phi(M_V)$. In Fig. 5.4.17, we have included the initial luminosity function $\Psi(M_V)$ as calculated by A. Sandage, who continued E.E. Salpeter's computations.

If our ideas are correct, then the luminosity function of *young galactic star clusters* should correspond to the initial luminosity function Ψ and not to the luminosity function Φ of our vicinity. Indeed, the investigation of various clusters confirms the concept that the division of an originally-present mass of gas into stars proceeds everywhere according to the same initial luminosity function $\Psi(M_V)$.

The difference $\Psi(M_V) - \Phi(M_V)$, summed over all M_V, corresponds to *those* stars which have evolved away from the main sequence since the formation of the galaxy. By far the major portion of these stars must at present be white dwarfs. In fact, the spatial density of white dwarfs calculated from Fig. 5.4.17 agrees with the observed density, as well as can be expected from the uncertainties of the data.

For the faint stars at the *lower end of the main sequence*, a relatively complete stellar census can be obtained only if we limit ourselves to the immediate vicinity of the Sun in the Milky Way. In Fig. 5.4.18, the luminosity function of the stars at a distance of ≤ 20 pc from the Sun according to R. Wielen et al. (1983) is drawn on the basis of the catalog of W. Gliese (1969), with the inclusion of some more recently discovered faint stars. The brighter stars up to about $M_V \simeq 8$ mag and out to that distance are all included; for the fainter stars, to about 13.5 mag, the number can be extrapolated from a smaller volume around the Sun. For still fainter stars, Φ represents only a lower limit for the number of stars. The maximum of the luminosity function at $M_V \simeq 13$ mag is, however, probably genuine.

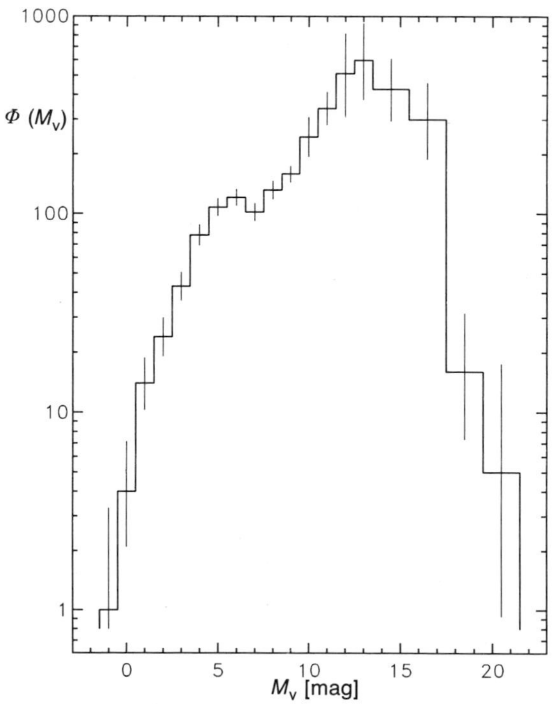

Fig. 5.4.18. The luminosity function $\Phi(M_V)$ of the stars within 20 pc distance from the Sun, after R. Wielen, H. Jahreiss, and R. Krüger (1983). Φ is the number of stars in a sphere of radius 20 pc with absolute magnitudes M_V per $\Delta M_V = 1$ mag

Equivalent to the luminosity function is the *mass function* $\phi(\mathcal{M})$, which gives the number of main sequence stars of mass $\mathcal{M}$ per pc^3 and per unit mass. It is obtained from the luminosity function $\Phi(M_V)$ by using the mass-luminosity relation (Fig. 4.6.2), whereby $\phi(\mathcal{M})\,d\mathcal{M} = \Phi(M_V)\,dM_V$. Particularly important for the theory of stellar genesis is the *initial mass function* $\psi(\mathcal{M})$ (or IMF), which is derived from the results of stellar evolution theory analogously to $\Psi(M_V)$ [cf. (5.4.10)].

The initial mass function is usually expressed in terms of $\xi(\mathcal{M}) = \mathcal{M}\psi(\mathcal{M})$, the *total mass* of all stars with individual masses $\mathcal{M}$ per pc^3 and per unit mass, where

$$\xi\,d\mathcal{M} = \mathcal{M}\psi\,d\mathcal{M} = \mathcal{M}\xi\,d\ln\mathcal{M}\ . \tag{5.4.11}$$

ξ can therefore also be regarded as the *number* of stars with $\mathcal{M}$ per pc^3 and per *logarithmic* mass interval, $\Delta\ln\mathcal{M} = 2.303\,\Delta\log\mathcal{M} = 1$.

According to E.E. Salpeter (1955), the *initial mass function* for the more massive stars ($\geq 1\,\mathcal{M}_\odot$) can be approximated by a power law:

$$\psi(\mathcal{M}) \propto \mathcal{M}^{-\beta-1} \quad \text{or} \quad \xi(\mathcal{M}) \propto \mathcal{M}^{-\beta}\ , \tag{5.4.12}$$

with $\beta = 1.35$. More recent investigations have indicated a steeper decrease at the more massive end, although the results of different determinations show considerable scatter ($1.4 \leq \beta \leq 2.0$). Towards smaller masses ($\leq 1\,\mathcal{M}_\odot$), the mass function flattens out sharply ($\beta \simeq 0.3$).

Finally, we obtain the *rates of formation* of the stars or *stellar genesis rates*, by dividing the initial mass function $\xi(\mathcal{M})$, which gives the mass of all the stars of mass $\mathcal{M}$ which have been formed during the existence of the Milky Way galaxy, by the age of the galaxy, T_0. This assumes that the conditions for the formation of stars have remained unchanged.

The *total rate of star formation:*

$$r = \frac{1}{T_0}\int_0^\infty \xi(\mathcal{M})\,d\mathcal{M} \tag{5.4.13}$$

is then found to be some $10^{-12}\,\mathcal{M}_\odot\,pc^{-3}\,yr^{-1}$. This corresponds to several $\mathcal{M}_\odot$ per year being formed into stars from the interstellar gas for the *entire* Milky Way system.

The preceding data and considerations are, more precisely, related to the disk and spiral arm Population I stars of the Milky Way. What is the situation for the *metal-poor halo Population II* stars? Of principal interest is the luminosity function of those stars which are still left over from the early period of the galaxy. We can best avoid the effects of making choices, which would be dangerous for such a statistical analysis, by carrying out a differential comparison of the luminosity functions of the halo and the disk (plus spiral arm) populations, using the homogeneous observational data for the stars within the immediate vicinity of the Sun.

The luminosity function of a *globular cluster* as a typical representative of Population II was first investigated, for the case of M 3, by A. Sandage in 1957. As expected, brighter stars ($M_V \leq +4$ mag) are lacking. Apart from a maximum at $M_V \simeq 0$ mag, which is produced by the cluster variables, the observed luminosity function of the globular cluster according to Sandage differs little from that of the stars of our neighborhood. If the unobservable region below $M_V = +6$ mag is therefore extrapolated, using the luminosity function for the vicinity of the Sun, we find for the whole cluster just under $6\cdot10^5$ luminous stars and $5\cdot10^4$ white dwarfs, with an overall mass of about $2.5\cdot10^5\,\mathcal{M}_\odot$. Only recently, with the use of CCD photometry, has it become possible to measure the main sequence of the globular clusters with sufficient accuracy down to considerably lower magnitudes. The first observations of M 13 indicate a steeper slope of its mass function as compared to that of the Sun's vicinity.

The question of the extent to which the *rate of star formation* is the same everywhere in the galaxy and at all times cannot be answered with certainty at the present time. The observed relatively small number of old, not yet evolved dwarf stars ($\geq$G) with low metal abundances gives, together with considerations of the formation of the elements in stars (Sect. 5.8.5), an indirect indication that in the early period of the Milky Way, roughly in the first 10^9 yr, the rate of formation of *massive* stars could have been considerably greater than it is today.

We shall take up the fundamental problems of the formation and evolution of our Milky Way galaxy, of its star population and the abundance distribution of the chemical elements, in more detail in Sect. 5.8, after we have dealt with other galaxies in the following Sects. 5.5 and 6.

5.5 Normal Galaxies. Infrared Galaxies

The leap into deep space outside our Milky Way galaxy, into the realm of the distant galaxies (or the extragalactic

Fig. 5.5.1. The Andromeda galaxy, M 31 = NGC 224, and its physical companions, the elliptical galaxies M 32 = NGC 221, and NGC 205 *lower left*. The inclination of the Andromeda galaxy is 78° and its distance from the Sun is 690 kpc. The picture was made with the Mt. Palomar 48″ Schmidt telescope

nebulas, as they were formerly called), and the beginnings of a cosmology based on observations, will be counted for all time as one of the most important achievements of our century.

We have already mentioned (in Sect. 5.2.3) the classical catalogue of Ch. Messier (M) dating from 1784, as well as the catalogues of J. L. E. Dreyer, which grew out of the work of W. and J. Herschel, the New General Catalogue (NGC) of 1890 and the Index Catalogue (IC) of 1895 and 1910, whose notation is still used today. The brightest galaxy, after the Magellanic Clouds in the southern sky, the *Andromeda Nebula* (Fig. 5.5.1), which was already observed in 1612 by S. Marius, has, for example, the catalogue numbers M 31 or NGC 224. In more recent times, we can list the Shapley-Ames Catalogue of 1932, which covers the entire sky and contains 1249 galaxies brighter than about 13.5 mag, and the Revised Shapley-Ames Catalogue by A. R. Sandage and G. A. Tammann (1981). The Revised New General Catalogue (RNGC) of J. W. Sulentic and W. G. Tifft (1973) is based on the Palomar Observatory Sky Survey. The Reference Catalogue of Bright Galaxies by G. and A. de Vaucouleurs (1964, and 1976 with A. Corwin) contains data for 4364 galaxies brighter than 16 mag. We also mention the Catalogue of Galaxies and Clusters of Galaxies by F. Zwicky et al. (1960/68) and the Morphological Catalogue of Galaxies by B. A. Vorontsov-Vel'yaminov et al. (1962/74). "*The Hubble Atlas of Galaxies*" by A.R. Sandage (1961) gives an unmatched survey of the subject.

We begin in Sect. 5.5.1 by treating the methods for determining the distances to the galaxies; in Sect. 5.5.2, we offer an overview of the manifold forms of galaxies, their classification and their absolute magnitudes. For "normal" galaxies, i.e. mainly spiral galaxies like our own Milky Way, and elliptical galaxies, we make use in Sect. 5.5.3 of the spectroscopic observations of stars and gas to discuss dynamic behavior and the mass distributions derived from it. The rotation curves of the spiral galaxies, which extend continuously far out into space, show the existence of extensive "dark" halos, which contain a considerable portion of the galactic mass and whose composition is still unknown; in any case, it is quite different from the mixture of stars and gas in the "visible" parts. The different star populations in the galaxies and their chemical compositions are described in Sect. 5.5.4, and the gas and dust components in Sect. 5.5.5. In connection with the dust in galaxies, we discuss the infrared galaxies, where intense processes of star formation took place recently. Ultraviolet radiation from the young OB stars is absorbed by the dust and re-emitted in the infrared.

5.5.1 Distance Determinations for Galaxies

The position in the cosmos of the "spiral nebulas" was the subject of heated debate among astronomers in the 1920's. We can refer here only briefly to the major contributions of H. Shapley, H. D. Curtis, K. Lundmark, and many others.

Then, in 1924, E. Hubble succeeded in resolving stars in the outer portions of the Andromeda galaxy, M 31, and in some other galaxies (we will, from the beginning, make use of the newer terminology); and, as a basis for photometric distance determinations, in identifying various objects with known absolute magnitudes:

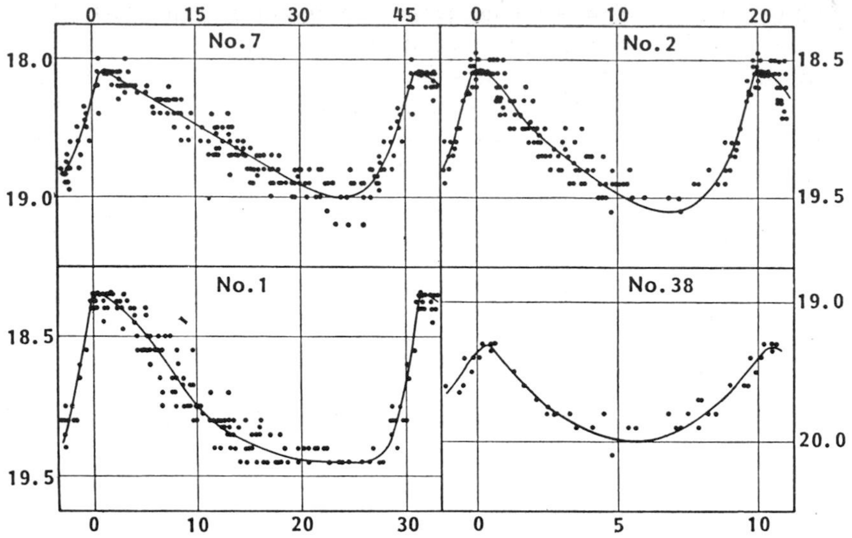

Fig. 5.5.2. Light curves from four cepheids in the Andromeda galaxy, after E. Hubble (1929). The abscissa is in days and the ordinate in photographic magnitude, m_{pg}, for each graph

Cepheids. Their light curves (Fig. 5.5.2) with periods of 10 to 48 days were at first naturally interpreted in terms of the "general" period-luminosity relation. Only in 1952 did W. Baade recognize, in part due to discrepancies concerning the absolute magnitudes of red giant stars, that the "classical" cepheids of Population I (δ Cep stars), thus in particular the cepheids with long periods in the Andromeda galaxy, are about 1.5 mag *brighter* than the corresponding cepheids of Population II (W Vir stars). This led to an "expansion" of all extragalactic distances by roughly a factor of 2. At present, the relation between the period P and the mean brightness $M_{\langle V \rangle}$ is well established, apart from small corrections which depend on the color B−V or on the amplitude of the star (Sect. 4.11.1). We can thus penetrate into deep space out to distance moduli of $m-M \simeq 29$ mag, i.e. to distances of about 6 Mpc (Megaparsec).

Along with the cepheids, Hubble was able to identify and use for distance determinations:

Novae, whose light curves are exactly similar to those of galactic novae. (The much brighter S And, with $m_{V,max} \simeq 8$ mag, which was observed in 1885 by E. Hartwig, was later recognized to have been a supernova.)

The nonvariable *brightest stars* with $M_V \simeq -10$ mag, or the brightest *globular clusters,* also with $M_V \simeq -10$ mag. Here, also, corrections later became necessary, after a portion of the "brightest stars" were found to be groups of stars or H II regions.

H II regions. The angular diameters of the bright H II regions determined from Hα plates were found after precise calibration to be an excellent aid to the measurement of great distances. They can be used out to distance moduli of $m-M \simeq 33$ mag, or 40 Mpc distance.

Because of their bright absolute magnitudes, *supernovae* offer the possibility of reaching out to much greater distances, up to about 1000 Mpc. They were recognized as a distinct phenomenon from the novae in 1934 by W. Baade and F. Zwicky. In particular, they allow distance determinations for clusters of galaxies, in which the scarcity of supernova explosions in each individual galaxy is balanced by the large number of galaxies in the cluster. The most suitable are the supernovae of type I in elliptical galaxies, since their light suffers only a slight extinction in the parent galaxy and their light curves are all very similar (Sect. 4.11.7). Their absolute magnitude at maximum is $M_{B,max} = -19.7$ mag, with a scatter of only about ± 0.3 mag. To be sure, this value is not independent of the calibration of the extragalactic distance scale; it is referred to a particular value of the Hubble constant, $H_0 = 50\ \mathrm{km\ s^{-1}\ Mpc^{-1}}$ (Sect. 5.9.1). We unfortunately cannot consider further the difficult question of the calibration of the distance scale here.

A distance determination using supernovae which is independent of the Hubble constant can be carried out by a method originally suggested by W. Baade and A. J. Wesselink for the measurement of the absolute radii of pulsating variables (Sect. 4.11.1). It requires a measurement of the magnitude at two times t_1 and t_2, and in addition, a determination of the radial velocity v_r on the basis of spectral data: the radiation flux F_λ emitted from the surface of a star of radius R is, from (4.2.11) and (4.2.18), diluted at a distance r by the factor $(R/r)^2$ and attenuated along the line of sight depending on the optical thickness τ_λ, so that a radiation flux

$$f_\lambda = F_\lambda \left(\frac{R}{r}\right)^2 e^{-\tau_\lambda} \tag{5.5.1}$$

is observed on the Earth. The ratio of radiation fluxes observed at the times t_1 and t_2 is then:

$$\frac{f_{\lambda,1}}{f_{\lambda,2}} = \frac{F_{\lambda,1}}{F_{\lambda,2}} \left(\frac{R_1}{R_2}\right)^2 . \tag{5.5.2}$$

In the neighborhood of the maximum of the light curve, we can estimate the F_λ to a sufficiently good approximation from the theory of stellar atmospheres using the spectral energy distribution or the color of the supernova, and thus obtain the ratio R_1/R_2. At the same time, we can calculate the difference of the radii by integration over the radial velocity (4.11.1):

$$R_2 - R_1 = \int_{t_1}^{t_2} v_\mathrm{r}(t)\,dt \simeq \frac{v_{\mathrm{r},1} + v_{\mathrm{r},2}}{2}(t_2 - t_1) . \tag{5.5.3}$$

We thus know both R_1 and R_2, and the distance r is then obtained from (5.5.1), providing that we can also estimate τ_λ.

In order to probe out further, past the range of the distance criteria calibrated for individual objects in the galaxies, we can, by relaxing our accuracy requirements somewhat, make use of the properties of *entire* galaxies such as their surface luminosities, the morphological luminosity classes (Sect. 5.5.2), the magnitude of the brightest galaxy in a cluster or the width of the 21 cm emission of neutral hydrogen from a galaxy (Sect. 5.5.3). Finally, we are left with only the cosmological red shift (Sect. 5.9.1) for estimating the greatest extragalactic distances.

For the *Andromeda galaxy*, M 31, we obtain, after correction of the (small) interstellar absorption, a true distance modulus $(m-M)_0 = 24.5$ mag, or a distance of 690 kpc. (In his pioneering publication in 1926, E. Hubble had given the distance as 263 kpc.)

The resolution of the central portion of the Andromeda galaxy, and of the neighboring smaller galaxies M 32 and NGC 205 (Fig. 5.5.1), was first achieved by W. Baade in 1944 using extremely refined photographic techniques. The brightest stars here are red giants with $M_\mathrm{V} \simeq -3$ mag, while the brightest blue stars of the spiral arm Population I are lacking.

With E. Hubble's results, it was definitively established that galaxies such as, among others, M 31 and our own galaxy, the Milky Way, are basically similar. Now, what is the situation regarding the apparent distribution of galaxies in the sky?

The zone around the galactic equator up to about $b = \pm 20°$, which appears to be nearly free of galaxies and is called the "zone of avoidance", is, as we have seen, caused by a thin layer of absorbing matter in the equatorial plane of the Milky Way. Thus, for example, the nearby large galaxy Maffei 1, at $b = -0.6°$ and a distance of only about 2 Mpc, remained hidden behind interstellar matter with an extinction of $A_\mathrm{V} \simeq 5$ mag and was discovered only in 1970.

Relatively often, *groups* of two, three, or more galaxies are observed, which clearly belong together physically. Our Milky Way forms the *local group* together with the Andromeda galaxy, M 31, its (physical) companions (Fig. 5.5.1), and about 20 additional galaxies. (The hypothesis that Maffei 1 also belongs to the local group could not be verified.)

However, at higher galactic latitudes, where galactic dark clouds are absent, the distribution of distant galaxies in the sky is quite inhomogeneous. They form in some cases *clusters of galaxies*, first noticed by Max Wolf, such as the Coma Cluster (in Coma Berenices), which contains several thousand galaxies (Sect. 5.7.1). The local group, together with a series of similar groups, appears to belong to the *Virgo Cluster*, whose center is at a distance of about 20 Mpc.

5.5.2 Classification of the Galaxies and Absolute Galactic Magnitudes

First, let us take a closer look at the great variety of forms in which galaxies occur! They were classified by E. Hubble, and his scheme, with some improvements, forms the basis of the Hubble Atlas of Galaxies (A. Sandage, 1961); it is also precisely defined there. It must be clear from the outset, as with the Harvard sequence of stellar spectral types, that such a purely descriptive scheme by no means necessarily represents an evolutionary sequence.

In the *Hubble-Sandage Scheme*, the galaxies are classified only according to their *shapes* as seen in blue-sensitive photographs (Table 5.5.1):

The *elliptical galaxies*, E0 to E7, have a rotationally symmetrical shape without noticeable structure (Fig.

Table 5.5.1. Classification of the Galaxies

	Lenticular galaxies	Normal spiral galaxies			Irregular galaxies
	S0	Sa	Sb	Sc	
Elliptical galaxies E0 . . . E7					Irr I
					Irr II
	SB0	SBa	SBb	SBc	
		Barred spiral galaxies			

5.5.3). The observed ellipticity is naturally in part a result of the projection of the true spheroid onto the sky as seen by the observer. The sub-classification of the E-galaxies depends on their *apparent* ellipticities, ranging from E0 (circular) to E7 (greatest ellipticity, $(a-b)/a \simeq 0.7$, where a and b are the apparent semimajor and semiminor axes). The true ellipticity is in general greater than the apparent ellipticity. The statistics of apparent ellipticities indicate that the true ellipticities of the E-galaxies are fairly evenly distributed. The surface brightness of these galaxies decreases uniformly on going outwards from the galactic center.

From the elliptical galaxies, there is a continuous transition through the *lenticular* or *lens-shaped galaxies* S0, which show indications of a disk and a circular absorption structure, to the *spiral galaxies* S. These are without exception more flattened.

The sequence branches at the S0 and S types: the *normal* spiral galaxies S have a central region or core from which the spiral arms grow out more or less symmetrically. In the *barred* spirals SB, first a straight "bar" emerges from the core, and the spiral arms are attached to it almost at right angles (Fig. 5.5.4). We also distinguish among the lenticular galaxies type SB0, in which a bar is

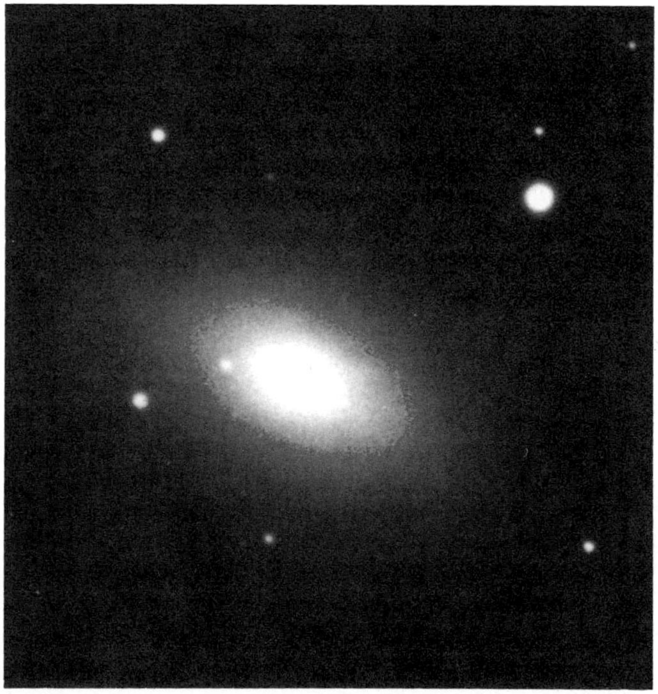

Fig. 5.5.3. The elliptical galaxy NGC 4697, of type E5. Photograph made with the 5 m Mt. Palomar telescope, from the Hubble Atlas of Galaxies

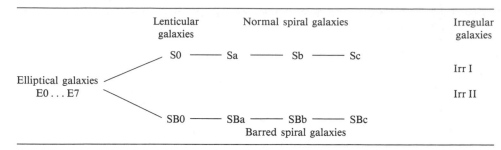

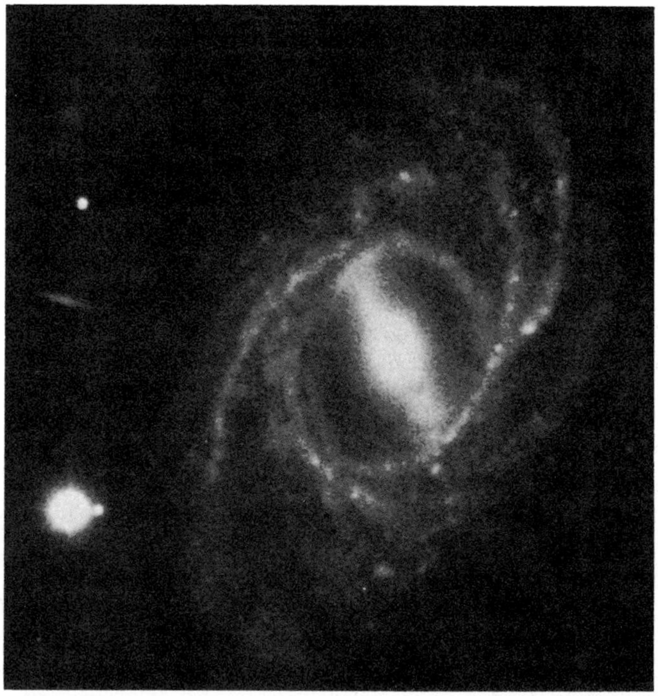

Fig. 5.5.4. The barred spiral galaxy NGC 2523, of type SBb(r), photographed with the 5 m Mt. Palomar telescope, from the Hubble Atlas of Galaxies

Fig. 5.5.5. The Large Magellanic Cloud (LMC). In this Hα photograph, the many H II regions can be clearly seen

recognizable, but usually does not extend out past the bright core region.

Between the spiral arms in both the normal and the barred galaxies, there are still large numbers of stars, so that the arms do not stand out strongly against the background on photometric recording curves.

The series of types S or SB 0, a, b, c is characterized by a relative decrease in the size of the *central region*, accompanied by more open curves of the *spiral arms*. For example, the Andromeda galaxy, M 31 (Fig. 5.5.1), is a typical Sb spiral; our Milky Way is on the border between Sb and Sc.

The Hubble Atlas divides the S and SB spirals into two further subclasses, depending on whether the spiral arms are attached directly to the central region or the ends of the bar (suffix s), or are tangential to an inner ring (suffix r).

At the centers of many galaxies, there is a clearly bounded, bright condensation region which is starlike in appearance, the *galactic nucleus*. Its angular diameter is e.g. only $1.6'' \cdot 2.8''$ in the case of M 31, corresponding to about 50 pc. The nuclei of galaxies and the "activity" which is associated with them will be discussed further in Sect. 5.6.

The Sc galaxies (such as M 33 in the local group) merge continuously into the *irregular galaxies*, Irr I. These relatively rare systems exhibit no rotational symmetry and no readily-visible spiral arms, etc. The best-known examples are our neighbors in the local system, the Large Magellanic Cloud (LMC; Fig. 5.5.5) and the Small Magellanic Cloud (SMC), in the southern sky at a distance of about 50 kpc. Precise observations have shown that the irregular distribution of luminosity in the Irr I systems is caused only by the bright blue stars which are particularly noticeable on blue-sensitive photographs, and the gas nebulas which surround them. The substrate of fainter red stars shows, in agreement with radioastronomical observations of the arrangement and velocity distribution of hydrogen based on the 21 cm line, that the major portion of the mass of, e.g., the Magellanic Clouds has a much more regular, flattened shape and undergoes considerable rotation.

The *extension of the Hubble sequence* introduced by G. de Vaucouleurs, continuing past Sc (or SBc) to the types Sd, Sm, and Im (or SBd, SBm, and IBm), takes into account the continuous transition from Sc to Irr I. In this scheme, for example, the Magellanic Clouds are classified as SBm (LMC) and IBm or Im (SMC).

A. Sandage and R. Brucato (1979) have redefined the group of *irregular* galaxies Irr II, introducing a type called *amorphous galaxies*, which is characterized (in blue-sensitive photographs) by a smooth, unresolved image without spiral structures, sometimes interrupted by dust absorption structures. Occasionally, weak filaments can also be seen. For example, M 82 (Fig. 5.5.16) belongs to this type. In recent times, it has become clear that the amorphous and other particular galaxies emit most of their radiation in the far infrared. We shall return to these *infrared galaxies* in Sect. 5.5.5.

The morphological classification of galaxies according to W. W. Morgan (1958) is based on their brightness concentrations. In this book, we use only the following types from this *Yerkes System*: *D galaxies* have a bright, elliptical core with an unusually extended outer shell, whose brightness decreases more slowly in going outwards than for the E-galaxies. The supergiants among them (see below), the *cD galaxies*, are found as the brightest objects at the centers of many clusters of galaxies (Sect. 5.5.1). The *N galaxies* are notable for their "star-shaped", very bright nuclei against the much fainter background of the rest of the galaxy (Sect. 5.6.4).

We discuss further possibilities for classifying the galaxies, in particular spectroscopic classification schemes, in Sects. 5.5.4 and 5.6.4.

We emphasize again that in the Hubble sequence, the galaxies are classified only according to their *shapes*.

Regarded physically, this is clearly a classification according to increasing angular momentum (rotational flattening!), as was pointed out by B. Lindblad.

The study of the distances to individual galaxies and their grouping into clusters has shown that the *absolute magnitudes* and diameters of the galaxies are spread over a wide range within a given Hubble type.

In the region of the spirals, barred spirals, and irregular galaxies, S. van den Bergh found in 1960 that a strong development of the spiral arms and analogous structures is correlated with increasing luminosity. He based a morphological *luminosity classification* on this observation (with notation analogous to that used for stars); it allows a unified calibration according to *absolute* magnitudes for the given region of Hubble types:

Luminosity Class	Mean Absolute Magnitude, M_{pg}
I Supergiants	−21.2 mag
II Bright Giants	−20.3 mag
III Normal Giants	−19.2 mag
IV Subgiants	−17.8 mag
V Dwarf Galaxies	−14.5 mag

Thus, for example, the Andromeda galaxy, M 31, is classified as Sb I-II, and M 33 as Sc II-III.

While absolutely bright systems are more common among the spiral galaxies, in the irregular galaxies, in contrast, dwarfs are predominant.

In the case of the elliptical galaxies, the *surface brightness* at the center of the image (proportional to the number of stars along the line of sight) is a measure of the luminosity. In particular, it is found that the brightest E-galaxy of a cluster (the socalled "first-ranked E-galaxy") always has the same absolute magnitude, $M_V = -23.3$ mag, within a very narrow range of variation.

The overall range of absolute magnitudes is considerably greater for the elliptical galaxies than for the spiral galaxies. From the giant E-galaxies, there is a decrease of luminosities amounting to nearly six orders of magnitude, passing through the normal E-galaxies and going to the faintest dwarf E-galaxies and the spheroidal dwarf galaxies, which are not sharply distinguished from the dwarf E-galaxies. An example of a spheroidal dwarf galaxy is the Draco system, with $M_V = -8.6$ mag, which can best be compared to a very large globular cluster; this also accords with its color-magnitude diagram (Fig. 5.5.6).

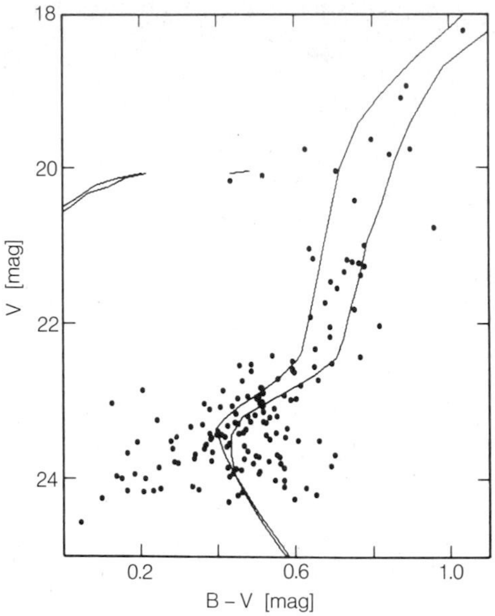

Fig. 5.5.6. The color-magnitude diagram of the Draco dwarf galaxy. Photometry was carried out using a CCD camera with the Canada-France-Hawaii 3.6 m telescope. The Draco system (distance 60 kpc, diameter 0.3 kpc) has many similarities to a metal-poor globular cluster, as is shown by the comparison to the color-magnitude diagrams of M 92 and M 3 (*fully-drawn lines*; the blue giant branch is shown for M 92, cf. Fig. 5.4.4). However, it is about 10 times larger. (From P. B. Stetson et al. (1985); courtesy of the *Publications* of the Astronomical Society of the Pacific)

In an analogous manner, we compare the *Large Magellanic Cloud*, with $M_V = -18.5$ mag (SBm III; Fig. 5.5.5) with the irregular dwarf galaxy IC 1613 (Im IV), which also belongs to the local system and has $M_V = -14.8$ mag (Fig. 5.5.7). It likewise contains a series of bright OB stars and H II regions.

Small dwarf galaxies often appear to be physically associated with neighboring giant galaxies. Along with the Magellanic Clouds, the systems in Sculptor, Draco, and Ursa Minor, which are all within 100 kpc from our galaxy, appear to be "satellites" of the Milky Way system. As companions to the Andromeda galaxy, the E-galaxies NGC 205 and M 32, with $M_V = -16.4$ mag (Fig. 5.5.1), have long been known. S. van den Bergh discovered in 1972 three other small spheroidal dwarf galaxies (And I-III) in the neighborhood of the Andromeda galaxy, with absolute magnitudes $M_V \simeq -11$ mag and diameters of 0.5 to 0.9 kpc; they are rather similar to "our" Sculptor system, for example. Recognizing such mini-galaxies is very difficult even in the local system; outside it, it is nearly hopeless.

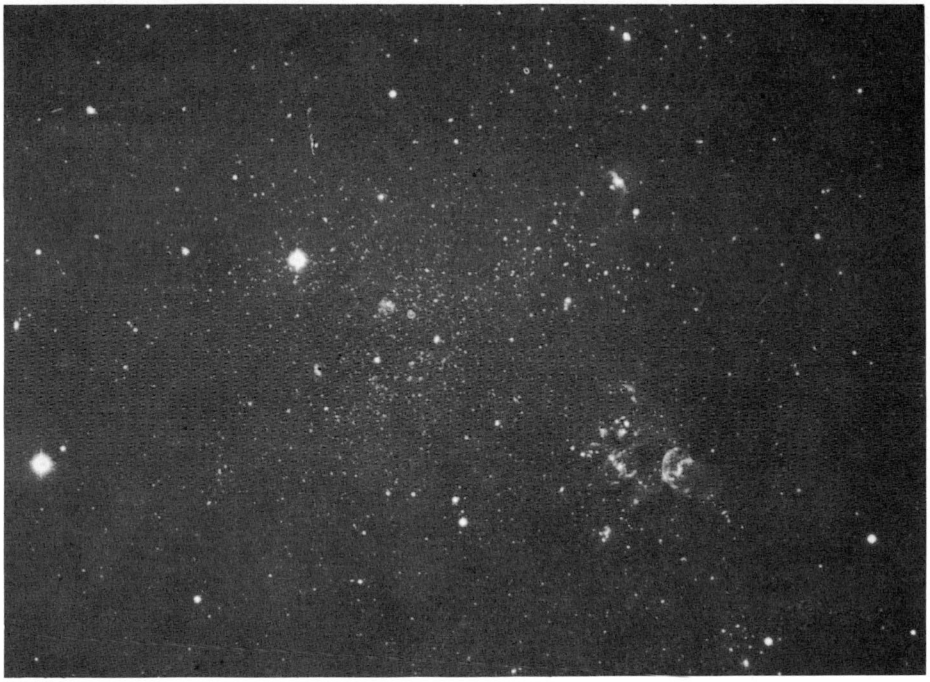

Fig. 5.5.7. The irregular dwarf galaxy IC 1613 (Im IV). A red-sensitive photograph made by W. Baade (1953) with the Hale Telescope. At the lower right, several H II regions may be seen. Compared to the Large Magellanic Cloud, IC 1613 has a visual luminosity about 30 times fainter

The *luminosity function* Φ (Fig. 5.5.8) contains information about the abundances of galaxies of different absolute magnitudes or luminosities. The distributions determined by E. Hubble (1936) and E. Holmberg (1950) are today of historical interest only. The luminosity function derived by F. Zwicky in 1957 from data on clusters of galaxies made it clear for the first time that the *dwarf galaxies* (with magnitudes fainter than $M_B \simeq -17$ mag) represent in numbers a considerable fraction of the population of the cosmos, although they have very little weight on the basis of their luminosities and masses.

The number Φ of galaxies (of all types) in a given volume of space with luminosities between L and $L+dL$ can be expressed empirically, following for example P. Schechter (1976), in the form:

$$\Phi(L)\,dL \propto \left(\frac{L}{L^*}\right)^{\alpha} e^{-L/L^*}\, d\left(\frac{L}{L^*}\right) . \qquad (5.5.4)$$

For $L \gg L^*$, Φ decreases exponentially, while for faint luminosities, $L \ll L^*$, it obeys a power law with the exponent α. If we relate the distribution function not to L, but instead to the absolute magnitude M, we find owing to $\Phi(L)\,dL = \Phi(M)\,dM$ and $dM = -2.5 \log(e)\,dL/L$

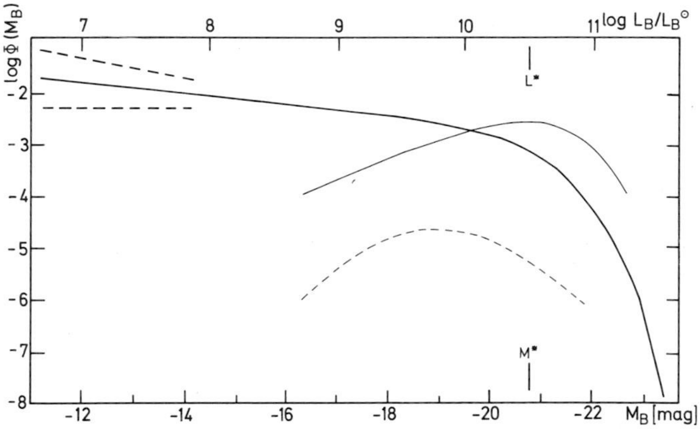

Fig. 5.5.8. The empirical luminosity function of the galaxies, $\Phi(M_B)$, after P. Schechter (5.5.4 and 5), normalized according to J.E. Felton (1978) for a Hubble constant of $H_0 = 50\,\mathrm{km\,s^{-1}\,Mpc^{-1}}$. The upper abscissa scale gives the monochromatic luminosity L_B of a galaxy in the blue, at $\lambda = 440$ nm, in solar units, $L_B^{\odot} = 5 \cdot 10^{23}\,\mathrm{W\,nm^{-1}} = 5 \cdot 10^{29}$ erg s^{-1} Å^{-1}. (———) Distribution of field galaxies and galaxies in non-compact clusters ($\alpha = -1.25$); dashed curves at the low-luminosity end: $\alpha = -1.5$ and -1.0. The upper curve, with $\alpha = -1.5$, corresponds to the luminosity function derived by F. Zwicky (1957), $\Phi(M) \propto 10^{+0.2M}$. (———) Distribution of galaxies in compact clusters ($\alpha = 0$). (– – –) The first, historical distribution function of E. Hubble (1936), corresponding roughly to a Gaussian distribution, with an arbitrary normalization, reduced approximately to the current value of H_0

(4.4.1) the following relation for the number of galaxies in the interval M to $M+dM$:

$$\Phi(M)\,dM \propto \Xi^{\alpha+1} e^{-\Xi}\,dM \ , \tag{5.5.5}$$

with

$$\Xi = 10^{-0.4(M-M^*)} \ . \tag{5.5.6}$$

The brightness distribution in the blue of both the field galaxies and of the galaxies in most of the large (not compact) clusters is well reproduced by a luminosity function of the form (5.5.5) or (5.5.4), with $\alpha = -1.25(\pm 0.25)$ and a characteristic absolute magnitude $M_B^* \simeq -20.8$ mag (corresponding to $M_V \simeq -21.7$ mag), or a characteristic monochromatic luminosity at $\lambda = 440$ nm of $L^* \simeq 3\cdot 10^{10} L_B^{\odot}$, where $L_B^{\odot} = 5\cdot 10^{23}$ W nm^{-1}. Only the number of the very brightest (cD-) galaxies in the centers of the clusters is not very well represented by this function.

The galaxies in *compact* clusters (Sect. 5.7.1) appear, in contrast to the field galaxies, to follow a Schechter distribution with $\alpha \simeq 0$.

The abundance distribution of the various Hubble types is *not* the same in different clusters of galaxies.

The main contribution to the luminosity of all galaxies together comes from galaxies with $L \simeq L^*$, i.e. in the blue, from systems which roughly correspond to our own Milky Way or to the Andromeda galaxy. The luminosity density $\mathscr{L}_B$ averaged over field and cluster galaxies probably lies within a factor of 3 of:

$$\mathscr{L}_B = \int_0^\infty L_B \Phi(L_B)\,dL_B \simeq 10^8 L_B^{\odot} \quad [\mathrm{Mpc}^{-3}] \ . \tag{5.5.7}$$

5.5.3 Dynamics and Masses of the Galaxies

In the *spectra of entire galaxies*, the Fraunhofer *absorption* lines verify that the galaxies consist in the main of *stars*. *Emission* lines from H II regions are observed in the arms of spiral galaxies and, still more noticeably, in the irregular systems. The very broad emission lines from the nuclei of Seyfert galaxies and the amorphous systems like M 82, as well as some giant elliptical galaxies (Sects. 5.5.5 and 5.6.4), clearly have quite a different origin.

The Doppler shifts of the absorption and emission lines (the latter can be more readily measured) give information about the *radial velocity* and the *rotation* of a galaxy. H.W. Babcock (1939) and N.U. Mayall (1950) carried out early radial velocity measurements on H II regions in the nearby Andromeda galaxy.

Considerable progress in extragalactic radioastronomy was made possible by the construction of the Dutch Synthesis-Radiotelescope in Westerbork, established on the initiative of J.H. Oort. Since 1971, observations of the 21 cm line of H I (and in the continuum) have been carried out with it, having an angular resolution of 24″·32″ in favorable cases; thus, the rotational velocities of a large number of galaxies could be determined radioastronomically, and their spiral arms could be resolved in some cases.

We consider first the rotation and the masses of the *spiral* galaxies, using as an example the well-investigated *Andromeda galaxy* (M 31). The uncorrected measurement of the radial velocity yields -300 km s^{-1} for the center of the galaxy, i.e. it is approaching us. Since M 31 has the galactic coordinates $l = 121°$ and $b = -21°$, most of this measured velocity is the reflection of our own galactic rotational velocity. After subtracting -300 km s^{-1} from the components of the radial velocity which are symmetric to the galactic center, taking into account the inclination of the rotation axis (78° with

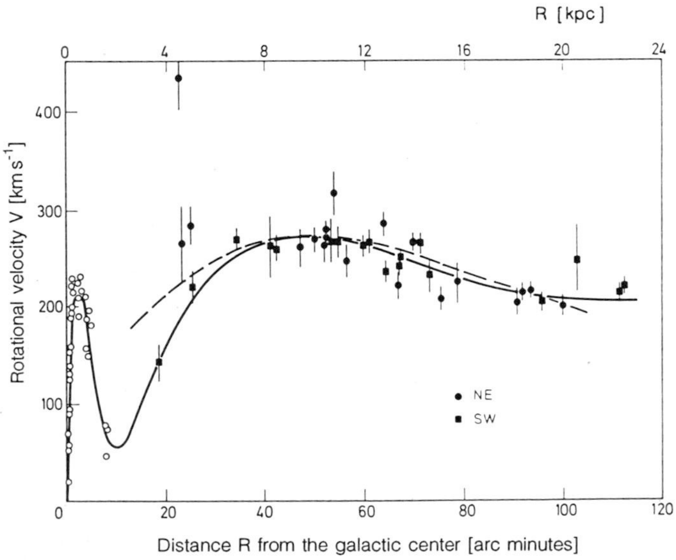

Fig. 5.5.9. Rotational velocities V in the Andromeda galaxy, M 31, as a function of the distance R from the galactic center (1′ corresponds to 0.2 kpc) from observations in the visible. (———) The rotation curve, after V.C. Rubin and W.K. Ford (1970), from 67 H II regions. The filled symbols denote measurements of the Hα line from OB associations, the open circles are measurements of the [N II] line at $\lambda = 658.3$ nm. (– – –) The rotation curve after J.M. Deharveng and A. Pellet (1975) from 1400 Hα measurements. The maximum at $V \simeq 87$ km s^{-1} in the galactic core at $R = 7$ pc could not be drawn in here

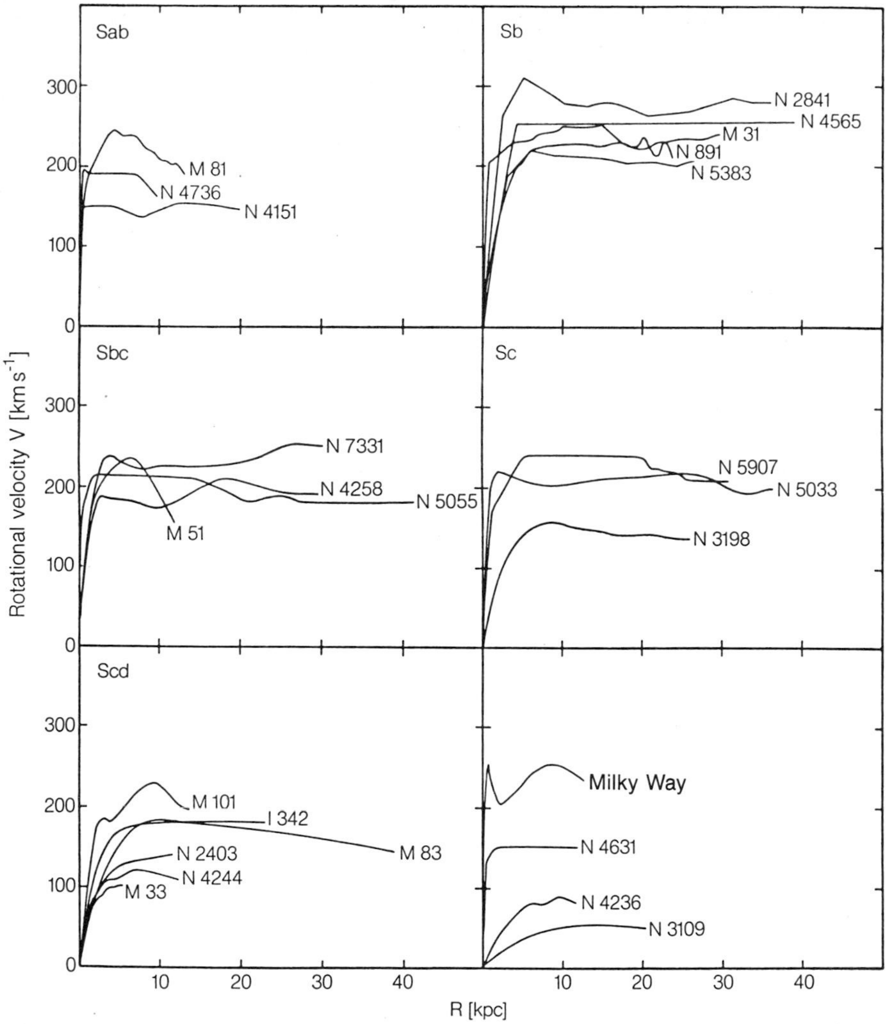

Fig. 5.5.10. Rotational velocities V in spiral galaxies of different Hubble types as a function of the distance R from the galactic center, taken from radioastronomical observations of the 21 cm neutral hydrogen line. From A. Bosma, Astron. J. **86**, 1825 (1981); reproduced with the kind permission of the author and the Astronomical Journal

respect to the line of sight), and assuming circular orbits, we arrive at the *rotational velocities*. The optical measurements from H II regions (Fig. 5.5.9) and the radioastronomical data using the 21 cm line (Fig. 5.5.10) show in the main good agreement. The latter are sensitive to parts of the galaxy which are farther out from the center than are the optical observations.

The rotational velocity $V(R)$ as a function of the distance R from the center exhibits a steep increase on going outwards, with a maximum of 225 km s^{-1} at $R = 0.4$ kpc, then decreases to a minimum at $R \simeq 2$ kpc, and finally increases again to a flat maximum of $\simeq 270$ km s^{-1} at $R \doteq 10$ kpc. Further out, $V(R)$ decreases only slightly, and remains essentially flat to the farthest point which can be measured with the 21 cm line, at $R_0 \simeq 30$ kpc.

In the region of the *starlike nucleus* of M 31, A. Lallemand, M. Duchesne, and M. F. Walker carried out measurements in 1960 using an electronic image converter and showed that the rotational velocity – going inwards towards the galactic center – has a deep minimum at $R \simeq 20$ pc and a steep maximum at $R \simeq 7$ pc (2.2″ from the center) of about 87 km s^{-1}, before finally decreasing to zero.

The velocity dispersion due to the statistical motions of the stars, both in the nucleus and in the central region, $R \leq 4$ kpc, is 100 to 200 km s^{-1} and thus of the same order of magnitude as the maximum rotational velocities.

Sufficiently far *outside* all attractive masses, we expect the Kepler orbit around a galaxy of total mass $\mathscr{M}$ to be given by (2.6.40):

$$V^2 = \frac{G\mathcal{M}}{R}\,, \tag{5.5.8}$$

i.e. the rotational velocity V decreases on going outwards as $R^{-1/2}$. Observations since the 1970's, especially those using the 21 cm line (Fig. 5.5.10), however indicate that for many galaxies, out to the farthest observable radii of 30 to 50 kpc the rotation curves are essentially flat, that is, the Keplerian case is *not* attained. The deviations of the rotation curves of some galaxies at smaller radii are probably due to perturbations by their companions.

Although we cannot directly determine the total masses of galaxies at the present state of the observational art, we can at least find their masses *within* a well-defined radial distance from the center with sufficient accuracy. For this purpose, we can use e.g. the radius R_0 out to which radial velocity measurements can be carried out, or a corresponding distance from normal photographic images in the visible region, e.g. the Holmberg radius, R_{Ho}.[15]

For a *spherically* symmetric mass distribution, the rotational velocity $V(R)$ at a radius R depends only on the mass $\mathcal{M}(R)$ *within R:*

$$V^2(R) = \frac{G\mathcal{M}(R)}{R}\,. \tag{5.5.9}$$

Using this expression, a flat rotation curve, $V(R) =$ const., corresponds to a linear increase of the mass distribution with radius, $\mathcal{M}(R) \propto R$, or a decrease of the mass density proportional to R^{-2}. A rigid rotation $V(R) \propto R$, as is observed to a good approximation in the inner regions of spiral galaxies, corresponds to an increase of $\mathcal{M}(R) \propto R^3$.

From (5.5.9) for M 31 within the Holmberg radius $R_{\mathrm{Ho}} \simeq 16$ kpc, with $V(R_{\mathrm{Ho}}) \simeq 230$ km s^{-1}, we obtain a mass $\mathcal{M}(R_{\mathrm{Ho}}) \simeq 2 \cdot 10^{11} \mathcal{M}_\odot$; and inside the furthest radioastronomical observation point (Fig. 5.5.10) at $R_0 \simeq 30$ kpc, we find a mass which is roughly twice as large, $\mathcal{M}(R_0) \simeq 4 \cdot 10^{11} \mathcal{M}_\odot$. According to the 21 cm observations, only about 3% of the mass consists of neutral hydrogen. The "lens" at the center of M 31, inside about 0.4 kpc, has a mass of about $6 \cdot 10^9 \mathcal{M}_\odot$. In the actual galactic nucleus, i.e. in a region of only about 7 pc radius, a mass of 10^7 to $10^8 \mathcal{M}_\odot$ is crowded together; the mass density here is 10^4 to 10^5 times that in the neighborhood of the Sun. The mass-luminosity ratio (which is known only to an order of magnitude) of about 20 solar units indicates that the nucleus of M 31 consists mainly of stars. For all of that portion of the Andromeda galaxy inside the Holmberg radius, the mass-luminosity ratio is roughly 8.

For an arbitrary (not spherically symmetric) mass distribution, in contrast to (5.5.9), the mass *outside* R also influences the rotational velocity $V(R)$. In order to determine the mass distribution, *models* are constructed which neglect the spiral structure, e.g. using concentric ellipsoids or disks. For these, the potential field and thus the distribution of rotational velocities, $V(R)$, can be calculated. By fitting these model calculations to the observational data (Figs. 5.5.9 and 5.5.10), essentially the *surface density* of the mass distribution can be obtained, i.e. the mass contained in a cylinder of cross-sectional area 1 pc^2 and perpendicular to the central plane.

The mass distribution of the spiral galaxies can be well reproduced, neglecting the nucleus itself, by superposing a thin *disk* and a *spheroidal component* which is only slightly flattened (Fig. 5.2.6). The inner portions of the spheroidal component correspond to the bright *central region* which can be seen in optical images (the central bulge); outside $R \geq 3 \ldots 5$ kpc, the disk dominates the mass distribution. The *surface density* decreases from 10^3 to $10^4 \mathcal{M}_\odot$ pc^{-2} in the central region to $\leq 10 \mathcal{M}_\odot$ pc^{-2} in the outermost zones which can still be observed by radioastronomical methods. The decrease of the mass density in going outwards is accompanied by a decrease of the *surface brightness*, e.g. in the visible region, which is different in the two components; within the disk, the intensity decreases exponentially with increasing radius.

This basic dynamic structure, consisting of a spheroid and a disk, is the same for all spiral galaxies, including our own Milky Way. Going along the Hubble sequence from Sa to Sd and Sm, the mass of the spheroidal component relative to that of the disk decreases, and the mass-luminosity ratio $\mathcal{M}/L_B$ (within the Holmberg radius) also decreases from about 10 to 2 solar units. Within a galaxy, the mass-luminosity ratio increases with increasing radius.

Comparing M 31 with our Milky Way system, we see that there is a considerable degree of similarity, which we shall not describe in great detail here.

Finally, let us turn to the parts of the spiral galaxies which are farthest out from their centers. The flat rotation curves out to the greatest distances which can still be observed ($R_0 \simeq 30-50$ kpc) forces us to the conclusion,

---

[15] The Holmberg radius is defined by the isophote of the photographic surface brightness being equal to 26.5 mag per square arc second.

practically independently of the details of the mass distributions, that a considerable portion, if not most of the galactic masses lie *outside* R_0. An acceptable model that would explain the observed $V(R)$ curves would be a roughly spherical mass distribution with a radius of the order of 100 kpc and an overall mass of $\geq 10^{12} \mathcal{M}_{\odot}$. This "*dark*" *halo*, also known as the *galactic corona*, is the continuation outwards of the spheroidal central region and of the (inner) halo of globular clusters. With a mass-luminosity ratio of 100–1000 solar units, it must have a completely different composition from e.g. the known mixture of stars, gas, and dust within the Holmberg radius. What sort of objects may account for the major part of the mass of this halo, whose existence has been inferred only from dynamical considerations, is at present still unknown; we can only speculate on their nature. Current observational techniques would not detect e.g. very cool dwarf stars, planet-like objects, or solid particles with diameters of more than a few cm. In addition, low-energy neutrinos (Sect. 5.9.5) or "exotic" particles postulated by modern elementary-particle theories would be possible carriers of the mass of a dark halo, assuming that they have nonzero rest masses (≥ 10 eV c^{-2}).

A massive galactic halo would also explain the relatively large masses of the clusters of galaxies, as derived from the velocity dispersion of their members (Sect. 5.7.1).

For more distant spiral galaxies whose rotation curves cannot be spatially resolved, the observed Doppler width ΔV_0 of the *profile of the 21 cm line*, integrated over the whole galaxy and corrected for its inclination, can be used to estimate their *masses*; this method was suggested by R. B. Tully and J. R. Fisher in 1977. ΔV_0 is roughly equal to twice the maximum rotational velocity, and therefore from (5.5.9),

$$(\Delta V_0)^2 \propto \mathcal{M}/R \ . \qquad (5.5.10)$$

When the distance r to the galaxy is known, its observable apparent angular diameter is $\alpha = R/r$, so that the overall mass is given by:

$$\mathcal{M} \propto (\Delta V_0)^2 \alpha r \ . \qquad (5.5.11)$$

Conversely, the *Tully-Fisher relation* is used to determine the *distances* to galaxies when the mass is known, e.g. from known mass-luminosity ratios and the measured luminosity or the absolute magnitude. In practice, correlations between ΔV_0 and the absolute magnitudes are established and calibrated empirically using data from nearby galaxies; for this purpose, infrared magnitudes are preferable to blue magnitudes owing to absorption by internal dust in the galaxies.

In the case of the *brighter elliptical galaxies* ($M_B \leq -21$ mag), the rotation is overshadowed by the statistical motions of the stars in the common gravitational field. Here, the total mass $\mathcal{M}$ of the galaxy can be estimated by determining the mean squared velocity $\langle v^2 \rangle$ or the velocity dispersion Δv from the width of the spectral lines and applying the virial theorem (2.6.25), which states that averaged over time, the kinetic energy of the galaxy is equal to $-\frac{1}{2}$ times its potential energy:

$$\mathcal{M}\langle v^2 \rangle = G \int_0^R \frac{\mathcal{M}(r)\, d\mathcal{M}}{r} \ . \qquad (5.5.12)$$

Here, R is the radius of the galaxy and $\mathcal{M}(r)$ is again the mass within a sphere of radius r. The total mass of a galaxy is thus of the order of:

$$(\Delta v)^2 \simeq \frac{G\mathcal{M}}{R} \ . \qquad (5.5.13)$$

The motions within binary galaxies and in clusters of galaxies can also be drawn upon to estimate the masses of elliptical galaxies (again using the virial theorem; Sect. 5.7.1).

While in the brighter E-galaxies, with a velocity dispersion of several hundred km s^{-1}, the maximum rotational velocities are not greater than 50 to 100 km s^{-1}, the absolutely *fainter elliptical galaxies* ($M_B \geq -21$ mag) rotate more rapidly; their maximum rotational velocities and velocity dispersion are of the same order of magnitude. The dynamics of these E-galaxies thus corresponds closely to those of the central regions (spheroidal components) of the spiral galaxies. The similarities of the two systems extends to their stellar populations (Sect. 5.5.4), so that in a first approximation, we can regard the central region of a spiral galaxy as a small elliptical galaxy.

The galaxy with the *largest known mass* is the elliptical giant galaxy M 87 (= NGC 4486 = Vir A), which is at a distance of about 15 Mpc from us. Out to an angular radius of 1′, corresponding to 4.4 kpc distance from the center, there are optical data for the velocity dispersion, from which $\mathcal{M}(4.4\ \text{kpc}) \simeq 2 \cdot 10^{11} \mathcal{M}_{\odot}$ is found. With the Einstein satellite, X-ray emission (in the 0.3–4 keV range) could be detected out to 100′ or 0.44 Mpc, resulting from hot gas at $T \lesssim 3 \cdot 10^7$ K (Sect. 5.7.2). Model calculations show that in order to bind this X-ray halo gravitationally to the galaxy, a mass of at least $3 \cdot 10^{13}$ to $10^{14} \mathcal{M}_{\odot}$ within the 0.44 Mpc radius would be necessary; out to this distance, no flattening of the $\mathcal{M}(r)$ curve can be discerned. The fraction of the total mass due to this hot gas detectable through its X-ray emissions is only 5%. Thus also, in the case of this elliptical galaxy M 87, as for many of the spiral galaxies, the problem of a "dark", massive halo of unknown composition arises.

5.5.4 Stellar Populations and Element Abundances in Galaxies

We shall now attempt to construct color-magnitude diagrams for other galaxies, as we did for the star clusters in our own galaxy, and then to study their different stars and stellar populations, in order finally to gain an insight into the formation and evolution of the galaxies.

In practice, we are faced with the problem of "light quantum economy". In the past 20 years, the construction of larger telescopes for the southern hemisphere and especially the development of better and better detectors (such as image converters and CCD's), as well as the extension of the accessible spectral regions, have all allowed considerable progress to be made in the study of galaxies. The 2.4 m Space Telescope promises a further improvement in observational possibilities in the coming years.

Individual stars even in the nearby Andromeda galaxy can at present be observed only down to absolute magnitudes of $M_V \simeq -2$ to -1 mag, using broadband photometry with linear detectors (e.g., CCD's); only a space telescope will allow us to extend this limit, and to observe, for example, the main sequences of older star clusters. Spectra with satisfactory dispersion are obtainable from the Andromeda galaxy at best for a few bright supergiants (and for gas nebulas). Only in the much nearer Magellanic Clouds can early main sequence stars still be analyzed.

In general, we must content ourselves with *integrated spectra* for the investigation of star populations and element abundances in other galaxies, particularly in their central regions.

The *classification of spectra* from the galaxies is based mostly on photographs of their central regions or of the whole galaxy with a dispersion of only 100 to 400 Å mm^{-1}. It represents an attempt to gain information about the stellar populations of the galactic system and to correlate these populations with the galaxy's classification in the Hubble sequence.

Since the spectrum of a galaxy is the superposition of many stellar spectra, it is dominated in the short wavelength region by the hot, blue stars, and in the long wavelength region by cool, red stars. The classification system of W.W. Morgan and N.U. Mayall (1957) is limited for the most part to the range $\lambda = 385$ to 410 nm:

A systems have broad Balmer lines; in the range 385–410 nm, the spectra correspond to spectral type *A*. A typical example is the galaxy NGC 4449, similar to the Magellanic Clouds; altogether, the Hubble types Irr I, Sc, and SBc occur in this group.

F systems correspond in the violet to spectral type F; the Sc galaxy M 33 is a typical representative. Sb galaxies also occur in this group.

K systems exhibit spectra which can be interpreted as a superposition of those of normal, not metal-poor G8 to early M giant stars (CN criterion) with fainter F8 to G5 stars. The prototype is M 31. In addition to the large Sb and Sa spirals, the corresponding barred spirals and elliptical giant galaxies such as M 87 are in this group.

Figure 5.5.11 summarizes the regions in the *Hertzsprung-Russell diagram* which give the main contributions to the spectra of A systems on the one hand and to K systems on the other, according to Morgan and Mayall.

Closely related to the spectral type is the *color index* B–V of the galaxies. Spiral galaxies which we see from their edges exhibit extinction and reddening due to their own interstellar matter, and this must be corrected for along with the extinction and reddening in our own galaxy. The *true color index* $(B-V)_0$ shows a close correlation with the Hubble type and likewise with the spectral type of the galaxies. $(B-V)_0$ is on the average 1.0 mag for S0 and giant E-galaxies, 0.8 for Sb, and

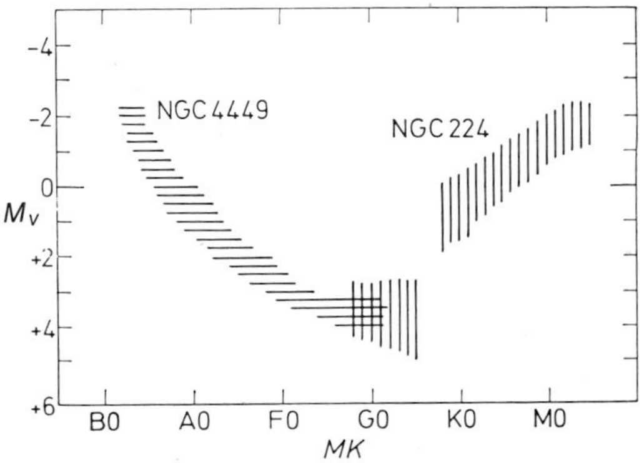

Fig. 5.5.11. The shaded regions of the Hertzsprung-Russell diagram make the major contributions to the light from the A system NGC 4449 (Irr I) ≡, and the K system M 31 = NGC 224 (Sb) ||||. The cooler portions of the main sequence do not play a noticeable role in the spectrum

0.4 mag for Irr galaxies. The corresponding color index $(U-B)_0$ is about $+1.0$, 0, and -0.2 mag, respectively. In the elliptical galaxies, $(B-V)_0$ is about 0.1 mag redder for the absolutely bright systems than for the dwarf systems. Within the spiral galaxies, we observe an increasingly blue color index on going outwards from the center.

Measurements of the energy distribution in a broad spectral range, e.g. using the extended Johnson color system or narrow-band spectrophotometry (Sect. 4.4.1), show that, corresponding to the spectral types according to Morgan and Mayall, or to the Hertzsprung-Russell diagram (Fig. 5.5.11), an important contribution to the integrated light from many galaxies falls in the range $\lambda \geq 800$ nm. Thus, the *red* and *infrared* spectral regions play an important role in the investigation of stellar populations in galaxies.

At the other end of the spectrum, the intensity below $\lambda \leq 400$ nm decreases strongly due to the large number of absorption lines. With decreasing wavelength, stars of earlier types become more and more predominant in the integrated spectra; finally, in the far ultraviolet ($\lambda \leq$ 160 nm), the integrated radiation comes almost exclusively from a relatively small number of OB stars. For nearby galaxies, the IUE satellite was able to register the energy distribution down to about 120 nm with a spectral resolution of 0.6 nm.

At this point we should mention that all measurements of the energy distribution, colors, etc., are referred to the (fixed) wavelength scale in the observer's system. However, due to the *red shift* z of the galaxies, the wavelength of their light is increased by the factor $(1+z)$, so that their radiation is, for one thing, "diluted" by this effect, and for another, a shorter-wavelength range (in the system of the emitting galaxy) enters the band chosen for the detector (Sect. 5.9.3). In the case of very distant galaxies, the energy distribution must be corrected for this effect (which was historically referred to as the "K term").

We shall attempt to bring some order into the initially rather confusing multiplicity of observational data from our own Milky Way system and from other galaxies, by introducing the concept of *stellar populations*.

In 1944, with the aid of the blackout during World War II, which made for favorable astronomical observing conditions, W. Baade at the 2.5 m telescope on Mt. Wilson was able to resolve individual stars in the central region of the Andromeda galaxy as well as in its elliptical companions M 32 and NGC 205, seen from their edges (Fig. 5.5.1). The brightest stars proved not to be blue OB stars, as in the spiral arms, but instead red giants, "similar" to those in the globular clusters of the Milky Way system. Baade recognized that different galaxies or parts of galaxies contain different star populations with quite distinct color-magnitude diagrams.

He initially distinguished the following star populations:

Population I with a color-magnitude diagram similar to that of our surroundings in the Milky Way (Fig. 4.5.2). The brightest stars are *blue OB stars* with $M_V \simeq -7$ mag.

Population II, with a color-magnitude diagram like that of the globular clusters (e.g., M 92, Fig. 5.4.3). The brightest stars here are *red giants*.

Furthermore, it was found that the stellar populations, or parts of them, also differ in terms of stellar dynamics and statistics: Population I contains *younger* stars, Population II *older* stars, and a portion of the latter, the Halo Population II, are *metal-poor* stars.

Thus, in the course of time, a clarification and refinement of the *classification scheme* of the stellar populations became necessary. This was worked out by J. H. Oort et al. in 1957 at a conference on stellar populations in Rome. Table 5.5.2 gives a summary of the classification of stellar populations, with some more recent additions; they are characterized primarily by the distribution of the orbits of their stars within the galaxy. The age and metal abundance are, to be sure, further parameters which are correlated with this one-dimensional sequence, but the correlation is not uniquely determined. On the contrary, in detailed investigations, they must be additionally measured. Following some methodological remarks about the empirical determination of these parameters, we will study the three *main populations* in more detail in the following paragraphs.

The *age* of a star cluster or any sort of genetically unified group of stars can be found, as we saw in Sect. 5.4, from its color-magnitude diagram by comparison with theoretically computed isochrones. It should be kept in mind, however, that such calculations depend on knowledge of the *chemical compositions* of the stars. Furthermore, the treatment of *mass loss*, the *mixing* of original matter with the products of nuclear fusion reactions, and also convection during stellar evolution, particularly in the region of the giant branches, is still fairly uncertain and limits the accuracy of age determinations.

The quantitative analysis of spectra with a high dispersion gives us knowledge of the *chemical composition* or metal abundance in the stars. This method is tedious and mainly limited to stars which are brighter than about 12

Table 5.5.2. Stellar Populations in the Milky Way System

	(Old) Halo Population II	Middle Population II	Disk Population	Older Population I	Extreme Population I
	Spheroidal Component		Disk Component		
Typical members	Subdwarfs Globular Clusters RR Lyr Variables with $P>0.4$ d Bright Red Giants		"Normal" stars in the central region and disk Planetary Nebulas Novae RR Lyr Variables with $P<0.4$ d		Bright blue OB Stars Open Clusters and associations Interstellar matter
Mean distance from the galactic plane, $\bar{z}$ [pc]	2000	700	400	160	120
Mean velocity perpendicular to the plane, $\bar{W}$ [km s^{-1}]	75	25	18	10	8
Concentration towards the galactic center	strong		considerable		weak
Metal abundance ε (relative to the Sun)	$\simeq 10^{-3}$ to $\simeq 1$		mainly $\simeq 1$ ($\frac{1}{3}$ to 3)		$\simeq 1$
Age of the stars [yr]	$\simeq 1.5 \cdot 10^{10}$		$\simeq 10^{10}$ to $\simeq 10^{9}$		$\leq 5 \cdot 10^{9}$

to 13 mag; however, it gives the most detailed information.

The results of such spectral analyses were already described in Sect. 4.9.4:

The *relative* abundance ratios of the *heavy elements* (nuclear charge $Z \geq 6$) for stars of the spiral arm, disk, and halo populations of our galaxy is *in the main* everywhere the *same* as in the Sun (Table 4.9.1). Even in the metal-poor, not very evolved stars of the Halo Population II (except for the most extremely metal-poor stars), the variations of the *relative* abundance ratios compared to those in the Sun are within a factor of 2 to 3. Only a few individual elements such as C, N, O, and Al, Ba, and Y show larger deviations.

It thus seems justified for many purposes to speak simply of *the* abundances of the heavy elements, or, frequently, *the metal abundance* ε (5.4.1). The metal abundance of fainter stars can be obtained from broadband *UBV photometry*, to be sure without the possibility of taking the anomalies of individual elements into account. As we saw from the example of the globular clusters (Sect. 5.4.3), the *UV excess* $\delta(U-B)$ in the two-color diagram is closely related to the metal abundance.

Before using the two-color diagram, we must correct the color indices for interstellar reddening. Since the slope of the interstellar reddening curve (Fig. 4.5.5) is known, *metal indices* which are nearly independent of interstellar reddening can be constructed, such as the Q index for UBV photometry, or the index Δm_1 in *Strömgren's* medium-band photometry (Sects. 4.4.1 and 5.3.1). The primary significance of the metal indices, aside from their use for faint stars, lies in the fact that they make it possible to extract at least a rough measure of the metal abundance from the *composite spectra* from distant globular clusters, galaxies, and portions of them.

A further refinement of the analysis of the composite spectra of galaxies has been achieved since the 1960's using the photoelectric scan technique. For example, H. Spinrad and B. J. Taylor (1969/71) have measured 36 regions of 1.6 to 3.2 nm width between 330 and 1070 nm covering important lines and bands, on the one hand for galactic spectra, and on the other for stars of known spectral type, luminosity class, metal abundance, etc. They then attempted to characterize a mixture of stars or of stellar populations in such a way that all 36 narrowband magnitudes would be correctly reproduced. The obvious question of how far such a decomposition can be carried in detail requires further work. More recently, due to the development of highly sensitive detector systems in the optical region, stellar populations in galaxies can be investigated on a large scale by *multichannel spectrophotometry*, at bandwidths of only 0.1 to 0.5 nm, i.e. with a spectral resolution corresponding to photographic spectra with about 100 Å mm^{-1} dispersion.

Important absorption lines or bands for the population analysis of galaxies are e.g. those of CN $\lambda = 388$ nm; Mg I $\lambda = 520$ nm; Na I $\lambda = 589$ nm; and the weaker groups of lines of Fe I $\lambda = 527$ and 533 nm, as well as in the infrared from CO $\lambda = 2.3$ µm; and H_2O $\lambda = 1.9$ µm.

After this introduction to methods, we return to the *stellar populations of our galaxy* (Table 5.5.2), then go on to those of other galactic systems. It appears expedient to begin with the two extreme main populations, the Halo Population II and the Spiral Arm Population I, and then to discuss the disk population. The transition populations will then require no further special description.

The *Halo Population II*. The halo[16] is, as we have seen, a nearly spherical subsystem of the spheroidal component of our Milky Way, with a radius of about 20 kpc. It contains the globular star clusters and numerous field stars, which are noticeable as high-velocity stars due to their high spatial velocities of up to $\simeq 300\,\mathrm{km\,s^{-1}}$ relative to the Sun. If their velocity components U and V in the galactic plane are plotted in a Bottlinger diagram (Fig. 5.2.12), it can be seen that these stars revolve around the galactic center on extended elliptical orbits, in some cases even in a retrograde sense. Their large velocity components W perpendicular to the galactic plane indicate that in contrast to the nearly coplanar orbits of the disk population and Population I, their orbital inclinations are distributed almost statistically.

The mass of the stellar halo is estimated to be 10 to 20% of the mass of the galactic disk, so that the gravitational field of the latter determines the present orbits of the halo stars for the most part.

The shape of the halo already shows that its *angular momentum* per unit of mass, or more precisely, its component $h = V \cdot R$ parallel to the axis of rotation of the galactic system, is small. We compare it to that of the Sun and to a mean value calculated by J.H. Oort (1970) for the entire Milky Way system:

Halo	Entire Milky Way System	Sun		(5.5.14)
$h \leq 50$	170	250	$[\mathrm{km\,s^{-1}} \cdot 10\,\mathrm{kpc}]$.	

These numbers are clearly important with respect to the *formation* and evolution of the Milky Way system (Sect. 5.8).

The halo population, globular clusters and field stars, is characterized by color-magnitude diagrams corresponding to those of the globular clusters (Fig. 5.4.4). Here, there are *no* bright OB stars; instead, the brightest stars are *red giants* with $M_V \simeq -3$ mag. Furthermore, the cluster variables or RR Lyrae stars with periods ≥ 0.4 d are also typical. The classical cepheids (like δ Cep) of Population I correspond here to the rarer W Virginis stars, with periods of 14 to 20 d.

All the stars and star clusters of the halo population are more or less *metal-poor*. Among the most metal-poor stars with the highest spatial velocities are the K giants HD 122 563, with a metal abundance of $\varepsilon \simeq 1/500$, and CD-38° 245, with the extremely low value $\varepsilon \simeq 5 \cdot 10^{-5}$, as well as the main sequence stars (subdwarfs) HD 140 283, with $\varepsilon \simeq 1/200$, and G 64-12 (= Wolf 1492), with $\varepsilon \simeq 1/3000$. Metal abundances of $\simeq 1/10$ (relative to the Sun) are found relatively frequently.

It would now seem obvious to ask if there is a correlation between these metal abundances and the galactic orbital elements or equivalent quantities. It is found that there *is* a strong correlation between very small metal abundances and a large eccentricity e, as well as a large inclination i of the galactic orbit, or a high velocity component $|W|$ perpendicular to the galactic plane. At larger metal abundances, above about 1/10 solar value, or for orbital eccentricities $e < 0.5$ or velocity components $|W| < 50\,\mathrm{km\,s^{-1}}$, the correlation is much weaker, or perhaps disappears entirely. There is thus, both in terms of metal abundance and in terms of galactic orbits, a rather numerous *intermediate population*, which forms the transition between the extreme halo population and the disk population (see below). Whether is is reasonable to define a special classification "Intermediate Population II" remains an open question.

Considering the globular clusters, the two-color diagrams and likewise the metal indices show that M 92 (Fig. 5.4.3) is just as metal-poor as the extreme subdwarfs (metal abundance $\leq 1/100$), while other globular clusters make a continuous transition to lower "metal defects". The color-magnitude diagrams (Fig. 5.4.4) become more similar going from M 92, with increasing metal abundance, to that of the old galactic cluster NGC 188, which has normal metal abundance. In this process, the giant branch moves downwards and ends earlier, and the horizontal branch gradually disappears.

[16] We consider here only the inner, stellar halo; the hypothetical dark population of the outer halo will not be discussed.

Regarding the positions of the globular clusters in the sky, it can be recognized that in general, their metal abundances increase with decreasing distance from the center of the galaxy. Some globular clusters near the galactic center appear to have only small metal deficiencies, even on the basis of their color indices.

The age of the halo population can be determined from the color-magnitude diagrams of the globular clusters (Figs. 5.4.3 and 4). From them (see Sect. 5.4), F. Hoyle and M. Schwarzschild concluded in 1955, by comparison with theoretical evolution diagrams, that the halo population forms the oldest component of the galactic system, with an age of $\simeq 10^{10}$ years. More recently, the color-magnitude diagrams of the globular clusters have attained an admirable precision and completeness. Thus, A. Sandage (1982) computes an *age* of the globular cluster system of about $17 \cdot 10^9$ years. By very exact determination of the "knee" on the $(B-V)_0$ scale of the color-magnitude diagrams of globular clusters and high-velocity stars, Sandage was also able to show that in any case, the ages of *all* members of the halo population must be the *same* within narrow limits. On the other hand, the precision of the *absolute* age determination should not be overestimated, owing to the uncertainties in some of the assumptions on which the stellar model calculations are based.

The *extreme* or *Spiral Arm Population I* is characterized by its bright ($M_V \simeq -6$ mag) blue O and B stars in *young* galactic clusters and associations (Fig. 5.4.2). All these formations are associated with (often dense) interstellar matter, from which they obviously were formed, in some cases rather recently. Our earliest ancestors could have observed the formation of the bright blue stars in Orion from their treetop-observatories! As is shown by numerous star analyses, the chemical composition of all these objects is quite similar to that of the Sun. In strongly compressed gas clouds, as was found in recent investigations in the infrared and the radiofrequency regions, a considerable portion of the condensible matter, especially the heavier elements, is, to be sure, separated from the gas, i.e. hydrogen, helium, etc., in the form of cosmic dust or "smoke". During stellar genesis, however, the solid and gaseous constituents are apparently united and thus the original element mixture is again formed.

The galactic orbits of the young stars of Population I and of the interstellar matter are, as we have already seen (Sect. 5.2.6), nearly circular. The theory of the spiral arms themselves will be discussed in Sect. 5.8.2.

The *disk population* of the Milky Way system consists on superficial examination of stars having roughly normal metal abundances with the well-known color-magnitude diagrams of our vicinity (Figs. 4.5.1 and 2) and roughly circular galactic orbits. Only on more detailed investigation do the relationships to the Halo Population II and the Spiral Arm Population I become apparent, along with other important indications of the evolution of the galaxy.

An initial estimate of the age of the disk population can be obtained from the color-magnitude diagrams of the galactic star clusters (Figs. 5.4.1b and 2). For the oldest, NGC 188, recent studies have yielded an age of 5 to $6 \cdot 10^9$ yr; M 67 is only slightly younger. There are also many *field stars* of the disk population with known absolute magnitudes and color indices, which fit into the color-magnitude diagrams of the oldest galactic clusters; however, the subgiants never fall far below NGC 188. Even though the maximum age of the disk population is not very precisely known, it must in any case be $\leq 10 \cdot 10^9$ yr and is thus clearly younger than the age of the globular clusters, about $17 \cdot 10^9$ yr. There are numerous younger star clusters (cf. for example Fig. 5.4.2); therefore, the evolution of the disk population covers the whole time period up to the formation of Spiral Arm Population I. We shall see in more detail how associations or clusters of young stars are continually forming in the spiral arms from interstellar matter. A portion of these groups "drifts apart" and thus continually replaces the evolutionary losses of the disk population.

A more exact investigation of the metal abundances of the disk stars shows that they differ only slightly among themselves, but that they do scatter within a factor of 3 to (at most) 5. In particular, the narrow-band photometric observations of B. Strömgren, B. Gustafsson, P. E. Nissen, and others, along with spectroscopic analyses, have shown that the metal abundance of the Hyades is about 1.5 times larger than that of the Sun, while the Pleiades have practically the solar mixture, with only a factor 1.1 overabundance.

The scatter of the metal abundances of the disk stars is on the one hand due to a dependence on the *age* of the stars, and on the other to *radial abundance gradients* within the disk. The age effect is difficult to see; it is in the range of $\Delta \log \varepsilon \simeq 0.5$ in 10^{10} yr. Since the formation of the disk, its metal content has increased only by about a factor of 3. Numerous investigations since about 1970 of our Milky Way and of other galaxies have shown that the metal abundance of the disk components decreases continuously from the center outwards; in the Milky Way, by about $\Delta \log \varepsilon/\Delta R \simeq -0.05\ \mathrm{kpc}^{-1}$, i.e. by a factor of 3 from the galactic center to the vicinity of the Sun. In the galactic central region, the metal abundance is about 2 to 3 times higher than the solar value, and we find "super metal rich" stars there.

Radial abundance gradients are also observed in the extreme Population I. The abundances of N/H and O/H in *H II regions* decrease from the center outwards on the order of $\Delta \log \varepsilon/\Delta R \simeq -0.1\ \mathrm{kpc}^{-1}$.

We now conclude our considerations of the Milky Way, at least temporarily, and turn to the *other galaxies:*

The *spiral galaxies* such as M 31 are basically similar to the Milky Way, as we might have expected. Along the main sequence from Sa to Sc and Sd, the proportion of gas, dust, and young stars increases, and the disk component becomes larger in comparison to the central region. The central regions, which are easier to observe in other galaxies than in our own, resemble small elliptical galaxies, judging by the stars they contain. As in the elliptical galaxies, the absolutely brighter (more massive) systems have, on the average, a redder color and a higher metal abundance; the similarities extend also to their dynamic behavior (Sect. 5.5.3).

In M 31, as in the Milky Way system, we find a *metal-rich* population (with stronger CN absorption) towards the center. Concerning the *globular clusters* in M 31, S. van den Bergh was able to verify that they are metal-poor by using their color indices; more precise investigations show, however, that they are somewhat less metal-poor than our own globular clusters.

Recently, it became possible to observe the central region of M 31 also in the ultraviolet (with the IUE satellite). The radiation flux at $\lambda \simeq 130$ nm could, for example, be explained by only *one* O6 V star or by five B0 III stars! However, since the emission region in the ultraviolet is spatially extended, this explanation in terms of young, blue stars must be abandoned; instead, the radiation no doubt originates from a much larger number of faint blue horizontal-branch stars in the old halo population.

Among the irregular galaxies (Sm, Im), our immediate neighbors, the *Magellanic Clouds* (LMC and SMC), have been most thoroughly investigated. Their most prominent feature is an extensive *Population I* with gaseous nebulas, blue OB stars, etc. There are, however, also *old* populations with red giants, etc. Furthermore, numerous globular clusters have been observed, and their color-magnitude diagrams, at least the upper parts, are known.

The *Globular Cluster System* in the Magellanic Clouds is quite different from that in our Milky Way galaxy, in terms of the types of its members and its kinematics. This shows that it is by no means a good assumption that galaxies of different Hubble types will contain only those populations which we already know from our own galaxy. There are only a very few "genuine" metal-poor globular clusters with an age similar to those in our galaxy. On the contrary, clusters of a medium age of several 10^9 yr predominate; all the globular clusters are more metal-poor than $\varepsilon \lesssim 1/3$. There are also *young*, massive globular clusters, only 10^7 to 10^8 yr old, with bright *blue* stars, which have no counterparts in our Milky Way system. In contrast to our system, in the Large Magellanic Cloud even the older globular clusters do not form a halo kinematically, but rather move within a disk. Remarkably, the rotational axis of the disk of the older clusters seems to be tilted by about 50° relative to that of the younger clusters and the interstellar gas (observations of the 21 cm line); this is perhaps a result of the gravitational interaction of the Cloud with our galaxy.

In the Magellanic Clouds, the chemical compositions of some *gaseous nebulas* and *bright stars* can be determined, in particular those of supergiants such as HD 33579 (B. Wolf, 1972). In the LMC, the metal abundances are a factor of 2 to 5 smaller than in the galactic Population I, while the relative abundances of the heavy elements are no different from those of Population I. In contrast, stars and nebulas in the SMC exhibit still lower metal abundances of around 1/3 to 1/10 of the solar values. (In both galaxies, there seem to be differences of about a factor of three between the metal abundances in gaseous nebulas and in B stars in blue globular clusters, as compared to the F supergiant field stars.) The formation of stars and their enrichment in heavier elements must have therefore followed rather different paths in the Magellanic Clouds and in the Milky Way.

The *elliptical galaxies* consist in the main of older stars, of which the brightest are *red giants.* In his early observations, W. Baade only rarely saw small filaments of dark matter, in which bright blue stars are also immediately found. Our main interest is naturally directed towards the *metal abundances* of the E-galaxies, as functions of the luminosities or the (roughly proportional) masses of the galaxies. The work of R. D. McClure and S. van den Bergh (1968), H. Spinrad and B. J. Taylor, as well as S. M. Faber (1972/73), and many more recent spectrophotometric investigations, have shown that the giant E-galaxies have normal to somewhat elevated metal abundances, but that with decreasing luminosity, the metal abundance also decreases down to values which approach those of the metal-poor globular clusters. In the dwarf E-galaxies, we see apparently (almost?) only metal-poor halo populations.

Although the dwarf elliptical galaxies represent "large globular clusters" on the basis of their stellar populations, there are differences in detail between their color-magnitude diagrams. For example, in the Draco dwarf system of the local group, the (first) giant branch is broader, and the asymptotic giant branch extends up to higher luminosities than in the diagram of the globular cluster M 92, which has a similar metal abundance. A somewhat larger range of ages of the stars and also different abundances of C, N, and O could explain these differences.

The *blue compact dwarf galaxies* are still a riddle. From their sizes and spectra, they resemble giant H II regions; but morphologically, they do not form a unified group. We find irregular and elliptical galaxies, tiny barred spirals, and galaxies interacting with each other, with two or more central regions. Their absolute magnitudes are fainter than $M_V \simeq -17$ mag, and their masses are probably at most several $10^9 \mathcal{M}_\odot$. From observations of the 21 cm line of H I, a high proportion of gas ($\leq 20\%$) can be derived. The spectroscopic studies of L. Searle and W. L. W. Sargent (1972), among others, show that some of these galaxies, for example II Zw 40 (notation from F. Zwicky's catalogue of compact galaxies), exhibit on the one hand low abundances of N, O, and Ne in their gaseous components, but on the other hand, they have a large population of blue stars. These galaxies are probably "at present" in a stage of intensive star formation, after having remained stable, without stellar genesis, for a long period of time.

To conclude this section, we consider the question (which, as we shall see, is very important for cosmology) of the abundance ratio of the two lightest elements, *helium* to *hydrogen*, in the different star populations. For the spiral arm population and the disk population, the He/H ratio is found initially by analysis of the spectra of O and B stars. The analyses of interstellar gas give results quite comparable to those for these young stars; here, the He/H ratio can be found from the recombination radiation of gaseous nebulas or H II regions, e.g. in the Orion Nebula, and from radiofrequency transitions between terms of very high principal quantum numbers (Sect. 5.3.5). All of these methods yield roughly He/H $\simeq 1/10$ (atom ratio). The same value is obtained for the Sun and for many planetary nebulas, although the latter represent a rather advanced stage of stellar evolution. During the formation of their gas shells, unburned matter from the outermost layers of the central star is often cast off. From data for the H II regions of our galaxy, a weak radial gradient of the helium abundance can be derived: He/H decreases from 0.11 in the inner parts of the disk to about 0.07 in the outer zones.

In the region of the halo population, the determination of the He/H ratio meets with the serious difficulty that no helium lines can be expected at the low temperatures of its main sequence and giant stars. We are thus forced to rely on data from the planetary nebulas and some other late evolutionary stages, whereby there is always a danger that products of nuclear fusion reactions could have gotten into the outer layers. Planetary nebula K 648, which is a member of the globular cluster M 15 and certainly belongs to Population II, shows that even in an object with a low metal abundance, the He/H ratio remains $\simeq 0.1$, similar to the stars of Population I and the disk. This and a some similar observations, along with the theory of the color-magnitude diagrams of the globular clusters, indicates that the two lightest elements H and He were formed in quite a different manner from the heavier elements, of $Z \geq 6$.

As we shall see in Sect. 5.9.4, helium was produced for the most part in the early phases of the cosmos, before the formation of the galaxies and stars; only a small amount was formed, like the heavier elements, within the stars. From observations of H II regions which are far removed from the center of our galaxy, of the gaseous nebulas in irregular galaxies, and of the (metal-poor) blue compact dwarf galaxies, we conclude that the *primordial helium* had an abundance in the range

$$\left(\frac{\mathrm{He}}{\mathrm{H}}\right)_0 = 0.07 - 0.08 \quad \text{or} \quad Y_0 = 0.22 - 0.25 \ , \qquad (5.5.15)$$

where Y_0 is the *mass* abundance of helium. Therefore, only a fraction $\Delta\,(\mathrm{He/H}) \leq 0.03$ or $\Delta Y \leq 0.06$ of the current helium abundance in the extreme Population I is of stellar origin.

5.5.5 Gas and Dust in Galaxies. Infrared Galaxies

We still know very little about the interstellar matter in the *elliptical galaxies*. We observe in them hardly any dust absorption, neither in the optical nor in the radiofrequency region of the 21 cm H I line, except for a few cases.

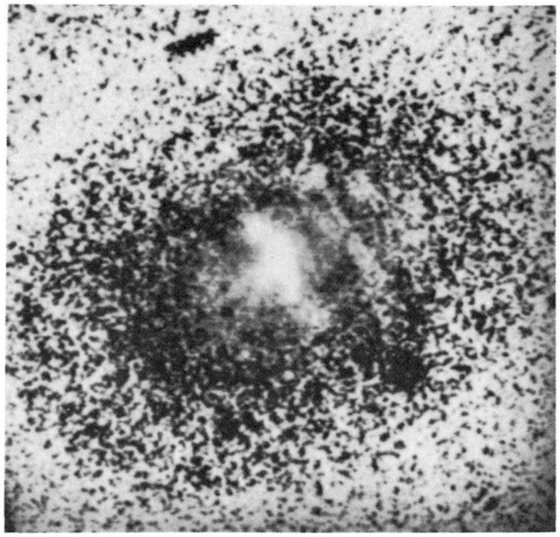

Fig. 5.5.12. Dust layers in the elliptical galaxy NGC 1052 (type E3, $M_B = -21.1$ mag). The *dust* appears *bright* in this (B – I) picture due to the image treatment. The picture was made from CCD images taken by W. B. Sparks et al. with the Danish 1.5 m telescope on La Silla (Chile). NGC 1052 contains about $10^8 \mathcal{M}_\odot$ of neutral hydrogen gas. Along with dust and H I, optical emission lines have also been observed in this "active" E-galaxy, for example from [O II], as well as continuum radiation in the radiofrequency region

This fact, together with the lack of young, blue OB stars, indicates that *no stellar formation* worthy of mention is now taking place in the E-galaxies. On the other hand, the observation of X-ray emissions from large elliptical galaxies shows that these systems are by no means free of gas, but instead contain very *hot, ionized gas* ($T \simeq 10^7$ K) with a mass fraction which is comparable to that of the interstellar matter in the spiral galaxies. To be sure, stars cannot be formed from such a diffuse, hot gas. Applying modern image enhancement methods to CCD pictures taken in different colors, it has finally proved possible to discern *dust* layers (with optical thicknesses in the visible of only ≤ 0.1) in some of the E-galaxies (Fig. 5.5.12).

The gas and dust content of the *spiral galaxies* in general increases on going from Hubble type Sa to Sm/Im; for example, the mass fraction of neutral hydrogen increases from about 0.02 to 0.10. Molecular hydrogen, which cannot be observed directly by radioastronomical methods, can still be measured indirectly by using the 2.6 mm line of carbon monoxide, as in our own galaxy. These extragalactic CO observations, however, are just beginning.

The *large-scale distribution* of gas and dust within the spiral galaxies is characterized by a strong concentration into a very flat disk, as in our Milky Way galaxy. In the Sa and Sb galaxies, as in the Milky Way, we find a minimum in the hydrogen density in the inner part ($R \leq 4$ kpc) of the disk; for later Hubble types, this "gap" is filled in. The layer of neutral hydrogen extends out past one to two Holmberg radii from the center in the early types, and considerably further in the later types.

Using the synthesis radiotelescope in Westerbork/ Netherlands, since 1971 it has been possible to resolve the H I density distribution in the *spiral arms* of galaxies outside the local group.

In Sect. 5.5.3, we have already described how rotation curves can be measured using the 21 cm line, and have discussed the determination of the masses of galaxies.

In galaxies with a suitable inclination, for example M 51 (Sbc I-II), M 101 (Sc I), or M 81 (Sb I-II), we can investigate the *large-scale structure of the spirals* by looking at them "from above" (Fig. 5.5.13); this is of course not possible for our own Milky Way system (Sect. 5.3.3). In those galaxies which we see by chance from the edge on, we can frequently observe a *deflection and fanning out of the H I layer*, similar to that known from our own galaxy (Fig. 5.5.14).

The *ionized* gas of the H II regions in the spiral galaxies, in particular the element abundances which can be derived from its emission lines, has already been treated in Sect. 5.5.4 in connection with the stellar populations.

The *dust* in spiral galaxies is warmed by the general radiation field, as we have seen from the example of our own system (Sect. 5.3.1), to between 15 and 50 K; furthermore, in and in the neighborhood of H II regions, higher temperatures up to the order of 100 K are attained through absorption of the ultraviolet radiation of *young OB stars*. Since the absorbed energy is re-emitted in the *far infrared*, we can make use of infrared astronomy to find *regions of star formation*, even when thick dust layers allow no optical radiation to pass through.

Infrared observations in spiral galaxies, especially by the IRAS satellite in its 60 μm band, have confirmed the close connection between the radiation of the blue OB stars, the dust zones which can be seen in the optical region, and the emissions of the warm dust in the far infrared. In the *Andromeda galaxy*, the infrared radiation comes mainly from a ring which is identifiable with gas/dust structures that can be seen in the visible, and from the central region (Fig. 5.5.15). In nearby spiral galaxies, we find dust temperatures averaged over the disk of about 25 K. The greater the luminosity of a galaxy in the infrared, the higher its dust temperature as a result of intense star formation. In about half of all observed

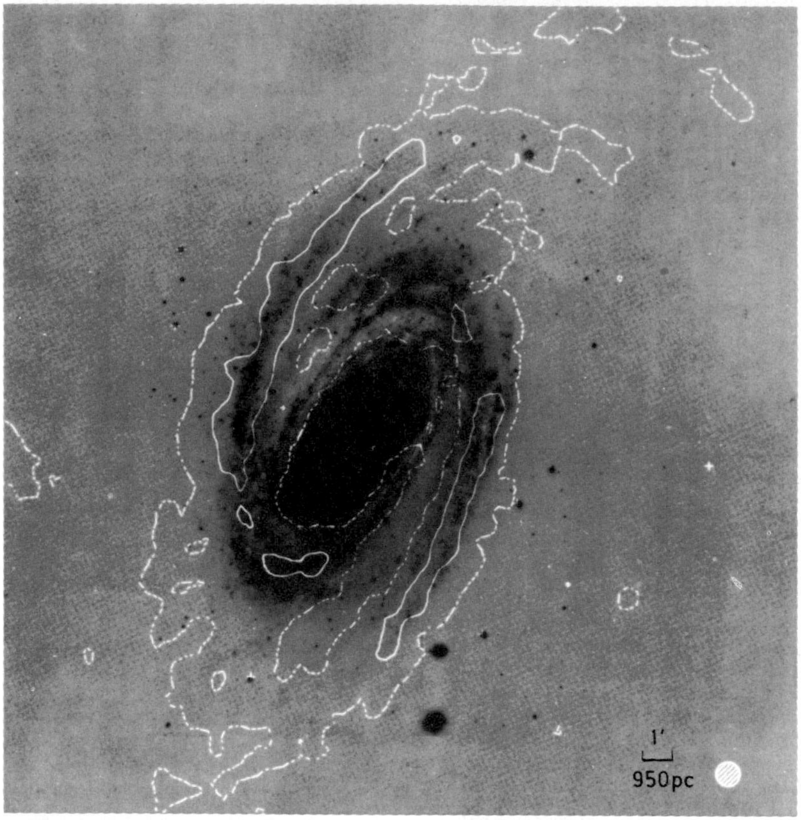

Fig. 5.5.13. Neutral hydrogen in the spiral galaxy M 81 = NGC 3031 (Sb I-II), from observations of the 21 cm line by A. H. Rots and W. W. Shane (1974) with the Westerbork Synthesis Radiotelescope. The isophotes for column densities of 3 and $10 \cdot 10^{24}$ H-atoms m^{-2} are overlaid onto the optical photograph. The galaxy is inclined at an angle of 35° to the line of sight; its distance from the Sun is 3.25 Mpc. The angular resolution of 50″ corresponds to 800 pc

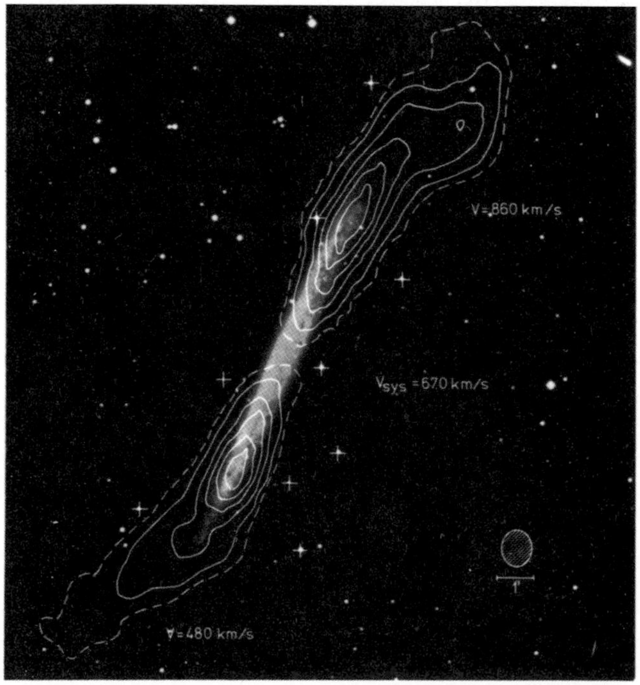

spiral galaxies, we find an infrared excess in the *nucleus* (compared to the stellar contribution of the K and M supergiants), which likewise has its origin in heated dust and thus from recently formed OB stars. Areas of the order of 100 pc across emit about $10^{10} L_{\odot}$ in the infrared. (These galaxies with intense star formation in their nuclei do not, however, fall into the class of galaxies with "active" nuclei, which we shall deal with in Sect. 5.6.)

The *luminosity* $L_{\rm IR}$ of the spiral galaxies in the far infrared is correlated with the blue luminosity $L_{\rm B}$ and increases on going from early to later Hubble types. The

Fig. 5.5.14. Bending of the layer of neutral hydrogen in the Sc galaxy NGC 5907, seen nearly on edge. Observations by R. Sancisi (1976) with the Westerbork Synthesis Radiotelescope at an angular resolution of 51″·61″. The isophotes from the 21 cm line of H I are overlaid on the optical image (a IIIa-J photograph by P. G. van der Kruit and A. Bosma with the 1.2 m telescope of the Hale Observatory). In order to emphasize the outer parts of the disk, with a radius of 50 kpc, only two velocity channels (bandwidth 27 km s^{-1}) are shown, centered approximately on the maximum rotational velocity (±190 km s^{-1}). The radial velocity of the galactic center is 670 km s^{-1}

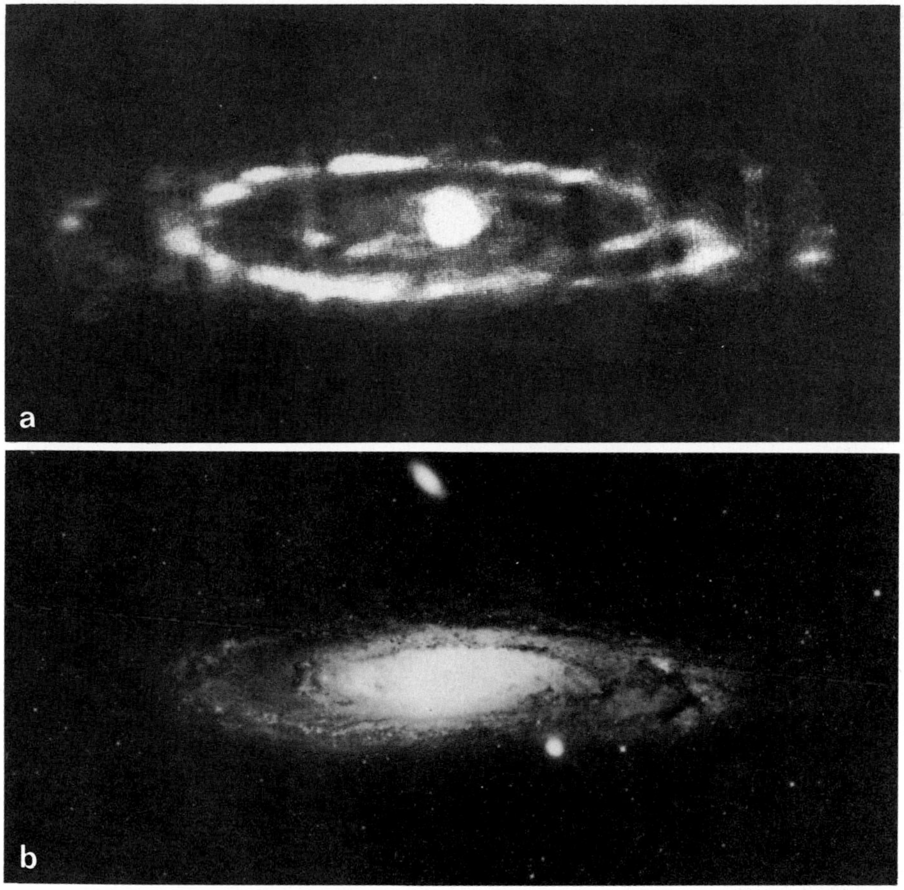

Fig. 5.5.15 a,b. The Andromeda galaxy, M 31. (**a**) Emission of the warm dust in the far infrared. Observations from the infrared satellite IRAS in the 60 μm wavelength band. (**b**) A blue-sensitive photograph with the Mt. Palomar 1.2 m Schmidt telescope for comparison; from H. J. Habing et al., Astrophys. J. **278**, L59 (1984). (Reprinted courtesy of the authors and *The Astrophysical Journal*, published by the University of Chicago Press; © 1984 The American Astronomical Society)

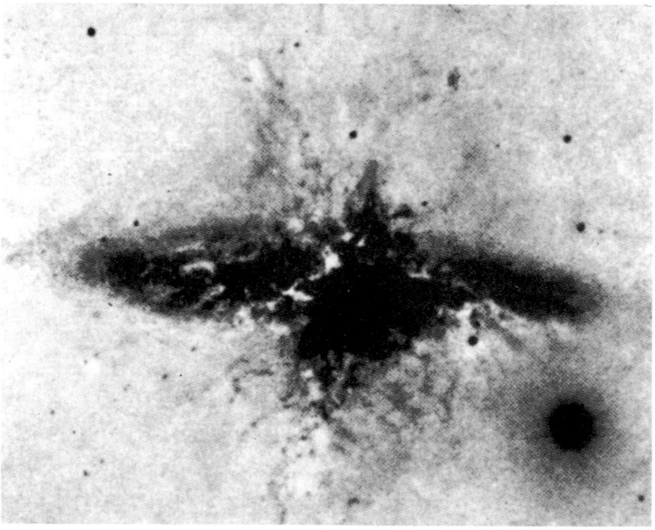

Fig. 5.5.16. The amorphous galaxy M 82 (Irr II). A (negative) Hα photograph taken by A. R. Sandage (1964) with the 5 m telescope on Mt. Palomar

strong variations of L_{IR}/L_B among galaxies of the same type are surprising; for example, in M 31, $L_{IR}/_B$ is only about 0.03, while the ratio is about 0.5 in our own galaxy. At the other end of the scale, we can observe spiral galaxies and especially irregular galaxies (see below) with values of $L_{IR}/L_B \gtrsim 1$.

In the case of the *irregular* (Irr II) or *amorphous galaxy* M 82 = NGC 3034, which along with the large spiral M 81 (Fig. 5.5.13) belongs to a relatively nearby group of galaxies at a distance of 3.3 Mpc, we can observe numerous details. On Hα photographs (Fig. 5.5.16), M 82 exhibits well-developed filaments out to distances of about 4 kpc from the disk, with outward velocities which increase with distance from the center.

This galaxy, which is also known as a radiofrequency source and whose optical continuum is partially polarized, was therefore originally interpreted as an *exploding* galaxy by C. R. Lynds and A. R. Sandage (1963): about 10^6 years ago, large masses of hydrogen were supposed to have been ejected from its center with velocities of several 1000 km s^{-1}. (These high velocities follow from the observed values of the order of 100 km s^{-1} on taking the small inclination of only about 8° into account.) This explosion model, however, proved *not* to be tenable, after N. Visvanathan and A. R. Sandage found in 1972 that the Hα emissions themselves are about 30% polarized. On the basis of this and other observations, we now interpret the Hα emissions as radiation from the disk which has been *scattered* by *dust particles* that are far out from the galactic center and have an extended motion field; Doppler shifts of the order of 100 km s^{-1} are caused by this scattering. According to observations by C. M. Telesco, D. A. Harper, G. H. Rieke, F. J. Low, and others (1979/1980), M 82 radiates its energy mostly in the far infrared around $\lambda = 100\ \mu m$, corresponding to a dust temperature of $\simeq 45$ K. The luminosity of about $3 \cdot 10^{10} L_\odot$ originates mainly from numerous OB stars in areas of stellar genesis in the central region, which are not accessible to optical observation because of the thick dust layers. The luminosity in the far infrared is considerably stronger than in the blue ($L_{IR}/L_B \simeq 4$), so that M 82 can be considered as the prototype of the *infrared galaxies*, whose optical appearance is completely dominated by strong dust absorption.

M 82 moves with its dust halo within an extended, thinner dust cloud in the M 81 group. The dust distribution, kinematic considerations, and the bridge of neutral hydrogen between M 81 and M 82 discovered with 21 cm observations, all indicate that about $2 \cdot 10^8$ yr ago, there was a near collision of the two galaxies, and the *tidal interactions* probably initiated the unusually high rate of star formation in M 82. Such a "*burst of star formation*" lasts for about 10^7 to 10^8 years.

In the course of the sky survey by IRAS, many distant galaxies were discovered in the 60 μm band, with L_{IR}/L_B ratios up to the order of 100; i.e., they are many times greater than that of M 82. With luminosities of up to several $10^{12} L_\odot$, these intense *infrared galaxies* represent the absolutely brightest stellar systems in the cosmos, along with the quasars (Sect. 5.6.5). Many of these galaxies have been found in the optical region using deep CCD images only after their discovery by IRAS. Their energy distribution in the infrared corresponds, as for M 82, to the radiation from warm dust, so that we can conclude that these distant galaxies have also "recently" experienced a burst of star formation at rates of several $100\,\mathcal{M}_\odot\ \mathrm{yr}^{-1}$.

In the nearby infrared galaxies, among those of highest luminosity are IC 4553 (= Arp 220, its number in H. Arp's Atlas of Peculiar Galaxies, 1966) and NGC 6240. For example, in NGC 6240 (Fig. 5.5.17) in the visible region two *separate nuclei* could be resolved; they have different radial velocities. This directly illustrates a very near collision of two galaxies which has caused a burst of star formation.

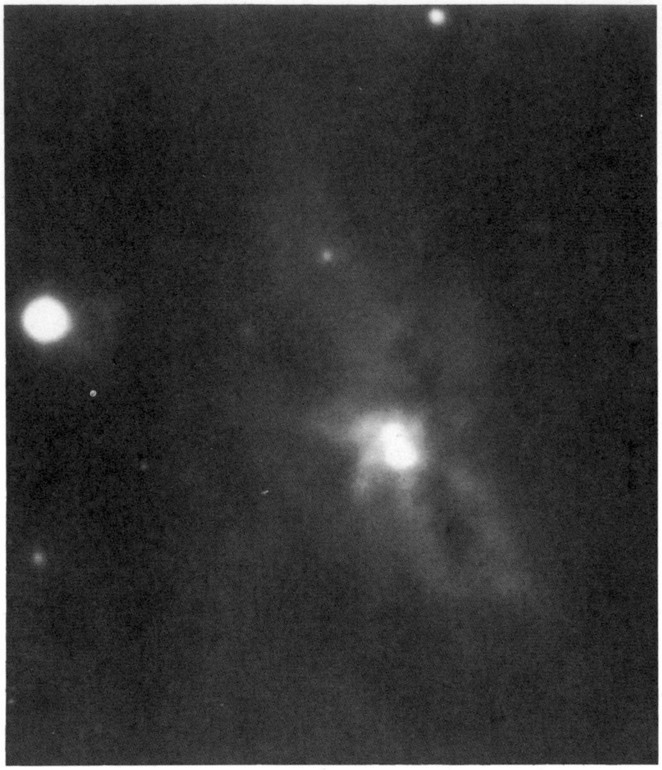

Fig. 5.5.17. The infrared galaxy NGC 6240. A photograph by J. Fried (1982) using the CCD camera with the 2.2 m telescope on the Calar Alto (red filter, 5 min. exposure). This galaxy, at a distance of 150 Mpc, has a radial velocity of 7350 km s^{-1} and shows two distinct nuclei at a separation of 1.8″ as well as noticeable "tidal streamers". The radial velocities of the nuclei differ by 150 km s^{-1}. NGC 6240 is an example of the collision of two galaxies, whose nuclei have not yet merged together

We shall return to the mutual interaction of galaxies in Sect. 5.7.3, in connection with the subject of groups and clusters of galaxies; here, we mention only that such interactions are by no means rare occurrences. In fact, perhaps *every* galaxy has been subjected (at least) once since its formation to a strong perturbation by tidal forces from another galaxy, and has passed through a phase of intense star formation as a result.

5.6 Radio Galaxies, Quasars, and Activity in Galactic Nuclei

A great number of galaxies, mainly large elliptical galaxies, are strong radio sources. In these radio galaxies, emissions in the radiofrequency range predominate over those in the optical region. Their nonthermal spectra and polarization indicate that they are synchrotron radiation, produced by the motion of highly energetic electrons in a magnetic field. We shall begin in Sect. 5.6.1 with the fundamentals of the theory of synchrotron radiation, giving a brief summary of the sources of nonthermal radiation in the Milky Way. Then, in Sect. 5.6.2, we treat the radio galaxies, discussing Cyg A and Cen A in detail. Our considerations of the radio galaxies lead us to postulate a compact energy source at the center of the galaxies. In general, an "activity" of varying strength can be observed in the centers of many galaxies; it expresses itself mainly through the production of nonthermal radiation in all wavelength ranges and of energetic particles. In Sect. 5.6.3, we interrupt the discussion of distant galaxies to give a description of the core of our own Milky Way, where we can observe many well-resolved details, although its activity is rather weak. Then, in Sect. 5.6.4, we turn to the Seyfert galaxies and to activity in the radio galaxies; and in Sect. 5.6.5, we discuss the quasars, the most luminous galaxies with the strongest activities. Finally, in Sect. 5.6.6, we attempt to show how all galactic activity phenomena have common underlying physical processes, and to discuss the origins of the enormous energy releases from quasars, etc.

5.6.1 Nonthermal Radiofrequency Radiation from the Milky Way and Normal Galaxies. The Theory of Synchrotron Radiation

In Sect. 5.3.5, we have already described the *thermal part* of the radiofrequency radiation in the Milky Way. It consists of *line emissions* from neutral hydrogen (21 cm line, H I regions) and from the CO molecule at $\lambda = 2.6$ mm, as well as from many other molecules. The 21 cm observations in other galaxies were treated in Sect. 5.5. We can observe also the *continuum* due to free-free radiation from ionized hydrogen, especially in the H II regions and gaseous nebulas, as well as in planetary nebulas.

Thermal radio continua can be recognized by the fact that when they are radiated from an *optically thin* layer, the intensity I_ν or the radiation flux per frequency interval, S_ν, is nearly independent of ν (compare Fig. 5.3.14).

To represent extended sources, the brightness temperature T_B is often used. This is the temperature which a *black body* would be required to have in order to emit just the observed radiation intensity I_ν in the radiofrequency region, according to the Rayleigh-Jeans law (4.2.24b) (which is sufficiently accurate in this spectral region):

$$T_B = \frac{c^2}{2k\nu^2} I_\nu$$

or (5.6.1)

$$I_\nu [\mathrm{W\,m^{-2}Hz^{-1}sr^{-1}}] = 3.08 \cdot 10^{-28} (\nu\,[\mathrm{MHz}])^2\, T_B [\mathrm{K}] \ .$$

For thermal radiation from optically thin sources, we thus find $T_B \propto \nu^{-2}$. The *spectral energy distribution* of the radio sources is usually approximated by a power law:

$$I_\nu \quad \text{or} \quad S_\nu \propto \nu^{-\alpha} \tag{5.6.2}$$

where α is called the spectral index (some authors prefer the opposite choice of sign for α). For thermal radiation from H II regions in optically thin layers, $\alpha \simeq 0$; in optically thick layers, $\alpha = -2$.

In contrast, the radiofrequency radiation of the Milky Way, which is spread over the whole sky, and likewise that of the strong *radio sources*, has a spectrum described approximately by:

$$I_\nu \propto \nu^{-0.8} \quad \text{or} \quad T_B \propto \nu^{-2.8} \ , \tag{5.6.3}$$

which, even after absorption and self-absorption are taken into account, can*not* be attributed to thermal emission. The measured radiation temperatures of $> 10^5$ K for long wavelengths are also hardly capable of being interpreted in terms of thermal radiation. Therefore, in 1950, H. Alfvén and N. Herlofson proposed the mechanism of *synchrotron radiation* or *magnetobremsstrahlung* to explain the nonthermal radio continua. This idea was then taken up and developed by I. S. Shklovsky, V. L. Ginzburg, J. H. Oort, and others. It was known to physicists that relativistic electrons (i.e. electrons with velocities $v \simeq c$, whose energy E is considerably greater than their rest energy $m_0 c^2 = 0.511$ MeV) moving on a circular orbit in the magnetic field of a synchrotron emit intense, continuous radiation in the direction of their motion; its spectrum extends into the far UV. This continuum can be distinguished from free-free radiation or bremsstrahlung by the fact that the acceleration of the electrons is due not to atomic electric fields, but instead to a macroscopic magnetic field B.

The *theory of synchrotron radiation* was developed in 1948/49 by V.V. Vladimirsky and J. Schwinger. It is based on the following considerations:

An electron which is orbiting around the field lines at the cyclotron frequency ω_c (5.3.30) emits radiation (according to the laws of electrodynamics) in a narrow cone of opening angle $\theta \simeq m_0c^2/E = \gamma^{-1}$, where γ is the Lorentz factor (Fig. 5.6.1). This radiation cone passes the observer rapidly, like the light beam from a lighthouse, so that he detects a series of light flashes, each of duration Δt. The spectral decomposition or, mathematically speaking, the Fourier analysis of this spectrum, taking the relativistic Doppler effect into account, yields a continuous distribution, with its maximum at the circular frequency $\omega_m \simeq 1/\Delta t$, proportional to $\gamma^3\omega_c$. Exact calculation gives for the frequency at the maximum:

$$\begin{aligned}\nu_m &= 0.29 \frac{3}{4\pi}\frac{eB_\perp}{m_0}\gamma^2 \\ &= 6.9\cdot 10^{-2}\frac{eB_\perp}{m_0}\left(\frac{E}{m_0c^2}\right)^2 ,\end{aligned} \qquad (5.6.4)$$

where $B_\perp$ is the component of the magnetic flux density perpendicular to the direction of motion of the electron having charge $-e$ and energy E. The total radiation at all frequencies is proportional to $B_\perp \nu_m \propto B_\perp^2 E^2$. Calculating in units which are appropriate for application to cosmic radio sources, e.g. ν_m in [GHz], $B_\perp$ in [10^{-10}T] or [10^{-6}G], and E in [GeV], we then find:

$$\nu_m\,[\mathrm{GHz}] = 4.7\cdot 10^{-3} B_\perp\,[10^{-10}\mathrm{T}]\,(E\,[\mathrm{GeV}])^2 \ . \qquad (5.6.5)$$

Thus, a characteristic galactic magnetic field of 10^{-10}T yields the following electron energies E at various observed characteristic frequencies ν_m or wavelengths λ_m

ν_m [GHz]	10^{-2}	1	10^2	10^4	10^6
λ_m	30 m	30 cm	3 mm	30 μm	300 nm
E [GeV]	1.5	15	$1.5\cdot 10^2$	$1.5\cdot 10^3$	$1.5\cdot 10^4$

(5.6.6)

Even for the production of radiation in the radiofrequency region, and quite certainly for visible light, electrons with energies in the range of those of *cosmic rays* (Sect. 5.3.8) are required. If the *energy distribution* of the electrons can be represented by a power law:

$$N(E)\,dE \propto E^{-p}\,dE \ , \qquad (5.6.7)$$

then the *emission coefficient* of the synchrotron radiation is found to be

$$j_\nu \propto B_\perp^{(p+1)/2}\nu^{-(p-1)/2} \ , \qquad (5.6.8)$$

and for the *intensity* I_ν, the same dependence on $B_\perp$ and ν is found for the optically thin case, which is usually realized. In some compact radio sources, we also have to take into account the *absorption* of the synchrotron radiation (synchrotron self-absorption and free-free absorption); it leads to a *flattening out* of the spectrum at the low-frequency end.

Changes in the energy distribution $N(E)$ of the electrons as a result of the *energy losses* on ionization (at low energies) or through synchrotron emission and inverse Compton scattering (at high energies) also lead to deviations from the simple power law (5.6.8).

Radio astronomy and likewise X-ray and gamma-ray astronomy give us the possibility of gaining information about the presence of *high-energy electrons* in cosmic objects. This is of great fundamental importance, since the original directional distribution of charged particles arriving at the Earth (cosmic-ray particles) can no longer be determined, due to the complex deflections caused by terrestrial and interplanetary magnetic fields. It is, to be sure, a serious drawback that we still know very little

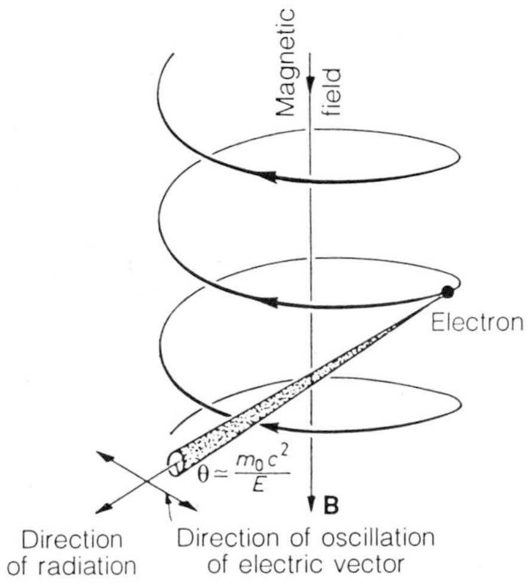

Fig. 5.6.1. The synchrotron radiation from a relativistic electron in a magnetic field B

about the magnetic fields which also enter into equation (5.6.8). We cannot decide without further information how to attribute the observed galactic yield of radiofrequency radiation to the factor $B^{(p+1)/2}$ and to the density of cosmic electrons $N(E)$ (in the appropriate energy range).

It is important for the interpretation of radio sources to know the "*lifetimes*" of the relativistic electrons. The time $t_{1/2}$ in which an electron loses half of its energy E to synchrotron radiation, $\dot{E} \propto B_{\perp} \nu_{\rm m}$, is proportional to $E/\dot{E}$. If we use (5.6.4) to eliminate the energy $E \propto \nu_{\rm m}^{1/2} B_{\perp}^{-1/2}$, we find $t_{1/2} \propto B_{\perp}^{-3/2} \nu_{\rm m}^{1/2}$. The precise value, again in the "cosmic" units used above, is:

$$t_{1/2}[\mathrm{yr}] = 5.7 \cdot 10^8 (B[10^{-10}\mathrm{T}])^{-3/2} (\nu_{\rm m}[\mathrm{GHz}])^{-1/2} \; . \tag{5.6.9}$$

If, for example, we observe synchrotron radiation at 1 GHz ($\lambda = 30$ cm), then, for a magnetic field of 10^{-10}T, on the one hand, electrons of energy 15 GeV would have been required (5.6.6); on the other hand, they could emit synchrotron radiation of this frequency only during $6 \cdot 10^8$ yr without being "refreshed". For emissions in the optical range, e.g. at 10^6 GHz (300 nm), the lifetime of the 10^4 GeV electrons is only $6 \cdot 10^5$ yr.

After these initial theoretical considerations, we turn to the *observation* of the nonthermal radiofrequency radiation of our *Milky Way* system and other *normal galaxies*, at first leaving their central regions and nuclei out of our discussion.

When G. Reber in 1939 investigated the distribution on the celestial sphere of radiofrequency radiation at 167 MHz or $\lambda = 1.8$ m with his radio telescope, which had only a moderate angular resolution, he immediately noticed the concentration of the radiation intensity towards the galactic plane and the center of the galaxy. Then, in 1950, M. Ryle, F. G. Smith, and B. Elsmore were able to show that several of the known brighter galaxies emit radiofrequency radiation with about the strength that would be expected from their similarity to our own galaxy. Shortly thereafter, R. Hanbury Brown and C. Hazard succeeded in determining the brightness distribution of the Andromeda galaxy in the radiofrequency range; it and the Milky Way galaxy were found to be very similar in this spectral region as well as in the optical region. Finally, the development of more powerful radiotelescopes with high angular resolution (interferometry, aperture synthesis; see Sect. 3.3.1) allowed rapid progress in the investigation of radio emissions from galaxies in the 1960's.

In our Milky Way system, we can observe a number of *discrete* nonthermal sources, whose more important types we have already met in previous chapters: the remnants of *supernovae*, on the one hand the *pulsars* and their surrounding nebulas (such as the Crab Nebula, M 1 = Tau A, Fig. 4.11.10, or the Vela supernova remnant, Fig. 4.11.12), and on the other the extended *ring* or *circular nebulas* (such as the remnant of Tycho's supernova, Fig. 4.11.13, or the Cygnus Loop); also, the *flare stars*, which are red dwarf stars with eruptions at irregular time intervals. After subtracting out these discrete sources, an essentially smooth, *diffuse* component remains, which in the range of m and dm wavelengths forms a *thick disk* of the order of several kpc deep. Along the galactic equator, several steps in the intensity distribution can be discerned; they are connected with the spiral arms. The spectrum of these components, $I_\nu \propto \nu^{-\alpha}$, has spectral indices α of 0.4 to 0.9 in the frequency range from 0.2 to 3 GHz. We interpret this component as due to synchrotron radiation, produced by the motion of electrons in the spatially-extended galactic magnetic field.

In addition to the disk component of nonthermal radiation, in the meter wavelength range we observe another component, which is distributed relatively uniformly over the whole celestial sphere; following J. E. Baldwin, it has been attributed to the galactic halo. Measurements with increasingly better angular resolution have, however, shown that more and more of this "halo component" can be ascribed to discrete galactic and particularly extragalactic radio sources, which are closely spaced in the sky, so that it is at present questionable whether anything will be "left over" for the galactic halo in between the many individual sources.

The determination of the spatial distribution of the radio sources in our Milky Way system is very difficult because we ourselves are *within* the system. We can gain more immediate information about the distribution of radio emissions from *neighboring* galaxies:

In those *spiral galaxies* which we observe on edge, there is frequently an ellipsoidal *disk*, which in the continuum, e.g. at 21.2 cm, is noticeably thicker than the intensity distribution of the 21 cm line from H I. Only rarely (Fig. 5.6.2) has a *halo* of nonthermal radiation been observed.

In some favorable cases, the *spiral arms* can also be investigated in the radio continuum:

The peak line, i.e. the intensity maximum of the spiral arms, does not exactly coincide with that seen in blue-sensitive photographs, which corresponds to the maxi-

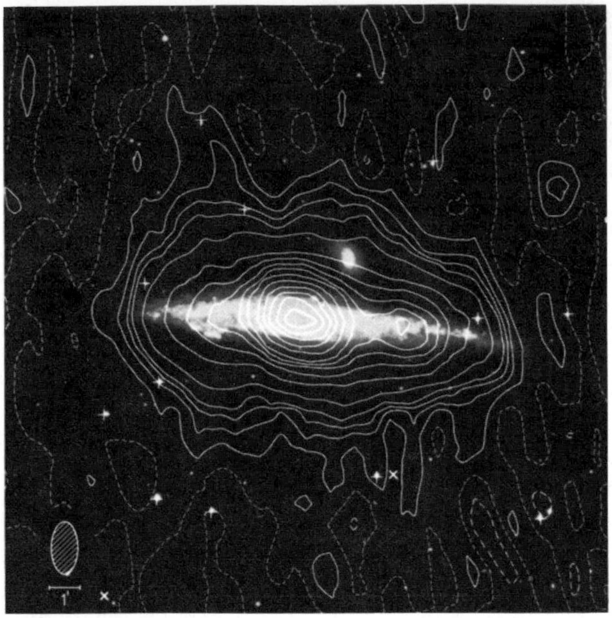

Fig. 5.6.2. The nonthermal radio halo of the spiral galaxy NGC 4631 (Sc/SBm); from observations by R. D. Ekers and R. Sancisi (1977) with the Westerbork Synthesis Radiotelescope. The isophotes at $\lambda = 0.49$ m (0.61 GHz) are superimposed onto the visible image

mum density of stars; this can be most clearly seen in the case of M 51 (Sbc I-II). Instead, the radio arms lie along the inner edges of the "optical" arms, in the area of the dust and gas arms which are marked by dark clouds and H II regions. D. D. Mathewson et al. have interpreted this, based on a suggestion of W. W. Roberts, in terms of the density-wave theory of the spiral arms (Sect. 5.8.2), according to which a compression of the gas and the magnetic field lines is to be expected in these regions. The resulting higher density of the synchrotron electrons and the intensification of the magnetic field would cause an increase in the strength of the radio emissions. The intensity of the radio arms decreases on going outwards from the galactic center.

We mention finally the surprising observations of NGC 4258 made by P. C. van der Kruit, J. H. Oort, and D. C. Mathewson in 1972 with the Westerbork Synthesis Radiotelescope. Although blue-sensitive photographs give the impression of a normal spiral galaxy (Sb II), the radio image at 1.4 GHz shows, in addition to the two "usual" optical spiral arms, two *radio arms*, which emerge from the central region and then stretch out along almost straight lines to between 5 and 15 kpc. These arms were identified optically only through interference filter photographs using Hα light, which, like the radio images, show an unusually flat distribution of intensity along the arms. The radio and Hα images, together with radial velocity measurements, can be interpreted with the help of the idea that two plasma clouds were ejected from the galactic core about $1.8 \cdot 10^7$ years ago, in opposite directions and with velocities of about 800 to 1600 km s^{-1}. These masses ($\simeq 10^7$ to $10^8 \mathscr{M}_\odot$) compressed the gas of the galactic disk and thus produced the "extra radio arms". The resulting smooth structures would then be affected, in the course of one or more galactic rotations, i.e. in $\simeq 10^8$ yr, by the differential rotation and thus gradually take on the appearance of ordinary spiral arms.

The distribution of synchrotron radiation from the spiral arms gives us information about the *origin of the synchrotron electrons*. The emission from a thick disk and the observed proportionality of the mean intensity to the average optical luminosity lead us to suspect that a major portion of the synchrotron radiation originates from the old disk population (Sect. 5.5.4). To what extent supernovae and their remnants can serve as sources and can explain the observed intensity is still not clear. The central regions of the galaxies probably are not the source of many of the synchrotron electrons, since no correlation between the intensities of the nonthermal disk components and those in the central regions is observed. Furthermore, it seems that the amplification of synchrotron radiation in the normal spiral arms (e.g., M 51) and also in the extra radio arms of NGC 4258 is due to compression of the plasma already present, including its magnetic field and synchrotron electrons.

5.6.2 Radio Galaxies

In 1946, J. S. Hey and his coworkers detected the first radio source in the Swan, *Cygnus A*, initially on the basis of its intensity fluctuations. Later, simultaneous measurements at widely separated points showed that these fluctuations result from scintillation in the ionosphere. Like optical scintillation, which is observed for stars but not for planets (with their larger angular diameters), radio scintillation is seen only in the case of sources with a sufficiently small angular diameter. In the socalled quasistellar radio sources or quasars, whose angular diameters are less than 1″, another kind of scintillation has been discovered; it originates in the interplanetary plasma.

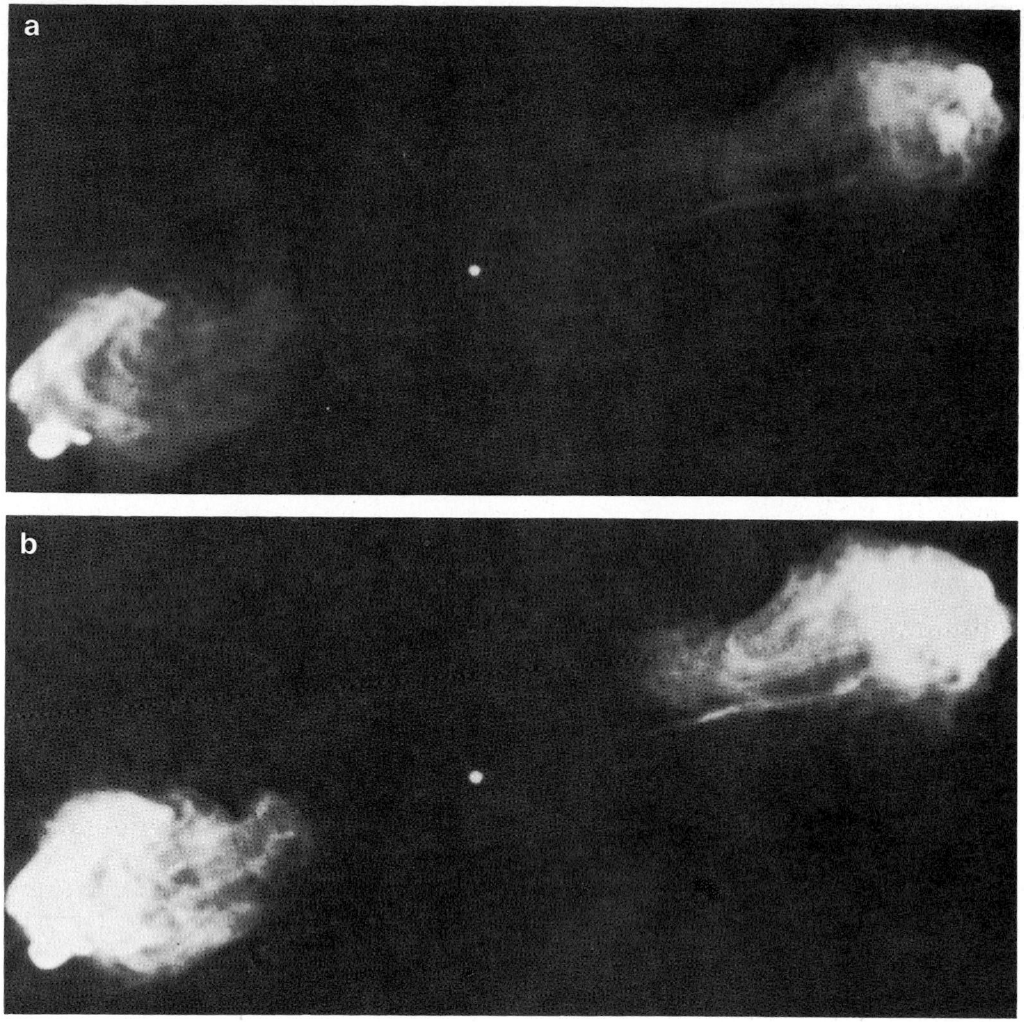

Fig. 5.6.3 a,b. The radio galaxy Cyg A = 3 C 405. Radio observations at 5 GHz (λ = 6 cm) with the VLA in a photographic representation, corresponding to two different "exposure times". The central source coincides with a cD-galaxy observed in the optical range, with a red shift of $z \simeq 0.06$. In (**a**), the structures in the outer emission regions, in particular the "*hot spots*", are easily distinguished; in (**b**), the *jet* which stretches from the center towards the NW (*right*) out to the outer emission region can be seen. The overall extension of the radio double structure is about 200 kpc; the angular resolution of 0.4″ corresponds to about 0.6 kpc. From R.A. Perley et al., Astrophysical J. **285**, L35 (1984). (Reprinted courtesy of the authors and *The Astrophysical Journal*, published by the University of Chicago Press; © 1984 The American Astronomical Society)

By 1954, it had become possible to determine the positions of several of the stronger radio sources with sufficient precision that W. Baade and R. Minkowski were able to identify them with visible objects. In particular, the second-strongest radio source in the northern hemisphere, Cyg A, could be attributed to a surprisingly faint object of photographic magnitude 17.9. Its spectrum exhibits Hα and several forbidden emission lines with a *red shift* $z = \Delta\lambda/\lambda \simeq 0.056$ (Sect. 5.9.1), corresponding to 16800 km s^{-1}, along with a weak continuum. It is therefore an *extragalactic object* at a distance of about 350 Mpc.

The optical image of Cyg A shows that it is a giant E-galaxy (cD galaxy) with a dark absorption band caused by dust.[17] The radiofrequency radiation from Cyg A = 3C405 (3C = 3rd Cambridge Catalogue of Discrete Radio Sources) originates for the most part not within the galaxy itself, but in two extended components which are located in nearly symmetric positions far out from its center. This was shown in 1953 by R. Hanbury Brown, R.C. Jennison, and M.K. Das Gupta using the correlation interferometer at the Jodrell Bank Observatory. A weaker radio source coincides with the position of the optical galaxy. In Fig. 5.6.3, we show a recent radio image of Cyg A, taken with the Very Large Array. The double structure has an overall length of about 0.2 Mpc. In the outer radio sources, we find smaller regions with very strong radio emission, which have relatively sharply defined outer boundaries. These "hot spots" were discovered in 1974 by P.J. Hargrave and M. Ryle with the Cambridge 5 km Aperture Synthesis Radiotelescope (at 5 GHz). Furthermore, a *jet* was found, stretching out from the center of one of the outer sources. The total *radio luminosity* of Cyg A is about 10^{38} W, corresponding to $2 \cdot 10^{11} L_\odot$.

Cyg A, with its characteristic double structure of radio emission lying symmetric to the central cD galaxy, is the prototype of a *radio galaxy*. In general, we define radio galaxies by the fact that their luminosities in the radiofrequency range are greater than those in the optical region. Their *radio spectra* are *nonthermal* ($I_\nu \propto \nu^{-\alpha}$) with an average spectral index (between $\lambda = 6$ and 11 cm) of $\langle\alpha\rangle \simeq 0.8$ in the extended double sources and $\langle\alpha\rangle \simeq 0$ in the compact sources. Here, again, we interpret the nonthermal radio emission as *synchrotron radiation* due to relativistic electrons, the spectrum being flattened out in compact sources due to absorption.

The *nearest radio galaxy* to us, at a distance of 5 Mpc, is Centaurus A = NGC 5128 in the southern hemisphere. Here, we can recognize a number of interesting details. The *optical* image initially shows NGC 5128 to be a giant elliptical galaxy (type E0 or S0), surrounded by a band of dark matter perpendicular to its axis (similar to that which originally made Cyg A appear to be two galaxies). Hα observations of H II regions in the band of dust show a rotation curve corresponding to a normal spiral galaxy with a mass of several $10^{11} \mathcal{M}_\odot$. NGC 5128 thus represents an unusual mixture of two forms of galaxies: a nearly spherical E-galaxy with a diameter of about 10 kpc, which is penetrated (or surrounded) by a rotating disk about 1 kpc thick and filled with dust. The first *radio map* of Cen A in the continuum at 1.4 GHz, made by B.F.C. Cooper, R.M. Price, and D.J. Cole in 1965, already allowed the identification of *two* pairs of plasma clouds, whose separations from the optical galaxy are about 1 Mpc and 10 kpc, respectively (Fig. 5.6.4a).

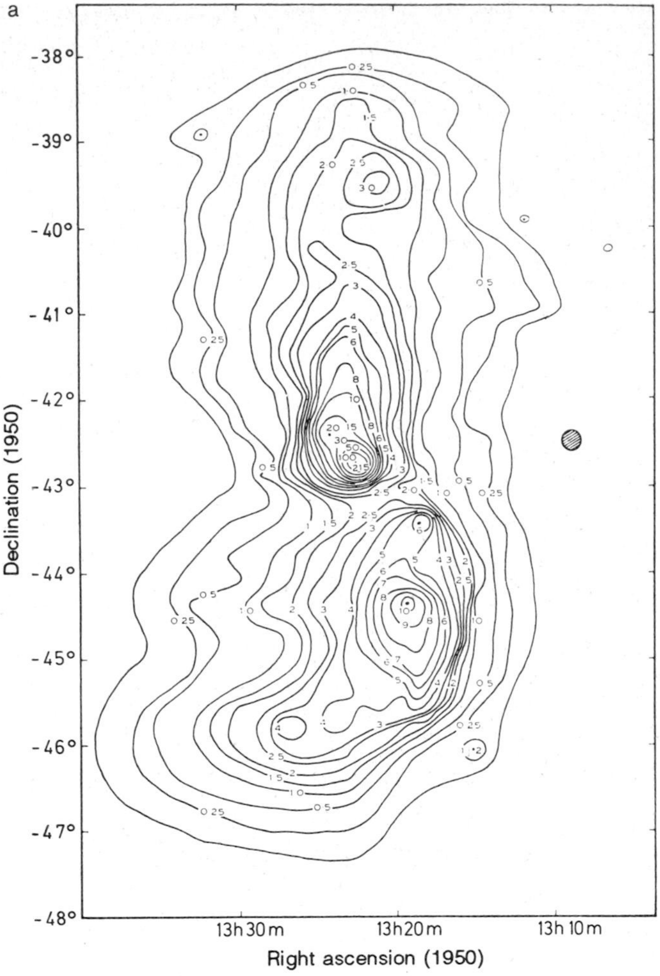

Fig. 5.6.4a. The radio source Cen A = NGC 5128: isophotes of the radio continuum at 1.42 GHz (λ = 21 cm), from B.F. Cooper et al. (1965). The brightness temperatures in [K] are shown. The shaded circle of 14′ diameter indicates the angular resolution of the antenna

Observations at high resolution with the Very Large Array at 4.9 and 1.5 GHz by J.O. Burns, E.D. Feigelson, and E.J. Schreier (1983) show additional structures in Cen A (Fig. 5.6.4b). The isophote picture of the inner

[17] W. Baade and R. Minkowski at first interpreted this optical image as two colliding galaxies. The stars would be only slightly disturbed by the collision, but interstellar gas would be swept out of both galaxies and excited to cause emission of radiofrequency radiation. Later investigations however failed to confirm this hypothesis.

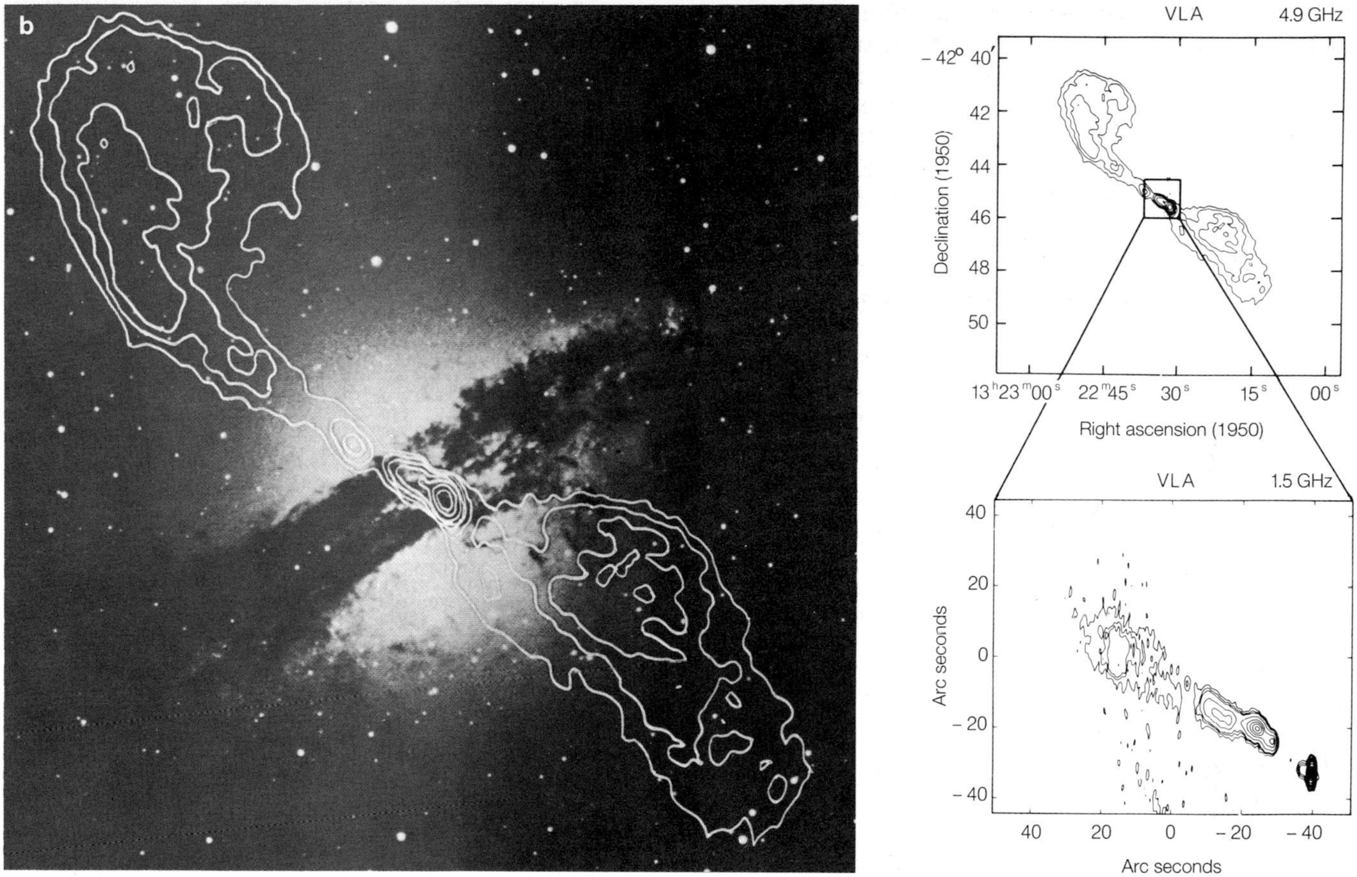

Fig. 5.6.4b. The inner portion of the radio source Cen A = NGC 5128; observations with the Very Large Array (VLA) at 4.9 GHz (λ = 6 cm) and a resolution of 15.6″·10.1″ (*upper right*), and at 1.5 GHz (λ = 20 cm) with 3.65″·1.05″ (*lower right*). The 4.9 GHz isophotes are superimposed at the left onto a blue photograph of the giant elliptical galaxy (E0 or S0+Spec) taken by J. Graham (1979) with the 4 m telescope of the Cerro Tololo Interamerican Observatory. Radio images from J. O. Burns et al., Astrophys. J. **273**, 128 (1983). (Reprinted courtesy of the authors and *The Astrophysical Journal*, published by the University of Chicago Press; © 1983 The American Astronomical Society)

pair of radio sources has an S-shaped structure, roughly symmetrical about a central point. This twisting continues in the outer pair. Near the center, we find a *jet* about 1.5 kpc long, consisting of several "nodes", which joins the center to the northern inner radio source across a gap of 0.4 kpc. It points in the direction of the axis of the optical galaxy, perpendicular to the dust band. A southern counterpart to this jet is not present. The jet was, incidentally, first discovered in the *X-ray region* using the Einstein Satellite. Since the nodes observed in the radiofrequency range coincide with those in the X-ray image, it is tempting to interpret the X-ray emission also as synchrotron radiation. To generate it, synchrotron electrons with up to about 10^5 GeV and certainly also heavier particles of comparable energies must be present. Some of the nodes are recognizable in the *optical region* as a nonthermal continuum and emission lines.[18] Within the central radio component, observations using long baseline interferometry at 2.3 GHz were finally able to discern an additional *inner jet* about 0.05″ or 1 pc long, in the same direction as the optical jet.

Sky surveys of all the optically identifiable radio galaxies have shown that roughly half of them (such as Cyg A

[18] The fact that the continuum in the *optical region*, with its UV excess (compared to normal galaxies), is also generated by the synchrotron mechanism was first shown by observations of the giant elliptical galaxy M 87 = NGC 4486, the radio source Virgo A (see also Sect. 5.5.3). A jet emerges from its center, emitting polarized blue light which can only be synchrotron radiation. This jet was later observed also in the radio and X-ray regions.

or Cen A) are giant E/S0 galaxies. The other half are quasars (see Sect. 5.6.5). The characteristic *double structure* of the radio emission is found in more than two-thirds of all the strong radio galaxies. Its extension or length lies typically in the range of 0.1 to 0.5 Mpc, rarely over 0.6 Mpc. The record is held by the giant radio galaxy 3C 236, at a distance of 600 Mpc from the Sun, with a length of 5.6 Mpc! As in the case of Cyg A, we frequently find a central, compact radio component, particularly in observations at high frequencies; it can usually be identified with the active core of the galaxy.

The *energy* contained in the synchrotron electrons and the magnetic field of a galaxy can be estimated using the theory of synchrotron radiation. The spectral index gives information about the *energy distribution* of the electrons, according to (5.6.7, 8). Then the emission per unit volume can be calculated as a function of the electron density *and* the magnetic field strength or its energy density. Finally, making the plausible assumption that the energy densities of the synchrotron electrons plus protons and that of the magnetic field are roughly equal to each other, the *total energy density* (or, nore precisely, a lower limit for it) can be obtained. Thus, for radio galaxies, energies in the range of 10^{49} to 10^{53} J (10^{56} to 10^{60} erg) and magnetic fields between 10^{-10} and $2 \cdot 10^{-8}$ T (10^{-6} to $2 \cdot 10^{-4}$ G) are estimated.

While jets have been found in the optical and X-ray regions in only a few radio galaxies and quasars, radio observations, in particular with the *Very Large Array*, have shown that *radio jets* are a widespread phenomenon and that relativistic particles are probably transported from the centers to the outer zones of radio galaxies by these jets. The fact that jets are often observed in only *one* direction, as e.g. in the case of Cen A, can probably be understood in terms of the relativistic Doppler effect, which intensifies the radiation from particles which are moving towards us (Sect. 5.6.5). When the supersonic plasma rays collide with a denser medium (gas in clusters of galaxies or intergalactic gas), a shock wave is formed, and it is presumably seen as a "hot spot" in the radiofrequency region. The effective collimation of the jets around their axes over great distances is surprising, as is their stability over periods of 10^6 to 10^7 yr; this is not yet well understood. In a few cases, e.g. for the radio galaxy NGC 6251, we observe a linear orientation of the jets over dimensions of 1 pc out to 1 Mpc. Crooked or "bent" jets such as in Cen A can perhaps be explained through perturbations by accompanying galaxies or through the precessional motions of the central galactic axis.

An important limitation of any model of a radio galaxy is the *lifetime* of the synchrotron electrons. We estimate for example in Cyg A a magnetic field of $1.5 \cdot 10^{-8}$ T (see above). Then, from (5.6.9), the electrons whose synchrotron emissions are observed at 5 GHz can emit this radiation only during a period of about 10^5 yr; however, their time of flight from the center to the edge of the galaxy, a distance of 0.1 Mpc, is in the most favorable case (propagation along straight lines at the velocity of light) $3 \cdot 10^5$ yr. In many cases, the previously suggested explanation of the radio galaxies in terms of a single explosion in the center 10^4 to 10^7 yr ago, causing ejection of clouds of plasma along the axis, meets with difficulties. Instead, there is probably a continuous "energy input" along the length of the jets, and an acceleration of the electrons to relativistic energies *in situ* within the shock waves of the "hot spots".

We shall postpone further discussion of the question of the energy source, and first consider the observations of the *nuclei* of the various types of galaxies. These are the locations of "activity", which varies strongly from galaxy to galaxy and also with time; the significance of this activity was pointed out by V. A. Ambartsumian as early as 1954, in a series of articles which were hardly noticed then. It is contained in the fact that enormous quantities of energy are released in the form of superthermal particles and photons. Here, we clearly are dealing with a source of synchrotron electrons and cosmic radiation, of nonthermal radiofrequency emissions, X-rays, and gamma rays, of nonthermal infrared emissions, etc.

In the following sections, we consider, in order of increasing activity, first the (more or less) *normal galaxies*, to which our Milky Way system belongs; then we turn to the *Seyfert galaxies*, the nuclei of the *radio galaxies*, and the *quasars*.

5.6.3 The Central Region of the Milky Way and Other Normal Galaxies

In the centers of many spiral galaxies, optical observations show a small *nucleus* with a high surface luminosity. In the Andromeda galaxy, M 31, where this nucleus was already discovered by E. Hubble, its size is $1.6'' \times 2.8''$, corresponding to dimensions of about 5 pc by 10 pc. The optical spectrum shows that the nucleus consists in the main of ordinary stars of later spectral types. The luminosity, as well as the internal motions, indicate an overall mass of 10^7 to $10^8 \mathscr{M}_\odot$ and thus a stellar density corresponding to about $2 \cdot 10^5$ pc^{-3}. The true significance of the galactic nuclei was, however, recognized only through radioastronomy. Even among the "normal" galaxies, many (but not all) of the nuclei are found to be strong, compact nonthermal radio sources. They show in principle the same phenomena which we shall meet with greater intensity in the Seyfert and N-galaxies, the quasars, etc.

The *central region* of our *Milky Way system* is not accessible to optical observations due to the dense dark clouds in the Scorpio-Sagittarius region. Only with radio antennas of high resolution in the centimeter wavelength range, and due to the development of infrared astronomy, has it been possible to investigate it since the late 1960's.

In the infrared, at shorter wavelengths ($\lambda \leq 10\,\mu m$), emissions from the stars dominate the overall radiation, while in the far infrared, we measure mainly the radiation from warm dust. Broadband infrared observations have been carried out with continually improving sensitivity and angular resolution following the pioneering investigations of E.E. Becklin, G. Neugebauer, F.J. Low, D.E. Kleinman, and others in 1968/69. From the surface luminosity at $\lambda = 2.2\,\mu m$, we can recognize a strong increase in the *stellar density* in the central region of the galaxy (within 1 to 2 pc from the center), the innermost part of the spheroidal component of the galaxy (Sect. 5.2.7). The diameter and the surface luminosity are similar to those observed in the Andromeda galaxy. It is reasonable to assume that both regions consist of stars and have the same spectral energy distributions at shorter wavelengths, also. Then the extinction in the near infrared can be determined to be $A_{2.2\,\mu m} \simeq 3$ mag. By extrapolation, a *visible extinction* A_V towards the galactic center in the range of 27 to 34 mag is obtained.

Observations with the infrared satellite IRAS have shown that the emissions between 12 and 100 µm wavelength also show an increased intensity towards the central region (with a radial extension of about 700 pc), along with a concentration in the plane of the Milky Way. The IRAS images show in particular dust-filled H II regions as discrete sources.

Before turning to the true galactic nucleus, in the innermost few pc, we discuss the radioastronomical observations of the gas in the general neighborhood of the center. We have already mentioned the *21 cm observations* of neutral hydrogen in the galactic plane (Sect. 5.3.3). At a distance of 3.7 kpc, we found strong deviations from the circular orbits typical of the outer disk in the *expansion* of the "3 kpc arm" and its counterpart on the other side of the galactic center. This was a first indication of the "unusual" processes in the central region of the Milky Way. Still further inwards, the 21 cm observations show complicated motions of the gas for which we still have no unique model; the explanation of the observed radial velocities is rendered difficult by the fact that there is no way to determine the distance to the H I regions. G.W. Rougoor and J.H. Oort originally (1960) interpreted the observations in terms of a thin *nuclear disk*, rotating at about $250\,km\,s^{-1}$ (and a ring), with a radius of ≤ 0.7 kpc. The discovery of H I regions having high velocities (by J.H. Oort and P.C. van der

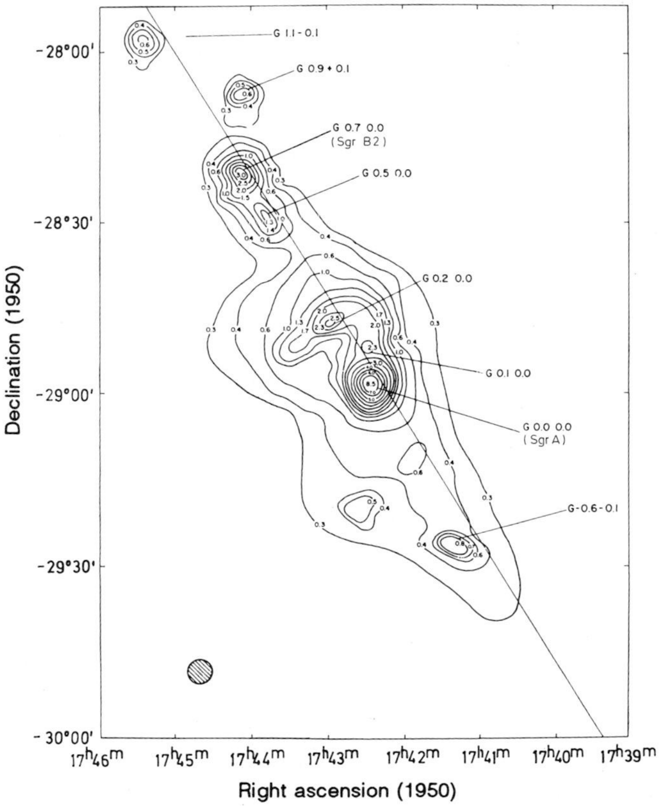

Fig. 5.6.5. The central region of the Milky Way at 8.0 GHz ($\lambda = 3.75$ cm) from D. Downes et al. (1966). The numbering of the isophotes corresponds to the antenna temperature. The radio sources G are denoted by their galactic coordinates l, b. Sagittarius A contains the nonthermal central source of our galaxy. The other sources and the extended source at $l \simeq \pm 0.5°$ emit thermal radiation. The "correct" galactic equator evidently lies 2′ or 3′ south of the position originally assumed in fixing the galactic coordinate system. The thermal source (giant H II region G 0.7 − 0.0 = Sagittarius B2) is associated with a large molecular cloud complex. The shaded circle of 4.2′ diameter indicates the angular resolution of the antenna. For a distance of 8.5 kpc to the galactic center, 1 minute of arc corresponds to 2.5 pc

Kruit, 1971) led to the idea that about 10^7 yr ago, around $10^7 \mathcal{M}_\odot$ of hydrogen was ejected from the galactic nucleus by an explosion. Another model (W. B. Burton and H. S. Liszt, 1978/80) describes the observations in terms of a *gaseous disk* which is rotating and expanding at the same time, *inclined* by about 25° to the galactic plane, and having a radius of 1.5 kpc, or also by a (likewise inclined) rotating "*bar*" of 0.6 kpc·1.9 kpc·0.1 kpc dimensions, in which the gas moves on strongly elliptical orbits. Independently of the precise model used, about $10^7 \mathcal{M}_\odot$ of H I with a density of the order of 0.3 to $1\cdot 10^6\,\mathrm{m}^{-3}$ must be present within a distance of 1 kpc from the galactic center.

Using the observations of different *molecular lines*, in particular that of CO at $\lambda = 2.6$ mm (11.5 GHz), which also indicates the distribution of H_2, as well as those of OH and H_2CO, a sharp minimum of molecular hydrogen is found between 2 and 4 kpc distance from the center; further inwards, the density then increases steeply. The kinematics of the molecular clouds presents on the whole a similar picture to that of the H I regions. Here, too, we probably are dealing with a rotating and expanding thin disk (of 1.5 kpc radius and 70 pc thickness) or a rotating bar. The inclination to the galactic equator seems to be smaller than that of the H I disk, about 7°. The density of H_2 is about a factor of 50 higher than that of H I; all together, about $10^9 \mathcal{M}_\odot$ of molecular hydrogen is found within 1.5 kpc from the center.

Near the center itself ($R \simeq 200$ pc), there is among other things a giant *molecular cloud complex*, which coincides with the radio source Sgr B2 and is characterized by the great variety of different molecules observed in it (Sect. 5.3.4). In general, the molecular clouds in the central region are associated with *giant H II regions* of several $10^4 \mathcal{M}_\odot$, which are noticeable in the cm range as strong *thermal* radio sources (Fig. 5.6.5). These radio sources are embedded in a less dense, ionized medium (electron densities of $\simeq 10^7\,\mathrm{m}^{-3}$), having the form of a disk with dimensions of about 100 pc·300 pc.

On the isophote chart of the central region (Fig. 5.6.5) at $\lambda = 3.75$ cm (8.0 GHz), recorded by D. Downes et al. in 1966 with a 36 m paraboloid, the radio source *Sagittarius A* can be recognized near the galactic center; it has a diameter of about 3.5′, corresponding to 9 pc. Its position relative to many other objects leaves no doubt that Sgr A represents the innermost central region or *nucleus* of our galaxy. With the higher resolution of the Very Large Array, four components can be distinguished (Fig. 5.6.6):

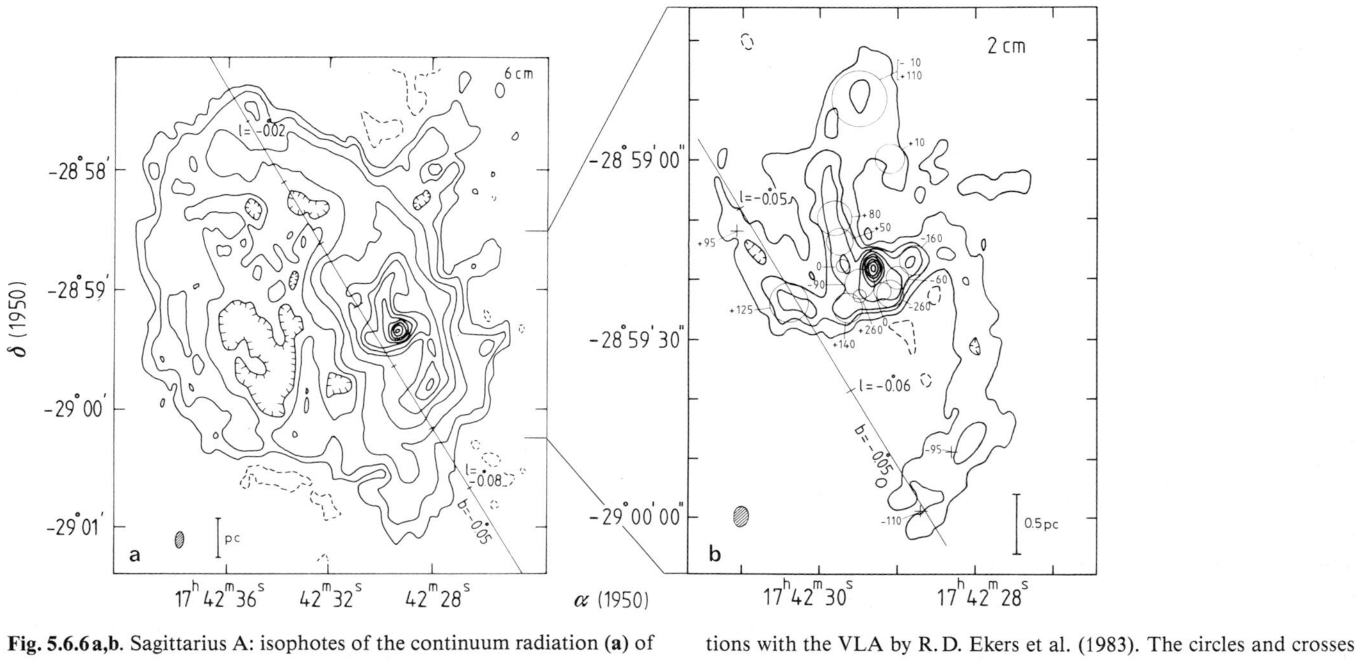

Fig. 5.6.6 a,b. Sagittarius A: isophotes of the continuum radiation (**a**) of the whole source at $\lambda = 6$ cm (5 GHz) with a resolution of 5″·8″; and (**b**) of the central region of the component Sgr A West at $\lambda = 2$ cm (15 GHz), with a resolution of 2″·3″. The images are based on observations with the VLA by R. D. Ekers et al. (1983). The circles and crosses in (b) mark the "[Ne II] clouds" observed by J.H. Lacy et al. (1980), with their radial velocities indicated in [km s^{-1}]

within a *halo* of weaker thermal and nonthermal radiation having about 20 pc diameter, we find the nonthermal, cup-shaped source *Sgr A East*, probably a supernova remnant. The halo also contains the thermal sources *Sgr A West*, an H II region of dimensions 2 pc·4 pc, and within it, in turn, the *ultracompact source Sgr A**, a strong nonthermal, elliptical source, whose semimajor axis could just be resolved using long-baseline interferometry in the cm wavelength range; it is $\simeq 10^{-3}$ seconds of arc or 10 AU. This corresponds roughly to the semiaxis of Saturn's orbit in our Solar System! Sgr A* is unique within the Milky Way; this source, at the position $l = -0.056°$ and $b = -0.046°$, can perhaps be regarded as *the* galactic center.

Let us now look somewhat more carefully at the *innermost few pc* around the center of the Milky Way! This region emits a luminosity of all together about $10^7 L_\odot$, almost exclusively in the *infrared*, due to dust which is heated by the ultraviolet radiation from hot stars. The radiofrequency and X-ray regions contribute only to a minor extent here. Within a few pc from the center, we find about $10^4 \mathcal{M}_\odot$ of neutral gas, based on observations of the 21 cm line of neutral hydrogen. The strongest emission line in the infrared is the fine structure line of [O I] at $\lambda = 63.2$ µm. Furthermore, radioastronomical observations of molecular lines and infrared observations of the radiation of the heated dust show that the center is surrounded between 2 and about 5 pc by a ring-shaped, slightly bent *disk of molecular gas and dust*. It rotates at around 110 km s^{-1} and contains about $10^4 \mathcal{M}_\odot$ of H_2 at temperatures $\lesssim 400$ K and particle densities around 10^7 m^{-3}. The molecular gas exhibits unusual excitation conditions; for example, strong emission lines of CO from high rotational levels (quantum numbers $J \simeq 7 \ldots 14$) are seen.

Dynamic considerations indicate a mass of $5 \cdot 10^6 \mathcal{M}_\odot$ within a distance of 2 pc from the center.

High-resolution VLA observations of the radio source Sgr A West, e.g. at $\lambda = 2$ cm (Fig. 5.6.6b), show that the thermal emission of the ionized hydrogen in this region comes from a small "spiral-shaped" structure. The ultracompact source Sgr A* is approximately at the center of symmetry of this structure, while the outer arcs of the spiral coincide with the inner edge of the molecule/dust disk. Matter presumably flows out of the disk, as a result of the galactic rotation, on spirals within the galactic nucleus (or out from it?).

In the *infrared*, the high-resolution *observations* at 2.2 µm by E.E. Becklin and G. Neugebauer (1975) indicate around 20 discrete sources in the innermost 1 to 2 pc, with luminosities of about $10^4 L_\odot$ each (in one case $10^5 L_\odot$); they are probably attributable to single bright supergiants of later spectral types or to clusters of early stars. One of these sources is only 0.5″ from the nonthermal source Sgr A*. Within the resolution of the Einstein Satellite (1′ or 3 pc), a discrete X-ray source is also at this position. The galactic center is furthermore a source of energetic *gamma radiation*: the emission at 0.511 MeV, due to positron annihilation, $e^+ + e^- \rightarrow \gamma + \gamma$, is strongly variable on a time scale of several months and therefore indicates a compact source with a diameter $\lesssim 0.1$ pc, i.e. $\lesssim$ several light months.

The observations of radio recombination lines in H II regions and especially the high-resolution measurements of *infrared fine-structure transitions*, such as [Ne II] $\lambda = 12.8$ µm, [Ar III] $\lambda = 9.0$ µm, [S III] $\lambda = 18.7$ µm, [O III] $\lambda = 51.8$ µm, and others, give us some insight into the *dynamics* of the near vicinity of the galactic center. For example, images made in the [Ne II] line show cloudlike structures in the emission regions, stretching along the "radio spirals" (Fig. 5.6.6b). The radial velocities of these "[Ne II] clouds" would be compatible with the rotation roughly in the galactic plane of a mass of 10^6 to $10^7 \mathcal{M}_\odot$ (black hole?, cf. Sect. 5.6.6). To be sure, this mass estimate is based on the assumption that the gas moves on approximately circular orbits around the galactic center. The dynamics near the center are, however, probably more complicated, as shown by the existence of the spiral structure as well as recent precise position measurements: the ultracompact radio source Sgr A* does not coincide either with one of the bright infrared sources (at $\lambda = 2.2$ µm) nor with the center of rotational symmetry of the [Ne II] clouds and the spiral-shaped radio emissions (deviations of 2.5″ or 0.1 pc).

A relatively weak *activity* as found in the nucleus of our Milky Way cannot at present be detected in other normal galaxies. For the central region of M 31, which has considerable similarities with that of our galaxy, spectral lines also in the *optical* region can be employed for the investigation of the dynamics (Sect. 5.5.3).

The development of extremely sensitive optical detectors with high spectral and spatial resolutions has in recent times permitted the discovery of *weak emission lines* in the nuclei of most spiral galaxies, including "normal" galaxies such as M 81 or M 51 (T.M. Heckman, 1980; W.C. Keel, 1983; and others). Their intensity ratios differ from those observed in H II regions, planetary nebulas, or supernova remnants. The occurrence of Hα lines with broad wings and of the neighboring [N II] ($\lambda = 658$ nm) line with comparable intensity is typical. In some of these emission regions, called "*liner*s" (*l*ow *i*onization *n*uclear *e*mission-line *r*egions), a weak nonthermal continuum can be distinguished, so that here, presumably, we are dealing with a somewhat stronger form of activity.

5.6.4 Seyfert Galaxies. The Active Nuclei of Radio Galaxies

Nuclear activity in the optical and radio spectral regions can be more readily detected in the class of (mostly spiral) galaxies which were discovered in 1943 by C.K. Seyfert; they were at first observed optically. The "*Seyfert nuclei*" exhibit a strong *emission* spectrum, in which the Balmer lines of hydrogen as well as He I and II and other allowed transitions are all characterized by large *linewidths*. The *forbidden* lines, mostly from higher excitation states, of N II, O II-III, Ne III-IV, S II etc. have the *same* widths in some Seyfert galaxies (Type Sy 2); in others (Type Sy 1), they are considerably *narrower* than the allowed lines (Fig. 5.6.7). The widths at half maximum for the Sy 1 galaxies correspond to a Doppler effect of about 5000 km s^{-1} in the allowed and 500 km s^{-1} in the forbidden lines; for the Sy 2 galaxies, all the linewidths lie in the range of 300 to 1000 km s^{-1}. We discuss the interpretation of the spectra in Sect. 5.6.6.

The apparently brightest Seyfert galaxies are NGC 4151 (Sy 1), with a nucleus of $m_V \simeq 12$ mag, and NGC 1068 (Sy 2) with $m_V = 10.5$ mag. The Sy 1 galaxies are absolutely brighter objects than those of type Sy 2; their nuclei attain absolute visible emissions of 10^{35} to $\geq 10^{38}$ W, or $2 \cdot 10^8$ to $\geq 2 \cdot 10^{11} L_\odot$.

The *optical continuum* contains a nonthermal contribution, which is superposed onto the radiation from the stars and increases towards the ultraviolet. The Sy 1 galaxies are strong *X-ray sources* and have a nonthermal spectrum even in the hard X-ray region (up to ≤ 50 keV), with $I_\nu \propto \nu^{-0.7}$, so that extremely energetic synchrotron electrons must be present in their nuclei. Extrapolation of the power law to higher energies leads us to suspect that the X-ray luminosities of the Sy 1 galaxies exceed their optical luminosities. NGC 4151 and the quasar 3 C 273 are at present the only discrete extragalactic *gamma-ray sources* which have been identified.

In the *infrared*, radiation from heated dust dominates the emissions of the Seyfert galaxies; their luminosities in the infrared are comparable to or somewhat greater than in the optical region. In the *radiofrequency region*, the

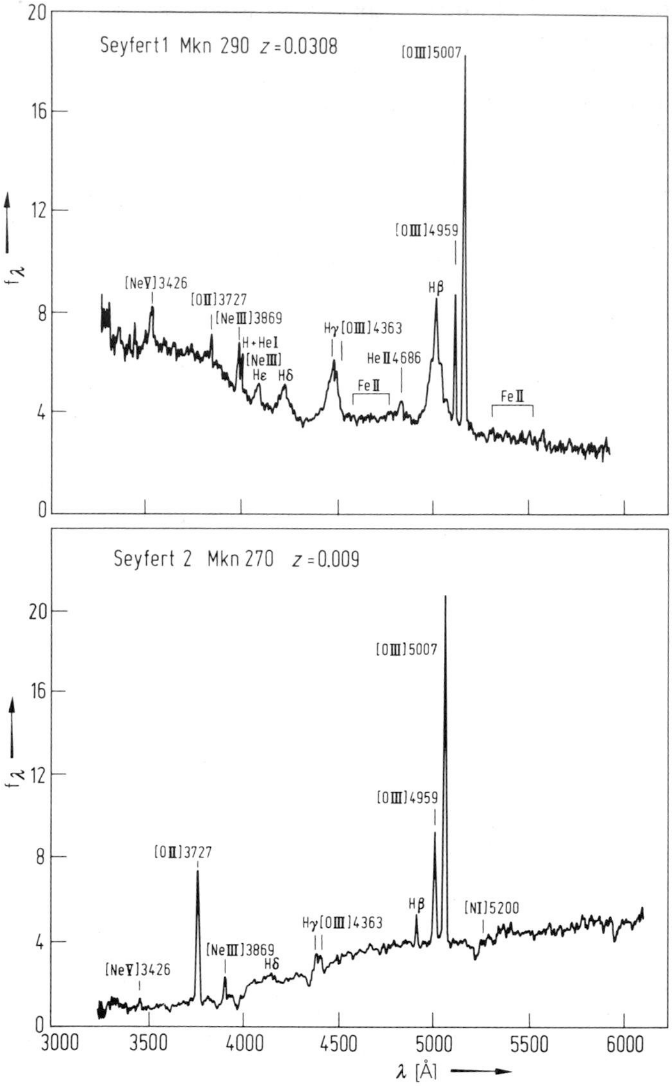

Fig. 5.6.7. The spectra of Seyfert galaxies (Type Sy 1 and Sy 2), after H. Netzer (1982). The radiation flux (outside the Earth's atmosphere) f_λ, in [10^{-17} W m^{-2} nm^{-1} = 10^{-15} erg$\cdot$s^{-1} cm^{-2} Å^{-1}], is plotted against the observed wavelength $\lambda = (1+z)\lambda_0$ in [Å] (z: red shift)

Seyfert galaxies represent compact sources, which emit nonthermal synchrotron radiation. Observations with the Very Large Array and long-baseline interferometry with angular resolutions of 0.001″ to $\leq 1''$ indicate that for some galaxies, the radio emissions show a double or triple structure, with some sources of only a few 0.1 pc diameter; in other galaxies, the radio source, which coincides with the optical nucleus, is lacking in noticeable structure. In comparison to the radio galaxies, the radio sources in the Seyfert galaxies are weaker and much less extended (≤ 0.1 to 1 kpc).

In the Seyfert galaxy NGC 1275 = 3 C 84, which was identified with the strong radio source Per A in 1954 by W. Baade and R. Minkowski, three collinear compact components (and some fainter structures) are observed in the cm wavelength region at a resolution of 10^{-3} seconds of arc; the emissions in the mm wavelength range seem to originate from a different source. The radiofrequency radiation varies strongly with time. In NGC 1275, in addition to the activity in the nucleus, numerous long, extended filaments are seen in Hα images (R. Lynds, 1970) out to about 75 kpc from the center, with velocities of several 1000 km s^{-1}; they show a certain superficial similarity to those of the Crab Nebula.

A *variability* of the emissions in all spectral ranges from the radio to the X-ray region on time scales of months to years is characteristic of the nuclei of all Seyfert galaxies; the amplitudes of the variation are in the range from 0.2 to 1 mag. In the Sy 1 galaxies, an increase in the (nonthermal) optical continuum is correlated with an increase in the intensity of the allowed emission lines. Recent observations of NGC 4151 also show variations in the line*widths*, so that possibly the type of a Seyfert galaxy also changes with time.

The phenomena already described for the "normal" galaxies give the impression of being somewhat fragmentary or poorly-developed counterparts of those seen in the Seyfert galaxies. One can thus conclude that all or at least most of the brighter galaxies are subject to the "Seyfert disease" at times. Since about 1% of all galaxies exhibit the Seyfert phenomenon, its total duration must be of the order of 10^8 yr (perhaps with interruptions).

The classification criterion described by Seyfert in 1943 is not the only possible one. Objects which have (optically) a bright, starlike nucleus surrounded by a fainter, nebulous shell are termed N-galaxies (W. W. Morgan, 1958). Probably all Seyfert galaxies are simultaneously N-galaxies, but certainly not *vice versa*. Related to the N-type, but still more generally defined, is the class of *compact galaxies* defined by F. Zwicky (1963); they have by definition an inner region whose surface luminosity is greater than 20 mag per square second of arc. Finally, galaxies which, for example in spectra with a low dispersion, show a strong *ultraviolet excess* (relative to normal objects), are termed *Markarian galaxies*. The relation of the criterion chosen by B. E. Markarian (1967) to the Seyfert criterion is evident. It remains a task for the future to find a classification which can present, in a clearcut way, as directly as possible from observations on the one hand the permanent characteristics of the galaxies, and on the other hand their temporary characteristics.

Many radio galaxies (Sect. 5.6.2) have compact nuclei with strong optical *emission lines*. We distinguish the "*broad-line*" radio galaxies with broad allowed lines of about 8000 km s^{-1} Doppler width and narrower forbidden lines from the "*narrow-line*" radio galaxies, whose allowed and forbidden lines have similar widths, around 500 km s^{-1}. The former are N-galaxies and exhibit similarities to the Sy 1 galaxies in many of their characteristics (emission lines, luminosity, X-ray emissions, variability); the latter are mostly E- or D-galaxies and are similar to the Sy 2 galaxies.

5.6.5 Quasars (Quasistellar Objects)

In the 1960's it was discovered that a number of by no means faint radio sources from the 3rd Cambridge Catalogue of Discrete Radio Sources (3C) could be identified with optical objects that could not be distinguished from stars on photographs taken with the Mt. Palomar 5 m-mirror. Their *diameters* of $<1''$ as determined optically and radioastronomically made it clear from the start that these new objects have unusually great surface brightnesses in both spectral regions. In 1962/63, M. Schmidt investigated their spectra; they show a *continuum* and *strong emission lines*, similar to those of the already known radio galaxies. The new factor was the enormous *red shift*, $z = \Delta\lambda/\lambda$, which, when interpreted in terms of the Hubble relation (5.9.2), indicate that these objects are very distant blue *galaxies*. Their visual absolute magnitudes surpass those of the normal giant galaxies by factors of up to 100; their radio luminosities correspond roughly to that of Cyg A. At this stage of the investigations, the name *Quasar* = *Quasistellar Radio Source* was coined.

In 1965, A. Sandage discovered that there are many more quasistellar galaxies, which are optically indistinguishable from the quasistellar radio sources in terms of their compact structures, their high surface brightnesses, and their blue color, but which emit no (or only very weak) radiofrequency radiation. A distinction was thus often made between QSR's = Quasistellar Radio Sources and QSO's = Quasistellar Objects (without radiofrequency emissions). Today, it is certain that both represent the same type of galaxy, of which a small fraction suffers from the "radio disease"; thus, *only* the name "quasars" or "quasistellar objects" is now used, and, as necessary, mention is made of whether or not they emit strong radiofrequency radiation.

In optical and radioastronomical sky surveys, several thousand quasars have since been discovered; for example, the catalogue of M.-P. Véron-Cetty and P. Véron (1989) contains about 4170 quasars, and that of A. Hewitt and G. Burbidge (1987) contains 3570 quasars with measured red shifts.

In two-color diagrams (U−B, B−V), the quasars with not too large z-values can be readily distinguished from stars by means of their blue color. They fall together with the nuclei of Seyfert galaxies and the N-galaxies in a narrow band roughly along the black body line (Fig. 4.5.5). Such a diagram emphasizes the close relationship of the objects in question.

The quasar which is nearest to us, discovered as the first quasar by M. Schmidt, is 3 C 273, with $m_V = 12.8$ mag and a red shift of $z = 0.158$ or a distance of about 950 Mpc. The remaining quasars are fainter than 16 mag. The *absolutely* brightest quasar (as of 1990) is S5 0014+81, with $M_V \simeq -33$ mag ($m_V = 16.5$ mag) and $z = 3.41$. The quasar with the largest known red shift was for several years PKS 2000-330 ($m_V = 19$ mag), with $z = 3.78$; then, since 1986, more and more quasars with larger red shifts were discovered in short order, among them about thirty with z greater than 4. The current record (1991) is held by PC 1247+3406 (PC = *P*alomar *C*CD Survey), with a red shift of $z = 4.90$.

In the spectrum of PKS 2000-330 (Fig. 5.6.8), for example the Lα resonance line of hydrogen is shifted from its rest wavelength at $\lambda_0 = 121.6$ nm, according to $\lambda = (1+z)\lambda_0$, to 581.2 nm. We can thus readily observe the strong ultraviolet emission lines from the distant quasars with $z \simeq 4$, including the Lyman line and C IV ($\lambda_0 = 155$ nm), in the optical region.

In the case of red shifts which are no longer small compared to one, we cannot use the simple Hubble relation, z proportional to distance, to determine the distance; instead, we need a generalized relation which is found from the theory of general relativity, given a model of the universe (Sect. 5.9.3).

The *absolute magnitudes* M_V of the quasars are in the range from −25 to −33 mag (for a Hubble constant $H_0 = 50\ \mathrm{km\,s^{-1}\,Mpc^{-1}}$, Sect. 5.9.1). This corresponds to visual luminosities of 10^{12} up to several $10^{14} L_\odot$ ($4 \cdot 10^{38}$ to $\simeq 10^{41}$ W). The quasars thus represent the *absolutely brightest* stellar systems in the universe. They consist of only the extremely bright *nuclei* of distant galaxies and are a continuation of the trend to strong activity shown by the Seyfert, N- and radio galaxies. Their *mother galaxies* are recognizable in normal optical images only in the cases of the nearest quasars, such as 3 C 273, as faint, extended nebulous spots; in general, they are washed out by the much brighter nucleus. Only recently, using extremely sensitive CCD detectors (with a large dynamic range!) and modern image-enhancement techniques, has it become possible to detect the associated galaxies of *all* the quasars with $z \lesssim 0.5$ (Fig. 5.6.9). From the luminosity distributions, after subtracting out the quasar itself, absolute magnitudes of about −21 to −23 mag and diameters of around 40 to 150 kpc are derived for the galaxies; these are typical values for large spiral or elliptical galaxies. The observations suggest that the quasars with strong radiofrequency emissions possibly belong to E-galaxies, and those with no radiofrequency emissions to S-galaxies.

The quasars are also among the strongest extragalactic *X-ray sources* and radiate in the range from 0.2 to 4.5 keV nearly as strongly as in the optical region; their *infrared* luminosities are of the same order of magnitude. The

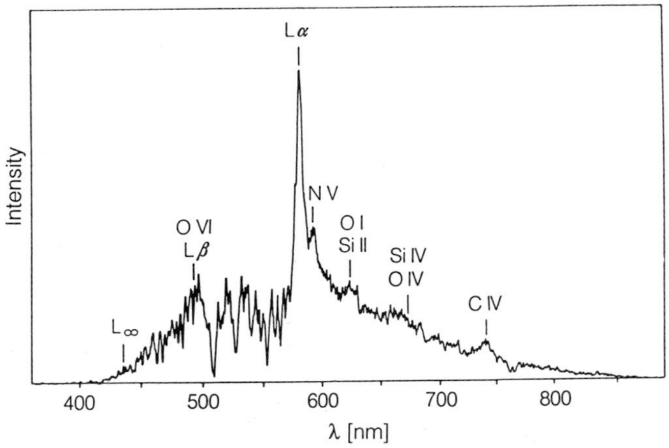

Fig. 5.6.8. The optical spectrum of the quasar PKS 2000−300 ($m_V \simeq 19$ mag) with a red shift $z = 3.78$; relative intensity vs. observed wavelength. The image was made with the Anglo-Australian 3.9 m telescope, resolution 1 nm. The identifications of the strongest lines and the Lyman edge are indicated; in the rest system, Lα is at 121.6 and C IV at 155 nm. The numerous absorption lines on the short-wavelength side of Lα result from various hydrogen clouds along the line of sight ("Lα forest"). From B. A. Peterson et al., Astrophys. J. **260**, L27 (1982). (Reprinted courtesy of the authors and *The Astrophysical Journal*, published by the University of Chicago Press; © 1982 The American Astronomical Society)

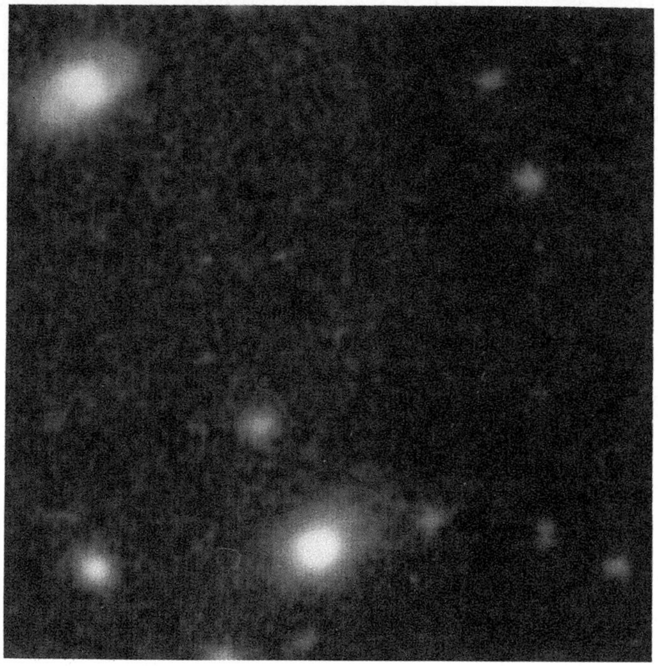

Fig. 5.6.9. A direct photograph of the quasar QSO 0054+144 ($z = 0.171$) with the CCD camera of the 2.2 m telescope on Calar Alto by J. Fried (red filter, 1 h exposure). The section is about 90′·90′; the quasar is the brightest object in the lower half of the picture. The bright quasar is surrounded by a nebulous spot, its mother galaxy; its luminosity distribution is typical of an elliptical galaxy. (Picture courtesy of J. Fried)

nearest quasar, 3 C 273, could also be identified as a discrete source in the gamma-ray region.

In the *radiofrequency region*, the quasars exhibit a nonthermal spectrum $I_\nu \propto \nu^{-\alpha}$, which without doubt can be interpreted as *synchrotron radiation*. In the case of 3 C 273, owing to its nearness almost the entire spectrum can be observed, in particular also the mm and infrared regions.[19] Between about 5 µm and 5 mm wavelengths, the spectral index is $\alpha \simeq 0.7$, similar to that of the radio galaxies (at longer wavelengths). Above 5 mm, we find a flat spectrum ($\alpha \simeq 0.1$), which probably results from the superposition of radiation from several compact components with different synchrotron self-absorption. In the near infrared, ($\lambda \lesssim$ 5 µm), the spectrum becomes steeper ($\alpha \simeq 1.6$). The continuum emission continues to obey essentially a power law beyond the optical region out to the X-ray and gamma-ray regions, so that it is tempting to interpret the quasar radiation in the entire electromagnetic spectrum as being due mainly to synchrotron radiation. To be sure, the energetic X-ray and gamma-ray photons are not generated directly by the synchrotron process, but instead by inverse Compton scattering (5.3.35) of less energetic photons on relativistic electrons. In many quasars, a broad emission maximum is observed in the optical and near ultraviolet around $\lambda \simeq 300$ nm (the "uv bump"), as *excess emission* relative to the synchrotron radiation continuum; it probably represents thermal radiation from an accretion disk around the central compact object (Sect. 5.6.6).

Observations in the radiofrequency range with *high angular resolution* show that many quasars consist of two or more components, usually a compact central source ($\ll 1''$) and a more extended, long, thin source (jet) directed away from the center, similar to the structure of the central region of Cyg A (Figs. 5.6.3 and 10).

For the nearest quasar, 3 C 273, optical and radio images both show that a thin jet (component 3 C 273 A) stretches out radially from the "actual" quasar to a distance of about 20″ (80 kpc). Since it cannot have propagated outwards at more than the velocity of light, its age must be at least of the order of 10^6 yr.

The actual radio quasar consists of a radio halo 80 pc across (3 C 273 B); within it, there are several smaller, variable components ($\lesssim 10$ pc) and a compact, unresolved component of $\lesssim 0.0004''$ or $\lesssim 1.5$ pc. In the frequency range $\gtrsim 10$ GHz, the compact component emits the greatest portion of the radiation flux; the surprising *smallness* of the central component of 3 C 273 and other quasars, which can evidently be regarded as the real sources of their energy and their activity, is verified by the *time variation* of the optical and radiofrequency radiation (with an amplitude of 0.2 to 1 mag); it has characteristic times of the order of weeks to years. The extension of the radiating region can thus be at most a few tenths of a light year.

Very-long-baseline interferometric observations, in 3 C 273 and some other more distant quasars and radio galaxies, of the time variations in the structure of the radio sources reveal motions which apparently occur at *faster-than-light velocities* (Fig. 5.6.10). These apparent velocities $v_\perp > c$ are explained by the relativistic Doppler effect: if a beam of relativistic particles is moving with $v \simeq c$ nearly directly towards the observer (at a small angle δ to the line of sight), its forward-directed synchrotron radiation (Fig. 5.6.1) is emitted by a source (as

[19] Simultaneous observations of 3 C 273 in different wavelength ranges from the radio to the X-ray region show a complex variability of the intensity and the spectral index, which we cannot discuss in detail here.

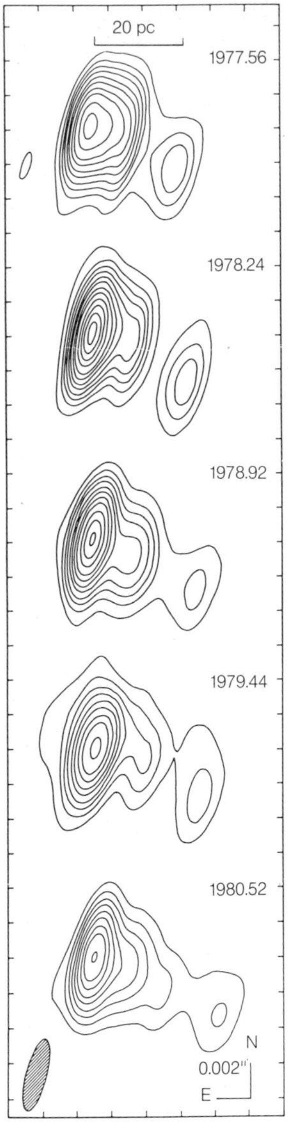

Fig. 5.6.10. The quasar 3 C 273, observations of the compact component B by very-long-baseline interferometry at 10.65 GHz (λ = 2.8 cm) in the years 1977 to 1980. The expansion of the source of $0.76 \cdot 10^{-3}{}'' \mathrm{yr}^{-1}$ corresponds to an apparent velocity of about 11 times the velocity of light ($H_0 = 50\,\mathrm{km\,s^{-1}\,Mpc^{-1}}$). The shaded ellipse indicates the resolution of $4.2 \cdot 1.2$ milliseconds of arc. From T. J. Pearson et al., Nature **290**, 365 (1981). (Reprinted by permission from Nature, **290**, 365. Copyright © 1981 Macmillan Magazines Ltd)

seen by the observer) which moves on the sphere with the apparent velocity:

$$v_\perp = \frac{v \sin\delta}{1 - \frac{v}{c}\cos\delta} \simeq \frac{v\delta}{\frac{\delta^2}{2}\frac{v}{c}} \simeq \frac{2c}{\delta} \tag{5.6.10}$$

(where $\cos\delta \simeq 1 - \delta^2/2$). In the case of 3 C 273, for example, the increase of the distance between two components observed in 1977/80 with $v_\perp/c \simeq 11$ corresponds to an observation angle of $\delta \simeq 2/11$ or about 10°. Furthermore, for a beam directed towards the observer, there is an amplification of the intensity relative to a source at rest by a factor of $8\gamma^3$, and for a beam perpendicular to the observer by a factor of γ^3, where $\gamma = E/m_0c^2 \gg 1$ (E = energy of the relativistic electrons). Conversely, the radiation from a jet ejected away from us is attenuated by a factor of $8\gamma^3$. We can thus understand, in principle, both the apparent faster-than-light velocities and also the frequent occurrence of only *one* jet in the radio sources, in terms of this relativistic effect.

The *total energy* of the synchrotron electrons, the associated heavy particles, and the magnetic field which must be present in a quasar can, again, as for the radio galaxies (Sect. 5.6.2), be estimated under the assumption of approximate equipartition; the result is about $10^{55}\,\mathrm{J} = 10^{62}\,\mathrm{erg}$. This quantity of energy corresponds to the entire relativistic rest energy m_0c^2 of a mass of $10^8\,\mathscr{M}_\odot$! The total *radiated power* of several $10^{39}\,\mathrm{W} = 10^{46}\,\mathrm{erg\,s^{-1}}$ (or $\simeq 10^{13}\,L_\odot$) typical of quasars could therefore be supplied by this reserve of energy for at most 10^8 yr.

As a result of this enormous energy requirement of the quasars, some astronomers have questioned the determination of their distances on the basis of Hubble's red shift relation. The large red shifts are interpreted by them in some other manner, and the quasars are taken to be "local" phenomena, which for example might have been formed by ejection of matter from other galaxies. In particular, these astronomers point out the frequent occurrence of quasar and galaxy pairs and groups with very small angular spacings, which thus apparently belong together, but which have very different red shifts. A satisfactory statistical investigation of such coincidences has yet to be performed.

On the other hand, a series of other observations speaks convincingly in favor of the *cosmological interpretation* of the red shifts of the quasars:

For one thing, many quasars also belong to groups and clusters in which the normal galaxies have the *same* red shifts as the quasars.

Then, especially, the *direct* observation of the mother galaxies of quasars with $z \lesssim 0.5$ clearly supports the cosmological interpretation of the red shift. Their luminosities agree with those of nearer large E- and S-galaxies precisely when the Hubble relation is applied to the distance determination of the quasars from their z values.

Finally, in every respect (energy flux and content, radio structure, activity in the nucleus, belonging to clusters of galaxies), there is a continuous transition from the

Seyfert and radio galaxies to the quasars. This is also apparent in the Hubble diagram (Fig. 5.9.3).

A further indication of the great distances to the quasars is given by the "*twin quasars*" QSO 0957+561 A and B discovered in 1979 by D. Walsh, R. F. Carswell, and R. J. Weymann, as well as similar objects observed more recently. The two quasars A and B, both with magnitudes of $m_V \simeq 17$ mag, are only 5.7″ apart and have *identical* red shifts ($z = 1.41$) and spectra. It is tempting to interpret them as images of a *single* quasar, whose light rays have passed through a "gravitational lens" on the way to us, i.e. they have been bent in such a way by the gravitational field of a massive galaxy that two images resulted. According to the theory of general relativity, (Sect. 4.12.9), a beam of light which passes a (pointlike) mass $\mathcal{M}$ at a distance a is deflected by an angle θ (in arc measure):

$$\theta = 2\frac{R_S}{a} = 4\frac{G\mathcal{M}}{ac^2} , \qquad (5.6.11)$$

where $R_S \simeq 3\,\mathcal{M}/\mathcal{M}_\odot$ [km] is the Schwarzschild radius. For example, for a galaxy with $10^{12}\mathcal{M}_\odot$ ($R_S \simeq 0.1$ pc) at a distance of 10 kpc, the deflection angle is $\theta \simeq 2\cdot 10^{-5}$ or 4″. Taking into account the *extended* mass distribution of the galaxy, we find that a gravitational lens produces three (or more) images, each having in general a different intensity. For QSO 0957+561, the component B has in the meantime been resolved into two images with a spacing of $\simeq 0.1''$, and furthermore the galaxy responsible for the deflection, with $z = 0.39$ (i.e. at a distance of $\simeq 2300$ Mpc) has been identified on the line connecting A and B, and its mass has been determined to be about $10^{12}\mathcal{M}_\odot$.

Deflection by a gravitational lens leads to *time-of-flight differences* in the different wavefronts which reach the observer from a quasar. A noticeable variation in the brightness of the quasar thus appears to the observer at different times in the different images. Now, if it is possible to identify the intensity variations in the different images as due to *one and the same* event in the source (and this is, in fact, the difficult part), then from the time difference and the angular spacing of the images, the *distance* to the quasar can be determined independently of the red shift z. This method is therefore independent of the Hubble relation (5.9.2) and can be used to determine H_0. In practice, the clear-cut identification of "associated" brightness variations in the different images is difficult. In 1989, for the first time, a time-of-flight difference of 1.1 yr was detected in the case of QSO 0957+561; the data are, however, still too imprecise to allow a determination of H_0.

In the spectra of many quasars, numerous faint, sharp *absorption lines* are observed, which for more distant quasars can be grouped according to their relative intensities into several systems with different absorption red shifts z_{abs}; as a rule, $z_{abs} \le z$, where z is the red shift of the quasar as determined from its emission lines. If z_{abs} differs only slightly from z, the absorption must be due to matter in the immediate neighborhood of the quasar, while differences $|z - z_{abs}|/z \ge 0.01$ can be attributed to absorbing matter in the outer halos of several galaxies along the line of sight between the observer and the quasar. For the more distant quasars (Fig. 5.6.8), there are numerous sharp absorption lines on the short-wavelength side of the Lα emission line. These are for the most part identifiable as Lα absorption at different red shifts ("Lα forest"), which originates in a large number of hydrogen concentrations in the intergalactic medium, and in the halos of galaxies along the line of sight.

Closely related to the quasars are the *BL Lacertae objects*, of which about 100 are known. They are named for BL Lac = 2200+420 ($m_V = 14.5$ mag, $z = 0.069$), which was for a long time "unrecognized" and was listed as a variable star in the catalogues. The characteristics of BL Lac objects are, for one thing, that emission lines in their spectra are unobservable or very faint, making a distance determination rather difficult. For another, the BL Lac objects exhibit a strong irregular variability of ≥ 1 mag over all wavelength ranges on a time scale of days to months. Their optical nonthermal continuum is steeper than that of the quasars and shows strong time-variable ($\le 30\%$) linear polarization.

5.6.6 Activity in the Nuclei of Galaxies

The observations of the nuclear regions of "normal" galaxies, Seyfert galaxies, radio galaxies, quasars, etc. described in the preceding sections allow us to conclude that although their activities differ quantitatively, they exhibit so many basic similarities that we are clearly dealing here with a single, *unified* phenomenon.

The following are characteristics of the *activity* in galactic nuclei: (a) the presence of a *compact* nucleus, which is brighter than in other galaxies of the same Hubble type; (b) *nonthermal* (optical) *continuum* radiation from the nucleus; (c) *emission lines* from the nucleus, which show signs of nonthermal excitation, having for example intensity ratios different from those in H II regions; (d) *nonthermal radio emissions;* and (e) *variability* of the continuum and/or of the emission lines.

While only some of these characteristics occur in the weaker forms of activity, we observe in the quasars, the radio galaxies, and many of the Seyfert galaxies the whole palette of activity phenomena.

Although the understanding of the precise physical nature of galactic activity and its causes is still fraught with unsolved problems, we can nevertheless attempt to put together a picture of the essential processes of activity in galactic nuclei.

The large-scale structure of a galaxy with an active nucleus is typified by its nonthermal *radio emissions.* Following a suggestion of P. A. G. Scheuer and A. C. S. Redhead (1972), the various forms of isophote pictures can be understood in a unified model: as we have already described for the case of the radio galaxies (Sect. 5.6.2), the energy of a *central source*, which coincides with the compact radio nucleus, is transported outwards to the outermost regions of the galaxy by two oppositely-directed *jets* of relativistic particles or plasma. In some cases, interaction of the jets with the surrounding medium produces intense radio emissions in "hot spots". The spectra from the radio to the X-ray regions are mainly ascribable to synchrotron radiation from relativistic electrons. The variety of appearances of the radio emissions is essentially determined by the *angle* δ of the jets to the line of sight, and by their *strengths.* At high energies and very small δ values, only *one* jet is observable (5.6.10). Presumably, δ decreases in going from the quasars without radio emission to the radio quasars to the BL Lac objects, while in the case of extended sources such as Cyg A, $\delta \simeq 90°$.

The gas in the neighborhood of the central sources is ionized by their *nonthermal continuum radiation* in the optical, ultraviolet, and X-ray regions; it can then be observed by means of its *emission lines.* As in the H II regions or in planetary nebulas, the degree of ionization and excitation depends on the strength of the ionizing UV flux ($\lambda \leq 91.1$ nm) and on the electron density n_e of the gas; however, here the ionization is due to *nonthermal* ultraviolet and X-ray radiation, $F_\nu \propto \nu^{-\alpha}$. The essential properties of the emission lines from the various active galaxies can be explained in terms of this *photoionization model* (whose details we cannot treat here), if the assumption is made that "gas clouds" move around the central source and cover only about one-tenth of its area. In this model, *narrow* lines are emitted in clouds having low electron densities ($n_e \simeq 10^{10}\,\mathrm{m}^{-3}$) which are relatively distant from the source (100 pc). In contrast, *broad* lines are emitted by clouds nearer to the source with diameters of about 10 to $10^3 R_\odot$, electron densities of 10^{13} to $10^{16}\,\mathrm{m}^{-3}$, and electron temperatures of around 10^4 K. In the Sy 1 galaxies and broad-line radio galaxies, these clouds are about 0.1 pc from the source; in quasars, they are ≥ 1 pc from the source. The central mass is estimated to be of the order of $10^9 \mathscr{M}_\odot$. In the BL Lac objects, the very weak emission lines can be explained as being due to a lack of gas in the neighborhood of the source.

The question of the *abundances* of the chemical elements in the gas clouds is of considerable interest. It is found, surprisingly, that Seyfert galaxies and quasars contain fairly normal cosmic matter; the derived abundances differ by no more than a factor of 3 from those of the solar mixture.

The primary question is of course that of the *origin* of the *quantity of energy* of $\leq 10^{55}$ J which must be contained in active galactic nuclei. We have already seen in connection with the quasars that nuclear energy is by no means sufficient to explain it. Even if, as a limiting possibility, the whole rest energy $\mathscr{M}c^2$ of a mass $\mathscr{M}$ could be made "usable", it would require about $10^8 \mathscr{M}_\odot$, i.e. a considerable portion of the mass of the galactic nucleus. The only process currently known to physics which can free a considerable fraction of the rest energy of a cosmic mass $\mathscr{M}$ is the release of (potential) *gravitational energy* by contraction or collapse. For example (this is to be understood only in the sense of a rough estimate!), if the originally thinly-spread matter contracts to a homogeneous sphere of radius R, then the gravitational energy released is

$$E_G \simeq \frac{G\mathscr{M}^2}{R} \, . \tag{5.6.12}$$

If we now require that $E_G \simeq \mathscr{M}c^2$, then (cf. Sect. 4.12.9) R must be of the order of the *Schwarzschild radius* R_S, i.e. the collapsed matter must assume an extremely compact structure. For $10^8 \mathscr{M}_\odot$, the radius $R_S \simeq 10^{-5}$ pc $\simeq$ 2 AU. Possibilities under discussion for the massive central objects are extremely dense star clusters, rapidly rotating massive disks, and black holes. For realistic estimates, an efficiency of less than 100% must be assumed for the release of the rest energy, so that the required masses would be more nearly in the range of 10^9 to $10^{10} \mathscr{M}_\odot$. From observations (very-long-baseline interferometry as well as the time variation of the radiation), the upper limit for the size of active galactic nuclei is found to be $\lesssim 10$ AU, in general agreement with the above estimate.

The *radiated energy* or luminosity L of several 10^{41} W ($3 \cdot 10^{14} L_\odot$) must finally be maintained by *accretion of matter* on the compact central object at a sufficiently high rate $\dot{\mathcal{M}}$. Where this matter comes from is still unexplained. Probably near collisions of galaxies play an essential role; the interaction between the galaxies could allow large amounts of gas to accumulate in the central regions (compare also Sects. 5.5.5 and 5.7.3). Finally, we estimate the required *accretion rate* $\dot{\mathcal{M}}$, whereby it is not important for our purposes to distinguish whether the material is falling in a spherical distribution or within an accretion disk:

$$L = \varepsilon \dot{\mathcal{M}} c^2 \ . \tag{5.6.13}$$

Here, ε is the efficiency ($0 \leq \varepsilon \leq 1$) for the conversion of mass to radiation energy. If we assume, somewhat arbitrarily, $\varepsilon = 0.1$, then matter would have to fall onto the central source at a rate of

$10 \ldots 100$	$\mathcal{M}_\odot\,\mathrm{yr}^{-1}$	in quasars,
≤ 0.1	$\mathcal{M}_\odot\,\mathrm{yr}^{-1}$	in Seyfert galaxies, and
$\leq 10^{-3}$	$\mathcal{M}_\odot\,\mathrm{yr}^{-1}$	in our Milky Way,

in order to supply the observed luminosity. In a time period of 10^8 yr, in the extreme case of the quasars, for example, 10^9 to $10^{10} \mathcal{M}_\odot$ would have to be accreted by the central source.

5.7 Clusters of Galaxies and Superclusters

Galaxies frequently belong to larger systems which are held together by the mutual gravitational attraction of their members. We can identify large or rich *clusters* containing up to several thousand galaxies, and smaller *groups* of decreasing size down to the *binary galaxies*. Among the galaxies brighter than 21 mag, some 10000 clusters and/or groups are known. A few noticeable concentrations of galaxies were studied as early as 1902/06 by M. Wolf, and later (around 1924) by C. Wirtz; systematic surveys of the distribution of galaxies on the celestial sphere were carried out from about 1930 on by H. Shapley, E. Hubble, and others. The catalogue of G.O. Abell (1958), with 2712 rich clusters of galaxies, and the Catalogue of Galaxies and Clusters by F. Zwicky et al. (1960/68) are both based on the Palomar Sky Survey.

In addition to the galaxies, many clusters also contain a tenuous, very hot gas, from which the thermal radiation in the X-ray region can be observed.

In Sect. 5.7.1, we first take up the classification of the various galaxy clusters, and their masses; then, in Sect. 5.7.2, we deal with the intergalactic cluster gas. Especially in the central regions of the denser clusters, "collisions" and near-misses of galaxies are frequent; the primary result of these events is a sweeping-out of the gas from the galaxies involved. Occasionally, two galaxies also merge to form a larger system. In Sect. 5.7.3, we give an overview of the interactions between galaxies and the evolution of the clusters to which they give rise.

As we shall see finally in Sect. 5.7.4, the clusters of galaxies themselves are not uniformly distributed in space, but instead form connected structures, mostly flat or long and filamentary: the *superclusters* of galaxies. These superclusters and clusters define the large-scale structure of the universe, making up a network of connected filaments and disks which surrounds vast, roughly spherical, nearly empty regions.

5.7.1 The Classification and the Masses of Clusters of Galaxies

It is not easy to distinguish clusters of galaxies, especially the more open ones, from chance density fluctuations in the general "field" of galaxies. G.O. Abell has defined a measure for the population of a cluster, its *richness*, by counting all the galaxies within an arbitrarily chosen interval of magnitudes (2 mag from the third-brightest galaxy) and within a fixed distance from the center of the cluster. For this distance, he chose $1.7/z$ minutes of arc (z: red shift), corresponding to 3 Mpc for a Hubble constant of $H_0 = 50\,\mathrm{km\,s^{-1}\,Mpc^{-1}}$ (Sect. 5.9.1).

Following Abell, we divide the clusters of galaxies into regular and irregular clusters. The *regular clusters* (corresponding approximately to the *compact* clusters in the scheme of F. Zwicky), contain large numbers of galaxies, have a roughly spherical shape, and exhibit a strong concentration of the galaxies towards the center of the cluster. They contain predominantly elliptical and lenticular galaxies (ca. 70-80%), of which the brightest (cD- and giant E-galaxies) are found mainly near the center. The prototype is the Coma cluster (= Abell 1656) at a distance of about 130 Mpc, with several thousand galaxies in a region of about 8 Mpc diameter.

The *irregular clusters* show no noticeable concentration towards their centers and do not have symmetric shapes; sometimes, they exhibit several density concentrations. These irregular clusters, which also include the *groups* of galaxies, having fewer members, contain galaxies of all the Hubble types; in particular, in contrast to the regular clusters, they have a high percentage (ca. 50%) of S-galaxies. In the inner region of the Virgo cluster with a diameter of about 3 Mpc, whose center is at a distance of around 20 Mpc from the Sun, we can find more than a thousand galaxies. The local group falls within the outer zone of this cluster.

There are three other classification schemes in use for clusters of galaxies, which on the whole are parallel. A. Oemler (1974) defines the *cD clusters*, with predominantly bright elliptical galaxies, corresponding roughly to Abell's regular clusters, and the *spiral-rich clusters* (corresponding to the irregular clusters); in addition, as an intermediate type, he introduced the *spiral-poor clusters*, dominated by S0 galaxies. The cluster types I to III according to L. P. Bautz and W. W. Morgan (1970) correspond to a series of clusters ranging from those dominated by a bright central galaxy (BM I) to the clusters without a predominant galaxy (BM III). H. J. Rood and G. N. Sastry (1971) give a finer classification scheme, which is based on the type and arrangement of the ten brightest galaxies of the cluster.

The diameters of the larger clusters of galaxies are in the range of 3 to 10 Mpc and are thus comparable with the sizes of the largest known radio galaxies such as 3 C 236 (Sect. 5.6.2). The *mass* $\mathcal{M}$ of a cluster of galaxies can be estimated with the help of the virial theorem, in a manner analogous to the mass determination of elliptical galaxies (5.5.13):

$$(\Delta v)^2 \simeq \frac{G\mathcal{M}}{R} . \tag{5.7.1}$$

Here, Δv is the dispersion of radial velocities within the cluster, and R is its radius. With typical values, $\Delta v \simeq 1000\ \mathrm{km\ s^{-1}}$ and $R \simeq 5$ Mpc, we obtain for the usual total mass:

$$\mathcal{M}(\text{cluster of galaxies}) \simeq 10^{15}\,\mathcal{M}_\odot . \tag{5.7.2}$$

This corresponds to an average mass-luminosity ratio for a cluster of galaxies of $\langle \mathcal{M}/L \rangle \simeq 200\,\mathcal{M}_\odot/L_\odot$. On the other hand, as was noted as early as 1933 by F. Zwicky, the summation of the "visible" masses of all the individual galaxies gives a much smaller total mass, only about one-tenth as large. The hot gas observed in clusters of galaxies by means of its X-ray emissions (see the following section) also contributes only about a tenth of the virial mass (5.7.2). This discrepancy can, however, be understood if the galaxies are assumed to have massive, dark halos, as is suggested e.g. by the observed rotation curves of many spiral galaxies (Sect. 5.5.3).

5.7.2 The Gas in Clusters of Galaxies

The clusters of galaxies form a class of strong extragalactic extended X-ray sources. Their spectra tell us that these X-ray emissions represent the *thermal emissions* of an extremely hot *gas* in the temperature range of 10^7 to 10^8 K. The optical thicknesses in the X-ray region are very small ($\leq 10^{-3}$), and the gas densities lie in the range of 10^2 to $10^3\ \mathrm{m^{-3}}$.

The X-ray emission exhibits complex behavior which varies from cluster to cluster. At the one extreme, we find widely scattered, irregular emission regions, often containing individual galaxies; at the other end of the scale, we observe a relatively flat emission curve, frequently with a stronger concentration around a bright galaxy near the center of the cluster (Fig. 5.7.1). The structure of the X-ray emission is clearly correlated to the cluster type: thus, Abell's regular clusters such as the Coma cluster and Abell 85, with a low content of spiral galaxies, have X-ray emissions with a smooth contour and a relatively high X-ray luminosity, $L_X \geq 10^{37}\ \mathrm{W} = 10^{44}\ \mathrm{erg\ s^{-1}}$. In contrast, the emissions from the irregular clusters which are rich in spiral galaxies, such as the Virgo cluster or Abell 1367, are less regular and have smaller luminosities ($L_X \leq 10^{37}$ W). The gas in the irregular clusters is somewhat cooler than the hot gas in the regular clusters, at 1 to $4\cdot 10^7$ K as opposed to $\geq 6\cdot 10^7$ K. Furthermore, the X-ray emissions from the clusters containing a bright cD-galaxy (such as Abell 85) are more strongly concentrated towards the center of the cluster (radius of the emission region ≤ 0.3 Mpc) than in the clusters which lack a dominant galaxy.

In a cluster of galaxies in dynamic equilibrium, both the individual galaxies and the atoms of hot gas move in the same common gravitational potential, and the spatial extent of each of the components is determined by its mean kinetic energy or temperature. From the observations in many clusters of the velocity dispersion Δv of the galaxies and the gas temperature T, it is known that the ratio β of the kinetic energy of the galaxies per unit mass, $\frac{1}{2}(\Delta v)^2$, to that of the gas, $3/2 \cdot kT/\mu m_H$,

$$\beta = \frac{\mu m_H (\Delta v)^2}{3kT} , \tag{5.7.3}$$

is of the order of magnitude of one ($\mu \simeq 0.6$: mean atomic mass of the completely ionized gas, m_H: mass of a hydrogen atom). This means that the hot gas and the "galaxy gas", in which we can regard the individual galaxies as "atoms", have about the same temperature and the same relative density curve. For example, $\Delta v = 1000\ \mathrm{km\ s^{-1}}$ corresponds to a temperature of about $T \simeq 10^8$ K.

To be sure, probably only the regular clusters are in dynamic equilibrium, but for all clusters down to the groups of galaxies, a correlation has been observed between the velocity dispersion Δv of the galaxies and both the gas temperature T and the X-ray luminosity L_X.

Observations of continuum and line emissions (e.g. Fe XXIV+XXV, 6.7 keV, and O VIII, 0.65 keV) in some nearby clusters of galaxies have shown that each of the individual galaxies is surrounded by its own X-ray halo. In the Virgo cluster, particularly the giant elliptical galaxies M 87 and M 86 are surrounded by gas at temperatures from several 10^6 to $3 \cdot 10^7$ K (see also Sect. 5.5.3); it is thus somewhat cooler than the general cluster gas.

An indirect method of detecting hot gas in clusters of galaxies was suggested by R. A. Sunyaev and Ya. B. Zeldovich (1970/72). A photon of the 3 K background radiation (Sect. 5.9.4) obtains a slightly increased energy, with a relative contribution on the order of kT/m_0c^2 ($m_0c^2 = 511$ keV, the rest energy of the electron; $T =$ temperature of the hot gas), through scattering from an electron in the cluster gas; for example, at $T = 10^8$ K or $kT = 10$ keV, this relative energy increase is 0.02. The scattering cross-section of this *inverse Compton scattering* is the Thomson scattering coefficient $\sigma_T = 6.65 \cdot 10^{-29}\ \mathrm{m}^2$, so that a photon moving through a distance l in a hot gas of electron density n_e experiences an energy or frequency increase given by:

$$\frac{\Delta \nu}{\nu_0} \simeq \left(\frac{kT}{m_0 c^2}\right) \sigma_T n_e l . \tag{5.7.4}$$

This energy shift corresponds to a decrease of the radiation temperature $T_0 \simeq 3$ K on the low-frequency side of the maximum of the 3 K radiation, according to the

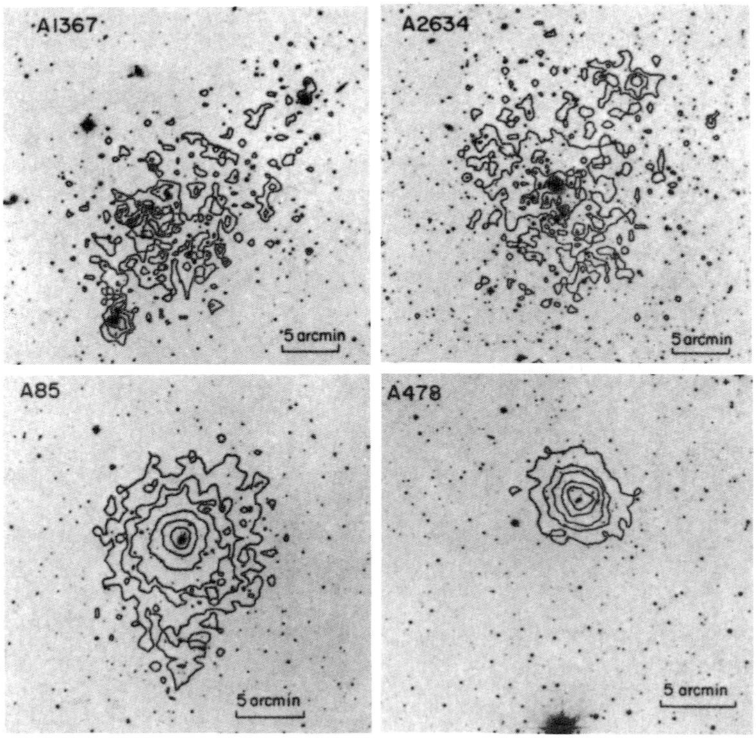

Fig. 5.7.1. X-ray emission from clusters of galaxies: observations with the Einstein Satellite by C. Jones et al. (1979). Lines of equal X-ray intensity are superposed onto the optical photographs from the Palomar Sky Survey. The clusters Abell 1367 and 2634 have irregular contours which often include individual galaxies, while Abell 85 and 478 show relatively smooth X-ray emissions, strongly concentrated around a bright galaxy in the center of the cluster

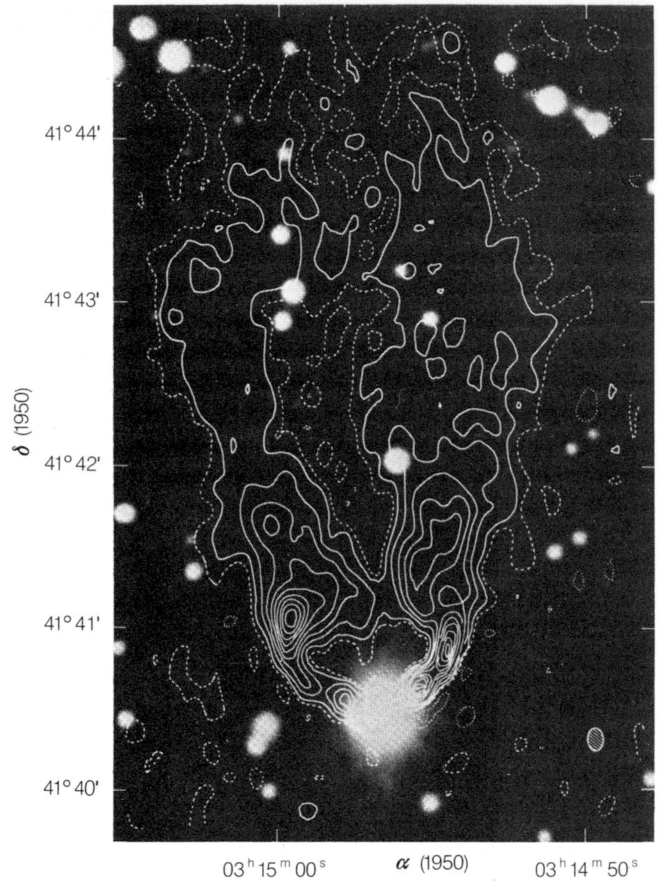

Fig. 5.7.2. The head-tail galaxy NGC 1265: observations at $\lambda = 6$ cm with the Westerbork Synthesis Radiotelescope. This overlay of the radio isophotes onto the optical image allows us to recognize that the radiofrequency emission regions were evidently ejected from the elliptical galaxy at the head. K. J. Wellington et al., Nature **244**, 502 (1973). (Reprinted by permission from Nature, **244**, 502. Copyright © 1973 Macmillan Magazines Ltd.)

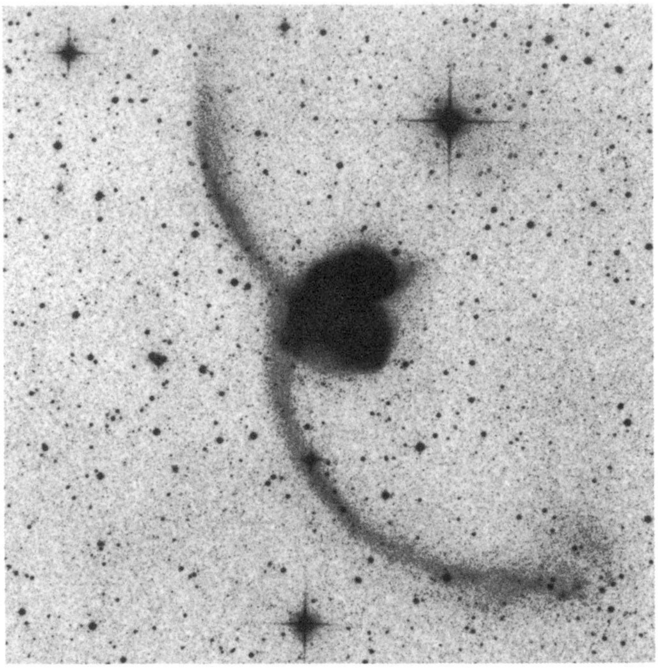

Fig. 5.7.3. The interacting galaxies NGC 4038/9 (= Arp 244 = Vorontsov-Velyaminov 245, the "feeler" galaxies or "antennas") in a photograph taken by O. Pizarro and H.-E. Schuster with the 1.62 m Schmidt camera of the European Southern Observatory (ESO). The projected distance between the ends of the "feelers" is about 160 kpc, and that between the centers of the two galaxies is about 9 kpc. (From S. Laustsen et al., 1987)

Rayleigh-Jeans law (4.2.24b), of $\Delta T_0/T_0 = -2\Delta\nu/\nu_0$. Therefore, the intensity of the background radiation in the center of a cluster of galaxies (with $n_e = 3\cdot 10^3\,\mathrm{m}^{-3}$, $L = 0.5$ Mpc) is decreased corresponding to $\Delta T \simeq 3\cdot 10^{-4}$ K. This extremely small effect could actually be observed for several clusters.

Finally, the existence of a hot gas in clusters of galaxies can also be concluded from the structure of a particular class of radio galaxies, the "*head-tail galaxies*" (Fig. 5.7.2), which occur in dense clusters. The supersonic motion of the galaxy through the cluster gas pushes the ejected relativistic electrons and magnetic fields backwards. From the observed geometry of the Mach cone, the velocities and pressures in the flow field can be estimated, and thus also the temperature and density of the cluster gas. The results agree with those obtained from X-ray emission.

5.7.3 Interacting Galaxies. The Evolution of Clusters of Galaxies

In near collisions of two galaxies, their mutual tidal interactions have a strong influence on the stars and especially on the interstellar gas within the galaxies. The atlases of peculiar or interacting galaxies by H. C. Arp (1966) and B. A. Vorontsov-Velyaminov (1959 and 1977) contain numerous examples of double and multiple systems of *interacting galaxies* with common shells, connecting bridges, or long, drawn-out tails of stars and glowing gas (see Fig. 5.7.3 and the picture opposite the title page). In radioastronomy, also, we find indications of interactions, for example from observations of the

21 cm line of neutral hydrogen, among other things in the form of connecting bridges or common gas shells: the "*Magellanic stream*", six H I clouds which follow the Large Magellanic Cloud on its orbit around the Milky Way, was presumably "drawn out" by tidal forces and is now falling into the Milky Way.

In Sect. 5.5.5, we have already interpreted the strong radiation in the far infrared shown by many galaxies as being due to tidal disturbances from near collisions of galaxies, which lead to an enormous increase in the rate of *star formation* within these *infrared galaxies* (Figs. 5.5.16 and 17).

Surveys of the galaxies searching for the various morphological indications of disturbances, i.e. for bridges, filaments, tails, deformations, ring-shaped galaxies, shell structures (in elliptical systems), double and multiple cores, etc., have shown in recent times that interactions are not rare occurrences, but instead represent a widespread phenomenon. Probably, in fact, every galaxy was at one point in its evolution subjected to strong tidal perturbations due to another galaxy passing nearby.

The interaction between galaxies in the *clusters of galaxies* is particularly important. Especially in the denser central regions of the regular clusters, within the Hubble time H_0^{-1} many near collisions of galaxies take place, and they have caused noticeable changes in the structures and galaxy contents of these clusters since they were originally formed. The various types of clusters of galaxies probably represent different phases in a *dynamic evolution*.

For one thing, a collision results in the loss of interstellar gas from the affected galaxies, and thus finally also in the loss of their spiral arms. The S-galaxies thus become lenticular galaxies, possibly via intermediate types such as the *anemic* (*A*-) galaxies, with very weak spiral arms, which were observed in the Virgo cluster by S. van den Bergh (1976).

The interstellar gas, heated by the collision, finally collects for the most part at the minimum of the gravitational potential at the center of the cluster; only particularly massive galaxies can retain the gas through their own gravitation. If a galaxy passes through this cluster gas at a supersonic velocity, its interstellar gas is likewise swept out, so that additional spiral galaxies in the cluster can be "annihilated" in this way. The denser regular clusters, with their stronger X-ray emissions and their small fraction of spiral galaxies, thus represent a dynamically further-developed stage than do the irregular galaxy clusters.

Analysis of the 6.7 keV X-ray emission line of Fe XXIV +XXV supports the above picture of the origin of the hot cluster gas. In nearly all clusters of galaxies, practically independently of the details of the model assumptions, an *iron abundance* is derived from this emission line which is about half the solar value. This relatively high abundance can be understood within the framework of our ideas about the formation of the chemical elements (Sect. 5.8.5) only if we assume that the gas came from the stars within the galaxies, i.e. is not of primordial origin.

Compared to the interstellar gas, the majority of the *stars* are considerably less influenced by a meeting between galaxies. However, in the case of an approximately central collision of two galaxies, the result can be that they merge into a single larger system. This "*galactic cannibalism*" was probably decisive for the formation of the cD-galaxies at the centers of clusters of galaxies; the occasional occurrence of two or more "nuclei" in these galaxies provides observational support for this hypothesis.

5.7.4 Superclusters of Galaxies

The very noticeable concentration of the apparently bright galaxies within the northern galactic hemisphere, especially in the constellation Virgo, led early to speculations that here, the clusters of galaxies belong to a still larger system, called the *local supercluster* or the *Virgo supercluster*; in particular, G. de Vaucouleurs studied this supercluster in detail in the 1950's. G. O. Abell noted that many of the rich clusters of galaxies in his catalogue seem to be members of superclusters. A quantitative indication of the occurrence of superclusters was given by the investigations by P. J. E. Peebles et al. of the two-point correlations of the distribution of galaxies projected onto the celestial sphere. A breakthrough in the observation of superclusters and a convincing demonstration of their existence was, however, made only in 1976, when the development of sensitive detectors made it possible to measure radial velocities from the spectra of many distant, faint galaxies and thus to investigate also the *radial* structure of the cluster distribution.

A typical supercluster contains, along with several groups of galaxies, also a few rich clusters (about 2 to 6), and forms a strongly flattened system with a diameter of around 100 Mpc, lacking in axial symmetry or a central concentration. The frequently-occurring filamentary or chainlike arrangement of the galaxies is quite noticeable.

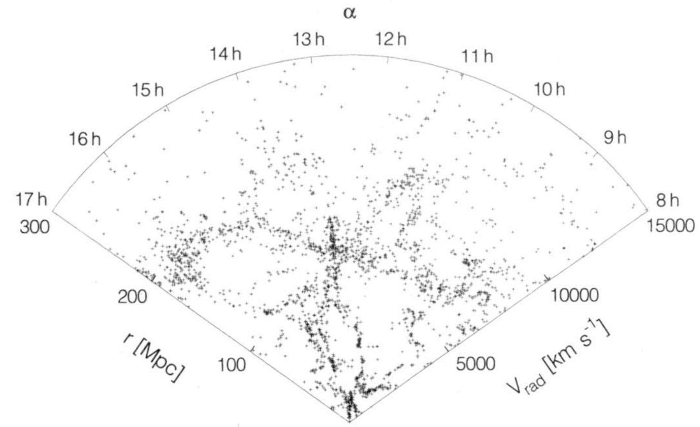

Fig. 5.7.4. The spatial distribution of 2500 galaxies out to a limiting blue magnitude of $m_B = 15.5$ mag, from the CfA Sky Survey (Harvard-Smithsonian Center for Astrophysics). The data are represented in a wedge diagram: the radial velocity $V_{rad} = cz$, the distance r, and right ascension (α between 8 h and 17 h) are plotted for a declination range between 26.5° and 44.5°. The distances are referred to a Hubble constant of $H_0 = 50$ km s^{-1} Mpc^{-1}. The group of galaxies at 13 h is the Coma cluster. The "Great Wall" is noticeable: it is a concentration of galaxies in a thin disk (≤ 10 Mpc thickness) at a distance of 140 to 200 Mpc, stretching over the whole range of right ascension. (Courtesy of Margaret J. Geller and John P. Huchra, Center for Astrophysics, Cambridge, MA)

Thus, the *Virgo supercluster* consists of the Virgo cluster itself, about 3 Mpc across and at a distance of 20 Mpc, of an irregularly-shaped ellipsoid of around 50 Mpc diameter with an axis ratio of 6 : 3 : 1, and of a "halo" of several galactic groups, including the *local group* at its edge. The *Coma supercluster* (Fig. 5.7.4), at a distance of about 140 Mpc, containing the Coma cluster and Abell 1367, among others, forms a flattened system (6 : 1) with a diameter of about 100 Mpc.

The characteristic time which a galaxy moving at a velocity of 1000 km s^{-1} requires to cross a supercluster of 100 Mpc diameter is 10^{11} yr and is thus longer than the age of the universe. The superclusters are therefore not relaxed systems, and the virial theorem cannot be used to estimate their masses. The direct summation of the masses of the individual members yields a typical mass for a supercluster of the order of magnitude of $10^{16} \mathcal{M}_\odot$; the overall luminosity is around $10^{14} L_\odot$.

A precise investigation of the structure of the superclusters is made difficult by the fact that their radial extension cannot be directly read off from the red shifts of their member galaxies. Instead, the nonuniform distribution of the galaxy clusters causes an acceleration of the galaxies towards the mass concentrations and thus invalidates the Hubble relation. For example, the local group is apparently falling towards the Virgo cluster at several 100 km s^{-1}.

The superclusters of galaxies are themselves not isolated systems in space, but instead touch each other at their edges and thus make up the *large-scale structure of the cosmos*, in the form of a network of connected flat disks and filaments (Fig. 5.8.4). These enclose roughly spherical regions which are nearly *free of galaxies* (voids), with a characteristic dimension of 50 Mpc. In the sky survey carried out by M. J. Geller and J. P. Huchra of the Harvard-Smithsonian Center for Astrophysics (1989), an extremely extended, contiguous structure was discovered, called the "Great Wall": it is a relatively thin (≤ 10 Mpc) layer which extends over at least 120 Mpc by 340 Mpc and contains a density of galaxies well above the average value (Fig. 5.7.4), at $\Delta\varrho/\varrho \simeq 5$. Its mass is estimated to be $2 \cdot 10^{16} \mathcal{M}_\odot$. Since this wall stretches out to the limits of surveys carried out up to the present, both in right ascension and in declination, its true extension and mass are probably still larger.

5.8 Galactic Evolution

The subject of the formation and evolution of galaxies is still fraught with riddles and unsolved problems. Even a superficial examination of the Hubble Atlas of Galaxies, by A. Sandage (1961), or in particular of H. Arp's Atlas of Peculiar Galaxies (1966), should serve as a "modest deterrent" to overeager theoreticians. We shall therefore limit ourselves here to an initial presentation of some mechanisms which seem to play an important role in the development of the galaxies: gravitational instability (Jeans) and the formation of star clusters and stars (Sect. 5.8.1); the dynamics of the spiral arms, especially the theory of density waves in the arms (Sect. 5.8.2); the formation of a galactic disk by collapse (Sect. 5.8.3); and some remarks about the development of galaxies (Sect. 5.8.4), insofar as this question is not dealt with under the topic of cosmology (Sect. 5.9). We then add in Sect. 5.8.5 some considerations of the explanation of the abundance distribution of the elements and of the chemical (nuclear) evolution of galaxies.

5.8.1 Jeans' Gravitational Instability. The Formation of Star Clusters and Stars

In Sect. 5.4, we have already shown that, and how, *stars* can be formed from interstellar matter in young galactic star clusters and associations. We now investigate the much deeper question of the conditions under which a mass of gas distributed over a region of space can become *unstable*, so that it is compressed under the influence of its own gravitational field. This process is followed, as we shall see, by a further splitting up of the matter, resulting in the formation of individual stars. Our question is answered by the criterion of *gravitational instability* discovered by J. Jeans (1902, 1928). We shall restrict ourselves to an estimate, which will allow us to recognize the important points; we first consider a roughly homogeneous sphere of radius R, density ϱ, and mass

$$\mathscr{M} = \frac{4\pi}{3}\varrho R^3 \ . \tag{5.8.1}$$

If this sphere is in equilibrium, then according to the *virial theorem* (2.6.25), the ratio of twice the kinetic energy, $2E_{\text{kin}}$, to the negative potential energy, $-E_{\text{pot}}$, is equal to *one*. If, in contrast, E_{kin} (i.e. the pressure in the interior of the sphere) is too small, or $-E_{\text{pot}}$ is too large, a gravitational instability is present and the mass collapses.

As we have already seen in Sect. 4.12.4, E_{kin} is equal to the thermal energy of the atoms or molecules (4.12.29); E_{pot} could be obtained by a simple integration (4.12.30). We thus find for the *stability limit*:

$$2\frac{E_{\text{kin}}}{-E_{\text{pot}}} = \frac{3kT}{\mu m_{\text{u}}}\mathscr{M} \Big/ \frac{3}{5}\frac{G\mathscr{M}^2}{R} = 1 \ . \tag{5.8.2}$$

As usual, G is the gravitational constant, k the Boltzmann constant, $m_{\text{u}} = 1.66 \cdot 10^{-27}$ kg is the atomic mass constant, T the absolute temperature, and μ is the molecular mass of the gas. A mass of gas can thus collapse only when its radius is *smaller* than the *Jeans radius*:

$$R_{\text{J}} = \frac{1}{5} G\mathscr{M}\frac{\mu m_{\text{u}}}{kT} \ . \tag{5.8.3}$$

With (5.8.1), we obtain immediately the result that its mass must be *larger* than the *Jeans mass*:

$$\mathscr{M}_{\text{J}} = 5.46\left(\frac{kT}{\mu m_{\text{u}} G}\right)^{3/2} \varrho^{-1/2} \ , \tag{5.8.4}$$

or, inserting the numerical values of the physical constants,

$$\frac{\mathscr{M}_{\text{J}}}{\mathscr{M}_{\odot}} = 3.82 \cdot 10^{-9}\left(\frac{T[\text{K}]}{\mu}\right)^{3/2} \varrho\,[\text{kg m}^{-3}]^{-1/2} \ . \tag{5.8.5}$$

If the gas is also subject to turbulent flow, we could insert simply the mean squared velocity $\langle v^2 \rangle$, obtained, for example, from the Doppler effect of a spectral line, instead of $3kT/\mu m_{\text{u}}$.

Our approach is still not very satisfying, because in reality we cannot assume an initially spherical distribution, but instead must consider a more or less inhomogeneous large mass of gas. This can be taken into account, according to W.H. McCrea (1957), by exerting an (imaginary) pressure on the surface of the sphere. This will favor its collapse and in many cases cause it to begin sooner. Equations (5.8.4) and (5.8.5) then are slightly modified, the values of the constants being somewhat changed.

As an important application, we investigate *instabilities of the interstellar gas*. We choose its density ϱ to be in the range 10^{-21} to 10^{-18} kg m^{-3} and its temperature T in the range 100 to 20 K (Sects. 5.3.3 and 4), and obtain from (5.8.5) a Jeans mass between about 10^5 and $10^2\,\mathscr{M}_{\odot}$. This means that the gas in the disk of a galaxy can initially form an object only of the order of a *star cluster*. Only when the matter is further compressed can the formation of individual stars begin (compare Sect. 5.4.7). The collapse of a mass of gas is favored by an external pressure, as is produced for example by the shock wave projected out from a supernova, or by a density wave in a spiral arm (Sect. 5.8.2).

Regarding many other problems, such as the formation of galaxies, of planetary systems, etc., we mention the following point: the *larger* the density ϱ of the original matter, the *smaller* will be the objects formed from it.

5.8.2 The Dynamics of the Spiral Arms. The Theory of Density Waves

Most galaxies which contain a rotating disk also have *spiral arms* embedded in the disk; the S0 galaxies are an exception. The spiral arms are highlighted by short-lived, bright OB stars, while the older red stars are more uniformly distributed throughout the whole disk.

The naive idea that a spiral arm always contains the *same* stars, gas clouds, etc., is refuted by considering that such a formation would be destroyed in the course of a few rotations of the disk, i.e. in several 10^8 yr, by the differential rotation.

B. Lindblad attempted many years ago to explain the persistence of the spiral structure over longer periods of time, in terms of a sectored system of *density waves.* In his model, the spiral arms of a galaxy are assumed to contain different stars, gas, etc. at different times, in a similar manner to an ocean wave whose crest is formed by constantly changing water particles. In order to justify the model, Lindblad and his coworkers carried out extensive calculations of the orbits of individual stars in the average gravitational field (potential field) of a galactic disk. These investigations were not well received, probably in part because Lindblad maintained for a long time that the spiral arms curve "forward", i.e. with the concave edge ahead. However, in many spiral galaxies, the "forward" and "backward" directions can be clearly distinguished, e.g. by observing the dark clouds, and it is found that the spiral arms in fact always "drag behind".

In the years following 1964, the *density wave theory* of the spiral structure has been revived by C.C. Lin et al., now however in the form of a continuum theory. If there is a location in the disk of a galaxy with, for example, a somewhat higher gas density, it will cause a perturbation (potential well) in the potential field describing the differential rotation of the disk. This in turn influences the velocities of the stars and thus the mass density of the "stellar gas". In order that the potential field, the gas density, and the stellar density all be mutually consistent, a (rather complicated) system of differential equations must be obeyed; it describes the structure and propagation of density waves in the galactic disk. From the manifold possible solutions, Lin and his coworkers selected *one*, corresponding to a quasistatic spiral structure. This structure moves, rigidly so to speak, with a constant angular velocity Ω_p. If this constant is fixed, then the form of the potential well can be calculated, and it then forms the basis for the spiral arm. For example, in our Milky Way system, observations give $\Omega_p \simeq 13.5\ \mathrm{km\ s^{-1} \cdot kpc^{-1}}$; i.e. in our neighborhood, the density wave moves around the galaxy with about *half* the velocity of the stars. The observable spiral arm (Fig. 5.8.1) is then produced, according to W.W. Roberts (1969), by the fact that the interstellar *gas* flows along the density wave from the concave edge (in our case at $\simeq 115\ \mathrm{km\ s^{-1}}$). A compression occurs in going towards the potential minimum, and it can be recognized initially by means of interstellar dust. W. Baade had already emphasized long ago that the dust arms provide the basis for the spiral structure. We have previously mentioned the radio images of M 51, which clearly show the corresponding amplification of radiofrequency emissions due to the compression of the magnetic field and to synchrotron electrons.

The compression of the gas leads furthermore to the formation of a *shock wave*, which in turn favors Jeans instabilities and therefore the development of *young stars* and *H II regions.* A narrow strip is indeed observed at the convex edge of the shock wave, containing bright blue stars and H II regions. It is followed by a broad, diffuse band with older stars and star clusters; finally, the old disk population is distributed nearly homogeneously. Quantitatively, Lin estimates the amplitudes of the gravitational fluctuations, and those of the gas and stellar densities, to be about 5% of the average values.

The density wave theory of the spiral structure has no doubt provided many new impulses, in particular for the evaluation of 21 cm observational data. Fundamental discrepancies between the theory and observations have so far not been found. On the other hand, it is readily apparent that the theory in its present form is still very incomplete. We mention several open *questions*, without

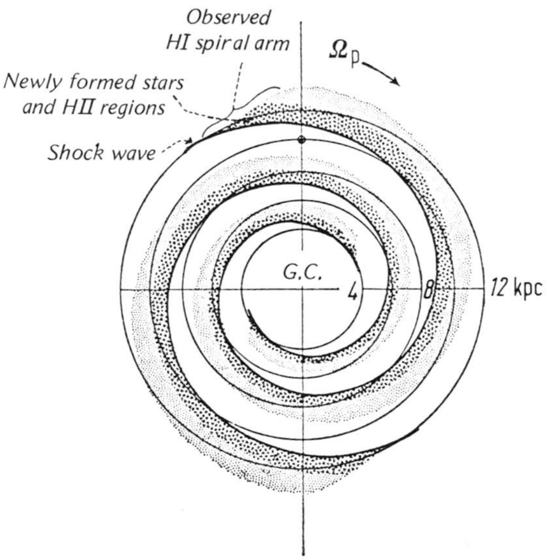

Fig. 5.8.1. The spiral structure of our galaxy (⊙ = Sun). The quasistationary spiral density wave according to C.C. Lin *et al.* (two arms) moves around at an angular velocity of $\Omega_p \simeq 13.5\ \mathrm{km\ s^{-1}\ kpc^{-1}}$. It is thus continually overtaken by the galactic matter, which moves about twice as fast (gas, stars, ...). This matter is compressed in the region of the potential well of the density wave. In the gas, dark clouds are produced by the compression, as well as an increased intensity of synchrotron radiation and a shock wave (fully-drawn curve). In the course of $\simeq 10^7$ yr young bright stars and H II regions are formed in this "shocked" gas. At a larger distance from the shock wave, older stars follow. The major portion of the potential well is filled with neutral hydrogen (H I; 21 cm radiation). From W.W. Roberts (1969)

going into the still provisional suggestions for their solution: why is *only one* quasistatic density wave observed? What determines its rotational frequency Ω_p? What is the origin of the initial *excitation* and of the *damping* of the density waves? In the case of the excitation, tidal interactions with neighboring galaxies (e.g. the Magellanic Clouds) or a rotating, "bar-shaped" mass in the central region of the galaxy have been suggested. Under what conditions does a galaxy *not* develop a spiral structure?

Given this list of questions, a quite different mathematical method is gaining in interest: using large computers, several research groups, beginning with R. H. Miller, K. H. Prendergast, and W. J. Quirk in 1968 and F. Hohl in 1970, have simulated the motions of several 10^5 stars under the influence of their mutual gravitation as an N-body problem, and have generated images of a series of the numerous successive states of such a system. The significance of this technique is to be found not least in the fact that it is now possible to *"experiment"* with various types of galaxies. In the cases thus far investigated, a spiral structure forms out of an initially homogeneous disk in the course of less than one rotation. It is not quasistationary, but instead changes and renews itself continually. The computer gives no answers to the question of why this happens, or the connection to Lin's theory. However, it can be safely assumed that while the hypothesis of the quasistatic spiral structure is not strictly correct, it is still a reasonable first approximation.

5.8.3 The Shapes of the Galaxies. The Formation of a Galactic Disk by Collapse

From the observed frequency distribution of the *apparent* axis ratios of galaxies of various Hubble types, the distribution of the *true* axis ratios q (= semiminor/semimajor axis) can be calculated, since the directions of the axes of rotation should be distributed randomly in space. A. Sandage, K. C. Freeman, and N. R. Stokes (1970), furthering the classic work of E. Hubble, found that the true axis ratios of the *elliptical* galaxies range between $q = 1$ to $q \simeq 0.3$, while *all* galaxies of the types S 0, SB 0 (without spiral arms) and Sa, Sb, Sc (with spiral arms) exhibit axis ratios in the limited range $q = 0.25 \pm 0.06$.

The radial distribution of *surface brightness*, $I(r)$, also differs for the two groups in a characteristic way.

The brightness distribution in *E-galaxies* and the globular clusters, which are similar in many respects, except for their masses, can be represented according to E. Hubble, G. de Vaucouleurs, and I. R. King by empirical formulas of the type:

$$\frac{I(r)}{I_0} = \frac{1}{1+(r/r_c)^2} \quad \text{or} \quad \log\frac{I(r)}{I_0} = -\left(\frac{r}{r_k}\right)^{1/4} \tag{5.8.6}$$

within certain limitations. Here, I_0 is the intensity at the center, and r_c and r_k denote the distance from the center at which the intensity has fallen to 1/2 or 1/10 of I_0, respectively.

This type of brightness distribution, or the corresponding density distribution, comes about due to the fact that for a high stellar density, the *relaxation time*, in which the stars reach equilibrium with a Maxwell-Boltzmann velocity distribution owing to interactions through near approaches, is relatively short. In the outer parts of the globular clusters and the dwarf E-galaxies, on the other hand, those stars which have too much energy and drift too far from the center are continually "plucked out" by the tidal forces from the nearby giant galaxy. Therefore, the cluster or the dwarf galaxy is cut off at a particular "tidal radius" r_t. From about $r > r_t/10$ outwards, equations such as (5.8.6) are thus no longer applicable. The approaches described here in schematic form can clearly also not describe the strong concentrations of mass near the nuclei of giant E-galaxies.

The spheroidal central components of the *spiral galaxies* can be approximated, like the elliptical galaxies, by (5.8.6). In contrast, in their *disks*, a brightness distribution given by

$$I(r) = I_0 e^{-r/r_0} \tag{5.8.7}$$

is observed. Since the mass-luminosity ratio does not vary much within a galactic disk (Sect. 5.5.3), we can conclude that a corresponding law holds for the radial distribution of the mass density per unit surface:

$$\mu(r) = \mu_0 e^{-r/r_0} \ . \tag{5.8.8}$$

The values of the *length scale factor* r_0 vary in the early galactic types (S0 to about Sbc) between 0.5 and 5 kpc; for galaxies of later types, the maximum r_0 values are around 1 kpc. According to K. C. Freeman (1970), all spiral galaxies exhibit a notably small scatter in their central surface brightnesses I_0 in the blue (21.65 ± 0.3 mag per square second of arc) and in their central mass densities μ_0. In this case, the exponential disk can be characterized by only *one* scale length r_0, and the luminosities ($\propto I_0 r_0^2$) would be determined by r_0 alone. It is, to be sure, not yet clear to what extent this result is affected by contributions from the spheroidal component.

While the more or less spherical systems evidently have little angular momentum, in the *disks*, in which all the stars rotate in the same sense around the center, there is a *large angular momentum*, as is to be expected from their degree of flattening. It is thus tempting to ask whether, and how, an (exponential) galactic disk can have formed from a more or less spherical cloud.

As the simplest model of such a *protogalaxy*, we consider a sphere of gas which is rotating rigidly at the constant angular velocity $\Omega = 2\pi/T$ (T = rotational period). Let $\mathcal{M}(r)$ be the mass within a sphere of radius r, R the overall radius, and $\mathcal{M} = \mathcal{M}(R)$ the overall mass. Then the ratio of the components of centrifugal force and gravitational force perpendicular to the axis of rotation is (Fig. 5.8.2):

$$\frac{\text{Centrifugal force}}{\text{Gravitation}} = \frac{\Omega^2 r \sin\theta}{G\mathcal{M}(r)\sin\theta/r^2} = \left[\frac{(\Omega r^2)^2}{G\mathcal{M}(r)}\right]\frac{1}{r}\;, \tag{5.8.9}$$

where G is again the gravitational constant.

We now allow our sphere to contract *radially*, in such a way that each mass element maintains its angular momentum. In particular, for a spherical shell of radius r, the angular momentum per unit mass, $\propto \Omega r^2$, and the mass $\mathcal{M}(r)$ inside the shell remain constant and the factor [...] in (5.8.9) does not change. But owing to the factor $1/r$, the ratio of centrifugal force to gravitational force increases until a new equilibrium is established between the force components perpendicular to the axis of rotation. The component of gravitational force parallel to the axis, however, remains unchanged. The protogalaxy must therefore become flattened, i.e. it must collapse in the direction of the rotational axis. The energy released in this process is mostly dissipated; in this way, a thin disk is finally formed. In order that a gravitational equilibrium (gravitation = centrifugal force) be maintained everywhere within the disk, a radial re-distribution of the mass must naturally take place. The rotation will then in general no longer be uniform.

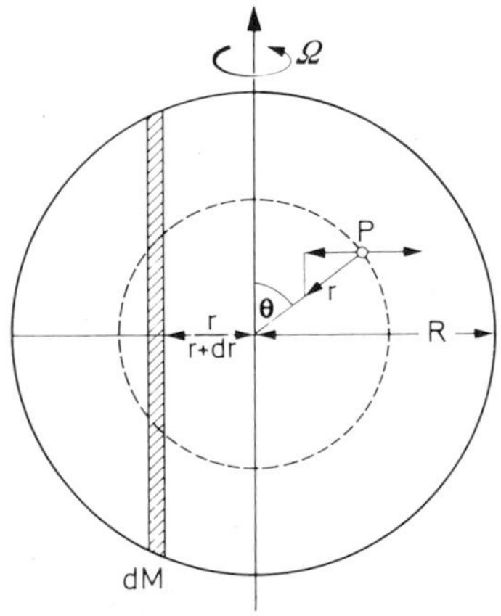

Fig. 5.8.2.
Model of a protogalaxy

We cannot follow these complex processes in detail, but we can expect that every mass element maintains its *angular momentum*. Thus, the fraction $d\mathcal{M}/\mathcal{M}$ of the mass whose angular momentum per unit mass lies in the interval h to $h+dh$ must be the same size in the finished galaxy as in the protogalaxy. Instead of $(d\mathcal{M}(h)/\mathcal{M})dh$, we could of course just as well consider the fraction of mass $\mathcal{M}(h)/\mathcal{M}$, after integration over h, in which the angular momentum per unit mass is less than a given value h.

In order to calculate these ratios, we take for lack of more precise knowledge, and as the simplest model of a protogalaxy, a homogeneous sphere of radius R, density ϱ, and mass $\mathcal{M}$. The angular velocity Ω can, from (5.8.9), and making use of

$$\mathcal{M}(r) = \frac{4\pi}{3}\varrho r^3 \quad\text{and}\quad \mathcal{M} = \frac{4\pi}{3}\varrho R^3 \tag{5.8.10}$$

be adjusted so that, at least within the entire equatorial plane, gravitational equilibrium holds, by setting

$$\Omega^2 = \frac{4\pi}{3}G\varrho = G\mathcal{M}/R^3\;. \tag{5.8.11}$$

The *angular momentum per unit mass* at a distance r from the axis is given by:

$$h(r) = r^2\Omega\;, \tag{5.8.12}$$

and the *total angular momentum* (moment of inertia $\times\,\Omega$) by:

$$H = \tfrac{2}{5}\mathcal{M}R^2\Omega = \tfrac{2}{5}(G\mathcal{M}^3 R)^{1/2}\;. \tag{5,8.13}$$

In order to calculate the distribution of the angular momentum, we cut a cylinder out of the sphere, parallel to the axis of rotation, with inner and outer radii r and $r+dr$. Its mass is then (see Fig. 5.8.2)

$$d\mathcal{M} = \varrho 2R\left(1-\frac{r^2}{R^2}\right)^{1/2} 2\pi r\,dr\ , \tag{5.8.14}$$

and its angular momentum is:

$$h\,d\mathcal{M} = r^2\Omega\,d\mathcal{M} = \Omega\varrho 2R\left(1-\frac{r^2}{R^2}\right)^{1/2} 2\pi r^3\,dr\ . \tag{5.8.15}$$

From (5.8.12), $dh = \Omega 2r\,dr$. Then with (5.8.10), we readily obtain the fraction of mass $d\mathcal{M}/\mathcal{M}$ whose angular momentum per unit mass lies in the interval from h to $h+dh$. For simplicity, we measure h in units of its maximum value ΩR^2 and write

$$h/\Omega R^2 = x\ ; \tag{5.8.16}$$

we then find

$$d\mathcal{M}/\mathcal{M} = \tfrac{3}{2}(1-x)^{1/2}\,dx\ . \tag{5.8.17}$$

By integration of this expression, we can obtain the fraction $\mathcal{M}(h)$ or $\mathcal{M}(x)$ of the mass for which the angular momentum per unit mass is $<x$:

$$\frac{\mathcal{M}(x)}{\mathcal{M}} = \frac{3}{2}\int_0^x (1-x)^{1/2}\,dx = 1-(1-x)^{3/2}\ . \tag{5.8.18}$$

For the *entire* sphere, by comparison, we have from (5.8.13) and (5.8.16):

$$\bar{x} = \frac{H}{\mathcal{M}\Omega R^2} = \frac{2}{5}\ . \tag{5.8.19}$$

These two distribution functions, $(d\mathcal{M}/\mathcal{M})/dx$ and $\mathcal{M}(x)/\mathcal{M}$, are plotted in Fig. 5.8.3. On the lower abscissa scale, we have plotted not the ratio of h to its maximum value ΩR^2, but rather the physically more meaningful ratio x' to the average value for the whole sphere, $\frac{2}{5}\Omega R^2$, with $x' = x/\bar{x} = 5x/2$.

It was shown in 1964 by D. J. Crampin and F. Hoyle, and then in 1970 by J. H. Oort, that some Sb and Sc galaxies, and also our Milky Way system, in fact do exhibit angular momentum distributions corresponding to (5.8.17 and 18); this indicates their possible formation from a uniformly rotating, homogeneous spherical protogalaxy. K. C. Freeman's above-mentioned investigations of exponential disks, whose surface densities μ obey the distribution law (5.8.8), lead considerably further in this direction.

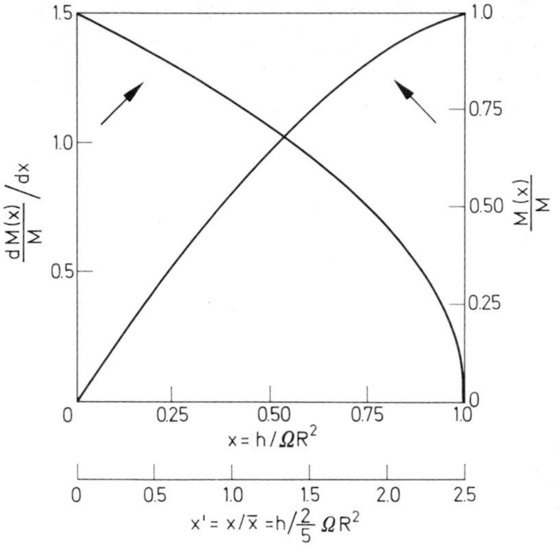

Fig. 5.8.3. The following quantities are plotted as functions of x = angular momentum per unit mass h, referred to its maximum value $h_{max} = \Omega R^2$, or of $x' = x/\bar{x}$, referred to the average value for the entire mass: (1) the fraction of the mass $d\mathcal{M}(x)/\mathcal{M}$ for which x lies between x and $x+dx$ (*left*), and (2) the fraction of the mass $\mathcal{M}(x)/\mathcal{M}$ for which x is smaller than the value on the abscissa (*right*). In the formation of a disk galaxy from a spherical protogalaxy with conservation of angular momentum, $d\mathcal{M}(x)/\mathcal{M}$ or $\mathcal{M}(x/\bar{x})/\mathcal{M}$ should be conserved as functions of $x/\bar{x}$ for each element of mass

From the distribution of the surface density μ, we first obtain the *total mass* $\mathcal{M}_E$ of an exponential disk:

$$\mathcal{M}_E = 2\pi\mu_0 r_0^2\ . \tag{5.8.20}$$

Note that e.g. 80% of the total mass is located within $r \leq 3r_0$.

We are now in a position to calculate the angular velocity of the rotation, $\Omega(r)$, and then the distribution of angular momentum per unit mass, $h_E(r)$. This then yields the fraction $\mathcal{M}_E(h_E)/\mathcal{M}_E$ of mass for which the angular momentum per unit mass is $\leq h_E$. By numerical integration, the total angular momentum,

$$H_E = 1.109(G\mathcal{M}_E^3 r_0)^{1/2}\ , \tag{5.8.21}$$

can also be calculated, and used to express h in units of its average value $H_E/\mathcal{M}_E$, with

$$x'_E = h_E/1.109(G\mathcal{M}_E r_0)^{1/2}\ . \tag{5.8.22}$$

We can now compare the function $\mathcal{M}_E(x'_E)/\mathcal{M}_E$ for exponential disk galaxies with that calculated for our "theoretical" protogalaxy, $\mathcal{M}(x')/\mathcal{M}$ (5.8.18 or Fig. 5.8.3, right-hand axis). According to K.C. Freeman, the agreement is excellent. This is a strong indication that such galaxies could in fact have been formed by *collapse* of spherical protogalaxies.

It is also interesting to compare the length scale factor r_0 of the disk from (5.8.7) with the radius R of our hypothetical protogalaxy for $\mathcal{M}_E = \mathcal{M}$ and $H_E = H$. From (5.8.21) and (5.8.13) we obtain immediately:

$$r_{0E}/R = 0.13 \ . \tag{5.8.23}$$

The radial contraction of the protogalaxy is thus considerable. The velocities in protogalaxies must therefore have been correspondingly smaller than the currently observed rotational velocities in the disk galaxies, owing to conservation of angular momentum.

The *total angular momentum* of a galactic disk, from (5.8.21) and (5.8.20), is

$$H_E \propto \mu_0^{-1/4} \mathcal{M}_E^{7/4} \ ; \tag{5.8.24}$$

it thus depends practically *only* on its mass $\mathcal{M}_E$, since μ_0 hardly varies.

The *time* required for a galactic collapse can be estimated in two ways: in a homogeneous-sphere protogalaxy, a point mass oscillates harmonically. The time τ required for it to plunge from the surface (radius R) to the center is equal to 1/4 of the oscillation period or the rotation period on a circular orbit of radius R. From (5.8.11), we see that

$$\tau_1 = \frac{\pi}{2}\sqrt{R^3/G\mathcal{M}} \ . \tag{5.8.25}$$

If the galaxy is already strongly condensed, τ will be about half the time required for a revolution around a Keplerian elliptical orbit having a semimajor axis of $R/2$, i.e.

$$\tau_2 = \frac{\pi}{2\sqrt{2}}\sqrt{R^3/G\mathcal{M}} \ . \tag{5.8.26}$$

With $\mathcal{M} = 1.5\cdot 10^{11}\,\mathcal{M}_\odot$ and $R = 25$ kpc, for example, we obtain:

$$\tau_1 = 2.5\cdot 10^8 \quad \text{and} \quad \tau_2 = 1.8\cdot 10^8\,\text{yr} \ . \tag{5.8.27}$$

Both times are, of course, roughly equal to the free fall time (4.12.39) and are of the order of magnitude of galactic rotational periods.

5.8.4 The Formation of Galaxies and Clusters of Galaxies

Galaxies contain mass concentrations of up to $10^{12}\,\mathcal{M}_\odot$; the characteristic masses of clusters of galaxies are of the order of $10^{15}\,\mathcal{M}_\odot$, and for superclusters of galaxies, $10^{16}\,\mathcal{M}_\odot$ (Sects. 5.5, 7). As we shall see, the average density of the matter in the cosmos is about $5\cdot 10^{-28}\,\text{kg m}^{-3}$ (5.9.35); the uncertainty in this figure is at least a factor of 3. The question of the extent to which *intergalactic* matter, i.e. matter outside the galaxy clusters, is distributed within the superclusters or in general in the universe, has not yet been settled. A density of $4\cdot 10^{-27}\,\text{kg m}^{-3}$, in the form of a very hot gas, would be compatible with observations.

The question of the formation of the galaxies and galaxy clusters is closely connected with the origin of the large-scale structure of the universe: the lattice-like arrangement of the superclusters and clusters of galaxies in filaments and flat disks, enclosing vast empty spaces (Sect. 5.7.4). According to recent cosmological theories (inflationary universe, Sect. 5.9.5), the *seeds* of this structure were planted as tiny density fluctuations at a very early evolutionary stage of the universe, about 10^{-35} s after the Big Bang, when the strong fundamental interaction separated from the electroweak interactions as a result of symmetry breaking (Table 5.9.3), thus causing an extremely rapid expansion (inflation) of the cosmos. The origin of the structure may even date to quantum fluctuations at a still earlier time. However, galaxies and clusters of galaxies can be formed only "much later", about 10^6 yr after the beginning of the expansion of the universe, at a red shift of $z \simeq 1000$; at that time, owing to recombination of the free electrons with protons (at $T \simeq 3000$ K), the strong coupling of the radiation field to matter was released and density fluctuations could rapidly increase. The fact that no sharp inhomogeneities were present in the cosmos before this time is demonstrated by the observed high degree of isotropy of the microwave background radiation (Sect. 5.9.4). After several 10^7 yr (corresponding to $z \simeq 100$), the density fluctuations must have become so intense that the "actual" formation of the galaxies or protogalaxies could take place. At this time, the *Jeans mass* (5.8.5) was about $2\cdot 10^5\,\mathcal{M}_\odot$; therefore, all larger masses would have been gravitationally unstable. We must thus date the formation of the galaxies back to an early phase of the expansion of the universe, when all the dimensions in the cosmos were a factor of $(1+z) \simeq 100$ smaller (5.9.29) and the average

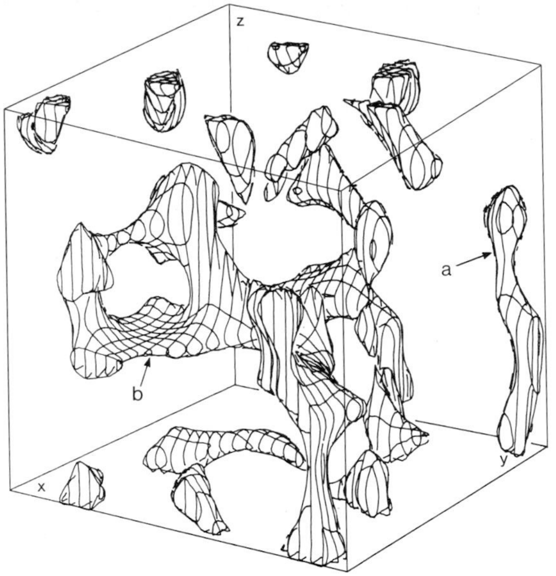

Fig. 5.8.4. The large-scale structure of the universe: surfaces of equal matter density in the early universe, based on numerical simulations of the increase of density fluctuations. Filaments (*a*) and flat disks ("pancakes", *b*) have formed and enclose large empty regions. The edge of the cube shown corresponds to 50 Mpc. From J. Centrella and A. L. Melott, Nature **305**, 196 (1983). (Reprinted by permission from Nature, **305**, 196. Copyright © 1983 Macmillan Magazines Ltd.)

density of matter was therefore $(1+z)^3 \simeq 10^6$ times larger (about $5 \cdot 10^{-22}$ kg m^{-3}) than today.

Numerical model calculations of the increase of density perturbations in the expanding cosmos, whose details we cannot treat further here, make it at least plausible how the characteristic framework of the superclusters and clusters of galaxies could have formed (Fig. 5.8.4). Ya. B. Zeldovich recognized as early as 1970 that it would be improbable for the original density fluctuations to have had spherical symmetry. He showed that density fluctuations of a general shape will collapse first in the direction of their shortest diameter and therefore form flat disks ("pancakes"). Further compression can lead to cigar-shaped formations.

A *lower limit* for the average matter density in the cosmos at the time of the formation of the galaxies can be obtained from the largest known red shift, $z = 4.9$, of the quasars, the most distant galaxies. At a time when galaxies obviously already existed, the average density was a factor of $(1+z)^3 \simeq 200$ times higher than today.

We can also estimate the density at the time of the formation of the galaxies in a quite different way: since it was shown in 1970 by J. H. Oort et al. that the exchange of angular momentum between the "finished" galaxies cannot have played a significant role, we have to assume that even the protogalaxies had the same angular momenta as the present-day galactic systems. The angular momentum of the galaxies must therefore be due to the *turbulence* of the galactic matter at the time of their formation. If we furthermore require that a *protogalaxy* can take on the necessary angular momentum, we are led back to the model of the previous section, at least in its essential features. A present-day galaxy with a scale length of $r_0 \simeq 2.5$ kpc must therefore, from (5.8.23), have been formed from a sphere of radius $R \simeq 20$ kpc. If this sphere contained $\simeq 1.5 \cdot 10^{11} \mathcal{M}_\odot$ (corresponding to the disk of the Milky Way system), then the average density of the sphere was about $3 \cdot 10^{-22}$ kg m^{-3}, i.e. of the same order of magnitude as the matter density in the cosmos at $z \simeq 100$ (and also the same as the average density in the dark halos of the spiral galaxies, with about $10^{12} \mathcal{M}_\odot$ and 50 kpc radius).

The theory of *gravitational instability* to be sure explains the "rough" distribution of matter in the galaxies, but not its finer details. If, as in (5.8.4), the masses of the galaxies were $\mathcal{M} \propto \varrho^{-1/2}$, then according to (5.8.11), their masses and therefore their luminosities L would be proportional to their characteristic lengths (e.g., r_0); but K. C. Freeman showed that $L \propto r_0^2$. A more reasonable approach might be to start with the origin of the *angular momentum* from turbulence of the pre-galactic matter. For disk-shaped galaxies, the specific angular momentum [H_E/M_E from (5.8.21) and (5.8.20) with $\mu_0 = \text{const.}$] is proportional to $r_0^{3/2}$. This relation is of course equivalent to the law for the luminosities mentioned above, $L \propto r_0^2$. If we start from the theory of isotropic and homogeneous turbulence, we find that a turbulence element of length r_0 has a characteristic velocity $v_0 \propto r_0^{1/3}$, so that the corresponding angular momentum per unit mass, $r_0 v_0$, becomes proportional to $r_0^{4/3}$. This leads to a mass or luminosity distribution $\propto r_0^{5/3}$, which agrees sufficiently well with the empirical relation $L \propto r_0^2$.

We must content ourselves here with these estimates and cannot delve deeper into the various approaches to the explanation of the formation of the galaxies, the clusters of galaxies, and the superclusters. A comprehensive theory has yet to be formulated. This is perhaps not astonishing, since we do not even understand the structure of *present-day* galaxies in detail, as our discussion of their rotation curves and dark halos has shown.

5.8.5 The Dynamic and Chemical Evolution of Galaxies. The Origin of the Abundance Distribution of the Elements

As we have seen, it must be assumed that the formation of the galaxies took place at an early phase of the universe, when its average density was still about 10^6 times higher than the present matter density. Under such conditions, density fluctuations could grow so rapidly that individual gas clouds with masses from $\leq 10^{11}$ to $10^{12} \mathcal{M}_\odot$ were formed.

The *evolution* of the galaxies is closely connected to the distribution of *angular momentum* and to the *formation of stars*. At a low rate of star formation, a gaseous protogalaxy with a high angular momentum forms a disk on contracting within a few 10^8 to 10^9 yr (Sect. 5.8.3). In the other extreme case, when all the gas is consumed right at the beginning by formation of stars, the protogalaxy changes its dimensions only slightly (by about a factor of 2), since the stars, in contrast to gas, cannot radiate away their kinetic energy, and, due to energy conservation, the sum of the kinetic and potential energies must remain constant.

The structure of the *disks* is, at least in their outer portions, surprisingly similar for all types of galaxies from S0 to Sm. The physical distinguishing feature of the Hubble types S0, Sa, Sb, etc. seems to be mainly (or entirely?) the fraction of their overall mass which is still present in *gaseous form*. This fraction varies from less than 1% in the regularly-shaped S0 disks to about 10% in irregular galaxies. The formation of spiral arms is evidently connected directly to the gas in the disk: the OB stars characteristic of spiral arms are formed from the gas and survive for only a small fraction of the lifetime of the galaxy.

The *spheroidal* components of the spiral galaxies were clearly formed from matter with *little angular momentum*. As the dynamic calculations of R.B. Larson and B.M. Tinsley (1974/78) demonstrate, the structure of our Milky Way and other spiral galaxies, consisting of a spheroid *and* a disk, must have been produced in a two-step process: in a first phase with very rapid star formation, the spheroidal component is formed within a few free-fall times, while in the following phase, with considerably slower star formation, the remaining gas can collapse to a disk before any more stars are formed.

The *giant E-galaxies* obviously evolved from protogalaxies with low angular momenta. Although there is at present no theory of the distribution of the originally turbulent gas into clouds having differing angular momenta, this assumption would not seem to meet with any insurmountable objections.

In dense *clusters* of galaxies, as we discussed in Sect. 5.7.3, the mutual interactions of the galaxies and interactions with the cluster gas can change the morphological type of the cluster in the course of time. Spiral galaxies lose their gas and presumably become S0 galaxies; in the centers of the clusters, the merging of several galaxies produces the giant cD-galaxies.

The tidal interaction of galaxies can evidently lead to a strong increase in the rate of star formation ("burst of star formation"), in particular in the denser central regions. As examples, we have met the "dusty" infrared galaxies M 82 and NGC 6240 (Sect. 5.5.5) and the blue compact dwarf galaxies (Sect. 5.5.4).

Following this short introduction to the dynamic aspects of the galaxies, we turn to their *nuclear* problems. The empirical basis for this topic was already presented in Sect. 5.5.4 in connection with the fundamental concept of stellar populations.

In relation to the relativistic theory of the expanding universe, G. Lemaître and G. Gamow in the 1940's developed the idea that the mixture of chemical elements, then regarded as universal, had been formed at the beginning of cosmic evolution during the Big Bang. In the course of the following years, nuclear physicists pointed out that the formation of the heavier elements according to Gamow's concept could have continued only up to mass number $A = 5$; on the other hand, it was discovered that the high-velocity stars and the subdwarfs are metal-poor to varying degrees.

The first attempt to combine the two problems of the *abundance distribution of the elements* and the *evolution of the galaxies* was then made by E.M. and G.R. Burbidge, W.A. Fowler, and F. Hoyle (1957); it is usually referred to for short as the B^2FH theory. Here, for the first time, the application on a broad basis of *nuclear physics* to astrophysics was undertaken. Especially W.A. Fowler's measurements of nuclear interaction cross-sections at low energies and in connection with stellar energy release were the first to place this area of research on a firm basis.

This theory presumes a cosmos consisting of *hydrogen* and *helium* (10:1) with traces of heavier elements, left over from the *Big Bang* (we adopt the modern terminology here), at the beginning of the universe from the astronomical point of view (Sect. 5.9.4).

A galaxy begins its life as a roughly spherical mass in which the first stars immediately begin to form. In their interiors, *nuclear* processes take place, leading to the formation of heavier elements. Some evolutionary processes lead to continuous or to explosive (supernovae) ejection of stellar matter into the interstellar medium. Thus, a second generation of metal-rich stars can be formed, etc. In the E-galaxies, depending on their masses, this process would sooner or later come to a stop; in the disk galaxies, the formation of heavier elements would have been essentially complete by the time of their collapse to form a disk.

We summarize, in schematic form, the nuclear processes originally considered by B^2FH, in particular since their terminology is still in widespread use today:

In Sect. 4.12.3, we have already discussed the processes which are important for *energy release* in stellar interiors: at $\simeq 10^7$ K, the nuclear burning of *hydrogen* to helium begins, at first through the p−p or fusion process and then at somewhat higher temperatures by means of the CNO cycle. Above $\simeq 10^8$ K, *helium* is then burned to carbon, and at still higher temperatures, the *carbon* itself is consumed.

Beginning with ^{12}C, the *α-process* forms heavier nuclei whose masses are multiples of 4, up to ^{40}Ca (at about 10^9 K) by combining α-particles.

The *e-process* generates the elements of the iron group, V, Cr, Mn, Fe, Co, and Ni, in thermal *e*quilibrium at $\simeq 4 \cdot 10^9$ K.

The heavier atomic nuclei are inaccessible to charged particles due to the high Coulomb barriers. The formation of still heavier elements can therefore take place only through *neutron* processes. We distinguish:

The *s-process*, which consists of neutron attachment to elements of the iron group, at a rate which is *s*low in comparison to the competing β-decays. The s-process generates for example Sr, Zr, Ba, Pb, . . . , and in general the stable nuclides in the valley of the nuclear binding-energy surface.[20] The fact that the s-process played a role in the formation of the solar abundance distribution, for example, was recognized by G. Gamow, who noted that the product of abundance times the interaction cross-section for $\simeq 25$ keV neutrons with the corresponding nuclides is a smooth function of the mass number A.

The *r-process* is the term used for the corresponding neutron attachment reactions when they are *r*apid in comparison to the competing β-decays. This process generates the neutron-rich isotopes of the heavy nuclei; B^2FH attribute the formation of in particular the radioactive elements, e.g. ^{235}U and ^{238}U (beginning with elements of the Fe group), to this process.

The *p-process* produces the neutron-poor or proton-rich isotopes of the heavy elements in a hydrogen-rich medium (*p*rotons).

The original theory of B^2FH has been modified in various detailed ways as a result of progress in nuclear physics and the theory of stellar evolution. It has been found especially that in the later phases of stellar evolution, e.g. in rapid collapse phases and within the shock waves of the supernovae, the hydrodynamic time scales lie well below 1 s and thus not all nuclear reactions can take place in thermodynamic equilibrium. Voluminous calculations of "*explosive nucleosynthesis*", beginning with the work of J. W. Truran, W. D. Arnett, et al. (1971), yield noticeably different nuclide distributions from those of equilibrium processes. The more recent results were already presented in Sect. 5.4.5 in connection with stellar evolution. Although many questions concerning the later evolutionary phases remain open, owing to the complex interactions of nuclear reactions and dynamic processes, e.g. the precise course of the r- and p-processes, the early evolutionary phases, with their reactions that also contribute to energy release, are fairly well understood.

A quite different view of the formation of the elements from that of B^2FH was developed by J. P. Amiet and H. D. Zeh (1968) following ideas of H. E. Suess and J. H. D. Jensen: it begins with a state having a high density of about $2 \cdot 10^{13}$ kg m^{-3} and a temperature of about $5 \cdot 10^9$ K. Under such conditions, free electrons are pressed into the nuclei, so to speak; the "valley" of stable nuclei is shifted on the N,Z-plane in the direction of the neutron-rich nuclei. When the pressure is reduced, this neutron-rich protomatter undergoes β-decay to form the more neutron-poor nuclei which are stable under normal conditions; however, relatively neutron-rich stable nuclei may (rarely) remain. This model has not been further pursued; in particular, the astrophysical explanation of the high density of the initial state has remained unclear.

Following this excursion into nuclear physics, we return to the problem of the *formation of the galaxies* and the *abundance distribution* of the elements found in them.

Initially, we assume that a *protogalaxy* containing nearly pure hydrogen (plus 10% helium) was present in the early universe, $\geq 10 \cdot 10^9$ yr ago; we then discuss

[20] If the binding energies of the atomic nuclei are plotted above a plane with the coordinates N (neutron number) and Z (nuclear charge or proton number), the resulting energy surface allows a clear representation of nuclear reaction processes including their energy balances. The stable nuclei are found on the valley floors of the energy surface.

along the lines of B^2FH the idea that the heavy elements were produced in connection with the birth and (to some extent explosive) passage of several generations of stars. In the dwarf E-galaxies, this process must have been arrested sooner than in the giant galaxies. In all the sufficiently large galaxies, we find that, independently of the formation of a disk and independently of the mass fraction of interstellar gas which is still present today, the order of magnitude of the extent to which the protomatter has been converted into heavy elements is the same. Furthermore, from the beginning of the formation of the heavy elements up to the time when they attained their current abundances, their mixing ratios remained more or less unchanged. Within many S galaxies, we can also observe an increase of the abundance of heavy elements in going from the edge of the disk towards the center.

In our *Milky Way system*, we can obtain a more precise picture of the chronology of formation of the heavy elements:

At present, stars are formed from the interstellar medium concentrated in a flat disk, with a metal content $Z \simeq Z_\odot = 0.02$ (5.4.2), at an overall rate of a few $\mathscr{M}_\odot$ per year and with a mass spectrum $\psi(\mathscr{M})$ (initial mass function, Sect. 5.4.8). In particular, the more massive stars enrich the interstellar gas with heavy elements at the end of their evolution, by e.g. supernova explosions. In this process, a part of their mass which has been modified by nuclear reactions remains in the remnant stars (e.g., neutron stars), and is thus removed from the cycle of interaction between the interstellar medium and many generations of stars.

The observed element abundances in the spheroidal component (halo) of, for example, the globular clusters and subdwarfs shows a maximum metal content of about $Z_\odot/6$, so that during a few free-fall times (5.8.27) of the protogalaxy of $\geq 10^{11}\,\mathscr{M}_\odot$, i.e. in $\leq 10^9$ yr, an increase of the metal abundance from practically zero ($Z < 10^{-8} Z_\odot$) to $Z_\odot/6$ must have occurred.

With the consolidation of the disk, of mass around $7 \cdot 10^{10}\,\mathscr{M}_\odot$, the formation of elements essentially came to its conclusion, since the oldest open star clusters already have roughly the normal composition $Z_\odot$. Although the dynamic considerations of O.J. Eggen, D. Lynden-Bell, and A.R. Sandage (1962) yielded a relatively short time (a few 10^8 yr) for the collapse into a disk, more recent age-dating of the globular clusters, giving about $16 \cdot 10^9$ yr (Sect. 5.4.4), indicates that the formation of the disk required several 10^9 yr. During this time, the enrichment of the metal content from about $Z_\odot/6$ to $Z_\odot$ occurred.

In this connection, we emphasize how small is the *nuclear energy release rate* in the Milky Way system and other galaxies in their present states. In (5.4.3), we related the hydrogen consumption of a star to its mass-luminosity ratio. If we take the latter to be $\simeq 10$ solar units, we find that our galaxy, in its *present* state, would have consumed only a few percent of its hydrogen in 1 to $2 \cdot 10^{10}$ yr. The production of a major portion of the heavy elements within the first $\leq 10^9$ yr by stellar evolution therefore implies that our Milky Way had previously a much greater luminosity than it does today.

According to Table 5.4.2, as a result of their lifetimes only relatively massive stars contribute to the formation of the heavy elements; e.g. for the enrichment of the halo, stars with $\geq \mathscr{M}_\odot$, or probably more likely with ≥ 10 to $20\,\mathscr{M}_\odot$ were important. On the other hand, stars with $\leq 1\,\mathscr{M}_\odot$ have not evolved away from the main sequence since the formation of the Milky Way, so that for example among the G-dwarfs, metal abundances in the *entire* range $0 \leq Z \leq Z_\odot$ should be found. If the formation of the halo stars had corresponded even roughly to the luminosity function of galactic star clusters (Sect. 5.4.8), then there should still be a large number of long-lived *metal-poor* dwarf stars present today, which contradicts observations. Instead, extremely metal-poor stars are very rare and stars with $Z \leq 10^{-4} Z_\odot$ are not observed at all. This "*G-dwarf problem*", and also the time evolution of the average metal abundance (with a rapid increase at the beginning and a flattening-out during the last $5 \cdot 10^9$ yr), as well as the radial abundance gradient within the disk, cannot be understood in terms of simple models based on a spatially and temporally constant rate of star formation in a closed system. Attempts to solve these problems therefore begin with hypotheses (in part *ad hoc* assumptions) such as (a) a radically different mass function $\psi(\mathscr{M})$ at the beginning, according to which in particular many more massive stars were formed; (b) a very rapid initial enrichment of the protogalaxy with heavy elements before the formation of the oldest halo stars, possibly through a "Population III" which no longer exists today; or (c) a transfer of original or nuclear-processed matter from the halo to the disk over a long period of time following the collapse which formed the disk. Recent observations by J. Norris, R. Zinn, G. Gilmore and others (1985) are interesting in this connection: they indicate a star component in our Milky Way with kinematics and metal abundances *between* those of the halo and the disk, i.e. a "thick disk" component.

The ideas about the formation of the heavy elements, especially in the Milky Way, make it possible from the nuclear physics point of view to determine a cosmic *time scale* (see also Sect. 5.9). In this socalled *cosmochronology*, long-lived radioactive nuclides of different half-lives are used as "chronometers", similar to radioactive dating in the Solar System (Sect. 2.8.5b). Examples are ^{187}Re ($T_{1/2} = 5 \cdot 10^{10}$ yr), ^{238}U ($T_{1/2} = 4.5 \cdot 10^9$ yr), ^{232}Th ($T_{1/2} = 1.4 \cdot 10^{10}$ yr) and ^{235}U ($T_{1/2} = 7.0 \cdot 10^8$ yr); also ^{244}Pu ($T_{1/2} = 8.3 \cdot 10^7$ yr) and ^{129}I ($T_{1/2} = 1.6 \cdot 10^7$ yr). A first guess is obtained by assuming that most of the heavy elements were produced within a short time during the formation of the Milky Way system: we then take two radionuclides, which are not replaced by the decay of any parent element and which decay with greatly differing half-lives, e.g. the uranium isotopes ^{235}U and ^{238}U, whose present abundance ratio is 1 : 138. In previous times, this ratio must have been much greater. If we extrapolate back to an initial ratio somewhat larger than one, we arrive at an age of formation of between 7 and $8 \cdot 10^9$ yr ago. For a more precise age determination, the *time dependence* of the production of elements during the evolution of the galaxy is required, in particular a model for the production of isotopes by the r-process, since all the nuclides used as chronometers were generated by this process. The uncertainties in the model, in some of the nuclear physical quantities, and in the observation and interpretation of the isotopic ratios of meteorites leave open a relatively wide range of values between about 13 and $25 \cdot 10^9$ yr for the *age of the Milky Way galaxy*, and therefore for all of the galaxies. These times are, in any case, compatible both with the age of the Milky Way as obtained from the color-luminosity diagrams of the globular clusters, some 14 to $18 \cdot 10^9$ yr (Sect. 5.4.4), and with the age of the universe as estimated from the expansion of the galaxies (Friedmann time, Table 5.9.1), in the range of 7 to $20 \cdot 10^9$ yr.

To conclude this section we now summarize briefly the fundamental open questions of this fascinating research area of the chemical evolution of the galaxies! Although we understand the essential features of the galactic cycle connecting the different generations of stars and the interstellar matter, and although we can reconstruct to a considerable extent the "history", of our Milky Way galaxy in particular, from present-day observations, we are still far from having a complete quantitative picture of element formation in the galaxies of different Hubble types. For one thing, we know nothing about the nature of the "invisible" major portion of the matter in many galaxies which indicates its presence in dynamic observations. The process of star formation in relation to the physico-chemical state of the interstellar medium is also still unclear to a great extent. In recent times it has become apparent, especially from infrared observations, that the rate of star formation is strongly increased by tidal interactions between the galaxies. Furthermore, our knowledge of the later phases of stellar evolution, and along with it the enrichment of the interstellar gas with heavy elements, still has serious gaps. Finally, the role in galactic evolution of the galactic nuclei and their activity, particularly during the early evolutionary stages, is still a riddle.

In contrast to the stars, the galaxies were apparently all formed at about the same time, around $2 \cdot 10^{10}$ yr ago. Young galaxies can therefore be observed only as very distant, faint objects (with large red shifts), whose light emitted in the early stages is just now reaching us. The observation of galaxies or quasars with $z \leq 4.9$ currently allows us to look back into the past up to about 95% of the age of the universe (Table 5.9.1), and there are indeed indications of evolutionary effects in the most distant systems (bluer color, greater luminosities). The rapid development of highly sensitive optical detectors in the past years promises, together with large telescopes, to allow enormous progress in the investigation of the early phases and the formation of the galaxies and galaxy clusters in the near future. According to theoretical estimates, we are not too far from being able to observe the birth of the galaxies.

5.9 Cosmology

Having studied our Milky Way galaxy, various other galaxies of the most diverse kinds, and the arrangement of the galaxies into clusters and superclusters, we now turn to the *cosmos as a whole*, its spatial structure and its evolution in time. We begin, in Sect. 5.9.1, with E. Hubble's discovery of the expansion of the universe. The proportionality of the red shifts of the galaxies to their distances opened up the possibility of determining cosmic distances. In Sect. 5.9.2, we then discuss the various models of the universe which are allowed within the framework of the theory of general relativity, and find that in the cosmos at present, matter dominates over radiation. In order to determine *the* correct model of our universe and its parameters, we must make use of a wide variety of different observations. To this end, we investigate somewhat more carefully the propagation of electromagnetic radiation over cosmic distances (Sect. 5.9.3). The discovery of an isotropic black-body radiation field at 3 K by A. A. Penzias and R. W. Wilson, which we interpret as a remnant of the Big Bang, indicates that the initial state of the universe was hot and was dominated by radiation, and that the lightest elements, in particular helium, were produced during this stage (Sect. 5.9.4). Finally, in Sect. 5.9.5, we summarize the most important relevant facts in order to arrive at a picture of the evolution of the universe from its earliest stages, which are closely connected with our knowledge of the fundamental physical interactions, through the production of the 3 K radiation, on to the formation of the galaxies and galaxy clusters, and finally to the present-day state of the cosmos.

5.9.1 The Expansion of the Universe. The Hubble Constant

Five years after his determination of the distances to faraway galaxies, E. Hubble in 1929 made a second discovery of enormous importance: the *red shift* of the lines in the spectra of distant galaxies:

$$z=\frac{\Delta\lambda}{\lambda_0}=\frac{\lambda-\lambda_0}{\lambda_0} \qquad (5.9.1)$$

(λ_0: laboratory wavelength or the wavelength in the rest system of the galaxy; λ: measured wavelength) increases *proportionally to their distances r*. If this red shift is interpreted as a Doppler effect, then for the *velocity* with which the galaxy is *receding*, $v=dr/dt$, we find the relation

$$v=c\Delta\lambda/\lambda_0=cz=H_0 r \ . \qquad (5.9.2)$$

(We limit ourselves initially to the case $z\ll 1$; otherwise, we would have to make a relativistic calculation of the Doppler effect).

Apparent exceptions for nearby galaxies (for example, the Andromeda galaxy is *approaching* us at 300 km s^{-1}) could be readily explained as resulting from the rotation of our own Milky Way galaxy. Recent measurements of the 21 cm line of hydrogen have confirmed to an excellent degree the wavelength dependence of the effect assumed in Hubble's relation (5.9.2).

For the *Hubble constant* H_0, Hubble himself in 1929 obtained the value 530 km $s^{-1}\cdot Mpc^{-1}$. Following W. Baade's revision of the cosmic distance scale (distinguishing between the Cepheids of Populations I and II), A. R. Sandage in 1958 calculated the most probable value to be 75 km $s^{-1}\cdot Mpc^{-1}$. A further revision of the extragalactic distance scale by A. R. Sandage and G. A. Tammann in 1974/75 led to a value of 50 km $s^{-1}\cdot Mpc^{-1}$. This result has been supported by more recent work of the same researchers. In contrast, some other investigations, especially those of G. de Vaucouleurs, have given numerical values around 100 km $s^{-1}\cdot Mpc^{-1}$, and, most recently, also values nearer to 75 km $s^{-1}\cdot Mpc^{-1}$. The differences in the individual determinations of H_0 lie outside the respective error limits (of about 10 to 15%), which have been estimated for the different methods. The resolution of this uncertainty in the determination of one of the most fundamental parameters is a high-priority goal of observational cosmology. Here, we shall assume the value:

$$H_0=50\ \mathrm{km\ s^{-1}\ Mpc^{-1}} \ . \qquad (5.9.3)$$

Conversely, (5.9.2) is often applied in order to estimate the *distance r* to a galaxy which cannot be telescopically resolved, by using the measured red shift of its spectrum. At the present state of the art, the value used for H_0 should always be indicated.

Relation (5.9.2) can be interpreted initially in a naive way as implying that an *expansion* of the universe from a relatively small volume began at a previous time τ_0.

If a particular galaxy, now at a distance r, was ejected with a velocity v in the course of the expansion, it would have required a time τ_0 to move through the distance r, this time is the same for all galaxies:

$$\tau_0=r/v=1/H_0 \ . \qquad (5.9.4)$$

The *Hubble time* τ_0 is thus equal to the reciprocal of the Hubble constant, $1/H_0$. If, as is usual, we quote H_0 in units of [km s^{-1} Mpc^{-1}] and τ_0 in [yr], then we find

$$\tau_0\,[\mathrm{yr}]=\frac{978\cdot 10^9}{H_0\,[\mathrm{km\ s^{-1}\ Mpc^{-1}}]} \ . \qquad (5.9.5)$$

While Hubble's older value of H_0 led to an age of the universe of $1.8\cdot 10^9$ yr, much too short in comparison to the age of the globular clusters, etc., the newer numerical value of 50 km s^{-1} Mpc^{-1} yields:

$$\tau_0=19.6\cdot 10^9\,\mathrm{yr}=6.2\cdot 10^{17}\,\mathrm{s} \ . \qquad (5.9.6)$$

5.9.2 Models of the Universe. The Matter Cosmos

The kinematics of the *expanding* universe contained in (5.9.2) would seem at first glance to represent a throwback to heliocentric concepts. That is however not the case! If, namely, (5.9.2) is written as a vector equation:

$$\boldsymbol{v} = H_0 \boldsymbol{r} \ , \tag{5.9.7}$$

where the origin of the coordinate system is taken to lie in our galaxy, then an observer in another galaxy at the distance $\boldsymbol{r}_1$ and with a velocity $\boldsymbol{v}_1$ relative to us (with $\boldsymbol{v}_1 = H_0 \boldsymbol{r}_1$) would find

$$\boldsymbol{v} - \boldsymbol{v}_1 = H_0(\boldsymbol{r} - \boldsymbol{r}_1) \ . \tag{5.9.8}$$

A universe expanding according to (5.9.7) thus offers the same aspect to all observers in different galaxies; i.e. our kinematic model of the universe is *homogeneous* and *isotropic*. It may be shown that (5.9.7) is the *only* velocity flow field which fulfills these conditions under the additional assumption of irrotational flow (curl $\boldsymbol{v} = 0$).

This initially purely kinematic model was extended by E. Milne and W.H. McCrea in 1934 into a *Newtonian cosmology*; they investigated the kinds of flow which can occur in a medium (the "galaxy gas") within the framework of Newtonian mechanics when homogeneity, isotropy, and irrotational flow are assumed. (The requirement of isotropy throughout the medium implies homogeneity, but not *vice versa*.)

If we consider a finite, expanding "cosmic sphere" with a radius $R(t)$ at a time t, then a galaxy which is e.g. on its surface will be attracted by the mass $\mathcal{M}$ within the sphere according to Newton's law of gravitation. The *equation of motion* is then:[21]

$$\frac{d^2R}{dt^2} = -\frac{G\mathcal{M}}{R^2} \tag{5.9.9}$$

with

$$\mathcal{M} = \frac{4\pi}{3} R(t)^3 \varrho(t) = \text{const.} \ , \tag{5.9.10}$$

where $\varrho(t)$ is the mass density at the time t considered.

If we multiply the above equation by $\dot{R} = dR/dt$, it can be integrated without difficulty, leading to[22]

$$\frac{1}{2}\left(\frac{dR}{dt}\right)^2 - \frac{G\mathcal{M}}{R} = h \tag{5.9.11}$$

with h = const., or

$$\left(\frac{\dot{R}}{R}\right)^2 = -\tilde{k}\frac{c^2}{R^2} + \frac{8\pi}{3} G\varrho(t) \ , \tag{5.9.12}$$

where we have written $h = -\tilde{k}c^2/2$ (c = velocity of light) in anticipation of a later comparison with relativistic calculations.

It is easy to convince oneself that at a particular time t, the same red shift law would be observed from every galaxy within our sphere. Since the distance r between two galaxies is proportional to the *scale factor* $R(t)$, the relative velocity is given by

$$v = \frac{\dot{R}}{R} r = H(t) r \ , \tag{5.9.13}$$

where we call $H(t) = \dot{R}/R$ the *Hubble function*. Denoting all quantities which refer to the *present* time $t = t_0$ by a subscript 0, we find for the Hubble constant in (5.9.2)

$$H_0 = \frac{\dot{R}_0}{R_0} \ . \tag{5.9.14}$$

In order to characterize fully a model of the universe, we require in addition to $H(t)$ or H_0 a second quantity describing the gravitational acceleration due to the mass $\mathcal{M}$, which acts inwards and opposes the expansion of the universe; this is the socalled *deceleration* or *delay parameter*

$$q(t) = -\left(\frac{\ddot{R}}{R}\right) \Big/ \left(\frac{\dot{R}}{R}\right)^2 = -\frac{1}{H(t)^2}\frac{\ddot{R}}{R} \tag{5.9.15}$$

or its current value,

$$q_0 = -\frac{1}{H_0^2}\frac{\ddot{R}_0}{R_0} = \frac{4\pi G\varrho_0}{3H_0^2} \tag{5.9.16}$$

according to (5.9.9). The delay parameter q_0 relates the acceleration $-\ddot{R}_0$ to a "unit acceleration", which would lead in the time $\tau_0 = H_0^{-1}$ starting from zero velocity to

---

[21] The use of Newtonian mechanics is justified as long as $G\mathcal{M}/Rc^2 \ll 1$; compare (4.12.55). The problems which arise for *infinitely* extended systems cannot be dealt with here.

[22] The pressure forces, which we have ignored here for simplicity, would not change the result obtained.

the expansion velocity R_0H_0 currently observed at a distance R_0.

The solution of our equations leads to *models of the universe* which, starting from a point (singularity) of infinitely high density, lead either to a *monotonic expansion* (total energy $\mathscr{M}h \geq 0$) or to a *periodic oscillation* between $R=0$ and $R_{max}(\mathscr{M}h<0)$. Static models are not possible within the framework of (5.9.9).

Generally speaking, Newtonian cosmology extends the purely kinematic cosmological model which we treated first, in the sense that it considers the Hubble constant H_0 to be a function of time t. In a periodic universe, for example, an age of red shifts would be followed by an age with violet shifts and conversely. We shall not discuss further here the question of the choice among the numerous different models possible within a Newtonian cosmology, nor the fundamental difficulties of such a cosmology; instead, we turn immediately to a *relativistic cosmology.*

After the formulation of the modern field theory of gravitation by A. Einstein (1916), in the form of his *Theory of General Relativity* (Sect. 4.12.9), Einstein himself and W. de Sitter in 1917 constructed some special models of the cosmos, in which the curvature of space was assumed to be independent of time. Einstein showed that his field equations (extended to include the Λ term, see below) have a static cosmological solution; de Sitter found a matter-free expanding cosmos as their solution. Referring to the older radial velocity measurements of V.M. Slipher ($\simeq$ 1912), C. Wirtz had noticed their increase with distance in 1924 and related it to de Sitter's relativistic cosmological model, five years before Hubble published his proportionality between the distance and the radial velocity. In 1922/24, A. Friedmann succeeded in formulating an important generalization to spaces with time-dependent curvatures. His publications went unheeded for a time, until about 1927/30 G. Lemaître, A.S. Eddington, and others took up the investigation of models of the *expanding universe* on the basis of the theory of general relativity.

If we assume from the beginning that the *cosmological postulate* holds, i.e. that the universe must be *homogeneous* and *isotropic* throughout, then following H.P. Robertson and A.G. Walker, we can cast the four-dimensional line element ds in the form:

$$ds^2 = c^2 dt^2 - R(t)^2 \left[\frac{dr^2}{1-kr^2} + r^2 (d\theta^2 + \sin^2\theta \, d\phi^2) \right] . \tag{5.9.17}$$

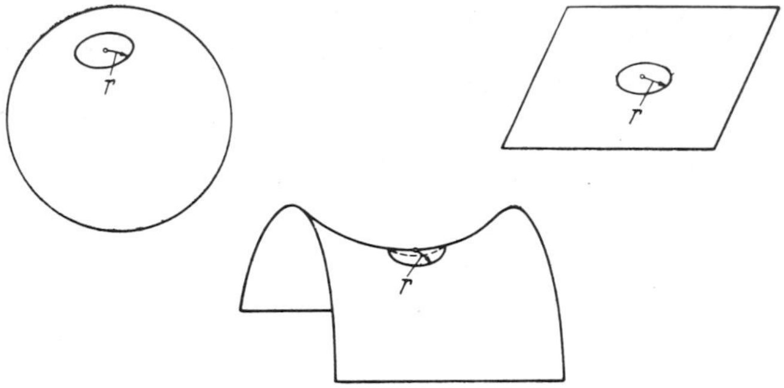

Fig. 5.9.1. Surfaces (i.e. two-dimensional spaces) with curvatures of $k>0$, $k=0$, and $k<0$ as models of curved spaces

Curvature k of the surface:	$k>0$	$k=0$	$k<0$
Geometry:	Spherical or elliptical	Euclidean	Hyperbolic
Circumference of a circle:	$<2\pi r$	$=2\pi r$	$>2\pi r$
Area of a circle:	$<\pi r^2$	$=\pi r^2$	$>\pi r^2$

Here, r, θ, and ϕ are dimensionless Lagrange coordinates, which remain constant for a galaxy that is following the expansion of the universe. The time dependence of the geometry thus depends exclusively on the expansion factor $R(t)$. $R(t)$ determines the *radius of curvature* of the three-dimensional space, defined analogously to the radius of curvature of a two-dimensional surface. The constant k, which can take on the values 0 or ± 1, fixes the sign of the *spatial curvature* which is the same everywhere at a given time; it corresponds to:

$k=0$ the well-known *Euclidian* space;
$k=+1$ a spherical or elliptical space; these spaces are *closed* and have a finite volume;
$k=-1$ hyberbolic space; it is *open.*

These three kinds of spaces or geometries are best imagined by considering their two-dimensional analogies (Fig. 5.9.1); in particular, a closed, expanding universe can be represented by the surface of a balloon which is being inflated.

In the *Robertson-Walker metric* (5.9.17), the number of Einsteinian field equations (4.12.59b), extended to include the term Λg_{ik} *with the cosmological constant* Λ, which relate the metric g_{ik} of the "cosmos" to its material contents, is reduced in a way which we cannot treat in detail here to two differential equations for $R(t)$; these are the *Friedmann-Lemaître equations*:

$$\left(\frac{\dot{R}}{R}\right)^2 = -k\frac{c^2}{R^2}+\frac{8\pi G}{3}\varrho+\frac{c^2}{3}\Lambda \qquad (5.9.18)$$

$$2\frac{\ddot{R}}{R}+\left(\frac{\dot{R}}{R}\right)^2 = -k\frac{c^2}{R^2}+\frac{8\pi G}{c^2}p+c^2\Lambda \ . \qquad (5.9.19)$$

For a *vanishing* cosmological constant, (5.9.18) is seen to be identical with the differential equation (5.9.12) from Newtonian cosmology. For systems with negligible pressure p, (5.9.19) also becomes identical to (5.9.9) after substitution by (5.9.18), so that we have exactly the same choice of cosmological models as in Newtonian cosmology. However, it is now *a priori* forbidden that velocities greater than that of light can occur; *relativistic* cosmology is the first theory to make possible a self-consistent and noncontradictory description of the *universe as a whole.*

A comparison of (5.9.18) with the Newtonian equations (5.9.9) and (5.9.12) further shows us that a *positive* constant Λ corresponds to a *repulsive* acceleration, which thus acts oppositely to gravitation. The Λ term was introduced by Einstein in 1917 into his gravitational equations; the purpose of this additional term was to make it possible to construct a *static* cosmological model, which Einstein originally held to be the only reasonable assumption. This point of view, however, became obsolete after Hubble's discovery of the red shift law. According to W.H. McCrea (1951) and Ya.B. Zeldovich (1968), the cosmological constant Λ (dimension m^{-2}) can be interpreted within the framework of quantum field theory in terms of the energy density of the vacuum, or the equivalent mass density ϱ_v, with $c^2\Lambda = 8\pi G\varrho_v$. However, up to the present time it has not been possible to determine the numerical value of Λ through physical or astronomical considerations; $|\Lambda| \leq 10^{-51}\,\mathrm{m}^{-2}$ is consistent with cosmological observations. For this reason, Λ is often set equal to zero. In the following discussion, we also assume $\Lambda = 0$. As we shall see however in Sect. 5.9.5, the cosmological constant may have played an important role in the very early evolutionary stages of the universe (inflationary expansion).

In relativistic cosmology, as in Newtonian cosmology, $R(t)$ is initially determined by the *Hubble constant* H_0 and the *deceleration parameter* q_0. The latter also determines the type of spatial curvature, i.e. the value of k or the basic structure of the spatial universe. If we introduce $H(t)$ and $q(t)$ (5.9.13) and (5.9.15) as well as the *critical density*:

$$\varrho_c(t) = \frac{3H(t)^2}{8\pi G} \qquad (5.9.20)$$

or the dimensionless *density parameter*:

$$\Omega(t) = \frac{\varrho(t)}{\varrho_c(t)} \qquad (5.9.21)$$

into the Friedmann-Lemaître equations for $p = 0$, we obtain:

$$2q-1 = \frac{\varrho}{\varrho_c}-1 = \Omega-1 = k\frac{c^2}{H^2R^2} \ . \qquad (5.9.22)$$

We can derive from this the following three possibilities, considering that c^2/H^2R^2 is always positive:

Deceleration parameter	Curvature	Density		Time dependence of $R(t)$
$0 \leq q < \frac{1}{2}$	$k = -1$	$\varrho < \varrho_c$	$\Omega < 1$	monotonically increasing
$q = \frac{1}{2}$	$k = 0$	$\varrho = \varrho_c$	$\Omega = 1$	monotonically increasing
$q > \frac{1}{2}$	$k = +1$	$\varrho > \varrho_c$	$\Omega > 1$	finite (cycloidal)

The present matter density ϱ_0 thus, together with H_0, determines the deceleration parameter q_0 and therefore the type of solution of the cosmological model.

For the present state of the universe, the mass density of the matter concentrated in the galaxies exceeds that of the radiation (3 K background radiation; see Sect. 5.9.4) by a large margin, so that the neglect of the pressure in the Friedmann-Lemaître equations is justified.

We speak generally of a *matter cosmos* when the density of *non*relativistic particles, i.e. practically the rest mass density, predominates, and their pressure is negligible. On the other hand, we speak of a *radiation cosmos* when the mass or energy density of *relativistic* particles, e.g. photons, is predominant.

If we disregard the early stages of the universe (up to about $2\cdot10^6$ yr after the singularity, Sect. 5.9.5), the matter cosmos represents a good approximation for a model of the universe. This is also true in particular for those epochs in which the most distant quasars with $z \simeq 4$ emitted their radiation, which we are just detecting today; thus we can employ models corresponding to (5.9.22) for comparison to the observations of galaxies, even the most distant ones.

For the *matter cosmos*, we expect conservation of mass, $\varrho(t)R(t)^3 = \text{const.} = \varrho_0 R_0^3$ (5.9.10), so that we can write (5.9.18) in the form:

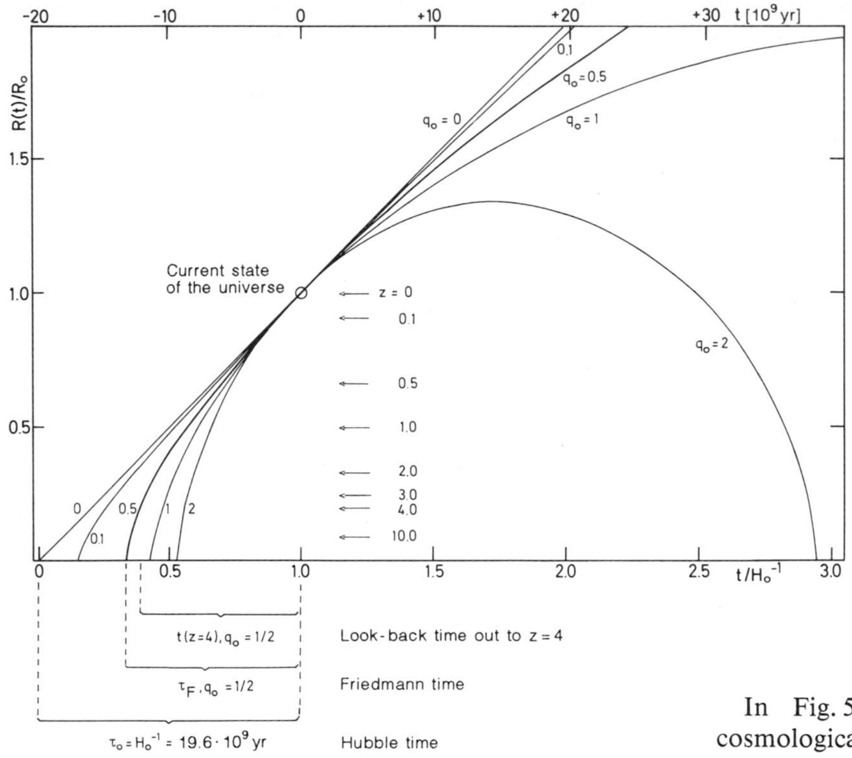

Fig. 5.9.2. Matter-dominated cosmological models ($q_0 \leq 1/2$ are open, $q_0 > 1/2$ are closed universes). The distance between two galaxies is proportional to the expansion factor $R(t)$. All models have *the same* Hubble constant $H_0 = 50\,\mathrm{km\,s^{-1}\,Mpc^{-1}}$, or the same tangent to the $R(t)$ curve at the present time. The time scale t is shifted here compared to that used in (5.9.23) and (5.9.24) by $(\tau_0 - \tau_F)$. The relationship between $R(t)$ and the red shift z is indicated by the arrows. The Hubble time τ_0 and, for the Einstein-de Sitter cosmos ($q_0 = 1/2$), the Friedmann time and the look-back time $t(z=4)$ are shown below

$$\dot{R}^2 = -kc^2 + \frac{A}{R}\,, \qquad (5.9.23)$$

where A is a constant. In the case $k = 0$ ($q = \frac{1}{2}$), we obtain by integration with $8\pi G\varrho_0/3 = H_0^2$ (5.9.20) the *Einstein-de Sitter cosmos* as solution:

$$\frac{R(t)}{R_0} = \left(\frac{3}{2}H_0 \cdot t\right)^{2/3}\,; \qquad (5.9.24)$$

this is a cosmological model of a monotonically expanding universe.

For sufficiently small R in (5.9.23), the term kc^2 is negligible compared to A/R, so that (5.9.24) represents a good approximation even for models with $k \neq 0$. In the cosmological models with negative curvature, $k = -1$ ($q_0 < \frac{1}{2}$), $R(t)$ increases continually and monotonically; for large R or t, $R(t)$ becomes proportional to t. However, the time dependence of the expansion factor in models with $k = +1$ ($q_0 > \frac{1}{2}$) is represented by a cycloid; in these models, the expansion is eventually followed by a contraction.

In Fig. 5.9.2, we have plotted $R(t)$ for several cosmological models with different delay parameters q_0; they are all characterized at the present time t_0 by the *same* Hubble constant H_0 and the same scale factor R_0. According to (5.9.14), H_0 determines the (common) tangent to the curves at $t = t_0$. Between t_0 and the intercept of this Hubble line with the abscissa is the time interval called the *Hubble time*, $\tau_0 = H_0^{-1}$, according to (5.9.4). The *beginning* of our cosmological epoch is determined by the intercept of the curve of $R(t)$ with the abscissa. It preceeds t_0 by a time interval which is termed the *Friedmann time*, τ_F (Table 5.9.1). There can be no galaxy, star cluster, etc. whose age is greater than τ_F. For

Table 5.9.1. The Friedmann time τ_F and the "look-back time" $t(z=4)$ for different values of the Hubble constant H_0 and the delay parameter q_0. The Hubble time $\tau_0 = H_0^{-1}$ is equal to τ_F for $q_0 = 0$

	τ_F [10^9 yr]			$t(z=4)/\tau_F$
H_0	50	75 [$\mathrm{km\,s^{-1}\,Mpc^{-1}}$]	100	
$q_0 = 0$	19.6	13.0	9.8	0.80
0.1	16.7	11.1	8.3	0.87
0.5	13.1	8.7	6.5	0.91
1.0	11.2	7.5	5.6	0.92
2.0	9.3	6.2	4.6	0.93

the Einstein-de Sitter cosmos, we obtain from (5.9.24) $\tau_F = \frac{2}{3}\tau_0$.

Before we discuss the observational data, in order to find out which model corresponds to the *real* universe, we shall deal with the propagation of light and radio signals within the framework of relativistic cosmology.

5.9.3 The Propagation of Radiation. A Comparison of Theory and Observations

According to the theory of general relativity, electromagnetic radiation propagates along *null geodesic* world lines, $ds^2 = 0$. In the line element of the Robertson-Walker metric (5.9.17), the Lagrange coordinates of galaxies which are moving apart, neglecting proper motions and simply as a result of the expansion of the universe, remain constant in time. If we consider a light signal which was emitted at a time t_e from a galaxy with the dimensionless radial coordinate r, and which reaches us (at $r = 0$) today (at $t = t_0$), then its propagation can be described by

$$c\frac{dt}{R(t)} = \text{const.} \qquad (5.9.25)$$

or

$$c\int_{t_e}^{t_0} \frac{dt}{R(t)} = \int_0^r \frac{dr'}{\sqrt{1-kr'^2}} \simeq r + \frac{k}{6}r^3 + \ldots , \qquad (5.9.26)$$

where we have for simplicity assumed that only the radial coordinates of the two galaxies are different.

The *frequency* ν_e of the radiation at the time of its emission can be defined by a particular number N of oscillations within a small time interval Δt_e. A present-day observer then measures the *same* number N in an interval Δt_0, or a frequency ν_0 given according to (5.9.25) by:

$$\frac{\nu_0}{\nu_e} = \frac{\Delta t_e}{\Delta t_0} = \frac{R(t_e)}{R_0} . \qquad (5.9.27)$$

The *red shift* z defined by (5.9.1) is now, due to $c = \nu\lambda$, given[23] by

$$1+z \equiv 1+z_e = \frac{\lambda_0}{\lambda_e} = \frac{\nu_e}{\nu_0} . \qquad (5.9.28)$$

From (5.9.27) and (5.9.28), we obtain the important result that the red shift is determined uniquely by the ratio of the expansion factor at the time of emission to the expansion factor at the time of detection of the light signal:

$$1+z = \frac{R_0}{R(t)} . \qquad (5.9.29)$$

We can now, for a given q_0, find the corresponding R for every z, and with the appropriate cosmological model the "look-back time" $t(z)$, i.e. the time that a light or radio signal would require to travel from the galaxy in question to the Earth (Fig. 5.9.2). In this way, we obtain among other things a lower limit $t(z \simeq 4)$ for the age of the most distant galaxies or quasars.

Now, in order to obtain a form of the Hubble relation (5.9.2) which is valid for *large* distances or red shifts, we must relate not only z but also the *distance* to the parameters of the cosmological model. The "proper distance" $r_E \simeq rR(t)$, which is given by the middle integral in (5.9.26), is unsuitable for this purpose, since it is not directly observable. On the other hand, we can introduce for example a "luminosity distance" r_L, which can be derived from observation of the apparent magnitude of the galaxy, by *defining* r_L for large distances using the usual relation (4.4.26) between the observed radiation flux f and the absolute luminosity of the galaxy, taken to be known; this relation is valid for small distances:

$$f = \frac{L}{4\pi r_L^2} . \qquad (5.9.30)$$

The luminosity, emitted at a time t_e, is equal to the radiation power and is therefore proportional to the energy $h\nu_e$ of the photons per time interval Δt_e. It is thus attenuated by a factor $(1+z)^2$ during its propagation until the time t_0 of its detection, owing to the cosmological red shift, since from (5.9.27), both the frequency ν_e decreases proportionally to $(1+z)$ and the time interval Δt_e increases proportionally to $(1+z)$. On the other hand, the radiation is diluted on a spherical surface with the emitting galaxy at its midpoint; the observer is located on this surface at the time t_0. Since this surface, from (5.9.17), is equal to $4\pi r^2 R_0^2$, where r is again the radial Lagrange coordinate of the galaxy, we finally obtain the result for the radiation flux referred to a unit surface:

[23] In contrast to (5.9.1), λ_0 denotes here the wavelength observed at the present time, and not the rest wavelength.

$$f(t_0) = \frac{L(t_e)}{(1+z)^2 4\pi r^2 R_0^2} , \tag{5.9.31}$$

and therefore

$$r_L = (1+z) r R_0 . \tag{5.9.32}$$

Other possibilities for measuring the distance by relating it to additional observable quantities such as the apparent angular diameter, the parallax, or the proper motion cannot be treated here.

The relationship between the red shift and the distance is now determined by equations (5.9.26), (5.9.29), and (5.9.32), together with the Friedmann-Lemaître equations. We give the result here for the *generalized Hubble relation* (in a universe dominated by matter):

$$cz \cdot \Psi(q_0, z) = H_0 r_L \quad \text{with}$$

$$\Psi(q_0, z) = 1 + \frac{1-q_0}{q_0}\left[1 - \frac{1}{q_0 z}(\sqrt{1+2q_0 z} - 1)\right]$$

$$= 1 + \tfrac{1}{2}(1-q_0)z + \dots \quad (q_0 z \ll 1) . \tag{5.9.33}$$

Corresponding to (5.9.30), r_L is given by the observation of galaxies of known absolute luminosities L. For $z \ll 1$, we obtain the original Hubble relation (5.9.2) again from (5.9.33).

If we express the brightness as usual in magnitudes (4.4.19), we find the following relation between the apparent bolometric magnitude m_{bol} and the red shift:

$$m_{bol} = 5 \log cz + 5 \log \Psi(q_0, z) + M_{bol} - 5 \log H_0 + 25 \quad [\text{mag}] , \tag{5.9.34}$$

where c is measured in $[\text{km s}^{-1}]$ and H_0 in $[\text{km s}^{-1}\,\text{Mpc}^{-1}]$. The absolute magnitude M_{bol} depends on L as given by (4.4.23). For small $q_0 z$, the term $5 \log \Psi(q_0 z) \simeq 1.086(1-q_0)z$.

In practice, instead of the bolometric magnitudes, the brightness integrated over a limited range of wavelengths is measured. We cannot deal here with all the required corrections to (5.9.34) in addition to z and q_0.

Now, which observations are available (some in fact, some only in principle) for determining the *actual parameters* of our cosmos?

We have already mentioned (in Sect. 5.9.1) the measurement of the *Hubble constant* H_0 from the red shift z of the spectra of those galaxies whose distance can be obtained independently (using cepheids, supernovae, etc.), up to $z = 0.1 \dots 0.15$. The large-scale expansion of the entire observable universe can be represented in a *Hubble diagram* (Fig. 5.9.3), in which, in accord with (5.9.34), the red shift z is plotted against the apparent magnitude, e.g. m_V. In order to determine from this diagram (luminosity) distances and thereby H_0, as well as the delay parameter q_0 for objects with the highest values of z, we require galaxies with *known, fixed*, and the largest possible absolute magnitudes, socalled "*standard candles*"; otherwise, the scatter in the Hubble diagram would be too great.

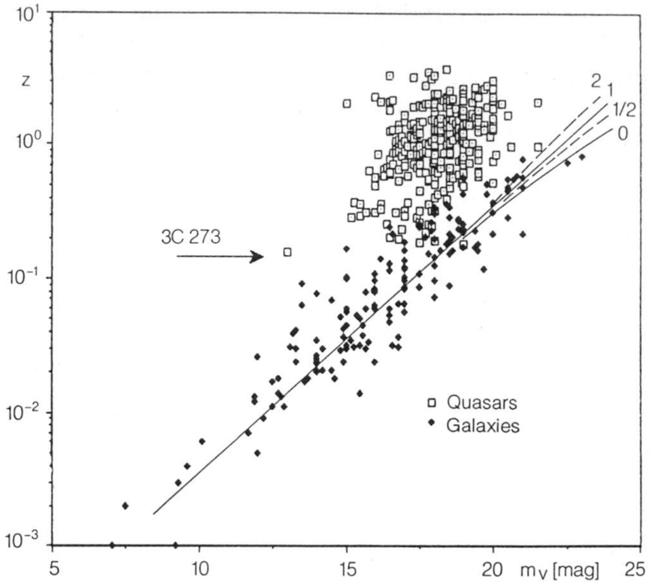

Fig. 5.9.3. A Hubble diagram for radio galaxies and quasars, taken mainly from the 5 GHz complete catalogue of H. Kühr et al. (1981). The red shift $z = \Delta\lambda/\lambda_0$ is plotted against the (uncorrected) apparent visual magnitude m_V. The Hubble relation (5.9.34) is adjusted to the observations for small z and is drawn for deceleration parameters of $q_0 = 0$, 1/2, 1, and 2. In this figure, a straight line of slope 0.2 at small values of z represents (5.9.34). The diagram is from H. Kühr (1987)

Observations of the brightest galaxies in clusters (Sect. 5.5.2) extend to about $z = 0.8 \dots 1.0$. The radio galaxies, whose prototype is Cyg A (Sect. 5.6.2), extend the relation between z and the apparent magnitudes of the normal giant galaxies. With $M_V \simeq -23$ mag, they are well suited to serve as "standard candles", since on the one hand, owing to their strong radio emissions, they can be recognized even at great distances ($z > 1$), and on the other hand, they can be investigated spectroscopically down to $m_V \simeq 23 \dots 24$ mag due to their spectral emission lines.

In recent years, several galaxies with extreme red shifts ($z > 2$) have been discovered; they, however, like the quasars (see below), cannot be used as "standard candles", because their intensity distributions in the optical and radiofrequency regions differ too much from those of galaxies with smaller z values.

Using direct photography through an interference filter, S. Djorgovski et al. in 1985 were able to observe a galaxy with $z = 3.2$ in the neighborhood of a quasar by detecting its strong Lα emission line. In order to photograph this line through a filter, they assumed that the known z of the quasar was approximately the same as that of the galaxy. S. J. Lilly (1988) identified a radio source with a radio galaxy (0902+34) having $z = 3.4$; it also exhibits strong Lα emissions. Then, in the course of a systematic search for strong emission lines from radio galaxies with sharply rising spectra in the radiofrequency region, K. Chambers et al. in 1988 discovered a galaxy (4C 41.17) with $z = 3.8$. The colors in the visible region of the two last-named galaxies indicate star populations more than 10^9 yr old, which therefore must have been formed at a very early time, corresponding to $z \gtrsim 5$.

The *quasars*, with or without radio emissions, attain larger values of z than the galaxies. The largest known red shift of a quasar at present (1991) is $z = 4.9$ (Sect. 5.6.5). The quasars and galaxies with active nuclei exhibit considerable *scatter* in the Hubble diagram, owing to their widely differing and variable core-region activities; they lie between the line representing ordinary giant galaxies and a region of up to 6 magnitudes greater brightness, and are therefore not usable as "standard candles". No quasar, however, has a *smaller* magnitude than would be indicated by the line for ordinary galaxies at its z value. The cosmological interpretation of the red shift of quasars is therefore supported by their positions in the Hubble diagram.

The determination of the *deceleration parameter* q_0 and thus the average density ϱ_0 from the Hubble diagram (Fig. 5.9.3) is very difficult. We must at the outset consider that for very distant galaxies of every type, in the long time interval which their light has required to reach the Earth, their brightnesses, colors, etc. will have changed markedly as a result of their *evolution*, so that a straightforward comparison with objects in our neighborhood is not possible. Taking the large uncertainties in the empirical determination of q_0 into account, we can conclude only that $-1.3 \leq q_0 \leq +2.0$.

The *statistics of the apparent magnitudes* of the galaxies promise further cosmological information: "How many galaxies per square degree have magnitudes (corrected for galactic absorption, etc.) in the region $m \pm 1/2$?". The limiting case of this distribution law for a Euclidean universe was already given in (5.2.1). Here, again, we cannot make much use of the quasars.

The direct measurement of the *average density* ϱ_0 or the density parameter Ω_0 also does not permit a very precise determination, via (5.9.22), of q_0. From the observed mean luminosities in the cosmos, $\mathcal{L}_B \simeq 10^8 L_B^{\odot}\,\mathrm{Mpc}^{-3}$ (5.5.7), together with the mass-luminosity ratios $\mathcal{M}/L_B$ of *galaxies* and *clusters* of galaxies derived from their dynamics (Sects. 5.5.3 and 5.7.1), we find a mean mass density of $\varrho_0 \simeq 10^{10} \mathcal{M}_\odot\,\mathrm{Mpc}^{-3} \simeq 6 \cdot 10^{-28}\,\mathrm{kg\,m}^{-3}$. With the critical density $\varrho_{c,0} = 4.7 \cdot 10^{-27}\,\mathrm{kg\,m}^{-3}$ (5.9.20), this corresponds to $\Omega_0 = \varrho_0/\varrho_{c,0} \simeq 0.14$ [24], although this value can be considered to be accurate only to within a factor of 2 to 3, due in particular to the unknown "dark matter" in the galaxies. We must furthermore take into account a possible contribution to ϱ_0 from *intergalactic* matter (the hot gas *within* the galaxy clusters was already included in the above value). Even a mean particle density of $\simeq 3\,\mathrm{m}^{-3}$ would give $\Omega_0 = 1$ and thus a closed universe. An intergalactic medium consisting of neutral or ionized, relatively cool ($T \leq 10^4$ K) hydrogen is not consistent with observations. In contrast, a hot intergalactic gas ($T \simeq 5 \cdot 10^8$ K), which would contribute to the very diffuse extragalactic X-ray emission and would lead to a density parameter of $\Omega_0 \leq 0.5 \ldots 0.9$, cannot at present be excluded.

Although the contribution of the 3 K background radiation, $\simeq 10^{-3} \varrho_0$, is negligible (Sect. 5.9.4), the background due to low-energy cosmic *neutrinos* (Sect. 5.9.5) may well play an essential role, in case the neutrinos turn out to have a nonzero rest mass. This possibility, which has been suggested by newer elementary-particle theories, has yet to be confirmed experimentally. Since the particle density of the background neutrinos, $\simeq 10^8\,\mathrm{m}^{-3}$, is similar to that of the photons of the 3 K radiation, even a very small neutrino mass of e.g. $10\,\mathrm{eV\,c}^{-2} \simeq 2 \cdot 10^{-35}$ kg would already amount to 0.3 of the critical density $\varrho_{c,0}$.

Finally, the observation of the abundance of *deuterium* in the cosmos offers another possibility for determining the density parameter indirectly (Sect. 5.9.4).

If we now attempt, in spite of all the uncertainties, to specify a cosmological model, we must take the large er-

[24] The uncertainty in the Hubble constant H_0 does *not* enter the density parameter Ω_0, since both ϱ_0 and $\varrho_{c,0}$ are proportional to H_0^2.

ror limits into account: those of the Hubble constant H_0, the delay parameter q_0, and the Friedmann time τ_F (Table 5.9.1), which gives the maximum age of cosmic formations. For this latter quantity, age determinations of globular clusters (Sect. 5.4.4) and radioactive dating (Sect. 5.8.5) are relevant. These methods lead to an age of 14 to $25 \cdot 10^9$ yr.

In summary, we estimate the most probable values for the parameters of our universe:

Hubble constant	$H_0 = 50\ \mathrm{km\,s^{-1}\,Mpc^{-1}}$
Deceleration parameter	$q_0 = 0.05$
or density parameter	$\Omega_0 = 0.1$
Hubble time	$\tau_0 = 19.6 \cdot 10^9$ yr
Friedmann time	$\tau_F = 17.6 \cdot 10^9$ yr
Critical density	$\varrho_{c,0} = 4.7 \cdot 10^{-27}\ \mathrm{kg\,m^{-3}}$
	$n_{c,0} = 3\ \mathrm{m^{-3}}$
Average density	$\varrho_0 = 4.7 \cdot 10^{-28}\ \mathrm{kg\,m^{-3}}$
	$n_0 = 0.3\ \mathrm{m^{-3}}$

(5.9.35)

(For a Hubble constant of $H_0 = 100\ \mathrm{km\,s^{-1}\,Mpc^{-1}}$ and $q_0 = 0.05$, we would find the Hubble time to be $\tau_0 = 9.8 \cdot 10^9$ yr, the Friedmann time $\tau_F = 8.8 \cdot 10^9$ yr, and the critical and average densities to be $n_{c,0} = 12\ \mathrm{m^{-3}}$ and $n_0 = 1.2\ \mathrm{m^{-3}}$). Therefore, we are probably in an *open*, continually expanding universe.

In contrast to this, the hypothesis of an inflationary universe suggested by recent unified elementary-particle theories (Sect. 5.9.5) would lead to values near $q_0 = 1/2$ and $\Omega_0 = 1.0$, corresponding to the limit of a Euclidean (likewise monotonically expanding) universe.

5.9.4 The Origin of the Elements and the Microwave Background Radiation. The Radiation Cosmos

Our interest now naturally turns to the initial stage of cosmic evolution, the *Big Bang*. Fundamental research on this topic (beginning about 1939) is inseparably connected with the names of G. Lemaître and G. Gamow.

At the enormously high temperatures of the *early universe*, we have to assume that for $kT \geq mc^2$, every elementary particle of mass m could be converted into every other such particle (owing to energy conservation). The theory of the initial stages of the expanding universe is therefore closely related to *elementary particle physics* (Sect. 5.9.5). Here, we first skip over the very earliest phases, and consider the universe from about 200 s after the singularity, when it had cooled off to around 10^9 K ($kT = 0.1$ MeV). At this temperature, the synthesis of the chemical elements (nuclides) from protons and neutrons could begin, since then *deuterium*, once formed, would no longer be destroyed by reactions with energetic particles, especially photons, in spite of its low binding energy of 2.2 MeV. Deuterium represents the starting nucleus for the gradual construction of heavier elements.

G. Gamow (1948) originally attempted to place the synthesis of *all* the heavier elements in this time period. This idea however proved to be untenable, since the formation of nuclides would come to a halt already at mass number A = 5, for which no stable nucleus exists. At this stage, practically only isotopes of *hydrogen* and of *helium* were formed; the synthesis of heavier nuclei occurred at considerably later times in the interiors of stars (Sect. 5.4.5). Measurement of the current abundance ratios of He/H and D/H, particularly in objects which contain original cosmic matter, thus gives us important information about the state of the early universe.

Gamow already noticed that we could also obtain *direct* information, so to speak, about this stage of the "Primeval Fireball", in addition to the element abundances. This is because shortly after this stage, the interaction between radiation and matter had become so weak that the *radiation field* of the cosmos could only expand *adiabatically* along with the rest of the universe. L. Boltzmann had shown long before that the radiation field of a black body radiation cavity remains black during adiabatic expansion, and that furthermore, the product $T^3 \times V$, the volume of the cavity, remains constant. However, after the completion of formation of the atoms (H+He), their particle number nV was also conserved, i.e. $T^3 \propto n$ would have to decrease. From nuclear-physics considerations, Gamow began with $T = 10^9$ K and $n \simeq 10^{24}\ \mathrm{m^{-3}}$ at the time of the formation of the elements. For the present-day universe, he used an average value of $n \simeq 1\ \mathrm{m^{-3}}$. After expansion to a 10^{24}-fold volume, he concluded, the present-day cosmos must be filled with *black-body radiation* at a temperature of the order of 10 K. The detection of this radiation at that time was, however, out of the question, given the state of the art of radioastronomical observational techniques.

Making use of the enormously improved technology in radioastronomy in the meantime, R.H. Dicke, P.J.E. Peebles, P.G. Roll, and D.T. Wilkinson, guided by newer calculations, began in 1964 to search for the cosmic background radiation. However, before they had completed their measurements, this microwave background radiation was discovered coincidentally in 1965 by A. A. Penzias and R. W. Wilson; it was immediately recognized as such by Dicke et al. Penzias and Wilson detected weak radiofrequency radiation at $\lambda = 7.35$ cm (4.08 GHz) using a large, low-noise horn antenna, which was originally constructed for purposes of communications via the Echo satellite. After subtraction of the contribution from the Earth's atmosphere and the receiver noise, an isotropic, unpolarized component remained, which was independent of the time of day or of the season and had an unexpectedly high excess-antenna temperature of (3.5 ± 1) K. It was later confirmed that this radiation was in fact the remnant of the Big Bang, through the observation that it follows *Planck's law* in the entire accessible wavelength region (Fig. 5.9.4), and is isotropic and unpolarized. For the temperature of the background radiation, the weighted average found from several measurements by G.F. Smoot et al. (1985) is $T_0 = (2.73 \pm 0.05)$ K. The intensity I_ν of this *3 K radiation* has its maximum at $\lambda = 1.7$ mm (180 GHz); on the short-wavelength side, it falls off rapidly. Below $\lambda \leq 3$ mm, the observations must be carried out from balloons or rockets. Above $\lambda \geq 30$ cm, the galactic nonthermal radiofrequency radiation predominates.

Since November, 1989, the COBE satellite (*Co*smic *B*ackground *E*xplorer), in a circular orbit at an altitude of 900 km, has been carrying out a survey in the microwave region with high sensitivity. During a period of one to two years, the isotropy of the 3 K radiation (at wavelengths of 3.3, 5.7, and 9.6 mm) and possible deviations from the Planck distribution (in the range from 0.1 to 10 mm) will be investigated. Also, the diffuse infrared radiation at 1 to 300 μm wavelength will be measured to search for indications of stars or galaxies from the early phases of the cosmos, whose radiation has been redshifted into this wavelength range.

Preliminary results of the observations by COBE confirm that the microwave radiation corresponds within 1% to a Planck distribution with a temperature $T = (2.735 \pm 0.06)$ K.

In fact, the background radiation had already been observed indirectly in 1941 by A. McKellar: from the intensity ratios of the interstellar absorption lines of CN at $\lambda = 387.46$ and 387.39 nm in the spectrum of ζ Oph, he derived an excitation temperature of 2.3 K for the first rotational level, lying $4.7 \cdot 10^{-4}$ eV (corresponding to $\lambda = 2.64$ mm) above the ground state; this was a completely incomprehensible result at the time. The modern value of "optical observations" of the 3 K radiation from lines of CN, CH, and CH^+ is 2.7 to 2.9 K.

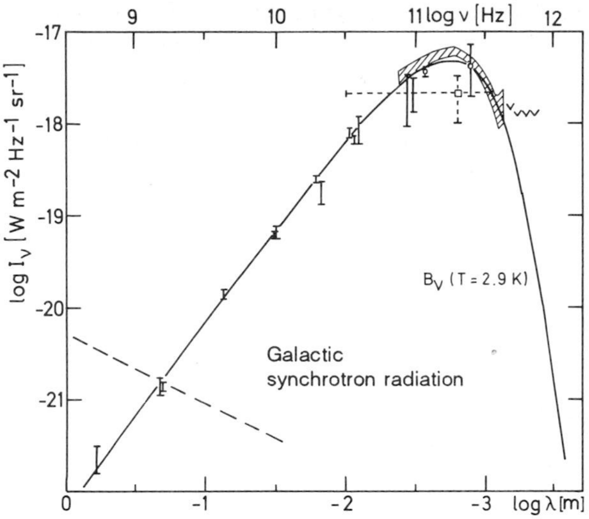

Fig. 5.9.4. The spectrum of the microwave background radiation, taken from observations in the microwave region (|), in the optical region (○), and in the infrared region (□ or shaded area); the upper limits of some infrared measurements are indicated by v's. The first observation of the background radiation by A. A. Penzias and R. W. Wilson (1965) was made at $\lambda = 7.35$ cm or 4.08 GHz. The observational results are compared to the spectrum of a black body radiator at $T = 2.9$ K. After D. P. Woody and P. L. Richards (1981) and G. F. Smoot (1983). According to newer measurements, T is 2.73 K. At long wavelengths, the nonthermal radio emissions of the Milky Way predominate

In recent times, interest has turned to possible *deviations* of the 3 K background radiation from the Planck distribution and from isotropy; these could give us information about "perturbations" in the density and velocity distributions during the early phases of the cosmos. "Wiggles" in the black-body spectrum, which up to the present time have been below the limit of detection, would give us indications of deviations from a completely adiabatic expansion of the radiation field, i.e. an energy input or dissipation. The high degree of isotropy of the background radiation confirms the cosmological postulate and gives a well-defined upper limit for possible temperature fluctuations in the early universe.

The only *anisotropy* in the intensity or the temperature of the 3 K radiation observed up to now has an angular dependence proportional to $\cos\theta$ ("dipole characteristic") with a relative amplitude of $\leq 10^{-3}$. It is a result of the *motion of the Earth* relative to the coordinate system fixed with respect to the background radiation. Due to the Lorentz transformation, an observer moving at a velocity v relative to this coordinate system measures a Planck distribution in every direction θ, but its temperature depends on θ:

$$\frac{T(\theta)-T_0}{T_0} \simeq \frac{v}{c}\cos\theta \quad (v \ll c) \ . \tag{5.9.36}$$

The maximum temperature rise is seen in the direction of the Earth's motion, or that of the Sun. The observed value is $\Delta T_{\max} = (3.2 \pm 0.2) \cdot 10^{-3}$ K, corresponding to $v \simeq 350$ km s^{-1} towards a point with the coordinates $\alpha \simeq 11$ h, $\delta \simeq -10°$ (D. J. Fixen, E. S. Chang, and D. T. Wilkinson, 1983). The main effect comes from the rotation of the Sun around the center of the Milky Way galaxy. If we relate the velocities to the center of mass of the *local group*, we find that it is moving at about 600 km s^{-1} relative to the cosmological coordinate system. The direction of the motion, which cannot be determined very precisely ($\alpha \simeq 10.4$ h, $\delta \simeq -27°$), is about 50° away from the center of the Virgo cluster.

The intensity variation of the 3 K radiation due to Compton scattering during transmission through the hot gas in the clusters of galaxies was already mentioned in Sect. 5.7.2.

The energy density of a black-body radiation field at a temperature T is given by the Stefan-Boltzmann law (4.2.27 and 39), $u = aT^4$, with $a = 7.56 \cdot 10^{-16}$ J·m^{-3} K^{-4}. For the microwave background radiation, at $T_0 = 2.7$ K, we find for the corresponding *mass density*:

$$\varrho_{\gamma,0} = \frac{u_0}{c^2} \simeq 4.5 \cdot 10^{-31} \text{ kg m}^{-3} \ . \tag{5.9.37}$$

(We again denote the present-day state of the cosmos by the index 0.) This is negligible compared to the present matter density, $\varrho_0 \simeq 5 \cdot 10^{-28}$ kg m^{-3} for a density parameter of $\Omega_0 = 0.1$ (5.9.35); i.e., the assumption of a *matter cosmos* (Sect. 5.9.2) for the present-day universe is clearly justified.

Although the mass density of the baryons[25], $\varrho_{b,0} \simeq \varrho_0 \simeq 10^3 \varrho_{\gamma,0}$, is strongly predominant today, the radiation field density was dominant during the early phases of the universe, since on expansion, the density of the radiation field decreases more rapidly than that of (nonrelativistic) matter. In order to describe this *radiation cosmos*, we first derive a conservation law:

$$\frac{d}{dt}(\varrho c^2 R^3) + p\frac{d}{dt}(R^3) = 0 \tag{5.9.38}$$

by modifying the two Friedmann-Lemaître equations; (5.9.18) is multiplied by $R(t)^3$ and differentiated with respect to time, and (5.9.19), after multiplication by $\dot{R}R^2 = \frac{1}{3}d(R^3)/dt$, is subtracted from it. To evaluate this conservation law, we still need the equation of state $p = p(\varrho)$. For the matter cosmos, with $p \simeq 0$ and $\varrho \simeq \varrho_b$, we obtain (5.9.10) once again:

$$\varrho_b(t)R(t)^3 = \text{const.} \tag{5.9.39}$$

For the *radiation cosmos*, the equation of state of a relativistic gas applies; in it, the pressure is equal to 1/3 times the energy density:

$$p_r = \tfrac{1}{3}u_r = \tfrac{1}{3}\varrho_r c^2 \ , \tag{5.9.40}$$

and therefore cannot be neglected. By integration of (5.9.38) we then obtain

$$\varrho_r(t)R(t)^4 = \text{const.} \tag{5.9.41}$$

In a radiation cosmos (index r), the density and the pressure are due not only to the photons (index γ), but rather to *all* relativistic particles. The fact that ϱ_r decreases on expansion of the universe by an additional factor of $R(t)$ relative to ϱ_b is a result of the energy decrease $\propto R^{-1}$ from the red shift, which must be considered in addition to the decrease from the increasing volume, $\propto R^3$, which affects both densities.

Now, starting from the present-day densities of matter and of the 3 K radiation, we can estimate the time t^* at which ϱ_r was equal to ϱ_b, i.e. the time before which the condition for a radiation cosmos was fulfilled. From (5.9.29), (5.9.39), and (5.9.41), t^* is given by:

$$\frac{R_0}{R(t^*)} = 1 + z^* = \frac{\varrho_{b,0}}{\varrho_{r,0}} = \alpha\frac{\varrho_{b,0}}{\varrho_{\gamma,0}} \ . \tag{5.9.42}$$

For this equation, we require $\alpha \equiv \varrho_{\gamma,0}/\varrho_{r,0}$. The ratio α is determined by the present-day energy density of the neutrino background radiation, which in turn depends on assumptions about the various types of neutrinos (Sect. 5.9.5). For the "standard cosmos", with three (massless) neutrino types (the electron neutrino ν_e, the muon neutrino or neutretto ν_μ, the tau neutrino ν_τ, and their respective antiparticles), we find $\alpha \simeq 0.6$. With $\varrho_{b,0}/\varrho_{\gamma,0} \simeq 10^3$, we obtain as a first estimate $z^* \simeq 600$.

[25] The term "baryons" includes neutrons, protons, and hyperons. The matter in the *present-day* cosmos consists predominantly of hydrogen, so that the baryon density is roughly equal to that of the H atoms.

The relationship between R and t in the *radiation cosmos*[26] (for $t \leq t^*$) is then obtained from the Friedmann-Lemaître equation (5.9.18), where the curvature term is negligible due to $R(t) \leq 10^{-3} R_0$:

$$\dot{R}^2 = \frac{8\pi G}{3} \varrho R^2 = \frac{8\pi G}{3} \frac{K}{R^2} \ . \tag{5.9.43}$$

Since, according to (5.9.41), $K = \varrho R^4 \simeq \alpha^{-1} \varrho_{\gamma,0} R_0^4$ is constant, the integration yields:

$$\frac{R(t)}{R_0} = \left(\frac{t}{t_r}\right)^{1/2}$$

with (5.9.44)

$$t_r \simeq \left(\frac{32\pi G}{3} \frac{\varrho_{\gamma,0}}{\alpha}\right)^{-1/2} \simeq 2 \cdot 10^{19} \mathrm{s} \ .$$

From this result, radiation or relativistic particles dominated the cosmos up to the time $t^* = t_r [R(t^*)/R_0]^2 \simeq 2 \cdot 10^6$ yr after the singularity.

Equation (5.9.41) also holds separately for each single type of relativistic particles, insofar as they have already "decoupled" from the remaining matter, i.e. they expand adiabatically along with the latter without significant interactions. For example, the photons interact with matter mainly through electrons. If the temperature in the universe sinks in the course of its evolution to about 3000 K, then the number of free electrons decreases drastically owing to recombination of hydrogen, and the radiation field decouples from the (nonrelativistic) matter. We can thus estimate the time t_γ after which the radiation field has expanded adiabatically, and finally cooled off to the present-day temperature of $T_0 = 2.7$ K: due to $\varrho_\gamma \propto T^4$, it follows from (5.9.41) and (5.9.29) for the time dependence of the temperature:

$$\frac{T(t)}{T_0} = \frac{R_0}{R(t)} = 1 + z \ . \tag{5.9.45}$$

This result indicates that our cosmos became practically transparent to radiation at about $z_\gamma \simeq 1000$. According to (5.9.44), this corresponds to $t_\gamma \simeq 8 \cdot 10^5$ yr. The observation of the microwave background radiation thus allows us a view into the past, back to the state of the universe about $8 \cdot 10^5$ yr after the singularity.

We now consider the *photon density* n_γ of the background radiation, which is obtained from the energy density u by dividing it by the average energy of the photons, $\langle h\nu \rangle \simeq 2.7\, kT$:

$$n_\gamma = \frac{u}{\langle h\nu \rangle} = 2.0 \cdot 10^7 (T[\mathrm{K}])^3 \quad [\text{photons m}^{-3}] \ . \tag{5.9.46}$$

At the present time, $n_{\gamma,0}$ is $\simeq 4 \cdot 10^8$ photons m^{-3}, so that for each baryon or hydrogen atom [$n_{\mathrm{b},0} \simeq 0.3$ m^{-3} for $\Omega_0 = 0.1$; (5.9.35)], there are about 10^9 photons from the 3 K radiation. In contrast to the mass or energy density of the radiation, the photon density is proportional to $R(t)^{-3}$. The ratio of photon to baryon *number* densities therefore remains constant during the expansion of the universe:

$$\frac{n_\gamma}{n_\mathrm{b}} = \frac{n_{\gamma,0}}{n_{\mathrm{b},0}} \simeq 10^9 \ . \tag{5.9.47}$$

The model chosen for the early, radiation-dominated cosmos, which is based essentially on the present-day temperature of the microwave background radiation, naturally also determines the course of the nuclear processes in the *synthesis of the chemical elements*; the background radiation was, in fact, predicted by G. Gamow et al. from theoretical considerations of just this process of element synthesis. Right after the discovery of the 3 K radiation, P. J. E. Peebles (1966) and R. V. Wagoner, W. A. Fowler, and F. Hoyle (1967) carried out detailed calculations of element synthesis. According to these and more recent calculations, for a ratio $n_\gamma/n_\mathrm{b} \simeq 10^9$, the formation of the elements from protons and neutrons began about 220 s after the singularity, when the temperature had fallen to $0.9 \cdot 10^9$ K. Mainly ^{4}He nuclei, but to a lesser extent also ^{2}D, ^{3}He, and ^{7}Li nuclei were formed. The free neutrons were almost entirely used up in the synthesis of ^{4}He. If we denote the densities of neutrons and protons immediately before the beginning of element synthesis by n_n and n_p, then $n_\mathrm{n}/2$ helium nuclei could have been formed, since a helium nucleus contains two neutrons, while $(n_\mathrm{p} - n_\mathrm{n})$ protons were left over. The *helium abundance* in numbers of atoms

$$\frac{n(^4\mathrm{He})}{n(\mathrm{H})} \simeq \frac{n_\mathrm{n}}{2(n_\mathrm{p} - n_\mathrm{n})} \ ,$$

or in terms of the mass ratio (5.9.48)

[26] Our estimate, which is based on (5.9.41) and not on the general (5.9.38), does not hold for the very early phases of the cosmos [$t \leq 1$ s; $R(t) \leq 10^{-9} R_0$]. Above $T \geq 10^{10}$ K, there are numerous kinds of relativistic particles and the ratio α is not constant in time.

$$\frac{Y}{X} \simeq \frac{2n_n}{n_p - n_n},$$

with $X+Y=1$, thus depends, for a wide range of values of the density parameter Ω_0, practically only on the ratio n_n/n_p.

Above $T \geq 10^{10}$ K ($kT \geq 1$ MeV), i.e. before the decoupling of the neutrinos (Sect. 5.9.5), the neutrons and protons were in thermodynamic equilibrium, and then according to the Boltzmann formula, the ratio n_n/n_p is given by

$$\frac{n_n}{n_p} = \exp\left(-\frac{\Delta m c^2}{kT}\right) \tag{5.9.49}$$

(where $\Delta m c^2 = 1.293$ MeV corresponds to the mass difference between the neutron and the proton). This ratio decreased monotonically during the expansion of the universe, starting from an initial value of 1. Below 10^{10} K, the reactions which convert neutrons into protons and *vice versa* became more and more rare, until finally only the β decay of neutrons into protons ($n \to p + e^- + \bar{\nu}_e$), with a half-life of 617 s, remained. The neutron to proton ratio then decreased further until, at the beginning of element synthesis, it had the value $n_n/n_p \simeq 0.14$. According to (5.9.48), this leads to a helium abundance of n(He) /n(H) $\simeq 0.08$ or $Y \simeq 0.25$, a result which agrees rather well with observations (5.5.15).

In contrast to ^{4}He, the formation of *deuterium* depends sensitively on the baryon density. If we regard the abundance of $n(^2\mathrm{D})/n(^1\mathrm{H}) \simeq 10^{-5}$ observed in the interstellar medium as primordial, then it follows that the density parameter $\Omega_0 \lesssim 0.1$. We therefore are living in an *open*, continually expanding cosmos. To be sure, for a trace element such as deuterium, the possibility of a significant rate of formation at considerably later times, and thus a larger Ω_0, cannot be completely excluded.

To conclude this section, we mention briefly the significance for the cosmological models of *stellar radiation* in the cosmos. It was already realized by J. Kepler (1610), E. Halley (1720), J.-P. Loy de Cheseaux (1744), and H. W. M. Olbers (1826), that the simple observation that the night sky is *dark* allows far-reaching conclusions to be drawn about the large-scale structure of the cosmos. The socalled *de Cheseaux-Olbers paradox* states the following: *if* the universe is spatially and temporally infinite and (more or less) uniformly filled with stars, then, in the absence of absorption, the entire sky would be illuminated with a brightness corresponding to the average surface brightness of the stars, i.e. about that of the Sun. The fact that this is not the case cannot be explained in terms of interstellar absorption alone, since the absorbed energy would not be lost and must reappear as radiation. In order to eliminate the paradox, however, it is sufficient to take into account the *limited* period of time ($\lesssim 10^{12}$ yr) during which the stars can maintain their luminosities. (The red shift of the radiation due to the expansion of the universe plays, in contrast, only a secondary role.) A simple energy estimate also argues against a cosmic radiation field with a temperature of the order of 10^4 K: if *all* the matter in the present-day cosmos were to be converted completely into radiation, then a temperature of only about $2.7 \cdot 10^{3/4} = 15$ K would be reached (owing to $\varrho_\gamma \propto T^4$, $\varrho_0 \simeq 10^3\ \varrho_{\gamma,0}$, and $T_0 = 2.7$ K); this temperature is obviously much lower than 10^4 K.

5.9.5 The Evolution of the Cosmos

After having reached some conclusions about the state of the universe in the distant past (except for the very earliest epochs) by making use of the observations of galaxies and of the 3 K radiation to construct cosmological models, we shall now describe the evolution of the cosmos as a continuous process from the initial extremely hot and dense phases up to the present time. We do this within the framework of the *standard model*, which is based on *general relativity* and the *cosmological postulate* of large-scale homogeneity and isotropy.

The fundamental parameters of our cosmos, which are in agreement with all the available empirical data, are summarized in (5.9.35 and 47):

In the present-day cosmos, the average matter density, for a Hubble constant of $H_0 = 50\ \mathrm{km\,s^{-1} \cdot Mpc^{-1}}$, is $\varrho_0 = 4.7 \cdot 10^{-28}\ \mathrm{kg\,m^{-3}}$ or $n_0 = 0.3\ \mathrm{m^{-3}}$, corresponding to a density parameter $\Omega_0 = 2q_0 = 0.1$; the temperature of the background radiation field is $T_0 = 2.7$ K and the ratio of photon to baryon densities is therefore 10^9. The oldest age of any object in the cosmos is given by the Friedmann time, $\tau_F = 17.6 \cdot 10^9$ yr.

We now first consider the *initial stages* of the universe.

The time dependence of the temperature T or the average thermal energy kT in the (radiation) cosmos is, according to (5.9.44) and (5.9.35), approximately equal to

$$t\,[\mathrm{s}] \simeq \left(\frac{10^{10}}{T[\mathrm{K}]}\right)^2 \simeq \left(\frac{10^{-3}}{kT[\mathrm{GeV}]}\right)^2 \quad \text{for} \quad t \lesssim 2 \cdot 10^6\,\mathrm{yr}\ , \tag{5.9.50}$$

if we leave out of consideration the details relating to the different types of relativistic particles.

However, we cannot use the standard model to describe the cosmos for arbitrarily short times after the singularity, since at very high mass concentrations and small dimensions, the theory of General Relativity, which treats space-time as a continuum, must be replaced by a quantum theory of gravitation. Such a theory has at present been only roughly formulated.

Following M. Planck, we can obtain the time τ_P or the associated length l_P and mass M_P, below which Einstein's theory of gravitation loses its validity, by forming natural units from the three constants G, c, and h or $\hbar$, which characterize the three fundamental theories of physics, general and special relativity and quantum mechanics:

$$\begin{aligned} \tau_P &= \left(\frac{G\hbar}{c^5}\right)^{1/2} \simeq 5.4\cdot 10^{-44}\,\mathrm{s} \ , \\ l_P &= c\tau_P \simeq 1.6\cdot 10^{-35}\,\mathrm{m} \ , \\ M_P &= \left(\frac{c\hbar}{G}\right)^{1/2} \simeq 2.2\cdot 10^{-8}\,\mathrm{kg} \ , \quad \text{and} \\ M_P c^2 &\simeq 1.2\cdot 10^{19}\,\mathrm{GeV} \ . \end{aligned} \tag{5.9.51}$$

In the framework of relativistic cosmological models, we can thus make no statements for example about times previous to the *Planck time* τ_P. The Planck mass and length can also be estimated to an order of magnitude by setting the size of a black hole of mass M_P, i.e. its Schwarzschild radius GM_P/c^2, equal to the length which is related to the fixed momentum $M_P c$ by Heisenberg's uncertainty relation, $M_P c l_P \simeq \hbar$.

In the early stages of the universe, the temperatures and densities were so great that the photons and a great variety of other relativistic particles were in *thermodynamic equilibrium*. Pair creation and annihilation of particles and antiparticles as well as their interactions with photons and other particles maintained an equilibrium for a particular species of particles of rest mass m, as long as the thermal energy $kT \gg mc^2$. When the energy dropped below mc^2 as a result of the expansion of the cosmos, particles of mass m which had decayed or been annihilated could no longer be reformed. The evolution of the early cosmos was therefore characterized by a sequential "dying out" of the various types of particles, beginning with the most massive, until after about 3 s, practically only photons and neutrinos remained and the fraction of baryons was only about 10^{-9} of the total.

The details of the cosmological model for the earliest phases depend decisively on the *physics of elementary particles*, and the great progress in this area in recent years has had a major effect on cosmology. We are concerned here for the most part with the *theory* of elementary particles and their interactions, since the energies involved, of 10^{19} GeV or more, are far above those accessible experimentally. Particle accelerators which are currently in operation or planned can reach energies of only 10^3 to 10^4 GeV.

Before we discuss the evolution of the cosmos after the first 10^{-43} s, it seems appropriate to give a brief overview of the more important concepts and results of elementary particle physics.

The *fundamental particles*, which show no indications of an internal structure, are currently considered to comprise the leptons and the quarks with their corresponding antiparticles.

The *leptons* are known to occur in three families: the first includes the *electron* e^-, its antiparticle, the positron e^+, and the associated electron *neutrino* ν_e with its antiparticle $\bar{\nu}_e$; the second family includes the muons μ^-, μ^+, and the muon neutrinos ν_μ and $\bar{\nu}_\mu$; and the third family consists of the tau particles τ^- and τ^+ (discovered only in 1975) and their neutrinos ν_τ and $\bar{\nu}_\tau$. Recent theories of elementary particles suggest that the neutrinos may have nonzero rest masses. Experimentally, up to now only upper limits could be determined, e.g. for the electron neutrino, $m(\nu_e) \lesssim 10\,\mathrm{eV}\,c^{-2}$; an "astrophysical" estimate from the neutrino emission accompanying the outburst of the bright supernova SN 1987 A (Sect. 4.11.7) yields $m(\nu_e) \lesssim 10 \dots 30\,\mathrm{eV}\,c^{-2}$.

The *quarks*, whose electric charges Q are found to be multiples of one-third of the elementary charge, and which were originally postulated theoretically by M. Gell-Mann and G. Zweig in 1963, include at present 5 species, termed *up* ($Q = +\frac{2}{3}$), *down* ($-\frac{1}{3}$), *strange* ($-\frac{1}{3}$), *charmed* ($+\frac{2}{3}$), and *bottom* ($-\frac{1}{3}$), and their corresponding antiquarks. An additional quark, the *top* quark, has also been predicted. The quarks are additionally distinguished by (three possible) *color* quantum numbers. Normally, quarks do not occur in the free state, but only as bound particles within the *hadrons: a baryon* (or antibaryon) consists of three quarks or antiquarks; a *meson* of a quark-antiquark pair. The structure is thus e.g. for the proton, $p = \{\mathrm{uud}\}$, for the neutron $n = \{\mathrm{ddu}\}$, for the hyperon $\Lambda^0 = \{uds\}$; for the pion $\pi^+ = \{u\bar{d}\}$, and for the kaon $K^+ = \{u\bar{s}\}$.

The *fundamental interactions* or forces between the various particles are summarized in Table 5.9.2. Through the theoretical works of G. Glashow, A. Salam, and S. Weinberg in the 1960's, it became clear that the *electromagnetic* interaction, which according to quantum electrodynamics is mediated by the exchange of massless *photons*, together with the *weak* interaction, which produces e.g. nuclear β-decay, can be regarded as different aspects of a single, "unified" *electroweak* interaction. The field quanta of the weak force predicted by this theory, the *vector bosons* $W^\pm$ and Z^0, with masses around 80 GeV c^{-2}, were observed in 1983 using the proton-antiproton accelerator SPS at the European nuclear research center CERN. Above $m(W^\pm, Z^0)\cdot c^2 \simeq 100$ GeV, the electromagnetic and the weak interactions are therefore of comparable strength; below this energy, a "symmetry breaking" takes place. The *strong* interaction, which according to quantum chromodynamics is mediated by the exchange of massless *gluons*, affects only the quarks, but not the leptons. The nuclear binding force is not an elementary force, but rather is an indirect result of the interactions between the quarks. The combination of the strong and the electroweak forces in *Grand Unified Theories* ("GUT theories") requires the introduction of additional field quanta, the *X bosons* with masses of the order of 10^{14} GeV c^{-2} (?). Above $m(X)\cdot c^2 \simeq 10^{14}$ GeV, the strong and the electroweak forces would thus also be similar. Finally, it is speculated that above 10^{19} GeV, a "superunification" of the gravitational force with the other elementary forces could occur ("supergravity").

The *masses* of the most important particles are listed in Table 5.9.3, along with the time dependence of the temperature and the thermal energy (5.9.50) during the expansion of the universe, so that we can follow the sequence of the annihilation of the different species of particles. We shall make use of this table in the following to describe the *evolution of the cosmos*, which is usually divided into different *eras* according to the predominant type of particles.

Between 10^{-43} and 10^{-35} s, we can expect from the *Grand Unified Theory* that a completely symmetrical state existed: all the interactions except for gravitation had the same strength; in the extremely hot and dense plasma, quarks, gluons, X-particles, leptons, photons, and their antiparticles were all present in about equal concentrations and were continually being interconverted. After the average energy in the cosmos decreased, as a result of its expansion, to a value less than the equivalent mass of the hypothetical X-particles, 10^{14} GeV, which as field quanta mediate the conversion of quarks into leptons and *vice versa*, the decaying X-particles were no longer replaced and the strong and electroweak interactions separated. The decay of the X-particles could have produced an extremely small relative *excess of quarks* over antiquarks of the order of 10^{-9} to 10^{-10}; in later phases of the cosmos, this excess then led to a corresponding ratio of baryons to photons, which finally can be observed today by comparing the present density of matter and of the 3 K radiation (5.9.47). The formation of the galaxies, and thus our own existence, was therefore a result of a small asymmetry in the decay of the X-particles only 10^{-35} s after the singularity. In this connection, the experimental confirmation of the theoretically predicted *proton decay*, $p \rightarrow e^+ + \pi^0 \rightarrow e^+ + 2\gamma$, would be of decisive importance; in spite of the extremely long predicted half-life of $\geq 10^{32}$ yr, such a confirmation may prove to be possible.

In the *quark era*, which followed the decay of the X-particles, a further symmetry breaking occurred at about

Table 5.9.2. The fundamental interactions. The strength of the interactions or the effective coupling constants are energy dependent; the values listed are for low energies ($\ll 100$ GeV). $\alpha = e^2/(2\varepsilon_0 hc) \simeq 1/137$ is the fine structure constant of atomic physics

Interaction	Strong		Electroweak		Gravitational
			Electromagnetic	Weak	
Strength	$\simeq 1$		$\alpha \simeq 10^{-2}$	$\simeq 10^{-5}$	$\simeq 10^{-39}$
Acts on	Quarks, indirectly on hadrons		Electrically charged particles	Leptons and quarks (or hadrons)	All particles
Quantum exchanged	Gluons, X-particles		Photon γ	Vector bosons	(Graviton)
Mass [GeV c^{-2}]	0	10^{14}?	0	83 ($W^\pm$), 91 (Z^0)	0

10^{-10} s or 100 GeV (corresponding to the masses of the $W^{\pm}$ and Z^0 bosons); at this time, the electromagnetic and the weak interactions separated. At average energies in the range around 1 GeV, the quarks, antiquarks, and gluons could finally no longer exist as free particles. The quarks and antiquarks combined to form hadrons, initiating, at about 10^{-6} s, the *hadron era*. The excess of quarks from the decay of the X-particles was transformed into an excess of baryons relative to antibaryons. When the energy of the cosmos decreased further, the hyperons and heavy mesons first annihilated with their corresponding antiparticles, then the neutrons, and finally the protons and the lighter mesons.

After the lightest hadrons, the π mesons, had decayed at about 10^{-4} s, the *lepton era* began. It was dominated by electrons, positrons, neutrinos, and photons. The muons were already annihilated at the beginning of this era, and the few (10^{-9}) baryons played no role in its energy density. At about 1 s, the interactions of the (electron) neutrinos with the other particles became so weak that they were no longer able to maintain thermal equilibrium. This decoupling of the neutrinos was the essential factor in determining the proton-neutron ratio, which in turn determined the abundance of the helium which was later formed. The neutrino temperature has continued to decrease up to the present time and is now 2 K. At about 3 s, the lepton era ended with the annihilation of the electrons and positrons except for a very small fraction of the electrons, which, together with the protons, guarantee charge neutrality of the universe.

Table 5.9.3. The various eras in the evolution of the cosmos. Particles and antiparticles of rest mass m annihilate each other when the average energy kT becomes $\leq mc^2$. The temperatures below $5 \cdot 10^9$ K refer to the photon gas. The mass values for the quarks, which do not exist as free particles at $kT \leq 1$ GeV, depend on the energy resolution (1 GeV for u, d, and s; mc^2 for c and b); for the lighter quarks, they differ strongly from the relevant constituent masses in the nonrelativistic quark model

Age	t [s]	T [K]	kT [GeV]	m [GeV c^{-2}]: Field quanta, quarks		Hadrons: Baryons		Hadrons: Mesons		Leptons		
"Grand Unification"	10^{-43}	10^{32}	10^{19}									Planck time
	10^{-35}	10^{27}	10^{14}	(10^{14}?)	X?							Symmetry breaking: strong – electroweak
	10^{-10}	10^{15}	10^{2}	91	Z							Symmetry breaking: electromagnetic – weak
				83	W							
Quark era								⋮				
				(>40)	t?							
				(4.8)	b			3.1	J/ψ			
				(1.5)	c	1.67	Ω			1.78	τ	
						⋮						
	10^{-6}	10^{13}	1			1.12	Λ	⋮				Transition: Quarks →Hadrons
						0.940	n					
Hadron era						0.938	p					
								0.50	K			
	10^{-4}	10^{12}	10^{-1}	(0.18)	s			0.14	π	0.11	μ	
				(0.009)	d							
Lepton era				(0.005)	u							
	1	10^{10}	10^{-3}									Decoupling of the neutrinos (→ 2 K)
	3	$5 \cdot 10^{9}$	$5 \cdot 10^{-4}$							$5 \cdot 10^{-4}$	e	
Photon era	$2 \cdot 10^{2}$	$9 \cdot 10^{8}$										Element synthesis (^{4}He)
	$2 \cdot 10^{13}$	$3 \cdot 10^{3}$	$z \simeq 1000$									Decoupling of the photons (→ 3 K)
	$6 \cdot 10^{13}$	$2 \cdot 10^{3}$	600									
	$1 \cdot 10^{15}$	$3 \cdot 10^{2}$	100									Formation of the galaxies
Matter era												
	$6 \cdot 10^{17}$	2.7	0									today 3 K 2 K

In the following *photon or radiation era*, after about 200 s, the synthesis of the light elements took place (Sect. 5.9.4); in this process, the free neutrons were almost completely used up in the formation of ^{4}He. Protons, helium nuclei, and electrons were about 10^9 times less abundant than photons (and the decoupled neutrinos). Near the end of the photon era, at about $8 \cdot 10^5$ yr, the photons decoupled from matter, after the free electrons had been removed by recombination with the protons to form hydrogen atoms. In contrast to the neutrinos, the photons could gain energy from the $e^{\pm}$ annihilation process, so that their temperature remained higher than that of the neutrinos and has the current value 2.7 K (Sect. 5.9.4).

Although there are about 10^9 times more photons than baryons in the cosmos, after around $2 \cdot 10^6$ yr, the mass or energy density of the photons became smaller than that of the baryons (5.9.42), and the *matter era* began. Only after the decoupling of radiation and matter could large-scale *density fluctuations* in the cosmos occur, beginning at "nucleation centers" which dated from a much earlier time (Sect. 5.8.4). After $3 \cdot 10^7$ yr, these fluctuations had grown so strong that the *formation of the galaxies and galaxy clusters* could take place, at a stage when the average density of the cosmos was still about 10^6 times higher than today. Finally, after $18 \cdot 10^9$ yr, the Friedmann time, we have reached the *present* state of the universe.

How will cosmic evolution continue in the *future*? This depends decisively on the average density ϱ_0 or the density parameter Ω_0 which, as we have seen, can be determined observationally only with a large uncertainty; we recall the problems of the "dark" matter in the galaxies, a possible intergalactic medium, and the neutrino mass. As the best value, compatible with all the observations, we have assumed Ω_0 to be $\simeq 0.1$. Accordingly, we are in an open universe which is continually expanding and whose density is decreasing. The temperatures of the radiation field and the neutrino background will continue to decrease. New stars will no longer be formed after the interstellar gas has been consumed, and the stars now present will have finished their evolution within $\leq 10^{12}$ yr. After still much longer times, of the order of $\geq 10^{32}$ yr, the nucleons will have also mostly decayed, according to the GUT theory; even black holes, which may have been formed in the early stages of the universe, will finally radiate away. We end up with a cold, dark, eternally expanding universe, containing only photons and neutrinos, out of which all the stars and galaxies have disappeared.

After this brief overview of cosmic evolution, we turn to some remaining open questions within the framework of the standard model of relativistic cosmology. Lacking a quantum theory of gravitation, we can only speculate about the first 10^{-43} s; and the age-old question, "What came *before* the creation of the world?" cannot be answered within the cosmological models that we have discussed. The high degree of isotropy of the 3 K radiation also remains unexplained within the standard model. It is, to be sure, taken into account in the cosmological postulate, but we have no physical explanation for it, since, for example, parts of the universe which lie on opposite sides of the sphere as seen from the Earth would have had no time to interact with each other (at most with the velocity of light) up to the time of the decoupling of the radiation at around $8 \cdot 10^5$ yr. Also, the fact that our universe does not deviate *very much* from the Euclidean case $k = 0$ or $\Omega_0 = 1$ is by no means self-evident. On the contrary, this fact requires an extraordinarily precise fixing of the parameter Ω in the early stages of the cosmos, so that it could remain in the neighborhood of one during about 10^{60} Planck times, unless we choose to regard its value as purely coincidental.

We cannot give here the details of extensions of the standard model, or of other cosmologies; instead, we must content ourselves with a few additional remarks:

The cosmological models which we have discussed so far start with the cosmological postulate of spatial homogeneity and isotropy. Should we not also require *homogeneity of the time scale* $-\infty < t < +\infty$? Should not the universe in principle be the same "from eternity to eternity?" This *complete cosmological postulate*, which has been suggested by numerous philosophers and theologians, was the basis of the *steady-state-universe theory* developed following 1948 by H. Bondi and T. Gold, then by F. Hoyle et al. In the sense of mechanics, this theory can also be considered to be an open relativistic model with $k = -1$ and $q_0 = -1$; in contrast to "conventional physics", however, it must assume the existence of a mechanism which allows the continual production of hydrogen within the cosmos, since this element is required to replace the matter which has "expanded away" and to "fuel" the stars. The whole problem of the creation of the universe is avoided in the steady-state model, but it also provides no explanation for the 3 K radiation. The observed indications of an evolution of the quasars and the galaxies are also not compatible with the complete cosmological postulate.

A. S. Eddington, P. A. M. Dirac, P. Jordan and others have attempted to approach the problem of "cosmological physics" from quite a different aspect. Taking the elementary constants of physics on the one hand:

e, h, c, the electron mass m_e and the proton mass m_P, and the gravitational constant G,

and those of cosmology on the other:

the Hubble time $\tau_0 \simeq 6 \cdot 10^{17}$ s and the average matter density $\varrho_0 \simeq 5 \cdot 10^{-28}$ kg m^{-3},

a number of *dimensionless constants* can be derived. A *first* group of dimensionless numbers having values near one is obtained, including the fine structure constant $\alpha = e^2/(2\varepsilon_0 hc) \simeq 1/137$, and $m_p/m_e = 1836$. A *second* group of the order of 10^{40} is found: e.g. the ratio of the electrostatic to the gravitational attraction of a proton for an electron:

$$\frac{e^2}{4\pi\varepsilon_0 G m_p m_e} \simeq 0.2 \cdot 10^{40} \,, \qquad (5.9.52)$$

or the ratio of the length $c\tau_0$ (roughly equal to the radius of the universe in a closed model) to the Compton length of the proton or to the classical electron radius:

$$c\tau_0 \Big/ \frac{h}{m_p c} \simeq 14 \cdot 10^{40}$$

or (5.9.53)

$$c\tau_0 \Big/ \frac{e^2}{4\pi\varepsilon_0 m_e c^2} \simeq 7 \cdot 10^{40} \,,$$

as well as the number of nucleons in a volume $(c\tau_0)^3$, of the order of magnitude of:

$$\varrho_0 \frac{c^3 \tau_0^3}{m_p} \simeq (0.13 \cdot 10^{40})^2 \,. \qquad (5.9.54)$$

These facts can be summarized by the statement that the deceleration parameter $q_0 \simeq G\varrho_0\tau_0^2$ of relativistic cosmology (5.9.16) is of the order of magnitude of 1. This means that the real world cannot be far removed from a Euclidean universe, which, as we have already remarked, is by no means self-evident.

If we consider the order-of-magnitude equality of the numbers in (5.9.52) on the one hand, to those in (5.9.53) and (5.9.54) on the other, to be an essential fact, then, since the age of the universe τ_0 occurs in the last three equations, we can speculate about a possible *time dependence* of the gravitational constant G and other physical constants. Recent experiments, however, leave almost no room for a variation in the value of G; the upper limit of the possible relative time variation is currently taken to be $|\Delta G/G| \leq 4 \cdot 10^{-12}$ yr^{-1}. It is perhaps more promising to search for an explanation of the relationship of the physical constants to the cosmological parameters within the framework of the new theories which unify the three fundamental interactions between the elementary particles, or in a "super unified" theory which includes gravitation (Table 5.9.2).

In this connection, the following thought also occurs: relativistic cosmology offers us an entire catalogue of *possible* models of the universe. Why, then, did precisely *our* universe, with its particular parameters, come into being? We can still not give an answer to this question. The *anthropic principle* (R. H. Dicke, 1961) represents an interesting attempt in this direction: our universe is as it is because we live in it. Simply because of our existence, because there are astronomers who can observe the cosmos, there was necessarily a limitation (or even determination) of the imaginable parameters: it must have been possible for the stars and the galaxies to have formed, for elements such as carbon to have been synthesized, and for planets to exist on which the conditions for the genesis and evolution of life were fulfilled for a sufficiently long period of time. Conversely, if the cosmos had been very different in its earlier periods, there would be no one today to observe it.

Finally, we discuss briefly the promising idea of A. Guth (1981), who attempted to solve several open problems within the framework of the standard model with his *inflationary universe:*

Within the first 10^{-35} s, during the "era of the grand unification", a "phase transition" occurred in the cosmos, accompanying the symmetry breaking of the fundamental interactions (Table 5.9.3). A brief supercooling followed by a re-heating to about 10^{14} GeV produced an extremely rapid exponential expansion of the universe, the *inflation*, during which its diameter increased by roughly a factor of 10^{50}. Following this process, i.e. after about 10^{-33} s, its evolution then continued in a manner corresponding to the standard model. This inflationary model can in principle explain the isotropy of the 3 K background radiation, the density fluctuations which later led to the formation of galaxies, and the current state of the cosmos, near to the Euclidean case with $\Omega = 1$; problems are still encountered in the description of the end of the inflationary phase and its transition to the standard model.

By the way, we obtain formally an *exponential expansion* as the solution to the Friedmann-Lemaître equation (5.9.18) if its dynamics are dominated by the term containing the cosmological constant Λ. It then follows from $(\dot{R}/R)^2 = c^2\Lambda/3$ that:

$$R(t) = R_0 \exp\left(\sqrt{\frac{c^3\Lambda}{3}}\,t\right) . \qquad (5.9.55)$$

Guth's hypothesis thus finally connects the cosmological constant to the physics of the fundamental interactions.

Since, during the initial phase transition, very small regions quickly attained enormous dimensions and could no longer interact with one another, it would be possible in the framework of the inflationary model that our observable universe was merely a "bubble" within a larger manifold, i.e. that it represents simply *one* universe among many.

6. The Origin of the Solar System. The Evolution of the Earth and of Life

We turn back now from the depths of interstellar space to our own Solar System, and the old question of how it came into existence. That this question cannot be answered by the handing-down of ancient myths, but only through our own probing: that daring thought was proposed in France already in 1644 by René Descartes in his whirlpool theory. In Germany, as late as 1755, Immanuel Kant had to publish the first edition of his "Allgemeine Naturgeschichte und Theorie des Himmels" anonymously, for fear of the (protestant) theologians; in it, he treated the origin of the Solar System for the first time "according to Newtonian principles". Kant assumed a rotating, flattened *primordial nebula* from which the planets and later their satellites were formed. The (independent) description offered somewhat later by S. Laplace in 1796 in his popular "Exposition du Système du Monde" was based on a similar hypothesis. We shall not deal with the details and differences among these historical approaches, but instead summarize briefly the most important *facts* (see also Table 2.5.1, Sect. 2.8, as well as Fig. 6.0.1) which are to be explained:

1) The *orbits of the planets* are nearly circular and coplanar. The direction of revolution is the same for all the planets (direct); it is also the same as the direction of rotation of the Sun. The orbital radii (the asteroids are taken together) represent *approximately* a geometric series

$$a_n = a_0 k^n \qquad (6.0.1)$$

with $a_0 = 1$ AU; for the Earth, $n = 0$ and $k \simeq 1.8$ (Fig. 6.0.1).

2) The *rotation* of the planets is for the most part in the direct sense (exceptions: Venus and Uranus), with the axis, i.e. the rotational angular momentum, essentially parallel to the orbital angular momentum and to the axis of rotation of the Sun (exceptions: Uranus and Pluto).

3) The *Sun* contains 99.87% of the mass, but only 0.54% of the *angular momentum of the overall system* ($\sum \mathscr{M} a v$), while, conversely, the planets (mainly Jupiter and Saturn) have only 0.135% of the mass, but 99.46% of the angular momentum[1].

4) The *earthlike planets* (Mercury, Venus, Earth, Mars, and the asteroids) have relatively high densities (3.9 to 5.5 g cm^{-3}); the *major planets* (Jupiter, Saturn, Uranus, and Neptune), in contrast, have low densities (0.7 to 1.6 g cm^{-3}). The former consist (like the Earth) in the main of metals and stone, and the latter of barely modified solar matter (hydrogen, helium, hydrides) with a solid, "earthlike" core. *Pluto*, with a density similar to that of the major planets, is distinguished from them by its small mass and its sharply inclined, eccentric orbit. Presumably, its unusual orbit is the result of perturbations which occurred after the formation of the Solar System. In the following, we shall therefore leave it out of consideration.

5) In the case of the *satellites*, we must distinguish between the inner satellites, which (and this must be explained by our cosmogony) have orbits with small eccentricities, slight inclinations to the equatorial plane, and revolution in the direct sense; and the outer satellites, with their considerably larger eccentricities and orbital inclinations. The latter were probably retroactively "captured" in the early period of the Solar System, while the former may be seen as the "original property" of their respective planets. The earthlike planets have longer rotational periods and few satellites, while the major planets exhibit relatively rapid rotation and, even neglecting the captured satellites, a large number of moons and in some cases *ring systems*.

In its composition and structure, our Moon follows the scheme of the earthlike planets, while the large satellites

[1] As can be readily calculated, the angular momentum (2.6.33) of the planetary orbits (Jupiter, Saturn) is $3.15 \cdot 10^{43}$ kg m^2 s^{-1}, and that of the Sun (with some uncertainty due to lack of knowledge of the angular velocity in the solar interior) is $1.7 \cdot 10^{41}$ kg m^2 s^{-1}.

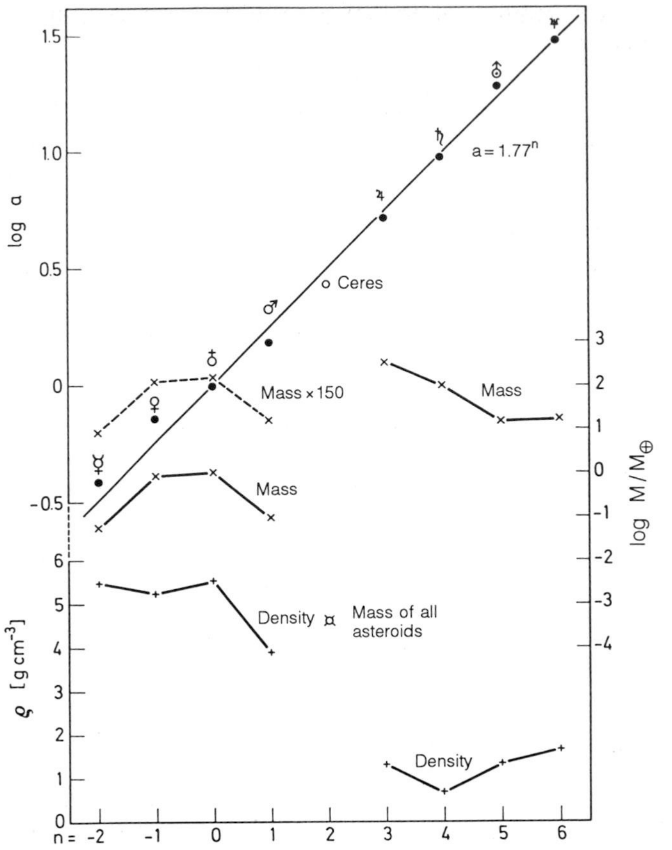

Fig. 6.0.1. The orbital semimajor axis a and the mass M of the planets (excepting Pluto) relative to those of the Earth (values = 1). The average densities ϱ (left-hand scale) are given in [g cm^{-3}]. *Lower part:* For the asteroids ($n = 2$), the orbital semimajor axis of Ceres and the estimated total mass are plotted. The line drawn in the *upper part* corresponds to the formula $\log a = 0.247n$ or $a = 1.77^n$.

of Jupiter and Saturn are in part like the Moon, but in part consist of a mixture of ice and silicates.

6) The numerous *impact craters* which are observed on the surfaces of planets and satellites from Mercury out to the Neptune system indicate a considerably larger number of "*planetesimals*" in the early period of the Solar System (small, solid objects up to a few km in size), than is present today.

7) From precise radioactive *age determinations*, we know that the Earth, the Moon, and the meteorites, and thus without doubt the Sun and the whole Solar System, were formed $4.5 \cdot 10^9$ yr ago during a relatively brief time interval.

The remarkable *regularities* in the structure of the Solar System support an evolution of the whole system in the sense of the theories of Kant, Laplace, and later of von Weizsäcker, ter Haar, Kuiper and others. In contrast, "catastrophe theories", which attribute the origin of the planets to the action of a star which passed close to the Sun (Jeans, Lyttleton, and others) appear to be excluded from the beginning. In addition to the improbability of such a near collision of two stars, these theories do not allow us to understand how a filament torn from the Sun by tidal forces could have formed the planets.

We shall therefore discuss the *cosmogony* of the Solar System directly in connection with the *formation of stars* from the interstellar medium as the result of a gravitational instability (Sect. 5.4.7). Here, the unusual distribution of angular momentum in the system makes it appealing to search for a common origin of the Sun and the planets out of a *rotating disk of gas and dust.*

The origin of our planetary system was very likely not a unique, accidental event; on the contrary, we can expect that numerous stars in the cosmos are circled by planets. Although we have a large amount of observational data for our own Solar System, the present observational techniques are *not* sufficiently sensitive to observe directly *other* planetary systems even around nearby stars. We have only sparse indications of the existence of other such systems or their possible precursors: *dark companions* of nearby stars can be detected indirectly, either astrometrically, from the orbital motions (Sect. 4.6.1), or spectroscopically, from periodic variations of the radial velocity of the principal star (Sect. 4.6.2). The latter method indicates companions for several stars having masses of the order of 1 to 10 Jupiter masses, e.g. for HD 114762. With the infrared satellite IRAS, infrared excesses were discovered in some stars such as α Lyr (A0V), α PsA (A3V), and β Pic (A5V), at $\lambda \gtrsim 25 \ldots 60$ µm; they result from circumstellar dust, which can be regarded as a *protoplanetary* system (Sect. 5.3.1). In the case of β Pic, CCD images in the near infrared later taken by B. A. Smith and R. J. Terrile (1984) allowed the observation of a flat, dust-filled gaseous disk with a diameter of ≥ 400 AU. The excess infrared radiation from the white dwarf ZZ Psc (B. Zuckerman and E. E. Becklin, 1987), which was initially taken to indicate an accompanying cool "brown dwarf" (Sect. 5.4.4), has furthermore, as a result of additional observations, been attributed to an extended disk of dust.

In the following sections, we discuss first, in Sect. 6.1, the common origin of the Sun and our planetary system from a rotating disk of gas and dust, and then deal with further details of the origin of the meteorites (Sect. 6.2) and of the Earth-Moon system (Sect. 6.3). In conclusion, we treat in Sect. 6.4 the especially interesting question of the evolution of the oceans and particularly of the atmosphere of the Earth, which is closely connected with the origin and evolution of life on our planet.

6.1 The Origin of the Sun and the Solar System

All stars are formed, as we have seen, from an interstellar gas and dust cloud of $\geq 10^3$ solar masses which contracts and then divides up into individual stars. Even single stars like our Sun originally belonged to such multiple systems, associations, or clusters.

The process of the formation of a star of 1 $\mathcal{M}_\odot$ was calculated by R. B. Larson in 1969, under the assumption of spherical symmetry, i.e. without angular momentum.

The initiation of the collapse of a cloud of radius R is determined by *Jeans' criterion* for gravitational instability (5.8.3):

$$R \leq 0.4 G \mathcal{M} \frac{\mu m_u}{kT} \quad . \tag{6.1.1}$$

(Larson chose a constant which was a factor of two larger).

With $\mathcal{M} = 1 \mathcal{M}_\odot = 2 \cdot 10^{30}$ kg, a molecular weight of $\mu = 2.5$, and an initial temperature estimated to be of the order of $T \simeq 10$ K, as well as the known values of the gravitational constant G, the Boltzmann constant k, and the atomic mass constant m_u, one finds that a cloud of $1 \mathcal{M}_\odot$ becomes unstable when its radius has reached a value:

$$R \simeq 1.6 \cdot 10^{15}\,\mathrm{m} \simeq 11\,000\,\mathrm{AU} \simeq 0.05\,\mathrm{pc} \quad . \tag{6.1.2}$$

Its average density is then

$$\varrho \simeq 1.1 \cdot 10^{-16}\,\mathrm{kg\,m^{-3}}$$

or

$$n \simeq 3 \cdot 10^{10}\,\text{molecules}\,\mathrm{m^{-3}} \quad . \tag{6.1.3}$$

This corresponds to the densities found in molecular clouds (Sect. 5.3.4). The collapse of such a cloud, up to the time of the formation of the Sun and the planetary system, requires several 10^5 yr, as can be estimated using (5.8.25) and (5.8.26).

Perhaps at this point we should first try to gain some insight into the problem of the *angular momentum* of cosmic masses and in particular of the Solar System, before taking up the results of numerical calculations of the collapse, including angular momentum. When our mass of gas (6.1.2) and (6.1.3) was decoupled from the rest of the Milky Way system as a result of the occurrence of the Jeans instability, it initially had a rotational period of the order of magnitude[2] of the galactic rotation, i.e. $\tau_{gal} = 2\pi/\omega_{gal} \simeq 10^9$ yr. If the mass of gas contracted to dimensions of the order of the radius of the Sun, $R_\odot = 7 \cdot 10^8$ m, we would obtain from conservation of angular momentum ($\propto R^2/\tau_{gal}$) a rotational period of about 0.07 d and an equatorial velocity of about 700 km $\mathrm{s^{-1}}$. This approaches the order of magnitude of the rotational velocities which are observed for young B stars. Although stars of later spectral types generally rotate much more slowly, such stars in young star clusters have been observed to rotate considerably faster than the Sun.

If we could transfer the angular momentum of the planetary orbits (essentially that of Jupiter) to the Sun, its equatorial velocity would increase from 2 km $\mathrm{s^{-1}}$ to about 370 km $\mathrm{s^{-1}}$. We thus arrive at the conclusion that the enormous angular momentum of the orbital motion of the planets is nothing other than a major fraction of the angular momentum which *any* object of around 1 $\mathcal{M}_\odot$ would have originally acquired from the rotation of the galaxy. The Sun, however, must have either transferred its own angular momentum to the interstellar medium later through magnetic coupling via the solar wind (R. Lüst and A. Schlüter, 1955), or, as numerical calculations suggest, may never have acquired it; instead, it may have been transferred outwards within an accretion disk by turbulent or magnetic friction already during the collapse process.

The flat *accretion disk* which surrounded the proto-Sun, the "*solar nebula*", consisting of a mixture of gas and dust similar to that known from the interstellar medium, must clearly have had the same chemical composition as the Sun itself (Table 4.9.1). On the other hand, its matter can never have been contained *within* the Sun, since, for example, the Earth and the meteorites contain

[2] More precisely stated: with rotation of the form $v \propto R^{-n}$ ($n = -1$ for rigid rotation, $n = 0.5$ for Keplerian rotation), the local angular velocity becomes $\omega = \frac{1}{2}|\mathrm{curl}\, v| = (1-n)\omega_{gal}/2$.

roughly a hundred times the concentration of *lithium* as does the Sun, where about 99% of this element has been eliminated in the course of time by nuclear processes. The solar nebula must thus have been formed about the same time as the Sun itself.

The fact that the formation of such a system is not an unusual coincidence is demonstrated by the observations mentioned at the beginning of this chapter of less massive companions and of dust rings or disks around some neighboring stars. Under just *which conditions* a planetary system or a double or multiple star system comes into being cannot be uniquely determined. In any case, the formation of a planetary system cannot presuppose any very special initial conditions for the gravitational collapse of an interstellar cloud.

Complex two-dimensional hydrodynamic model calculations of the gravitational collapse of a cloud taking into account its *angular momentum*, performed by P. Bodenheimer, G.E. Morfill, W. Tscharnuter, H.J. Völk, and others with the goal of explaining consistently the origin of the Sun and the planetary system "at one stroke", initially indicated that even a very small original overall angular momentum (ratio of rotational to gravitational energy $\simeq 10^{-5}$) would prevent the formation of a single star. Instead, a rotating *ring* is formed, which presumably later becomes unstable and breaks up into a double or multiple star system. If, however, a sufficiently strong *turbulent friction* is assumed within the collapsing cloud, then a single *central star* is formed with gradually increasing mass, surrounded by a flat, rotating *accretion disk* (Fig. 6.1.1). Within the disk, matter flows inwards, while angular momentum is transported outwards as a result of viscosity. The *temperature* of the disk is relatively low, and its radial dependence varies only slowly at first: from about 1000 K in the neighborhood of Mercury's orbit, it decreases on going outwards to several tens of Kelvins at the edge of the disk (at around 40 AU). At the radius of the Earth's orbit, for example, the temperature is about 300 K, and it is near 150 K at the orbit of Jupiter.

A central question now remains: how were the earthlike and the major planets, the meteorites, etc. formed out of the accretion disk? There is as yet no generally accepted theory to describe this process. A first possibility is formation as a result of *gravitational instabilities* within the disk. The Jeans condition (5.8.5) for the beginning of the condensation of, for example, one Jupiter mass at 150 K requires a density of about 10^{-6} kg m^{-3}, which appears plausible for an evolutionary phase somewhat later than that depicted in Fig. 6.1.1. While the formation of the

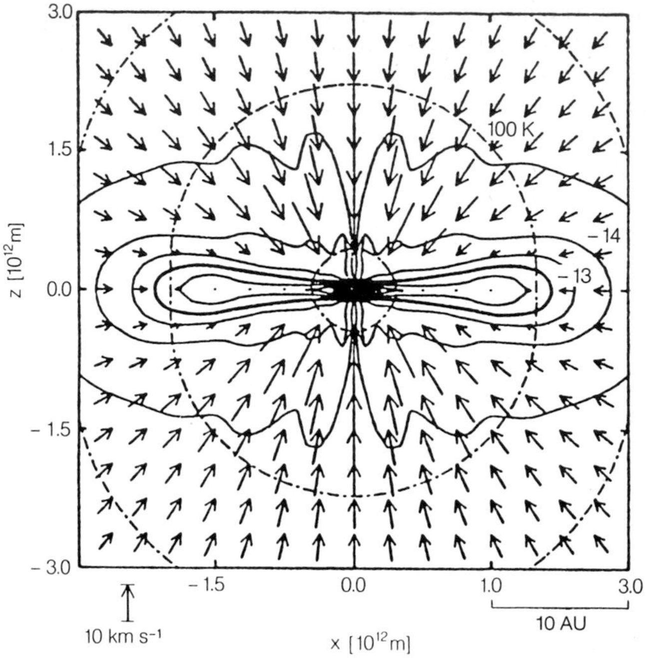

Fig. 6.1.1. The gravitational collapse of a rotating protostellar cloud (ratio of rotational to gravitational energy $1.2 \cdot 10^{-4}$) of mass $3\mathscr{M}_\odot$ with turbulent viscosity, after G.E. Morfill et al. (1985). The time depicted is about $3.3 \cdot 10^4$ yr after the beginning of the collapse, when the central mass has increased to $0.5\mathscr{M}_\odot$. A meridional section is shown, with the axis of rotation in the z-direction: the contours represent lines of constant density ($\log \varrho$; ϱ in [g cm^{-3}]), the dot-dashed lines are lines of constant temperature, and the arrows indicate the velocity field. (Reproduced with the kind permission of the University of Arizona Press, Tucson AZ)

large outer planets, which consist of nearly unchanged solar matter, could have been caused or at least favored by such an instability, for the inner planets it remains to be explained how, following the production of a solid core by sedimentation of the condensed matter, the light elements such as hydrogen and helium, which make up about 99% of the overall mass, could have escaped. We can only speculate at present about this problem (considerably higher temperatures in the inner parts of the solar nebula? tidal forces from the Sun?). The asteroids and other small bodies in the planetary system, finally, have masses which are too small for them to have been formed by gravitational instabilities; they could, however, have resulted from the breakup of larger objects.

A further development of the theory of the origin of the planets and other objects in the Solar System begins with the *dust component* in the accretion disk, which, as we observed in the interstellar medium, makes up about 1% of the overall mass. Mutual inelastic collisions cause the dust particles within the turbulent gaseous disk to accrete into larger and larger clumps and to form the *planetesimals*, irregularly shaped solid bodies with dimensions of the order of 1 to 10 km. Later, over periods of the order of 10^8 yr, stepwise collision processes between planetesimals of varying sizes led to *solid planets*. Depending on the relative velocities of the planetesimals, we expect them also to have been broken up again by some of the collisions. All of these processes release large quantities of energy, so that melting of surface layers can occur. The *meteorites* contain important documents about these processes; we shall discuss their origin in more detail in Sect. 6.2.

In the outer, cooler region of the planetary system, the solid cores could collect and bind the surrounding *gas*, which consisted mainly of hydrogen and helium, through gravitational forces; this gave rise to the typical structure of the *Jupiter-like* planets.

The density of planetesimals in space was considerably higher in the first 10^9 yr of the Solar System than it is today, as we can see from the numerous impact craters on the Moon, Mercury, Mars and many satellites of the outer planets. The large objects in the planetary system thus moved for a long period of time after their formation through a medium with strong *frictional forces* caused by numerous clumps of stone of varying sizes. This was probably the source of the small eccentricities and inclinations to the plane of the ecliptic of the planetary orbits, as well as the corresponding orbital inclinations of the (regular) satellites of the major planets. Older, purely celestial-mechanical calculations concerning the origin and evolution of the Solar System are not accurate, due to their neglect of this friction. With the end of the intense bombardment by planetesimals, the end of the accretion phase of our Solar System was also reached.

We can today regard the majority of the asteroids and meteorites as the "remains" of the planetesimals, together with the two moons of Mars, the many small satellites of Jupiter, Saturn, and Uranus, and the comets which come from the outermost parts of the Solar System.

6.2 The Origin of the Meteorites

The point of view held previously, that the meteorites represent *the* cosmic material with *the* cosmic element abundance distribution has, as we have already seen, long since been supplanted by a thorough study of their mineralogical, chemical, and isotopic structures.

In Fig. 2.9.4, we summarized on the left side the main types of meteorites and on the right their finer classification. The iron meteorites are clearly the most differentiated; the most far-reaching conclusions can be expected from the investigation of the *chondrites*. As we have seen, these contain the chondrules which are spherules of millimeter dimensions consisting of various silicates, along with iron (especially the enstatites), embedded in a more fine-grained matrix of similar composition.

A starting point for the interpretation of the extraordinary variety of meteoric structures can be taken to be the *condensation and separation of solids* from the gas phase of solar composition, a very effective chemical fractionation process. It was the great achievement of H. C. Urey (1952) to apply the experience and methods of physical chemistry for this process to the cosmogony of the Solar System. Calculations were carried out by J. W. Larimer and E. Anders (1967/68), who discussed two limiting cases for the time dependence: (1) Thermodynamic equilibrium; the cooling process takes place so slowly that formation of solid solutions in diffusion equilibrium is possible. (2) Rapid cooling, so that the condensed elements and compounds cannot interdiffuse.

At a pressure of about 10 Pa (10^{-4} bar), the condensation of solar matter takes place (under both assumptions) according to the following steps:

$T<2000$ K:	barely-volatile compounds of Ca, Al, Mg, Ti ...
1350 to 1200 K:	magnesium silicates, nickel-iron.
1100 to 1000 K:	alkali silicates.

At this point, about 90% of the chondritic material has consensed:

680 to 620 K:	iron sulfide (FeS, troilite) and other sulfides, then Pb, Bi, Tl, In ...
400 K:	Fe_3O_4 is formed from iron and water vapor.
400 to 250 K:	hydrated silicates condense.

The condensation of many trace elements is more complex and depends on the precise assumptions made; according to H. C. Urey, it can be used as a "cosmic thermometer".

Analysis of the "ordinary" chondrites reveals several processes involving different components which took place at differing temperatures between about 1300 K and 400 K:

Material with a high content of Ca, Al, Ti, ... was separated out partially at $T \geq 1300$ K, at least in the region of formation of the meteorites and the earthlike planets. Its abundance is relatively enriched in the Earth-Moon system, and in contrast is reduced in ordinary and enstatitic chondrites.

A metal-silicate fractionation influenced many types of meteorites and probably also the earthlike planets at temperatures from about 1050 to 700 K. Remelting and degassing at around 600 to 450 K led to a reduced abundance of volatile elements in some meteorites and in the Earth-Moon system.

The accretion of meteoric matter took place in the main within a relatively narrow range of temperatures around 450 K and at low pressures ($\simeq 1$ Pa). The C1 chondrites were formed at about 360 K and 0.2 Pa.

We cannot treat here the further details of these processes, and in particular their causes, especially since it has not yet proved possible to fit together those processes which have been recognized to be important into a self-consistent astronomical picture. The still somewhat mysterious reheating and remelting processes are presumed to have taken place within larger objects in which a metallic core and a silicate-rich mantle had formed.

The rare *carbonaceous chondrites* are of special interest with regard to the origin of the Solar System and especially the beginning of life on the Earth. They are classified according to their content of volatile elements (H, C, S, O, ...) into the types C1, C2, and C3. These differences correspond on the one hand to the relative amounts of the high-temperature fractions of the chondrites and of similar iron particles which were formed at around 1200 K, and on the other hand to the low-temperature fractions of the fine-grained matrix in which they are embedded, which condensed at only about 450 to 300 K.

The *abundance distribution* of the chemical elements in this matrix and in the C1 meteorites agrees, apart from the highly volatile elements, with that of the *Sun* (Table 4.9.1). The cosmic abundance of extremely rare elements can thus be determined from them, even when these elements can hardly be detected spectroscopically in the Sun. Many of these elements are of great importance for the understanding of nucleosynthesis. The relative abundances of the isotopes are essentially the same in the meteorites, on the Earth, and in the Sun (where however only a few are precisely known). The C1 chondrites, consisting entirely of this dark matrix, contain up to 4% carbon, mainly in the form of organic compounds. They can be most readily compared with terrestrial coals or humic acids.

Some researchers let themselves be tempted at first by these carbonaceous chondrites into making the assumption of some sort of extraterrestrial life. This problem, along with the related question of terrestrial contamination, was finally resolved with the aid of the C2 chondrite which fell to Earth near Murchison in Australia in 1969; it was possible to collect its fragments within a few months. The *Murchison meteorite* contains among other compounds the amino acids, which play an essential role in terrestrial life forms. But although in terrestrial organisms practically only the optically active (levorotating or dextrorotating) L-configuration occurs, both the mirror-image isomers (L- and D-forms) are roughly equally represented in the meteorite: it contains an optically inactive *racemic mixture*. The meteorite also contains numerous additional amino acids which do not occur in living organisms. The investigation of these amino acids, as well as of the (mostly linear) hydrocarbons from carbonaceous chondrites, has shown that these compounds must have been formed in the early phases of the Solar System, without the intervention of mysterious life forms.

How can we understand this process? One possibility is offered by the *Fischer-Tropsch synthesis* (1923), in which carbon monoxide, CO, and hydrogen, H_2 can be combined in the presence of suitable catalysts to yield *hydrocarbons*, mainly of the linear type C_nH_{2n+2}. It can be assumed that in the solar nebula, not all of the CO would have been converted into CH_4 (as would be expected in thermal equilibrium below 650 K), since this reaction proceeds extremely slowly. The remaining CO could have combined with H_2 at 380 to 400 K, with catalysis by the Fe_3O_4 and hydrated silicates which were then present, in a kind of Fischer-Tropsch synthesis. In the presence of NH_3, according to E. Anders, the biologically important amino acids and the many other organic compounds found in carbonaceous chondrites would also have been synthesized.

We shall return to the exciting question of the origin of life in Sect. 6.4. First, however, we consider further the origin of the meteorites.

A precise *temporal* investigation made possible the discovery that many parts of meteorites contain Xe isotopes which (as is confirmed by laboratory experiments) can only have been formed by decay of the "extinct" radioisotopes ^{129}I (with a half-life of $1.6 \cdot 10^7$ yr) and ^{244}Pu (half-life $8.3 \cdot 10^7$ yr, fission). These investigations, very complex in detail, support the idea that the formation of all meteors, including the carbonaceous chondrites, took place within at most $\simeq 10^7$ yr.

It was then possible to determine the socalled *irradiation age* of numerous meteorites, i.e. the time during which the meteorite was subjected to irradiation by cosmic rays. Their main components penetrate about 1 m into the meteoric material and leave a trail of in part stable, in part radioactive reaction products, whose relative abundances allow calculation of the time during which the corresponding piece of meteoric material was irradiated. In many cases, the irradiation age gives us the time at which the piece was broken out of a larger object (protometeorite) and thus exposed to cosmic irradiation. The irradiation ages of the tough iron meteorites lie mostly between a few 10^8 and 10^9 yr, while the much more frangible stone meteorites attained their present size in the main only 10^6 to $4 \cdot 10^7$ yr ago. According to mineralogical indications, which permit conclusions to be drawn about the gravitational forces in the protometeorites, their sizes can be estimated to have been between 50 and 250 km, with iron cores of $\simeq 10$ km diameter.

These results, along with the "formation temperatures" and the surface reflectivities (Sect. 2.9.2), indicate that the major portion of the meteorites was formed in the *asteroid belt*. Only for a portion of the carbonaceous chondrites can an origin in the neighborhood of Jupiter's orbit or even further out in the Solar System be considered probable. (In the cases of a few individual meteorites, their chemical and isotopic compositions indicate an origin presumably on the Moon or Mars.)

In some carbonaceous chondrites, particularly the *Allende meteorite* which fell in Mexico in 1969 (C3), and the previously mentioned Murchison meteorite (C2), *inclusions* with extremely *unusual isotopic compositions* of many elements have been in found recent times; their compositions differ markedly from the ratios found in solar matter, and cannot be explained through physical processes during the origin and evolution of the Solar System. Excesses of ^{16}O, ^{22}Ne, ^{26}Mg (from the radioactive decay of short-lived ^{26}Al), ^{14}N, ^{50}Ti, and of several Xe isotopes, among others, have been determined. In some other inclusions, radioactive dating gives ages of up to $4.9 \cdot 10^9$ yr, an age which is considerably greater than that of the Solar System itself! These exciting results indicate the possible existence of *presolar matter*, whose isotopic composition dates from the time *before* the formation of the Solar System. Perhaps our planetary system was not formed from a well-mixed interstellar gas-dust cloud, so that small solid particles could somehow survive its formation process. Another possibility would be that a nearby supernova explosion $4.5 \cdot 10^9$ yr ago set off the gravitational instability which led to the origin of the Solar System, and in the process could have "contaminated" at least the outer portions of the solar nebula.

6.3 The Earth-Moon System

Following our discussion of the origin of the planets and the meteorites, we now turn to the especially interesting and enlightening question of the origin and evolution of the Earth-Moon system. The results of older astronomical and geophysical research are complemented here by the fascinating findings from the manned and non-manned landings on the Moon (1966–1973).

Our Earth-Moon system must have been formed in a quite different way from the satellite systems of the other planets: in all the other systems, with the exception of Pluto/Charon, the ratios of *mass* and *angular momentum* of the satellite to those of the planet are namely $\ll 1$, while the mass ratio Moon : Earth is 1 : 81.3, and the orbital motion of the Moon takes up 83% of the angular momentum of the whole system.

In order to learn something about the earlier states of the Earth-Moon system, we begin with terrestrial observations and, following the direction indicated by the classic works of G. Darwin (1897), take up the phenomenon of *tidal friction* which we have already mentioned briefly (Sect. 2.6.6). The two tidal maxima which the Moon continually drags around the Earth, both in the oceans and in the solid material of the planet, cause a braking action on the Earth's rotation. After subtracting all the other effects, a consideration in particular of past eclipses leads us to an estimate of the resulting increase of the length of the day, amounting to 1.64 ms per century. The angular momentum released by the Earth can only be taken up by the orbital motion of the Moon, i.e. the period of revolution and the orbital radius of the Moon are increasing.

This effect, which from an astronomical viewpoint is vanishingly small, becomes however considerable over the course of geological times. In 1963, J. W. Wells and C. T. Scrutton noted that the calcium carbonate shells of corals (and other lifeforms) living in waters with strong tides show fine bands, which reflect the periods of the year, the synodic month, and the days. While recent corals confirm the well-known astronomical data, the investigation of petrified corals has shown that for example in the middle Devonian, i.e. about $370 \cdot 10^6$ yr ago, 1 yr had $\simeq 400$ (then current) days and 1 synodic month had $\simeq 30.6$ d, in sufficiently good agreement with an extrapolation of the present values. But evidently even the geological data offer us only the "recent history" of the Moon; for the distant past, we must rely on theoretical extrapolations. These have been carried out several times following G. Darwin's first efforts, with the result that the Moon must have come very near to the Earth (within a few Earth radii) about 1.5 to $2.5 \cdot 10^9$ yr ago, which would have caused a tidal wave of genuinely apocalyptic proportions. The geological evidence strongly suggests that this event never occurred, at least not in the past $4 \cdot 10^9$ yr. This is, however, not an argument against the theory of tidal friction as such, but only against its extrapolation using the present-day value of the friction. In fact, we know hardly anything about the structure of the oceans and thus about the strength of the tidal friction over the course of geologic time. Along with the actual tidal friction, a braking action due to the impacts of planetesimals or meteors, which were much more numerous in the early period of the Solar System, could have been significant.

We thus attack our problem of the evolution of the Earth-Moon system from the opposite direction and attempt to evaluate the results of the lunar landings in these terms.

The entire lunar surface is covered with a layer of fine dust and fragments of varying sizes, in part baked into breccia; it is called the regolith. This material was clearly formed by the impacts of numerous small and large meteorites. The most outstanding features of the current lunar surface (apart from a few volcanic structures), the maria and the craters, were created by the impacts of meteorites (the smaller features also by secondary impacts) ranging up to the size of asteroids, onto the originally solid crust. Only later were they covered for the most part by enormous basalt flows. In the Mare Imbrium, which has a diameter of 1150 km and originally about 50 km depth, we can distinguish three stages of this lava flooding process, which occurred about $3.9 \cdot 10^9$ yr ago. Both the points of origin of the lava flows, and also the "drowned craters" which emerge here and there from the otherwise smooth surface of the hardened lava, indicate that the lava flooding generally had nothing to do with the meteoric impacts themselves. Radioactive age determinations confirm that the lava flows are several 100 million years younger than the meteoric impacts. Local melting on impact played a minor role. Instead, the mare basalt ($\varrho \simeq 3300$ kg m^{-3}) later flowed up through cracks in the stone crust, from which we can conclude that at that time, the Moon itself was still partially molten at a depth of 200-400 km. In comparison to the rocks of the highlands and several intermediate levels, the mare basalt shows clear indications of an additional chemical differentiation; in particular, it shows a lower content of Al_2O_3.

If we compare the numbers of craters of differing diameters in the highlands (Fig. 2.8.5), the surfaces of maria of differing ages, and that of the largest craters themselves, we find that the intensity of the cosmic bombardment decreased by several orders of magnitude during the first 10^9 yr following the formation of the Moon (we could just as well say the following the origin of the Solar System), and that from then on ($-3.8 \cdot 10^9$ yr), the decrease continued much more slowly. The fact that there are so few meteoritic craters on the Earth is now readily understandable: the present crust of the Earth was formed only after the supply of meteors had been nearly exhausted! The observations on the Moon, as well as those of Mercury, Mars, Venus, and many satellites of the outer planets, which show craters quite comparable to those on the Moon, indicate that in the early period of the Solar System, out to the orbit of Neptune, the plane of the planetary orbits was filled rather densely with objects of the order of ≤ 100 km in size, the planetesimals.

When the first pictures of the *dark* side of the Moon became available, it caused some surprise that large impact craters are apparently more common on the hemisphere facing the Earth. From this we can conclude that the rotation and revolution of the Moon were already synchronous about $3.8 \cdot 10^9$ yr ago, just as they are now.

In contrast to the Earth, no iron core has been detected in the Moon. The average density of the Moon, 3340 kg m^{-3}, is very similar to that of the crust and the upper mantle of the Earth, although clear differences are found in their detailed chemical compositions; in particular, readily volatile elements are less common on the Moon. We cannot take up the details of lunar petrology and of selenology here. Instead, we turn to the cardinal question of the *origin* of the Moon: was the Moon formed by *splitting off* from a rapidly rotating proto-Earth, or was it *captured*?

It is clear from the outset that the capture of a "finished" Moon is relatively unlikely on the basis of age determinations of the lunar rocks. We need also not take seriously the popular version of the splitting-off theory, according to which the Pacific Ocean is the scar left by the departing Moon; this is contradicted by the results of plate tectonics.

The *splitting-off theory* (G. Darwin) is supported by the agreement of the average density of the Moon with that of the upper layers of the Earth. On the other hand, the objection was raised early that the present value of the angular momentum of the Earth-Moon system would not have sufficed to make the proto-Earth rotate rapidly enough to cause the centrifugal force to be greater than the gravitational force at any point (corresponding length of the day $\simeq 2.7$ h). We can reply that the whole process of formation of the Earth-Moon system took place during a period of intense bombardment by meteorites or planetesimals, and that a considerable *exchange of angular momentum* must have occurred during this time. Since we have no information about the original orbits of the planetesimals, we can carry out no quantitative calculations of this effect. A collision with one or more large planetesimals could have initiated the separation of the Moon. In addition, as pointed out by E. A. Ringwood, the sinking of the heavy nickel-iron component into the Earth's core would have decreased its moment of inertia and as a result, from conservation of angular momentum, would have increased its rotational velocity.

On the other hand, the *capture* of the Moon is supported by its underabundance of volatile elements relative to that in the Earth's mantle, which indicates that the Moon was formed at a different place in the Solar System, somewhat nearer to the Sun. Another point favoring this theory is the mass ratio of the satellite to the planet, which is unusual within the Solar System. However, we still have the difficulty of explaining the relatively low global abundance of iron on the Moon. It would seem very improbable that the lunar matter was *previously* fractionated in a gravitational field. A further problem of the capture theory, that the Moon would have to have been slowed down drastically in the course of a near collision with the Earth in order to have been captured at all, does not seem to be too serious: this braking could have occurred as the result of a collision with a large planetesimal while the Moon was in the vicinity of the Earth.

At present, we cannot decide between these two alternative explanations of the origin of our Moon. The two theories, which were originally rather schematically formulated, appear today under consideration of the high density of planetesimals in the early period of the Solar System to be not so very divergent.

At the conclusion of our considerations of the origin and evolution of the objects in the Solar System, we shall, with a certain amount of local patriotism, regard the *evolution of the Earth* in somewhat more detail. The Earth's inner structure, the series of its geological periods, and the continental drifts were already discussed in Sect. 2.8.5.

6.4 The Evolution of the Earth and of Life

We first turn to the question of the *formation* of the *oceans* (the hydrosphere) and of the Earth's *atmosphere*, which is particularly interesting with regard to the origin and evolution of life on the Earth. Both can*not* have belonged to the original structure of the Earth; we have seen, indeed, that in the early stages of the Earth-Moon system (as in the meteorites), the light elements were expelled from the planets and other objects, at least in the inner regions of the Solar System. The well-known fact that the noble gases are very rare in the Earth's atmosphere, although helium and neon are among the most common elements on the Sun, shows that our atmosphere cannot be a direct remnant of the solar nebula. Instead, the atmosphere and the oceans were formed in a secondary process[3], probably from *volcanic exhalations*, which yield in particular H_2O and CO_2 as well as traces of SO_2, CO, H_2, N_2, etc. The most abundant argon isotope, ^{40}Ar, was produced in contrast by the decay of ^{40}K in the Earth's crust (potassium feldspars), and helium was formed as a result of the α-decay of the known radioactive elements.

The Earth is the only planet whose surface temperature allowed the formation of a *hydrosphere*, i.e. of large quantities of liquid H_2O. Through reactions with silicates in the oceans, this in turn permitted most of the carbon dioxide to be removed from the atmosphere in the course of time. Today, we find the major portion of the original CO_2 in the form of potassium and magnesium carbonates in sedimentary rocks.

The *protoatmosphere* of the Earth contained as yet *no oxygen*, since that element was completely bound up in compounds as oxides, silicates, etc. The formation of O_2, and thereby also of ozone (O_3) in the optically thin atmosphere began when water vapor, H_2O, was decomposed by solar ultraviolet radiation (photodissociation) into $2H+O$. But this process could have produced at

[3] The atmospheres of Venus and Mars, consisting mainly of carbon dioxide, are also of secondary origin. In the atmospheres of the major planets, in contrast, ammonia (NH_3) and methane (CH_4) were formed directly in the course of chemical fractionation from the original solar matter.

most about 10^{-3} of the present oxygen concentration: a thicker layer of oxygen (along with the ozone which accompanies it) namely absorbs the short-wavelength solar radiation, so that the amount of irradiated gas is self-limiting.

The additional oxygen in our atmosphere can only have been produced by *photosynthesis by lifeforms*, i.e. in connection with their evolution. The production capacity of the *present-day* flora can be made clear by considering the estimate that all the oxygen in the Earth's atmosphere passes through the process of photosynthesis or carbon dioxide assimilation in a period of only 2000 yr! In this process, with the aid of the energy $h\nu$ from solar radiation, energy-rich glucose is produced and oxygen is released according to the overall reaction:

$$6CO_2+6H_2O+h\nu \rightarrow C_6H_{12}O_6+6O_2 \ . \qquad (6.4.1)$$

In the complex intermediate reactions, a decisive role is played by *chlorophyll*. A further idea of the carbon metabolism of the Earth's plants is given by the enormous mass of the fossil fuel (hydrocarbon) deposits.

The history of the Earth's atmosphere and of the Earth itself is thus closely connected with the *origin of life*. Like cosmogony, this problem was also the domain of mythological fantasies for a long period of time. After the sucessful synthesis of urea in 1828 by Friedrich Wöhler removed the boundary between inorganic and organic chemical compounds, contributions from the fields of astrophysics, geology, and biochemistry have, to be sure, not yet solved the problem of the origin of life, but have at least placed it firmly within the realm of scientific research.

Before we turn to the questions of the origin and evolution of life on the Earth, we first give a brief summary of the *molecular-biological basis* of (present-day) life.

Living organisms are extremely complex physico-chemical structures. Even the simplest lifeform makes up a *system*, in which two essential functions interact:

1) The ability to *reproduce*, from complex molecules to entire organisms, is anchored in the *genetic material*, which contains the *information* and the *control mechanisms* that, similarly to the memory of a computer, allow a molecule or organism of the same kind to be "copied".

"Errors" in this replication mechanism, which occur as a result of chemical influences or ionizing radiation, or even through thermal motions of the molecules, can produce *mutations*. The origin of a mutation is thus simply a problem in probability theory. *Once* it has occurred, then the modified molecule or organism will continue to be *exactly* reproduced.

The mutations, and furthermore the many possibilities of genetic recombination, form the basis for the *evolution* of lifeforms, according to the mechanism of *natural selection* discovered by Charles Darwin in 1859. Many mutations lead, as might be expected, to nonfunctional structures which again quickly "disappear". The structures which are at least capable of functioning within themselves are immediately exposed to the effects of their environment and (later) of other lifeforms. In this struggle for existence, to use the somewhat unfortunate catchphrase coined by H. Spencer, that molecular complex or organism will win, which has the highest *rate of reproduction*. The latter thus provides a clearcut measure of the selection value of a particular mutation.

2) A mechanism for the delivery of the necessary *energy* and of course the necessary *materials* must be associated with the control system. The construction of a complex form with particular structures, which is thus "improbable" or not likely to occur through coincidence alone, presupposes a certain amount of *information* or "know how". It is thus connected according to the fundamentals of thermodynamics with a decrease in the *entropy* (entropy is proportional to the logarithm of the probability of a state), or an increase in the negative entropy or negentropy of the system. The latter can however only occur through an input of energy, whereby in the environment (from which the energy is taken), a corresponding increase in entropy, according to the 2nd Law of Thermodynamics, must be taken into account.

Modern biochemistry has succeeded in the past decades in obtaining a clear overview of the most important groups of substances and processes which make possible the wonder of life. Here, again, we meet up with the interaction of *information* and *function*, analogous to the *legislative* and the *executive* branches of a government, in a hierarchy of reaction cycles, which altogether make possible the formation and reproduction of functional macromolecular or living systems.

In all organisms, the *nucleic acids* occur as carriers of *information* or of the genetics of the molecules, while the *structure* and *function* of the organisms are determined by *proteins*, which, with varying degrees of specialization, carry out their functions according to the "instructions" of the nucleic acids. The fundamental structures and functions are basically the same throughout the whole organism.

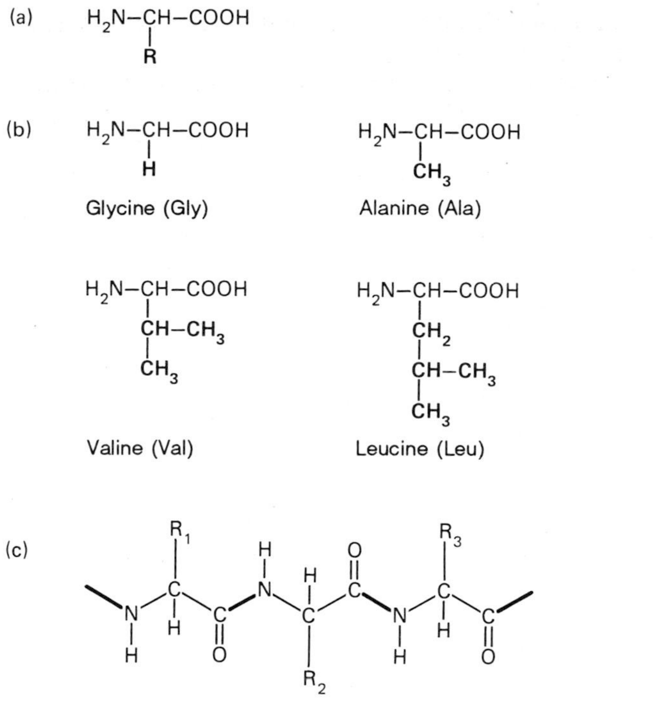

Fig. 6.4.1 a–c. Amino acids and proteins. (**a**) The basic structure of the majority of natural amino acids, which differ only in the nature of the side group R. (**b**) Examples of the protein-component amino acids with their symbols. (**c**) A protein molecule (polypeptide) is formed by joining from about 50 up to several 1000 amino acids in a linear chain, by rather rigid peptide bonds (indicated as dark lines). $R_1, R_2, \ldots$ are the amino acid side groups, e.g. H in glycyl, CH_3 in alanyl

We first consider the proteins somewhat more closely, and then turn to the nucleic acids.

The *proteins*, the most important carriers of function, e.g. the recognition and processing of particular substances, catalysis, regulation of reaction rates, etc., are macromolecules in which up to several thousand *amino acids* are bound together in a certain sequence by rather rigid peptide bonds into a *peptide chain* (Fig. 6.4.1). It is very remarkable that in the proteins of *all* lifeforms on Earth, only about 20 particular amino acids occur, although chemically, many more are possible.

The polypeptides can be folded in a variety of ways and can, in particular, be rolled up into balls (globulines). The shape of the ball is determined uniquely by non-covalent bonds through the sequence of amino acids. The resulting shape of the molecule and its distribution of interactive forces allow its activity to be *specialized* with respect to certain substances or chemical reactions. An especially important group of proteins in this sense are the *enzymes*.

As can be readily calculated, various arrangements of altogether 20 amino acids in chains of about 1000 links can *theoretically* yield so many different proteins, that even a set of samples would fill the entire known universe! We can thus readily understand the enormous *variety* of organic lifeforms; it must, in fact, be limited by some ordering principle.

The carriers of this information, "how and what is to be produced", are the *nucleic acids*. The discovery of their role in the replication of organisms was the beginning of modern molecular biology. O. T. Avery and coworkers recognized in 1944 that the *deoxyribonucleic acids* (abbreviated DNA) are the material carriers of genetic information. J. D. Watson and F. H. Crick in 1953 then succeeded in constructing a spatial molecular model of DNA, the famous *double helix*, and thus attained some insight into the close relationship between the information localized in the DNA and the mechanism of replication.

The deoxyribonucleic acids are macromolecules formed by joining *nucleotides* in a particular sequence. These in turn consist of a sugar molecule (2-deoxyribose), a phosphoric acid group, and a heterocyclic nitrogen base. Clearly with regard to the above-mentioned limitation of the variety in protein formation, only *four* different

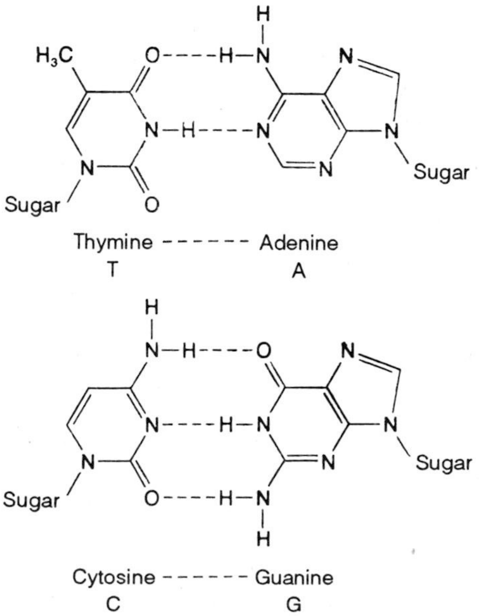

Fig. 6.4.2. Base pairing in deoxyribonucleic acid. A pyrimidine base (*left*) and a purine base (*right*) are joined by hydrogen bonding (*dashed lines*) and thus fit into the DNA helix (Fig. 6.4.3). (Reproduced with the kind permission of the G. Fischer Verlag, Stuttgart)

nucleotides occur, containing the purine bases *adenine* and *guanine* and the pyrimidine bases *thymine* and *cytosine*, which are denoted for short by their initials A, G, T, and C (Fig. 6.4.2).

The *DNA molecule* consists of two strands of nucleotides, which together form a *double helix*. The bases are joined pairwise as in Fig. 6.4.2 by hydrogen bonds. There are thus four *base pairs* A–T, G–C, T–A, and C–G. The information about the production of a protein is now coded in the sequence of the bases in a DNA strand, so to speak in a "genetic language", using the four symbols A, G, T, C. According to the principle of base pairing, the structure of *one* strand in the double helix completely determines that of the *other* strand and thus forms the basis for the *replication* of a DNA molecule, the most important "elementary process" of biology: the double helix is opened up like a zipper by breaking of the relatively weak hydrogen bonds in the base pairs, then the corresponding free bases attach themselves to each base in the strand, resulting in two identical double helices (Fig. 6.4.3).

The genetic information is in fact not read off directly from a DNA strand itself, but instead is first transferred to a corresponding molecule of a ribonucleic acid (RNA) and transported by it to the place where protein synthesis occurs. RNA has a similar structure to DNA, except that in the nucleotides, deoxyribose is replaced by ribose and the base uracil (U) takes the place of thymine (T)[4]. Thus, the "DNA symbols" AGTC are transcribed as UCAG in the "RNA code". Interesting as they are, we cannot go into the details of the transcription process here; instead, we turn directly to the deciphering of the *genetic code:*

In order to specify *one* of the 20 amino acids uniquely using the 4-symbol code, at least 3 symbols must be combined into a "word" (triplet). In fact, each amino acid is specified by a sequence of three nucleotides in the DNA or in the corresponding RNA after transcription. For example, the sequence G–C–U specifies alanine. The sequence of triplets along the DNA chain thus serves so to speak as a "blueprint" for the production of a corresponding protein with a particular series of amino acids. Since the genetic code could spell altogether $4^3 = 64$ words, while only 20 amino acids need to be specified, nature "allows" some amino acids to be specified by several different words; however, some of the words are used as "punctuation marks".

A section of DNA strand consisting of about 1000 nucleotide pairs, containing the information required to produce a particular protein, is called a *gene*, the carrier of inheritance. The genetic code is *universal*, i.e. it is the same for all living organisms on Earth. Since chemical processes such as DNA replication and protein synthesis

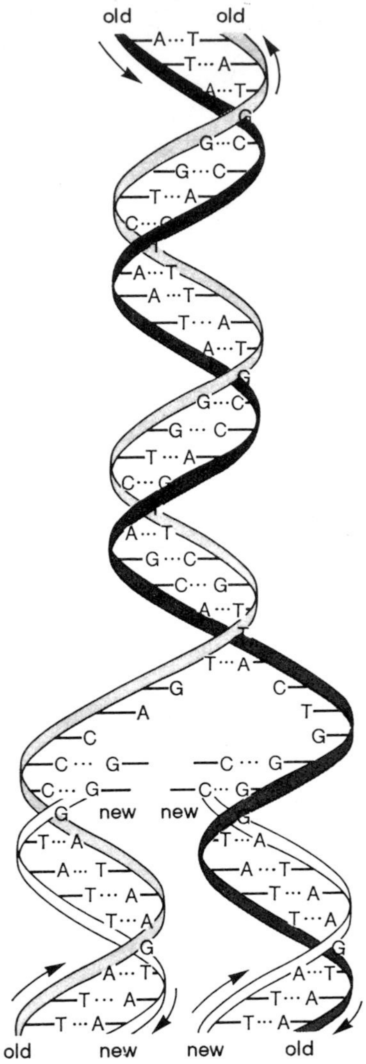

Fig. 6.4.3. The DNA molecule is a double helix, i.e. it has the form of a spiral staircase. The "balustrades" of the staircase are formed by the phosphate and deoxyribose (saccharide) groups (not shown). The steps of the staircase are formed by the base pairs A–T and G–C (Fig. 6.4.2). In the replication process, the two complementary strands of the double helix are separated like an opened zipper (*upper part*). Then each base attaches itself to a new base according to the pairing scheme described, and the result (*below*) is two identical DNA double helices

[4] In uracil, the CH_3 group on the benzene ring of thymine (Fig. 6.4.2) is replaced by H.

must be accomplished in the organism rapidly and with the least possible energy consumption, they are catalyzed by special proteins, the *enzymes*, which themselves are again produced using the "know how" of the DNA. It is, to be sure, a still-unsolved problem to identify the *beginnings* of these closed cycles, but the mechanism of the genetic code, which we have indicated only schematically here, makes the replication of particular systems of molecules and thus of particular species of organisms fundamentally understandable. It also explains the origin of *mutations* through disturbances in the word sequence along the DNA chain.

Having described the molecular-biological fundamentals of the structure, the functions, and the mechanisms of self-reproduction in present-day living organisms, we turn to the question of the *origin of life* on Earth and its *evolution*, first investigating the *material preconditions* for this process.

The complex molecules which are typical of the structure of living matter, especially the nucleic acids and the proteins, are not stable in the presence of oxygen. Their formation from inorganic substances is therefore not possible on the Earth under present conditions without the intervention of living organisms. In the "beginning", they could only have been formed in an *oxygen free* atmosphere. On the other hand, such molecules could not have survived for long on the surface of the proto-Earth without the protective oxygen and ozone layers in the atmosphere, owing to the very strong irradiation in the far ultraviolet ($\lambda < 290$ nm). However, they could have collected for example in the pores of minerals or on the bottoms of shallow bodies of water of ≥ 10 m depth.

The similarity of the genetic code and its application in essentially the same organic molecules for the DNA and for the proteins in lifeforms from the most primitive bacteria to human beings indicate that the origin of life occurred at a single time and place, where the most important macromolecules were already present and could be used as "building blocks". The formation of such high-molecular weight *prebiotic molecules* "all by themselves" is not so extremely improbable as one might at first think. Beginning in 1953, S. L. Miller, led by the considerations of H. C. Urey, was able to prepare complex organic molecules, in particular amino acids, by subjecting a mixture of methane (CH_4), ammonia (NH_3), and water vapor (H_2O), a kind of protoatmosphere, to electrical discharges or irradiation with UV or gamma rays. According to current ideas (see above), the early atmosphere was, to be sure, not a strongly reducing methane atmosphere, but rather consisted mainly of CO_2 and H_2O, with only traces of CH_4 and NH_3; but experiments have shown that even under these conditions, small amounts of the relevant organic molecules can be formed. However, it is questionable whether concentrations would result which were sufficiently high for life to have come about.

A quite different, interesting possibility for the formation of complex molecules is indicated by the investigation of the interstellar medium and of meteorites. As we have seen in Sect. 5.3.4, radio astronomers have observed surprisingly complex organic molecules in the dense *interstellar clouds*, from which stars are later formed. On the other hand, a rich variety of amino acids and other substances which are necessary for the origin of living organisms are found in the *carbonaceous chondrites*. These meteorites, with their organic matter, were formed about $4.5 \cdot 10^9$ yr ago, along with the Sun and all the other objects in the Solar System, from the "solar nebula", a rotating disk of gas and dust (Sect. 6.2). Today, we also know that the Earth, along with the Moon and the other planets, was subjected to an intense bombardment by *meteorites* for about 10^9 yr after its formation; among them, as among present-day meteorites, there must have been a considerable fraction of carbonaceous chondrites. Since many of them were slowed down by the Earth's atmosphere and also fell into water, sometimes after fragmentation, it is quite probable that the prebiotic molecules came for the most part from these *carbonaceous chondrites*.

Organic molecules, however, are still far from being living organisms! We cannot say how the first lifeforms were structured and how they came into being. Even the *simplest living organisms* or "*biomolecules*" must, in contrast to lifeless matter, be able to carry out a *metabolism* and especially to *reproduce themselves*; in this process, occasional "errors" occur, or *mutations*. From the palette of different mutations, the best-adapted species were *selected* in the process of *evolution*. Behind the question of the origin of life and of its evolution stands the fundamental problem of whether our current physics and natural sciences contain a sufficient basis for understanding the origin of biomolecules capable of self organization, or whether we need to assume a mysterious "vis vitalis" or something similar. In 1971, M. Eigen succeeded in explaining the Darwinian principle of survival of the fittest in terms of purely *physico-chemical fundamentals*. The actual origin of life, as well as all the mutations, must be regarded as "coincidences"; all the rest is physics. Eigen first determined that no solution can be hoped for on the basis of classical (equilibrium or quasi-equilibrium) thermodynamics. The interactions of proteins and nucleic acids necessary for every life-process are so complex, even in the most primitive forms imaginable, that their occurrence under such conditions must be regarded as out of the question. "Life" is in fact only possible with the aid of an energy input, metabolism, etc.; i.e. as a (quasi-) stationary irreversible thermodynamic process. In a series of fundamental publications during the 1970's, I. Prigogine, M. Eigen, H. Kuhn, H. Haken, and others were able to describe the essential features of the origin and evolution of self-organizing molecules as *necessary* properties of systems of nonlinear reaction kinetics equations, which are *far* from the state of thermodynamic equilibrium and are subjected to strong energy currents. The origin of biomolecules can be attributed to characteristic *instabilities* of these systems. The further development of the living organisms takes place necessarily in discrete steps, which each begin with a new "invention": the production of a *membrane* around the nucleus of a cell, the principle of sexuality with its greater security in reproduction, the combination of different types of cells with particular cooperative functions, etc. We can unfortunately not go into these very interesting considerations further here or justify them in more detail.

Making use of the current status of research in the fields of palaeontology, geology, geochronology, etc., we shall now attempt to draw a consistent picture of the *early history of the Earth and its atmosphere in connection with the development of life*, based on the works of E. S. Barghoorn, J. W. Schopf, P. E. Cloud and others. For most of the details, we must of course refer the reader to the specialized literature on this subject.

We first ask the question: what do we know about the earliest lifeforms? In sediments as old as 3.5 to $3.6 \cdot 10^9$ yr from South Africa and Australia, we find definite traces of living organisms: microfossiles such as the *Ramsay spheres* (yeast-like organisms) and *stromatolites*, layers built up from the calcium carbonate deposits of blue algae (cf. Table 2.8.2). The currently oldest known, radioactively dated rock deposits are the *Isua formations* in Southwest Greenland, with an age of $3.8 \cdot 10^9$ yr. Their sediments contain carbonaceous inclusions, which were interpreted by H. D. Pflug and M. Schidlowski as fossile structures of blue bacteria and other microorganisms which have been modified by later metamorphosis of the rocks containing them. According to this interpretation, which is to be sure not uncontroversial, there was already a widespread, differentiated world of bacterial flora on the Earth as long as $3.8 \cdot 10^9$ yr ago, and with certainty about $3.5 \cdot 10^9$ yr ago.

As we have seen, the Earth-Moon system originated about $4.5 \cdot 10^9$ yr ago. However, while the rocks of the Lunar Highlands or the Terrae on the Moon date back to this time, the oldest rocks on the Earth are considerably younger. The Isua sediments verify the existence of a widespread hydrosphere as early as $3.8 \cdot 10^9$ yr ago, and they themselves are probably derived from ancient rock strata dating back to about 4.0 to $4.1 \cdot 10^9$ yr ago. Since, on the one hand, we can hardly assume that the Earth's crust was solid earlier than roughly $4.2 \cdot 10^9$ yr ago, and on the other, the first traces of life date from 3.8 to $3.5 \cdot 10^9$ yr ago, only the amazingly short period of around 400 to at most 800 million years remains for the decisive steps in the origin of the first lifeforms and their evolution to the stage of bacteria and blue algae to take place! The structural materials for this process, amino acids and other prebiotic molecules, were probably "delivered" for the most part by the impacts of carbonaceous chondrites from space. It may also have been important for the origin of life that around 10^9 yr after the formation of the Solar System, the intense bombardment by meteorites was drastically reduced, and that the oceans and the protoatmosphere, consisting mainly of CO_2 and H_2O, had already begun to develop.

An *oxygen-free atmosphere* was a prerequisite for the origin of organic life; many substances essential for life are attacked by oxygen. However, there was originally also no ozone layer which would have protected the early lifeforms from lethal solar ultraviolet radiation. The most primitive organisms must therefore have developed either under a protective layer of water, or else in pores in rocks which let in some light but not too much of the deadly UV radiation.

These first organisms were certainly heterotrophic, i.e. they lived on the organic substances already present. However, *autotrophic* organisms of different kinds, which make use of inorganic starting materials, developed very early from bacteria and expanded rapidly owing to the "invention" of *photosynthesis* by the early plants, making possible the chemical dissociation of water (6.4.1). Parallel to the development of the plants, a general enrichment of the hydrosphere and the atmosphere in *oxygen* took place.

In the period about $2.0 \cdot 10^9$ yr ago, there were, according to the remains that we can study today, only *prokaryotic lifeforms*, i.e. single-celled organisms lacking a well-defined nuclear membrane or chromosomes. These include on the one hand the *eubacteria*, the true bacteria, and the blue algae; and on the other, the *archaebacteria*, some of whose representatives, e.g. the sulfur-, salt-, or methane-bacteria, are still living today, often as autotrophes, under extreme conditions (in hot sulfur-containing springs, concentrated brine, or sewage wastes). The small amount of oxygen produced by the early prokaryotic lifeforms remained in the hydrosphere and was mostly used up in converting ferrous to ferric oxides. The *banded iron ore deposits* are characteristic of this geological epoch. Their formation can be interpreted (P. E. Cloud, 1973) to be a result of the formation of Fe^{2+} solutions under an oxygen-free atmosphere, which were then transported into the seas and there, by the action of oxygen-producing microscopic algae, were periodically precipitated as Fe^{3+} compounds.

We find the first *eukaryotic lifeforms*, i.e. those *with* nuclear membranes, about 1.4 to $1.8 \cdot 10^9$yr ago; their origins and their connection to the prokaryotic organisms are still largely unexplained. The first eukaryotic organisms were perhaps prokaryotic organisms with a "built-in" parasite.

Only about $2 \cdot 10^9$ yr ago did the production of *excess oxygen* begin. Geologically, this epoch is characterized by the sediments called *red beds*, which are colored red by Fe_2O_3. Free oxygen made it possible for O_2-processing enzymes to develop and quickly provided the *animals* with a high-yield energy source (respiration!).

There was a fundamental change when the *oxygen content* of our atmosphere reached *a few percent* of the present value, 0.6 to $0.7 \cdot 10^9$ yr ago. This oxygen sufficed to allow the formation of an ozone layer, which absorbed the lethal UV radiation so strongly that now the *surface of the water* and soon thereafter the *land* became habitable.

Shortly before the transition to the Cambrian, 0.6 to $0.7 \cdot 10^9$ yr ago, in the phanerozoic era (the period of "appearing life"), the *metazoa* arose, i.e. *multi-celled* animals with complex internal organization; initially, they had soft bodies. Folling the Cambrian, i.e. less than $0.6 \cdot 10^9$ yr ago, a great variety of shell-bearing metazoa developed. At this point, the series of *geological periods* described in Table 2.8.2 began; their petrified lifeforms (fossiles) have provided us with a detailed record of the further evolution of the plant and animal kingdoms. About 410 million years ago, from the Silurian to the Devonian eras, the first widespread forests developed, and in the Carboniferous era, whose fossilized plants are still serving us as fuel today, the present-day level of oxygen was no doubt reached, if not even surpassed for periods of time.

There have been repeated critical *periods of change* in the evolution of the plant and animal kingdoms, each lasting several million years, in which at first the variety of different species was drastically reduced, and then numerous new species appeared; for example, the transition from the Permian to the Triassic eras, or near the end of the Cretaceous (extinction of the dinosaurs). The causes of such periods, in which the rate of mutations was evidently much higher than before and after, are not clear; speculations include increased UV irradiation, a stronger intensity of cosmic rays as a result of the polarity reversals of the Earth's magnetic field, or the impacts of large meteorites. Some believe that there is a connection between the extinction of the dinosaurs and the observation of a high concentration of iridium and osmium in a thin layer of clay which is found between the Cretaceous and the Tertiary deposits and is spread over nearly the whole Earth. An enrichment in these metals, which are normally quite rare in the Earth's crust but are common in certain meteorites, could in fact be an indication of the impact of an enormous meteorite.

Finally, in the most recent past, geologically speaking, the *evolution of humans* has taken place, as with other organisms in a series of steps and under the influence of natural selection. Beginning with common ancestors, the evolution of human beings (hominids) branched off from that of the hominoid apes, probably in the Tertiary, 10 to $15 \cdot 10^6$ yr ago. The finding of *Australopithecus afarensis*, about $3 \cdot 10^6$ yr old, by D. C. Johanson and T. Gray in 1974 in the Afar Region of Ethiopia, can be regarded as the oldest definite trace of the hominids to come to light thus far (cf. Table 2.8.2). *Ancient man* of the species Australopithecus already had an upright gait and could make and use tools, but had a brain volume only about as great as that of the hominoid apes, and had no speech as we understand it. The further evolution of humans was characterized by an enormous development of the brain, in particular of the forebrain, leading from *early man*, Homo habilis (about 2 to $1.6 \cdot 10^6$ yr ago) and Homo erectus (up to about 300 000 yr ago) finally to *modern man*, Homo sapiens.

The whole evolution of life, if we may be allowed to reduce it to its simplest elements, is evidently based on the *principle* of producing more and more complex systems by the storage and passing-on of more and more *information*. These systems are then increasingly successful in applying their ordering principle (production of negentropy) as against the "natural" tendency of the 2nd Law of Thermodynamics in the establishment of a statistical-thermodynamical equilibrium, i.e. to maximum disorder (increase of entropy). This evolution leads even in the prokaryotic organisms away from the fixation of genetic information towards the development of instinctive, i.e. program-controlled behavior schemes, then further to memory and the beginnings of intelligent behavior. The development of humans (just a few million years ago) was driven essentially by their enormously improved techniques for storing and using information, in comparison to simpler organisms, at first through language, then through writing (of all kinds), and most recently through the invention of the electronic computer and information storage devices.

This ever-increasing "struggle against the law of entropy" is *necessarily* coupled with an increasing *energy consumption*. A species is, as one says, in a dynamic equilibrium. That is, it continually takes in energy from its food supplies, in the case of plants from sunlight, which is in part necessary for the (re)production of the system and in part leads to waste heat (as in any thermal energy source), which must be discarded. Man can only continue his development along the lines described above if he succeeds in making more and more energy "usable" from natural sources. Along with the muscle power of humans and animals, the forces of wind and water were used. The next great step was the application of the fossile fuels, coal and petroleum, in thermal power plants. It is certainly no accident that our age of automation has also brought with it the application of nuclear power.

To the question of the *future* of the Sun, and thus of life on the Earth, the theory of stellar evolution permits us to give a quite specific answer. The Sun is moving (over a period of a few billion years) towards the upper right on the Hertzsprung-Russell diagram; it will become a red giant star. Its radius will grow enormously in this process, its bolometric magnitude will increase by several factors of ten, and the temperature on the Earth will rise to well above the boiling point of water. This will doubtless mean the *end* of all organic life on the Earth.

On the other hand, the question has often been raised as to whether there is life on other celestial bodies. This question is at present only reasonable if we understand life to mean the occurrence of organisms whose structures have a certain similarity to those of terrestrial lifeforms. We need not belabor the point that the environmental conditions permitting such life are fixed within rather narrow limits.

In our Solar System, only Mars can be considered as a possibility. Its thin atmosphere contains in fact small amounts of the required gases, but it offers no "UV protection"; the temperature lies somewhat lower than that of the Earth. It therefore appears possible that at most extremely primitive organisms might be found there, but even this is not very probable according to the present state of our knowledge.

In our Milky Way and in other galaxies, there are numberless stars which are similar to our Sun. There is nothing to contradict the assumption that some of these stars also have planetary systems, and it seems quite plausible that here and there in such a system, a planet offers similar conditions on its surface to those found on the Earth. Why should not life have evolved there also?

A first survey of the radiofrequency emissions from nearby star systems to search for signals from "extraterrestrial civilizations" (Project SETI: Search for Extraterrestrial Intelligence) was begun in 1960 by F. D. Drake. Up to now, this socalled *bioastronomy* has been equally unsuccessful in finding evidence for planets around other stars and in detecting "artificial" radio signals from outer space.

* * *

From the study of cosmic structures and evolutionary processes, we have returned to the problems of the Earth, of life and of our existence. The circle of our considerations is thus completed. Equally fascinating as the insight into cosmic events is the process of the acquisition of human knowledge itself. The boundary of our knowledge is constantly moving out into regions which were previously unknown or at least not understood. This process seems to be connected with a kind of adjustment of the human intellect and its thought processes to the areas of being which have only become accessible through just this intellect. We owe in fact to just a few great figures of history the finding of new forms of thinking and of dealing with problems which only became clear through their own genius. All great discoveries and deeds contain a certain illogical element.

Our categories of knowledge are arranged, so to speak, in layers, which evidently are the result of the basic areas of interaction of humanity with its environment. Even the most primitive human uses different concepts and ways of thinking when he deals with nonliving natural objects, with his animals and other living creatures (which are beneath him in the evolutionary chain), with his own emotional life and that of his fellow humans, or finally with overreaching intellectual or especially religious subjects. These different areas of experience, which at first were almost heedlessly allowed to have their parallel but separate existences, have come into closer and closer contact in the course of time, and mankind has long since suffered inner and outer struggles in an effort to unite them into a whole, out of the unprovable conviction that this must indeed be possible.

No one can finally refute the view held by many primitive religions that on a certain level, the universe is simply a result of the arbitrary decisions of some sort of gods and demons. In fact, however, the "other" hypothesis has proved repeatedly to be more fruitful.

The problems of human life and living with each other are interrelated with those of the understanding of natural phenomena. We see them on the one hand from our *subjective point of view*, and speak of desire and duty, of love and hate, of justice and injustice. On the other hand, they are a part of the *objective framework* of the Possible, governed by the known (or still unknown) laws of nature. The great conflicts of guilt and fate, of human goals and our ability to attain them, or however they may be termed, arise out of this remarkable duality of the subjective and the objective worlds.

It has been the conviction of truly religious men and women in all periods of history that these tensions could gradually, bit by bit, be resolved, presuming that humanity could lay aside its old prejudices and "renew itself", a process which is perhaps not so different from the renewal and reformation of natural science in its great epochs.

Whether humankind turns its gaze *outwards* and probes ever deeper into the further reaches of nature, or looks *inwards* and finds there new realms for humanity, over and over again we view with amazement and joy a

"New Cosmos".

Appendix

A.1 Various Units in the International System of Units (SI) and the Gaussian System[1]

We use the *International System of Units* (*S*ystème *I*nternational d'*U*nités) throughout this book; in parallel, the more important quantities are also given in the Gaussian System. It is expedient to introduce "astronomical units" as well (see the back inside cover).

SI Base Units:

Meter	m	(length)
Kilogram	kg	(mass)
Second	s	(time)
Ampère	A	(electric current)
Kelvin	K	(temperature)
Mole	mol	(quantity of matter)
Candela	cd	(light intensity)[2]

Prefixes for Orders of Magnitude:

			10^{-1}	deci	d
10^2	hecto	h	10^{-2}	centi	c
10^3	kilo	k	10^{-3}	milli	m
10^6	mega	M	10^{-6}	micro	μ
10^9	giga	G	10^{-9}	nano	n
10^{12}	tera	T	10^{-12}	pico	p

Some Relations Between the SI and the Gaussian System of Units in the Area of Electromagnetism:

The Gaussian System of "mixed" cgs units employs both electrostatic units (esu) and electromagnetic units (emu) in the corresponding applications. The permeability and permittivity of vacuum are dimensionless constants having the value 1 in this system.

X = quantity in SI units
$\tilde{X}$ = quantity in Gaussian units

[1] See Symbols, Units, and Nomenclature in Physics, Document U.I.P. 20 (1978), from Physica **93A**, 1 (1978).
[2] Not used in this book.

We limit ourselves here to the *vacuum*, so that the magnetic induction $\boldsymbol{B}$ and the magnetic field strength $\boldsymbol{H}$ are related by the equations:

$$\boldsymbol{B} = \mu_0 \boldsymbol{H} \quad \text{or} \quad \tilde{\boldsymbol{B}} = \tilde{\boldsymbol{H}} \; .$$

Electric charge: $\tilde{e} = \dfrac{e}{\sqrt{4\pi\varepsilon_0}}$

Electric field strength: $\tilde{\boldsymbol{E}} = \sqrt{4\pi\varepsilon_0}\,\boldsymbol{E}$

Magnetic flux density: $\tilde{\boldsymbol{B}} = \sqrt{\dfrac{4\pi}{\mu_0}}\,\boldsymbol{B} = \sqrt{4\pi\varepsilon_0}\,c\boldsymbol{B}$

ε_0 and μ_0 are the permittivity and permeability constants of the vacuum (see the back inside cover), with $\varepsilon_0\mu_0 = 1/c^2$, c = velocity of light.

Force on a moving charge e: $\boldsymbol{F} = e\boldsymbol{E} + e\boldsymbol{v}\times\boldsymbol{B} = \tilde{e}\tilde{\boldsymbol{E}} + \dfrac{\tilde{e}}{c}\boldsymbol{v}\times\tilde{\boldsymbol{B}}$

Energy density in vacuum: $w = \dfrac{1}{2}\left(\varepsilon_0 E^2 + \dfrac{B^2}{\mu_0}\right) = \dfrac{1}{8\pi}(\tilde{E}^2 + \tilde{B}^2)$

Poynting vector: $\boldsymbol{S} = \boldsymbol{E}\times\boldsymbol{H} = \dfrac{1}{\mu_0}\boldsymbol{E}\times\boldsymbol{B} = \dfrac{c}{4\pi}\tilde{\boldsymbol{E}}\times\tilde{\boldsymbol{B}}$

Larmor frequency: $\omega_{\mathrm{L}} = \dfrac{eB}{2m} = \dfrac{\tilde{e}\tilde{B}}{2mc}$

Cyclotron frequency: $\omega_{\mathrm{c}} = \dfrac{eB}{m} = \dfrac{\tilde{e}\tilde{B}}{mc}$

1st Bohr radius: $a_0 = 4\pi\varepsilon_0\dfrac{\hbar^2}{me^2} = \dfrac{\hbar^2}{m\tilde{e}^2}$

Classical electron radius: $r_{\mathrm{e}} = \dfrac{1}{4\pi\varepsilon_0}\dfrac{e^2}{mc^2} = \dfrac{\tilde{e}^2}{mc^2}$

Various Derived Units:

Length		
Ångström	1 Å	$= 10^{-10}$ m $= 10^{-8}$ cm
Mass:		
Metric ton	1 t	$= 10^3$ kg
Atomic mass unit m_{u}	1 u	$= 1.6605 \cdot 10^{-27}$ kg
Time:		
Minute	1 min	= 60 s
Hour	1 h	= 60 min = 3600 s
Day	1 d	= 24 h = 86400 s
Year	1 yr	$\simeq 3.156 \cdot 10^7$ s

Frequency:		
Hertz:	1 Hz	$= 1\ \mathrm{s}^{-1}$
Angle:		
Planar radian	1 rad	$= 1\ \mathrm{m\ m}^{-1}$ (dimensionless)
		$= 57.2958° = 3437.74' = 206\,264.81''$
Degree	1°	$= \pi/180$ rad $= 1.7453 \cdot 10^{-2}$ rad
Minute of arc	1′	$= (1/60)° = 2.9089 \cdot 10^{-4}$ rad
Second of arc	1″	$= (1/60)' = 4.8481 \cdot 10^{-6}$ rad
Solid Angle:		
Steradian	1 sr	$= 1\ \mathrm{m}^2\ \mathrm{m}^{-2}$ (dimensionless)
		$= (180/\pi)^2$ degrees squared
		$= 3282.8$ degrees squared
Force:		
Newton	1 N	$= 1\ \mathrm{m\ kg\ s}^{-2} = 10^5$ dyn $= 10^5\ \mathrm{cm\ g\ s}^{-2}$
Pressure:		
Pascal	1 Pa	$= 1\ \mathrm{m}^{-1}\ \mathrm{kg\ s}^{-2} = 1\ \mathrm{N\ m}^{-2} = 10\ \mathrm{dyn\ cm}^{-2}$
Bar	1 bar	$= 10^5$ Pa
Energy:		
Joule	1 J	$= 1\ \mathrm{m}^2\ \mathrm{kg\ s}^{-2} = 1\ \mathrm{N\ m} = 1\ \mathrm{W\ s}$
		$= 10^7\ \mathrm{erg} = 10^7\ \mathrm{cm}^2\ \mathrm{g\ s}^{-2}$
Electron volt	1 eV	$= 1.6022 \cdot 10^{-19}\ \mathrm{J} = 1.6022 \cdot 10^{-12}$ erg
Energy equivalents:		
kT for $T = 1$ K	1 K·k	$= 1.3807 \cdot 10^{-23}\ \mathrm{J} = 8.6174 \cdot 10^{-5}$ eV
Kilogram	$1\ \mathrm{kg} \cdot c^2$	$= 8.9876 \cdot 10^{16}$ J
Atomic mass unit	$m_u c^2$	$= 1.4924 \cdot 10^{-10}\ \mathrm{J} = 931.49$ MeV
Proton mass	$m_p c^2$	$= 1.5033 \cdot 10^{-10}\ \mathrm{J} = 938.27$ MeV
Electron mass	mc^2	$= 8.1872 \cdot 10^{-14}\ \mathrm{J} = 0.5110$ MeV
Power:		
Watt	1 W	$= 1\ \mathrm{m}^2\ \mathrm{kg\ s}^{-3} = 1\ \mathrm{J\ s}^{-1} = 1$ VA
		$= 10^7\ \mathrm{erg\ s}^{-1}$
Temperature:		
Celsius scale	t [°C]	= absolute temperature T [K] $- 273.15$ K
Temperature equivalent of the electron volt	$1\ \mathrm{eV\ k}^{-1}$	$= 11\,605$ K
Electric Charge:		
Coulomb	1 C	$= 1\ \mathrm{A\ s} = 2.9979 \cdot 10^9$ esu
Electrostatic Potential, Voltage:		
Volt	1 V	$= 1\ \mathrm{m}^2\ \mathrm{kg\ s}^{-3}\ \mathrm{A}^{-1} = 3.3356 \cdot 10^{-3}$ esu
Magnetic Flux Density (in Vacuum):		
Tesla	1 T	$= 1\ \mathrm{kg\ s}^{-2}\ \mathrm{A}^{-1} = 1\ \mathrm{V\ s\ m}^{-2} = 10^4$ G (Gauss)

A.2 Names of the Constellations

Standard abbreviation and latin names (nominative and genitive forms) of the constellations

And	Andromeda	Andromedae
Ant	Antlia	Antliae
Aps	Apus	Apodis
Aql	Aquila	Aquilae
Aqr	Aquarius	Aquarii
Ara	Ara	Arae
Ari	Aries	Arietis
Aur	Auriga	Aurigae
Boo	Bootes	Bootis
Cae	Caelum	Caeli
Cam	Camelopardalis	Camelopardalis
Cap	Capricornus	Capricorni
Car	Carina	Carinae
Cas	Cassiopeia	Cassiopeiae
Cen	Centaurus	Centauri
Cep	Cepheus	Cephei
Cet	Cetus	Ceti
Cha	Chamaeleon	Chamaeleontis
Cir	Circinus	Circini
CMa	Canis Major	Canis Majoris
CMi	Canis Minor	Canis Minoris
Cnc	Cancer	Cancri
Col	Columba	Columbae
Com	Coma Berenices	Comae Berenices
CrA	Corona Austrina	Coronae Austrinae
CrB	Corona Borealis	Coronae Borealis
Crt	Crater	Crateris
Cru	Crux	Crucis
Crv	Corvus	Corvi
CVn	Canes Venatici	Canum Venaticorum
Cyg	Cygnus	Cygni
Del	Delphinus	Delphini
Dor	Dorado	Doradus
Dra	Draco	Draconis
Equ	Equuleus	Equulei
Eri	Eridanus	Eridani
For	Fornax	Fornacis
Gem	Gemini	Geminorum
Gru	Grus	Gruis
Her	Hercules	Herculis
Hor	Horologium	Horologii
Hya	Hydra	Hydrae
Hyi	Hydrus	Hydri

Ind	Indus	Indi
Lac	Lacerta	Lacertae
Leo	Leo	Leonis
Lep	Lepus	Leporis
Lib	Libra	Librae
LMi	Leo Minor	Leonis Minoris
Lup	Lupus	Lupi
Lyn	Lynx	Lyncis
Lyr	Lyra	Lyrae
Men	Mensa	Mensae
Mic	Microscopium	Microscopii
Mon	Monoceros	Monocerotis
Mus	Musca	Muscae
Nor	Norma	Normae
Oct	Octans	Octantis
Oph	Ophiuchus	Ophiuchi
Ori	Orion	Orionis
Pav	Pavo	Pavonis
Peg	Pegasus	Pegasi
Per	Perseus	Persei
Phe	Phoenix	Phoenicis
Pic	Pictor	Pictoris
PsA	Piscis Austrinus	Piscis Austrini
Psc	Pisces	Piscium
Pup	Puppis	Puppis
Pyx	Pyxis	Pyxidis
Ret	Reticulum	Reticuli
Scl	Sculptor	Sculptoris
Sco	Scorpius	Scorpii
Sct	Scutum	Scuti
Ser	Serpens	Serpentis
Sex	Sextans	Sextantis
Sge	Sagitta	Sagittae
Sgr	Sagittarius	Sagittarii
Tau	Taurus	Tauri
Tel	Telescopium	Telescopii
TrA	Triangulum Australe	Trianguli Australis
Tri	Triangulum	Trianguli
Tuc	Tucana	Tucanae
UMa	Ursa Major	Ursae Majoris
UMi	Ursa Minor	Ursae Minoris
Vel	Vela	Velorum
Vir	Virgo	Virginis
Vol	Volans	Volantis
Vul	Vulpecula	Vulpeculae

Usage of the names of constellations: nonvariable *stars* are denoted by Greek (in some cases Roman) letters or by numbers together with the genitive of the Latin name of their constellation, usually in the standard abbreviation with three letters; for example,

β UMa = Beta Ursae Majoris, or 48 UMa = 48 Ursae Majoris, ι Her = Iota Herculis, l Car = l Carinae, a Cen = a Centauri.

Variable stars are denoted by capital letters, R, S, . . . , Z, RR, RS, . . . ZZ, AA . . . AZ, BB . . . QZ, together with the genitive form of their constellation (334 combinations; J is not used); additional variables in a constellation are denoted by e.g. V 335, etc: RR Lyr, W Vir, SS Cyg, V 1057 Cyg.

For *strong radio* and *X-ray sources*, the latin name of constellation in the nominative is used together with capital letters and with numbers; e.g. Tau A = Taurus A, Her X-1 = Hercules X-1, Sco X-3 = Scorpius X-3.

Bibliography

The references listed here are restricted to the most important reference works, periodicals, books, etc., which should be helpful in attaining a deeper understanding of individual problems and fields of work.

Literature Survey

Information about individual papers that have been published in specialist journals etc. can be found in

Astronomy and Astrophysics Abstracts (Springer, Berlin Heidelberg)

which appears every six months (first appeared in 1969).

Conference Proceedings

Papers presented at a conference devoted to a particular subject are usually collected in proceedings. In general we do not list the various volumes among the references for each chapter of this book. Instead, we just mention here the following series:

Reviews in Modern Astronomy (Springer, Berlin Heidelberg)
Publications of the International Astronomical Union. Proceedings of Symposia Series (Reidel, Dordrecht)
Highlights of Astronomy. International Astronomical Union (Reidel, Dordrecht)
Reports on Astronomy. Transactions of the International Astronomical Union (Reidel, Dordrecht)
Proceedings of IAU Colloquia (various publishers)
Proceedings NATO Advanced Study Institute Series
Proceedings Liège International Astrophysical Colloquia
Astronomy and Astrophysics Library (Springer, Berlin Heidelberg)
Astrophysics and Space Science Library (Reidel, Dordrecht)

Handbooks, Review Articles, Collections of Numerical Data and Formulas, etc.

Annual Review of Astronomy and Astrophysics (Annual Reviews, Palo Alto, CA)
Flügge, S. (ed.): *Encyclopedia of Physics,* Vols. 50–54, *Astrophysics I–V* (Springer, Berlin Göttingen Heidelberg 1958–1962)
Middlehurst, B.M., Kuiper, G.P. (eds.): *The Solar System,* 5 vols. (University of Chicago Press, Chicago 1953–1966)
Kuiper, G.P., Middlehurst, B.M. (eds.): *Stars and Stellar Systems,* 8 vols. (University of Chicago Press, Chicago 1960–1975)

Schaifers, K., Voigt, H. H. (eds.): *Landolt-Börnstein, Numerical Data and Functional Relationships in Science and Technology*, New Series, Group VI, Volume 2, *Astronomy and Astrophysics* (Springer, Berlin Heidelberg 1981 [Subvolume a], 1982 [Subvolume b und c])
Allen, C. W.: *Astrophysical Quantities* (Athlone, London 1973)
Lang, K. R.: *Astrophysical Formulae* (Springer, Berlin Heidelberg 1980)
Zombeck, M.: *Handbook of Space Astronomy and Astrophysics* (Cambridge University Press, Cambridge 1990)

Sky Atlases

Bečvár, A.: *Atlas Coeli 1950.0* (Czechoslovakian Academy of Sciences, Prague 1962)
Marx, S., Pfau, W.: *Sternatlas* (1975,0) (Barth, Leipzig 1983)
Schurig, R., Götz, P., Schaifers, K.: *Himmelsatlas* (*Tabulae caelestes*) (Bibliographisches Institut, Mannheim 1980)
Tirion, W.: *Sky Atlas 2000.0* (Cambridge University Press, Cambridge 1981)
Ridpath, I. (ed.): *Norton's 2000.0 Star Atlas and Reference Handbook* (Longman, Harlow 1989)

Important Journals

a) Popular Magazines

New Scientist (PC Magazines)
Sky and Telescope (Sky Publishing Corp., Cambridge, MA)
Scientific American (Scientific American Inc., New York)

b) Specialist Journals

The Astronomical Journal (American Institute of Physics, New York)
Astronomische Nachrichten (Akademie-Verlag, Berlin)
Astronomy and Astrophysics (A European Journal) (Springer, Berlin Heidelberg)
The Astrophysical Journal (University of Chicago Press, Chicago)
Astrophysics and Space Science (Reidel, Dordrecht)
Celestial Mechanics (Reidel, Dordrecht)
Icarus. International Journal of Solar System Studies (Academic, New York)
Mercury (The Journal of the Astronomical Society of the Pacific, San Francisco, CA)
Monthly Notices of the Royal Astronomical Society (Blackwell Scientific Publications, Oxford)
Nature (Macmillan Journals, London)
Publications of the Astronomical Society of Japan (Tokyo)
Publications of the Astronomical Society of the Pacific (San Francisco, CA)
Science (American Association for the Advancement of Science, Washington, DC)
Solar Physics (Reidel, Dordrecht)
Soviet Astronomy (Translation of Astron. Zhurnal) (American Institute of Physics, New York)
Space Science Reviews (Reidel, Dordrecht)

Introductions to General Astronomy, Popular Reference Works and Reviews

Audouze, J., Israel, G. (eds.): *The Cambridge Atlas of Astronomy* (Cambridge University Press, Cambridge 1988)
Harwit, M.: *Astrophysical Concepts.* Astronomy and Astrophysics Library (Springer, Berlin Heidelberg 1988)
Karttunen, H., Kröger, P., Oja, H., Poutanen, M., Donner, K. J. (eds.): *Fundamental Astronomy* (Springer, Berlin Heidelberg 1987)
Kaufmann III, W. J.: *Universe* (Freeman, New York 1988)
Kippenhahn, R.: *Light from the Depths of Time* (Springer, Berlin Heidelberg 1987)
Longair, M. S.: *Theoretical Concepts in Physics* (Cambridge University Press, Cambridge 1984)
Pasachoff, J. M.: *Contemporary Astronomy* (Saunders, Philadelphia 1977)
Pasachoff, J. M.: *Astronomy: From the Earth to the Universe* (Saunders, Philadelphia 1983)
Roy, A. E., Clarke, D.: *Astronomy* (*Principles and Practice*); *Astronomy* (*Structure of the Universe*) (Hilger, Bristol 1988, 1989)
Shu, F. H.: *The Physical Universe* (University Science Books, Mill Valley 1982)

Bibliography of the Individual Sections

References are given in lists for groups of sections that deal with related subject matter. In general, each book is listed under only one group, so in borderline cases it is worth looking under groups on related topics.

History of Astronomy (Sects. 2.1, 3.1, 4.1 and 5.1)

Abetti, G.: *History of Astronomy* (Sidgwick and Jackson, London 1954)
Ashbrook, J.: *The Astronomical Scrapbook: Skywatchers, Pioneers, and Seekers in Astronomy* (Sky Publishing, Cambridge 1984)
Berry, A.: *A Short History of Astronomy, 1898* (Dover, New York 1961)
Dreyer, J. L. E.: *A History of Astronomy from Thales to Kepler* (Dover, New York 1953)
Gingerich, O. (ed.): *Astrophysics and Twentieth-Century Astronomy to 1950* (Cambridge University Press, New York 1984)
Hearnshaw, J. B.: *The Analysis of Starlight* (Cambridge University Press, Cambridge 1987)
Herrmann, D. B.: *The History of Astronomy from Herschel to Hertzsprung* (Cambridge University Press, New York 1984)
King, H. C.: *The History of the Telescope* (Dover, New York 1979)
Lang, K. R., Gingerich, O.: *A Source Book in Astronomy and Astrophysics. 1900–1975* (Harvard University Press, Cambridge, MA 1979)
Neugebauer, O.: *Astronomy and History* (Springer, Berlin Heidelberg 1983)
Pannekoek, I.: *A History of Astronomy* (Dover, New York 1989)
Parker, B.: *Creation. The Story of the Origin and Evolution of the Universe* (Plenum, New York 1988)
Smith, R.: *The Expanding Universe: Astronomy's Great Debate 1900–1931 Cambridge University Press, Cambridge 1982)*
Verschuur, G. L.: Interstellar Matters (Springer, Berlin Heidelberg 1989)

Coordinates and Time. Planetary Orbits. Celestial Mechanics and Space Travel (Sects. 2.3–7)

Brouwer, D., Clemence, G. M.: *Methods of Celestial Mechanics* (Academic, New York 1961)
Brown, E. W.: *An Introductory Treatise on the Lunar Theory* (Dover, New York 1960)
Brown, E. W., Shook, C. A.: *Planetary Theory* (Dover, New York 1964)
Green, R.: *Spherical Astronomy* (Cambridge University Press, Cambridge 1985)
Leech, J. W.: *Classical Mechanics* (Methuen, London 1958)
Podobed, V. V.: *Fundamental Astrometry* (University of Chicago Press, Chicago 1965)
Roy, A. E.: *Orbital Motion* (Hilger, Bristol 1982)
Smart, W. M.: *Textbook on Spherical Astronomy* (Cambridge University Press, Cambridge 1960)
Smart, W. M.: *Celestial Mechanics* (Longmans, London 1960)
Taff, L.: *Celestial Mechanics* (Wiley, New York 1985)

Physical Nature of the Planets and Their Satellites (Sect. 2.8)

Atreya, S. K.: *Atmospheres and Ionospheres of the Outer Planets and Their Satellites* (Springer, Berlin Heidelberg 1986)
Beatty, J. K., Chaikin, A. (eds.): *The New Solar System* (Cambridge University Press, Cambridge 1990)
Briggs, G., Taylor, F.: *The Cambridge Photographic Atlas of the Planets* (Cambridge University Press, Cambridge 1982)
Burns, J. A. (ed.): *Planetary Satellites* (University of Arizona Press, Tucson 1977)
Carr, M. (ed.): *The Geology of the Terrestrial Planets*. NASA SP-466 (1984)
Chamberlain, J. W., Hunten, D. M.: *Theory of Planetary Atmospheres: An Introduction to Their Physics and Chemistry*, 2nd ed. (Academic, San Diego, CA 1987)
Cole, G. H. A.: *Physics of Planetary Interiors* (Hilger, Bristol 1984)
Cook, A. H.: *Interiors of the Planets* (Cambridge University Press, Cambridge 1980)
Dressler, A. J.: *Physics of the Jovian Magnetosphere* (Cambridge University Press, Cambridge 1983)
Gehrels, T. (ed.): *Protostars and Planets* (University of Arizona Press, Tucson 1978)
Gehrels, T. (ed.): *Jupiter* (University of Arizona Press, Tucson 1976)
Gehrels, T., Matthews, M. (eds.): *Saturn* (University of Arizona Press, Tucson 1984)
Greenberg, R., Brahic, A. (eds.): *Planetary Rings* (University of Arizona Press, Tucson 1984)
Guest, J. E., Butterworth, P., Murray, J., O'Donnell, W.: *Planetary Geology* (David & Charles, London 1979)
Guest, J. E., Greeley, R.: *Geology on the Moon* (Wykeham, London 1977)
Houghton, J. T.: *The Physics of Atmospheres* (Cambridge University Press, Cambridge 1986)
Hubbard, W.: *Planetary Interiors* (Van Nostrand Reinhold, New York 1984)
Hunt, G., Moore, P.: *The Planet Venus* (Faber and Faber, London 1982)
Hunt, G., Moore, P.: *Atlas of Jupiter* (Mitchell Beazley, London 1981)
Hunt, G., Moore, P.: *Atlas of Saturn* (Mitchell Beazley, London 1982)
Hunt, G., Moore, P.: *Atlas of Uranus* (Cambridge University Press, Cambridge 1989)
Hunten, D. M., Colin, L., Donahue, T. M., Moroz, V. I. (eds.): *Venus* (University of Arizona Press, Tucson 1983)

Kopal, Z.: *Physics and Astronomy of the Moon* (Academic, New York 1971)
Lewis, J., Prinn, R.: *Planets and Their Atmospheres: Origin and Evolution* (Academic, New York 1983)
Melchior, P.: *The Physics of the Earth's Core* (Pergamon, Oxford 1986)
Moore, P., Cross, M.A.: *The Moon* (Mitchell Beazley, London 1981)
Moore, P., Hunt, G.: *The Atlas of the Solar System* (Mitchell Beazley, London 1983)
Morrison, D. (ed.): *Satellites of Jupiter* (University of Arizona Press, Tucson 1982)
Voigt, A., Giebler, H.: *Berliner Mondatlas* (Wilhelm-Foerster-Sternwarte, Berlin 1974)

Comets, Meteors and Meteorites. Interplanetary Matter (Sect. 2.9)

Akasofu, S.-I., Kamide, Y. (eds.): *The Solar Wind and the Earth* (Terra Scientific, Tokyo; Reidel, Dordrecht 1987)
Alpert, Ya.L.: *The Near-Earth and Interplanetary Plasma,* Vols. 1 and 2 (Cambridge University Press, Cambridge 1983)
Brandt, J.C., Chapman, R.C.: *Introduction to Comets* (Cambridge University Press, Cambridge 1983)
Bronshten, V.: *Physics of Meteoritic Phenomena* (Reidel, Dordrecht 1983)
Dodd, R.T.: *Meteorites* (Cambridge University Press, Cambridge 1982)
Gehrels, T. (ed.): *Asteroids* (University of Arizona Press, Tucson 1979)
Hawkins, G.S.: *Meteors, Comets, and Meteorites* (McGraw-Hill, New York 1964)
Kowal, Ch.T.: *Asteroids: Their Nature and Utilization* (Ellis Horwood, Chichester 1988)
Lovell, A.C.B.: *Meteor Astronomy* (Clarendon, Oxford 1954)
Mackin, R., Jr., Neugebauer, M.: *The Solar Wind* (Pergamon, Oxford 1964)
Mason, B.: *Meteorites* (Wiley, New York 1962)
Roth, G.D.: *The System of Minor Planets* (Faber and Faber, London 1962)
Whipple, F.L.: *The Mystery of Comets* (Cambridge University Press, Cambridge 1986)
Wilkening, L.L. (ed.): *Comets* (University of Arizona Press, Tucson 1982)

Astronomical Instrumentation and Observational Methods (Sects. 3.2–5)

Beckman, J.E., Phillips, J.P.: *Submillimetre Wave Astronomy* (Cambridge University Press, Cambridge 1982)
Born, M., Wolf, E.: *Principles of Optics* (Pergamon, Oxford 1980)
Christiansen, W.N., Högbom, J.A.: *Radiotelescopes* (Cambridge University Press, Cambridge 1985)
Chupp, E.L.: *Gamma Ray Astronomy* (Reidel, Dordrecht 1976)
Culver, R.B.: *An Introduction to Experimental Astronomy. An Observational Workbook* (Freeman, Oxford 1984)
Eccles, M.J., Sim, M.E., Tritton, K.P.: *Low Light Level Detectors in Astronomy* (Cambridge University Press, Cambridge 1983)
Hanbury Brown, R.: *The Intensity Interferometer* (Taylor and Francis, London 1974)
Hillier, R.: *Gamma Ray Astronomy* (Oxford University Press, Oxford 1984)
Hirsh, R.: *Glimpsing an Invisible Universe. The Emergence of X-Ray Astronomy* (Cambridge University Press, Cambridge 1983)
Jaschek, C.: *Data in Astronomy* (Cambridge University Press, Cambridge 1989)
Jones, R., Wykes, C.: *Holographic and Speckle Interferometry* (Cambridge University Press, Cambridge 1983)

Kingston, R.H.: *Detection of Optical and Infrared Radiation* (Springer, Berlin Heidelberg 1978)

Kitchin, C.R.: *Astrophysical Techniques* (Hilger, Bristol 1984)

Kleinknecht, K.: *Detectors for Particle Radiation* (Cambridge University Press, Cambridge 1986)

Kraus, J.D.: *Radio Astronomy* (Cygnus-Quasar Books, Powell, OH 1986)

Lena, P.: *Observational Astrophysics* (Springer, Berlin Heidelberg 1988)

Longair, M.S.: *High Energy Astrophysics* (Cambridge University Press, Cambridge 1981)

Malin, D., Murdin, P.: *Colours of the Stars* (Cambridge University Press, Cambridge 1984)

Meaburn, J.: *Detection and Spectrometry of Faint Light* (Reidel, Dordrecht 1976)

Pecker, J.C.: *Experimental Astronomy* (Reidel, Dordrecht 1970)

Rohlfs, K.: *Tools of Radio Astronomy.* Astronomy and Astrophysics Library (Springer, Berlin Heidelberg 1986)

Steel, W.: *Interferometry* (Cambridge University Press, Cambridge 1984)

Sullivan, W.T.: *The Early Years of Radio Astronomy* (Cambridge University Press, Cambridge 1984)

Vaucouleurs, G. de: *Astronomical Photography* (Faber and Faber, London 1961)

Verschuur, G.L.: *The Invisible Universe Revealed* (Springer, New York 1987)

Walker, G.: *Astronomical Observations – An Optical Perspective* (Cambridge University Press, Cambridge 1987)

Theory of Radiation. Spectra and Atoms. Stellar Atmospheres and Fraunhofer Lines (Sects. 4.2 and 4.7–9)

Athay, R.G.: *Radiation Transport in Spectral Lines* (Reidel, Dordrecht 1972)

Bekefi, G.: *Radiation Processes in Plasmas* (Wiley, New York 1966)

Böhm-Vitense, E.: *Introduction to Stellar Astrophysics.* Vol. 1. *Basic stellar observations and data.* Vol. 2. *Stellar atmospheres* (Cambridge University Press, Cambridge 1989)

Cannon, C.J.: *The Transfer of Spectral Line Radiation* (Cambridge University Press, Cambridge 1985)

Chandrasekhar, S.: *Radiative Transfer* (Dover, New York 1950, 1960)

Condon, E.U., Shortley, G.H.: *The Theory of Atomic Spectra* (Cambridge University Press, Cambridge 1963)

Cowan, R.D.: *The Theory of Atomic Structure and Spectra* (University of California Press, Berkeley 1981)

Dalgarno, A., Layzer, D.: *Spectroscopy of Astrophysical Plasmas* (Cambridge University Press, Cambridge 1987)

Gray, D.F.: *The Observation and Analysis of Stellar Photospheres* (Wiley, New York 1976)

Haken, H., Wolf, H.C.: *Atomic and Quantum Physics* (Springer, Berlin Heidelberg 1987)

Kaler, J.B.: *Stars and Their Spectra. An Introduction to the Spectral Sequence* (Cambridge University Press, Cambridge 1989)

Kalkofen, W. (ed.): *Methods in Radiative Transfer* (Cambridge University Press, Cambridge 1984)

Mihalas, D.: *Stellar Atmospheres* (Freeman, San Francisco 1978)

Mihalas, D., Mihalas, B. W.: *Foundations of Radiation Hydrodynamics* (Oxford University Press, New York 1984)

Novotny, E.: *Introduction to Stellar Atmospheres and Interiors* (Oxford University Press, New York 1973)

Oxenius, J.: *Kinetic Theory of Particles and Photons.* Springer Ser. Electrophys., Vol. 20 (Springer, Berlin Heidelberg 1986)

Rybicki, G. B., Lightman, A. P.: *Radiative Processes in Astrophysics* (Wiley, New York 1985)

Thomas, R.: *Stellar Atmospheric Structural Patterns* NASA SP-471 (1983)

Unsöld, A.: *Physik der Sternatmosphären. Mit besonderer Berücksichtigung der Sonne* (Springer, Berlin Göttingen Heidelberg 1955, 1968)

White, H. E.: *Introduction to Atomic Spectra* (McGraw-Hill, New York 1934)

Spectroscopic Data

Bashkin, S., Stoner, Jr., J. O.: *Atomic Energy-Level and Grotrian Diagrams* (North-Holland, Amsterdam, Vol. I (H I-P XV) (1975), Addenda (1978); Vol. II (S I-Ti XXII) (1978); Vol. III (V I-Cr XXIV) (1981); Vol. IV (Mn I-XXV) (1982)

Moore, C. E.: *A Multiplet Table of Astrophysical Interest*, Revised Ed. NSRDS-NBS 40 (National Bureau of Standards, Washington, DC 1972)

Moore, C. E.: *An Ultraviolet Multiplet Table.* NBS Circular 488 (National Bureau of Standards, Washington, DC 1950–1962)

Merrill, P. W., Moore, C. E.: *Partial Grotrian Diagrams of Astrophysical Interest*, NSRDS-NBS 23 (National Bureau of Standards, Washington, DC 1968)

Reader, J., Corliss, C. H., Wiese, W. L, Martin, G. A.: *Wavelengths and Transition Probabilities for Atoms and Atomic Ions*, NSRDS-NBS 68 (National Bureau of Standards, Washington, DC 1980)

The Sun (Sects. 4.3 and 4.10)

Athay, R. G.: *The Solar Chromosphere and Corona – Quiet Sun* (Reidel, Dordrecht 1976)

Bahcall, J. N.: *Neutrino Astrophysics* (Cambridge University Press, Cambridge 1989)

Billings, D. E.: *A Guide to the Solar Corona* (Academic, New York 1966)

Brandt, J. C.: *Introduction to the Solar Wind* (Freeman, San Francisco 1970)

Bray, R. J., Loughhead, R. E.: *Sunspots* (Chapman and Hall, London 1964)

Bray, R. J., Loughhead, R. E., Durrant, C. J.: *The Solar Granulation* (Cambridge University Press, Cambridge 1984)

Bruzek, A., Durrant, C. J.: *Illustrated Glossary for Solar and Solar-Terrestrial Physics* (Reidel, Dordrecht 1977)

Durrant, C. J.: *The Atmosphere of the Sun* (Hilger, Bristol 1988)

Gibson, E. G.: *The Quiet Sun.* NASA SP-303 (NASA, Washington, DC 1973)

Hundhausen, A. J.: *Coronal Expansion and Solar Wind* (Springer, Berlin Heidelberg 1972)

Jordan, S.: *The Sun as a Star.* NASA SP-450 (NASA, Washington, DC 1981)

Krüger, A.: *Introduction to Solar Radioastronomy and Radiophysics* (Reidel, Dordrecht 1979)

McLean, D. J., Labrum, N. R. (eds.): *Solar Radiophysics* (Cambridge University Press, Cambridge 1985)

Noyes, R. W.: *The Sun, Our Star* (Harvard University Press, Cambridge, MA 1982)

Priest, E.R.: *Solar Magnetohydrodynamics* (Reidel, Dordrecht 1984)
Priest, E.R. (ed.): *Solar System Magnetic Fields* (Reidel, Dordrecht 1985)
Smith, H.J., Smith, E.V.P.: *Solar Flares* (Macmillan, New York 1963)
Stix, M.: *The Sun – An Introduction* (Springer, Berlin Heidelberg 1989)
Sturrock, P.A. (ed.): *Physics of the Sun;* Vol. 1, *The Solar Interior;* Vol. 2, *The Solar Atmosphere;* Vol. 3, *Astrophysics and Solar-Terrestrial Relations* (Reidel, Dordrecht 1985)
Švestka, Z.: *Solar Flares* (Reidel, Dordrecht 1976)
Tandberg-Hanssen, E.: *Solar Activity* (Blaisdell, Waltham, MA 1967)
Tandberg-Hanssen, E., Emslie, A.G.: *The Physics of Solar Flares* (Cambridge University Press, Cambridge 1988)
Thomas, R.N., Athay, R.G.: *Physics of the Solar Chromosphere* (Interscience, New York 1961)
White, O.R.: *The Solar Output and Its Variation* (Colorado Ass. University Press, Boulder 1977)
Zirin, H.: *Astrophysics of the Sun* (Cambridge University Press, Cambridge 1988)

Stars: Basic Parameters, Spectra, Binary Stars, Variable Stars (Sects. 4.4–6 and 4.11)

Adams, D.J.: *Cosmic X-ray Astronomy* (Hilger, Bristol 1979)
Aitken, R.G.: *The Binary Stars* (Dover, New York 1964)
Bode, M.F., Evans, A. (eds.): *Classical Novae* (Wiley, Chichester 1989)
Clark, D.H., Stephenson, F.R.: *The Historical Supernovae* (Pergamon, Oxford 1977)
De Jager, C.: *The Brightest Stars* (Reidel, Dordrecht 1980)
Garrison, R.F. (ed.): *The MK Process and Stellar Classification* (David Dunlap Obs., Ontario, Canada 1984)
Glasby, J.S.: *The Dwarf Novae* (Constable, London 1970)
Glasby, J.S.: *The Nebular Variables* (Pergamon, Oxford 1974)
Gursky, H., Ruffini, R.: *Neutron Stars, Black Holes and Binary X-Ray Sources* (Reidel, Dordrecht 1975)
Gurzadyan, G.A.: *Flare Stars* (Pergamon, Oxford 1980)
Heintz, W.D.: *Double Stars* (Reidel, Dordrecht 1978)
Hoffmeister, C., Richter, G., Wenzel, W.: *Variable Stars* (Springer, Berlin Heidelberg 1985)
Jaschek, C., Jaschek, M.: *The Classification of Stars* (Cambridge University Press, Cambridge 1987)
Kenyon, S.J.: *The Symbiotic Stars* (Cambridge University Press, Cambridge 1986)
Kitchin, C.R.: *Stars, Nebulae and the Interstellar Medium* (Hilger, Bristol 1987)
Kopal, Z.: *Close Binary Systems* (Chapman and Hall, London 1959)
Kopal, Z.: *Language of the Stars* (Reidel, Dordrecht 1979)
Lewin, W., Heuvel, E. van den (eds.): *Accretion Driven Stellar X-Ray Sources* (Cambridge University Press, Cambridge 1983)
Petit, M.: *Variable Stars* (Wiley, Chichester 1987)
Petschek, A.G. (ed.): *Supernovae* (Springer, Berlin Heidelberg 1990)
Pringle, J.E., Wade, R.A.: *Interacting Binary Stars* (Cambridge University Press, Cambridge 1985)
Shlovsky, I.S.: *Supernovae* (Wiley, London 1968)

Sexl, R., Sexl, H.: *White Dwarfs – Black Holes: An Introduction to Relativistic Astrophysics* (Academic, New York 1979)
Smith, F. G.: *Pulsars* (Cambridge University Press, Cambridge 1977)
Strohmeier, W.: *Variable Stars* (Pergamon, Oxford 1972)
Underhill, A.: *The Early Type Stars* (Reidel, Dordrecht 1966)
Wolff, S.: *The A-Stars: Problems and Perspectives.* NASA and CNRS (Nat. Techn. Inf. Service, Springfield 1983)

Atlases of Stellar Spectra
Keenan, P. C., McNeil, R. C.: *An Atlas of Spectra of the Cooler Stars* (Ohio State University Press, Columbus 1976)
Morgan, W. W., Keenan, P. C., Kellman, E.: *An Atlas of Stellar Spectra* (University of Chicago Press, Chicago 1943)
Morgan, W. W., Abt, H. A., Tapscott, J. W.: *Revised MK Atlas for Stars Earlier than the Sun* (Kitt Peak National Observatory, Tucson, AZ 1978)
Seitter, W. C.: *Atlas für Objectiv Prismen Spektren. Bonner Spektral Atlas I, II.* (Dümmlers, Bonn 1970, 1975)
Yamashita, Y., Nariai, K., Norimoto, Y.: *An Atlas of Representative Stellar Spectra* (University of Tokyo Press, Tokyo 1977)

Stellar Structure and Evolution. General Theory of Relativity (Sects. 4.12 and 5.4)

Audouze, J., Vauclair, S.: An Introduction to Nuclear Astrophysics (Reidel, Dordrecht 1980)
Bowers, R., Deeming, T.: *Astrophysics I: Stars* (Jones and Bartlett, Boston 1984)
Chandrasekhar, S.: *An Introduction to the Study of Stellar Structure* (1939) (Dover, New York 1957)
Chandrasekhar, S.: *The Mathematical Theory of Black Holes* (Oxford University Press, Oxford 1983)
Clayton, D. D.: *Principles of Stellar Evolution and Nucleosynthesis* (University of Chicago Press, Chicago1983)
Collins, II, G. W.: *The Fundamentals of Stellar Astrophysics* (Freeman, San Francisco 1989)
Cox, J. P.: *Theory of Stellar Pulsation* (Princeton University Press, Princeton 1980)
Cox, J. P., Giuli, R. T.: *Principles of Stellar Structures,* Vol. 1: *Physical Principles;* Vol. 2: *Applications to Stars* (Gordon and Breach, New York 1968)
Eddington, A. S.: *The Internal Constitution of the Stars* (1926) (Dover, New York 1959)
Goldberg, H., Scadron, M.: *Physics of Stellar Evolution and Cosmology* (Gordon and Breach, New York 1981)
Kaplan, S. A.: *The Physics of Stars* (Wiley, New York 1982)
Kaufmann III, W. J.: *Black Holes and Warped Spacetime* (Freeman, San Francisco 1979)
Katz, J. I.: *High Energy Astrophysics. Frontiers in Physics*, Vol. 63 (Addison-Wesley, Menlo Park, CA 1987)
Kippenhahn, R., Weigert, A.: *Stellar Structure and Evolution* (Springer, Berlin Heidelberg 1990)
Meadows, A. J.: *Stellar Evolution* (Pergamon, Oxford 1978)
Novotny, E.: *Introduction to Stellar Atmospheres and Interiors* (Oxford University Press, New York 1973)

Reddish, V.C.: *Stellar Formation* (Pergamon, Oxford 1978)
Rindler, W.: *Essential Relativity* (Springer, New York Heidelberg Berlin 1977)
Rolfs, C.E., Rodney, W.S.: *Cauldrons in the Cosmos. Nuclear Astrophysics* (University of Chicago Press, Chicago 1988)
Schwarzschild, M.: *Structure and Evolution of the Stars* (Princeton University Press, Princeton 1958 and Dover, New York 1965)
Shapiro, S.L., Teukolsky, S.A.: *Black Holes, White Dwarfs, and Neutron Stars: The Physics of Compact Objects* (Wiley, New York 1983)
Shklovskii, I.S.: *Stars: Their Birth, Life and Death* (Freeman, San Francisco 1978)
Tayler, R.J.: *The Origin of the Chemical Elements* (Wykeham, London 1972)
Tayler, R.J.: *The Stars: Their Structure and Evolution* (Wykeham, London 1970)
Wald, R.: *General Relativity* (University of Chicago Press, Chicago 1984)
Will, C.M.: *Theory and Experiment in Gravitational Physics* (Cambridge University Press, Cambridge 1985)

Structure and Dynamics of the Milky Way (Sect. 5.2)

Binney, J., Tremaine, S.: *Galactic Astronomy* (Princeton University Press, Princeton, NJ 1987)
Bok, B.J., Bok, P.F.: *The Milky Way* (Harvard University Press, Cambridge, MA 1981)
Chandrasekhar, S.: *Principles of Stellar Dynamics* (Dover, New York 1960)
Mihalas, D., Binney, J.: *Galactic Astronomy – Structure and Kinematics of Galaxies* (Freeman, San Francisco 1981)
Mihalas, D., Routly, P.: *Galactic Astronomy* (Freeman, San Francisco 1968)
Scheffler, H., Elsässer, H.: *Physics of the Galaxy and Interstellar Matter,* Astronomy and Astrophysics Library (Springer, Berlin Heidelberg 1987)
Spitzer, L., Jr.: *Dynamical Evolution of Globular Clusters* (Princeton University Press, Princeton, NJ 1987)

Interstellar Matter. Cosmic Rays (Sect. 5.3)

Alfvén, H.: *Cosmic Plasma* (Reidel, Dordrecht 1981)
Aller, L.H.: *Physics of Thermal Gaseous Nebulae* (Reidel, Dordrecht 1984)
Allkofer, O.C.: *Introduction to Cosmic Radiation* (Thieme, München 1975)
Bohren, C.F., Huffman, D.R.: *Absorption and Scattering of Light by Small Particles* (Wiley, New York 1983)
Bowers, R., Deeming, T.: *Astrophysics II: Interstellar Matter and Galaxies* (Jones and Bartlett, Boston 1984)
Dyson, J.E., Williams, D.A.: *The Physics of the Interstellar Medium* (Manchester University Press, Manchester 1980)
Ginzburg, V.L., Syrovatskii, S.I.: *The Origin of Cosmic Rays* (Pergamon, Oxford 1964)
Ginzburg, V.L.: *Elementary Processes in Cosmic Ray Astrophysics* (Gordon and Breach, New York 1969)
Goudis, C.: *The Orion Complex: A Case Study of Interstellar Matter* (Reidel, Dordrecht 1982)
Greisen, K.: *The Physics of Cosmic X-ray, γ-ray, and Particle Sources* (Gordon and Breach, New York 1971)
Gurzadyan, G.A.: *Planetary Nebulae* (Reidel, Dordrecht 1970)
Hulst, H.C. van de: *Light Scattering by Small Particles* (Wiley, New York 1957)
Kaplan, S.A.: *Interstellar Gas Dynamics* (Pergamon, Oxford 1966)

Kaplan, S. A., Pikelner, S. B.: *The Interstellar Medium* (Harvard University Press, Cambridge, MA 1970)
Longair, M. S.: *High Energy Astrophysics* (Cambridge University Press, Cambridge 1981)
Osterbrock, D. E.: *Astrophysics of Gaseous Nebulae and Active Galactic Nuclei* (Univ. Science Books, Mill Valley, CA 1989)
Pottasch, S. R.: *Planetary Nebulae* (Reidel, Dordrecht 1983)
Ramana Murthy, P. V., Wolfendale, A.: *Gamma-Ray Astronomy* (Cambridge University Press, Cambridge 1986)
Spitzer, L., Jr.: *Physical Processes in the Interstellar Medium* (Wiley, New York 1978)
Wolfendale, A. W.: *Cosmic Rays* (Lewnes, London 1963)

Galaxies and Clusters of Galaxies (Sects. 5.5 – 8)

Arp, H.: *Atlas of Peculiar Galaxies* (California Institute of Technology, Pasadena 1966)
Frank, J., King, A. P., Raine, D. J.: *Accretion Power in Astrophysics* (Cambridge University Press, Cambridge 1985)
Hey, J. S.: *The Radio Universe* (Pergamon, London 1971)
Kaufmann III, W. J.: *Galaxies and Quasars* (Freeman, San Francisco 1979)
Mitton, S.: *Exploring the Galaxies* (Faber and Faber, London 1976)
Pacholczyk, A. G.: *Radio Galaxies* (Pergamon, Oxford 1977)
Sandage, A.: *The Hubble Atlas of Galaxies* (Carnegie Institution of Washington 1961)
Shipman, H. L.: *Black Holes, Quasars, and the Universe* (Houghton Mifflin, Boston 1980)
Takase, B., Kodaira, K., Okamura, S. (eds.): *An Atlas of Selected Galaxies* (University of Tokyo Press, Tokyo; VNU Science Press, Utrecht 1984)
Tayler, R. J.: *Galaxies: Structure and Evolution* (Wykeham, London 1978)
Verschuur, G. L., Kellermann, K. L. (eds.): *Galactic and Extragalactic Radio Astronomy*, Astronomy and Astrophysics Library (Springer, Berlin Heidelberg 1988)
Weedman, D. W.: *Quasar Astronomy* (Cambridge University Press, Cambridge 1986)

Cosmology (Sect. 5.9)

Abbott, L. F., Pi, S.-Y.: *Inflationary Cosmology* (World Scientific, Singapore 1986)
Barrow, J. D., Tipler, F. J.: *The Anthropic Cosmological Principle* (Oxford University Press, Oxford 1986)
Börner, G.: *The Early Universe. Facts and Fiction* (Springer, Berlin Heidelberg 1988)
Grotz, K., Klapdor, H. V.: *The Weak Interaction in Nuclear, Particle and Astrophysics* (Hilger, Bristol 1990)
Harrison, E. R.: *Cosmology, the Science of the Universe* (Cambridge University Press, Cambridge 1981)
Hawking, S. W.: *A Brief History of Time: From the Big Bang to Black Holes* (Bantam Books, New York 1988)
Misner, C. W., Thorne, K. S., Wheeler, J. A.: *Gravitation* (Freeman, San Francisco 1973)
Narlikar, J.: *Introduction to Cosmology* (Jones and Bartlett, Boston 1983)
Peebles, P. J. E.: *Physical Cosmology* (Princeton University Press, Princeton 1971)
Peebles, P. J. E.: *The Large-Scale Structure of the Universe* (Princeton University Press, Princeton, NJ 1980)
Rowan-Robinson, M.: *The Cosmological Distance Ladder* (Freeman, Oxford 1985)

Sciama, D. W.: *Modern Cosmology* (Cambridge University Press, Cambridge 1971)
Silk, J.: *The Big Bang* (Freeman, San Francisco 1988)
Weinberg, S.: *Gravitation and Cosmology* (Wiley, New York 1972)
Weinberg, S.: *The First Three Minutes. A Modern View of the Origin of the Universe* (Basic Books, New York 1977)
Zee, A.: *The Unity of Forces in the Universe*, Vols. I, II (World Science Press, Singapore 1982)
Zeldovich, Ya. B., Novikov, I. D.: *The Structure and Evolution of the Universe* (University Chicago Press, Chicago 1983)

Formation of the Solar System – Evolution of the Earth and of Life (Sects. 6.1–4)

Alfvén, H., Arrhenius, G.: *Structure and Evolutionary History of the Solar System* (Reidel, Dordrecht 1975)
Black, D. C., Matthews, M. S. (eds.): *Protostars and Planets II* (University of Arizona Press, Tucson 1985)
Encrenaz, T., Bibring, J.-P.: *The Solar System*. Astronomy and Astrophysics Library (Springer, Berlin Heidelberg 1990)
Henderson-Sellers, A.: *The Origin and Evolution of Planetary Atmospheres* (Hilger, Bristol 1983)
Monod, J.: *Chance and Necessity* (Random House, New York 1972)
Prigogine, I., Stengers, I.: *Order out of Chaos* (Fontana, London 1985)
Rutten, M. G.: *The Origin of Life* (Elsevier, Amsterdam 1971)
Suess, H. E.: *Chemistry of the Solar System. An Elementary Introduction to Cosmochemistry* (Wiley, New York 1987)
Taube, M.: *Evolution of Matter and Energy on a Cosmic and Planetary Scale* (Springer, Berlin Heidelberg 1985)
Unsöld, A.: *Evolution kosmischer, biologischer und geistiger Strukturen* (Wissenschaftliche Verlagsgesellschaft, Stuttgart 1983)
Wasson, J.: *Meteorites: Their Record of Early Solar-System History* (Freeman, Oxford 1985)
Wood, J. A.: *Meteorites and the Origin of the Planets* (McGraw-Hill, New York 1968)

Figure Acknowledgements

Frontispiece		S. Laustsen, C. Madsen, R.M. West: *Entdeckungen am Südhimmel. Ein Bildatlas der Europäischen Südsternwarte (ESO)* (Springer-Verlag Berlin, Heidelberg and Birkhäuser Verlag, Basel 1987) Fig. 78
Figs.	2.2.3, 2.3.1–3, 2.4.1–4, 2.5.1, 2.5.5, 2.8.1	*Seydlitz: Allgemeine Erdkunde,* 5. Teil, 7. Aufl. (Schroedel, Hannover 1961)
Fig.	2.6.7	After A. Unsöld: Phys. Bl. **5**, 205 (1964)
Fig.	2.7.1	NASA: D. Mondey (ed.): *The International Encyclopedia of Aviation* (Crown, New York 1977), p. 414
Fig.	2.7.2	After H.M. Schurmeier, R.L. Heacock, A.E. Wolfe: Sci. Am. **214**, 57 (Jan. 1966)
Figs.	2.7.3, 2.8.5	Photo. NASA
Fig.	2.7.4	G. Hunt, P. Moore: *Saturn* (Herder, Freiburg 1983) p. 16 (Fig. 2)
Table	2.8.2	K. Krömmelbein: Private communication (1967); R. Kraatz: Private communication (1984); W.B. Harland, A.V. Cox, P.G. Llewellyn, C.A.G. Pickton, A.G. Smith, R. Walters: *A Geologic Time Scale* (Cambridge University Press, Cambridge 1982)
Fig.	2.8.2	Drawn after W.-H. Ip, W.I. Axford: In Landolt-Börnstein, Neue Serie VI/2a (Springer, Berlin, Heidelberg 1981)
Fig.	2.8.3	A. Unsöld: *Evolution kosmischer, biologischer und geistiger Strukturen* (Wissenschaftliche Verlagsgesellschaft Stuttgart 1983) and J.D. Phillips: Oceanus **17**, 24 (1973/74)
Fig.	2.8.4	Photo. Lick Observatory, from Sky Telesc. **26**, 342 (1963)
Fig.	2.8.6	E. Brüche, E. Dick: Phys. Bl. **26**, 351 (1970) Fig. 7
Fig.	2.8.7	H. Wänke, F. Wlotzka: Universitas **26**, 850 (1971)
Fig.	2.8.8	G.E. McGill: Nature **296**, 14 (1982) Fig. 2
Fig.	2.8.9	NASA: Naturwissenschaften **59**, 395 (1972) Fig. 5
Figs.	2.8.10, 13, 14b, 15, 18	Photo. NASA/Raumfahrt-Bildarchiv H.W. Köhler, Augsburg
Fig.	2.8.11	NASA: Naturwissenschaften **59**, No. 4, titlepage (1972)
Fig.	2.8.12	Drawn after V.I. Moroz: Space Sci. Rev. **29**, 3 (1981); A. Seif, D.B. Kirk: J. Geophys. Res. **82**, 4364 (1977); W.B. Hanson, S. Sanatani, D.R. Zuccaro: J. Geophys. Res. **82**, 4351 (1977); J.T. Houghton: *The Physics of Atmospheres* (Cambridge University Press, Cambridge 1977)
Fig.	2.8.14a	Photo. B. Lyot and H. Camichel, Observatoire Pic du Midi
Fig.	2.8.16	Photo. H. Camichel, Observatoire Pic du Midi
Fig.	2.8.17	NASA: G. Briggs, F. Taylor: *The Cambridge Photographic Atlas of the Planets* (Cambridge University Press, Cambridge 1982) p. 228

Figs.	2.8.19, 20	Photo. Jet Propulsion Laboratory, California Institute of Technology
Fig.	2.9.1	Photo. Hale Observatories
Fig.	2.9.2	P. Swings, L. Haser: *Atlas of Representative Cometary Spectra* (Université de Liège, 1956) Plate IV
Fig.	2.9.3	W. Gentner: Naturwissenschaften **50**, 192 (1963) Fig. 1
Fig.	2.9.4	E. Anders: Acc. Chem. Res. (1968)
Fig.	2.9.5	H. Fechtig, C. Leinert, E. Grün: In Landolt-Börnstein, Neue Serie VI/2a (Springer, Berlin Heidelberg 1981)
Fig.	3.1.1	After B. Rossi: In *Electromagnetic Radiation in Space*, ed. by. J.G. Emming (Reidel, Dordrecht 1967) p. 164
Fig.	3.2.4	S.W. Burnham: Publ. Yerkes Observatory of the University of Chicago **1** (1900)
Fig.	3.2.5	Photo. Mt. Wilson and Palomar Observatories
Fig.	3.2.6	*Das Weltall* (Time-Life, Munich 1964) p. 37
Fig.	3.2.9b	H.N. Russell, R.S. Dugan, J.W. Stewart: *Astronomy II* (Ginn, New York 1927) Fig. 254
Fig.	3.2.10	A. Lohmann, M. Reinecke, H. Ruder, G. Weigelt: Sterne Weltraum **16**, 284 (1977)
Figs.	3.2.11, 12	H. Tüg: Sterne Weltraum **16**, 366 (1977)
Fig.	3.2.14	Photograph from the IUE satellites, image no. SWP 5176, observer: R. Wehrse (1979)
Fig.	3.2.15	J.C. Brandt and the HRS Investigation Definition and Experiment Development Teams: In: *The Space Telescope Observatory*, ed. by D.N.B. Hall, NASA CP-2244 (1982) p. 76
Fig.	3.2.16	N.P. Carleton, W.F. Hoffmann: Phys. Today **31**, No. 9, 30 (1978)
Fig.	3.3.1	Photo. G. Hutschenreiter; Max-Planck-Institut für Radioastronomie, Bonn
Fig.	3.3.2	P. Thaddeus: Phys. Today **35**, No. 11, 36 (1982)
Fig.	3.4.1	M.S. Longair: *High Energy Astrophysics* (Cambridge University Press, Cambridge 1981) Fig. 4.2
Fig.	3.4.2	M. Garcia-Munoz, G.M. Mason, J.A. Simpson: Astrophys. J. **217**, 859 (1977)
Fig.	3.4.3	M. Simon: Private communication; see also J.A. Esposito et al.: 19th International Cosmic Ray Conference, Vol. 3, ed. by F.C. Jones, J. Adams, G.M. Mason (NASA Conf. Publ. 2376, 1985) p. 278
Fig.	3.4.4	M. Gottwald: Sterne Weltraum **22**, 466 (1983)
Fig.	3.5.1	B. Rossi: In *Electromagnetic Radiation in Space*, ed. by J.G. Emming (Reidel, Dordrecht 1967) p. 164 (Fig. 9)
Fig.	4.2.1	A. Unsöld: *Physik der Sternatmosphären*, 2nd ed. (Springer, Berlin Göttingen Heidelberg 1955)
Fig.	4.3.1	E.H. Schröter, E. Wiehr: Sterne Weltraum **24**, 319 (1985)
Fig.	4.3.2	M. Minnaert, G.F.W. Mulders, J. Houtgast: *Photometric Atlas of the Solar Spectrum* (Schnabel, Kampfert, Helm, Amsterdam 1940) Detail
Fig.	4.3.4	H. Neckel: Space Sci. Rev. **38**, 187 (1984) and Sterne Weltraum **23**, 297 (1984)
Fig.	4.4.1	H.L. Johnson, W.W. Morgan: Astrophys. J. **114**, 523 (1951)
Fig.	4.4.2	Drawn after H. Tüg, N.M. White, G.W. Lockwood: Astron. Astrophys. **61**, 679 (1977); C. Jamar, D. Macau-Hercot, A. Monfils,

G.I. Thompson, L. Houziaux, R. Wilson: ESA SR-27 (1976); A.D. Code, M. Meade: Astrophys. J. Suppl. **39**, 195 (1979); R.L. Kurucz: Astrophys. J. Suppl. **40**, 1 (1979)

Fig. 4.5.1 H.N. Russell, R.S. Dugan, J.W. Stewart: *Astronomy II* (Ginn, New York 1927)

Fig. 4.5.2 H.L. Johnson, W.W. Morgan: Astrophys. J. **117**, 338 (1953)

Fig. 4.5.3 W.W. Morgan, P.C. Keenan, E. Kellman: *An Atlas of Stellar Spectra* (University of Chicago Press, Chicago 1942) Detail

Fig. 4.5.5 W. Becker: In *Stars and Stellar Systems III* (University of Chicago Press, Chicago 1963) p. 254; Effective temperatures after I. Bues

Fig. 4.6.1 R.H. Baker: *Astronomy,* 6th ed. (Van Nostrand, New York 1955)

Fig. 4.6.2 Drawn after D.M. Popper: Annu. Rev. Astron. Astrophys. **18**, 115 (1980)

Fig. 4.6.4a J. Adam: Private communication (1987)

Fig. 4.6.4b C. Hoffmeister, G. Richter, W. Wenzel: *Veränderliche Sterne,* 2nd ed. (Springer, Berlin Heidelberg 1984) Fig. 124

Fig. 4.6.5 R.A. Hulse, F.H. Taylor: Astrophys. J. **195**, L51 (1975)

Fig. 4.7.3 P.W. Merrill: Pap. Mt. Wilson Observ. **IX**, 118 (1965)

Fig. 4.8.1 A. Unsöld: *Physik der Sternatmosphären*, 2nd ed. (Springer, Berlin Göttingen Heidelberg 1955) p. 106

Figs. 4.9.1–3 A. Unsöld: Angew. Chem. **76**, 281 (1964)

Tables 4.9.1, 2 Sun: H. Holweger: Private compilation (1985); Meteorites: E. Anders, M. Ebihara: Geochim. Cosmochim. Acta **46**, 2263 (1982)

Fig. 4.10.1 A. Unsöld: *Physik der Sternatmosphären*, 2nd ed. (Springer, Berlin Göttingen Heidelberg 1955)

Fig. 4.10.2 R.E. Danielson: Astrophys. J. **134**, 280 (1961)

Fig. 4.10.3 R. Howard: In *Illustrated Glossary for Solar and Solar-Terrestrial Physics*, ed. by A. Bruzek, C.I. Durrant (Reidel, Dordrecht 1977) p. 7

Fig. 4.10.4 J. Houtgast: Rech. Astron. Utrecht **13**, 3 (Utrecht 1957)

Fig. 4.10.5 C. de Jager: *Handbuch der Physik*, Vol. 52, *Astrophysik III* (Springer, Berlin Göttingen Heidelberg 1959) p. 136

Fig. 4.10.6 G. van Biesbroeck: In *The Sun*, Vol. 1, ed. by G.P. Kuiper (University of Chicago Press, Chicago 1953) p. 604

Figs. 4.10.7, 10 J.A. Eddy: *A New Sun. The Solar Results from Skylab* (NASA, Washington, DC 1979) pp. 97, 162

Fig. 4.10.8 D.H. Menzel, J.G. Wolbach: Sky Telesc. **20**, 252 (1960) Fig. 5

Fig. 4.10.9 T. Royds: Mon. Not. R. Astron. Soc. **89**, 255 (1929)

Fig. 4.10.11 Cape Observatory: Proc. R. Inst. G.B. **38**, No. 175, Plate I (1961)

Fig. 4.10.12 L. Golub, M. Herant, K. Kalata, I. Lovas, G. Nystrom, F. Pardo, E. Spiller, J. Wilczynski: Nature **344**, 842 (1990)

Fig. 4.10.13 H.P. Palmer, R.D. Davies, M.I. Large: *Radio Astronomy Today* (Manchester University Press, Manchester 1963) p. 19

Fig. 4.10.14 J.M. Wilcox: Space Science Lab., Univ. of Calif., Berkeley **12**, 53 (1971) Fig. 2

Fig. 4.11.1 C. Hoffmeister, G. Richter, W. Wenzel: *Veränderliche Sterne*, 2nd ed. (Springer, Berlin Heidelberg 1984) Fig. 10

Fig. 4.11.2 After J.L. Linsky, B.M. Haisch: Astrophys. J. (Lett.) **299**, L27 (1979) Fig. 1

Fig. 4.11.3 T.P. Snow: In *Mass Loss and Evolution of O-Type Stars,* IAU Symp.

No. 83, ed. by P.S. Conti, C.W.H. de Loore (Reidel, Dordrecht 1979) p. 65

Fig. 4.11.4 E.L. Robinson: Annu. Rev. Astron. Astrophys. **14**, 119 (1976)

Fig. 4.11.5 J. Trümper, W. Pietsch, C. Reppin, W. Voges, R. Staubert, E. Kendziorra: Astrophys. J. (Lett.) **219**, L 109 (1978) Fig. 2

Fig. 4.11.6 W.H.G. Lewin, P.C. Joss: Nature **270**, 211 (1977) Fig. 2

Fig. 4.11.7 G.W. Clark: Sci. Am. **237**, 42 (Oct. 1977)

Fig. 4.11.8 R. Minkowski: Annu. Rev. Astron. Astrophys. **2**, 248 (1964)

Figs. 4.11.9, 12, 5.3.1, 10, 12 Photo. European Southern Observatory

Fig. 4.11.10 Photo. Hale Observatories

Fig. 4.11.11 F.R. Harnden Jr., F.D. Seward: Astrophys. J. **283**, 279 (1984)

Fig. 4.11.13 R.M. Duin, R.G. Strom: Astron. Astrophys. **39**, 33 (1975)

Fig. 4.11.14a R.H. Becker, S.S. Holt, B.W. Smith, N.E. White, E.A. Boldt, R.F. Mushotzky, P.J. Serlemitsos: Astrophys. J. (Lett.) **234**, L 73 (1979) Fig. 2

Fig. 4.11.14b S.H. Pravdo, R.H. Becker, E.A. Boldt, S.S. Holt, R.E. Rothschild, P.J. Serlemitsos, J.H. Swank: Astrophys. J. (Lett.) **206**, L 41 (1976) Fig. 1

Fig. 4.11.15 See also S.S. Murray, G. Fabbiano, A.C. Fabian, A. Epstein, R. Giaccioni: Astrophys. J. (Lett.) **234**, L 69 (1979)

Tab. 4.12.1 Z. Abraham, I. Iben Jr.: Astrophys. J. **170**, 157 (1971)

Fig. 4.12.2 E. Meyer-Hofmeister: In Landolt-Börnstein, Neue Serie VI/2b (Springer, Berlin Heidelberg 1982) p. 152, Fig. 3

Fig. 4.12.4 T. Kirsten: Phys. Bl. **39**, 313 (1983)

Fig. 5.2.4 Photo. Mt. Wilson and Palomar Observatories in O. Struve: *Astronomie* (de Gruyter, Berlin 1962) p. 326, Fig. 26.3

Fig. 5.2.5 J.C. Duncan: *Astronomy* (Harper, New York 1950) p. 408

Fig. 5.2.6 J.H. Oort: In *Stars and Stellar Systems*, Vol. 5, ed. by A. Blaauw, M. Schmidt (University of Chicago Press, Chicago 1965) p. 484

Fig. 5.2.7 H. Scheffler: In Landolt-Börnstein, Neue Serie VI/2c (Springer, Berlin Heidelberg 1982) p. 191

Fig. 5.2.8 G. Westerhout: University of Maryland, USA

Fig. 5.2.11 R. Wielen: In Landolt-Börnstein, Neue Serie VI/2c (Springer, Berlin Heidelberg 1982) p. 208, Fig. 4

Fig. 5.2.12 O.J. Eggen: R. Obs. Bull. **84**, 114 (1964)

Fig. 5.3.4 Drawn after B.D. Savage, J.S. Mathis: Annu. Rev. Astron. Astrophys. **17**, 73 (1979) Table 2

Fig. 5.3.5 D.S. Mathewson, V.L. Ford: Mem. R. Astron. Soc. **74**, 143 (1970)

Fig. 5.3.6 After D.C. Morton: Astrophys. J. **197**, 85 (1975)

Fig. 5.3.8 J.H. Oort: In *Interstellar Matter in Galaxies*, ed. by L. Woltjer (Benjamin, New York 1962)

Fig. 5.3.9 M.A. Gordon, W.B. Burton: Astrophys. J. **208**, 346 (1976) Fig. 4

Fig. 5.3.11 Photo. Mt. Wilson and Palomar Observatories, in P.W. Merrill: *Space Chemistry* (University of Michigan Press, Ann Arbor 1963) p. 122

Fig. 5.3.13 L. Goldberg, L.H. Aller: *Atoms, Stars and Nebulae* (Blackiston, Philadelphia 1946) p. 182

Fig. 5.3.14 C.G. Wynn-Williams, E.E. Becklin: Publ. Astron. Soc. Pac. **86**, 5 (1974) Fig. 4

Fig.	5.3.15	R.E. Lingenfelter: Astrophys. Space Sci. **24**, 89 (1973)
Fig.	5.3.16	M. Grewing: In Landolt-Börnstein, Neue Serie VI/2c (Springer, Berlin Heidelberg 1982) p. 134, Fig. 4 and P. Meyer: *Origin of Cosmic Rays*, IAU Symp. No. 94, ed. by G. Setti, G. Spada, A.W. Wolfendale (Reidel, Dordrecht 1981) p. 7
Fig.	5.3.17	H.A. Mayer-Hasselwander, K. Bennett, G.F. Bignami, R. Buccheri, P.A. Caraveo, W. Hermsen, G. Kanbach et al.: Astron. Astrophys. **105**, 164 (1982)
Fig.	5.3.18	K. Pinkau: In *X-Ray Astronomy,* Proc. 21st Plenary Meeting of COSPAR, Innsbruck, ed. by W.A. Baity, L.E. Peterson (Pergamon, Oxford 1979) p. 523, Fig. 9
Fig.	5.4.1 a	H.L. Johnson: Astrophys. J. **116**, 646 (1952)
Fig.	5.4.1 b	O.J. Eggen, A. Sandage: Astrophys. J. **158**, 672 (1969)
Fig.	5.4.2	After A. Sandage, O.J. Eggen: Astrophys. J. **158**, 697 (1969)
Fig.	5.4.3	A. Sandage: Astrophys. J. **162**, 852 (1970) Figs. 13, 14
Fig.	5.4.4	A. Sandage: Astrophys. J. **252**, 574 (1982) Fig. 5
Fig.	5.4.5	Drawn after P.M. Hejlesen: Astron. Astrophys., Suppl. **39**, 347 (1980) Table 1
Fig.	5.4.6	R. Kippenhahn, H.C. Thomas, A. Weigert: Z. Astrophys. **61**, 246 (1965)
Fig.	5.4.7 a,b	E. Meyer-Hofmeister: In Landolt-Börnstein, Neue Serie VI/2b (Springer, Berlin Heidelberg 1982) p. 152 and M. Patenaude: Astron. Astrophys. **66**, 225 (1978) Figs. 1, 3
Fig.	5.4.8	Drawn after A.V. Sweigart, P.G. Gross: Astrophys. J., Suppl. **36**, 405 (1978); D. Schönberner: Astron. Astrophys. **79**, 108 (1979)
Fig.	5.4.9	Drawn after J.R. Wilson, R. Mayle, S.E. Woosley, T.A. Weaver: Proc. 12th Texas Symp. on Relativistic Astrophysics, ed. by M. Livio, G. Shaviv, Ann. N.Y. Acad. Sci. **470**, 267 (1985)
Fig.	5.4.10	R. Kippenhahn, K. Kohl, A. Weigert: Z. Astrophys. **66**, 58 (1967)
Fig.	5.4.11	I. Iben Jr.: Astrophys. J. **141**, 1010 (1965)
Fig.	5.4.12	M. Walker: Astrophys. J., Suppl. **2**, 376 (1956)
Fig.	5.4.13 a	M.L. Kutner, K.D. Tucker, G. Chin, P. Thaddeus: Astrophys. J. **215**, 521 (1977)
Fig.	5.4.13 b	P. Thaddeus: In *Symposium on the Orion Nebula to Honor Henry Draper*, ed. by A.E. Glassgold, P.J. Huggins, E.L. Schucking, Ann. N.Y. Acad. Sci. **395**, 9 (1982)
Fig.	5.4.14	B. Zuckerman: Nature **309**, 403 (1984)
Fig.	5.4.15	I. Appenzeller: In Landolt-Börnstein, Neue Serie VI/2b (Springer, Berlin Heidelberg 1982) p. 357 [after I. Appenzeller, W. Tscharnuter: Astron. Astrophys. **40**, 397 (1975)]
Fig.	5.4.16	R. Mundt: In *Nearby Molecular Clouds*, ed. by G. Serra, Lecture Notes in Physics, Vol. 217 (Springer, Berlin Heidelberg 1985) p. 160
Fig.	5.4.18	R. Wielen, H. Jahreiss, R. Krüger: In *The Nearby Stars and the Stellar Luminosity Function.* IAU Coll. No. 76, ed. by A.G.D. Philip, A.R. Upgren (L. Davis, Schenectady, NY 1983) p. 163
Figs.	5.5.1, 3–5	Photo. Mt. Wilson and Palomar Observatories, from A. Sandage: *The Hubble Atlas of Galaxies* (Carnegie Institution of Washington, DC 1961)
Fig.	5.5.2	E. Hubble: Astrophys. J. **69**, 120 (1929)

Fig. 5.5.6 P.B. Stetson, D.A. Vandenberg, R.D. McClure: Publ. Astron. Soc. Pac. **97**, 908 (1985) Fig. 7b
Fig. 5.5.7 Photo. Hale Observatories
Fig. 5.5.8 Drawn after P. Schechter: Astrophys. J. **203**, 297 (1976); J.E. Felten: Astron. J. **82**, 861 (1978)
Fig. 5.5.9 V.C. Rubin, W.K. Ford Jr.: Astrophys. J. **159**, 390 (1970); J.M. Deharveng, A. Pellet: Astron. Astrophys. **38**, 15 (1975)
Fig. 5.5.10 A. Bosma: Astron. J. **86**, 1825 (1981) Fig. 3
Fig. 5.5.11 W.W. Morgan, N.U. Mayall: Publ. Astron. Soc. Pac. **69**, 295 (1957)
Fig. 5.5.12 W.B. Sparks, J.V. Wall, D.J. Thorne, P.R. Jorden, I.G. van Breda, P.J. Rudd, H.E. Jorgensen: Mon. Not. R. Astron. Soc. **217**, 87 (1985)
Fig. 5.5.13 A.H. Rots, W.W. Shane: Astron. Astrophys. **31**, 245 (1974) Fig. 1b
Fig. 5.5.14 R. Sancisi: Astron. Astrophys. **53**, 159 (1976) Fig. 1
Fig. 5.5.15 H.J. Habing, G. Miley, E. Young, B. Baud, N. Boggess, P.E. Clegg, T. de Jong, S. Harris, E. Raimond, M. Rowan-Robinson, B.T. Soifer: Astrophys. J. (Lett.) **278**, L 59 (1984)
Fig. 5.5.16 A.R. Sandage: Sci. Am. **211**, 39 (Nov. 1964)
Fig. 5.5.17 J.W. Fried, H. Schulz: Astron. Astrophys. **118**, 166 (1983)
Fig. 5.6.2 R.D. Ekers, R. Sancisi: Astron. Astrophys. **54**, 973 (1977) Fig. 1
Fig. 5.6.3 R.A. Perley, J.W. Dreher, J.J. Cowan: Astrophys. J. (Lett.) **285**, L 35 (1984)
Fig. 5.6.4a B.F.C. Cooper, R.M. Price, D.J. Cole: Aust. J. Phys. **18**, 602 (1965)
Fig. 5.6.4b J.O. Burns, E.D. Feigelson, E.J. Schreier: Astrophys. J. **273**, 128 (1983) Figs. 2, 11
Fig. 5.6.5 D. Downes, A. Maxwell, M. Meeks: Astrophys. J. **146**, 657 (1966) Fig. 4
Fig. 5.6.6 R.D. Ekers, J.T. van Gorkom, U.J. Schwarz, W.M. Goss: Astron. Astrophys. **122**, 143 (1983) Figs. 1, 5
Fig. 5.6.7 H. Netzer: In Landolt-Börnstein: New Series, VI/2c (Springer, Berlin Heidelberg 1982) p. 300, Fig. 1
Fig. 5.6.8 B.A. Peterson, A. Savage, D.L. Jauncey, A.E. Wright: Astrophys. J. (Lett.) **260**, L 27 (1982)
Fig. 5.6.9 T. Gehren, J. Fried, P.A. Wehinger, S. Wyckoff: Astrophys. J. **278**, 11 (1984)
Fig. 5.6.10 T.J. Pearson, S.C. Unwin, M.H. Cohen, R.P. Linfield, A.C.S. Readhead, G.H. Seielstad, R.S. Simon, R.C. Walker: Nature **290**, 365 (1981) Fig. 1
Fig. 5.7.1 C. Jones, E. Mandel, J. Schwarz, W. Forman, S.S. Murray, F.R. Harnden Jr.: Astrophys. J. (Lett.) **234**, L 21 (1979) Fig. 2
Fig. 5.7.2 K.J. Wellington, G.K. Miley, H. van der Laan: Nature **244**, 502 (1973)
Fig. 5.7.3 S. Laustsen, C. Madsen, R.M. West: *Entdeckungen am Südhimmel. Ein Bildatlas der Europäischen Südsternwarte (ESO)* (Springer, Berlin, Heidelberg and Birkhäuser Verlag, Basel Boston 1987) Fig. 92
Fig. 5.7.4 M.J. Geller, J.P. Huchra: Science **246**, 897 (1989)
Fig. 5.8.1 W.W. Roberts: Astrophys. J. **158**, 132 (1969) Fig. 7
Fig. 5.8.4 J. Centrella, A.L. Melott: Nature **305**, 196 (1983)
Fig. 5.9.3 H. Kühr: Dissertation, Universität Bonn (1980); H. Kühr, A. Witzel,

I. K. Pauliny-Toth, U. Nauber: Astron. Astrophys. Suppl. **45**, 367 (1981)

Fig. 5.9.4 After D. P. Woody, P. L. Richards: Astrophys. J. **248**, 18 (1981) and G. F. Smoot: In *Early Evolution of the Universe and Its Present Structure,* IAU Symposium No. 104, ed. by G. O. Abell, G. Chincarini (Reidel, Dordrecht 1983) p. 153

Fig. 6.0.1 Partly after W. Gentner: Naturwissenschaften **56**, 174 (1969) Fig. 3

Fig. 6.1.1 G. E. Morfill, W. Tscharnuter, H. J. Völk: In *Protostars and Planets II*, ed. by D. C. Black, M. S. Matthews (University of Arizona Press, Tucson 1985) p. 493

Fig. 6.4.2 S. Fittkau: *Kompendium der organischen Chemie* (G. Fischer, Stuttgart 1980)

Fig. 6.4.3 Th. Wieland, G. Pfleiderer (Ed.): *Molekularbiologie* (Umschau, Frankfurt/M. 1969) p. 46, Fig. 3

Selected Exercises

1. **Coordinate Systems** (Sects. 2.2, 2.3, and 5.2)

1.1 The relationship between the horizontal axis system (azimuth A and altitude h or zenith distance $z = 90° - h$) and the equatorial system (declination δ and hour angle t or right ascension α = sidereal time − t) is contained in the *polar triangle* (or nautical triangle): pole P − zenith Z − celestial object O, with the sides z, $90° - \phi$, and $90° - \delta$ (ϕ = altitude of the pole).

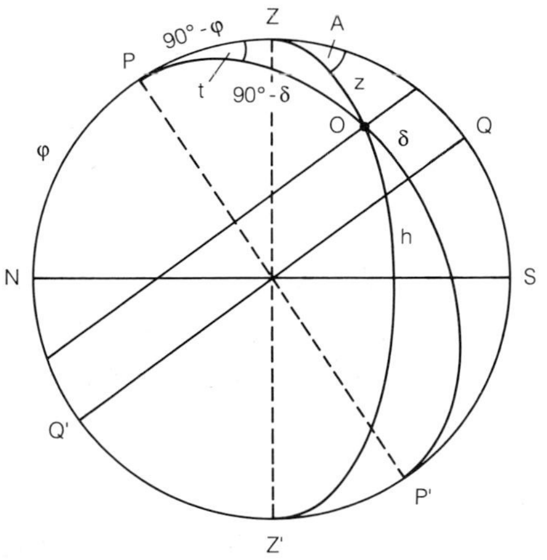

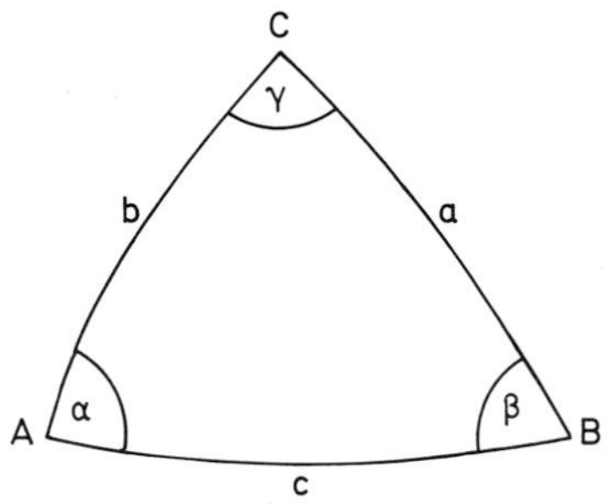

Give formulas for the calculation of z given ϕ, δ, and t and for δ given ϕ, z, and A. What are the corresponding equations for going from the equatorial system (α, δ) to (a) ecliptic coordinates and to (b) galactic coordinates?

1.2 Where and when does the Sun rise in Heidelberg (geographic latitude $\phi = 49.41°$) on the longest day of the year? What are the azimuth and the sidereal time of rising and setting of Arcturus (α Boo, α = 14 h 15 min 39.6 s, $\delta = +19°10'57''$) in New York ($\phi = 40.7°$)? How long does the star remain above the horizon?

Solve exercise (a) without considering refraction in the Earth's atmosphere, and (b) with a refraction of $34'50''$ at the horizon (Sect. 2.2). We define the instant of sunrise by the time at which its upper rim becomes visible.

Hint: From *spherical trigonometry*, we know that in a triangle ABC with sides a, b, c (expressed as angles) and the opposite angles α, β, γ, the following relations hold:

$$\sin a : \sin b : \sin c = \sin \alpha : \sin \beta : \sin \gamma$$

$$\cos a = \cos b \cos c + \sin b \sin c \cdot \cos \alpha$$

$$\cos \alpha = -\cos \beta \cos \gamma + \sin \beta \sin \gamma \cdot \cos a \ .$$

2. Resolving Power, Magnitude (Sects. 2.8 and 3.2)

The apparent visual magnitude of Jupiter's moon J1 Io (at an average opposition of the planet) is $V = +4.8$ mag. Stars out to about 6 mag can be seen with the unaided eye under favorable conditions. Why was Io only discovered after the invention of the telescope by G. Galilei? How much brighter is Jupiter than Io?

Hint: Calculate Io's maximum angular distance from Jupiter and the apparent radius of the planet (Table 2.8.4) and compare them with the resolving power of the human eye (3.2.2). The difference in brightness between Jupiter and Io can be roughly estimated using the ratio of the sizes of the disks of the planet and its satellite.

3. Production of Heat by Radioactivity (Sects. 2.5, 2.8, 2.9, and 4.9)

The production of heat in rocks by radioactivity takes place mostly through the decays of ^{238}U, ^{232}Th, and ^{40}K. The specific energy production q for natural uranium (99.3% ^{238}U) is $9.4 \cdot 10^{-5}$, for thorium ^{232}Th $2.6 \cdot 10^{-5}$, and for natural potassium (0.02% ^{40}K) $3.6 \cdot 10^{-9}$ J kg^{-1} s^{-1}.

3.1 Calculate q from the cosmic element abundances (Table 4.9.1) for (a) cosmic matter containing all the elements, and (b) chondritic matter, assuming for simplicity that it consists practically of olivine $(Mg, Fe)_2SiO_4$ and traces of K, Th, and U in the cosmic element ratio. Compare the result of (b) with the actual value for chondrites, $q = 5.2 \cdot 10^{-12}$ J kg^{-1} s^{-1}. Which radioactive decay chain dominates the production of heat?

3.2 Calculate the heat flux Q at the surface of a homogeneous sphere of radius R and density ϱ using q. How large would Q be for the Earth if it consisted entirely of chondritic matter? How large would Q be for a layer of granite ($q = 1.1 \cdot 10^{-9}$ J kg^{-1} s^{-1}) 10 km thick spread over the whole surface of the Earth?

3.3 The energy balance of an object in the Solar System at a distance r from the Sun is given by (2.8.2). How does the radioactive flux Q compare to the radiation flux $S(r)$ from the Sun for an albedo $A = 0.5$ for various values of r and R (radius of the object)? How large must the radius R be for Q to be larger than $S(r)$? Are there objects in the Solar System for which $Q \geq S(r)$? Discuss examples for objects in the asteroid belt, at the distance of Saturn's radius, and at the edge of the Oort Cloud of comets (Sect. 2.5).

3.4 Calculate the element abundances of ^{238}U, ^{232}Th, and ^{40}K at the time of the formation of the Solar System, $4.5 \cdot 10^9$ yr ago, using the half-lives given in Sect. 2.8.5b. How large were the relative contributions of these radionuclides to heat production in rocks at that time?

4. Atmospheres of Planets and Moons (Sects. 2.6 and 2.8)

A celestial object can hold its own atmosphere when the thermal velocity $\bar{v}$ of the molecules (2.8.23) is very small compared to the escape velocity $v_e = \sqrt{(2G\mathcal{M})/R}$ (2.6.42). Show that, assuming the acceleration of gravity $g = G\mathcal{M}/R^2$ to be independent of the altitude in the atmosphere, the following relation holds:

$$\left(\frac{\bar{v}}{v_e}\right)^2 = \frac{H}{R}$$

where H is the scale height (2.8.17) in the atmosphere. Make up a list of the planets and their moons ranked according to v_e and $v_e/\bar{v}$, using the data given in Sect. 2.8, and compare it with the observed existence of atmospheres. What is the resulting minimum value of the ratio $v_e/\bar{v}$ which would allow the object to retain its own atmosphere?

5. "Cosmic Collisions" (Sects. 2.9 and 6.4)

5.1 Employ the present flux of particles at 1 AU distance from the Sun from Fig. 2.9.5 to estimate by extrapolation how large is the probability of collision of the Earth with a meteorite or asteroid of radius 1, 10, and 100 km.

5.2 An asteroid with a radius of 10 km and a density of $\varrho = 3000\ \mathrm{kg\,m^{-3}}$ is assumed to collide with the Earth at a relative velocity of $50\ \mathrm{km\,s^{-1}}$. Assuming that its mass remains "together" during the collision, estimate:

a) the magnitude of the momentum of the asteroid in comparison to the orbital momentum of the Earth,
b) the maximum angular momentum transferred to the Earth in comparison to its rotational angular momentum. How much of a change in the rotational period would be produced by such a collision? How long would it take for tidal forces (Sect. 6.3) to cause the same change?
c) Assume that all the kinetic energy of the asteroid could be converted into thermal energy of the molecules of the Earth's atmosphere. How high would the temperature of the atmosphere rise? Compare this kinetic energy with the total thermal energy of the atmosphere, with the total dissociation energy of its molecules (N_2: 9.8 eV, O_2: 5.1 eV), with the rotational energy of the Earth, and with the energy radiated to the Earth from the Sun in one year.

5.3 Assume that the mass of the asteroid were distributed evenly over the entire surface of the Earth. How thick would the resulting layer be?

Assume that the mass of the asteroid were distributed homogeneously throughout the Earth's atmosphere up to an altitude of 10 km in the form of dust particles of 1 µm radius. Estimate the resulting extinction coefficient (5.3.9, 10 with $Q_e \simeq Q_{sca} \simeq 1$) and the change in the Earth's albedo and its atmospheric temperature. Compare the mass of the asteroid with that of the Earth's atmosphere.

6. Asteroids (Sects. 2.6, 2.8, and 4.3)

6.1 A small asteroid at a distance 3 AU from the Sun is assumed to have a radius of 10 km and an albedo $A = 0.03$ (similar to that of the carbonaceous chondrites). How high is its effective temperature for very rapid and for very slow rotation? Where is its radiation maximum, assuming it to be a black body radiator?

Compare the effective temperature with the values given in Table 2.8.1 for the planets.

6.2 Is a *binary system* of two asteroids bound by gravitation possible? Assume both asteroids to have the same mass $\mathcal{M}$ and calculate numerically the relationship between their distance and their orbital period (a) for $\mathcal{M} = 10^{21}$ kg (corresponding to the largest asteroids); (b) for $\mathcal{M}$ of an asteroid having 10 km radius and a density of 3000 kg m^{-3}; and (c) for $\mathcal{M} = 1$ kg (!).

How large are the forces on such "binary asteroids" at 3 AU distance from the Sun due to other bodies in the Solar System?

6.3 In the story "The Little Prince" by A. de Saint-Exupéry (1943), the prince visits a tiny planet which has just enough room for a street lamp and a lamplighter. The planet has been rotating faster and faster from year to year, so that the lamp now has to be lit and extinguished once every minute.

Assume the planetoid to be a sphere of radius 1.25 m with a density of 3000 kg m^{-3}. How large are the gravitational and the centrifugal accelerations at its equator? How long would it take for a stone to fall from an altitude of 1 m onto its surface? Is the centrifugal force sufficiently strong to overcome the frictional force between the standing lamplighter (mass 70 kg) and the surface of the planetoid? What is the situation concerning the stability of the planetoid at a rotational period of 1 min, if the tensile strength of its material is 10^8 Pa?

7. Distances and Space Velocities of the Stars (Sects. 2.3, 4.4, and 5.2)

7.1 How large is the angular distance on the sphere between the two brightest stars in Gemini, Castor (α Gem) and Pollux (β Gem)? How large is their true spatial distance in [pc]? The coordinates (α, δ) for the Epoch 2000 and the parallaxes p of the two stars are

α Gem: $\alpha = 7\,\mathrm{h}\ 34\,\mathrm{min}\ 35.9\,\mathrm{s}$, $\delta = +31°53'18''$, $p = 0.067''$

β Gem: $\alpha = 7\,\mathrm{h}\ 45\,\mathrm{min}\ 18.9\,\mathrm{s}$, $\delta = +28°1'34''$, $p = 0.094''$.

7.2 In a star catalogue, the coordinates of Sirius (α CMa) for the Epoch 1900 are given as $\alpha = 6\,\mathrm{h}\ 40\,\mathrm{min}\ 44.6\,\mathrm{s}$ and $\delta = -16°34'44''$. Calculate its position for the year 2000, taking into consideration the precession from Table 2.3.1, and compare the result with the entry in the Bright Star Catalogue. How much of a contribution does the proper motion of Sirius make (it is $-0.545''\,\mathrm{yr}^{-1}$ in right ascension, $-1.211''\,\mathrm{yr}^{-1}$ in declination)? The radial velocity of Sirius is $-8\,\mathrm{km\,s}^{-1}$ and its parallax is $0.378''$. By how much does its distance from the Sun change in 100 years due to the radial velocity? Calculate the galactic coordinates (l, b) of Sirius and verify the result using Fig. 5.2.8. How large are the components of the space velocity in the Milky Way?

8. Sirius A and B – The Fundamental Parameters of Stars (Sects. 4.4, 4.6, and 4.12)

The visual binary system α CMa = Sirius has a parallax $p = 0.378''$. Relative to the brighter Sirius A ($m_V = -1.46$ mag), its fainter companion, the white dwarf Sirius B ($m_V = +8.44$ mag) moves on an elliptical orbit with a semimajor axis $a = 7.5''$ and a period of $P = 50.1$ yr. The ratio of the semimajor axes of the orbits of the two components around their common center of gravity is $a_B/a_A = 2.1$.

8.1 How large are the masses $\mathscr{M}$ of the two stars? What are the greatest and the least distances of approach of the two components in AU? How large are the average orbital velocities in the center of mass system?

8.2 Analysis of the spectra of the two stars yields the following effective temperatures and surface gravities:

$$T_{\text{eff, A}} = 9980\ \text{K} \quad T_{\text{eff, B}} = 22\,500\ \text{K}$$

$$g_A = 204\ \text{m s}^{-2} \quad g_B = 7.1 \cdot 10^6\ \text{m s}^{-2}\ ,$$

and thus, as bolometric corrections (4.4.21), B.C.(A) $\simeq$ 0.2 mag and B.C.(B) $\simeq$ 2.4 mag. Calculate the luminosities L, the radii R, and the average densities $\bar{\varrho}$ of the two stars. To what extent is the mass-luminosity relation (4.6.5) fulfilled by the two components?

8.3 Interferometric measurements yield an apparent angular diameter for Sirius A of $0.0059''$. Compare this value with the spectroscopically-determined diameter. For white dwarfs, the mass-radius relation (4.12.51) is given by $R/R_\odot = 0.0128\ (\mathscr{M}/\mathscr{M}_\odot)^{-1/3}$. What radius does this relation yield for Sirius B?

9. Radiation Force and Maximum Luminosity of the Stars (Sects. 2.6, 4.2, and 4.6)

A star reaches its maximum luminosity, the *Eddington luminosity* L_E, when the outwardly-directed radiation acceleration g_{rad} at its surface is equal to its gravitational acceleration $g = G\mathscr{M}/R^2$ (4.6.7); for $g_{\text{rad}} > g$, there are thus no stable stars (A.S. Eddington, 1921).

9.1 The radiation acceleration is equal to the momentum which is transferred to a unit mass per second through the absorption of radiation; cf. (2.6.9), where for radiation the momentum is equal to energy/c, c = velocity of light. Show that the following relation holds:

$$g_{\text{rad}} = \frac{1}{c} \int_0^\infty k_{\nu,M} F_\nu \, d\nu$$

where F_ν is the monochromatic radiation flux (4.2.6) and $k_{\nu,M}$ is the mass extinction coefficient (4.2.15). In a completely ionized hydrogen plasma, the radiation force is transferred to the free electrons mainly by Thomson scattering with the frequency-independent cross section σ_T (3.4.6). Show that in this case, the Eddington luminosity is given by

$$L_{\mathrm{E}} = \frac{4\pi c G \mathscr{M}}{\sigma_{\mathrm{T},M}}$$

(where $\sigma_{\mathrm{T},M} = \sigma_{\mathrm{T}} \cdot N_{\mathrm{e}}/\varrho$). How large is L_{E} in units of the solar luminosity for $\mathscr{M} = 1\ \mathscr{M}_{\odot}$?

9.2 In the brightest supergiants, absolute bolometric magnitudes of $M_{\mathrm{bol}} \simeq -10$ mag are observed. Calculate their maximum masses under the assumption that M_{bol} corresponds to the Eddington luminosity, and discuss them in connection with the empirical mass-luminosity relation (4.6.5).

10. Excitation, Ionization, and Continuous Absorption in the Solar Atmosphere (Sects. 4.2, 4.7, and 4.8)

In the photosphere of the Sun, in the layer $\bar{\tau} = 0.1$ the temperature is $T = 5070$ K, the electron pressure is $P_{\mathrm{e}} = 0.38$ Pa, and the gas pressure is $P_g = 4360$ Pa (Table 4.8.1). How large are the electron density N_{e} and the total particle density N_{tot} in this layer?

10.1 Calculate, using the Boltzmann formula (4.7.6), the occupation numbers of the first two excited energy states (principal quantum numbers $n = 2$ and 3) of neutral hydrogen relative to the ground state ($n = 1$). The energies are given by

$$\chi_n\,[\mathrm{eV}] = 13.60 \left(1 - \frac{1}{n^2}\right)$$

(4.7.3), and the associated statistical weights are $g_n = 2n^2$. How great is the degree of ionization $N(\mathrm{H\,II})/N(\mathrm{H\,I})$ of hydrogen and how large is the relative fraction of negative hydrogen ions, $N(\mathrm{H}^-)/N(\mathrm{H\,I})$? Use the Saha formula (4.7.14, 4.7.17) and take only the states $n = 1$ to 3 into account in the partition function Q_0 of H I. For H II, Q_1 is equal to 1. The H^- ion has an "ionization energy" (electron affinity) of 0.75 eV and only one bound state (ground state), $Q(\mathrm{H}^-) \equiv g_0(\mathrm{H}^-) = 1$.

10.2 Calculate the contributions of H I and H^- to the continuous absorption κ_λ of the solar photosphere in the layer $\bar{\tau} = 0.1$ at the wavelength $\lambda = 364$ nm, i.e. on the short wavelength side of the Balmer series limit (compare Fig. 4.8.2). The atomic absorption or photoionization cross sections of H I are given by (4.7.36), where $\kappa_\lambda = \alpha_{n\kappa}(\lambda) \cdot N_n = \kappa_{\lambda,\mathrm{at}} \cdot N_n$ (4.2.15). Which energy states n of H I need to be taken into account for calculating the photoionization cross section at the wavelength 364 nm? For the H^- ion, $\alpha_{0\kappa}(364\ \mathrm{nm}) \simeq 2.4 \cdot 10^{-21}\ \mathrm{m}^2$.

How strong is Thomson scattering from free electrons in comparison to κ_λ?

11. Radiation Transport: Temperature Stratification of a Stellar Atmosphere (Sects. 4.2 and 4.8)

The radiation transport equation for an atmosphere of parallel plane layers (4.8.4) in the "grey case", i.e. for an absorption coefficient which is independent of the frequency or the wavelength, has the form:

$$\cos\theta \frac{dI}{d\tau} = I - B \ ,$$

where τ is the frequency-independent optical depth, $I = \int_0^\infty I_\nu d\nu$ and $B = \int_0^\infty B_\nu d\nu = \sigma T^4/\pi$.

Consider only two directions of radiation propagation, one outwards, $\cos\theta = +1$, with the intensity I^+, and one inwards, $\cos\theta = -1$, with I^-. In this greatly simplified *two-ray approximation*, the mean intensity (4.2.34) is given by $J = \frac{1}{2}(I^+ + I^-)$ and the radiation flux (4.2.6) by $F = I^+ - I^-$.

11.1 Show that on adding the two radiation-transport equations for the two directions, $\cos\theta = \pm 1$, one obtains energy conservation, i.e. that in radiation equilibrium $J = B$, the radiation flux F is independent of the depth τ ($F = \sigma T_{\mathrm{eff}}^4$).

11.2 Subtracting the two equations gives a differential equation for J. Solve this equation with the boundary condition that no radiation goes inwards at the surface of the star (i.e. $I^-(\tau = 0) = 0$) and compare the resulting temperature stratification $T(\tau)$ with the grey stratification (4.8.7).

12. Supernovae and Pulsars (Sects. 2.6, 3.5, 4.2, 4.4, 4.11, and 5.4)

12.1 The *bolometric* magnitude $L(t)$ of the bright supernova SN 1987 A of February 23rd, 1987 in the Large Magellanic Cloud decreased exponentially with time during the period from about 120 d to 800 d after the outburst.

For this time interval, determine $L(t)$ approximately from the figure. How large is the decrease in magnitudes per day? Give the half-life $\tau_{1/2}$(SN) of the light curve and compare it with the half-life of the radioactive decay of ^{56}Co to ^{56}Fe, $\tau_{1/2} = 77.28$ d.

12.2 How many solar masses of ^{56}Co would be required to explain the observed luminosity $L(t)$ in the exponential part of the curve as due to the decay of ^{56}Co? In the decay of a ^{56}Co nucleus, on the average 3.6 MeV of energy is released.

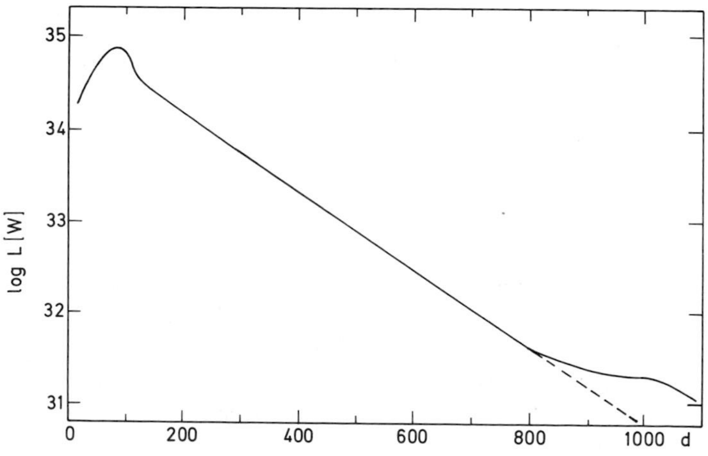

12.3 After about 800 d, the bolometric light curve flattens out and, e.g. 1030 d following the explosion, it has a value of $2\cdot 10^{31}$ W, well above the extrapolated exponential decrease. A possible explanation for this would be an additional energy input from the pulsar or rotating neutron star (not yet directly observable) at the cost of its rotational energy $E_{\text{rot}} = \frac{1}{2} J\omega^2$.

Calculate the moment of inertia J of the neutron star assuming that it is a homogeneous sphere (2.6.45) with a radius $R = 10$ km and a mass $\mathcal{M} = 1.4\,\mathcal{M}_{\odot}$. The pulsar in the Crab Nebula (Sect. 4.11.7) changes its period $P = 2\pi/\omega$ according to the relation $P/\dot{P} = 2500$ yr ($\dot{P} = dP/dt$); about 1/100 of the decrease in its rotational energy E_{rot} emerges as pulsar radiation. Calculate $E_{\text{rot}}/100$ and compare the result with the observed luminosity of SN 1987 A, 1030 d after its explosion.

12.4 The neutron star formed by a supernova explosion has a surface temperature T of over $2\cdot 10^6$ K for up to 100 yr after the outburst.

At what wavelength is the maximum of the radiation from a black body at $T = 2\cdot 10^6$ K? How large is the corresponding luminosity if the radius of the neutron star is 10 km?

The X-ray satellite ROSAT can just detect a "point source" at a flux of $2\cdot 10^{-17}$ W m^{-2} (in the energy range from 0.1 to 0.3 keV), or $6\cdot 10^{-17}$ W m^{-2} (0.5 to 2 keV). Could the thermal X-radiation from a neutron star formed by the explosion of SN 1987 A with $R = 10$ km and $T = 2\cdot 10^6$ K be observed by ROSAT (distance to the Large Magellanic Cloud: 50 kpc)?

13. Thermonuclear Reactions (Sects. 4.7 and 4.12)

13.1 Two positively-charged nuclei (with charges $Z_1 e$ and $Z_2 e$) separated by a distance r repel each other by a force corresponding to the *Coulomb potential*:

$$V(r) = \frac{1}{4\pi\varepsilon_0}\frac{Z_1 Z_2 e^2}{r} \;.$$

A nuclear reaction is possible only if the two nuclei approach each other to within $r_0 = 10^{-15}$ m. How high is the Coulomb barrier $B = V(r_0)$? How great is its thickness $\Delta = \delta - r_0 \simeq \delta$ for a particle with the kinetic energy $E = \frac{1}{2} m v^2 = \frac{3}{2} kT$, where δ is given by $V(\delta) = E$? Express δ in units of the de Broglie wavelength $\Lambda = \hbar/(mv)$. (For simplicity, we assume that the nucleus is at rest.)

According to G. Gamow, the probability for *tunnelling* through a Coulomb barrier is $P(v) = \exp(-2\pi\eta)$ where η is given by:

$$\eta = \frac{1}{4\pi\varepsilon_0}\frac{Z_1 Z_2 e^2}{\hbar v} \;.$$

How does η depend on δ? Calculate B, δ, and $P(v)$ for the reaction ^{12}C (p, γ) ^{13}N from the CNO triple cycle for $T = 10^7$ and $2\cdot 10^7$ K.

13.2 For more precise calculations, the *screening* of the nuclear charges by the electrons in the plasma must be taken into account. In the framework of the Debye theorie, the electrostatic potential then becomes:

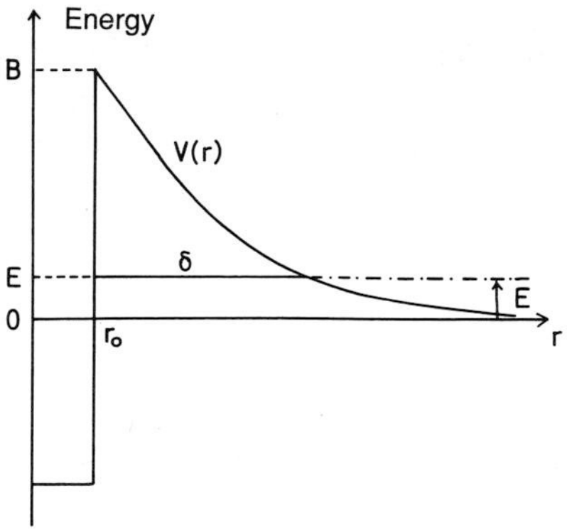

$$V_s(r) = V(r)e^{-r/r_D}$$

where the Debye length r_D for a completely ionized hydrogen plasma of particle density N is given by:

$$r_D = \sqrt{\frac{\varepsilon_0 kT}{e^2 N}} .$$

How do B, δ, and $P(v)$ change when the screened potential $V_s(r)$ is used? Take N to be 10^{32} and $10^{35}\,\mathrm{m}^{-3}$.

13.3 The *rate* of a thermonuclear reaction is essentially determined by the velocity distribution $\Phi(v)$ of the nuclei, as well as by $P(v)$. We can take a Maxwell-Boltzmann distribution (4.7.33) for $\Phi(v)$. At what value of v are the two opposed exponential functions $\exp(-b/v)$ and $\exp[-(v/v_0)^2]$ equal (where b and v_0 are constants)? For what v does the maximum of the integrand $\sigma(v)v\Phi(v)$ in (4.7.33), the socalled "Gamow peak", occur? Assume the interaction cross-section to be given by $\sigma(v) = \mathrm{const}\cdot P(v)/v^2$.

14. Energy Release through Accretion (Sects. 2.6, 4.11, 4.12, and 5.6)

If a mass m is brought from a great distance onto the surface of a sphere having a mass $\mathcal{M}$ and radius R, then the gravitational energy

$$\Delta E = -\frac{G\mathcal{M}m}{R}$$

is released (4.12.26). The accretion of matter at a rate $\dot{\mathcal{M}} = dm/dt$ can, if ΔE is completely converted to radiation, therefore give rise to an *accretion luminosity* (4.11.9):

$$L_A = \left|\frac{dE}{dt}\right| = \frac{G\mathcal{M}\dot{\mathcal{M}}}{R} .$$

14.1 Calculate L_A in units of the solar luminosity for a mass flux $\dot{\mathcal{M}} = 10^{-10}\,\mathcal{M}_\odot\,\mathrm{yr}^{-1}$, typical of close binary systems (Sects. 4.11.5 and 6), onto (a) a white dwarf, and (b) a neutron star, each having $\mathcal{M} = 1\,\mathcal{M}_\odot$.

14.2 In cataclysmic variables, the mass flux to the white dwarf is not directed radially onto the dwarf, but instead into an accretion disk, whereby each mass particle moves inwards on a spiral orbit. Assume that its orbit at each moment can be approximated by a circular orbit with a Kepler velocity according to (2.6.40). Show using (2.6.35) that when the "innermost" orbit, close to the surface of the white dwarf, is reached, only the luminosity $\frac{1}{2}L_A$ is obtained. How can the full accretion luminosity L_A be produced?

14.3 How large is L_A when a mass $\mathscr{M}$ of matter is accreted up to its Schwarzschild radius R_S (4.12.62)? (Here, make the greatly simplified assumption that Newtonian gravitational theory can be applied down to R_S.) How does L_A depend upon $\mathscr{M}$? In Sect. 5.6.6, the radiation from the nuclei of active galaxies is estimated to be $L = \varepsilon \dot{\mathscr{M}} c^2$ where ε is the efficiency (5.6.13). How does this expression relate to L_A?

14.4 The Eddington luminosity L_E (see Exercise 9) gives an upper limit for the rate of continuous matter fall, $\dot{\mathscr{M}}$, since when $L_A \geq L_E$, radiation pressure prevents further accretion. How large is the maximum rate for accretion onto a white dwarf of $1\,\mathscr{M}_\odot$, a neutron star of $1\,\mathscr{M}_\odot$, and onto a galactic nucleus (Schwarzschild radius R_S) of $10^9\,\mathscr{M}_\odot$?

15. Microwave Background Radiation and Interstellar Absorption Lines

(Sects. 4.7, 4.9, 5.3, and 5.9)

In the spectrum of the bright O star ζ Oph, the following weak absorption lines of the CN radical have been observed, which originate from the two lowest levels of CN with the rotational quantum numbers $J = 0$ and 1 (see figure):

Line	Wavelength λ [nm]	Equivalent width W_λ [pm]	Relative f-value
$R(0)$	387.46	0.73	1
$R(1)$	387.39	0.21	$\frac{2}{3}$
$P(1)$	387.57	0.11	$\frac{1}{3}$

15.1 Calculate using (5.3.22) the energy of the level $J = 1$ and the wavelength λ_0 and frequency ν_0 which correspond to the transition $(J = 0) \leftrightarrow (J = 1)$. The rotational constant is $B = 1.8910\,\mathrm{cm}^{-1}$.

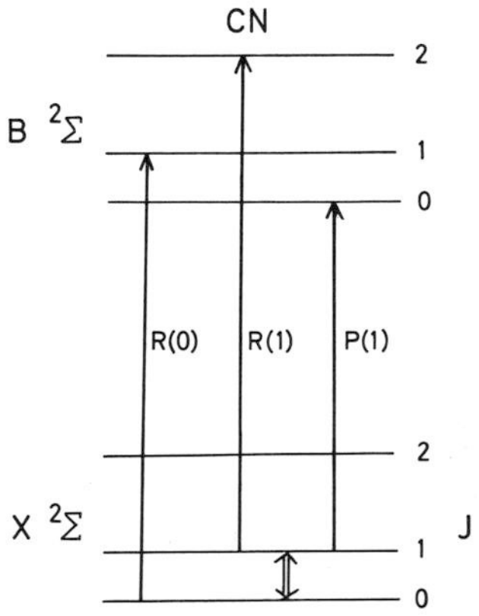

15.2 Determine the excitation temperature ("rotational temperature") from the ratio of the occupation numbers $N_1/N_0 = N(J = 1)/N(J = 0)$ according to the Boltzmann formula (4.7.6). The lines can be assumed to be unsaturated, so that W_λ is proportional to the absorption coefficient κ_ν and thus to the product $Nf\lambda_0^2$ (4.9.20a).

15.3 Is the excitation into the level $J = 1$ caused by the microwave background radiation at $T = 2.7$ K? Verify the answer by solving the kinetic equilibrium (4.7.21) for a 2-level atom with $J = 0$ and $J = 1$ (cf. Fig. 4.7.6), taking the following processes into account:

a) spontaneous emission with the Einstein coefficient $A_{10} = 1.2 \cdot 10^{-5}\,\mathrm{s}^{-1}$,
b) induced emission $B_{10}J_\nu$ and absorption $B_{01}J_\nu$ (4.7.23), where J_ν is given by the intensity of the 3 K radiation (Kirchhoff-Planck function) at the frequency ν_0, and
c) excitation and deexcitation by electron impact (4.7.34), with $C_{10} = \alpha N_e$ and $\alpha = 2.1 \cdot 10^{-12}\,\mathrm{m}^3\,\mathrm{s}^{-1}$.

Assume $N_e = 10^5\,\mathrm{m}^{-3}$ for the electron density in the interstellar medium (H I clouds), corresponding to a degree of ionization $N_e/N_H = 5 \cdot 10^{-4}$. What would the equilibrium be like for an electron density 10 times higher?

15.4 Compare the radiation field of the microwave background with the general stellar radiation in the Milky Way, when the latter is assumed to be diluted black-body radiation at $T = 10^4$ K and a dilution factor (5.3.13) of 10^{-15}. How large would the average distance between stars have to be (at a mean stellar radius of $2R_\odot$) to make the intensity of the stellar radiation equal to that of the 3 K radiation, (a) at λ_0 and (b) at the maximum of the Planck distribution?

16. Identification of Spectral Lines and Red Shifts (Sects. 4.7, 5.6, and 5.9)

The optical spectrum of a radio source exhibits six broad emission lines, ordered in decreasing intensity, at $\lambda = 563.2$, 323.9, 579.2, 503.2, 475.3, and 459.5 nm wavelengths. Try to identify these lines as Balmer lines from neutral hydrogen. What red shift z or radial velocity is found from the Hubble relation (5.9.2), and to what distance does this correspond? What would one predict for the wavelengths of Hα and Lα with this z? Are there plausible identifications for the remaining lines? Is the identification definite, or could the observed lines also be attributed to e.g. the Lyman series?

Hint: The laboratory wavelengths of the hydrogen lines can be found in spectroscopic tables, or (4.7.3) can be used. The principal quantum number of the lower level is $n = 1$ for the Lyman series and $n = 2$ for the Balmer series. For further identifications, consider the emission lines of other objects, e.g. the Orion Nebula, Fig. 5.3.11. (Here, we are dealing with 3C 273, the first quasar, discovered by Maarten Schmidt.)

17. The Distribution of Galaxies (Sects. 5.7 and 5.9)

In the galaxy survey carried out by the Center for Astrophysics in Cambridge, Mass., a large structure, spread out over (at least) 7.5 h in right ascension and 36° in declination (the "Great Wall") was found; in it, the density of galaxies is about 5 times higher than their average density in the cosmos (Fig. 5.7.4). This structure has an average radial velocity of 8700 km s^{-1} with a variation of only about 500 km s^{-1} in the radial direction.

17.1 Calculate the volume and the mass of the "Great Wall" as functions of the Hubble constant H_0 and the density parameter $\Omega = \varrho/\varrho_c$ (5.9.21). How many galaxies like our Milky Way would correspond to this mass?

17.2 Estimate the volume of the region which is free of galaxies in Fig. 5.7.4, assuming it to be spherical with its center at $\alpha \simeq 15$ h and $V_{rad} \simeq 7400$ km s^{-1}.

Subject Index

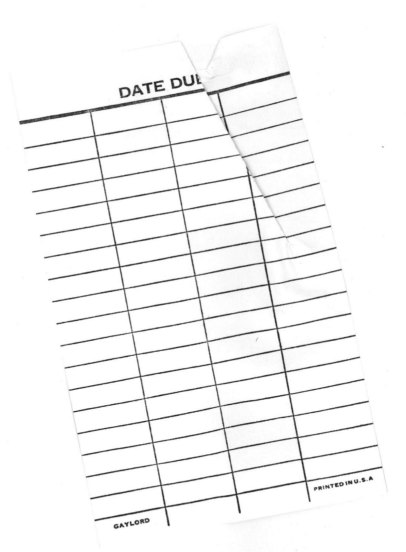

Fundamental Physical Constants *

Constant	SI units	Gaussian units
Gravitational constant, G	$6.6726 \cdot 10^{-11}$ $m^3\,kg^{-1}\,s^{-2}$	$6.6726 \cdot 10^{-8}$ $cm^3\,g^{-1}\,s^{-2}$
Velocity of light in vacuum, c	$2.9979 \cdot 10^{8}$ $m\,s^{-1}$	$2.9979 \cdot 10^{10}$ $cm\,s^{-1}$
Permeability constant of vacuum, μ_0	$1.2566 \cdot 10^{-6}$ $m\,kg\,s^{-2}\,A^{-2}$ $= V\,s\,A^{-1}\,m^{-1}$	–
Permittivity constant of vacuum, ε_0	$8.8542 \cdot 10^{-12}$ $m^{-3}\,kg^{-1}\,s^4\,A^2$ $= A\,s\,V^{-1}\,m^{-1}$	–
Planck's constant, h	$6.6261 \cdot 10^{-34}$ $m^3\,kg\,s^{-1} = J\,s$	$6.6261 \cdot 10^{-27}$ erg s
$\hbar = h/2\pi$	$1.0546 \cdot 10^{-34}$ J s	$1.0546 \cdot 10^{-27}$ erg s
Elementary (electron's) charge e	$1.6022 \cdot 10^{-19}$ A s = C	$4.8032 \cdot 10^{-10}$ esu
Atomic mass constant $m_u = u$	$1.6605 \cdot 10^{-27}$ kg	$1.6605 \cdot 10^{-24}$ g
Rest masses:		
Proton, m_p	$1.6726 \cdot 10^{-27}$ kg	$1.6726 \cdot 10^{-24}$ g
	1.0073 u	
Neutron, m_n	$1.6749 \cdot 10^{-27}$ kg	$1.6749 \cdot 10^{-24}$ g
	1.0087 u	
Electron, $m_e = m$	$9.1094 \cdot 10^{-31}$ kg	$9.1094 \cdot 10^{-28}$ g
	$5.4858 \cdot 10^{-4}$ u	
Hydrogen atom, m_H	$1.6736 \cdot 10^{-27}$ kg	$1.6736 \cdot 10^{-24}$ g
	1.0078 u	
m_p/m_e	$1.8362 \cdot 10^{3}$	
Specific charge of electron, e/m	$1.7588 \cdot 10^{11}$ $C\,kg^{-1}$	$5.2728 \cdot 10^{17}$ $esu\,g^{-1}$
Avogadro's constant, N_A	$6.0221 \cdot 10^{23}$ mol^{-1}	
Universal gas constant $\mathscr{R}$	8.3145 $J\,mol^{-1}\,K^{-1}$	$8.3145 \cdot 10^{7}$ $erg\,mol^{-1}\,K^{-1}$
Boltzmann's constant k	$1.3807 \cdot 10^{-23}$ $J\,K^{-1}$	$1.3807 \cdot 10^{-16}$ $erg\,K^{-1}$
Fine structure constant α	$7.2974 \cdot 10^{-3}$ = 1/137.036	
Compton wavelengths:		
Proton, Λ_p	$1.3214 \cdot 10^{-15}$ m	$1.3214 \cdot 10^{-13}$ cm
Electron, Λ_e	$2.4263 \cdot 10^{-12}$ m	$2.4263 \cdot 10^{-10}$ cm
	$2.4263 \cdot 10^{-2}$ Å	
Classical electron radius, r_e	$2.8179 \cdot 10^{-15}$ m	$2.8179 \cdot 10^{-13}$ cm
Thomson cross section, σ_T	$6.6525 \cdot 10^{-29}$ m^2	$6.6525 \cdot 10^{-25}$ cm^{-2}
Rydberg constant, R_∞	$1.0974 \cdot 10^{7}$ m^{-1}	$1.0974 \cdot 10^{5}$ cm^{-1}
	1/911.27 $Å^{-1}$	
1st Bohr radius, a_0	$5.2918 \cdot 10^{-11}$ m	$5.2918 \cdot 10^{-9}$ cm
Planck's radiation constant, c_2	$1.4388 \cdot 10^{-2}$ m K	1.4388 cm K
Stefan-Boltzmann radiation constant, σ	$5.6705 \cdot 10^{-8}$ $W\,m^{-2}\,K^{-4}$	$5.6705 \cdot 10^{-5}$ $erg\,cm^{-2}\,s^{-1}\,K^{-4}$

* All numerical values have been rounded to 5 figures. For more precise values (with error limits), see E. R. Cohen and B. N. Taylor: The 1986 Adjustment of the Fundamental Physical Constants, report of the CODATA Task Group on Fundamental Constants, CODATA, Bull. **63**, Pergamon, Elmsford, NY (1986), or Physics Today, Part 2, BG11 (August 1987).